Ocean Engineering & Oceanography

Volume 14

The Ocean engineering & Oceanography (OEO) series publishes state-of-art research and applications in the two related and interdependent areas of ocean engineering and oceanography. The series contains monographs, lecture notes, edited volumes and selected PhD theses focusing on all different theoretical as well as applied aspects and subfields of ocean engineering and oceanography research. Topics in the series include marine electronic engineering, concerned with the design of electronic devices for use in the marine environment, offshore engineering (or offshore construction), concerned with the design of fixed and floating marine structures, pipelines and other underwater equipment, and oceanography, concerned with physical, chemical, geological and biological processes throughout the marine environment.

All books published in the series are submitted for consideration in Web of Science.

More information about this series at http://www.springer.com/series/10524

Abdus Samad · S. A. Sannasiraj · V. Sundar ·
Paresh Halder
Editors

Ocean Wave Energy Systems

Hydrodynamics, Power Takeoff and Control Systems

Springer

Editors
Abdus Samad
Department of Ocean Engineering
Indian Institute of Technology Madras
Chennai, Tamil Nadu, India

S. A. Sannasiraj
Department of Ocean Engineering
Indian Institute of Technology Madras
Chennai, Tamil Nadu, India

V. Sundar
Department of Ocean Engineering
Indian Institute of Technology Madras
Chennai, Tamil Nadu, India

Paresh Halder
Institute of Ocean Energy
Saga University
Saga, Japan

The funder for the Chapter 17 is HORIZON 2020 (Project SETITAN, ID: 764014).

ISSN 2194-6396 ISSN 2194-640X (electronic)
Ocean Engineering & Oceanography
ISBN 978-3-030-78715-8 ISBN 978-3-030-78716-5 (eBook)
https://doi.org/10.1007/978-3-030-78716-5

This Springer imprint is published by the registered company Springer Nature Switzerland AG
The registered company address is: Gewerbestrasse 11, 6330 Cham, Switzerland

We dedicate this book to all researchers who contributed to this important field that has currently gained enormous momentum. The authors have derived a lot from the past world carried out and reported in the literature.

Preface

The three most important and challenging dimensions of sustainable development are the economic, social, and environmental ones. Energy is one of the major factors governing the nature of the environment. The depletion of conventional energy systems has been forcing researchers toward finding renewable sources for energy. Among the renewable energy sources, ocean energy, despite its large availability, has been only a little-explored till now.

Marine energy harvesting systems face challenges because of cyclone, tsunami, gust, ocean current, corrosive water, marine growth, cost, maintenance, and other issues which include societal and environmental issues. A bunch of parameters dictates the system designs for their performance and survival in the ocean. Among the ocean energy technologies, wave energy is at the forefront with technology reaching a readiness level, gradually but steadily, and several countries have installed either fully operating commercial power plants or pilot plants. Several new concepts are being tested in laboratories and through field experiments. Among different types of wave energy extraction devices, the oscillating water column is undergoing extensive investigations around the globe, with a focus on the enhancement of its operating efficiency.

It is at this juncture that we realized the need for a book offering timely insights into ocean wave energy, which will inform engineers and researchers about the most recent technological development and know-how gained by the use of such technologies. Furthermore, such a book should also serve as an additional source of information for graduate students attending ocean wave energy courses in colleges and universities and wish to get a broad and in-depth insight into current technologies. Therefore, the chapters were chosen to cover a whole range of topics and disciplines related to ocean wave energy conversions, such as mechanical design, electrical design, hydrodynamics, and report findings from physical and numerical modeling. About 50% of the book deals with hydrodynamics analysis of different wave energy converters; the remaining chapters cover control and power takeoff mechanisms.

We wish to thank the students Mr. Hithaish, Ms. Kathyayani, Mr. Suman, Mr. Amit, Dr. Tapas Das, Dr. John Ashlin, Dr. R. Wilbert, Ms. Sukanya, and Mr. Danny, who have assisted in checking the chapters including their formatting, for their direct or indirect support at every stage of the preparation of this book. We also acknowledge the reviewers for their comments and suggestions for improvement of every single chapter of the book. The editors also thank the authors for the timely submission of their respective manuscripts.

The editors thank the Springer supporting staff and IIT Madras colleagues during the development of the book. The editors also acknowledge their respective family members for their untiring support in the development of the book.

Chennai, India
March 2021

Abdus Samad
S. A. Sannasiraj
V. Sundar
Paresh Halder

Contents

Contributors

S. J. Ashlin COAST Engineering Research Group of University of Plymouth, Plymouth, UK

Rameez Badhurshah IIT Bombay, Powai, Mumbai, India

Marcos Blanco CIEMAT, Madrid, Spain

Sergio M. Camporeale Department of Mechanics, Mathematics and Management, Polytechnic University of Bari, Bari, BA, Italy

Tapas K. Das Wave Energy and Fluids Engineering Laboratory (WEFEL), Department of Ocean Engineering, Indian Institute of Technology Madras, Chennai, Tamil Nadu, India

Daniel Raj David University of Western Australia, Perth, Australia

Pasquale G. F. Filianoti Department of Civil, Energy, Environment and Materials Engineering, Mediterranean University of Reggio Calabria, Reggio Calabria, RC, Italy

Luis García-Tabarés CIEMAT, Madrid, Spain

Paresh Halder Institute of Ocean Energy, Saga University, Saga, Japan

P. Madhan Kumar Wave Energy and Fluids Engineering Laboratory (WEFEL), Department of Ocean Engineering, Indian Institute of Technology Madras, Chennai, TamilNadu, India

Marcos Lafoz CIEMAT, Madrid, Spain

M. H. Mohamed Mechanical Power Engineering Department, Faculty of Engineering EL-Mattaria, Helwan University, Cairo, Egypt;
Department of Mechanical Engineering, College of Engineering and Islamic Architecture, Umm Al-Qura University, Makkah, Kingdom of Saudi Arabia

Jorge Nájera CIEMAT, Madrid, Spain

Gustavo Navarro CIEMAT, Madrid, Spain

Tirumaleswara Reddy Nemalidinne DHI (India) Water and Environment Pvt Ltd, Delhi, India

Krisna Adi Pawitan Quantum Wave Microscopy Unit, Okinawa Institute of Science and Technology Graduate University, Okinawa, Japan

Juan Ignacio Pérez ETSICCP-UPM, Campus Ciudad Universitaria, Madrid, Spain

A. Ramadan College of Engineering and Technology, Arab Academy for Science and Technology and Maritime Transport (AAST), Cairo, Egypt

Dionisio Ramírez ETSII-UPM, Madrid, Spain

Abdus Samad Wave Energy and Fluids Engineering Laboratory (WEFEL), Department of Ocean Engineering, Indian Institute of Technology Madras, Chennai, Tamil Nadu, India

S. A. Sannasiraj Department of Ocean Engineering, Indian Institute of Technology Madras, Chennai, India

Álvaro Santiago CIEMAT, Madrid, Spain

Miguel Santos-Herrán CIEMAT, Madrid, Spain

R. Suchithra Wave Energy and Fluids Engineering Laboratory (WEFEL), Department of Ocean Engineering, Indian Institute of Technology Madras, Chennai, Tamil Nadu, India

V. Sundar Department of Ocean Engineering, Indian Institute of Technology Madras, Chennai, India

Zhi Yung Tay Engineering Cluster, Singapore Institute of Technology, Singapore, Singapore

Jorge Torres CIEMAT, Madrid, Spain

Marco Torresi Department of Mechanics, Mathematics and Management, Polytechnic University of Bari, Bari, BA, Italy

Vengatesan Venugopal Institute for Energy Systems, School of Engineering, The University of Edinburgh, Edinburgh, UK

Vishnu Vijayasankar Wave Energy and Fluids Engineering Laboratory (WEFEL), Department of Ocean Engineering, Indian Institute of Technology Madras, Chennai, Tamil Nadu, India

Chapter 1
Wave Energy Potential

V. Sundar and S. A. Sannasiraj

Abstract A brief introduction to Ocean energy followed by an outlook on the wave energy potential around the globe and a comprehensive discussion on the distribution of wave energy potential along the coast of Indian peninsula are presented in this chapter. The wave energy distribution along the coastal region is expected to facilitate the understanding and planning process of installation of new energy devices. Further the importance of wave energy and its impact on the environment are reported.

1.1 World Energy Outlook

The conventional sources of energy have been depleting, forcing researchers to look for alternate green energy sources, of which Ocean energy could be a viable option. The techniques to optimize energy extraction and harness the energy from the ocean gain a great deal of attention worldwide. This shall be complemented by pilot projects focusing on demonstration and environmental monitoring. The World energy Outlook (WEO) has three main scenarios, "The Current Policies Scenario" (CPS), "The New Policies Scenario" (NPS), and the "450 parts per million of carbon dioxide equivalent" (ppm CO_2 eq.). Figure 1.1 shows the shares of primary energy sources in global energy demand from the year 2010–2035. Renewable energy sources continue to increase from 13 to 27%, while nuclear energy will almost remain constant.

1.2 Ocean Energy

Ocean energy can be broadly harnessed from either the thermal energy transferred to the water by the sun or the physical power driven by waves and tides. The method of using thermal energy is otherwise known as Ocean Thermal Energy Conversion

V. Sundar (✉) · S. A. Sannasiraj
Department of Ocean Engineering, Indian Institute of Technology Madras, Chennai, India
e-mail: vsundar@iitm.ac.in

A. Samad et al. (eds.), *Ocean Wave Energy Systems*, Ocean Engineering & Oceanography 14, https://doi.org/10.1007/978-3-030-78716-5_1

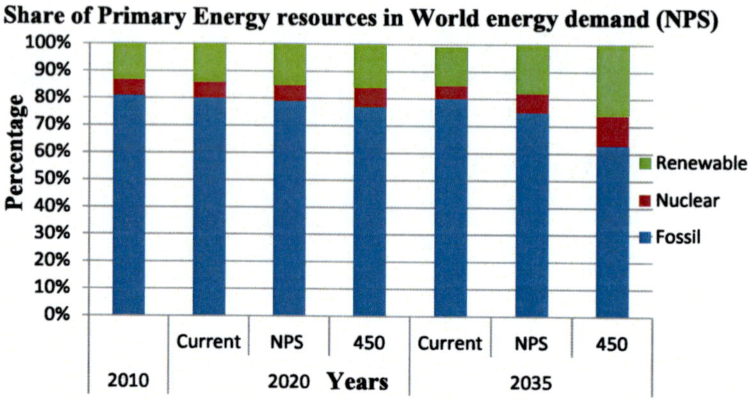

Fig. 1.1 World energy outlook—key global projections [5]

(OTEC). About 70% of the earth's surface area is covered by ocean waters. Thus it absorbs most of the solar energy, which in turn can be converted to electricity.

The three major conversion systems under OTEC are namely closed-cycle, open-cycle, and hybrid systems. In closed-cycle system a liquid (working fluid) with a low boiling point, which can be vaporized by the warmth of the ocean surface water, is employed. The expansion due to liquid vaporization drives a turbine to generate electricity. The seawater is brought to boiling temperature by operating at low pressures. The steam thus generated drives a turbine in an open system. A hybrid system is a combination of both open and closed systems.

The energy obtained from ocean waters is drastically different from that of the OTEC. The tides are caused due to the gravitational pull of the celestial bodies, whereas the waves are generated by wind. Thus, it is understood that tidal and wave energies are intermittent and abundant in nature, whereas the source of thermal energy is limited, and it varies significantly across different locations.

The tidal energy is conventionally harvested by creating a reservoir through a dam-like structure that allows passage of water with force through turbines, which runs an electrical generator. Instead, the tide-induced currents, which is a major underwater current could also be utilized. The three fundamental ways to harness wave energy are i) Wave overtopping systems, where the potential head of the overtopped volume of water head in the reservoir is higher than that of the ocean still water level, ii) The rise and fall or oscillatory motion of floating objects can drive hydraulic pumps to generate electricity by induced electromagnetic field theory, iii) oscillating water column adopts the rise and fall of water level within an enclosed chamber which in turn compresses the air above the water level that drives the turbine to generate electricity.

Marine renewable energy is the most sought-after resource to replace conventional, depleting fossil fuels. The 7000 km of the Indian coastline is not explored sufficiently to use marine renewable energy. The study at Vizhinjam, remains as a

Fig. 1.2 Tapping of ocean energy – efforts in India

pilot plant without any significant breakthroughs in terms of structural optimization and Power take off (PTO) arrangements. Some comprehensive efforts to exploit marine energy in India are shown in Fig. 1.2. It is to be noted that there could be an increase in research and development in this front among the developing countries. In contrast, a few of the developed nations have explored the ocean energy resources vastly and highly successful pilot plants have been commissioned. The contents of this chapter explore the various wave and tidal energy conversion technologies. This is done with the view of new knowledge gained in other parts of the world and in India in terms of technology advancement and improved knowledge on the coastal waters of India. The Department of Ocean Engineering, IIT Madras, being the original investigators of the Vizhinjam plant are in the right position with their extensive contributions to the coastal zone development and management, including disaster mitigation.

1.3 Environmental Impacts

Rapid development in the field of ocean energy is witnessed worldwide through innovative technologies and modern age devices to harness the abundant energy resource. A few of the said technologies are tested out in real-field with the aid of pilot plants. In the modern-day and age striving to achieve sustainability, these technologies should come hand in hand with environmental conservation.

Owing to the lack of commissioning of pilot plants, it is difficult to quantitatively assess the environmental impacts of wave energy conversion technologies. The parameters such as shore processes dictating the shoreline stability, marine pollution, water quality, navigation, fisheries, flora and fauna, ports and harbour structures, coastal population, recreation, tourism are to be thoroughly investigated prior to any such project implementation.

1.4 Tidal Datum

The water level in the ocean rises and falls and oscillates over the mean Sea level, MSL, due to the gravitational attraction between the celestial bodies. It would take a few hours to reach the high tide (highest level) and fall towards low tide (lowest level) in a day. Therefore, to plan any activity such as data recording, construction/commissioning of structures, maintenance/repairs of structures, vessel maneuvering etc., it is essential to understand the water level changes as per tidal fluctuations. More often, planners and engineers refer to the tidal charts and survey datum to plan engineering works. The tidal datum is exclusively site-specific information that serves as a reference to measure the localized water level variations. All tidal datum is established with respect to fixed benchmarks, which is easy to recover. The coastal zone regulation authority/agency adopts the tide datum to demarcate high seas boundaries, territorial sea, exclusive economic zones, state- and privately-owned land. A Schematic representation of the different tidal datum levels viz. HAT (Highest Astronomical Tide), LAT (Lowest Astronomical Tide) and MSL (Mean Sea Level) are shown in Fig. 1.3. In addition to the above levels, the storm surge and wave run-up need to be considered for the survival of any structure as well as its impact on the coastal environment. The comprehensive definitions of tidal datum are maintained by the National Hydrographic office of India.

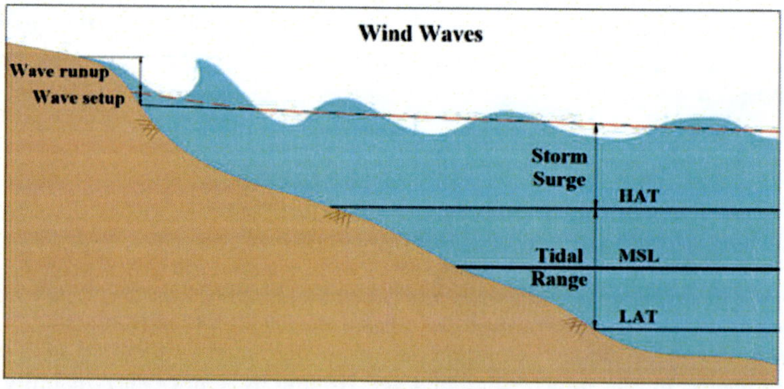

Fig. 1.3 A sketch of tide induced water levels at shoreline [4]

1.5 Importance of Wave Energy

The most interesting and promising among the forms of ocean energy is from wave motion. During the process of wave power extraction, it does not create any waste; does not emit CO_2; is without any noise pollution; and is also environmentally friendly. Unlike most of the renewable energy resources, wave energy can produce power throughout the year. The wave energy that is highly concentrated near the ocean surface is stored in the oceans worldwide. Out of the abundance of energy stored in waves, only a small part of it is used for commercial electricity generation today. The largely untapped resource could play an important role not only in compensating for the depleting energy sources but also will be an answer to the ever-increasing demand for electricity. It is clean, predictable than other renewable resources such as wind and solar.

Prior to planning installation of wave energy convertors, a detailed investigation needs to be carried out to understand the available location-wise wave power potential. The available wave power depends on the wave period and its height. Further, the seasonal variation is also important. One should not be carried away if the wave power potential is high only over a short duration within a year. An ideal site should have consistent power availability over the entire year. The available wave power has been evaluated from the wave characteristics for different locations along the Indian coastline. The wave power potential thus obtained is used to identify possible sites for the installation of wave energy conversion plants/devices to generate electricity from waves [7].

1.6 Wave Power Potential

1.6.1 *Methods of Evaluation*

It is mandatory to have in-depth knowledge on the wave climate and wave power potential. The design of a wave energy converter should consider the wave climate for its operation and efficiency and the extreme wave characteristics for its survival in a harsh environment. This chapter deals with the wave power potential.

The wave power per unit wave front in deep water for a regular sine wave is given by,

$$J = \frac{\rho T (gH)^2}{64\pi} = \alpha H^2 T \tag{1.1}$$

In this expression 'ρ' is the mass density of water, g, the acceleration of gravity, T, the period of the wave, H, the wave height and the coefficient, $\alpha = 0.98\ \text{kW/m}^2$.The wavelength, L in terms of its period is given by the equation,

$$L = \frac{gT^2}{2\pi} = 1.56T^2 \quad (1.2)$$

For irregular waves, the wave power per unit frontage is,

$$J = \alpha_J H_{mo}^2 T_J \quad (1.3)$$

where, $\alpha_J = \alpha/2 = 0.49$ kW/m^2 in deep water; H_{mo} is the significant wave height; and T_J is the energy transport period. H_{mo} and T_J are defined by means of the so-called wave spectrum. It is observed that in most cases, H_{mo} is close to the average wave height of the largest 1/3 of observed waves in a record while T_J is approximately 20 percent longer than the zero-up crossing period T_z which is defined as the average time interval between successive crossings of the mean water level as the fluid moves in the upward direction.

The maximum power, which may be absorbed by a heaving axisymmetric point absorber, is

$$p_{\max} = \frac{\rho}{2}\left(\frac{gT}{4\pi}\right)^3 H^2 = \left(\frac{L}{2\pi}\right)J \quad (1.4)$$

P_{max} is the incident power of a wavefront of width equal to the wavelength divided by 2π and this width is termed as the absorption width or capture width. The device is capable of maximum power absorption, while the device produces a wave displacement oscillation of optimum phase and amplitude. The resonance of the oscillating system with the incident waves will lead to the optimum phase; when the magnitude of absorbed and radiated power is almost similar, the optimum amplitude is achieved.

Another theoretical upper limit, derived for a heaving point absorber as reported by Budal and Falnes [1], relates the maximum power output to the physical dimension of the devices. Based on their analysis, it is reported that the ratio of the absorbed power to the oscillating volume displacement amplitude should be less than $\pi\rho g H/4T$.

$$\left(\frac{P}{V}\right)_{\max} = \left(\frac{\pi}{4}\right)\frac{\rho g H}{T} \quad (1.5)$$

The above expression implies that the wave energy converters can be economical if the size of the point absorbers are reduced for a given power level, thereby effectively harnessing only a portion of the available energy.

1.6.2 *Estimated Wave Power Potential*

1.6.2.1 Across the Globe

There is approximately 8000–80,000 TW/year of wave energy in the entire ocean and on average about 1–10 TW, with each wave crest transmitting 10–50 kW per meter. The energy levels in different parts of the world are depicted in Fig. 1.4. From an investigation of global wave energy resources, Cornett [2] derived from analysis of wave climate predictions generated by the WAVEWATCH-III (NWW3) wind-wave model of Tolman [8, 9] spanning a 10-year period from 1997 to 2006. The global distribution of annual mean wave power thus derived is presented in Fig. 1.5. The annual mean wave power is highest in the higher latitudes of the southern hemisphere (between 40°S and 60°S), particularly in the southern Indian Ocean around the Kerguelen Archipelago and near the southern coasts of Australia, New Zealand, South Africa, and Chile; in the North Atlantic south of Greenland and Iceland and west of the U.K. and Ireland; and in the North Pacific south of the Aleutian Islands and near the west coast of Canada and the U.S. states of Washington and Oregon. The maximum annual mean wave power in the southern hemisphere is ~ 125 kW/m, found southwest of Australia near 48°S, 94°E. In the northern hemisphere, the annual mean wave power south of Iceland exceeds 80 kW/m around 56°N, 19°W, while the maximum in the North Pacific is ~ 75 kW/m, found near 41°N,174°W. It is important to note that wave power estimates presented herein describe the energy flux due to wave propagation and that only a fraction of the energy flux available at any site can be captured and converted into more useful forms of energy. It is also important to recognize that these estimates are less reliable in shallow coastal waters.

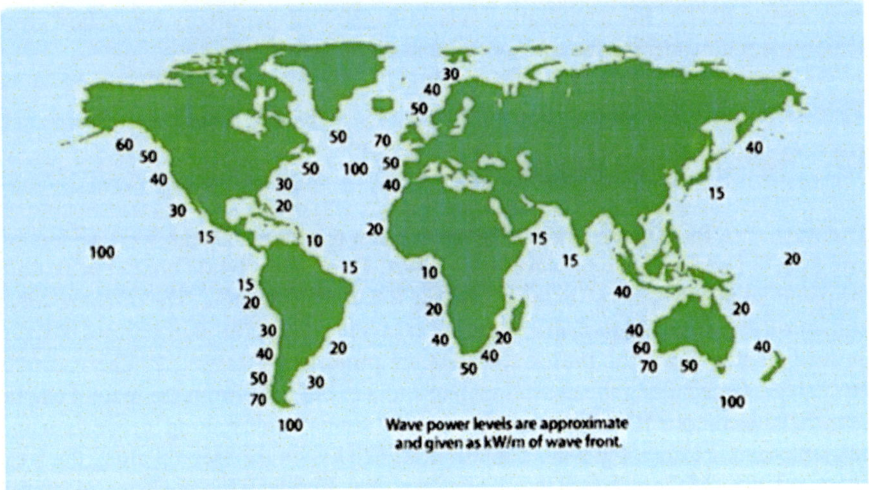

Fig. 1.4 Energy levels around the globe [10]

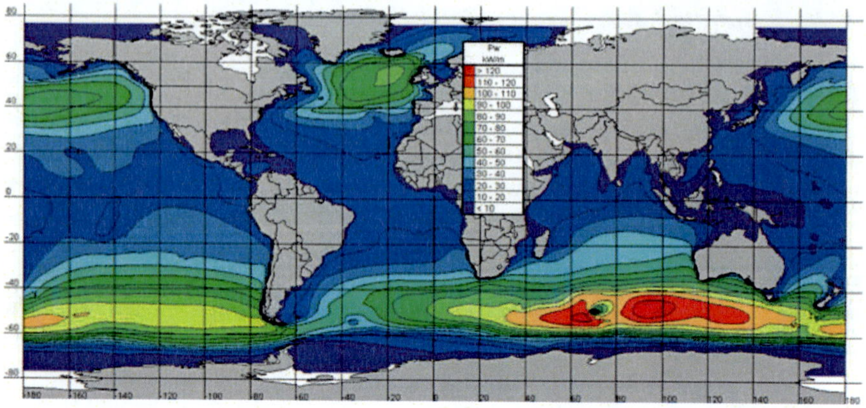

Fig. 1.5 Global annual mean wave energy levels predicted with WAVE WATCH III [2]

1.6.2.2 Indian Scenario

One of the earliest works reported on the distribution of wave power potential along the Indian coast was due to Narasimha Rao and Sundar [3]. They employed the data from the National Institute of Oceanography, NIO. The data was collected from ships and Indian daily weather reports covering the period 1968–1973. For understanding the wave statistics, this amount of data would suffice. The season-wise distribution of mean wave height, wave period, and the power potential along the Indian coast are reported in Table. 1.1.

Wave energy is certainly one such alternative, with an estimated worldwide power of 20,000 TW/year. In India, the potential wave power harvested is approximately estimated as 40 GW. Even harvesting at least 10–20% of this energy would be a great achievement considering the energy demand that is persisting.

1.7 Wave Energy Map for INDIA

The wave data for 10 years from 1993 to 2002 in the Indian Ocean were generated using the 3rd generation win-wave model (WAM) [6]. This data is crucial to explore the potential of wave energy extraction along the Indian coastline. The offshore wave climate off Indian coasts has been extracted at salient points. The distributions of wave power potential along the Indian coastline are projected in Table 1.2. These details are also presented along the coastal map of India in Fig. 1.6. From the map, it can be seen that the contour 10–15 kW/m is distributed almost evenly along the western and eastern coasts. Further, the wave contour of 15–20 kW/m is observed along the west coast, off viz., Maharashtra, Goa, Karnataka, and Kerala. This presence of higher power along the west coast could probably be due to the strong waves during SW

Table 1.1 Wave power potential along the Indian coastline [3]

Location	South-West Monsoon			North-East Monsoon			Non-monsoon		
	Mean wave Ht (m)	Mean wave period (s)	Wave power kW/m	Mean wave Ht (m)	Mean wave period (s	Wave power kW/m	Mean wave Ht (m)	Mean wave period (s)	Wave power kW/m
10–15°N and coast-85°E (off madras)	1.71	0.80	16.62	1.53	5.86	13.44	1.14	5.5	7.00
15–20°N and coast-85 E (off Visakhapatnam)	2.04	8.25	33.65	1.60	6.28	15.75	1.24	7.10	10.70
20–25°N and coast-85–95E(off Kolkata)	1.96	7.66	28.84	1.33	8.01	13.88	1.72	6.51	18.87
5–10°N and coast-75–80°E (off cape comorin)	1.78	6.29	19.52	1.22	5.35	7.80	1.29	5.46	8.90
10–15°N and coast-70°E (off cochin)	2.03	6.77	27.34	1.03	5.05	5.25	1.01	5.38	5.37
15–25°N and coast-70°E (off Bombay)	2.63	6.93	46.98	1.00	5.00	4.90	1.01	5.25	5.24

Table 1.2 Wave energy potential

Sr. No.	Places	State	Lat	Lon	H_{max} (m)	H_{av} (m)	Hs (m)	T_{mean} (s)	q_{mean} (°)	Power (kW/m)
1	Madhi	Gujarat	22.00	69.00	3.57	1.25	1.98	7.21	232	13.93
2	Gojines	Gujarat	22.00	69.13	1.35	0.47	0.75	3.5	233	0.97
3	Porbandar	Gujarat	21.50	69.50	3.40	1.19	1.89	7.33	228	12.83
4	Mangrol	Gujarat	21.00	70.00	3.46	1.21	1.92	6.89	242	12.47
5	Khambhaliya	Gujarat	21.01	70.12	1.35	0.47	0.75	3.5	233	0.97
6	Jafrabed	Gujarat	20.42	71.50	1.35	0.47	0.75	3.5	248	0.97
7	Narpad	Gujarat	20.00	71.51	1.35	0.47	0.75	3.5	248	0.97
8	Haveli	Gujarat	20.00	72.50	2.43	0.85	1.35	5.73	232	5.12
9	Virar	Gujarat	19.50	72.50	2.91	1.02	1.62	5.97	243	7.68
13	Bandra	Mumbai	19.00	72.57	1.35	0.47	0.75	3.5	248	0.97
10	Pune	Maharastra	18.50	72.50	3.51	1.23	1.95	6.99	252	13.07
11	Guhagar	Maharastra	17.50	73.00	3.51	1.23	1.95	7.68	254	14.36
12	Ratnagiri	Maharastra	17.00	73.00	3.89	1.36	2.16	7.61	258	17.40
14	Shrivardhan	Maharastra	18.00	72.84	1.35	0.47	0.75	3.5	248	0.97
15	Jakimiraya	Maharastra	17.00	73.18	1.35	0.47	0.75	3.5	263	0.97
16	Rajapur	Maharastra	16.50	73.00	4.11	1.44	2.29	7.78	259	19.94
17	Kudal	Maharastra	16.00	73.00	4.26	1.49	2.37	8	260	21.95
18	Tarkarli	Maharastra	16.00	73.39	1.35	0.47	0.75	4.5	263	1.24
19	Panaji	Goa	15.50	73.50	3.91	1.37	2.17	8.24	251	19.12
20	Agonda	Goa	15.00	73.87	1.35	0.47	0.75	4.5	263	1.24
21	Gokarna	Karnataka	14.50	74.00	3.89	1.36	2.16	8.38	247	19.16

(continued)

Table 1.2 (continued)

Sr. No.	Places	State	Lat	Lon	H_{max} (m)	H_{av} (m)	Hs (m)	T_{mean} (s)	q_{mean} (°)	Power (kW/m)
22	Tenginagundi	Karnataka	14.00	74.39	1.35	0.47	0.75	4.5	263	1.24
23	Barkur	Karnataka	13.50	74.50	3.60	1.26	2.00	8.33	249	16.35
24	Surathkal	Karnataka	13.00	74.50	4.03	1.41	2.24	8.32	253	20.45
25	Sasihithlu	Karnataka	13.00	74.68	1.35	0.47	0.75	4.5	263	1.24
26	Kasaragod	Kerala	12.50	74.50	4.26	1.49	2.37	8.43	253	23.13
27	Taliparambha	Kerala	12.00	75.00	3.86	1.35	2.14	8.49	247	19.13
28	Ettikulam	Kerala	12.00	75.10	1.35	0.47	0.75	4.5	263	1.24
29	Payyoli	Kerala	11.50	75.50	3.34	1.17	1.86	8.27	246	13.99
30	Parappanangadi	Kerala	11.00	75.73	1.35	0.47	0.75	4.5	263	1.24
31	Calicut	Kerala	11.00	75.50	4.06	1.42	2.25	8.52	247	21.24
32	Thrissur	Kerala	10.50	76.00	0.74	0.26	0.41	9.26	241	0.77
33	Vypin	Kerala	10.00	76.07	1.35	0.47	0.75	4.5	278	1.24
34	Cochin	Kerala	10.00	76.00	3.94	1.38	2.19	8.68	239	20.43
35	Allepey	Kerala	9.50	76.00	4.14	1.45	2.30	8.79	242	22.84
36	Thurayilkunnu	Kerala	9.00	76.42	1.35	0.47	0.75	5.5	263	1.52
37	Kollam	Kerala	9.00	76.50	2.86	1.00	1.59	8.49	233	10.49
38	Trivandrum	Kerala	8.50	76.50	4.31	1.51	2.40	8.9	230	25.08
39	Nagerkoil	Tamil Nadu	8.00	77.00	4.63	1.62	2.57	8.73	207	28.32
40	Kanniyakumari	Tamil Nadu	8.00	77.50	4.31	1.51	2.40	8.3	197	23.39
41	Koodankulam	Tamil Nadu	8.00	78.00	4.74	1.66	2.63	8.38	191	28.54
42	Tiruchendur	Tamil Nadu	8.50	78.50	3.89	1.36	2.16	8.05	164	18.40

(continued)

Table 1.2 (continued)

Sr. No.	Places	State	Lat	Lon	H_{max} (m)	H_{av} (m)	Hs (m)	T_{mean} (s)	q_{mean} (°)	Power (kW/m)
43	Tuticorin	Tamil Nadu	9.00	78.50	3.26	1.14	1.81	7.91	167	12.71
44	Devipatnam	Tamil Nadu	9.50	79.00	2.34	0.82	1.30	6.69	168	5.56
45	Keraikudi	Tamil Nadu	10.00	79.50	0.46	0.16	0.25	8.5	161	0.27
46	Velakkani	Tamil Nadu	10.50	80.00	2.91	1.02	1.62	6.79	133	8.73
47	Kottuchery	Tamil Nadu	11.00	79.94	1.35	0.47	0.75	3.5	158	0.97
48	Karaikkal	Tamil Nadu	11.00	80.00	3.03	1.06	1.68	7.16	131	9.94
49	Pitchavaram	Tamil Nadu	11.50	80.00	3.23	1.13	1.79	7.5	131	11.84
50	Puducherry	Tamil Nadu	12.00	80.00	3.00	1.05	1.67	7.77	135	10.59
51	Kalpakkam	Tamil Nadu	12.50	80.50	3.54	1.24	1.97	7.83	134	14.88
52	Chennai	Tamil Nadu	13.00	80.50	3.37	1.18	1.87	7.71	136	13.27
53	Pulicat	Andhra Pradesh	13.50	80.50	3.40	1.19	1.89	7.88	138	13.79
54	Pambali	Andhra Pradesh	14.00	80.33	1.35	0.47	0.75	3.5	128	0.97
55	Durgarajupatnam	Andhra Pradesh	14.00	80.50	3.40	1.19	1.89	7.92	140	13.86
56	Krishnapatnam	Andhra Pradesh	14.50	80.50	3.43	1.20	1.90	7.95	144	14.15
57	Mulampeta	Andhra Pradesh	15.00	80.19	1.35	0.47	0.75	3.5	143	0.97
58	Chakicherla	Andhra Pradesh	15.00	80.00	1.83	0.64	1.02	6.7	159	3.39
59	Ammanabrole	Andhra Pradesh	15.50	80.50	2.31	0.81	1.29	7.5	156	6.08
60	Palakayatippa	Andhra Pradesh	16.00	81.32	1.35	0.47	0.75	3.5	158	0.97
61	Machilipattinam	Andhra Pradesh	16.00	81.50	3.29	1.15	1.83	8.08	155	13.21
62	Yanam	Andhra Pradesh	16.50	82.50	3.66	1.28	2.03	8.13	159	16.47
63	Chintapalle	Andhra Pradesh	17.00	82.43	0.45	0.16	0.25	3.5	143	0.11

(continued)

Table 1.2 (continued)

Sr. No.	Places	State	Lat	Lon	H_{max} (m)	H_{av} (m)	Hs (m)	T_{mean} (s)	q_{mean} (°)	Power (kW/m)
64	Kakinada	Andhra Pradesh	17.00	82.50	3.26	1.14	1.81	8.08	160	12.98
65	Pudimadaka	Andhra Pradesh	17.50	83.00	2.86	1.00	1.59	7.68	165	9.49
66	Konada	Andhra Pradesh	18.00	83.50	2.46	0.86	1.37	7.55	169	6.90
67	Penumatti	Andhra Pradesh	18.00	83.76	1.35	0.47	0.75	4.5	173	1.24
68	Naupada	Andhra Pradesh	18.50	84.50	3.51	1.23	1.95	7.92	169	14.81
69	Pukkalapalyam	Andhra Pradesh	18.98	1.35	1.35	0.47	0.75	4.5	188	1.24
70	Kaviti	Andhra Pradesh	19.00	85.00	3.54	1.24	1.97	7.87	171	14.96
71	Malud	Orissa	19.50	85.50	3.40	1.19	1.89	7.77	173	13.60
72	Nuliasahi	Orissa	20.00	86.50	2.91	1.02	1.62	7.3	174	9.39
73	Mankadakhia	Orissa	20.00	86.53	1.35	0.47	0.75	3.5	188	0.97
74	Kendrapara	Orissa	20.50	87.00	2.37	0.83	1.32	7.14	173	6.08
75	Bideipur	Orissa	21.00	87.00	1.49	0.52	0.83	5.62	175	1.88
76	Chandipur	Orissa	21.00	87.41	1.35	0.47	0.75	3.5	188	0.97
77	Dantan	West Bengal	21.50	87.50	1.03	0.36	0.57	6.95	181	1.11
78	Nandigram	West Bengal	22.00	88.00	1.23	0.43	0.68	4.36	196	1.00
79	Bakhkhali	West Bengal	21.50	88.50	2.31	0.81	1.29	6.38	177	5.17
80	Sundarban	West Bengal	22.00	89.00	1.94	0.68	1.08	5.78	186	3.30
81	Khulna District	Bangladesh	22.00	89.50	0.51	0.18	0.29	8.53	195	0.34

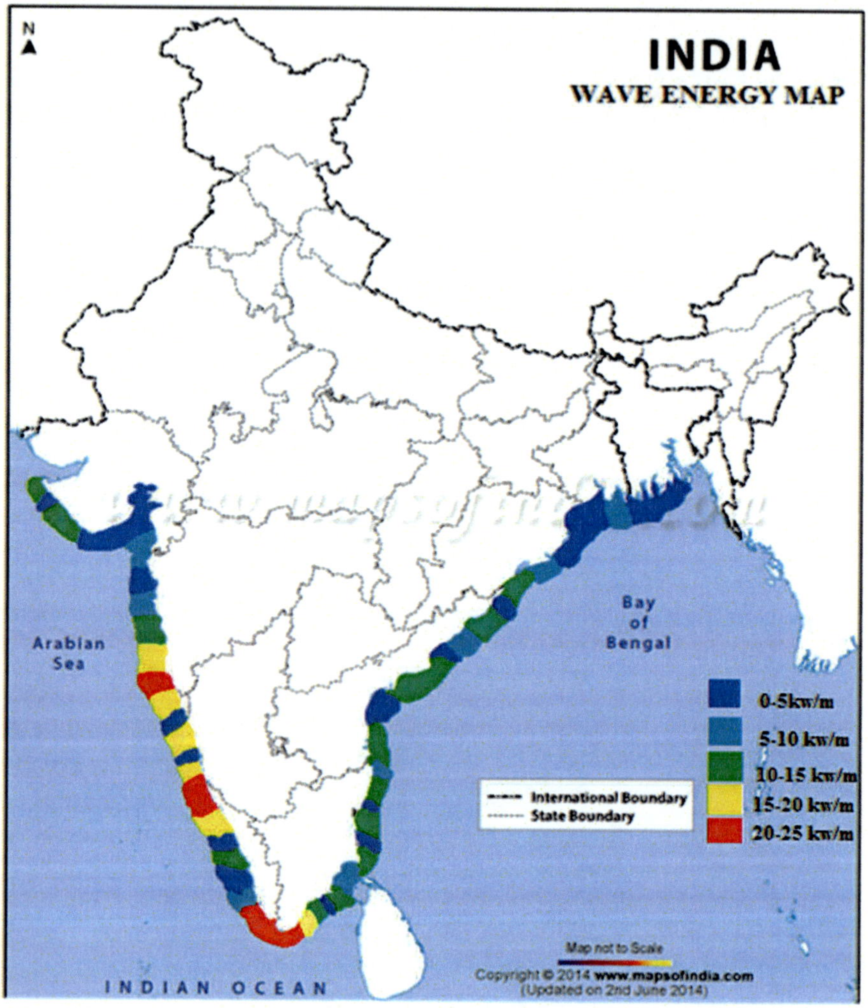

Fig. 1.6 Wave energy map for India

monsoon. Maximum wave power is obtained at the southern tip of Indian peninsula (Kanyakumari, Nagercoil district, Koodankulam), which could be due to the effect of refraction and the presence of strong winds prevailing in the region.

The distribution of wave power in a finite range and the length of coastline over which it is spread are grouped and presented in Table 1.3. It is also projected as pictorial representation, i.e., the distribution along the length of the coast and total power available in each power contours in Figs. 1.7 and 1.8 respectively. The wave energy concentration along the coastal stretch of different maritime states is projected in Table 1.4. It is to be noted that the revised estimate shows the total power available

Table 1.3 Wave power contour

Contour power level (kW/m)	Contour length (km)	Total power flux crossing contour (GW)
0–5	1530	3.825
5–10	822	6.165
10–15	1634	20.425
15–20	665	11.64
20–25	400	9

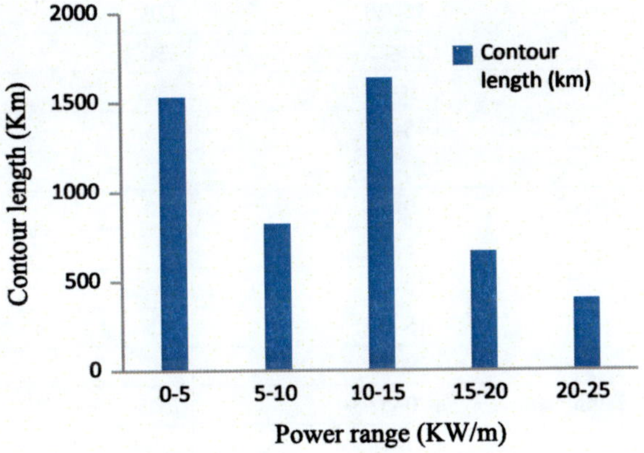

Fig. 1.7 Distribution of wave energy along different latitudes of Indian coast

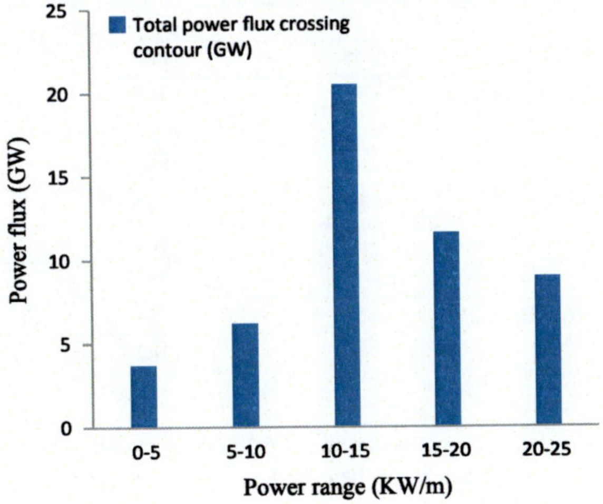

Fig. 1.8 Total power flux available along different latitudes of Indian coast

Table 1.4 Wave power contour along different maritime states

State	Contour length (km)	Total power flux crossing contour (GW)	Contour power level (kW/m)
Gujarat	0–5	465	1.2
	5–10	110	0.8
	10–15	325	4.1
Maharashtra	0–5	90	0.2
	5–10	115	0.9
	10–15	120	1.5
	15–20	210	3.7
	20–25	130	2.9
Karnataka	0–5	130	0.3
	15–20	215	3.8
	20–25	95	2.2
Kerala	0–5	130	0.3
	5–10	60	0.5
	10–15	65	0.8
	15–20	125	2.2
	20–25	85	1.9
Tamil Nadu	0–5	75	0.1
	5–10	110	0.8
	10–15	529	6.6
	15–20	115	2.0
	20–25	90	2
Andhra Pradesh	0–5	150	0.4
	5–10	85	0.6
	10–15	550	6.9
Orissa	0–5	235	0.6
	5–10	160	1.2
	10–15	45	0.6
West Bengal	0–5	255	0.6
	5–10	182	1.4

along the coastline is about 50 GW. However, by considering wave power above 10 kW/m which can be efficiently harvested, the total power available would be 41 GW which is matching with the wave estimate in the eighties of 40 GW.

It should further be noted that the entire 41 GW could not be harvested due to many other physical morpho-dynamic factors and site conditions such as water depth etc. A realistic estimate at each site can be made based on a detailed survey along a particular coastal stretch.

References

1. Budal, K., & Falnes, J. (1980). Interacting point absorbers with controlled motion. In B. Count (Ed.), *Power from sea waves* (pp. 381–399). Academic Press.
2. Cornett, A. M. (2008). A global wave energy resource Assessment ISOPE 2008, Vancouver, Canada, paper NO: ISOPE-2008-579.
3. Narasimha Rao, T. V. S., & Sundar, V. (1982). Estimation of wave power potential along the Indian coastline. 0360-5442/82/100839-07503.00/0 1982 Pergamon Press Ltd. Ocean Engineering, 29, Issue 11, 1343–1355.
4. CSIRO & BoM. (2007). Climate change in Australia, CSIRO, Bureau of Meteorology, and the Australian Greenhouse Office in partnership with the Australian Climate Change Science Program, Canberra. www.climatechangeinaustralia.gov.au/technical_report.php.
5. REN21. Renewable Energy Policy Network & Institute for Sustainable Energy Policies.
6. Sannasiraj, S. A. (2007, July 16–17). Wave prediction off Indian coasts. In *Proceedings of Indo-Japan workshop on Coastal Problems and Mitigation measures–Including the effects of Tsunami*, IIT Madras, India.
7. Sannasiraj, S. A., & Sundar, V. (2016). Assessment of wave energy potential and its harvesting approach along the Indian coast. Renewable Energy.
8. Tolman, H. L. (2002a). User manual and system documentation of WAVE-WATCH III version 2.22. Technical Note, U.S. Department of Commerce, NOAA, NWS, NCEP, Washington, DC.
9. Tolman, H. L., (2002b). The 2002 release of WAVEWATCH III. In *7th International Workshop on Wave Hindcasting and Forecasting, Meteorological Service of Canada, Wave Power: Moving towards Commercial Viability*, 30 November 1999, Broadway House, Westminster, London.
10. Thorpe, T. W. (1992). *A review of wave energy* (Vols. 1 and 2). ETSU Report Number R-72, December 1992.

Chapter 2
Wave Energy Convertors

V. Sundar and S. A. Sannasiraj

Abstract A detailed review of the works carried out in connection with wave energy extraction devices are presented in this chapter. Although, there are a few methods of extracting the energy from waves, the emphasis is laid on Oscillating water column (OWC) devices due to its' more attractive, simple, and proven technology. The classification of the OWC devices based on location, its principle and the different options for its installation is made. If one works out the cost-benefit ratio, it is rather not encouraging to plan for a device to extract energy. Hence, the possibilities of merging OWC with breakwaters would be more viable and cost-effective too. Here again, one should consider the purpose of the breakwater as it may be either for development harbor or coastal protection. With this background, the different options of multi-purposed wave energy devices and, in specific, the conceptual design of integrating the breakwater and OWC are presented in this chapter.

2.1 General

The energy demand is increasing globally and ocean energy is one of the most promising long-term resource. The untapped energy that is being held by the world oceans is tremendous, wherein untiring efforts towards perfecting the technology towards an increase in the efficiency of the system will certainly be rewarding. The impact of wave energy converter technologies on the environment as per Thrope [68], is summarized under Table 2.1. The present analysis is limited to wave energy.

The Advantages of a Wave Energy Converter Are

- An unlimited natural energy source is utilized.
- The energy is resourcefully consumed, which facilitates protecting the coastline from high-impact wave forces.

V. Sundar (✉) · S. A. Sannasiraj
Department of Ocean Engineering, Indian Institute of Technology Madras, Chennai, India
e-mail: vsundar@iitm.ac.in

A. Samad et al. (eds.), *Ocean Wave Energy Systems*, Ocean Engineering & Oceanography 14, https://doi.org/10.1007/978-3-030-78716-5_2

Table 2.1 Environmental impacts of wave energy conversion technologies [68]

Environmental effects	Wave energy devices (Location-wise)		
	Shoreline	Nearshore	Offshore
Land use/sterilization	W		
Construction/Maintenance sites	W		
Recreation	W	W	
Coastal erosion	W	W-M	W-M
Sedimentary flow patterns		W	W
Navigation hazard		W	W
Fish and marine biota	W	W	W
Acoustic noise	W		
Working fluid losses		W	W
Endangered species	W	W	
Device/mooring damage		W-M	W-M

W Weak effect; *M* Medium effect.

- On commissioning the system, several reports state that the investment could be recovered within a short time.
- By incorporating an energy device in the breakwater for the development of artificial harbors, its capital cost could easily be absorbed as it will serve a dual purpose.
- Small scale plants can serve as lighter buoys and in the event of large-scale production of mini plants, they can certainly serve remote coastal villages.
- Apart from integrating with harbor breakwater, it can also be coupled with desalination, an idea being explored by few countries.
- The aesthetic value of the coast is unaffected
- Enhances the growth of flora and fauna.
- Job opportunities as well uplift of the local public and economic development
- Oxygen and hydrogen can be generated by the electrolysis of water.

2.2 Harnessing of Wave Energy

Levels of energy conversion

Primary Conversion The first level for the extraction of energy is through the oscillation induced by the waves on a floating body or inside a chamber of a fixed structure. The oscillating systems should be capable of storing the energy extracted from waves for further conversion.

Secondary Conversion This is the second level of converting the energy stored from primary conversion to electricity. This level involves the utilization of drives and control systems, including controllable values, hydraulic, pneumatic, and electronic components. The secondary conversion is obtained by converting the kinetic/potential energy to rotational energy using a turbine, thereby resulting in the rotation of a shaft.

Tertiary Conversion This is the last and final level in which the rotary motion power is transferred to electric generators which converts the harnessed power into electricity or the rotary motion is directly applied to pump the fluid like in a desalination system.

Classification of wave energy converters

As Per Their Horizontal Size and Orientation
Point absorbers are devices that are small compared to a typical wavelength. This type of systems has a typical power rating of the order of a few hundred kilowatts. Hence, a large power plant would consist of hundreds or perhaps thousands of such units, which need to be dispersed in a very long and relatively narrow array along the coast.

Line absorbers are the counterpart of point absorbers and are essentially elongated floating structures. These are systems of a length which is comparable to or larger than one typical wavelength. They are further classified as a terminator and attenuator. Line absorbers act as terminators if they are aligned along the direction of the wave crests and act as an attenuator if they are aligned normal to the wave crests.

As Per Different Location with Respect to the Coastline
Wave energy devices may be located onshore, nearshore or offshore. Onshore wave energy devices are normally placed on the coastline, over a cliff, while the nearshore wave energy devices are located in the shallow waters, which are normally within 15 km distance from the shoreline, whereas, offshore systems were installed in deep waters.

As Per Their Locations with Respect to the Mean Water Level
In general, most of the wave energy converters are partially submerged. Some wave energy converters are fully submerged, freely floating or resting on the seabed and piercing the free surface.

2.3 Conversion Process

The different process by which energy in waves can be converted to electrical energy was discussed by Hagerman [21], and later updated by Brooke [9]. The variation from one concept to another depends on the mode of oscillation for the energy absorption, and type of reaction point. These oscillations are utilized to absorb the energy as shown in shore based, nearshore and offshore devices in Fig. 2.1a–c, respectively.

Wave Energy Conversion Process	Concept
Fixed oscillating water column	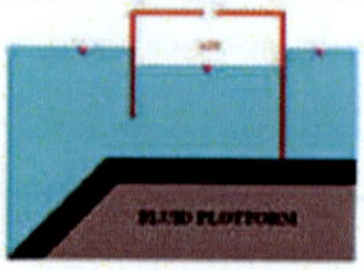
Reservoir filled by wave surge	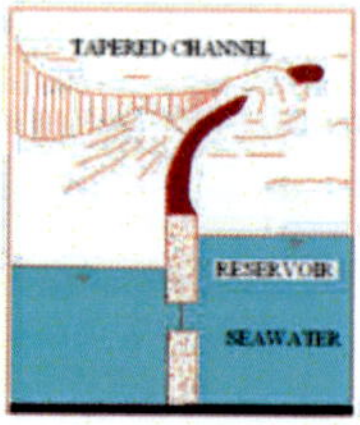
Pivoting flaps	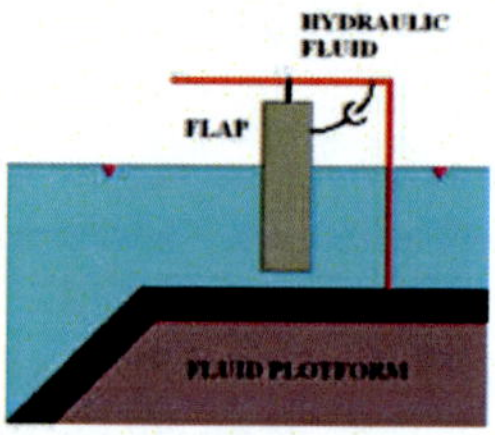

Fig. 2.1 **a** Classifications of wave energy devices (Onshore). **b** Classifications of wave energy devices (Nearshore). **c** Classifications of wave energy devices (Offshore)

The generic types of wave energy harnessing schemes are illustrated in the above said figures.

One of the earliest floating devices is the backward bent duct buoy (BBDB). The BBDB consists of a long water filled duct, which floats in a horizontal position over the water surface such that the duct opening is positioned away from the incident waves. The other end of the duct is connected to a vertical chamber, the oscillations within the chamber due to the rise and fall of water column, drives a turbine. Lighter buoys operating in shallower waters adopt the BBDB design. The details of the characteristics of BBDB have been discussed by Seymour [49].

2.4 Wave Energy Devices

Depending on the type and their location with respect to the coast and offshore, several devices have been developed to extract the wave energy for conversion into electricity. The devices are referred to as the oscillating water column, hinged contour device,

Freely floating oscillating water column (**Backward bent duct buoy, BBDB**)	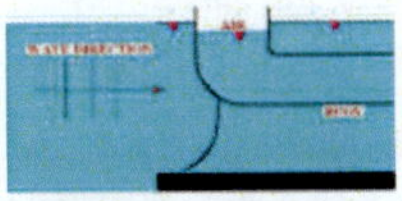
Moored floating oscillating water column (**Backward bent duct buoy, BBDB**)	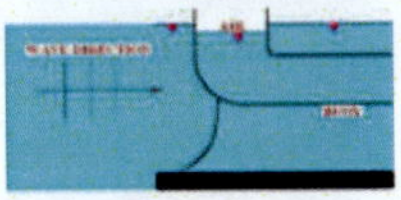
Bottom mounted oscillating water column	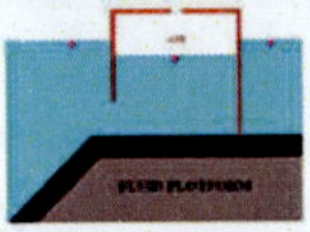
Reservoir filled by direct wave action	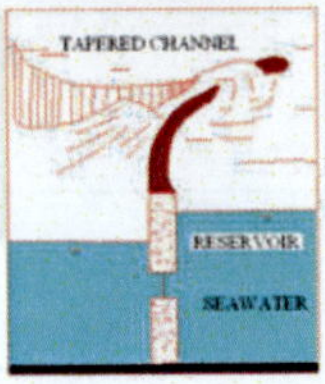
Flexible pressure device	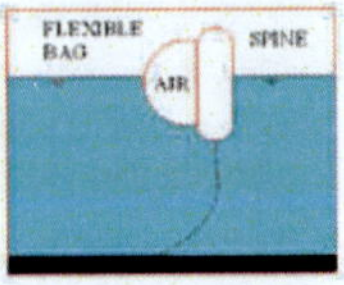
Submerged buoyant absorber with sea floor reaction	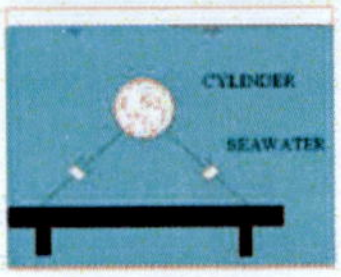
Heaving float in bottom mounted or moored floating caisson	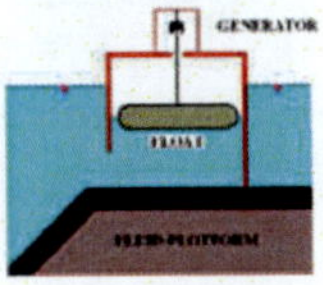

Fig. 2.1 (continued)

buoyant moored device, and overtopping device. Recent innovations of different conversion devices are following the above process.

The oscillating water column device has a chamber with its one end open in a submerged state to the sea and the other end vented to an air turbine. A propagating wave crest acts like a piston that compresses the air inside the chamber, and thus, forcing the air to drive the turbine. When the wave trough is present in the chamber,

<table>
<tr><td>Freely heaving float with sea floor reaction point</td><td></td><td>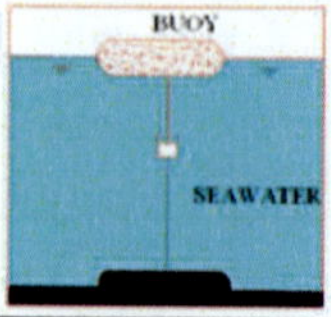</td></tr>
<tr><td>Freely heaving float with mutual force reaction</td><td></td><td>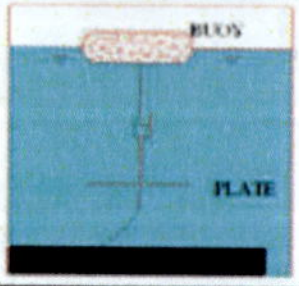</td></tr>
<tr><td>Countering float with mutual force reaction</td><td></td><td>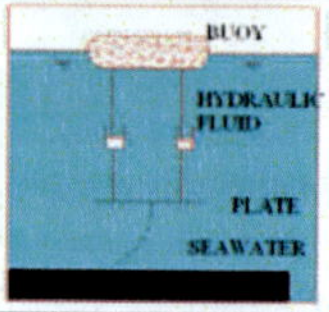</td></tr>
<tr><td>Countering float with sea floor reaction point</td><td></td><td>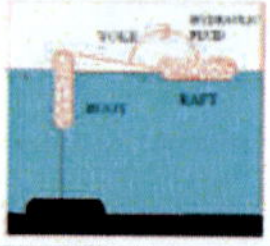</td></tr>
<tr><td>Pitching float with mutual force reaction</td><td></td><td></td></tr>
<tr><td>Flexible bag with spine reaction point</td><td></td><td>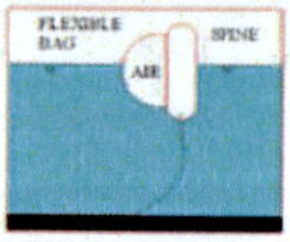</td></tr>
<tr><td>Reservoir filled by direct wave action</td><td></td><td>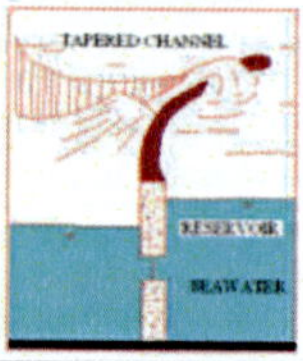</td></tr>
</table>

Fig. 2.1 (continued)

the air is sucked back through the turbine that can rotate in the same direction irrespective of the direction of the air column that is essential to drive an alternator thus, generating *electricity*. Wells turbine is one such turbine for an efficient power extraction from an OWC.

The hinged contour device comprises a series of articulated floating sections moored to the seafloor that is set in motion due to the waves. The hinges in turn linked to hydraulic rams, drive hydraulic motors to produce electricity.

The buoyant moored device is a floating type with the main structure responsible for the conversion of energy from the ocean waves to electrical following the wave motion. The device is anchored to the seafloor. Although, different mechanisms are employed electricity generation through turbines, the basic mechanism is the application of hydraulic pumps or pumps supplying seawater under pressure to drive the turbines.

The overtopping device has a ramp over which the waves propagate and overtop, creating a head of water that could probably be used to run a turbine. These devices are usually floating types but can also be of fixed shore-based type.

It is to be mentioned herein that the biggest advantage of OWC devices is that the turbine alone is the only moving part, which makes it more attractive than all other systems. Hence, it is envisaged to provide more emphasis on OWC in this chapter.

Oscillating Water Column (OWC) Device

The OWC wave energy device consists of a chamber of the pyramidal top, through which the air column above the water surface oscillates due to the waves penetrating the chamber, as shown in Fig. 2.2. The lip wall facing the waves allows them to propagate under it into the chamber. When the crest is inside the chamber, the air column above is pushed upwards through the duct to drive a turbine placed in the cylindrical duct. When the trough is present inside the chamber, the air column inside the chamber is pushed down and hence there is a reversal in the direction of airflow above the water column. This sets an oscillation of the air column depending on the characteristics of the incoming ocean waves which dictates the magnitude of energy that can be captured by the system. A special turbine-like Wells turbine mounted on top of the duct needs to be designed to rotate in the same direction irrespective of the direction of airflow. The power is thus continuously generated irrespective of the airflow direction. A generator is coupled to the turbine shaft that produces electricity by rotating its armature shaft.

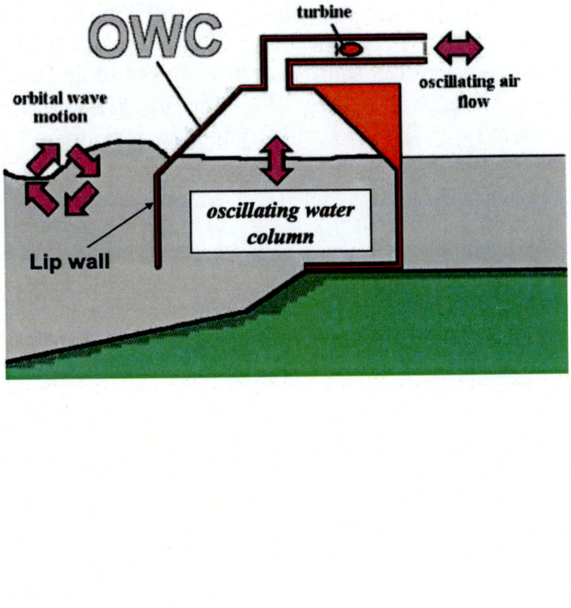

Fig. 2.2 Oscillating water column (OWC) device

OWC Design

The site-specific incident wave characteristics and local water depths dictate the design parameters for the OWC's air chamber. If the size of the chamber is incorrectly sized, it could lead to waves resonance inside the air chamber, this effect results in net-zero passage of air through the turbine; preferably, the air chamber dimensions are to be designed to maximize energy capture from the incoming wave climate and must be conducive to air through the turbine. This outcome can be achieved with an aid of a funnel-shaped design, as the chamber narrows down from the water surface towards the turbine, which aids in streamlining the airflow towards the turbine. Numerous concepts and techniques were put forth and tested over the past few decades, amongst which the OWC principle is found to be most promising for efficient wave energy extraction. Most of the wave energy devices have been proved in the field and most of the devices work on the OWC principle. Most of the basic configurations based on this principle are shown in Fig. 2.3.

2.5 Wave Energy Developments and Activities

2.5.1 General

The effort in research and development on this topic has been ever increasing. Several new ocean energy companies have evinced keen interest in the development of new devices/technologies. Few examples of the devices are the Pelamis, the Archimedes Wave Swing, and the Limpet. The classification, along with the installed version of the different wave energy converters, is shown in Fig. 2.4. A brief overview of the energy demonstration projects and prototypes are listed below.

Prototype OWC prototypes were built in Norway (Toftestallen, near Bergen, 1985), Japan (Sakata port, 1990) [42], India (Vizhinjam, near Trivandrum, 1990) [44], Portugal [14], and UK (the LIMPET plant in Islay Island, Scotland) [67]. The largest of the above, a near-shore bottom standing 2 MW plant (named Osprey), was destroyed in 1995 shortly after being towed and sunk into place near the Scottish coast. Smaller shoreline OWC prototypes (also equipped with Wells turbine) were built in Islay, UK (1991) [58], and more recently in China. The Australian company, Energetech developed a technology using a large parabolic-shaped collector to concentrate the incident wave energy (a prototype was tested at Port Kembla, Australia, in 2005) [26]. Falcão [17] has reported that more than 1000 wave energy converters have been designed and developed around the globe. A summary of the OWC devices in the open ocean is provided in Table 2.2.

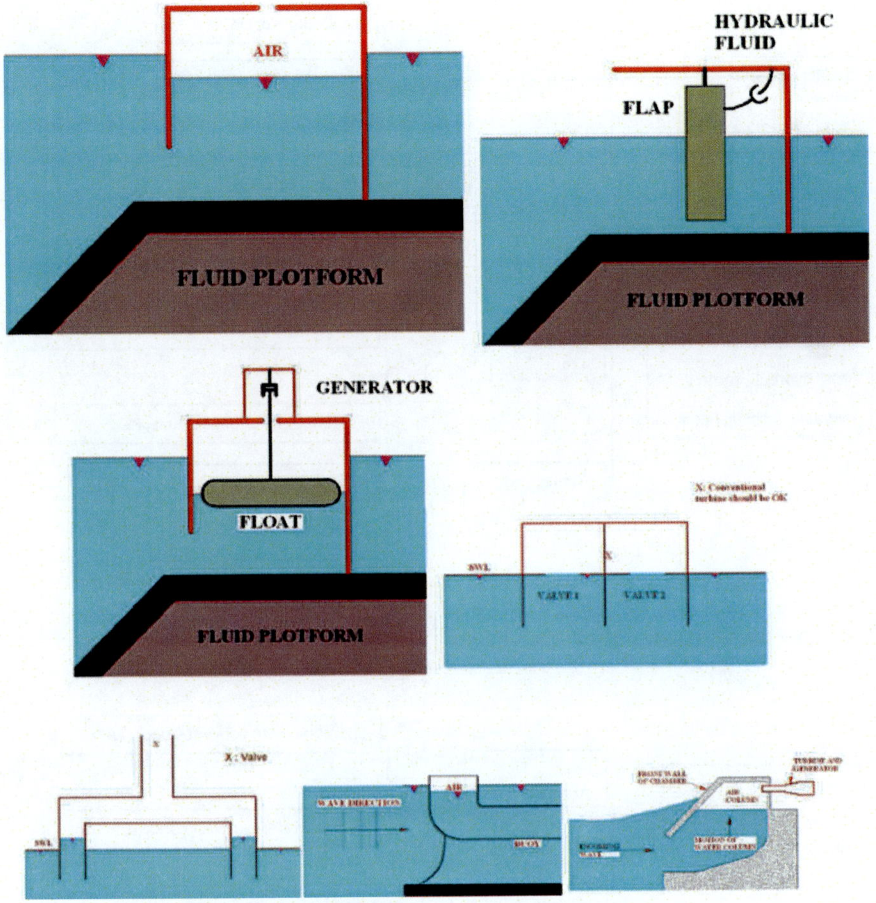

Fig. 2.3 Concepts on the principle of OWC

2.5.2 Shoreline Wave Energy System

These devices do not require complicated installation techniques and are easy to commission and easy to maintain. The two main classifications of devices located on the shoreline are the oscillating water column (OWC) and the tapered channel (TAPCHAN). Another type of shoreline (or bottom-fixed nearshore) based device is the Pendular. It essentially comprises a pendulum-like paddle structure that would oscillate by the incoming waves and thus, rotate the pump. The duration of the pilot project **LIMPET**, Islay, Scotland was 36 months from November 1998 [10].

In 1985, the TAPCHAN (TAPered-CHANnel) was the first overtopping device built in the real field of Toftestallen, Norway with a power capacity of 350 kW. By the principle of operation, TAPCHAN is a fixed type overtopping device. The

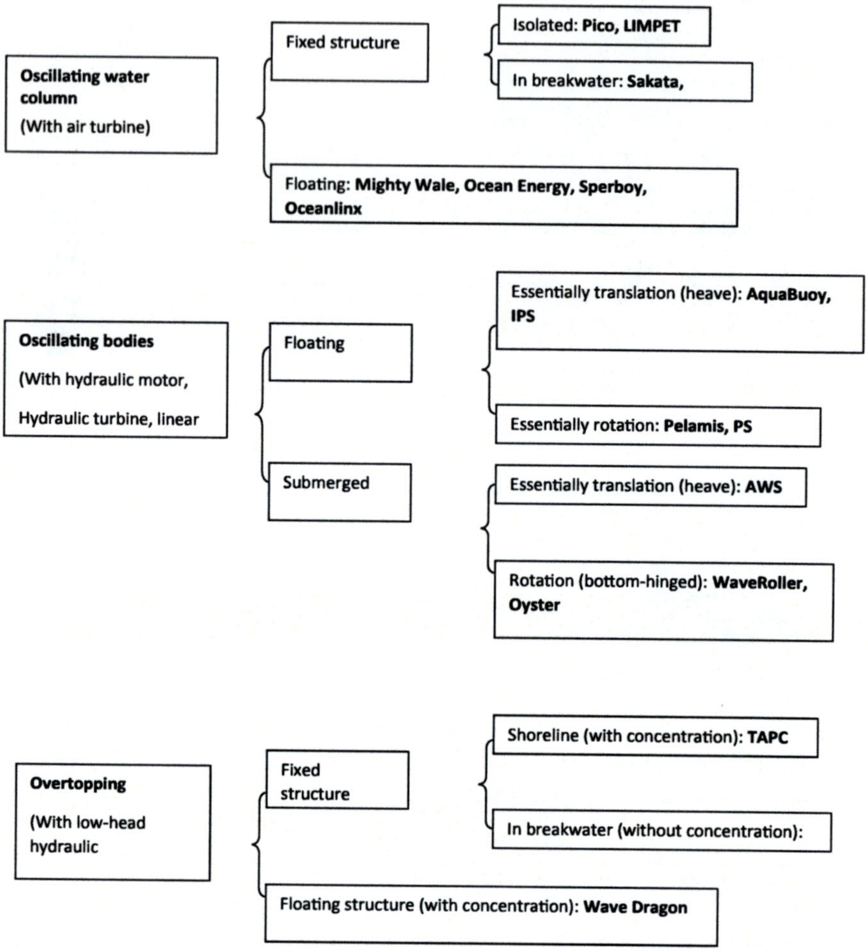

Fig. 2.4 Classification of installed wave energy devices based on working principle [26]

details regarding the device and mode of operation and an aerial view of the real field TAPCHAN device are shown in Fig. 2.5a, b, respectively. The gradually narrowing channel causes wave amplification and this causes water to spill over and fall into the reservoir. From the reservoir, water is led to the turbine for generating electricity. A conventional low head Kaplan axial flow turbine was used to convert the hydraulic energy into electric energy. The only disadvantage of TAPCHAN is that it will work only in a specific environment characterized by a high wave energy flux.

Table 2.2 Summary of OWC devices around the Globe

Location	Type	Rated Output	Operation period
Shanwei, Guangdong, China	Coastal OWC	100 kW	Since 2005
Dawanshan, China	Coastal OWC	3 kW	Since 1990
Isle of Islay, Scotland	Coastal OWC (Islay 1)	75 kW	1988–1999
Kujukuri, Japan	OWC with pressure storage	30 kW	Since 1987
Niigata, Japan	Breakwater OWC	40 kW	1986–1988
Toftestallen, Norway	Coastline OWC	500 kW	1985–1988
Sanze, Japan	Coastal OWC	40 kW	1983–1984
Vizhinjam, Kerala, India	Near shore close to a breakwater	50 kw	Since 1996
Pico, Portugal	Coastal	1 MWh	Since 2005
Islay, U.K	Coastal	500 kw	Since 2000

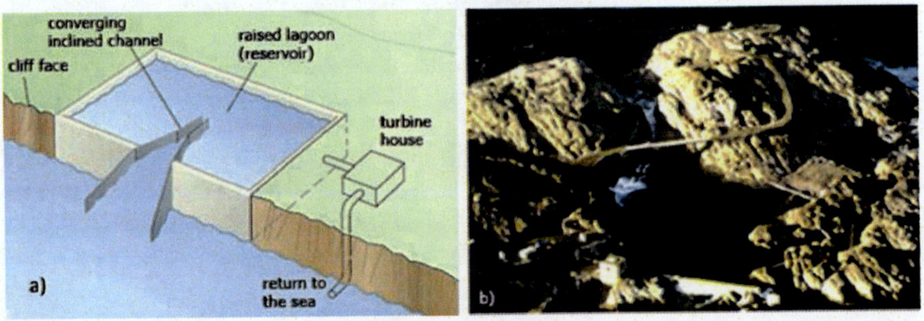

Fig. 2.5 **a** Schematic sketch of the TAPCHAN overtopping device. **b** Aerial photographic view of Norwegian TAPCHAN

2.5.3 Near Shore Wave Energy System

The OWC is the most popular fixed-bottom nearshore structure, which is ideal to operate in shallow water depths between 10 and 25 m. OWC can be incorporated as a part of existing or new breakwaters, whose primary purpose is to serve as a harbor or coastal protection structure. This approach was adopted in the port of Sakata, Japan, and in Vizhinjam Harbor, Trivandrum, India, the views of which are presented in Fig. 2.6a, b, respectively.

Fig. 2.6 Views of OWC plants off **a** Sakata Port, Japan, **b** Vizhinjam Harbour, India

2.5.4 *Offshore Wave Energy Systems*

Overtopping Devices

These devices operate in deep water conditions, where the water depth is usually above 50 m. To potentially harness energy from deep waters, offshore devices are mostly found in the vicinity of the surface waters. In most of the designs proposed and tested at laboratory/prototype sea trials, the primary components of the energy conversion device either floats over the water surface or it is found immediately below the water surface. The generally used methods of converting the oscillating motion of the body into mechanical energy are obtained by involving hydraulic pumps or rams incorporated into, or acting on, the moorings. Offshore devices are working on the principle of a point absorber, as described by Sjostrom [50]. The principle of fully submerged Archimedes Wave Swing [46] that the wave swing consists of a bottom-fixed, air-filled cylindrical chamber, the volume of which oscillates due to the reciprocating motion of a piston located on its top or 'floater's heave motion. This heave motion is transmitted to a linear electrical motor. The floater is sealed against a fixed vertical sleeve and the device is tuned by adjusting the air pressure inside the float. The Pelamis is a floating device composed of cylindrical sections which are linked by hinged joints and these cylindrical section moves with respect to each other in the waves. Energy is extracted at the joints by hydraulic rams which in turn drives the electrical generators [11]. Another form of the floating device involves a gyroscope to provide an inertial reference frame for a hydraulic operated power take-off mechanism. The Edinburgh Duck is one of the most popular energy conversion devices proposed by Salter in 1974, which has been evolved into various systems over the years and is being incorporated into a broad range of novel concepts with widespread applications.

Wave Dragon is the first floating offshore platform deployed in Nissum Bredning, Denmark, and had the grid-connected in 2003. For positional stability, slack mooring is used. The plan and sectional view of the prototype are detailed in Fig. 2.7. The inlet by virtue of its shape concentrates the wave energy. As the wave rides over the steep ramp, it gets amplified and flows into the reservoir. The water thus collected is used for rotating the turbine for electricity production. The advantage of this technology is that it can capture a high magnitude of wave energy available in offshore areas. The Wave dragon can be deployed either as a single unit or as an array. Initially, there were proposals for the prototype deployment in the U.K., Spain, Portugal, France, Germany, and Norway by 2006. The issues related to survivability in severe environmental conditions remain an unsolved problem. So, technology is still in the formative stage.

The Sea-wave Slot-cone Generator (SSG) is the recently developed onshore overtopping device [55]. The operating principle of SSG (Fig. 2.8) is like Wave Dragon and consists of the integrated breakwater with multiple reservoirs that use the run-up concept to collect the water from ocean waves. The wave amplification on the front side ramp of the SSG structure spills water into a reservoir that drives a multiple vertical axis turbine.

The ramp and its inclination are the important parameters for the optimal design of overtopping type devices since the wave amplification and run-up of the incident waves depend on them. Herein, it is to be noticed that the rate of overtopping into the reservoir must be maximized to generate a high degree of electricity, which is vulnerable to the stability of the lee side of the coastal protection structure. Since the waves in the ocean are random in nature, the overtopping devices' angle of the

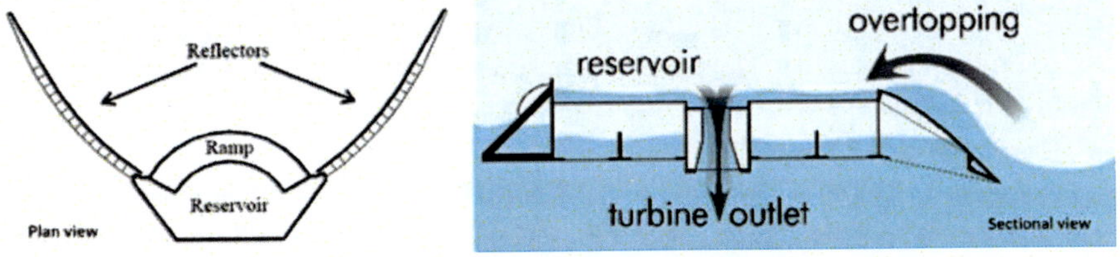

Fig. 2.7 Plan and sectional view of the wave dragon overtopping device [64]

Fig. 2.8 The artistic view of a 3-level SSG with multistage turbine [55]

ramp must be designed for the average wave conditions. The optimum angle of the ramp is reported [28] to be between 30° and 40°. Considering the stability of the lee side of the coastal protection structure and the maximized requirement of wave overtopping, the feasibility of the multipurpose option of harnessing wave energy and coastal protection through the use of overtopping type devices is limited.

Oscillating Bodies

The oscillating bodies (OB) devices, which are also frequently called third-generation wave energy devices, are mainly meant to be deployed in the offshore regions for exploiting wave energy from more powerful wave regimes. The OB devices are complex in nature with different operating principles compared to both overtopping devices and oscillating water columns. In general, OB devices are partially or completely submerged with respect to mean sea level. Furthermore, to maximize electricity production, the OB devices are predominantly designed to operate in an array of devices. The general operating mechanism of OB devices is explained below.

The hydrodynamic wave force on bodies induces vibratory motion depending on the parameters such as wave period, the inertia of the body, damping, and stiffness. The WEC undergoes oscillatory motion about a fixed reference or relative to other parts of the body. Generally, the wave-activated body is connected to a piston, which pumps water on a wheel to cause rotary motion, and this rotary motion is connected to the generator for power generation. Presently, direct linear converters are widely used to produce electricity. The units can be smaller and modular, which are the main advantages of this system of conversion. The typical example of the floating type OB device, namely, Pelamis is discussed below.

The Pelamis is an offshore WEC for power generation from the waves developed by Pelamis Wave Power Ltd. of Scotland. The device consists of cylindrical sections linked by hinged joints. The device is laid perpendicular to the wave crest and is kept in position by a compliant mooring system. Being compliant, it can get over

Fig. 2.9 Pelamis Wave energy converter [63]

the adverse effects of severe environmental conditions for its survivability. On wave activation, the relative motion between adjacent units is captured by a hydraulic ram to generate electricity. The length of the device depends on the predominant wave conditions of the region. The first prototype tested in Scotland between 2004 and 2007 had a length of 120 m and the diameter was 3.5 m. Figure 2.9 shows the first prototype deployed at sea. The sea trial test paved the way for further research on the structural configuration and the power module system of the device.

Oscillating Water Column

Oscillating water column (OWC) is one of the wave energy converters, which is the first kind of device developed in the early stage of research [38]. As discussed earlier in the introduction section, the principle of the OWC device is like an air pump due to the up and down oscillation of the water column inside the chamber. Suction and expulsion of air would occur in the chamber through an air hole (power take-off (PTO) device) that is connected to a turbine. The turbine can convert the bidirectional airflow into a unidirectional movement of the turbine. The rotation of the turbine can be connected with an electric generator for electricity production. The turbine is the only moving component of the system and it is placed above the water surface. This is the main advantage of the OWC system. Hence, installation, operation, and maintenance are easier in an OWC system. Further, it can be either bottom mounted or floating type. This helps the installation at the shoreline, nearshore, and offshore regions.

Masuda [37] of Japan Marine Science and Technology Centre (JAMSTEC) has tried many wave power devices to light navigational buoys and has come out successful with a buoy in which an air chamber is subjected to variations of pressure caused by the oscillating water column height of the waves. In principle, this device is a circular symmetric tube half immersed in the sea. The mass of water enclosed by the tube is forced to oscillate due to wave action. This water column acts as a piston, forcing the above air to be exhaled and inhaled into the system. The airflow from and to the chamber caused by the oscillating water column is used to drive an air turbine. Encouraged by the success of the system in small buoys, Masuda has designed larger systems for tapping the energy of the waves.

The JAMSTEC in Japan has undertaken numerous research and development activities for ocean wave energy extraction technology, one of the earliest in the 1970s is a floating prototype structure known as "Kaimei". The 'Kaimei' is a barge 80 m long 12 m wide weighing 800 tons, moored by chains. Nine sets of turbines and generators of capacity 125 kW capacities each, working on the OWC principle were attached. The prototype device was subjected to field tests in the Sea of Japan, adjacent to the coast of Yamagata Prefecture. The overall efficiency of the system is found to be very low since the barge itself was heaving up and down with waves.

JAMETEC through several theoretical and experimental studies developed the floating wave power device called "Mighty Whale". The open sea test was conducted in 1987 at 40 m water depth in Japan. The side view of the device in the real field is shown in Fig. 2.10. The device was 50 m long, 30 m wide, and 12 m deep, incorporating three air chambers for wave energy conversion. The energy absorption units were placed side by side on the front side of the device facing the wave direction. The power rating of the individual units was 110 kW. The operational experience showed the prospect of wave energy plant as multi-purpose like a breakwater, water purification, aeration etc.

Fig. 2.10 The side view of the Mighty Whale in the real field (*Source* http://www.jamstec.go.jp/gallery/e/research/system/002.html)

Fig. 2.11 Photographic view of the Pico –power plant

The typical example for the fixed onshore oscillating water column type wave energy converter, PICO power plant, is discussed as follows. This plant is built at Pico of Azores to study the feasibility of supplying power to the local grid. Figure 2.11 gives the photographic view of the installed plant. The natural gully available at the location was made use of informing a major part of the construction. The envisaged power capacity of the device was 400 kW, and the diameter of the well turbine was 2.3 m. Shortly after its commissioning in 2003, some technical problems were noticed and after its rectification, it was functioning efficiently during the period 2006–2008. In 2008, there were attempts to increase the capacity of the project, but it was suspended considering the uncertainty in the structural longevity of the plant. In 2010, additional facilities were incorporated for enhancing the device's performance. Presently the plant is being used for testing of air turbines for formulating concepts in optimized design.

While oscillating water columns, fixed to the seabed were discarded during earlier considerations, fixed structures are under active consideration because of the following advantages.

- Lesser space between devices
- Lower survival wave heights
- Higher energy conversion efficiency
- No flexible power cables
- Closer proximity to shore
- Higher load factor.

The main disadvantages are:

- Lower incident wave energy
- Interference from seaweed and
- Installation difficulties.

Since the cost of moorings and their maintenance is expected to be about 30% of the floating structure, the fixed structure will be cheaper to install and maintain.

2.6 Onshore/Nearshore OWC Wave Energy Devices

The concept of adding two harbor walls in front of the OWC wave energy device and defining the system as a multi-resonant oscillating water column device was introduced by Ambli et al. [1]. The harbor concept was tested at the National Hydrodynamic Laboratories (NHL) in Trondheim, Norway, and it was found that the performance of an OWC in the presence of harbor walls substantially increases the efficiency of the device. The multi-resonant OWC device was theoretically modeled by Evans [13] on the assumption that; an incident wave propagating in a direction parallel to the harbor walls will enter undisturbed. The efficiency of multi-resonant OWC was reported to improve as the length of the harbor walls matches an odd multiple of a quarter wavelength of the predominant incident wave was demonstrated by Hunter [25] and claimed it is due to the formation of standing wave resonance that enhances the column excitation.

Takahashi [52] has described the development of a wave power extracting caisson breakwater in Japan. Zheng [60] through a detailed experimental program on parametric optimization of the prototype Multi OWC wave power device proved that the flared rather than the rectangular harbor could significantly increase the wave energy extraction for a wide range of wave period. Koola [29] carried out detailed investigations to arrive at the dimensions of the OWC device for India's first wave energy prototype type system located at Vizhinjiam in the state of Kerala, India. Ravindran et al. [45] have described in detail the prototype demonstration bottom seated OWC plant of 150 kW peak capacity that was erected in Vizhinjam, the Arabian Sea during 1991.

Takahashi et al. [53] presented a report on the wave power conversion prototype wave the waves ranging between 1 and 5 m. The wave power also varied significantly and exceeded values as high as 100 kW/m. The annual average electric power in the area of Sakata Port was reported to be about 27 kW.

The performance of wave energy devices installed in shallow water regions with high power densities using shallow-water, nonlinear wave theory was investigated theoretically by Korde [30]. Falcao [15] applied a stochastic model to devise an optimal algorithm for the rotational speed control of an oscillating–water column (OWC) the wave power output of the plant with known hydrodynamic coefficients for the OWC as well as the turbine performance curves. Falcao [16] reported a theoretical analysis of the ocean wave energy absorption by a periodic linear array of oscillating water columns (OWCs) of the arbitrary platform.

Boccoti [5] proposed a new kind of absorber that showed the possibility of obtaining an impressive natural resonance in wind-generated waves. The same was also proved by carrying out a small-scale field experiment. It was concluded that his wave absorber would enable him to defend coasts with a very low environmental impact. Also, this device can be used as breakwaters for converting large quantities of wave energy into electric power.

The performance characteristics of a Double Chamber OWC proposed by Boccotti [6] were evaluated through a well-controlled experimental investigation in a wave flume by Wilbert et al. [59]. The efficiency of the system was assessed by considering the variations in the geometrical dimensions of the system along with the different characteristics of incident waves. For a relative bottom opening, of 0.45, the energy conversion capacity of this system with a maximum efficiency of about 80% was reported to be nearly 10% higher than the conventional single chamber OWC.

Other important contributions towards the enhancement of the efficiency of OWC are due to Malmo and Reitan [40, 66], Falcão and Rodrigues [18], McIver and Evans [29], Whittaker and Stewart [58], Korde [31–33], Mavrakos and McIver [41], Tseng et al. [54], Wang et al. [57].

2.7 Offshore OWC Wave Energy Devices

Hotta et al. [24] reported on the floating wave power device "Mighty Whale" developed by JAMSTEC, Japan discussed earlier. Hong et al. [22] studied the motion and the drift force of a floating OWC (oscillating water column) water energy device in regular sea waves taking into account the oscillating surface-pressure due to the pressure drop across the duct of the OWC air chamber. Hong et al. [23] developed an analytical model to analyze the hydro-elastic responses of a very long floating structure (VLFS) positioned behind a reverse T-shaped freely floating breakwater coupled with an OWC device as shown in Figs. 2.12 and 2.15.

All OWC principle is based on a large opening facing the waves so that maximum energy can be trapped. However, a caisson breakwater with an OWC with a small opening was considered by Boccotti et al. [7] in their theoretical model through which it was concluded that with some swell waves, the wave heights in front of Breakwater was greater than in front of a conventional breakwater-OWC system. The opening was provided at the upper end of a vertical duct, as shown in Fig. 2.13. Boccotti et al. [8] proceeded to prove their theoretical predictions through a small-scale field experiment on a breakwater incorporated with an OWC with a small opening in the natural laboratory of Reggio Calabria. One of the nine caissons is shown in Fig. 2.14. The concept shows that each caisson consists of three independent cells. Hong et al. [23] theoretically investigated the hydro-elastic responses of a very-long floating structure (VLFS) placed behind a reverse T-shape freely floating breakwater with an oscillating water column (OWC) device as shown in Fig. 2.15.

Using a fully nonlinear numerical model, Koo [34] investigated the motion responses of a floating box-shaped breakwater to serve as a wave energy converter as shown in Fig. 2.16. The pneumatic damping system in the breakwater is provided with a nozzle for airflow on the top of the chamber and the energy can be absorbed as the airflow passes through the outlet. The relative mean vertical velocity between the body and the wave elevation inside the chamber is known to be a key factor affecting air pressure and wave damping [20, 22, 23, 34, 35, 51].

To improve the efficiency of the OWC system, Ashlin et al. [2] investigated the effect of the air orifice opening ratio, *a/A* [ratio of air hole area (*a*) to plan area of OWC device (*A*)], by varying from 0.1 to 1.5% in incremental steps of 0.1% along

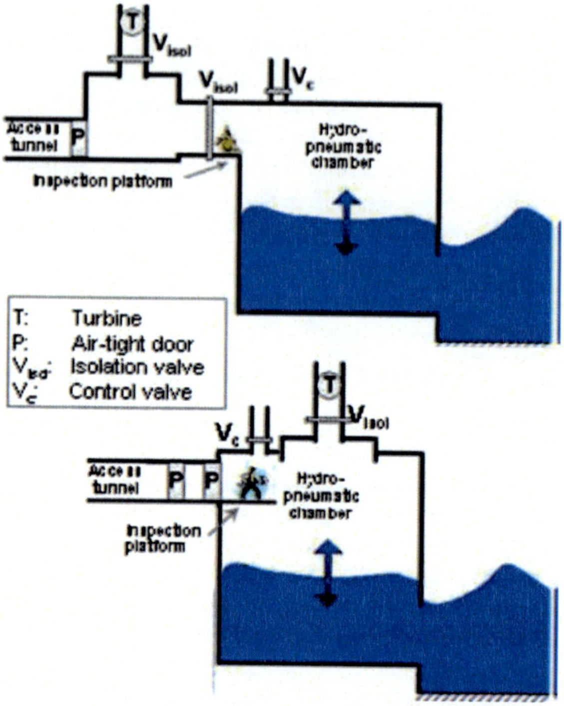

Fig. 2.12 Two alternate configurations for CEODOURO port [62]

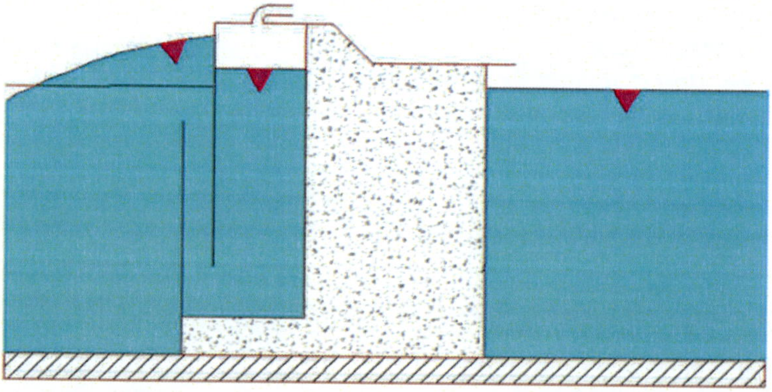

Fig. 2.13 Caisson breakwater merged with OWC with a smaller opening towards the waves [7]

with 2.1%. The optimum air damping is predicted in between $a/A = 0.6$ and 0.7% and has been further adopted in investigating the influence of different bottom profile shapes (characterized by four different bottom profiles: flat bottom; circular curve bottom of radius 300 mm; and, bottom with a linear slope of 1:5 and 1:1) inside the OWC chamber on its hydrodynamic characteristics through physical model testing

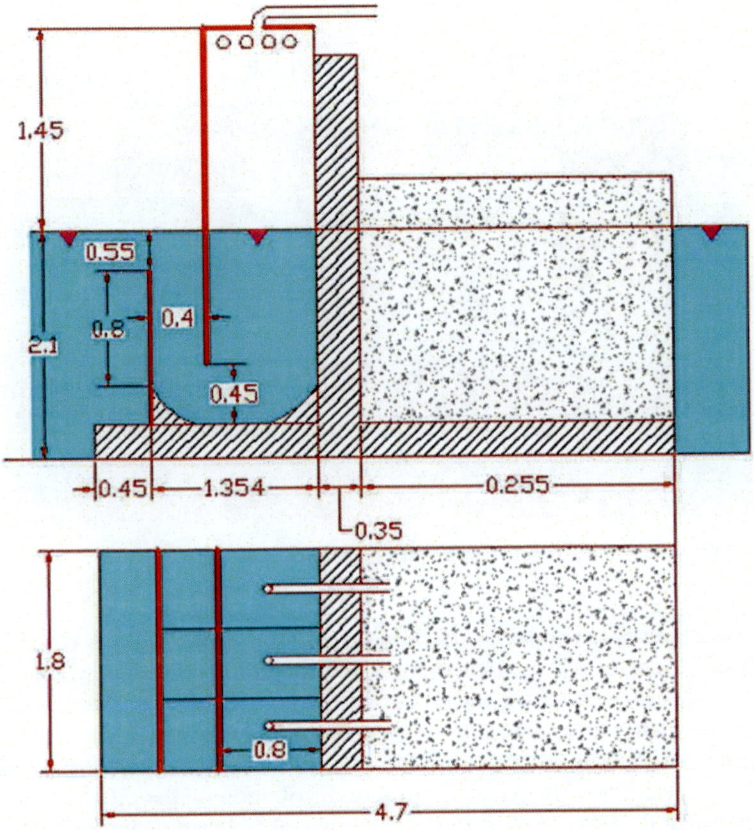

Fig. 2.14 Caisson breakwater merged with OWC with a smaller opening towards the waves [8]

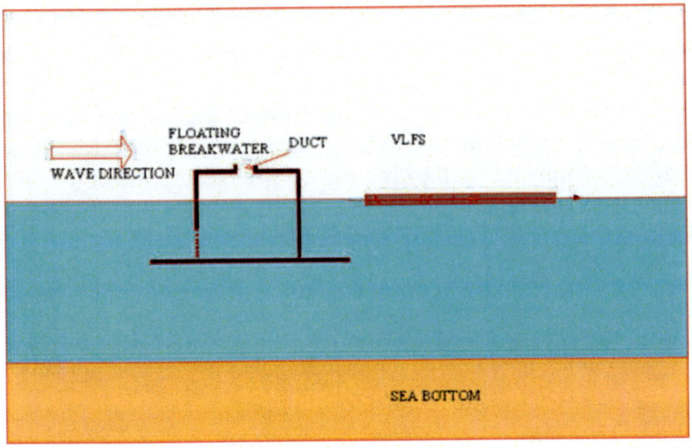

Fig. 2.15 Combination of inverse T breakwater with OWC and VLFS [23]

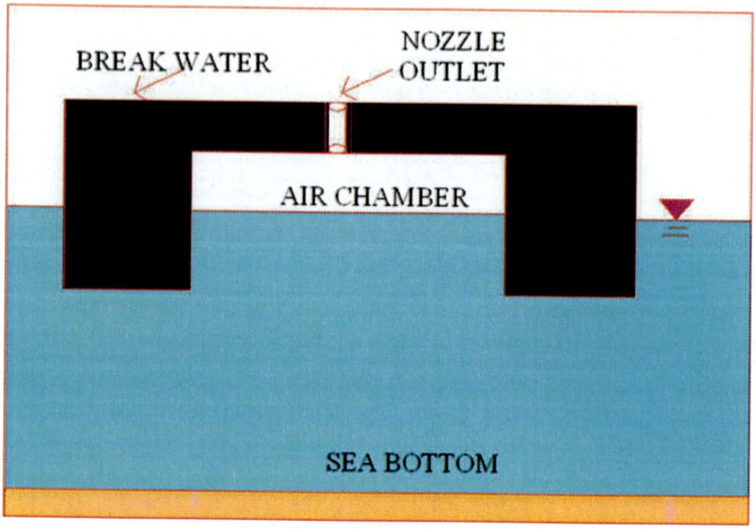

Fig. 2.16 Concept of a pneumatic-type floating breakwater [34]

in a flume subjected to regular and random waves. From these results, it was observed that the OWC device with a circular curve bottom profile proved to exhibit a better efficiency than the other three shapes because of higher wave amplification, higher air pressure, higher wave power absorption, and less reflection.

A series of experiments was carried out to evaluate the design wave loads on the OWC system [3]. The dynamic pressure distribution on the different structural components of the OWC device, the total horizontal and vertical wave forces acting on the OWC model due to both regular and random wave incidences were measured through a well-controlled experimental program. From these results, it was observed that in the high-frequency region, the seaward pressure on the lip wall is found to be larger than the shoreward pressure. Thus, the seaward pressure should be considered as a critical parameter, while designing the lip wall of the OWC device as it governs the stability toward the seaside. It is also ascertained that the measured wave loads on the OWC device are determined to be lesser than the theoretical wave load on a vertical wall. It is mainly due to the high wave power absorption of the OWC device. The reduction in the wave load on the OWC device shows the advantages of using them in the construction of a caisson-type breakwater.

2.8 Special Types of Breakwaters with WEC

In this section, special types of breakwaters that could be adopted for incorporating WEC are discussed.

Floating Types

The curtain wall breakwaters are commonly used as secondary breakwaters to protect small craft harbors and locations where a certain degree of wave transmission is allowed in small fishing harbors and marinas. These types of breakwaters are either supported on piles or sometimes over sheet piles. Depending on the seabed conditions, they may sometimes be fixed over a rubble mound although this is not so common. These types of structures are projected in Fig. 2.17. The T-type and inverted T type of breakwaters were discussed in the previous chapter. The floating T-shaped breakwater moored to the seafloor that can easily be transported from one location to another probably to carry out a specific operation over a certain period as shown in Fig. 2.18 can be converted to a floating OWC breakwater as projected in Fig. 2.19 to fulfill the requirement of tranquility as well as for wave energy conversion. These types of devices can be to serve for a relatively shorter period. As Boccotti et al. [7] have proved that smaller opening is quite efficient in the wave energy absorption for conversion; the device suggested has been modified to a floating structure as shown in Fig. 2.20. The illustration of a floating OWC with a berthing facility on its lee side is shown in Fig. 2.21. It is to be noticed that the system incorporates the harbor walls that is claimed to enhance energy from the waves. These walls may also be made as converging from the seaward side towards the OWC. Further, the air column chamber is funnel-shaped so as allow more air to penetrate for effective driving of the turbine. A modification to the above concept is to provide a sloping bottom to force the waves to amplify inside the chamber. This concept is shown in Fig. 2.22.

Caisson Types

Majority of the systems, if not all of those that are installed either as commercial or pilot plants fall under this category. There has been substantial progress made in terms of research and development which is still being refined. A typical concept

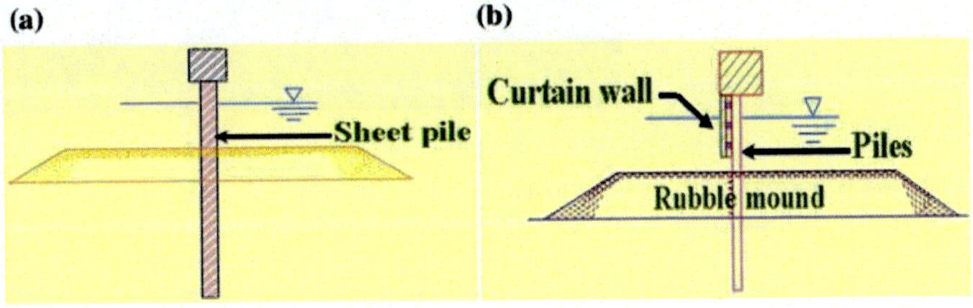

Fig. 2.17 **a** Simple sheet pile breakwater, **b** Simple skirt pile-supported

Fig. 2.18 A view of floating T-Shaped/skirt breakwater

Fig. 2.19 Simple floating skirt breakwater with OWC

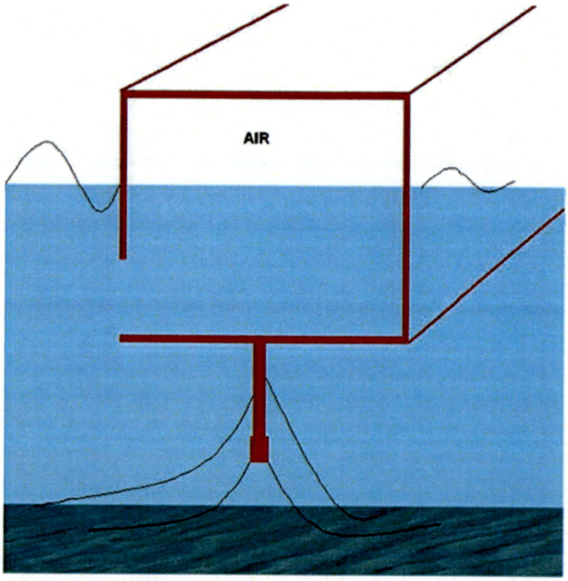

wherein the bottom slope can also play an important role in the wave energy inside the chamber is illustrated in Fig. 2.23.

A view of dual cylindrical caisson breakwater at Nagashima Port, Japan [65] is shown in Fig. 2.24. The dual cylinder caisson breakwater consists of inner and outer cylinders. The outer part of the cylinder is perforated and the section in between the outer and inner forms a wave dissipating chamber and hence can withstand

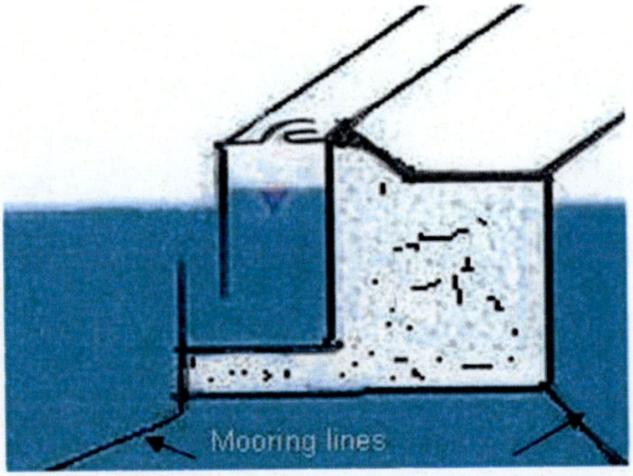

Fig. 2.20 Simple floating skirt breakwater with OWC as modified by Boccotti et al. [7]

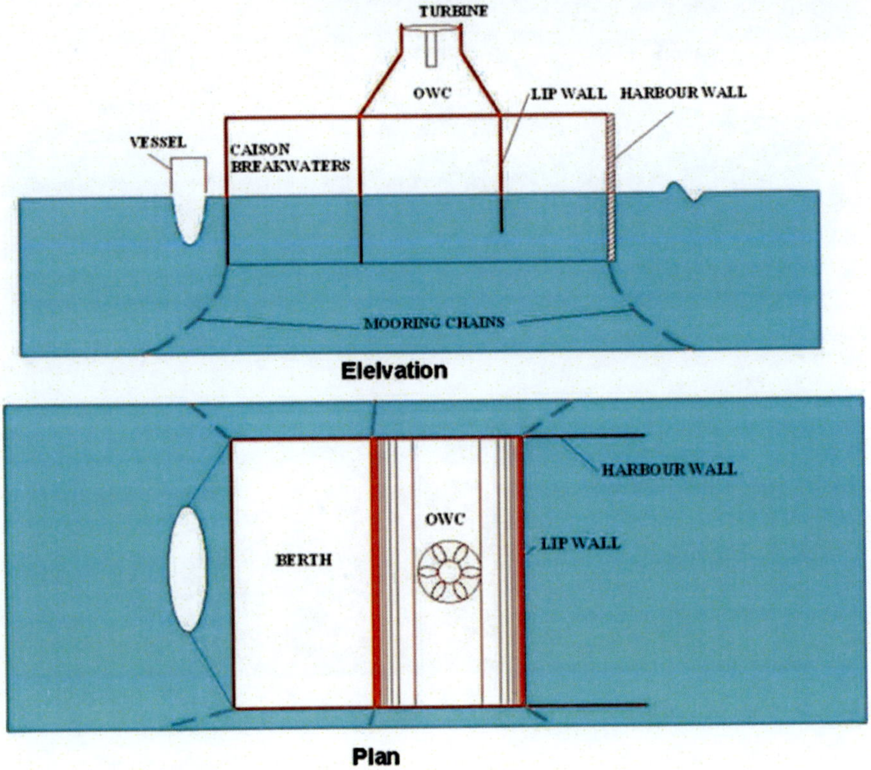

Fig. 2.21 Floating OWC with berthing facility on its lee side

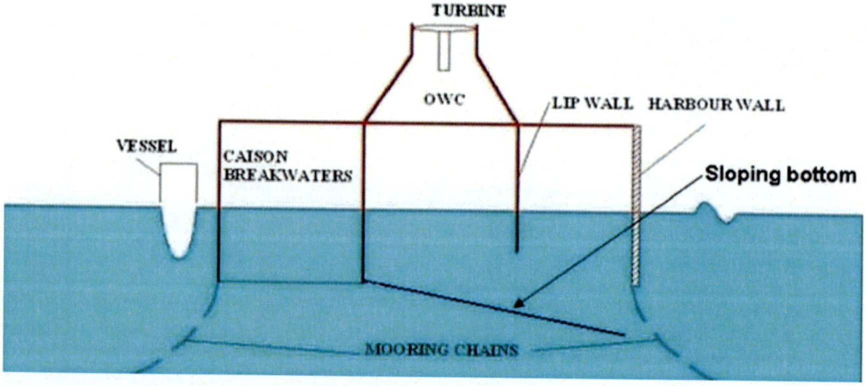

Fig. 2.22 Floating OWC with a sloping bottom and with berthing facility on its lee side

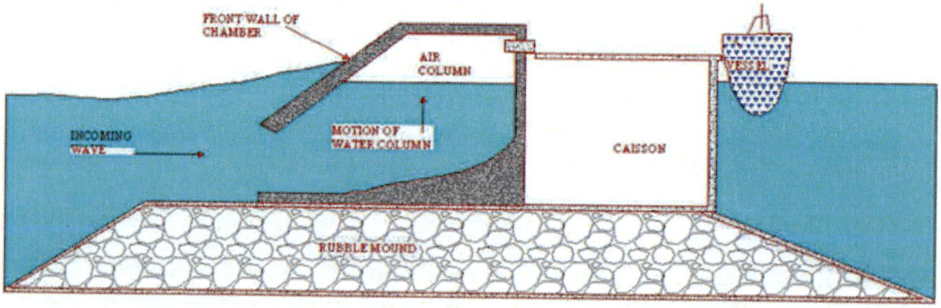

Fig. 2.23 Caisson type OWC with a sloping bottom and with berthing facility on its lee side

Fig. 2.24 View of dual cylindrical caisson breakwater at Nagashima Port, Japan [65]

large forces with relatively small cross-section leading to reduction of construction material. This configuration may be considered for investigation for caisson type OWC. The front face instead of being perforated can be of slit type or just open or it can have an arrangement of smaller opening for oscillating water column as suggested by Boccotti et al. [7]. The idea is projected in Fig. 2.25.

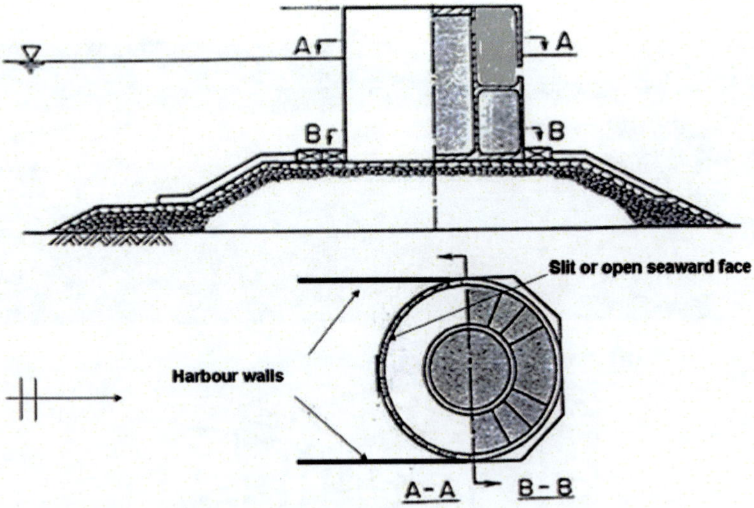

Fig. 2.25 Slit or open front face OWC circular caisson breakwater

A curvilinear plan shape of the front wall as obtained with circular caissons used in Hanstholm, Denmark (1965) reduces the forces on the wall compared to if it had been a plane vertical wall. Such a breakwater existing in Naples, Italy, and its illustration are shown in Fig. 2.26a, b, respectively. Each of the circular fronts may be a single unit of an OWC and a sloping bottom as stated earlier, which could probably enhance the magnification of incident wave energy. Another form of caisson breakwater that could be adopted incorporating OWC is the seaside front slit type breakwater as shown in Fig. 2.27. The total force and moment on such a breakwater will be less which is an added advantage.

Pile Supported Types

These types of structures are quite common for water depth up to about 30 m. A few concepts are presented below. If the OWC needs to be integrated, it is essentially of a fixed type. However, pile-supported chamber-type floating body with a proper damping system is more efficient than other types of floating breakwaters in attenuating long-period waves (this is an important requirement for the tranquility of harbors, wherein, the problems arise only due to such waves, like the collision of ships, snapping of mooring lines, etc.). This has been also highlighted by Koo [34]. The concepts of fixed OWC (flat floor and sloping floor) are shown in Fig. 2.28a, b, whereas, the concepts of heaving OWC (flat floor and sloping floor) are shown in Fig. 2.29a, b respectively. The efficiency of the system is governed by a proper damping system apart from the capture factor. Herein, the shape of the walls (OWC chamber) needs to be investigated in depth apart from other factors like the harbor walls. The four systems will also have harbor walls although explicitly not shown in the illustrations. A typical deep-water berthing facility is shown in Fig. 2.30. This

(a)

(b)

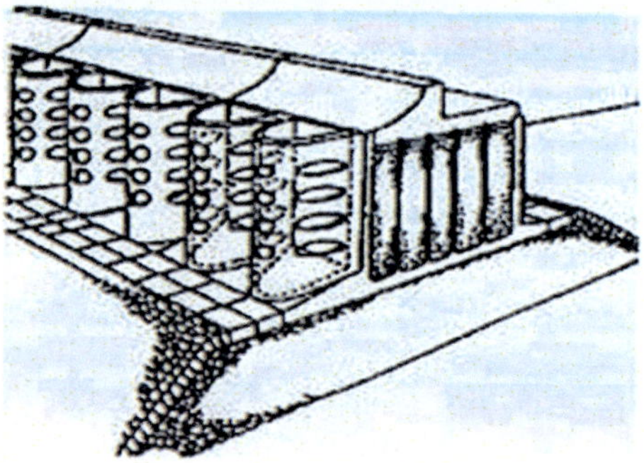

Fig. 2.26 **a** Photograph of Circular front Slit or open front face OWC circular caisson breakwater. **b** Circular front Slit or open front face circular caisson breakwater

Fig. 2.27 Seaside front slit type caisson breakwater

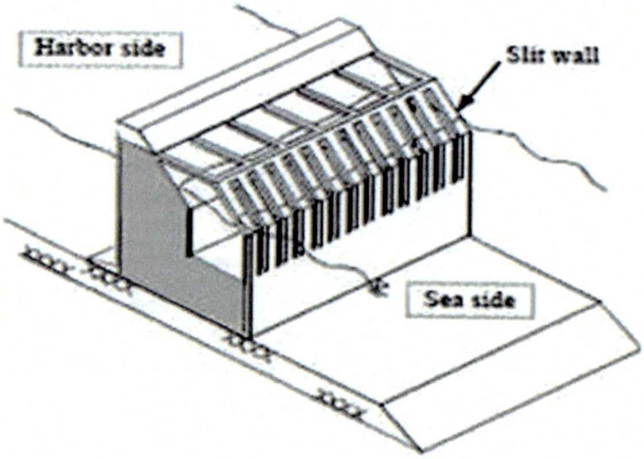

encompasses an approach trestle (an open jetty supported on piles), a breakwater (maybe of caisson type or rubble mound type). Any of the aforesaid OWC systems can easily be integrated into breakwater or even with the approach trestle. This however needs a careful and detailed investigation. Although it is not too ambitious and is a proposition that could be achieved soon is a harbor (minor or fishing or commercial) with OWC.

A new concept namely WAVERUSH for incorporating wave energy converter in a conventional sloping breakwater (no such system found in literature) that employ rows of flexible tubes mounted over the breakwater was introduced by Roberts and Shepherd [47] which is claimed to have advantages of lesser space required than that for bottom seated OWC device. Lópeza et al. [61] through the implementation of a novel methodology reported the details of a case study in Galicia (NW Spain), in which three deployment sites of OWC were considered, and it was claimed that the turbine-induced damping must be matched to the wave climate of the site for its maximum efficiency. Vicinanza et al. [56] have reported a comprehensive review on the integration of OWC with breakwaters.

Rubble Mound Breakwaters

The other possibility could be to consider it as part of detached breakwaters that can in addition to wave energy extraction can serve as a coastal protection measure, i.e., a mitigation measure for coastal erosion [43]. Offshore detached breakwaters as a shore defense measure are increasing at a remarkable pace, since; they effectively reduce and absorb the incident wave energy. The ends of the breakwaters diffract the waves resulting in a calm region in between them. They are usually of mound type, aligned parallel to the shoreline to serve as a shore protection measure. These prevent the destructive waves from reaching the shore and protect it against the encroachment of the sea. Salient formations are witnessed when the shore advances towards the breakwater; when the advancement reaches the breakwater, it is termed as a tombolo. Offshore breakwaters are expensive and should be used after good model tests as sometimes they bring in precisely the opposite effects as against what is desired. The shoreline variation due to the construction of a single offshore breakwater and due to segmented offshore breakwaters are shown in Fig. 2.31a, b, respectively.

If each of these breakwaters is fixed with one or a series of OWC, the results are expected to quite fruitful as the costs will be shared and the benefits are more than just coastal protection or energy from ocean waves. A typical concept of incorporating OWC with rubble mound breakwater for this purpose is shown in Fig. 2.32a–c. Any of the aforesaid OWC systems can easily be integrated into breakwater or even with the approach trestle. This however needs a careful and detailed investigation. Although it is not too ambitious and is a proposition that could be achieved soon is a harbor (minor or fishing or commercial) with OWC.

To optimize the spacing between an array of OWC devices embedded in a rubble mound breakwater Ashlin et al. [4] subjected OWC integrated with a rubble mound breakwater for multi-purpose options, termed it as 'OWCBW system' to regular and random waves in normal and oblique wave incidences. Three different spacings between the centers, *S*, i.e., single, twice, and thrice the width of the OWC model, were considered. It was observed that the OWCBW system with thrice the width of the model spacing exhibited better performance. The front-line of an array of OWC devices are subjected to a higher amount of wave energy concentration due to the convergence of water waves, and thus, increasing the absorbed power to a maximum

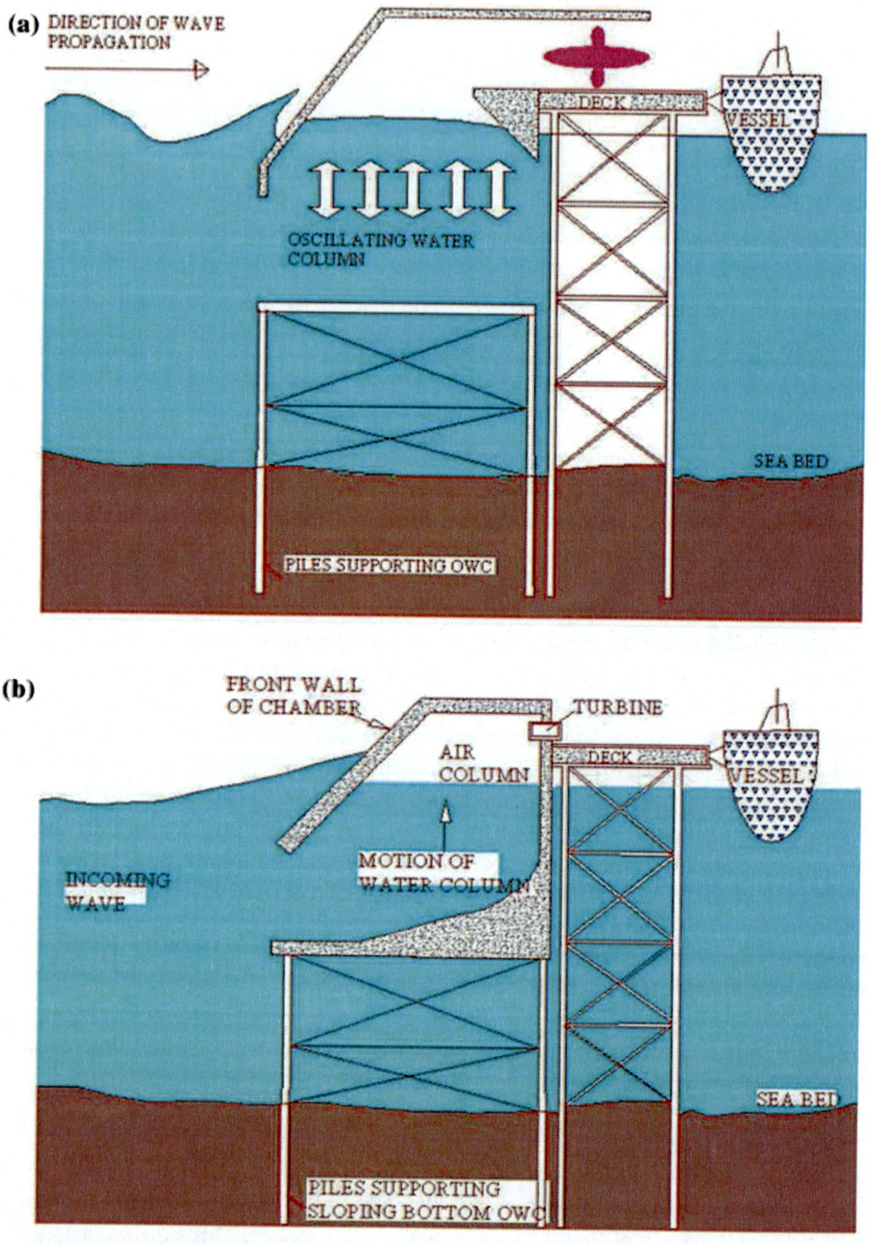

Fig. 2.28 **a** Pile supported fixed OWC (flat floor) berthing structure. **b** Pile supported fixed OWC (sloping bottom) berthing structure

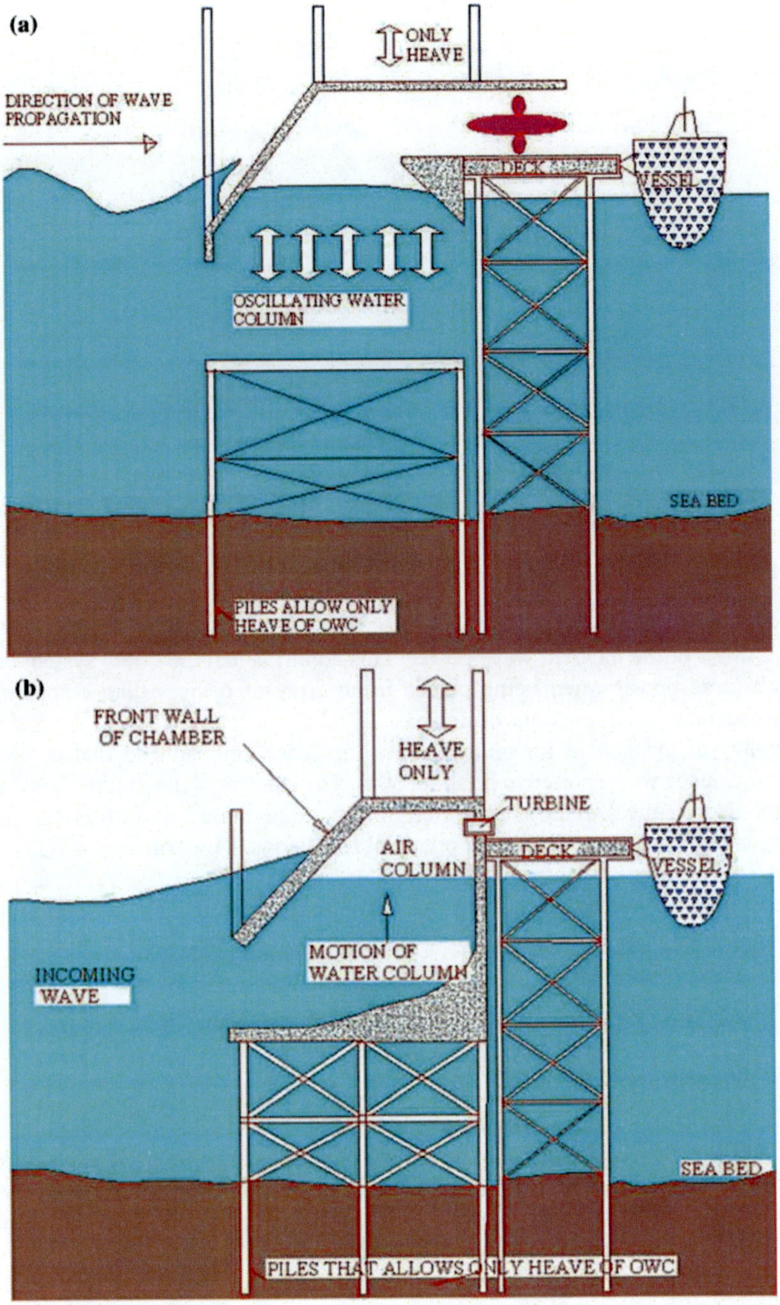

Fig. 2.29 **a** Pile supported heaving OWC (sloping bottom) berthing structure. **b** Pile supported heaving OWC (sloping bottom) berthing structure

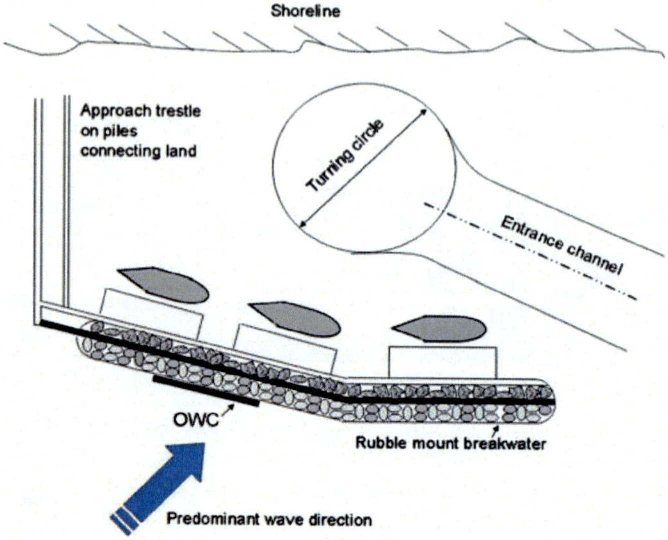

Fig. 2.30 a Typical deeper water direct berthing facility

of 2.2 times of the incident wave power. This confirms that the OWC system could absorb more power when being placed in an array of devices than compared to an isolated device. The extent of decrease in relative capture width at the resonant frequency of the system for oblique wave incidence compared to that of normal wave incidence was reported to be about 30%. The concept of integrating OWC with a semi-circular breakwater as projected in Fig. 2.32d could be worth investigating which is expected to have a greater potential for Islands. Dhinakaran et al. [12] have carried out a comprehensive review of literature on the works carried out on the wave hydrodynamic characteristics of semi-circular breakwaters, that could be considered for the formation of harbors as well as offshore detached breakwaters to serve as a coastal protection measure.

2.9 Summary

Most of the proposed devices for conversion of wave energy are oscillating systems with a frequency-dependent response showing the phenomenon of resonance. When the wave period is at resonance with the natural (Eigen period) of the system, maximum energy conversion takes place. In the case of a narrow bandwidth resonance, the energy conversion is less rapid. Larger dimensional wave energy converters such as terminators and attenuators have a wider bandwidth to absorb the energy compared to that of point absorbers. Conversely, the smaller structural volume of point absorbers leads to a larger ratio between structural volume and power converted.

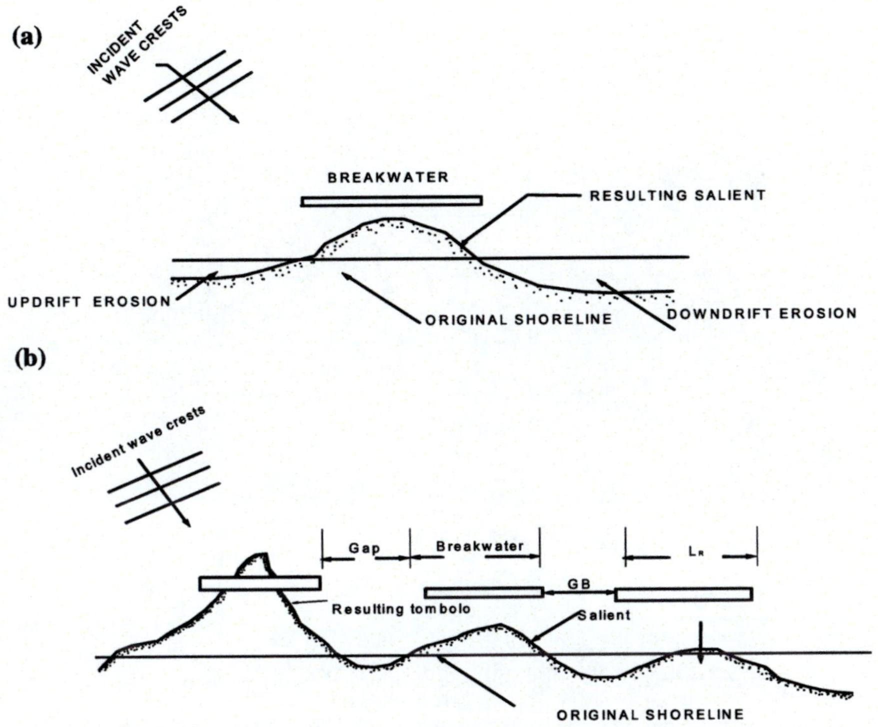

Fig. 2.31 **a** Definition sketch of the single-detached breakwater. **b** Definition sketch of Segmented detached breakwater

The early studies investigated the effect of regular waves to arrive at the optimum control parameters. The target was to achieve the optimum phase and optimum amplitude of the oscillation to maximize the efficiency of the power conversion. Successively, optimum control parameters were arrived at for real-time wave climate, and thus the requirement to predict the incident wave characteristics became more apparent. Thus, a clear resolution of the problem was needed to arrive at optimum characteristics with the regulated design parameters to achieve optimum control infield practice.

All the above aspects have been clearly discussed in detail by Falnes [19].

As wave energy converter systems can undergo six degrees of freedom of motion particularly in the case of freely floating converters exposed to the dynamic behavior of ocean waves, with a lesser degree of motions for restrained structures it is worthwhile to refine systems with just one degree of motion and perfect it with optimum energy extraction efficiency.

To summarize the factors to be considered for wave energy devices.

- The available wave power to be extracted is higher for long waves and hence sites with long waves should be ideal. However, for the development of harbors, the

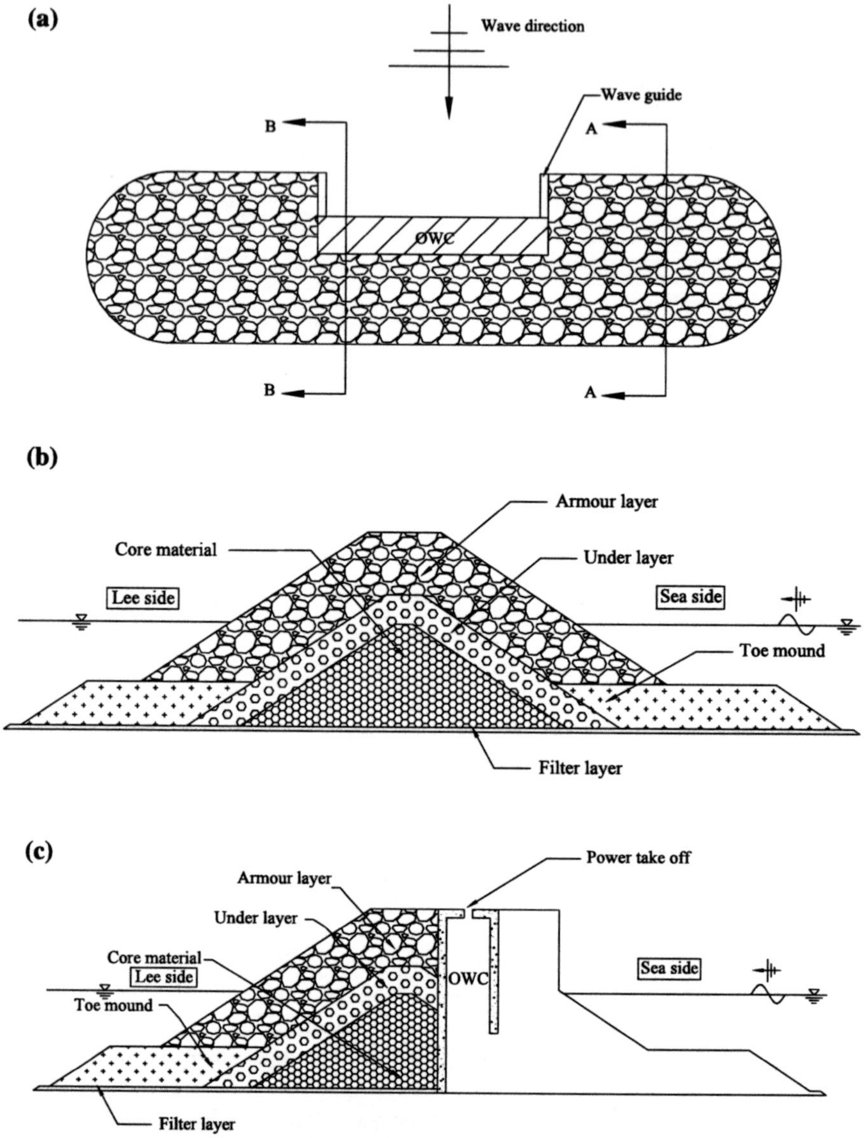

Fig. 2.32 **a** Plan view of OWC integrated with Offshore Detached Breakwater. **b** Cross section of rubble mound breakwater at A-A. **c** Cross section of OWC integrated with rubble mound breakwater at B-B. **d** Cross section of OWC integrated with semicircular breakwater

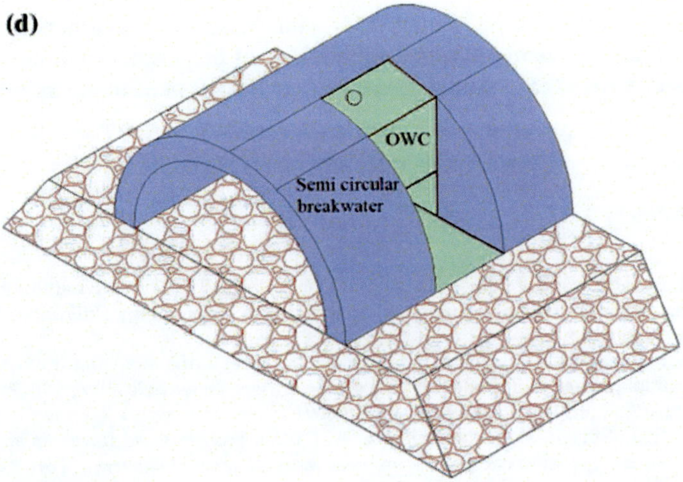

Fig. 2.32 (continued)

criteria being just the reverse and hence it is more meaningful to combine both structures.

- Maximum wave power exists in deep waters which gradually decreases towards the shore. The information on bathymetry in addition to wave characteristics is important.
- Waves in an open ocean are extremely dynamic in nature with varying height, length, and direction. Locations with well-defined predominant wave directions (narrow range) could be ideal.
- The system should be designed for both operational wave characteristics for its efficiency as well as for extreme wave characteristics for its stability and survival.
- A great deal of attention is needed for ensuring a minimum fluctuation in power, while it is being extended to the grid system.
- The rules and regulations imposed by the International, National, and local bodies/agencies in regard to Navigation, fishing, coastal highway, tourism, etc., need to be addressed and considered.
- Locations of huge tidal variations to be avoided as they can affect mooring and the efficiency of the device. Further, if coupled with harbor breakwaters the location may pose problems for the berthing of vessels.
- Some of the special type breakwaters will later become standard breakwaters. E.g, The perforated caisson breakwater has become popular that it is considered as a standard breakwater now. Common special type breakwaters are generally non-gravity types like piled, floating, or pneumatic types. Similarly, it is expected that a few concepts on the integration of WEC with breakwaters may become a standard one in the future.

- Although, several OWC integrated with caisson breakwaters have been constructed, integrating it with a rubble mound breakwater or similar ones for shallower waters that would be beneficial for Islands need to be explored.

References

1. Ambli, N., Bonke, K., Malmo, O., & Reitan. H. (1982). The Kvaener miltiresonant OWC. In *Proceedings of The 2nd International Symposium on Wave Energy Utilisation*, Trondheim, Norway, Tapir, (pp. 275–295).
2. Ashlin, S. J., Sundar, V., & Sannasiraj, S. A. (2016). Effects of bottom profile of an oscillating water column device on its hydrodynamic characteristics. *Renewable Energy, 96*(Part A), 341–353. https://doi.org/10.1016/j.renene.2016.04.091
3. Ashlin, S. J., Sundar, V., & Sannasiraj, S. A. (2017). Pressures and forces on an oscillating water column type wave energy caisson breakwater. *Journal of Waterway, Port, Coastal, and Ocean Engineering, 143*, 1–18. https://doi.org/10.1061/(ASCE)WW.1943-5460.0000405
4. Ashlin, S. J., Sannasiraj, S. A., & Sundar, V. (2018). Performance of an array of oscillating water column devices integrated with an offshore detached breakwater. *Ocean Engineering, 163*, 518–532. https://doi.org/10.1016/j.oceaneng.2018.05.043
5. Boccotti, P. (2003). On a new wave energy absorber. *Ocean Engineering, 30*, 1191.
6. Boccotti, P. (2007). Caisson breakwaters embodying an OWC with a small opening—Part I: Theory. *Ocean Engineering, 34*, 806–819.
7. Boccotti, P., Filianoti, P., Fiamma, V., & Arena, F. (2007a). Caisson breakwaters embodying an OWC with a small opening—Part I: Theory. *Ocean Engineering, 34*(5–6).
8. Boccotti, P., Filianoti, P., Fiamma, V., & Arena, F. (2007b). Caisson breakwaters embodying an OWC with a small opening—Part II: A small-scale field experiment. *Ocean Engineering, 34*(5–6).
9. Brooke, J. (2003). *(2003) Wave energy conversion*. New York: Elsevier Ocean Engineering.
10. Cuan B. B., Trevor J. T., & Whittaker, M. F. (2002). Overview and initial operational experience of the LIMPET wave energy plant. In *Proceedings of 12th ISOPE*, Kitakyushu, Japan, May 26–31 (pp 586–594).
11. Dalton, G. J., Alcorn, R., & Lewis, T. (2009). Case study feasibility analysis of the Pelamis wave energy convertor in Ireland, Portugal, and North America. *Renewable Energy* (article in press), 1–13.
12. Dhinakaran. G., Sundar. V., & Sundaravadivelu. R. (2012). Review of the research on emerged and submerged semicircular breakwaters. *Journal of Engineering for the Maritime Environment, Proceedings of IMechE, Part M, 226*(4). 397–409.
13. Evans, D. V. (1982). Wave power absorption within a resonant harbour. In *Proceedings of the 2nd International Symposium on Wave Energy Utilization*, Trondheim, Norway (pp. 371–378).
14. Falcão, AF de O. (2000). The shoreline OWC wave power plant at the Azores. In *Proceedings of 4th European Wave Energy Conference*, Aalborg, Denmark (pp. 42–47).
15. Falcão, A. F. O. (2002). Control of an oscillating-water-column wave power plant for maximum, energy production. *Applied Ocean Research, 24*(2), 73–82.
16. Falcão, A. F. O. (2002). Wave-power absorption by a periodic linear array of oscillating water columns. *Ocean Engineering, 29*(10), 1163–1186.
17. Falcão, A. F. de O. (2010). Wave energy utilization: A review of the technologies. *Renewable Sustainable Energy Review. 14*(3), 899–918. https://doi.org/10.1016/j.rser.2009.11.003
18. Falcão, A. F. O., & Rodrigues, R. J. A. (2002). Stochastic modeling of OWC plant for maximum energy production. *Applied Ocean Research, 24*(2), 59–71.
19. Falnes, J. (1993). Wave energy converters: generic technical evaluation study. Final report for the B-study of the DG XII Joule Wave Energy Initiative. Commission of the European Communities under contract number JOU2-0003-DK.

20. Falnes, J., & McIver, P. (1985). Surface wave interactions with systems of oscillating bodies and pressure distributions. *Applied Ocean Research, 7*(4), 225–234.
21. Hagerman, G. M. (1985). Oceanographic criteria and site selection for ocean wave energy conversion. In D. V. Even & A. F. O. Falcão (Eds.), Hydrodynamics of ocean wave energy utilization (pp. 169–178). Berlin, Heidelberg: Springer
22. Hong, D. C., Hong, S. Y., & Hong, S. W. (2004). Numerical study of the motions and drift force of a floating OWC device. *Ocean Engineering, 31*(2), 139–164.
23. Hong, D. C., Hong, S. Y., & Hong, S. W. (2006). Reduction of hydroelastic responses of a very-long floating structure by a floating oscillating-water –column breakwater system. *Ocean Engineeering, 33*, 610–634.
24. Hotta, H., Washio, Y., Yokozawa, H., & Miyazaki, T. (1996). Research and development on the development of the wave power device "Mighty Whale." *Renewable Energy, 9*, 1223–1226.
25. Hunter, R. S. (1982). Future possibilities for the NEL Oscillating water column wave energy converter—Experimental measurements and theoretical predictions of the phase control in regular waves. Report Pr39: WAVE/0R0 for Department of Energy, National Engineering Laboratory, East Kilbride, Glasgow.
26. IEA-OES. (2008). International energy agency-Implementing agreement on ocean energy systems. Annual report.
27. Kim, D., & Iwata, K. (1991). Dynamic behavior of tautly moored semi-submerged structure with pressurized air-chamber and resulting wave transformation. *Coastal Engineering in Japan, 34*(2), 223–242.
28. Kofoed J. P. (2002). *Wave overtopping of marine structures—Utilization of wave energy* (Ph.D. thesis). Hydraulics & Coastal Engineering Laboratory, Department of Civil Engineering, Aalborg University.
29. Koola, P. M. (1990). *Investigations on the performance behavior of oscillating water column wave energy device* (Ph.D. thesis). Ocean Engineering Centre, Indian Institute of Technology Madras, Chennai, India.
30. Korde, U. A. (1997a) Performance of a wave energy device in shallow-water nonlinear waves: Part I. *Applied Ocean Research, 19*(1), 1–11.
31. Korde, U. A. (1997). Performance of a wave energy device in shallow-water nonlinear waves: Part II. *Applied Ocean Research, 1*, 13–20.
32. Korde, U. A. (1999). On providing a reaction for efficient wave energy absorption by floating devices. *Applied Ocean Research, 19*(5), 235–248.
33. Korde, U. A. (2002). Latching control of deep-water energy devices using an active reference. *Ocean Engineering, 29*(11), 1343–1355.
34. Koo, W. (2009). Nonlinear time-domain analysis of motion-restrained pneumatic floating breakwater. *Ocean Engineering, 36*, 723–731.
35. Koo, W. C., Kim, M. H., Lee, D. H., & Hong, S. A. (2006). Nonlinear time–domain simulation of pneumatic floating breakwater. *International Journal of Offshore and Polar Engineering, 16*(1), 25–32.
36. Koo, W. (2009). Numerical analysis of pneumatic damping effect on a motion-constraint floating breakwater. In *Proceedings of the ASME 2009 28th International Conference on Ocean, Offshore and Arctic Engineering. Volume 4: Ocean Engineering; Ocean Renewable Energy; Ocean Space Utilization, Parts A and B*. Honolulu, Hawaii, USA. May 31–June 5, 2009 (pp. 177–181).
37. Masuda. Y. (1971). Wave-activated generator. In *International Colloq exposition oceans*. Bordeaux, France.
38. McCormick. M. E. (1981). *Ocean wave energy conversion*. New York: Wiley.
39. McIver, P., & Evans, D. V. (1988) An approximate theory for the performance of a number of wave energy devices set into a reflecting wall. *Applied Ocean Research, 10*(2) 58–65.
40. Malmo, O., & Reitan, A. (1985). wave power absorption by an oscillating water column in a channel. *Journal of Fluid Mechanics, 158*, 153–175.
41. Mavrakos, S. A., & McIver, P. (1997). Comparison of methods for computing hydrodynamic characteristics of arrays of wave power devices. *Applied Ocean Reserch, 19*(5–6), 283–291.

42. Ohneda, H., Igarashi, S., Shinbo, O., Sekihara, S., Suzuki, K, Kubota, H, Ogino, H, Morita, H. (1991). Construction procedure of a wave power extracting caisson breakwater, Proc. 3rd Symp. *Ocean Energy utilization,* Tokyo, 171–179 .
43. Pilarczyk, K. W., & Zeidler, R. B. (1996). *Offshore breakwaters and shore evolution control.* Rotterdam: A.A. Balkema (balkema@balkema.nl).
44. Ravindran, M., & Koola, P. M. (1991). Energy from sea waves—The Indian wave energy program. *Current Science, 60*, 676–680.
45. Ravindran, M., Pathak, A. G., Koola, P. M., & Latha, G. (1995). Indian wave energy programme: Progress and future plans. In *Proceedings of the Second European Wave Power Conference,* Lisvon, Portugal, November, 14–19.
46. Rademakers, L. W. M. M., Schie, R. G., Schuitema, R., Vriesema, B. Gardner, F. (1998). Physical model testing for characterizing the AWS. In *Proceedings of the Third European Wave Power Conference*, Partras, Greece (pp. 31–38).
47. Roberts, I., & Shepherd. K. (2009). WaveRush: A new concept for a breakwater wave energy converter. In *Proceedings of Coasts, Marine Structures and Breakwaters*, September 16–18
48. Salter, S. H. (1974). Wave Power. *Nature, 249*, 720–724.
49. Seymour, R. J. (1992). *Ocean energy recovery: The state of the art*. New York: American Society of Civil Engineers.
50. Sjostrom, B. O. (1993). The past, present, and furture of the hose pump wave energy converter. In *Proceedings of the European Wave Energy Symposium*, Edinburgh.
51. Suzuki, M., & Arakawa, C. (2003). Numerical methods to predict characteristics of oscillating water column for terminator type of wave energy converter. In *Proceedings of 13th International Offshore and Polar Engineering Conference,* (Vol. 1, pp. 333–340). Honolulu, HI: ISOPE
52. Takahashi, S. A. (1988). *Study on design of a wave power extracting caisson breakwater*. Wave Power Laboratory, Port and Harbour Research Institute Japan.
53. Takahashi, S., Nakada, H., Ohneda, H., & Shikamori, M. (1992) Wave power conversion by a prototype wave power extracting Occasion in Sakata Port. In *Proceedings of International Conference on Coastal Engineering.*
54. Tseng, R.-S., Wu, Z. S., & Huang, C. C. (2000). Model study of a shoreline. *Wave Power System, Ocean Engineering, 27*, 801–821.
55. Vicinanza, D., Ciardulli, F., Buccino, M., Calabrese, M., & Kofoed, J. P. (2011). Wave loadings acting on innovative breakwaters for energy production. *The Journal of Coastal Research, 64*, 608–612 (17) (PDF) The SSG Wave Energy Converter: Performance, Status and Recent Developments. Available from https://www.researchgate.net/publication/233420375_The_SSG_Wave_Energy_Converter_Performance_Status_and_Recent_Developments. Accessed September 20, 2020.
56. Vicinanza, D., Lauro, E. D., Contestabile, P., Gisonni. C., Lara. J. L., & Losada. I. J. (2019). Review of innovative harbor breakwaters for wave-energy conversion. *Journal of Waterway, Port, Coastal, Ocean Engineering, 145*(4), 03119001, 1–18.
57. Wang, D. J., Katory, M., & Li, Y. S. (2002). Analytical and experimental investigation on the hydrodynamic performance of onshore wave-power devices. *Ocean Engineering, 29*(8), 871–885.
58. Whittaker, T. J. T., & Stewart, T. P. (1993). An experimental study of nearshore and shorline oscillating water columns with harbours. In *Proceedings of the European Wave Energy Symposium*, Edinburgh, Scotland, July (pp. 151–156).
59. Wilbert, R., Sundar, V., & Sannasiraj, S. A. (2013). Wave interaction with a double chamber oscillating water column device. *International Journal of Ocean and Climate Systems, 4*(1), 21–40.
60. Zheng, W. (1989). Experimental research and parameters optimization of a prototype OWC wave power device. In *Proceedings of the International Conference on Ocean Energy Recovery (ICOER'89)* (pp. 43–50).

Bibliography

61. Lópeza, I., Carballoa, R., & Iglesiasb, G. (2019). Site-specific wave energy conversion performance of an oscillating water column device. *Energy Conversion and Management, 195*, 457–465.
62. Martins, E., Carrilho, L., Neumann, F., Ramos, F. S., Justino, P. A., Gato, L., & Trigo, L. (2005). Ceodouro project: Overall design of an OWC in the new OPorto break water. EWTEC—European Wave and Tidal Energy Conference, Glasgow (United Kingdom), 29 Aug–2 Sep.
63. Retzler, C. (2006). Measurements of the slow drift dynamics of a model Pelamis wave energy converter. *Renewable Energy, 31*(2), 257–269.
64. Kofoed, J. P., Frigaard, P., Madsen, E. F. & Sørensen. H. C. (2006). Prototype testing of the wave energy converter wave dragon. *Renewable Energy, 31*(2), 181–189.
65. Tanimoto, K., & Takahashi, S. (1994). Design and construction of caisson breakwaters—the Japanese experience. In *Coastal Engineering, Special Issue on 'Vertical Breakwaters'*, Oumeraci, H. et al. (eds.), Amsterdam, Holland: Elsevier Science Publishers B.V., Bd. 22, Nr. 1/2, S. 57–77.
66. Malmo, O., & Reitan, A. (1986) Wave power absorption by an oscillating water column in a reflecting wall. *Applied Ocean Research, 8*(1) 42–48.
67. Heath, T., Whittaker, T. J. T., Boake, C. B. (2000). The design, construction and operation of the LIMPET wave energy converter (Islay, Scotland). In *Proc. 4th European Wave Energy Conf*., Aalborg, Denmark, 2000, pp. 49–55.
68. Thorpe, T. W.(1992). A review of wave energy. ETSU-R-72.

Chapter 3
Direct Absorber for Wave Energy Conversion

Check for updates

A. Ramadan and M. H. Mohamed

Abstract In this chapter, considerable and promising designs have been investigated as wave energy conversion systems. A new design is introduced to the wave energy conversion system to produce power from regular and irregular waves. Laboratory experiments have been conducted for the Wave Hunter system in the wave tank of the Fluid Dynamics and Ship Theory Institute (FDS) of the Hamburg University of Technology. The experimental tests are done for two floats with different designs. The first one with 50 cm in diameter with a conical shape and inverted cup, the second one is 40 cm with the same design of 50 cm with adding a baffle in the lower part. The results showed that that float indicates a significant efficiency especially the 40 cm buoy with 40% in regular waves instead of 33% for 50 cm float, this is because of the design of the float. This float is characterized by baffles touching the surface of the water as wings. This is leading to the increase of Froude-Krylov and Drag forces in the vertical direction. Finally testing the system in irregular waves to indicate how it is efficient in the real sea condition. The experiments of the irregular waves are done for different wave characteristics of two models JONSWAP and P-M models. The float of 50 cm is tested in these waves and gets an efficiency of 38% as average for all conditions which are higher than any system with the same configurations around the world. These results encourage us to use this system to be installed in real sea conditions for wave energy conversion.

A. Ramadan
College of Engineering and Technology, Arab Academy for Science and Technology and Maritime Transport (AAST), Cairo, Egypt

M. H. Mohamed (✉)
Mechanical Power Engineering Department, Faculty of Engineering EL-Mattaria, Helwan University, P.O Box: 11718, Cairo, Egypt
e-mail: mhmohamed@uqu.edu.sa

Department of Mechanical Engineering, College of Engineering and Islamic Architecture, Umm Al-Qura University, P.O. 5555, Makkah, Kingdom of Saudi Arabia

A. Samad et al. (eds.), *Ocean Wave Energy Systems*, Ocean Engineering & Oceanography 14, https://doi.org/10.1007/978-3-030-78716-5_3

3.1 Introduction

The world is going through a major crisis in energy sources, where most of the world depends mainly on fossil fuels for the production of electric power requirement. The growing number of the population and increasing rate of well-being with a decrease in the energy resources required are the major problems which must be faced now. However, consumption of fossil fuels has become a destructive force, because of emissions, spills and leaks pollutants' dispersion, acid rains, global warming, climate changes, sea level rise as well as strip mining.

The most appropriate solution to meet the energy deficit is the search for alternative sources of renewable energy to replace fossil fuels for the production of electricity. More clean energy is expected to blossom for satisfy the energy requirement for civilization, which is inflating. Wave energy is one of the green energies available and has enormous energy stored. However, the problem is that type of energy has not been effectively used so far.

All wave energy converters absorb power contained in small waves through hydrostatic forces (that is buoyancy versus weight or versus hydrostatic pressure). However, extreme energy is contained in waves arise from hydrodynamic forces. In the present chapter, considerable and promising designs have been investigated as wave energy conversion system. A new design is introduced to the wave energy conversion system to produce power from regular and irregular waves.

3.1.1 Wave Energy Physics and Resource

Waves are caused by the action of wind on the sea or by earth movement. Waves can be in many meters in height and stores a great deal of energy. Sea water waves are derived from the winds as they blow across the oceans. This energy transfer provides a natural storage of wind energy in the water near the free surface. Figure 3.1 indicates briefly the physics of wave generation in open sea. This type of renewable energy encourages the researcher according to the following advantages:

1. The power produced from waves is steady and more predictable than wind or solar energy it is available up to 90%.
2. It is continuous and the waves can travel long distances without considerable energy loss.
3. Waves contain about 1000 times more kinetic energy than the wind.
4. Ocean wildlife will remain undisturbed by the devices.
5. The cost per kilowatt is already in the same range as wind as and less than solar devices.
6. Ocean waves encompass two forms of energy: the kinetic energy and the potential energy of elevated water particles.

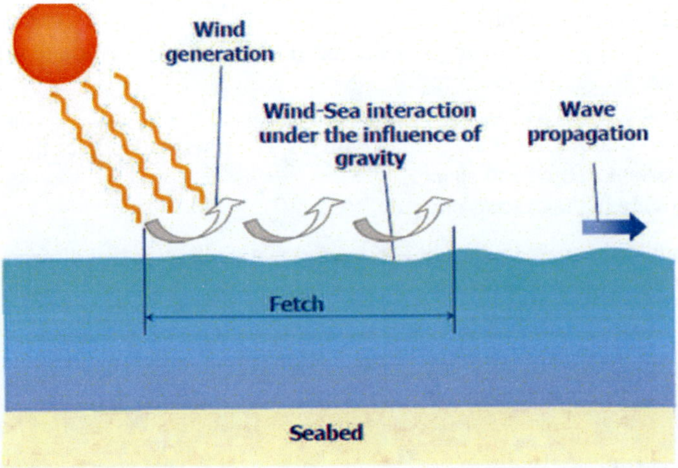

Fig. 3.1 Physics of wave generation

3.2 Theoretical Background and Governing Equations

There are some theories used to explain and solve a lot of problems in water waves. These theories are also used in laboratory experiments and to establish numerical models for wave energy conversion system. One of these theories is the Linear Wave Theory (LWT) is selected to describe wave motion. This theory is the simplest and suitable theory for our case of study. In this chapter, the Linear Wave theory is explained in details to derive the governing equations of float motion.

3.2.1 Linear Wave Theory of Ocean Surface (LWT)

The solution of the equation of motion for water depends on the surface boundary conditions, but the surface boundary conditions are waves. LWT is the simplest form of mathematical representation for the regular wave. The model and simulation of the movement for floats in the wave must be known and the mathematical equations related to that, especially the hydrodynamics forces affecting on floats due to the incident waves.

LWT assumes that, waves are small in amplitude, sinusoidal and progressive in a constant water depth. Once the waves became larger, this theory may not be accurate for representation the wave. The main postulates of this theory are [1]:

1. The fluid is homogeneous.
2. The fluid is incompressible, ideal and inviscid.
3. Surface tension is negligible.
4. Pressure at the surface is uniform and constant.

5. The flow is irrotational.
6. The particular wave being considered doesn't interact with any other water motion.
7. The bottom boundary, or bed, is horizontal, fixed impermeable boundary.
8. The wave amplitude is very small with respect to water depth for the wave form is invariant in time and space.
9. Waves are plane or long-crested.

To simplify the mathematics, it is assumed that the flow is two dimensional with waves traveling in the z-direction.

3.2.2 *Dispersion Relation*

The potential theory of two-dimensional potential flows is used to get the dispersion relation. The potential theory describes the water waves with small amplitude compared to its wave length. Equations of this theory are derived basically from conservation of mass and momentum. This relation is connecting the frequency of water wave to the wave number. The flow of water wave can be described as a potential flow with velocity potential φ (y, z, t). The potential theory assumes that the flow is ideal flow, irrotational and incompressible. Irrotational according to the shear stress, which is relatively small and the sign of shear stress changing due to periodic motion of wave, therefore it will cancel this shear stress. This is meaning that ($\nabla^2\varphi = 0$ (Laplace equation)) where (φ) is potential flow [2]. The coordinate system of wave propagation and float motion can be seen in Fig. 3.2.

$$\frac{\partial\varphi}{\partial y} = v\text{: Velocity component in horizontal direction.} \tag{3.1}$$

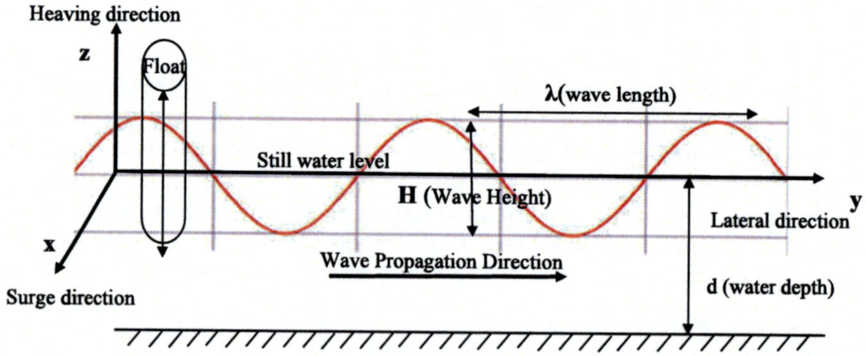

Fig. 3.2 Wave coordinates system for LWT

$$\frac{\partial \varphi}{\partial z} = w\text{: Velocity component in vertical direction.} \tag{3.2}$$

Fluid flow in wavy motion must satisfy the continuity equation, which is written for two dimensional flows as in Eq. (3.3) after substituting Eqs. (3.1) and (3.2) in it.

$$\frac{\partial^2 \varphi}{\partial y^2} + \frac{\partial^2 \varphi}{\partial z^2} = 0 \tag{3.3}$$

This linear partial differential equation can be solved for wave motion using the following boundary conditions:

Horizontal bottom at z = d, this is meaning no flow across this boundary:

$$\frac{\partial \varphi}{\partial z} = 0 \text{ at } z = d \tag{3.4}$$

Kinematic condition at water surface: the flow velocity in the z direction $\frac{\partial \varphi}{\partial z}$ at this point is equal to the total derivative of the water surface with respect to time in the vertical direction. This means $z - \zeta(y, z, t)$ must be a constant at the surface, where ζ is the wave elevation:

$$\frac{d}{dt}[z - \zeta(y, t)] = 0 \text{ at the surface } \frac{d}{dt} = \frac{\partial}{\partial t} + v\frac{\partial}{\partial y} \tag{3.5}$$

$$\text{Then } \frac{\partial \varphi}{\partial z} = \frac{\partial \zeta}{\partial t} + v\frac{\partial \zeta}{\partial y} \text{ at water surface} \tag{3.6}$$

Assume that, the particle velocity components are small compared to the wave velocity and the derivatives of the wave elevation in z direction, since the wave elevation is assumed small compared to the wave length, this results in the linearized kinematic boundary condition:

$$\frac{\partial \varphi}{\partial z} = \frac{\partial \zeta}{\partial t} \text{ at } z = \zeta \tag{3.7}$$

Dynamic free surface condition: at the water surface the pressure is atmospheric pressure which is constant p_o. The Bernoulli's equation at the surface is [2]:

$$\left(\rho\frac{\partial \varphi}{\partial t} + \rho g\zeta + p_o\right) + \frac{\rho}{2}\underbrace{\left[\left(\frac{\partial \varphi}{\partial y}\right)^2 + \left(\frac{\partial \varphi}{\partial z}\right)^2\right]}_{\text{Continuity equation}} = \text{Constant} \tag{3.8}$$

The additional condition can be used for wave propagation as follow:

$$\varphi(y, z, t) = \varphi(y - ct, z) \tag{3.9}$$

$$\zeta(y,t) = z\,(y - ct) \tag{3.10}$$

The limiting case of small wave height is used by assuming that the product of wave heights and potential is of small order of magnitude than corresponding linear occurrence.

$$\frac{\partial\varphi}{\partial z} = \frac{\partial\zeta}{\partial t} \text{ at z} = \zeta \tag{3.11}$$

$$\frac{\partial\varphi}{\partial t} - g\zeta = \text{Constant} \tag{3.12}$$

Using Taylor expansion in z direction, the required derivatives at free surface are:

$$\frac{\partial\varphi(y,\zeta,t)}{\partial z} = \frac{\partial\varphi(y,0,t)}{\partial z} + \frac{\partial\varphi^2(y,0,t)}{\partial z^2}\zeta + \ldots \tag{3.13}$$

$$\frac{\partial\varphi(y,\zeta,t)}{\partial t} = \frac{\partial\varphi(y,0,t)}{\partial t} + \frac{\partial\varphi^2(y,0,t)}{\partial t\partial z}\zeta + \ldots \tag{3.14}$$

Combining the kinematic condition with linearity and only the first element of the expansion is retained as the others are of higher order, then the continuity equation is written after derivative with respect to time as follow:

$$\frac{\partial^2\varphi}{\partial t^2} = g\frac{\partial\varphi}{\partial z} \text{ at z} = 0 \tag{3.15}$$

Finally the solution of this equation leads to the dispersion relation for both shallow and deep water waves as follows [3]

$$\omega^2 = g\,k\tanh(kd) \text{ Shallow Water} \tag{3.16}$$

$$\omega^2 = g\,k \text{ Deep Water} \tag{3.17}$$

d is the water depth
g gravity acceleration and k is wave number.

From this relation, the water wave is divided into three types according to the depth and wave length ratio; this is summarized in Table 3.1. The deep water portion (red part) is the type of wave equations which are used in the present chapter due to the condition of wave tank and experiments conditions. From these equations the speed of wave can be determined to get the available power in water according to the condition of water wave. For the sinusoidal gravity wave and the coordinates system of wave propagation could be seen in Fig. 3.2. The stated limit for $\frac{d}{\lambda}$ gives a dispersion relation accurate within 10%. Because wave properties can be measured

Table 3.1 Wave equations

	Shallow water $\frac{d}{\lambda} \le \frac{1}{20}$	Transition water $\frac{1}{20} < \frac{d}{\lambda} < \frac{1}{2}$	Deep water $\frac{d}{\lambda} \le \frac{1}{2}$
1. Wave profile	$h_\omega = \frac{H}{2}\cos\left(\frac{2\pi x}{\lambda} - \frac{2\pi t}{T}\right) = \frac{H}{2}\cos\theta$		
2. Wave velocity	$C = \frac{\lambda}{T} = \sqrt{gd}$	$C = \frac{\lambda}{T} = \sqrt{\frac{gT}{2\pi}\tan h\left(\frac{2\pi d}{\lambda}\right)}$	$C_0 = \frac{\lambda}{T} = \frac{gT}{2\pi}$
3. Wave length	$\lambda = CT = T\sqrt{gd}$	$\lambda = \frac{gT}{2\pi}\tan \mathrm{h}\left(\frac{2\pi d}{\lambda}\right)$	$\lambda_0 = \frac{gT}{2\pi} = C_0 T$
4. Group velocity	$C_g = C = T\sqrt{gd}$	$C_g = \frac{1}{2} = \left(1 + \frac{4\pi d/\lambda}{\sin \mathrm{h}(4\pi d/\lambda)}\right)C$	$C_g = \frac{C_0}{2} = \frac{gT}{4\pi}$
5. Water particle velocity a. Horizontal b. Vertical	$u = \frac{H}{2}\sqrt{gT}\cos\theta$ $v = \frac{H}{2}\left(1 + \frac{\pi}{\alpha}\right)\cos\theta$	$u = \frac{H}{2}\frac{gT}{\lambda}\frac{\cos \mathrm{h}(2\pi(z+d)\lambda)}{\sin \mathrm{h}(2\pi d/\lambda)}\cos\theta$ $\zeta = \frac{H}{2}\frac{gT}{\lambda}\frac{\sin \mathrm{h}(2\pi(z+d)\lambda)}{\sin \mathrm{h}(2\pi d/\lambda)}\cos\theta$	$u = \frac{\pi H}{T}e^{2\pi z/\lambda}\cos\theta v = \frac{\pi H}{T}e^{2\pi z/\lambda}\sin\theta$
6. Water particle acceleation a. Horizontal b. Vertical	$\dot{u} = \frac{\pi H}{T}\sqrt{\frac{g}{d}}\sin\theta$ $\dot{v} = -2H\left(\frac{\pi}{T}\right)^2\left(1 + \frac{\pi}{d}\right)\cos\theta$	$\dot{u} = \frac{gH\pi}{\lambda}\frac{\cos \mathrm{h}(2\pi(z+d)/\lambda)}{\sin \mathrm{h}(z+d/\lambda)}\sin\theta\dot{v} = \frac{gH\pi}{\lambda}\frac{\sin \mathrm{h}(2\pi(z+d)/\lambda)}{\sin \mathrm{h}(z+d/\lambda)}\cos\theta$	$a_u = 2H\left(\frac{\pi}{T}\right)^2 e^{2\pi z/\lambda}\sin\theta a_v = 2H\left(\frac{\pi}{T}\right)^2 e^{2\pi z/\lambda}\cos\theta$
7. Water particle displacement a. Horizontal b. Vertical	$\zeta = \frac{TH}{4\pi}\sqrt{\frac{g}{d}}\sin\theta$ $\zeta = \frac{H}{2}\left(1 + \frac{z}{d}\right)\cos\theta$	$\zeta = \frac{H}{2}\frac{\cos \mathrm{h}(2\pi(z+d)\lambda)}{\sin \mathrm{h}(2\pi d/\lambda)}\sin\theta$ $\zeta = \frac{H}{2}\frac{\sin \mathrm{h}(2\pi(z+d)\lambda)}{\sin \mathrm{h}(2\pi d/\lambda)}\cos\theta$	$\zeta = \frac{H}{2}e^{2\pi z/\lambda}\cos\theta$ $\zeta = \frac{H}{2}e^{2\pi z/\lambda}\cos\theta$
8. Subsurface pressure	$p = pg(h_\omega - z)$	$p = pgh_\omega\frac{\cos \mathrm{h}(2\pi(z+d)/\lambda)}{\sin \mathrm{h}(2\pi/\lambda)} - pgz$	$p = pgh_\omega e^{2\pi z/\lambda} - pgz$

with accuracy 5–10%, the approximations are useful for calculating wave properties including water particles displacement, water particles velocity and acceleration, their kinematics and dynamics including wave pressure, their resultant forces and moments.

3.2.3 *Energy in Water Wave*

The energy available in the water wave was explored and studied for decades since 1940. Salter published a paper on the wave energy Duck [4] for ground breaking. The rate of wave transport of energy across some line in the sea is known as the power and, therefore the power of a wave can be calculated from the stored energy per unit width of the wave. The mass per unit area of sea surface; m_{wave} in half wave above the still water level is [5]

$$m_{wave} = \left(\frac{\rho}{2}\right)\left(\frac{H}{2\sqrt{2}}\right) \text{ kg/m}^2 \quad (3.18)$$

According to the potential theory, we can state the energies as follows.

1. The potential energy per unit area of sea surface, E_p, can then be

$$E_p = m_{wave} g \Delta h = \left(\frac{\rho}{2}\right)\left(\frac{H}{2\sqrt{2}}\right) g \left(\frac{H}{2\sqrt{2}}\right) = \frac{\rho g H^2}{16} \text{ J/m}^2 \quad (3.19)$$

2. The kinetic energy per unit area of sea surface, E_k of a wave is calculated from the moving part of wave using the Airy's linear wave theory velocity v as follow:

$$E_k = \frac{m_{wave} v^2}{2} = \frac{\rho}{2} \int_0^{\infty} a^2 \omega^2 e^{-2k_o z} dz \quad (3.20)$$

The total energy $E = E_k + E_p$.

From previous equations, it can be got the wave power per unit width of a wave [5].

$$P_w = E\, c_g = \frac{1}{2}\, \rho\, g\, a^2 c_g \text{ W/m} \quad (3.21)$$

Where

a is the amplitude of a wave = (H/2).

(c_g) is the Group velocity and defined as a series of wave's velocity with varying amplitude which is called wave packet.

$$c_g = \frac{gT}{4\pi} \text{ m/s (deep water)} \quad (3.22)$$

$$c_g = C = \sqrt{gd} \text{ (shallow water)} \quad (3.23)$$

Another formula for the wave power per unit width was stated by Orer and Ozdamar [6]

The velocity of wave is $c_g = C/2\,[1 + (4\pi d/\lambda)/\sinh(4\pi d/\lambda)]$ where $c = \frac{\lambda}{T}$, then the power is written as follow:

$$P_w = \frac{1}{16} \rho g H^2 \frac{\lambda}{T} [1 + (4\pi d/\lambda)/\sinh(4\pi d/\lambda)] \text{ W/m} \quad (3.24)$$

The description of the energy available in real sea wave is using the wave energy spectrum S(f) is the most commonly used method for accurate estimation of this power [7]. Real sea waves can be described in terms of their significant wave height (H_s). First it is needed to know the definition of (H_s) for looking at a wind driven sea. A practical definition that is often used the height of the highest 1/3 of the waves, $H_{1/3}$. Therefore, [7] described the total stored energy in a wave per unit area of sea surface in terms of (H_s) and in terms of (H_s) the wave energy spectrum S(f), which is detailed as follows [8]:

$$\mathrm{E} = \frac{\rho g}{2} \int_0^\infty S(f) df = \frac{\left(\rho g H_s^2\right)}{16} \tag{3.25}$$

$$\mathrm{H_s} = \sqrt{2}\,\mathrm{H} \tag{3.26}$$

The wave power level per unit width in a wave is:

$$\mathrm{P_w} = \rho g \int_0^\infty c_g S(f) df \tag{3.27}$$

In deep water when depth is higher than half wave length, the power of a wave can be written in terms of significant wave height and periodic time [9, 10]:

$$\mathrm{P_w} = 0.3\,\mathrm{H_s^2 T_e k\ W/m} \tag{3.28}$$

where $\mathrm{T_e} = 0.86\,\mathrm{T_p}$ and $\mathrm{H_s} = 4\sqrt{h_{w.dv.}}$.

$h_{w.dv.}$ Standard deviation of water displacement
T_p The average peak periods of waves.

3.2.4 *Wave Energy Spectrum*

These spectrums are used mainly in this chapter to optimize the frequency range for the irregular wave used. This frequency range is used mainly in the experiments work to safe effort and time. From these calculations, the significant wave heights can be determined with these optimized frequency ranges Eq. (3.29).

$$\mathrm{H_s} = 4\sqrt{\mathrm{S}_{(\omega)} \Delta\omega} \tag{3.29}$$

Figure 3.3 illustrates these spectrums and optimum frequencies should be used. The maximum wave spectrum means maximum energy could be produced from wave. In this study, the Pierson- Moskowitz and JONSWAP spectrums are selected as the wave energy spectrums model to be used as they are suitable for describing fully developed seas.

Fig. 3.3 Wave spectrums

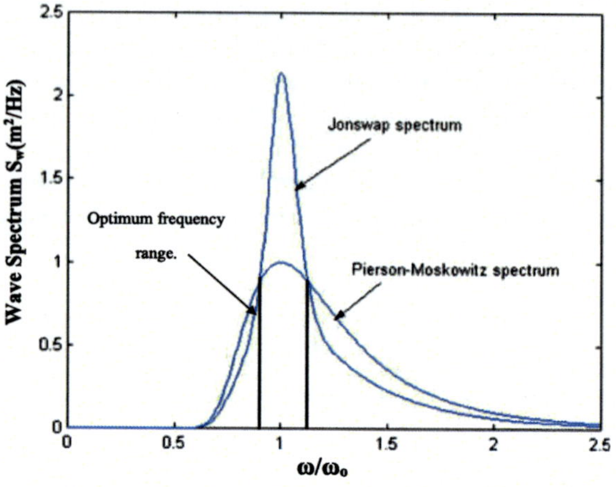

3.2.4.1 Pierson-Moskowitz Spectrum

It is the simplest spectrum used for describing wave energy spectrum [11]. They assumed that if the wind blows steadily for a long time over a large area, waves would come into equilibrium with the wind. This is the concept of a fully developed sea. Here, a long time is roughly ten thousand wave periods, and a large area is roughly five-thousand wave lengths on a side. It is the simplest spectrum used for describing wave energy spectrum [11]. They assumed that if the wind blows steadily for a long time over a large area, waves would come into equilibrium with the wind. This is the concept of a fully developed sea. Here, a long time is roughly ten thousand wave periods, and a large area is roughly five-thousand wave lengths on a side.

After some measurements were done in the north Atlantic for waves, they calculated the wave spectra for various wind speeds, and they found that the spectrum was [10].

$$S_w = \frac{\alpha^2 g}{\omega^5} \exp.\left[- \beta \left(\frac{\omega_o}{\omega}\right)^4 \right] \tag{3.30}$$

where $\alpha = 8.1 \times 10^{-3}$, $\beta = 0.74$

ω_o natural angular frequency,
ω Angular frequency.

The modified Pierson- Moskowitz spectrum in frequency domain is:

$$S_f = \frac{A_s}{(2\pi)^{4f^5}} \exp.\left[\frac{-B_s}{(2\pi)^{4f^4}} \right] \tag{3.31}$$

where $A_s = \frac{173H_s}{T_{av.}}$ and $B_s = \frac{691}{T^4_{av.}}$.

$T_{av.}$ is the average of peak periodic times of wave records.

3.2.4.2 JONSWAP Spectrum

After analyzing data collected during the Joint North Sea Wave Observation Project-JONSWAP [12], found that the wave spectrum was never fully developed. It continues to develop through non-linear, wave-wave interactions even for very long times and distances. They therefore proposed a spectrum in the form:

$$S_w = \frac{\alpha^2 g}{\omega_o^5} \exp.\left[\frac{-5}{4}\beta\left(\frac{\omega_p}{\omega_o}\right)^4\right]\gamma^r \tag{3.32}$$

$$r = \exp.\left[-\frac{(\omega_o - \omega_p)^2}{2\sigma^2\omega_p^2}\right] \tag{3.33}$$

ω_p Peak angular frequency of wave records.

Wave data collected during the JONSWAP experiment were used to determine the values for the constants.

$$\alpha = 0.0081,\ \gamma = 3.3$$

$$\sigma = \begin{cases} 0.07 & \omega_o \le \omega_p \\ 0.09 & \omega_o > \omega_p \end{cases} \tag{3.34}$$

The JONSWAP spectrum is similar to the Pierson-Moskowitz spectrum except that it continues to grow gradient or time as specified by the α term, and the peak in the spectrum is more pronounced as specified by the γ term. The latter turns out to be particularly important because it leads to enhanced non-linear interactions and a spectrum those changes in time according to the theory [12].

3.2.5 Forces on Floating Bodies

Newton equation of motion is used to describe the oscillating motions of the floating body as it responds to the incident waves, causing the hydrodynamic loading or forces on this body. These forces could be divided into two types according to the procedures of work. The first type is the lateral force, which is produced from the incident wave on the body. This force is calculated from the Morison equation [13], which represents the design point of the system to withstand this force on the body. The second is Froude-Krylov force, which depends mainly on the water wave particles acceleration vertically and it is converted to lift force for moving the system. In the following paragraphs all forces will be described in details.

3.2.5.1 Morison Equation

The acceleration of water particles and velocity horizontally between fluid and float is the main source for this force (Fig. 3.4). The general Morison equation are obtained or developed from the hydrodynamic pressure forces corresponding only to the velocity and acceleration component normal to the float axis. The axial components are not considered in this equation [13]. First of all, the dynamic pressure on the float can be seen in Eq. (3.35)

$$dF = dF_m + dF_d = C_m \rho \frac{\pi D^2}{4} dz \frac{\partial u}{\partial t} + C_d \frac{\rho}{2} D \, dz \, |u| \, u \tag{3.35}$$

u and $\frac{\partial u}{\partial t}$ are the horizontal wave particle velocity and acceleration respectively.

φ Velocity potential.
$C_m = 1 + C_a$ Inertia coefficient for cylinder and added mass consideration.

C_a Radiation coefficient.
C_d The drag coefficient.

In this equation the hydrodynamic transparent and interference of incident waves are negligible. For this reason, the new equation of acceleration can be written $\frac{\partial u}{\partial t} = \frac{du}{dt}$ which is obey to the linear wave theory.

Then the Morison equation rewrite as:

$$dF_m = C_m \rho \frac{\pi D^2}{4} dz \frac{du}{dt} \tag{3.36}$$

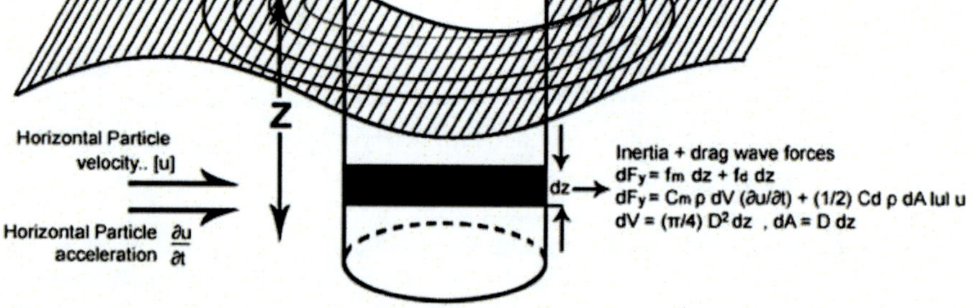

Fig. 3.4 Lateral wave forces on a vertical cylinder

This is (F_m) Froud-Krylov and hydrodynamic mass forces which is called linear inertia force. The nonlinear (F_d) drag force is caused by viscous effects and the associated downstream.

$$dF_d = C_d \frac{\rho}{2} D\, dz\, |u|\, u \tag{3.37}$$

So, for moving the cylinder with water wave, Froude-Krylov is remaining without changing, but the drag force or hydrodynamic mass force is taking into consideration the relative velocity between cylinder and water. The new Morison equation now is:

$$dF = \rho dV \frac{du}{dt} + C_a \rho dV \left(\frac{du}{dt} - \dot{u}_c \right) + C_d \frac{\rho}{2} dA\, |u - u_c|\, (u - u_c) \tag{3.38}$$

u_c is the velocity of the cylinder in water wave.
C_a,C_d These coefficients are known according to experiments were done on cylinders and they equal to unity for cylindrical shapes.

3.2.5.2 Vertical Forces of Float

The vertical forces on the float are responsible for the motion vertically of float with water wave, which is coming from the pressure distribution on the surface draft of the float, drag force and the inertia force. To develop the equation for this type of force it should be known first the basic equations of pressure distribution on float. From Eq. (3.39), Froude-Krylov [14] follows from the pressure field of the undisturbed incident wave.

$$F = -\int_s p\, n\, ds\; = \rho \int_s \frac{\partial \emptyset}{\partial t} n\, ds \tag{3.39}$$

The mass of float in a harmonic wave field is

$$m_b = \rho\, V \tag{3.40}$$

For the force effects the float performs are:

1. Heave motion (vertical motion)

$$h_f = H_f e^{-i(wt-\delta)} = \; -i\, \omega\, h_f \tag{3.41}$$

h_f : float displacementH_f : float displacement amplitude

2. Vertical velocity

$$\underset{f}{\dot{h}} = H_f(-i\, \omega) e^{-i(wt-\delta)} = -\omega\, h_f \tag{3.42}$$

3. Hydrostatic force

$$F_R = -\rho g A_o h_f \tag{3.43}$$

A_o is the water line area.

4. Inertia force

$$F_T = -\rho V h_f \tag{3.44}$$

5. Hydrodynamic inertia force (added mass force)

$$F_H = -m\left(h_f - \underset{w}{\dot{v}}\right) \tag{3.45}$$

$$m = m_b + m_{add} \tag{3.46}$$

$\underset{w}{\dot{v}}$ Water particle acceleration in vertical direction.
m_{add} The added mass of water due the geometry [15].

$$m_{add} = \frac{\rho D^3}{6} + \mathrm{m_{inverted\ cup}} \tag{3.47}$$

$\mathrm{m_{inverted\ cup}}$ mass of water inside the inverted cup.
D Buoy diameter

$$\mathrm{a} = -2H\left(\frac{\pi}{T}\right)^2 e^{\frac{2\pi h}{\lambda}} \cos\theta \tag{3.48}$$

6. Linear drag force

$$F_D = -C_D\frac{\rho}{2}A_o\left(\underset{f}{\dot{h}} - v_w\right) \tag{3.49}$$

v_w Water particle velocity in vertical direction.

From the previous equations of forces in details, the equilibrium equation is written for forces with Froude-Krylov due to Eq. (3.50).

$$\rho\int_S \frac{d\phi}{dt}n_z dS - \rho g A_o h_f - \rho V\ddot{h}_f - m\left(\ddot{h}_f - \dot{v}_w\right) - C_D\frac{\rho}{2}A_o\left(\ddot{h}_f - v_w\right) = 0 \tag{3.50}$$

By separation of the variables; the linear equation of motion is obtained:

$$m\,\ddot{h}_f - C_D\frac{\rho}{2}A_o\ddot{h}_f + \rho g A_o h_f - \rho\int\limits_S \frac{dv_w}{dt}n_z dS + m\,\dot{v}_w$$
$$+C_D\frac{\rho}{2}A_o v_w = F_{ex.}e^{-i\omega t} \qquad (3.51)$$

$F_{ex.}$ The excitation force of the water wave.

From these equations, the natural frequency of any floated buoy is as follow:

$$\omega_o = \sqrt{\frac{\rho A_o g}{m_b + m_{add}}} \qquad (3.52)$$

The natural frequency is the first parameter must be calculated to know the resonance case for any float used in the WEC system. This resonance is happen according to the incoming wave frequency, it is used to predict the maximum power may be captured at this case.

The analytical and numerical simulations are depend mainly on solving the equation of motion and get the power captured theoretically.

The first procedure in theoretical part is the calculation of natural frequency, then the horizontal forces that influence on the motion of float in waves. This is performed by using the Morison equations form Eqs. (3.35) to (3.38). Secondly the equation of motion in the vertical direction is solved. These equations from (3.39) to (3.52) are used in the analytical and simulation to know the optimum float shapes whose gives the maximum power captured.

3.3 Wave Energy Conversion Systems

Waves had the potential to provide a completely sustainable source of energy which could be captured and converted into electricity by wave energy converters. These WEC had been developed to extract energy from shoreline out to the deeper water offshore. These devices were generally categorized by the installation location and the Power Take-Off (PTO) system. Locations were shoreline, near shore and offshore. Most devices could be characterized as belonging to six types [16]; Attenuator; Point absorber; Oscillating wave surge converter; Oscillating water column; Overtopping device; and submerged pressure differential.

In the present section the available data on wave energy conversion projects around the world is reviewed.

3.3.1 *Attenuator*

An attenuator is a floating device which works parallel to the wave direction and rides the waves. Movements along its length could be constrained to produce energy. This was the case of the PELAMIS WEC. PELAMIS project [17].

This was the first commercial scale wave power station in the world. The sea snake-like device was made up of a series of three semi-submerged cylinders linked by hinged joints. The power conversion units in each 140-m long Pelamis P-750 s generated up to 750-kW with efficiency 16% for offshore heaving body.

3.3.2 *Oscillating Wave Surge Converter*

This device extracts both wave surges and its particle motion energies. The WEC arm oscillated as a pendulum mounted on a pivoted joint in response to the water motion due to waves. This was the case of the OYSTER WEC [18].

OYSTER (1) is currently undergoing sea trials with full-scale 315 kW Oyster wave energy device.

3.3.3 *Oscillating Water Column*

An Oscillating Water Column (OWC) is a partially submerged, hollow structure. It is open to the sea below the water line, enclosing an air column on top of a water column. Waves cause the water column to rise and fall, which in turn compresses and decompresses the air column. This trapped air is allowed to flow to and from the atmosphere via a turbine, which usually had the ability to rotate regardless of the direction of the airflow. The rotation of the turbine was used to generate electricity [19].

The OWC WEC most used air turbines were the Wells and the Impulse type. The Denniss-Auld type was also used [20]. Examples of this type are CEANLINX project [21], LIMPET project [22] and PICO plant [23].

3.3.4 *Overtopping Device*

This type of WEC relies on physical capture of water from waves which is held in a reservoir above sea level, before being returned to the sea through conventional low-head turbines which generates power. An overtopping device might use collectors to concentrate the wave energy [24]. The WAVE DRAGON project was a good illustration of this type of WEC. WAVE DRAGON project is an offshore floating overtopping WEC device. This WEC is slack moored and can orient itself to face into the dominant wave direction. Since 2003, a scale-model of the WAVE DRAGON device had been deployed in Denmark and connected to a public power grid. The rated power of the Nissum Bredning prototype is 20 kW resembling 4 MW when deployed in a relatively low-energy (24 kW/m) wave climate and 7 MW when deployed in a 36 kW/m climate; this mean average efficiency is 20%.

3.3.5 Submerged Pressure

These WECs, also called Direct Drive, are typically located near shore and attached to the seabed. The motion of the waves causes the sea level to rise and fall above the device, inducing a pressure differential in the WEC. The alternating pressure could then pump fluid through a system to generate electricity. The ARCHIMEDES WAVE SWING (AWS) is the typical project in this case. The AWS consists of a hollow cylinder and a lid, called floater. The cylinder is filled with air, and is fixed on the sea bottom [25]. Energy is extracted from the floater reciprocating motion. The AWS is designed for deployment in exposed ocean of water depths of between 70 and 150 m. The model predicts capacity factors in the region of 25% for a 2.5 MW device deployed in a 44 kW/m resource indicating an average annual energy production of around 5.1 GWh per device.

3.3.6 Point Absorber

Point absorbers, often simply called a buoy, are relatively small devices in comparison to other WEC types. They are typically cylindrical in shape and constrained to one major motion degree, usually up-and-down. They are generally significantly smaller in diameter than a wavelength.

A point absorber captured energy in all directions through its movements at near the water surface. The PTO system might take a number of forms as the following main projects.

1. CETO project [26]. The CETO system distinguished itself from other traditional WEC being a fully submerged, pumping technology with the power generated onshore by a standard hydroelectric turbine system. The high-pressure seawater was also used to supply a reverse osmosis desalination plant.
2. SEAREV project [27] and Clément et al.[28]

The SEAREV WEC is a floating device enclosing a heavy horizontal axis wheel serving as an internal gravity reference The efficiency is reached theoretically to 80%. Two major advantages of this arrangement are;

1. All the moving parts (mechanic, hydraulic, electric components) were sheltered from the sea action inside a closed and waterproof shell.

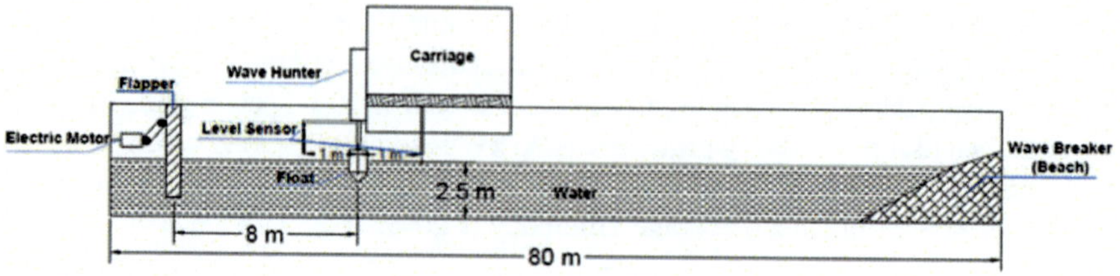

Fig. 3.5 The wave tank dimension with position of the wave hunter

2. The wheel pendulum involved neither end stop nor any security system limiting the stroke.
3. Wave Hunter.

Laboratory experiments have been conducted for the Wave Hunter system in the wave tank of the Fluid Dynamics and Ship Theory Institute (FDS) of Hamburg University of Technology (TUHH) [29]. The wave tank has the dimension of 80 m long, 5 m width and 2.5 m water depth. The test rig is installed at a distance of 8 m away from the wave generator to avoid the strong effect of reflected waves as shown in Fig. 3.5. The float is connected to the WAVE HUNTER system which is mounted on the towing carriage as in Fig. 3.6. Float consists of an inverted cup and conical shape in lower end. The inverted cup has two main advantages:

1. Preventing the slamming of float with wave, therefore it increases the extracted power.
2. The large amount of mass required for synchronize the float with wave is not needed.

3.3.6.1 Power Extraction Experimentally from Regular Wave.

The Power Capture of Float 50-C

The first experiment performed for power captured is done on the float 50-C with regular wave due to the results coming out from RAO's experiments. As it is illustrated before in work scenario, the reason of measuring water level behind the float is to see if those types of floats have the ability to work as a wave breaker. The installations of whole instruments are discussed previous paragraph and in water tank. The natural frequency of this float is $\omega_o = 3.34$ rad/s ($T_o = 2$ s), where at this value the maximum RAO's is reached.

Figure 3.7 shows the wave and float displacement for wave characteristic of H = 20 cm & T = 2 s. From this figure where mc = 0, the flat portion is inside the water with larger value than other cases. Therefore, it is subjected to high laterals force from wave motion. Then, the float motion is much delayed than the water motion with high damping in this motion. This means less power could be gained from wave and of course less efficiency of conversion due to reduction in float's height which

Fig. 3.6 Float designs with dimension in (mm)

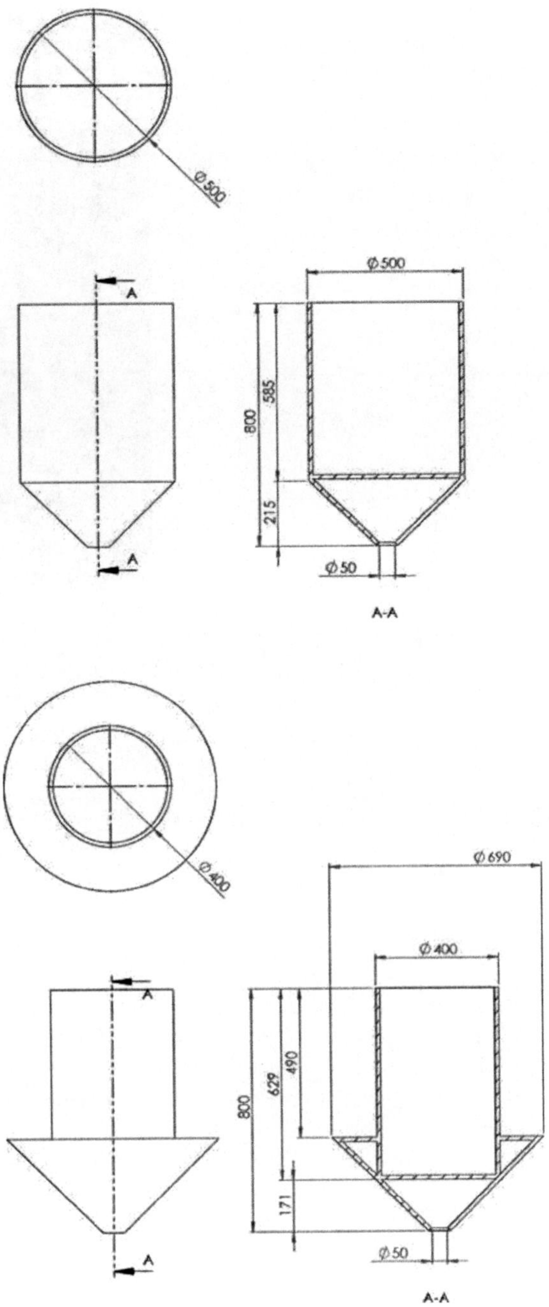

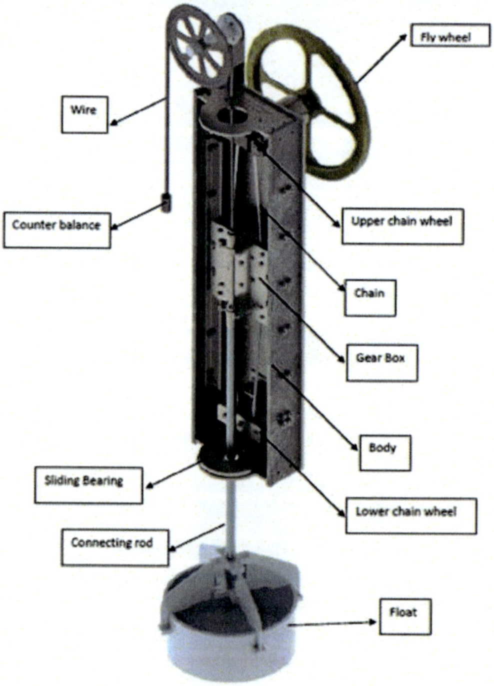

Fig. 3.7 Pictures of the connection method of float in the water tank to the system

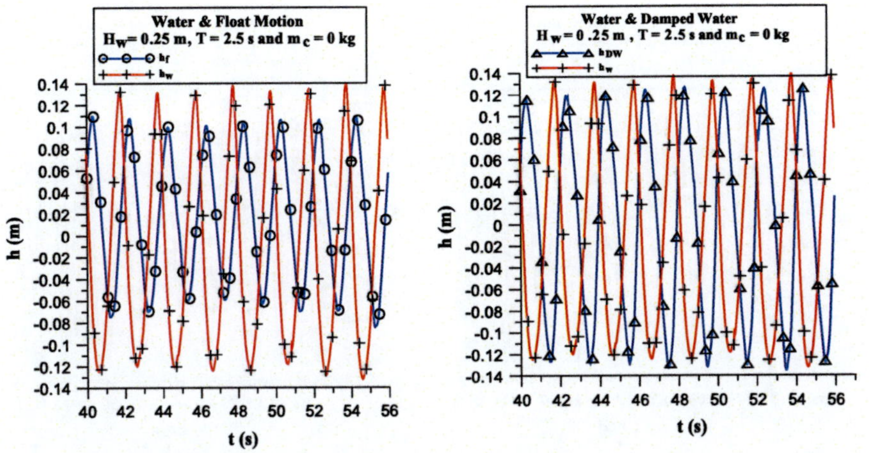

Fig. 3.8 Water displacement with float and damped water for H = 20 cm, $m_c = 0$, T = 2 s, F_{load} = 3.15 N, and RPM = 60

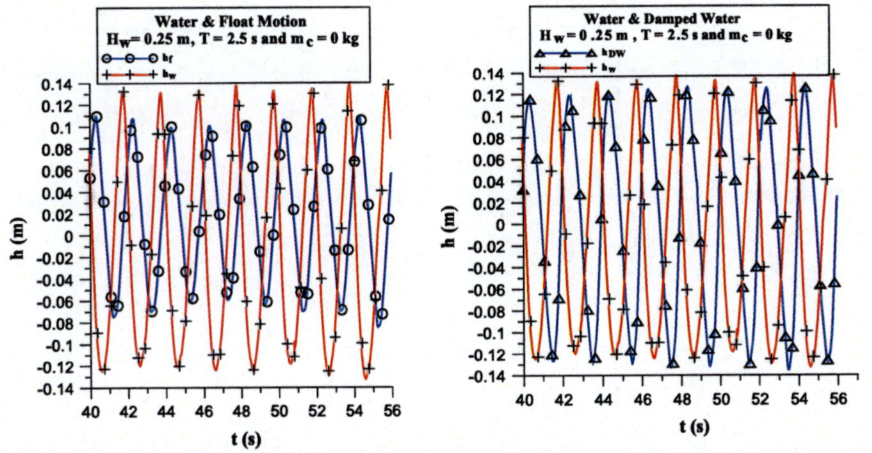

Fig. 3.9 Water displacement with float and damped water for H = 25 cm, $m_c = 0$, T = 2.5 s, F_{load} = 3.85 N, and RPM = 66

is acquired. The reduction in power gained also is coming from the high vibration of the connected rod and it is noticed during the experiments in this case with the system. Furthermore, it increases the friction in the movable parts. The same trend can be seen in other wave characteristics as in Figs. 3.8, 3.9, 3.10, 3.11, 3.12, and 3.13. The damped water behind the float is about 20% only of the incident water wave heights; this due to the diameter of float is only 0.1 of water tank width.

After adding mass of 10 kg to the counter weight (m_c = 10 kg) and same wave characteristics incident as in first case of 0 kg, it is found in Fig. 3.14 an improvement

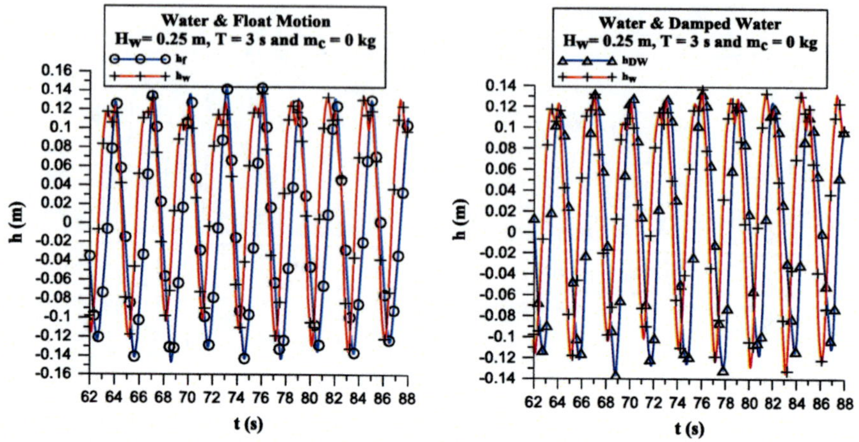

Fig. 3.10 Water displacement with float and damped water for H = 25 cm, $m_c = 0$, T = 3 s, $F_{load} = 4.68$ N, and RPM = 61.5

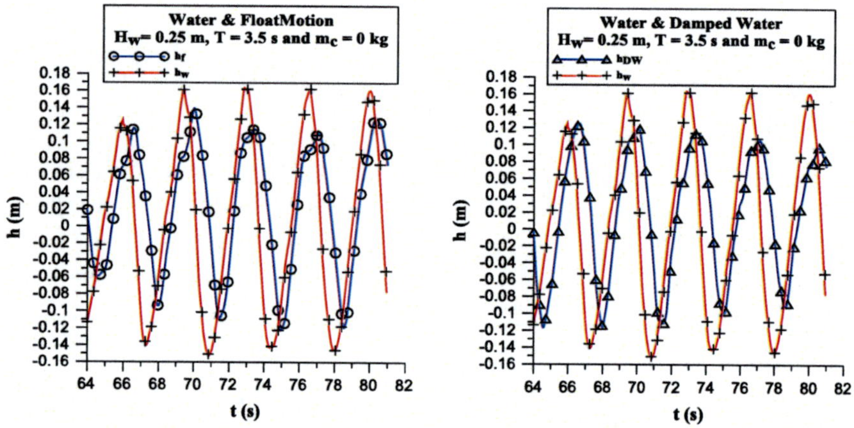

Fig. 3.11 Water displacement with float and damped water for H = 25 cm, $m_c = 0$, T = 3.5 s, $F_{load} = 4.08$ N, and RPM = 54.15

is obtained in the motion of float related to the motion of water wave. This also leads to more damping in water behind the float. That is because more energy acquired by float from water wave than first case. Figures 3.14, 3.15, 3.16, 3.17, 3.18 and 3.19 show what explained previously for other water wave generation with different characteristics of wave heights and period. Even for this case of work ($m_c = 10$ kg), there is a flat portion exposed to the lateral wave force with small value compared to the first case. The vibration and friction of connecting rod and movable parts are less than before. In this case the damped water behind the float is increased with

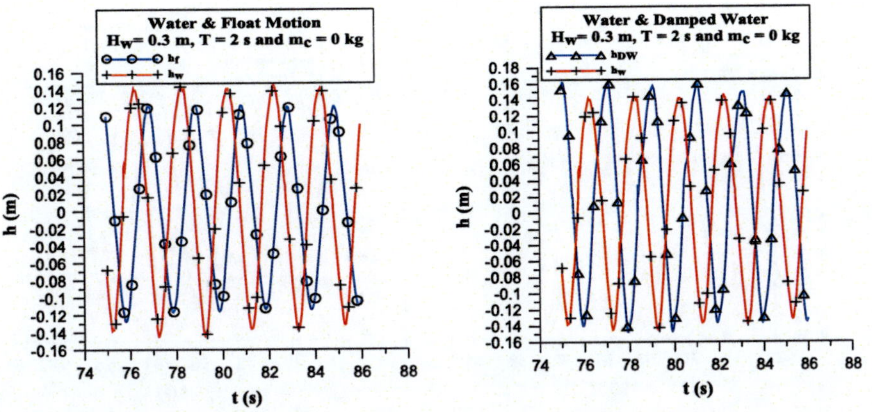

Fig. 3.12 Water displacement with float and damped water for $H = 30$ cm, $m_c = 0$, $T = 2$ s, $F_{load} = 6.17$ N, and RPM = 75

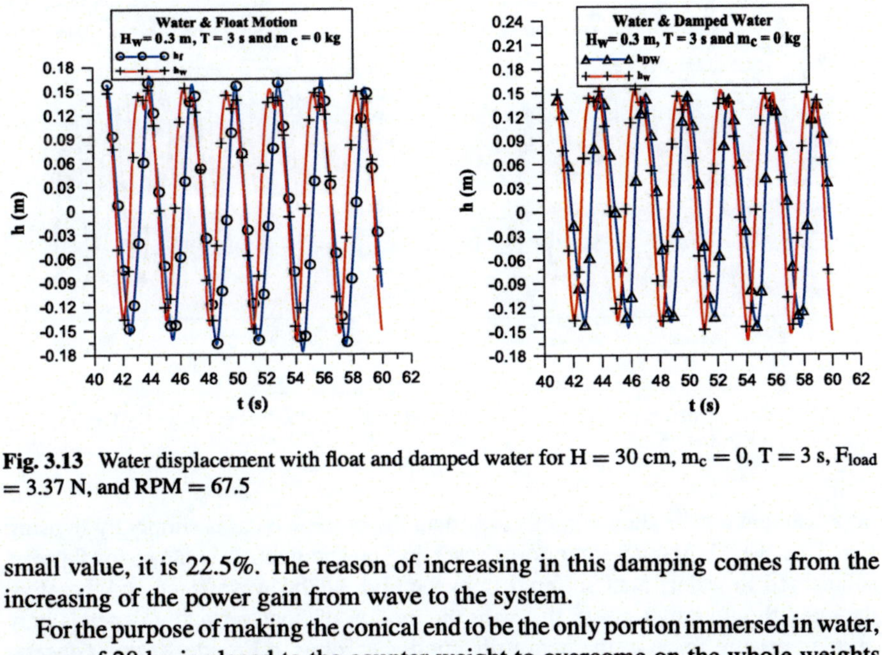

Fig. 3.13 Water displacement with float and damped water for $H = 30$ cm, $m_c = 0$, $T = 3$ s, $F_{load} = 3.37$ N, and RPM = 67.5

small value, it is 22.5%. The reason of increasing in this damping comes from the increasing of the power gain from wave to the system.

For the purpose of making the conical end to be the only portion immersed in water, a mass of 20 kg is placed to the counter weight to overcome on the whole weights of gearbox, connection rod, and the float with holders. Then it is tested, this case for the same conditions of water waves as before to know the optimum part should be immersed from the float in water. Figure 3.20, 3.21, 3.22, 3.23, 3.24, 3.25, 3.26, and 3.27 describe the whole experiment conditions. The float height and wave height are matched with reasonable value. This means, much power can be captured from those waves and directly increases the efficiency of system for energy conversion. This

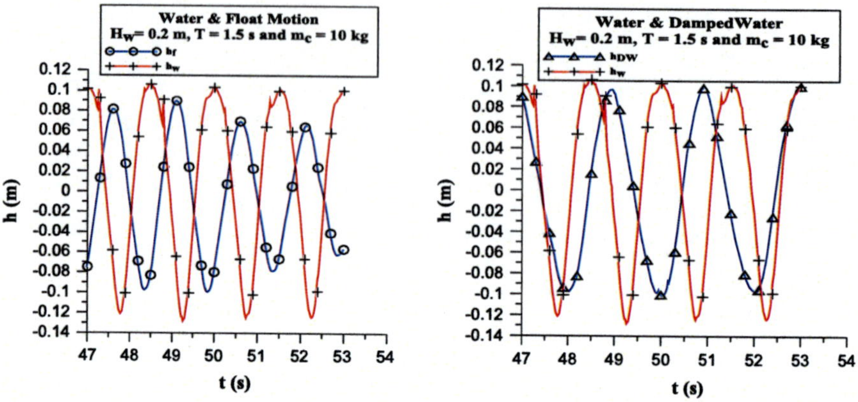

Fig. 3.14 Water displacement with float and damped water for H = 20 cm, m_c = 10, T = 1.5 s, F_{load} = 5.2 N, and RPM = 55.5

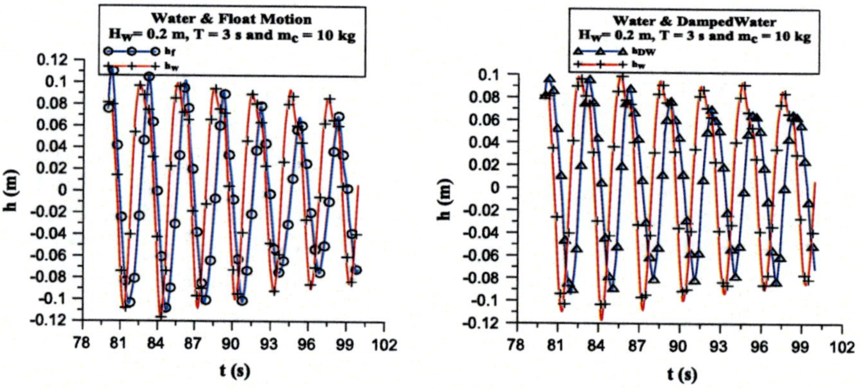

Fig. 3.15 Water displacement with float and damped water for H = 20 cm, m_c = 10, T = 3 s, F_{load} = 3.2 N, and RPM = 45

improvement can be shown in this case than the other two cases simply by revising Fig. 3.12, which describes the force resolving on the float. It is clear, for flat part submerged in water; float suffered from the high lateral wave force and therefore damping the float motion. In this case the motion of float vertically depends only on the pressure force in the heave direction according to Froude-Krylov concept, without any assisting of any other lift force. In addition, the friction is increased in the movable parts due to the force in the horizontal direction and perpendicular to the heave motion. This is the reason behind it is needed the conical end of float to be the submerged part and to be against the lateral force of wave to convert a part of this lateral force to be a lift force. This force is added to the pressure force for smooth vertically motion and decreases the influence of lateral force on the float. Therefore, the float height is increased and automatically increases the captured power from

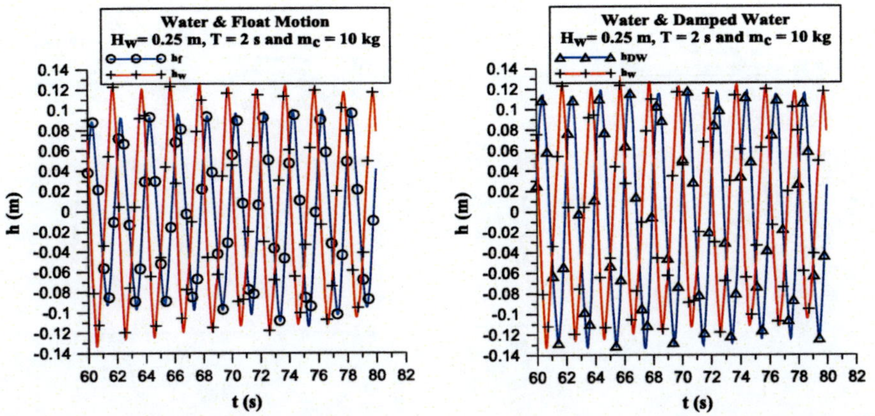

Fig. 3.16 Water displacement with float and damped water for H = 25 cm, m_c = 10, T = 2 s, F_{load} = 5.35 N, and RPM = 64.5

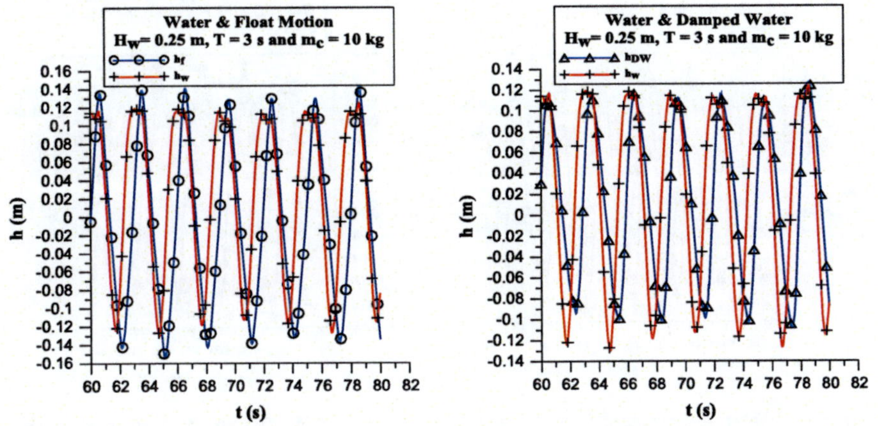

Fig. 3.17 Water displacement with float and damped water for H = 25 cm, m_c = 10, T = 3 s, F_{load} = 5.075 N, and RPM = 52.4

waves. The damped water behind float does not exceed than before of case m_c = 10 kg which is 22.5%.

From the previous discussion of float motion with water wave motion, it is needed to know what is happen for the RAO's under the load condition of this float 50-C. Figure 3.28 shows the RAO's of float 50-C at waves of H = 20, 25 cm and different frequencies conditions at m_c = 20 kg. Obviously from the results output experimentally; it seems that the RAO's slightly differ than that without load and approximately same value of 1. This is the most advantage and powerful point for the WAVE HUNTER system, which lead to highly efficient in wave energy conversion to mechanical energy. The captured power from the system can be seen averagely in

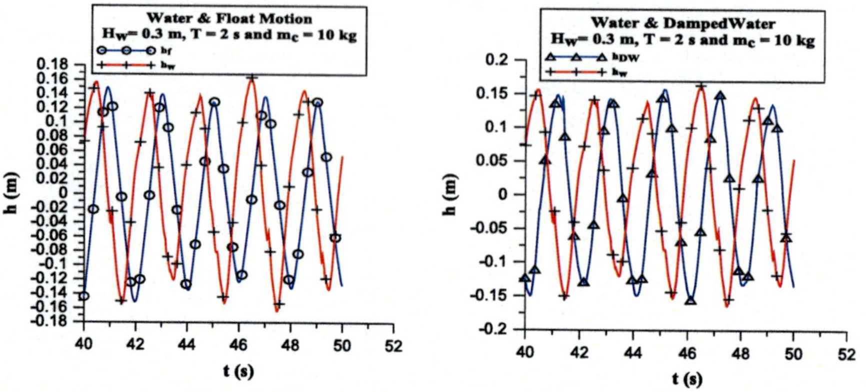

Fig. 3.18 Water displacement with float and damped water for H = 30 cm, $m_c = 10$, T = 2 s, $F_{load} = 8.88$ N, and RPM = 79.5

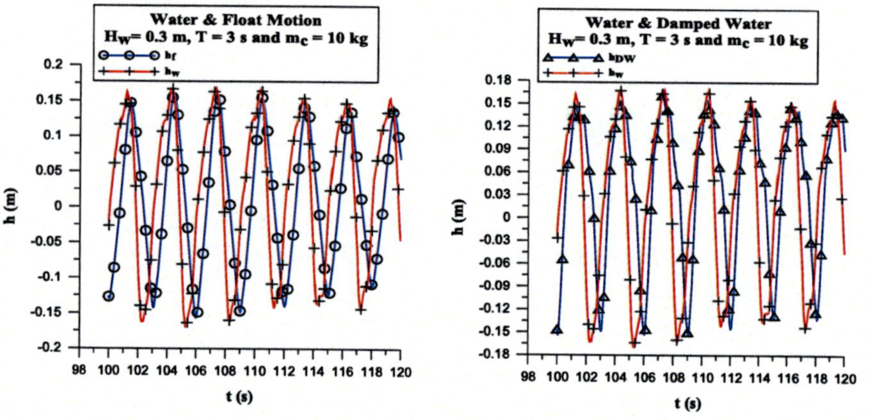

Fig. 3.19 Water displacement with float and damped water for H = 30 cm, $m_c = 10$, T = 3 s, $F_{load} = 8.91$ N, and RPM = 55.5

all cases of experiments in Fig. 3.29. The power captured is increased by increasing the water wave height incident on float, and this means that the friction is constant. This gives the possibility of using the system at many conditions of wave without fearing from the friction force affecting on the motion of float.

It is determined the efficiency of conversion under all conditions of wave incident on float as discussed before. The efficiency is calculated according to the available power in the incident wave; this is done with two different equations, Eqs. (3.53) and (3.54) from two different sources [5, 6] respectively.

$$P_w = \frac{1}{2}\rho\, g\, a^2 c_g \text{ W/m} \tag{3.53}$$

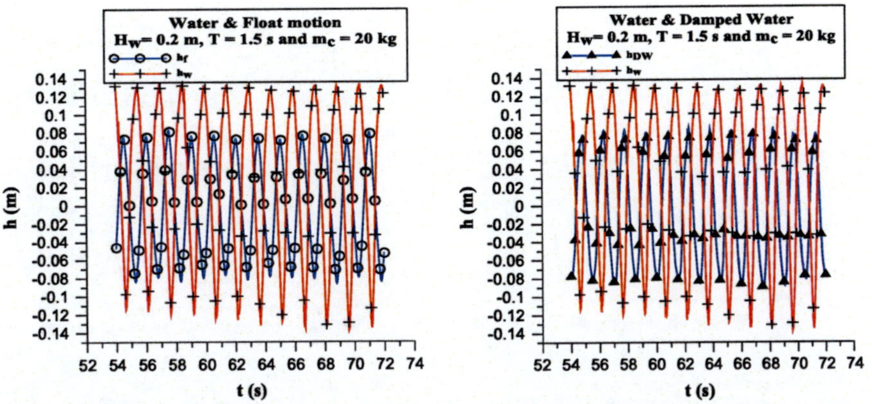

Fig. 3.20 Water displacement with float and damped water for H = 20 cm, $m_c = 20$, T = 1.5 s, $F_{load} = 4.03$ N, and RPM = 69.15

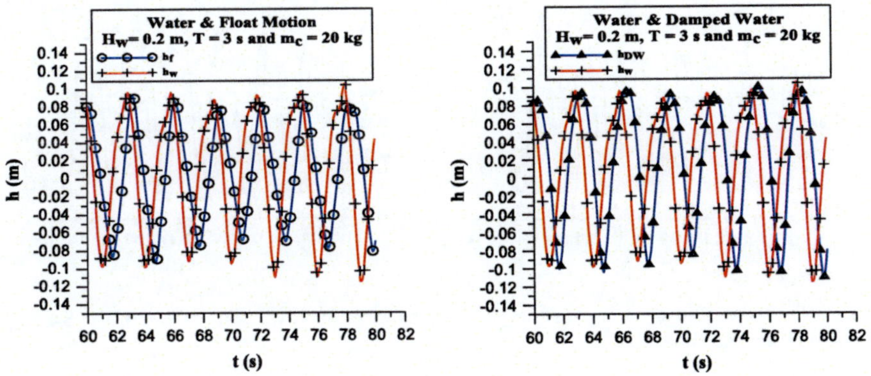

Fig. 3.21 Water displacement with float and damped water for H = 20 cm, $m_c = 20$, T = 3 s, $F_{load} = 4.3$ N, and RPM = 45

$$P_w = \frac{1}{16}\rho g H^2 \frac{\lambda}{T}\,[1 + (4\pi d/\lambda)/\sinh(4\pi d/\lambda)\,]\ \text{W/m} \tag{3.54}$$

$$\eta = \frac{P_{cap.}}{P_w} \tag{3.55}$$

Figure 3.30 presents the efficiency of conversion for whole conditions are discussed before. It is confirming that the maximum efficiency could be reached 35% when using $m_c = 20$ kg, according to the conical end as the only submerged part in water. It is clear also from the float motion and RAO's in this case. The higher efficiency is reached at 2 s (resonance case). The radiated waves from float in small periods are much higher than larger period, which is related to the time of wave

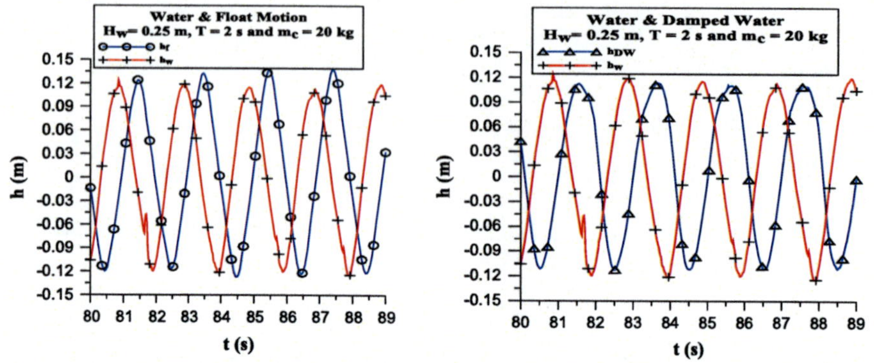

Fig. 3.22 Water displacement with float and damped water for H = 25 cm, $m_c = 20$, T = 2 s, F_{load} = 6.15 N, and RPM = 83

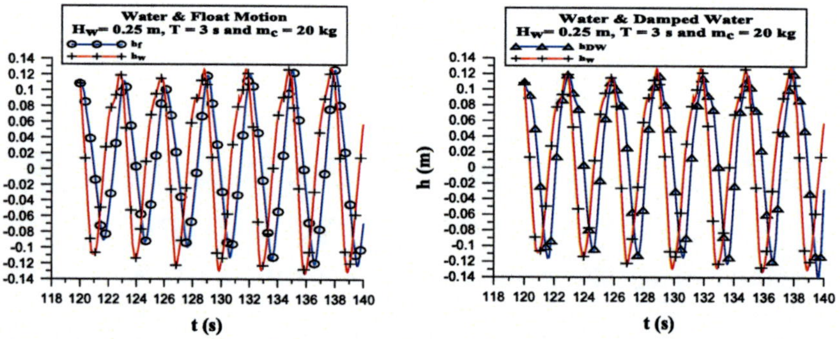

Fig. 3.23 Water displacement with float and damped water for H = 25 cm, $m_c = 20$, T = 3 s, F_{load} = 7.34 N, and RPM = 48.3

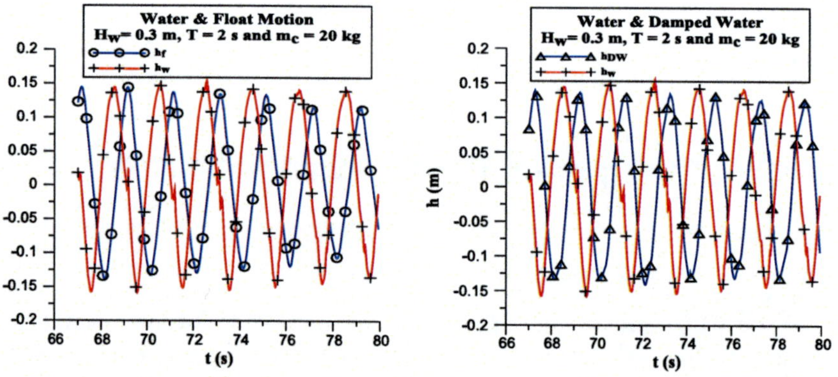

Fig. 3.24 Water displacement with float and damped water for H = 30 cm, $m_c = 20$, T = 2 s, F_{load} = 10.52 N, and RPM = 83.5

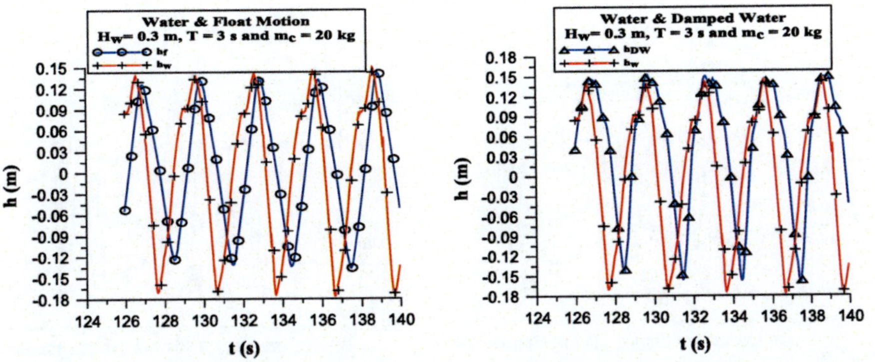

Fig. 3.25 Water displacement with float and damped water for H = 30 cm, m_c = 20, T = 3 s, F_{load} = 9.42 N, and RPM = 54

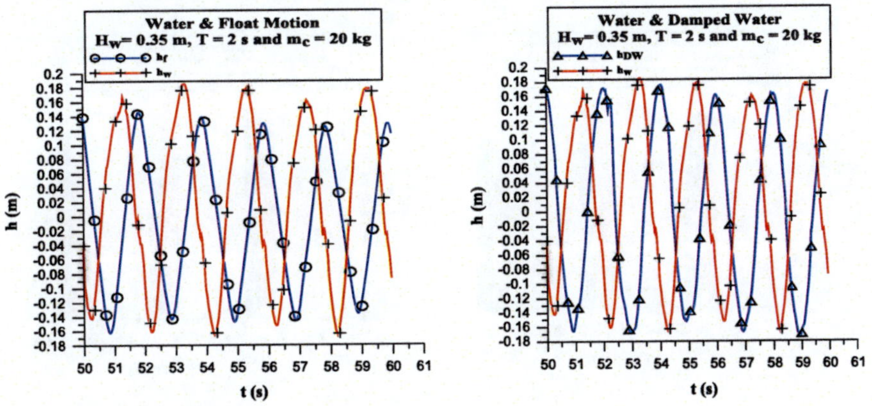

Fig. 3.26 Water displacement with float and damped water for H = 35 cm, m_c = 20, T = 2 s, F_{load} = 10.38 N, and RPM = 85

coming. The radiated wave numbers effect on the wave height incident and also on float motion. Average efficiency for the whole system of float 50-C can be shown in Fig. 3.31). Figure 3.32 illustrates the average efficiency with dimensionless parameter T_o/T for selecting the suitable natural frequency of float according to the incident wave period. This average efficiency is the average under all condition of work. The lowest value of efficiency can be seen as 12%, but the highest one is 28%. Therefore, from the survey for other systems with same configuration of point absorber system for wave energy conversion, this system has high and reasonable efficiency where the maximum efficiency could be reached by other systems not more than 30% under any condition of wave incident.

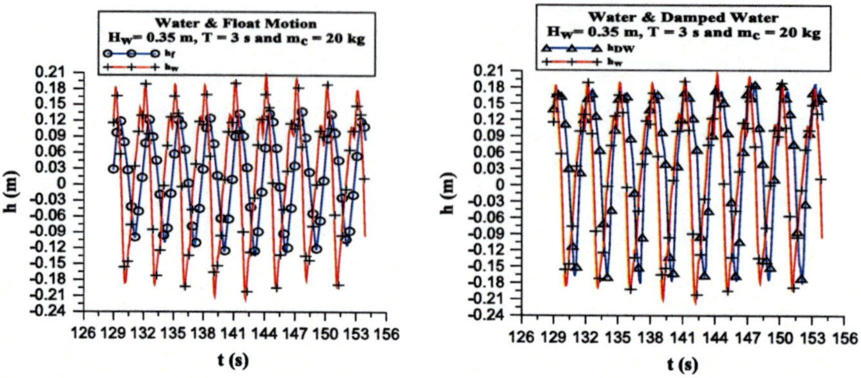

Fig. 3.27 Water displacement with float and damped water for H = 35 cm, m_c = 20, T = 3 s, F_{load} = 11.68 N, and RPM = 57

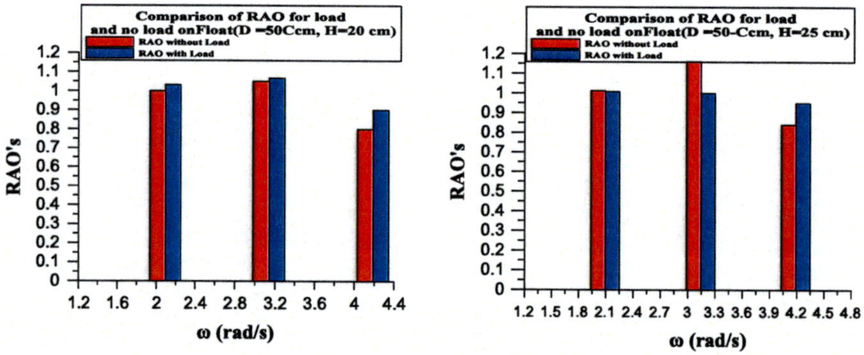

Fig. 3.28 RAO comparison of float 50-C in load and without load for H = 20, 25 cm, m_c = 20

The Power Captured of Float 40-C

The second float 40-C is examined according to the results coming out from RAO's experiments. The natural frequency of this float is ω_o = 3.34 rad/s (T_o = 1.85 s), where at this value the optimum RAO's is reached. Figure 3.33 shows the float and damped wave condition in case of incident wave with H = 20 cm, T= 1.5s and m_c = 0 kg. The float as it can be seen doesn't gain any energy from wave and doesn't move, it is vibrating only. Otherwise the damped water height behind the float is increased than the incident wave due to the radiated and separation of water from the float.

In Fig. 3.34 the wave incident is changed to H = 20 cm with T = 2s and m_c = 0 kg, there is an improvement in float movement and also in damping of water according to the energy gain by float from the incident wave compared with the previous wave

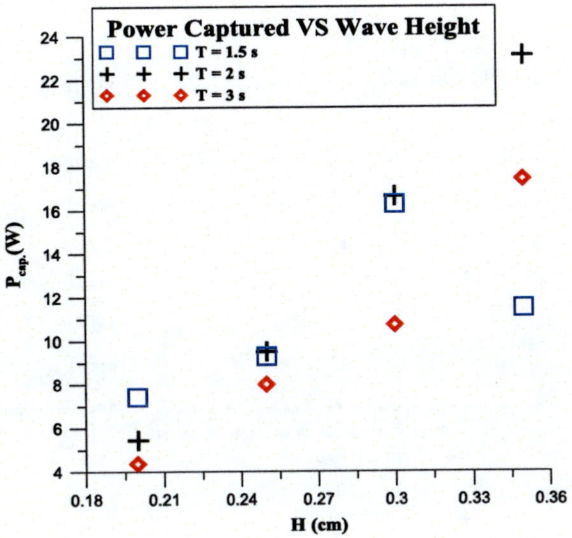

Fig. 3.29 Average power captured from the system VS H

of T = 1.5s. This is because the natural frequency of this float is close to 2 s not to 1.5 s, and the no. of wave cycle incident in T =1.5s is too many.

Figure 3.35 describes the motion of float and damped water with incident wave of properties H = 20 cm and T = 2 s but the counter weight is 10 kg. In this case the damped water is reasonable compared to the float motion, where the float approximately not moved.

Figure 3.36 shows the motion of the float and damped water in case of wave incident as before in Fig. 3.35 with changing in period T = 3 s. The phase angle is getting small than other cases. The motions of float and incident wave are approximately in one phase.

The same conditions of what happen in case of T = 3 s is repeated but with T = 4 s Fig. 6.48 with little enhancement than in 3 s period. In the other side, the damped water is not efficient enough compared to the float 50-C where the percent of damping didn't not exceed 18% from the incident wave (Fig. 3.37).

Figure 3.38 indicates the increasing in power gain and improves the damped water behind the float where H = 25 cm, T = 2 s. The damped water is reached 20% approximately. The water wave and float is not in phase as before in 20 cm. This means more power could be gained due to the resonance case of float with water wave. The float displacement and power captured are increased in height as it can be seen in Fig. 3.38. Moreover, the improvement is happened in the phase of motion between float and water wave. The damping of water is reached to be 16% of wave incident height.

In Fig. 3.39 shows the maximum displacement of float could be gained from wave incident on it. The displacement of float is higher than the incident wave displacement due to high radiated waves with long time period and permits the float to move smoothly without any effect coming from the wave incident. In other hand

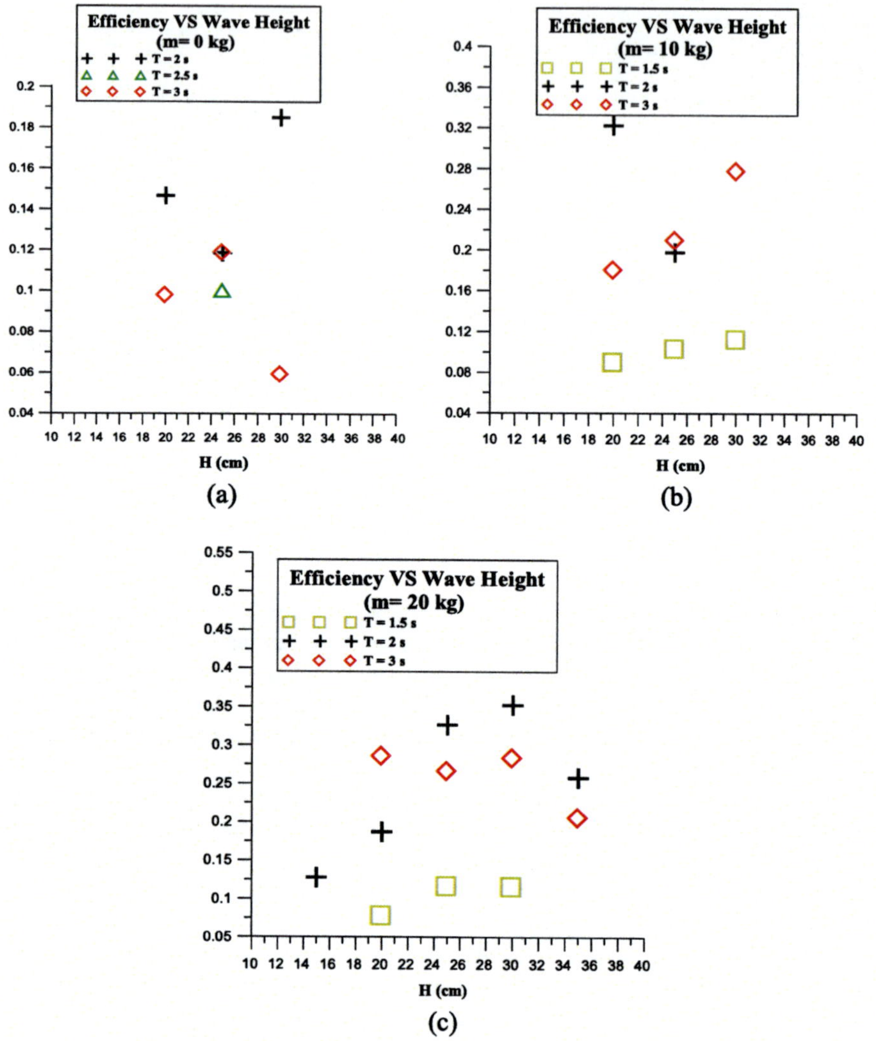

Fig. 3.30 Efficiency of float 50-C against wave height for $m_c = 0$, 10 and 20 kg

the damped is lowered according to the same reason. Figure 3.40 illustrates the condition of H = 30 cm and T = 2 s with more improvement than other cases from RAO's and phase condition. Figures 3.41 and 3.42 the motion is improved in phase and gain more energy from incident wave by float at T = 3 s with $m_c = 10$ kg. Moreover, the damping of water is reached to 10%. Figures 3.43 and 3.44 show the wave characteristics of period and height with float motion and damping water. The difference is only in $m_c = 20$ kg, this is used to less the draft of flat part in water and make the conical part to be the only contact part with water. This is lead to

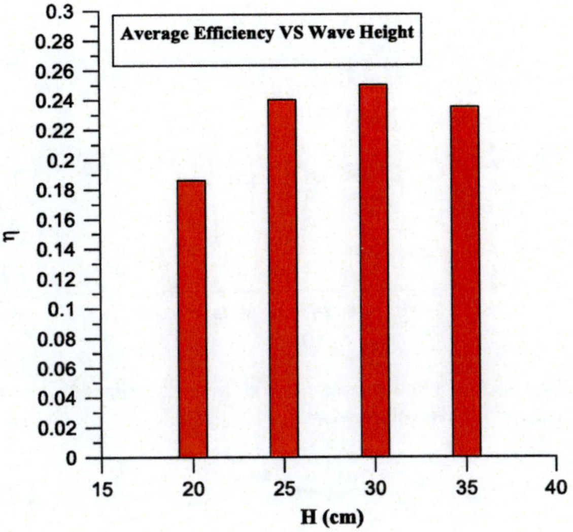

Fig. 3.31 Average Efficiency of float 50-C VS H

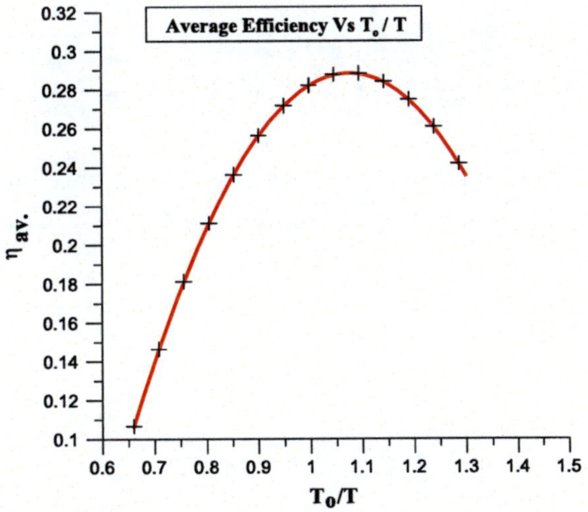

Fig. 3.32 Average Efficiency of float 50-C against dimensionless parameter T_o/T

improvement in motion of float with water, but may be not increased in efficiency according to the weight of float 40-C is less than that in float 50-C. Therefore, the weight of 20 kg is not the optimized due to the draft of flat part (Figs. 3.45, 3.46, 3.47, 3.48, 3.49, 3.50, 3.51, 3.52, 3.53and 3.54).

The power captured averagely by the system can be seen in Fig. 3.55 for different periods at different wave heights for cases of $m_c = 10$ and 20 kg. Figures 3.56 and 3.57 present the average efficiency of float 40-C for different periods (1.5, 2 and 3 s) with wave heights for two cases 10 kg and 20 kg. As it is expected from previous

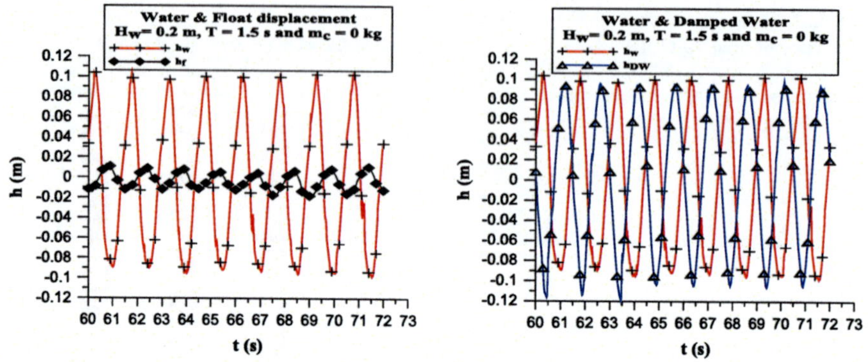

Fig. 3.33 Water displacement with float and damped water for H = 20 cm, m_c = 0, T = 1.5 s, F_{load} = 1.32 N, and RPM = 9

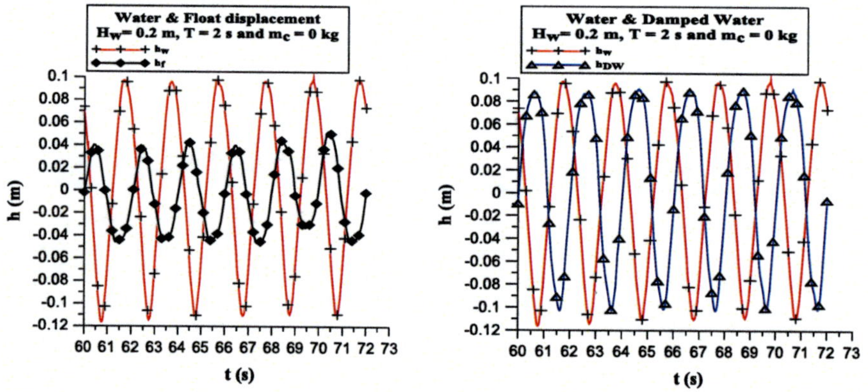

Fig. 3.34 Water displacement with float and damped water for H = 20 cm, m_c = 0, T = 2 s, F_{load} = 0.86 N, and RPM = 30

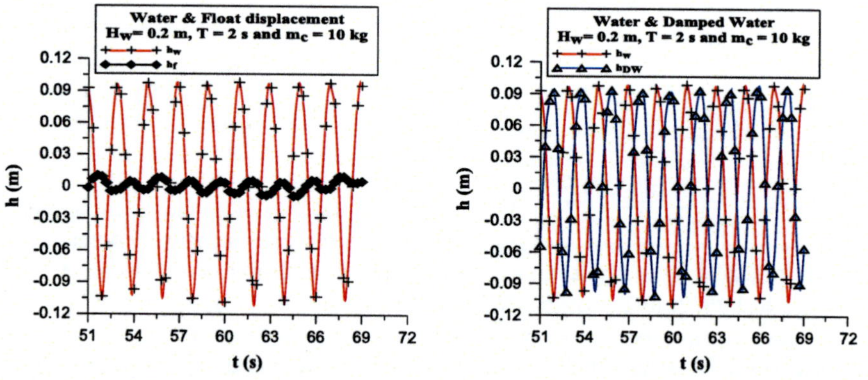

Fig. 3.35 Water displacement with float and damped water for H = 20 cm, m_c = 10, T = 1.5 s, F_{load} = 1.08 N, and RPM = 22.5

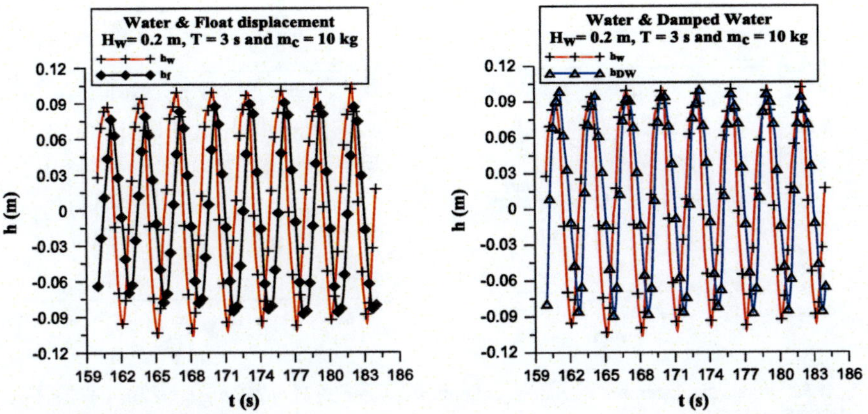

Fig. 3.36 Water displacement with float and damped water for H = 20 cm, m_c = 10, T = 3 s, F_{load} = 2.23 N, and RPM = 38.1

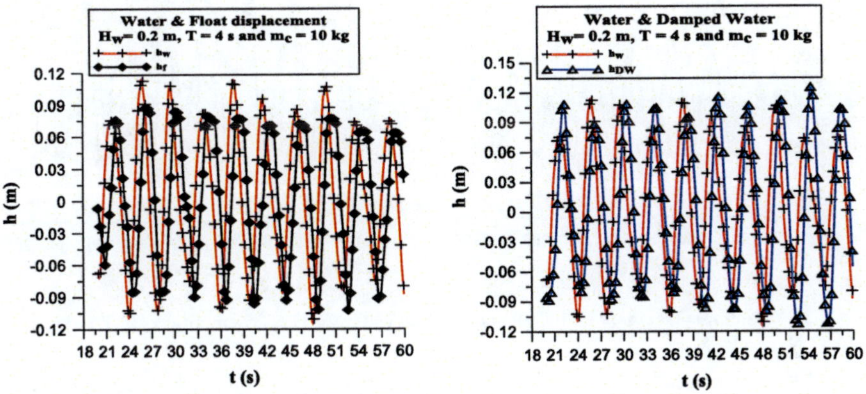

Fig. 3.37 Water displacement with float and damped water for H = 20 cm, m_c = 10, T = 4 s, F_{load} = 1.54 N, and RPM = 45

paragraph, the maximum efficiency is 40.38% at m_c = 10 kg and resonance period of 2 s and H = 25 cm. From the whole view of efficiencies in both 10 kg and 20 kg, beside that; mass of 10 kg is suitable for 40-C float than 50-C float, due to the difference in mass between them. This is clear from Figs. 3.56 and 3.57. The average efficiency in all cases of work as maximum is 38% at H = 25 cm Figs. 3.58 and 3.59. The comparison now illustrates that increase in diameter of float increase the efficiency for our system, according to the same value of friction in all cases.

In other point of view, the efficiency in 40-C is higher than in 50-C because the design of float itself which is the major reason. The float 40-C is conical end as 50-C but with diameter 70 cm in the upper cone and 5 cm in the lower cone, this is mean

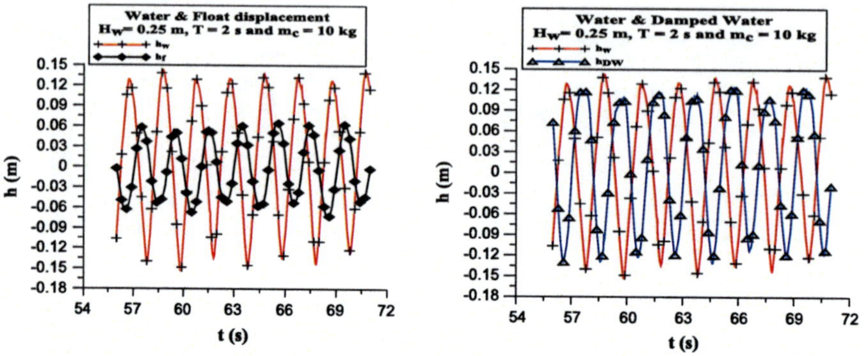

Fig. 3.38 Water displacement with float and damped water for H = 25 cm, $m_c = 10$, T = 2 s, F_{load} = 1.27 N, and RPM = 46.1

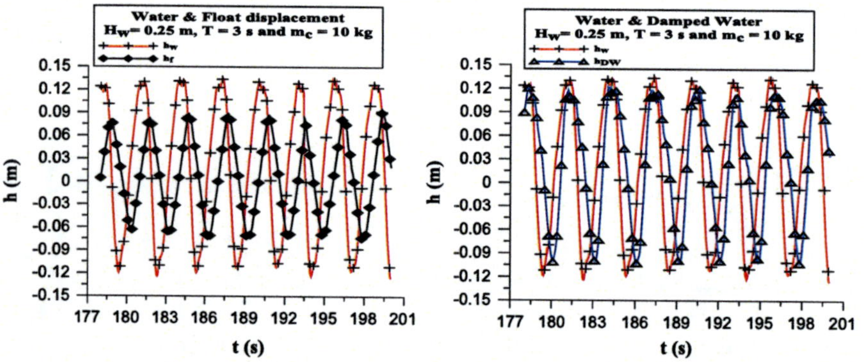

Fig. 3.39 Water displacement with float and damped water for H = 25 cm, $m_c = 10$, T = 3 s, F_{load} = 4.81 N, and RPM = 32.55

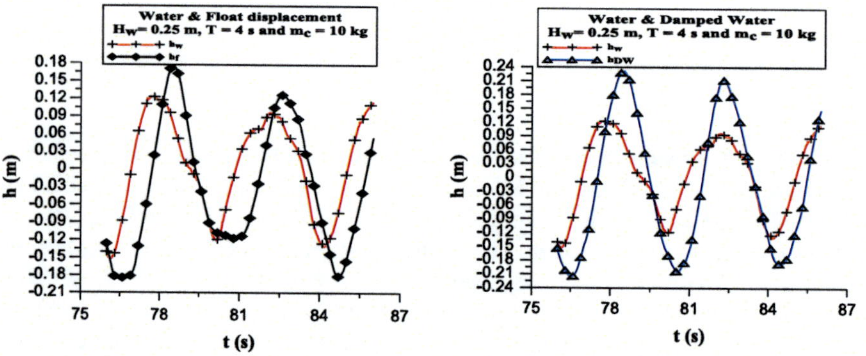

Fig. 3.40 Water displacement with float and damped water for H = 25 cm, $m_c = 10$, T = 4 s, F_{load} = 2.74 N, and RPM = 61.5

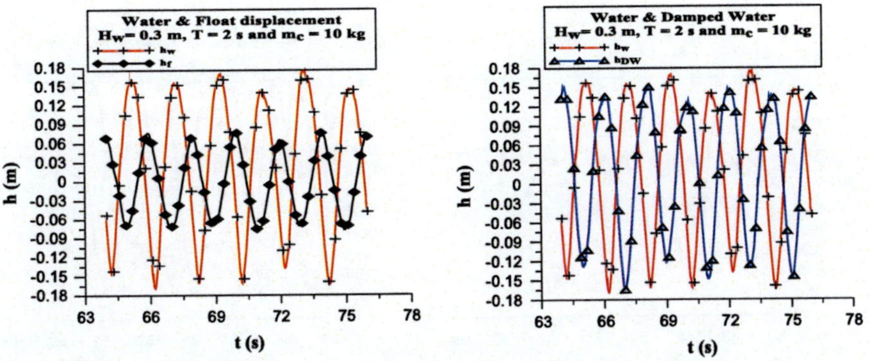

Fig. 3.41 Water displacement with float and damped water for H = 25 cm, m_c = 10, T = 4 s, F_{load} = 2.76 N, and RPM = 49.05

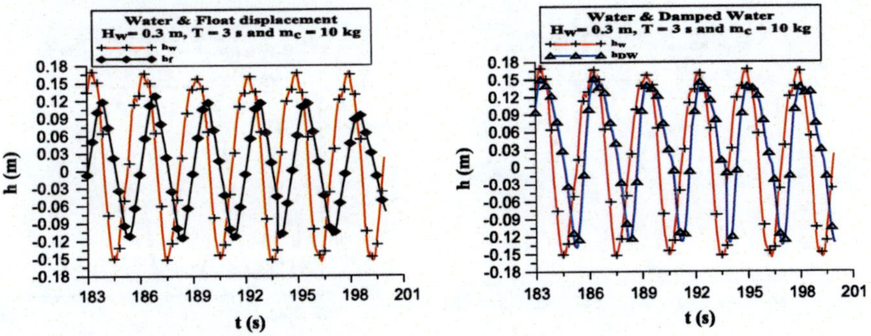

Fig. 3.42 Water displacement with float and damped water for H = 25 cm, m_c = 10, T = 4 s, F_{load} = 4.62 N, and RPM = 49.05

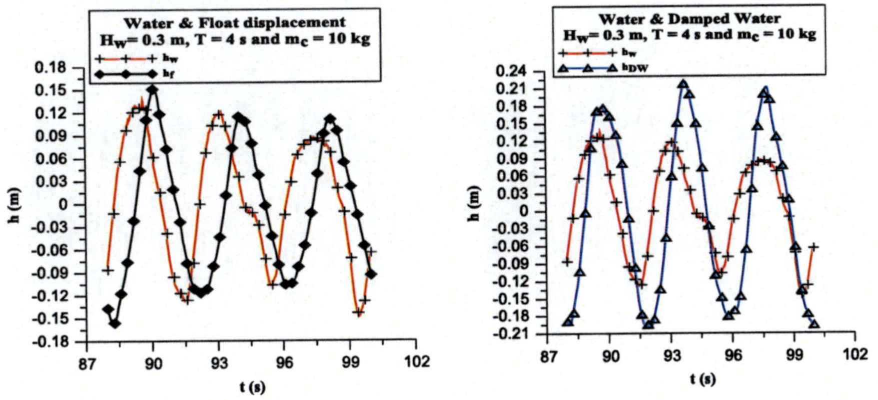

Fig. 3.43 Water displacement with float and damped water for H = 25 cm, m_c = 10, T = 4 s, F_{load} = 6.55 N, and RPM = 45

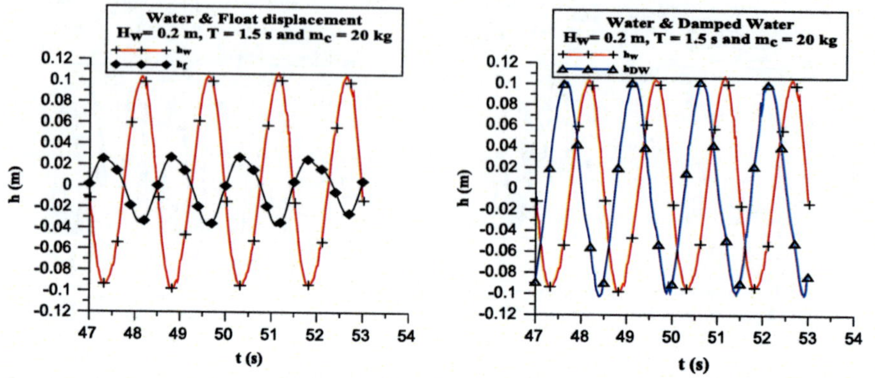

Fig. 3.44 Water displacement with float and damped water for H = 25 cm, $m_c = 20$, T = 4 s, $F_{load} = 1$ N, and RPM = 28.5

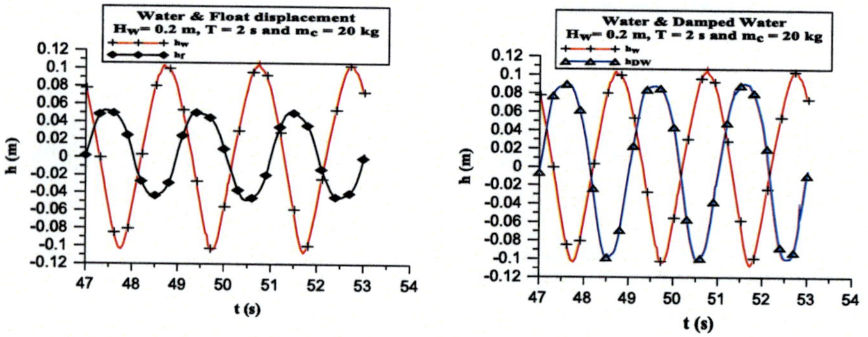

Fig. 3.45 Water displacement with float and damped water for H = 25 cm, $m_c = 20$, T = 4 s, $F_{load} = 0.9$ N, and RPM = 45

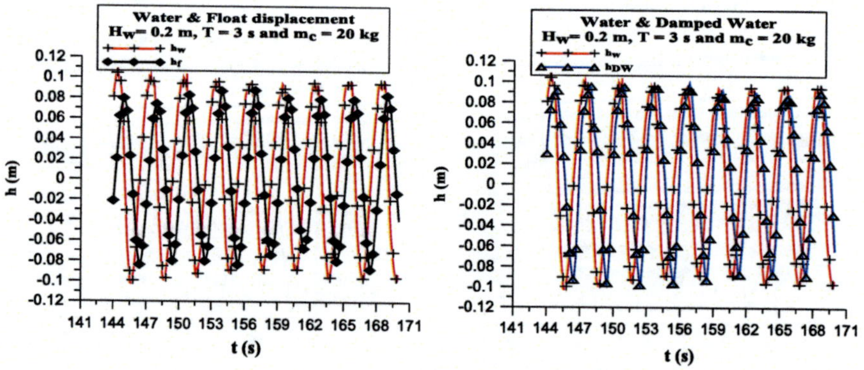

Fig. 3.46 Water displacement with float and damped water for H = 25 cm, $m_c = 20$, T = 4 s, $F_{load} = 2.1$ N, and RPM = 36.6

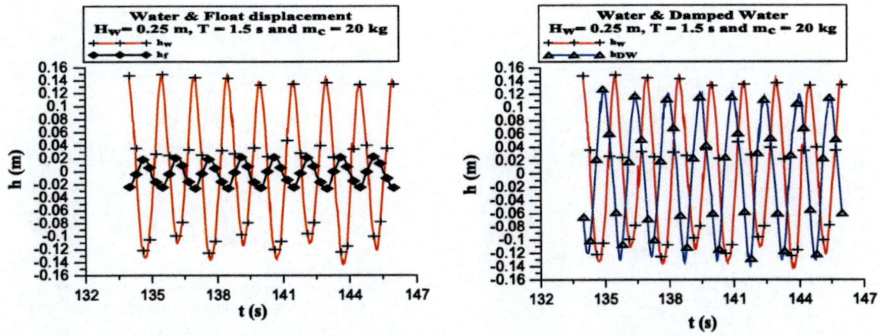

Fig. 3.47 Water displacement with float and damped water for H = 25 cm, $m_c = 20$, T = 4 s, F_{load} = 2.7 N, and RPM = 20

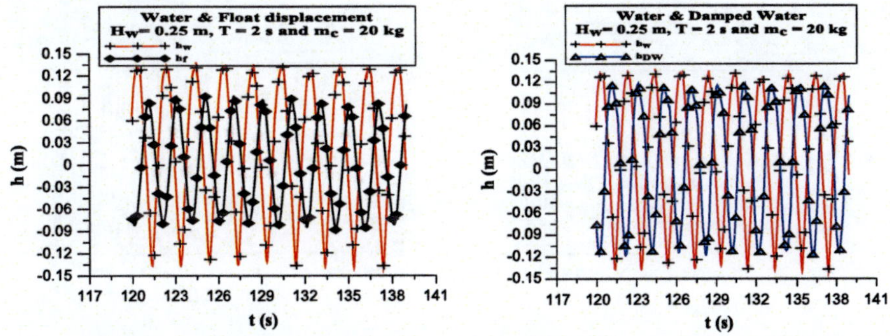

Fig. 3.48 Water displacement with float and damped water for H = 25 cm, $m_c = 20$, T = 4 s, F_{load} = 2.6 N, and RPM = 57.75

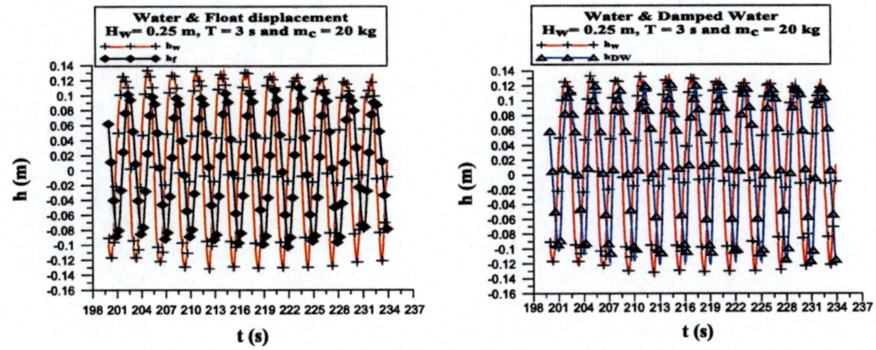

Fig. 3.49 Water displacement with float and damped water for H = 25 cm, $m_c = 20$, T = 4 s, F_{load} = 3.6 N, and RPM = 45

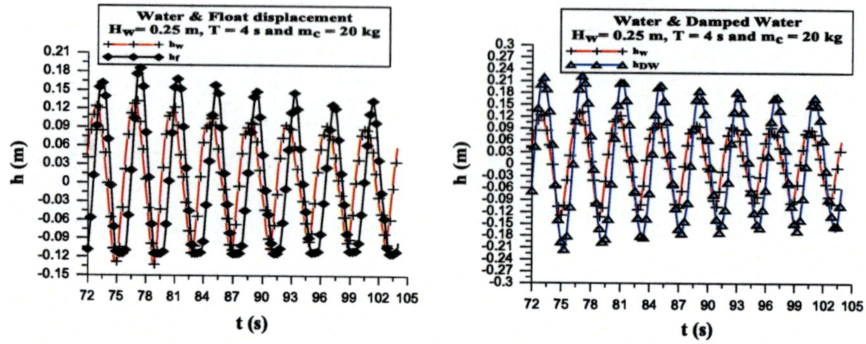

Fig. 3.50 Water displacement with float and damped water for H = 25 cm, m_c = 20, T = 4 s, F_{load} = 2.04 N, and RPM = 56.1

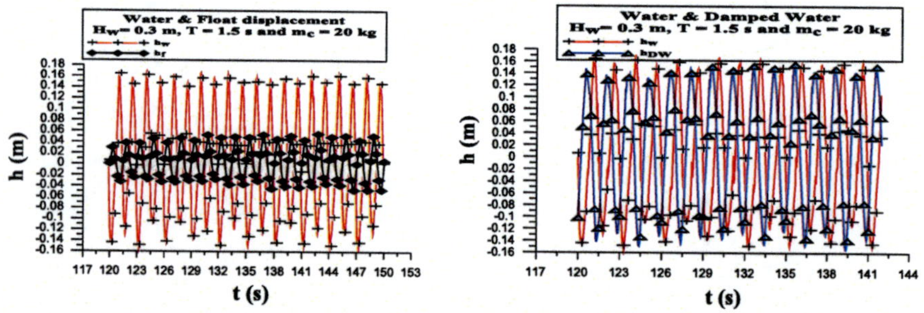

Fig. 3.51 Water displacement with float and damped water for H = 25 cm, m_c = 20, T = 4 s, F_{load} = 2.89 N, and RPM = 76.5

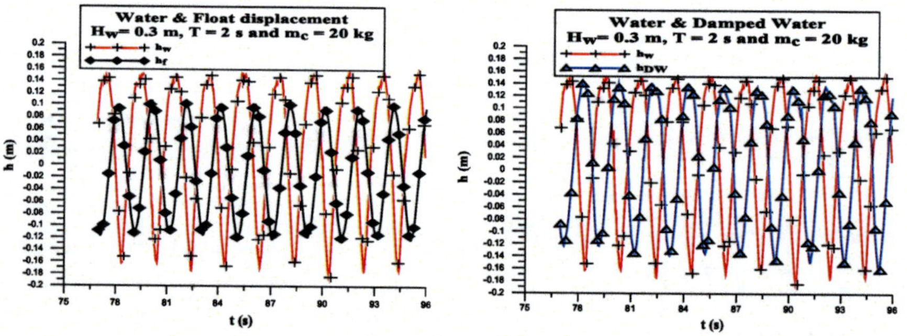

Fig. 3.52 Water displacement with float and damped water for H = 25 cm, m_c = 20, T = 4 s, F_{load} = 1.4 N, and RPM = 41.73

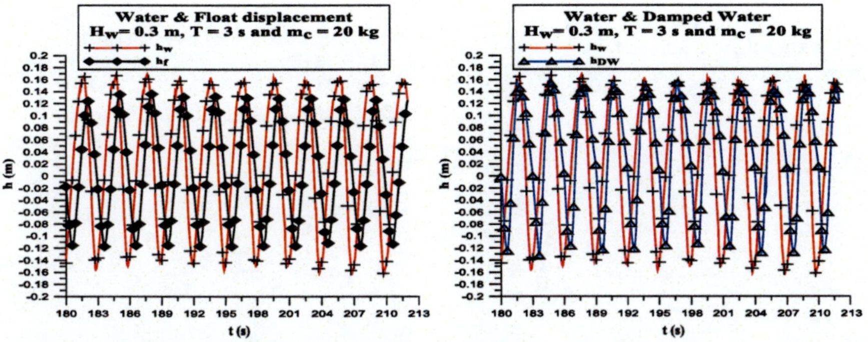

Fig. 3.53 Water displacement with float and damped water for $H = 25$ cm, $m_c = 20$, $T = 4$ s, $F_{load} = 4.08$ N, and RPM $= 51.255$

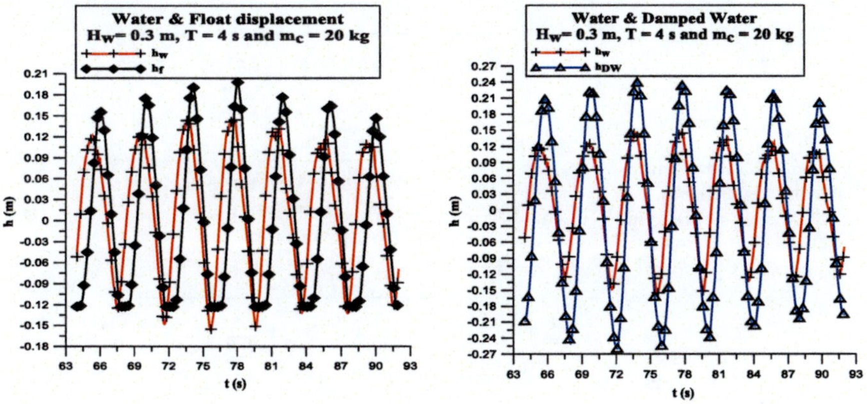

Fig. 3.54 Water displacement with float and damped water for $H = 25$ cm, $m_c = 20$, $T = 4$ s, $F_{load} = 3.58$ N, and RPM $= 65.1$

that the drag force from water on float is increased than 50-C due to the excess of diameter in upper cone of 40-C float, which working as wings for the float to lift it.

Power Extraction Experimentally from Irregular Wave

After the results coming out from regular waves, the float of 50-C is tested in irregular waves. This is according to the average efficiency reached.

Figure 3.60 presents the water wave and float motion in case of irregular waves for different wave characteristics for JONSWAP model as $H_s = 20.7$, 21.4, 25 cm with $T_p = 2.5$, 2.1, 1.5 s, respectively. The wave height is ranged from 5 to 30 cm with different periodic times and it is changed rapidly for a few seconds. According

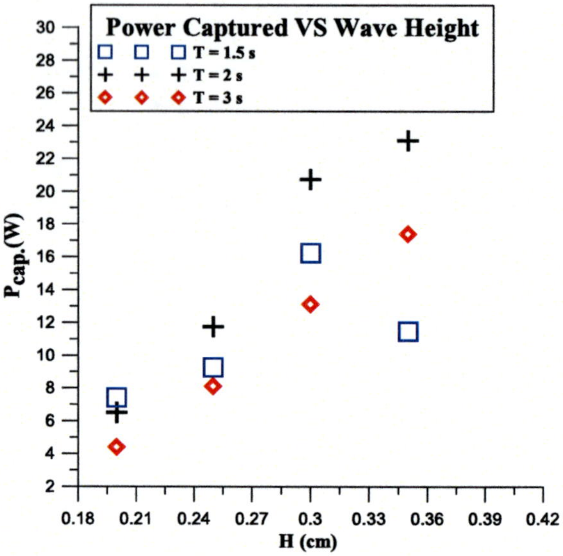

Fig. 3.55 Power captured with wave height at different periods float 40-C

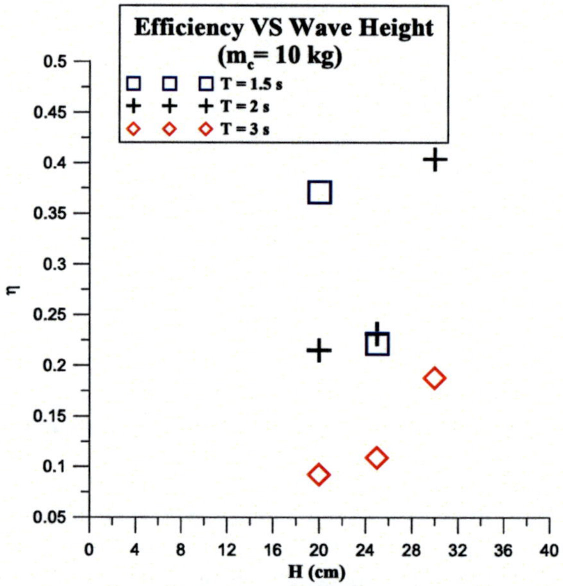

Fig. 3.56 Efficiency with wave height at different periods with $m_c = 10$ kg for float 40-C

to this fast change, it is not appeared a fine matching of float with water comparing to the regular wave.

Figure 3.61 shows the water wave and float with same characteristics of H_s&T_p but this time with different model where the new model is Pierson-Moskowitz. The influence of forces, power captured and finally the efficiency of the system is presented

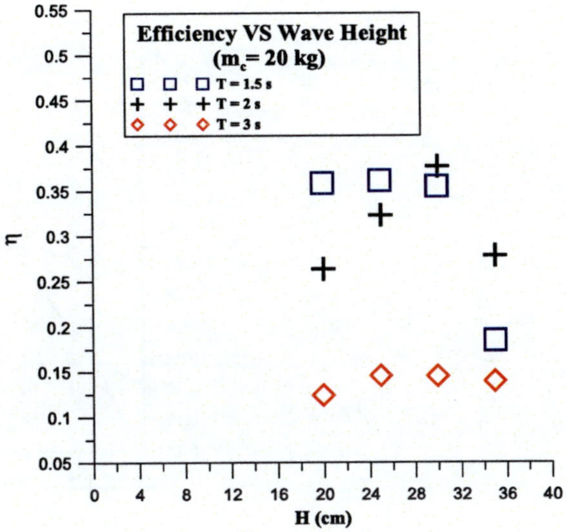

Fig. 3.57 Efficiency with wave height at different periods $m_c = 20$ kg for float 40-C

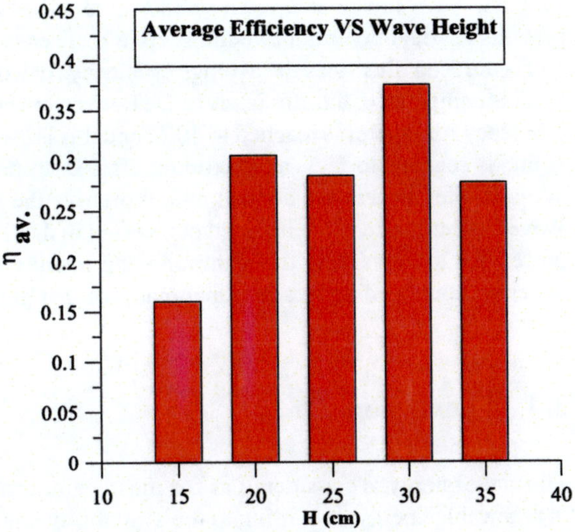

Fig. 3.58 Average efficiency for float 40-C for all cases

in Figs. 3.62, 3.63, and 3.64, respectively. From these figures, it is found that the maximum efficiency is reached to 34.8% in the JONSWAP model and 38.4% in Pierson-Moskowitz model averagely in all cases of work. In Pierson-Moskowitz model the efficiency is reached to 54.6% for $H_s = 20.7$ cm and $T_p = 2.5$ s case. The irregular wave generation and extract power from these waves are the real test of our system for power extraction and efficiency of conversion. The average efficiency in

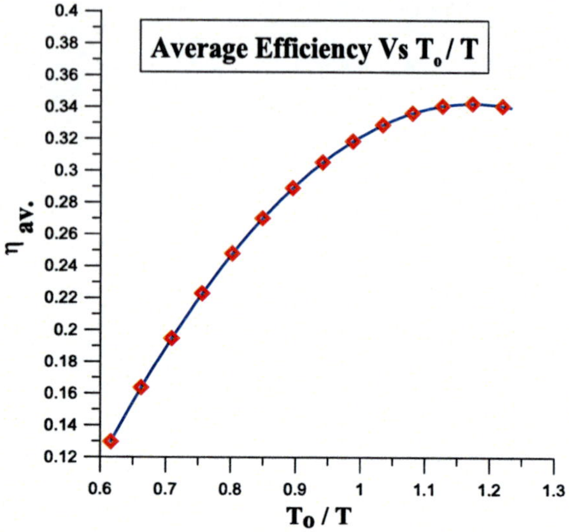

Fig. 3.59 Average efficiency for float 40-C for all cases with dimensionless parameter T_0/T

irregular wave can be estimated as 38% in all cases of work. This value means, the system has great potential to build it up in real sea condition.

Figure 3.65 illustrates the average efficiency for irregular waves with two different models compared to the efficiency of De Backer and et al. system [30]. The maximum efficiency for Backer is reached to 40%, and this is low than The Wave Hunter system which is reached to 57% as maximum also. Both floats in Backer system and The Wave Hunter system are conical end, moreover the increased in efficiency for The Wave Hunter is due to the inverted cup design in the float which increase the captured power due to preventing the slamming and making synchronize between float and water without needing to additional mass for that purpose.

3.4 Conclusion

The wave energy is considered as the most significant and promising energy source in renewable energy, according to the availability annually. The point absorber wave energy conversion system is considered as an optimum solution for wave energy conversion due to its performance improvements during the experimental work for different designs. The point absorber includes a float as the hydrodynamic system with a mechanical part as a power take-off (PTO). By logic, in any new design, the float should smaller relatively than the other systems in the high-efficiency systems. Its shape and design are normally cylindrical; therefore it is constrained to one essential degree of motion, typically up-and-down and generally. The point absorber (Float) captures the wave energy in all directions during its motion.

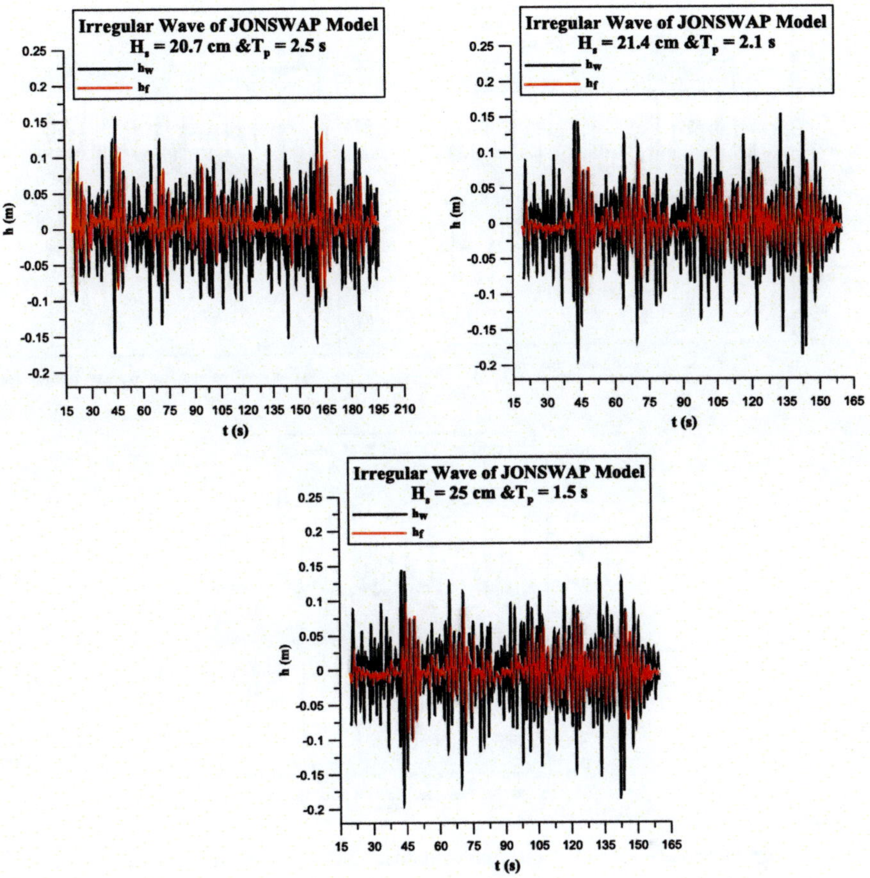

Fig. 3.60 Irregular wave JONSWAP Model motion with float 50-C motion at $m_c = 20$ kg

The performance of the wave energy conversion WAVE HUNTER system is investigated regarding its efficiency and motion behavior. The performance improvements during the experimental and simulation work are included in this report. The efficiency of the system in regular waves is around 30% for float of 50 cm with conical end and inverted cup, in some wave conditions and for all cases of regular wave incident. With such values the WAVE HUNTER is comparable with other state of the art WECs. For the irregular wave incident with the models of JONSWAP and Pierson-Moskowitz the maximum efficiency is 33.4%. The WAVE HUNTER system can work with floats of higher diameter to produce more power from regular and irregular waves. Increasing efficiency is then possible because the captured power will increase strongly while the friction losses of the movable parts will increase only slightly.

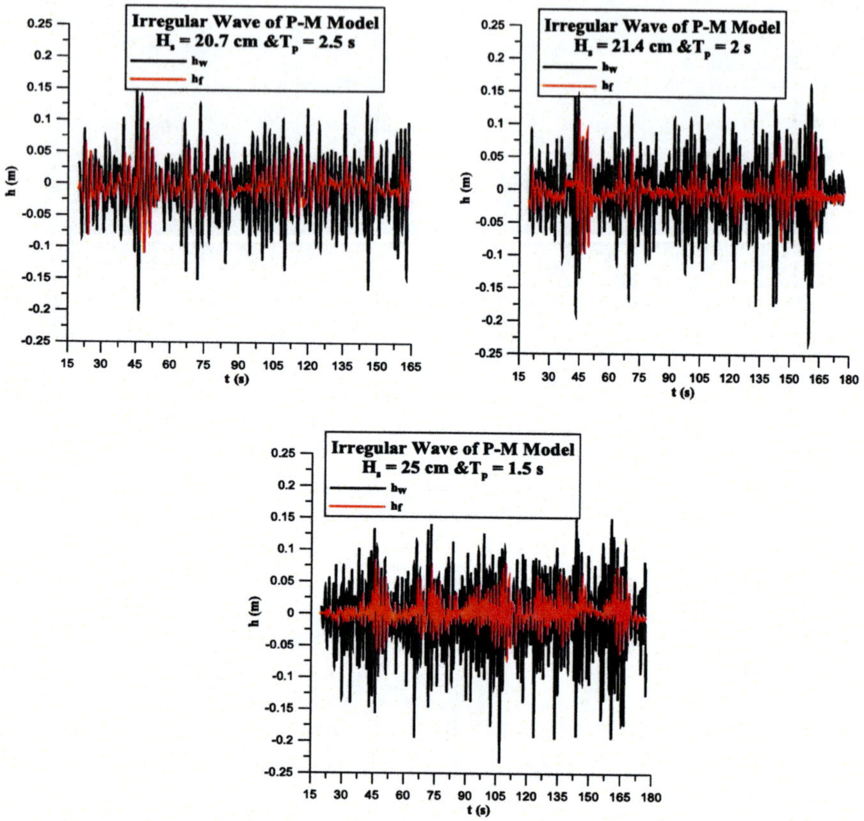

Fig. 3.61 Irregular wave P-M model motion with float 50-C motion at $m_c = 20$ kg

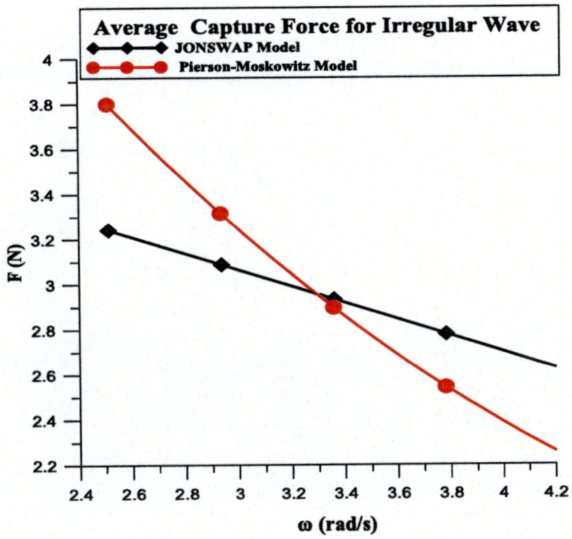

Fig. 3.62 Average capture forces for both models JONSWAP and P-M of float 50-C motion at $m_c = 20$ kg

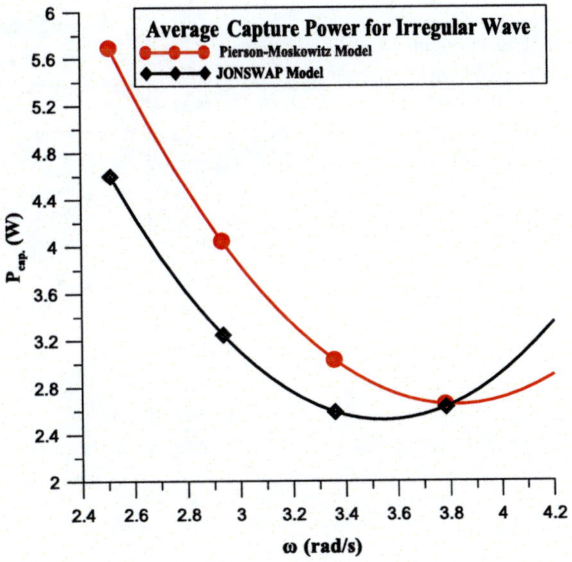

Fig. 3.63 Average capture power for both models JONSWAP and P-M of float 50-C motion at $m_c = 20$ kg

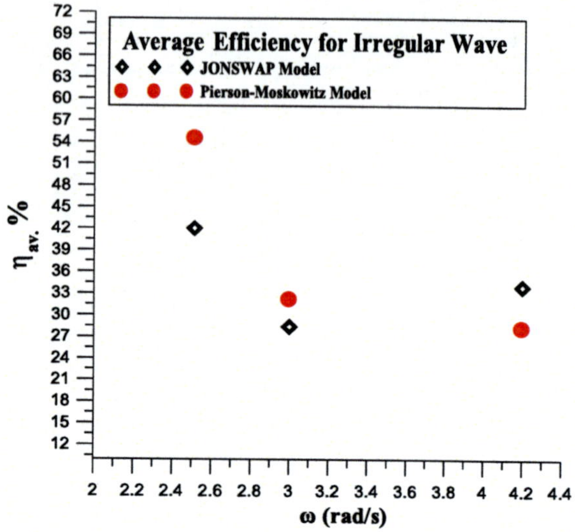

Fig. 3.64 Average efficiency for both models JONSWAP and P-M with frequencies

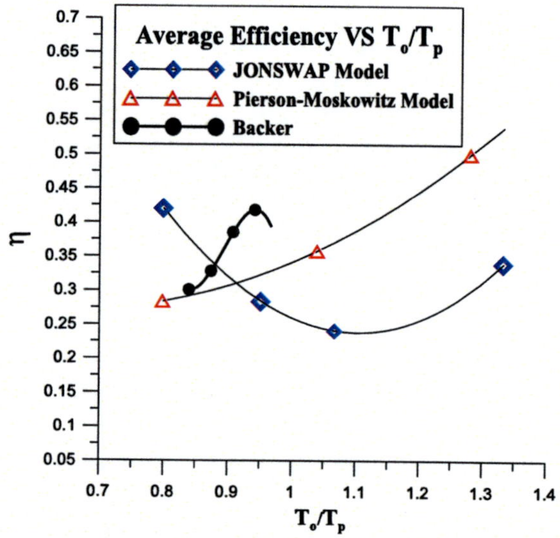

Fig. 3.65 Average efficiency for both models JONSWAP and P-M compared with De Backer data [30] dimensionless parameter T_O/T_P

References

1. Newell, C. (2010). *A mathematical and numerical examination of wave-current interaction and wave-driven hydrodynamics*. IRIS NUI Galway's New Institutional Research.
2. Singh, S. P., & Sen, D. (2007). A comparative linear and nonlinear ship motion study using 3-D time domain methods. *Ocean Engineering, 34*(13), 1863–1881.
3. Salter, S. H. (1974). Wave power. *Nature, 249*(5459), 720–724.

4. Linton, C. M., & McIver, P. (2010). *Handbook of mathematical techniques for wave/structure interactions*. CRC Press.
5. Chaplin, R. V. (2013). Seaweaver: A new surge-resonant wave energy converter. *Renewable Energy, 57*(September), 662–670.
6. Orer, G., & Ozdamar, A. (2007). An experimental study on the efficiency of the submerged plate wave energy converter. *Renewable Energy, 32*(8), 1317–1327.
7. Falnes, J. (2007). A review of wave-energy extraction. *Marine Structures, 20*(4), 185–201.
8. McCormick, M. E. (2013). Ocean wave energy conversion. Courier Corporation.
9. Filippo, M., Bonamano, S., Stella, G., Peviani, M., & Marcelli, M. (2012). Italian off-shore wave energy map, using gauges and numerical model data. In *Proceedings of the European Seminar OWEMES 2012 Offshore Wind and Other Marine Renewable Energies in Mediterranean and European Seas* (pp. 467–76). Rome-Italy.
10. Finnegan, W. (2013). Wave-structure interaction of offshore wave energy converters. Doctor of Philosophy, Ireland: The College of Engineering and Informatics, NUI Galway.
11. Pierson, W. J., & Moskowitz, L. (1964). A Proposed Spectral Form for Fully Developed Wind Seas Based on the Similarity Theory of SA Kitaigorodskii. *Journal of Geophysical Research, 69*(24), 5181–5190.
12. Hasselmann, K., Barnett, T. P., Bouws, E., Carlson, H., Cartwright, D. E., Enke, K., Ewing, J. A., et al. (1973). *Measurements of wind-wave growth and swell decay during the joint north sea wave project (JONSWAP)*. TLJ200100 · TLJ200510. Germany: Deutches Hydrographisches Institut.
13. Morison, J. R., Johnson, J. W., & O'Brien, M. P. (1953). Experimental studies of forces on piles. *Coastal Engineering Proceedings, 1*(4), 25.
14. Clauss, G. F., Klein, M., & Onorato, M. (2011). Formation of extraordinarily high waves in space and time. In *ASME 2011 30th International Conference on Ocean, Offshore and Arctic Engineering* (Vol. 2, pp. 417–29). Rotterdam, The Netherlands: The American Society of Mechanical Engineers. https://doi.org/10.1115/OMAE2011-49545.
15. Wiegel, R. L. (2005). *Oceanographical engineering* (Vol. 4). New Jersey-USA: Courier Dover Publications.
16. "EMEC: European Marine Energy Centre." 2015. Accessed June 17. http://www.emec.org.uk/.
17. Power, Pelamis Wave. 2014. Bernera Wave Farm. [WWW] Available from http://www.pelamiswave.com/our-projects/project/4/Bernera-Wave-Farm. Accessed 6/4/2021.
18. "Oyster 1, Orkney." 2015. Accessed January 7. http://www.aquamarinepower.com/projects/oyster-1-orkney.aspx.
19. Mueller, M. A., Polinder, H., & Bakerm N. (2007). Current and novel electrical generator technology for wave energy converters. In *Electric Machines & Drives Conference, 2007. IEMDC'07. IEEE International* (Vol. 2. pp. 1401–1406). Antalya: IEEE.
20. Cruz, J. (2008). Ocean wave energy. Springer Series in Green Energy and Technology.
21. "OCEANLINX Project." 2015. Accessed January 8. http://www.oceanlinx.com/.
22. Marleaux, P. Voith Hydro Region Asia Pacific. "Voith | Wave Power Plants." Accessed 17/01/2021. http://voith.com/en/products-services/hydro-power/ocean-energies/wave-power-plants-590.html.
23. Le Crom, I., Bermejo, H. C., Pecher, A., Azevedo, E. B., Reis, F. V. (2011). Incident wave climate at the OWC Pico plant: Validation of a feed-forward based propagation method (ANN) and a numerical simulation (SWAN) with measured data. In *9th Ewtec 2011*. University of Southampton.
24. Muetze, A., & Vining, J. G. (2006). Ocean wave energy conversion-a survey. In: *41st IAS Annual Meeting. Conference Record of the 2006 IEEE, Industry Applications Conference, 2006* (Vol. 3, pp 1410–1417). Tampa, FL: IEEE.
25. Wu, F., Zhang, X.-P., Ping, Ju., & Sterling, M. J. H. (2008). Modeling and control of AWS-based wave energy conversion system integrated into power grid. *IEEE Transactions on Power Systems , 23*(3), 1196–1204.
26. Mann, L. D., Burns, A. R., & Ottaviano, M. E. (2007) CETO, a Carbon free wave power energy provider of the future. In *Proceedings of the 7th European Wave and Tidal Energy Conference (EWTEC)*.

27. Babarit, A., Guglielmi, M., & Clément, A. H. (2009). Declutching control of a wave energy converter. *Ocean Engineering, 36*(12–13), 1015–1024.
28. Clément, A., McCullen, P., Falcão, A., Fiorentino, A., Gardner, F., Hammarlund, K., Lemonis, G., et al. (2002). Wave energy in Europe: current status and perspectives. *Renewable and Sustainable Energy Reviews, 6*(5), 405–431.
29. Ramadan, A., Mohamed, M. H., Abdien, S. M., Marzouk, S. Y., El Feky, A., & El Baz, A. R. (2014). Analytical investigation and experimental validation of an inverted cup float used for wave energy conversion. *Energy, 70*(June), 539–546.
30. Backer, De., Griet, M. V., Frigaard, P., Beels, C., & De Rouck, J. (2010). Bottom slamming on heaving point absorber wave energy devices. *Journal of Marine Science and Technology, 15*(2), 119–130.

Chapter 4
Development of Oscillating Water Column and Wave Overtopping—Wave Energy Converters in Europe Over the Years

Krisna Adi Pawitan

Abstract The latest Intergovernmental Panel on Climate Change (IPCC) in 2018 gave the clearest yet call for immediate action to reduce the amount of human-related activity CO_2 emission. This fueled researchers to explore more sustainable ways of harvesting energy over the years, with ocean waves being one of the most attractive sources of renewable energy. This chapter will initially explore multiple ways of devices to harvest the energy from an ocean wave. The exploration then focuses on the development of Oscillating Water Column (OWC) type Wave Energy Converter (WEC), especially after the introduction of perforated vertical breakwater back in 1961 which open the possibility to combine both energy generation capability and coastal protection in the same structure. Examples of several projects which have been done in the past and are currently under construction in Europe will be given as an illustration of the current progression in the utilization of wave energy using OWC technology. Also, a Wave Overtopping Device—Wave Energy Converter will also be explored in a similar manner. A couple of representative project examples will also be given to similarly illustrate current progress on the development of such devices until now.

4.1 The Importance of Wave Energy Resources Utilisation

The intergovernmental Panel on Climate Change (IPCC) 5th assessment report (AR5) [1] indicates a clear involvement of human activities in the increase of average global temperature. It also stated that global warming leads to an increase in sea level. According to the AR5 IPPC report in 2014, the increase in global temperature is extremely likely to have been caused by anthropogenic gasses which increase the concentration of CO_2 in the atmosphere. The latest IPCC [2] report further indicates an increase in temperature in excess of 1.5 °C above the pre-industrial level. This

K. A. Pawitan (✉)
Quantum Wave Microscopy Unit, Okinawa Institute of Science and Technology Graduate University, Okinawa, Japan
e-mail: krisna.pawitan@oist.jp

A. Samad et al. (eds.), *Ocean Wave Energy Systems*, Ocean Engineering & Oceanography 14, https://doi.org/10.1007/978-3-030-78716-5_4

is the clearest yet call for immediate action to reduce the amount of CO_2 emission. To reduce the amount of greenhouse gasses in the UK, Department of Energy and Climate Change releases the Carbon Reduction Commitment (CRC) which pursues the target of cutting the CO_2 emission by 60% from 1990 levels by 2050 in the region. This plan pushed the development of greener energy generation ahead of "dirtier" energy sources, such as coal and natural gas. It is estimated that for every 1 kWh produced by marine renewables, the CO_2 emission can be reduced by 394 g when compared to Combined Cycle Gas Turbine (CCGT), 120 g when compared to Biomass, 937 g when compared to Coal powered energy generator, and about 1000 g when compared to Diesel power with 25% efficiency [3]. Fueled by the initiative, Scotland's renewable energy generation for consumer electricity in the last 10 years tripled the production to about 10GW by 2018 or accounted for over 70% of the total electricity used in Scotland. Wind energy, which accounted for about 71% of the total renewable energy generation, remains the biggest contributor by 2019. The wave and tidal energy sector, on the other hand, only responsible for about 0.18% [4]. This gives room for improvement for the wave energy sector to be developed and to have a bigger contribution to the world's total energy generation.

Among many renewable resources, the ocean remains as the biggest energy resource waiting to be harvested. It is calculated to be as much as 1000 GWh of wave energy reaching the British Isles every day on average [5]. The world energy council approximated no less than 29,500 TWh/yr of wave energy available, with the densest energy potential located in Europe and the west coast of America [6–8]. This source of energy, contrary to the theoretical energy potential suggested, remains underused compared to other renewable energy resources such as wind and solar [9, 10], with conversion technologies at a relatively early stage of development and with consequent high energy cost. In order to harvest, one needs to understand how a sea-wave forms. To put it simply, (wind-induced) wave is basically energy passing through the surface of the water. This energy mainly comes from the wind and is then transmitted across the ocean through the water. A stronger wind, which often occurs in the middle of the ocean, blows the surface of the ocean water and creates waves. The wave travels toward the coastline transmitting the energy to an accessible location. This form of energy, along with offshore wind and sea tidal and current, is commonly known as marine renewables. Deployment of wave energy converter technologies, unfortunately, is lagging behind the other two, due to its high cost of about EUR 330–630/MWh, which quite far compared to the target set by the European Union of USD 220/MWh by 2025 (IRENA [8]). This is worsen by poor performance of several pilot projects to date. This, at the same time provides a wide opportunity for a creative mind to fill in the gap in wave energy utilization and figure out the best way to harvest the energy. The following subchapter will explore the development of the wave energy utilisation attempt that has been done over the years, including a more in-depth overview of the development on the Oscillating Water Column (OWC) research with several examples of such projects, particularly in Europe, given as a representation of the general development of the device. Furthermore, a brief introduction of wave overtopping devices will be given similarly, including several examples of the projects which had been done in the past and are currently running, particularly in Europe.

4.2 A Brief Introduction to Wave Energy Harvesting Mechanism

Even though wave energy has been established as one of the biggest potential sustainable energy resources over the last 70 years, the mechanism for which the energy is taken from the ocean has not been converged. Many types of WECs have been developed in the past to harvest some of this ocean energy while surviving the harsh operation condition on-site at the same time. There are many types of wave energy converter developed to date, among them, there are four types of mechanism that stand out called wave attenuator, point absorber, wave overtopping, and Oscillating Water Column (OWC)—Wave Energy Converters (WEC). A wave attenuator is a floating Wave Energy Converter (WEC) of which the energy is generated by means of relative motion between two arms of the device. Because the device is floating, it can be rotated to face the direction of the incoming wave to guarantee maximum wave absorption in accordance with the design characteristics. Point absorber is also a floating WEC that took advantage of floating buoy(s) which oscillate up and down as the wave passes and connected to an energy extraction mechanism from such motion. In general, this type of WEC can absorb wave energy from multiple wave directions. On the down side, these types of WECs need to be positioned in a difficult-to-reach location and require additional effort for maintenance, such as towing to near-shore position, furthermore because these devices only serve as energy generation, the cost then wholly intended for energy generation. Wave Overtopping and Oscillating Water Column (OWC) devices, on the other hand, have advantages due to their simplicity in energy harvesting mechanism (often leads to low maintenance cost) and their ability to be combined with coastal protection building, such as breakwater or seawall, thus allows the cost to be split between energy generation and coastal protection. Without discounting the importance of other types of WECs mechanism, the following sub-chapters will be focused only on the development of Oscillating Water Column (OWC) and Wave overtopping types of Wave Energy Converter, especially in Europe.

4.3 Oscillating Water Column (OWC) Type Wave Energy Converter

4.3.1 General Introduction of OWC

The general principle of an OWC can be seen in Fig. 4.1. In general, an OWC device utilises wave crests and troughs to drive the water column trapped inside the OWC chamber to oscillate up and down. This oscillation occurs due to hydrodynamic pressure variation at the column base. These water movements, furthermore, force the air inside the chamber to move out and in via an air channel. The power generation can then be done by placing a power take-off mechanism in this air channel.

The development of an Oscillating Water Column started in the 1940s with the development of a floating OWC device to power a navigation buoy [11–13]. It was designed and developed due to its simplicity and potential for having relatively low maintenance and therefore the initial development was only focused on energy generation. Only after the idea of a perforated vertical breakwater was established in 1961 by Jarlan [14] that showed the advantage of having a wave absorption chamber to reduce the wave loading experienced by the front vertical wall breakwater, engineers then began to study the possibility of combining OWC concept into a vertical breakwater [15] thus making the possibility of higher survivability in the coastal defense structure while producing energy at the same time. These studies found that integrating an OWC into a vertical breakwater is not only possible but also has the theoretical potential of 100% energy absorption [16]. Apart from the energy absorption capability, by definition, a coastal protection structure should be exposed to a highly energetic portion of the ocean. This, in turn, will enable the WEC device to be exposed to those energetic portions of the ocean as well. This scheme, furthermore, enables the ability to share the cost between the energy generation and coastal protection constructions.

Fig. 4.1 Simplified version of an Oscillating Water Column schematics

4.3.2 Working Principle and Design Analysis of OWC

Due to the working principle of the device, minimum submerged moving parts are needed for the device to generate power, thus the possibility of mechanical failure due to salt-water and metal interaction is estimated to be minimal. Although it is important to be noted that the air flowing out of and into the chamber may be very humid and saturated with salt-water spray. Over the years, many researchers have explored the optimum design of an OWC. In general, the studies focused on the front wall wave loads, OWC caisson dimensioning, and on power take-off optimisation. Small-scale physical model studies found that an inclined-slope front wall will reduce the wave loads experienced by the front wall but will have a higher wave transmission coefficient when compared to an ordinary vertical wall [15, 17]. Subsequent studies also found that the front wall geometry and the front penetration—defined as the depth of the submerged part of the front wall divided by the distance between the still water level (*swl*) and the OWC's chamber floor—also play a significant role in determining the overall efficiency of the device, with typical efficiencies of up to 70% reached. Furthermore, [18, 19] hypothesized that the total forces acting on the front wall will be the sum of wave pressure distribution and hydrostatic pressure distribution and are reduced by the opposing pressure distribution acting on the in-chamber side of the front wall. Unfortunately, hitherto, there have been no experimental or numerical results to support this argument. In addition to the wave loading, [15] published guidance to select the dimension of the OWC caisson design based on the wave characteristics. For an OWC design, there are three most influential geometries: chamber width (B_c), submerged front wall (w), and chambers height (crest freeboard) (h_c) (Fig. 4.2). Takahashi [15] also recommends several values for the OWC installed breakwater design based on the hydrodynamic efficiency and structural stability as follows: B_c equal to about 0.13L (wavelength), w equal to about $0.25H_{max}$ (highest incident wave height), and h_c equal to about $1.0H_{m0}$. Here H_{max} refers to the highest

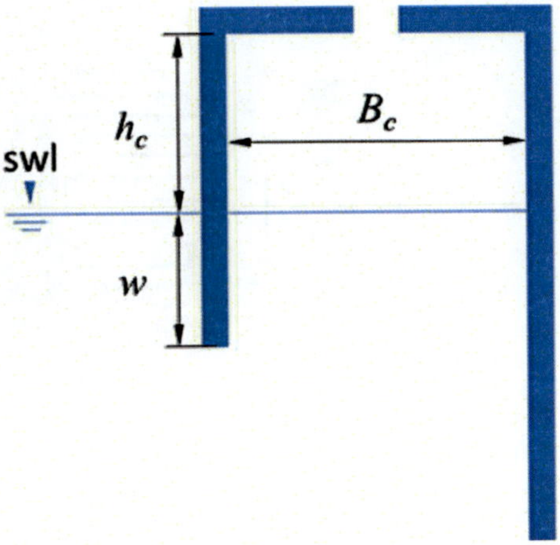

Fig. 4.2 Most influential geometries in designing an OWC chamber after Takahashi [15]: chamber width (B_c), submerged front wall (w), and chamber height (h_c)

individual wave height in a record, while H_{m0} or the significant wave height is defined as an imaginary higher one-third wave and calculated as the mean of the top 1/3 wave height in descending order in a given record.

After the dimensions related to the OWC chambers have been selected, it is important to continue the exploration into the loads acting on the in-chamber walls. For the front wall facing seaward, a lot can be borrowed from the coastal engineering research on the wave loadings on a vertical wall breakwater. One of the most prominent and commonly used predictions in that field is the trapezoidal pressure distribution with the maximum pressure occurred on the still water level proposed by Goda [20, 21]. The prediction tools themselves enables the engineers to calculate the pressure at the still water level (refer to as p_1), the pressure at the bottom of the breakwater foundation (p_2), at the toe of the breakwater (p_3), and finally the uplift force experienced by the breakwater due to wave loadings (p_4). The trapezoidal shape mentioned and the location of p_1-p_4 summary are illustrated in Fig. 4.3. Here η indicates the height to which the wave pressure is exerted, and d denotes the distance between the still water level and the toe of the breakwater.

Because the wave loads experienced by the front wall are assumed to be similar to a conventional breakwater, now in-chamber wall's wave load is important to be discussed. Early exploration on the rear wall wave pressure was done leading to the then proposed 4 MW wave generator at Siadar Bay, Scotland. For the sake of preliminary design, [22] then postulates that the wave pressure distribution acting on the rear wall of the air chamber may come from the "missing" part of the front wall and the pressure distribution should be hydrostatic as illustrated in Fig. 4.4 by red arrows.

However, more recent small-scale physical model results, using a regular wave experiment, suggested that the pressure acting on the rear wall is closer to a Goda-like pressure distribution [23]. Building upon large-scale experimental results, [24] expanded the in-chamber wave loading prediction to include both the chamber's rear

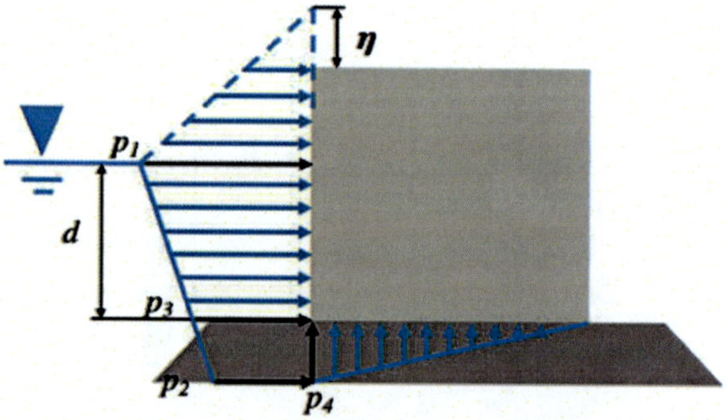

Fig. 4.3 Simplified illustration of the trapezoidal wave loading proposed by Goda [21]. Here p_1 denotes the pressure acting at still water level, p_2 at the bottom of the breakwater foundation, p_3 at the toe of the breakwater, and p_4 the uplift pressure due to incident wave. η indicates the height at which the wave pressure is exerted to

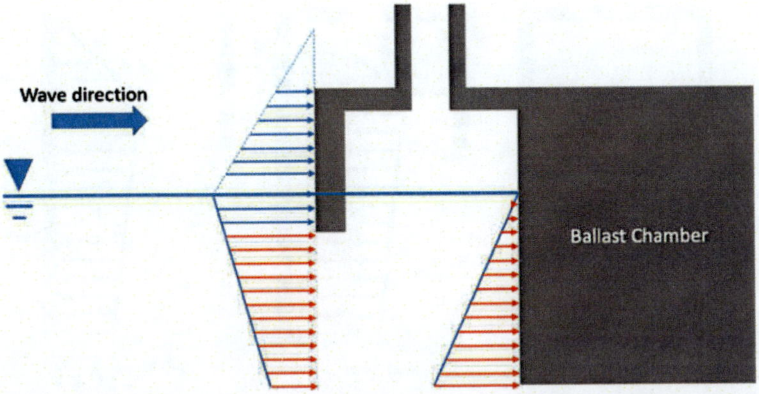

Fig. 4.4 Illustration of hypothesised pressure distribution on the rear wall after Patterson et al. [22]

wall and the chamber's ceiling by modifying the incoming Goda-like trapezoidal pressure from the incident wave. The expanded method incorporates a transfer coefficient (K_t) to the integrated pressure calculation introduced by Goda [20], hereafter referred to as F_{GODA}. The transfer coefficient (K_t) itself is based upon the front wall submersion relative to the water depth at the toe of the device and can be calculated as follows.

$$0 \leq \frac{w}{d} \leq 0.3 \rightarrow K_t = 1$$

$$0.3 < \frac{w}{d} \leq 0.9 \rightarrow K_t = \frac{0.81 - 0.7(w/d)}{0.6}$$

The "transferred" in-chamber integrated pressure (F_t) then can be calculated as follows,

$$F_t = F_{GODA}.K_t$$

The authors suggested that the rear wall pressure changes depending on the chamber's condition and separate the estimation method into closed chamber condition indicating zero air escape thus zero energy production, fully open chamber condition indicating zero chamber pressure buildup thus zero energy production, and operating chamber condition, which is defined as the intermediate condition between the first two extremes mentioned, demonstrating that the pressure acting on the in-chamber rear wall probably closer to the one illustrated in Fig. 4.3 instead of the one shown in Fig. 4.4. A very simplified concept of this can be seen in Fig. 4.5. To indicate the transition state of the operating condition, the authors also introduce *proportion coefficient* (P) which indicates the condition of the chamber with $P = 1$ meaningfully closed and $P = 0$ meaningfully open. The author proposed the following empirical formula to calculate the *proportion coefficient* (P),

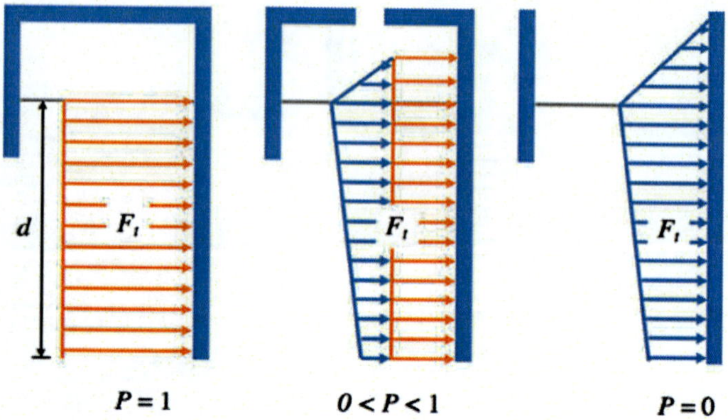

Fig. 4.5 The pressure distribution schematics for the operating condition at various openings are represented by P values. The opening illustration is exaggerated in this figure, subfigure **a** closed chamber condition ($P = 1$), **b** intermediate cases, and **c** open chamber condition ($P = 0$)

$$P = f\left(\frac{A_o}{A_c}\right) = 1 - 45.55\left(\frac{A_o}{A_c}\right)$$

Here the A_o and A_c represent the ratio between the opening orifice area ratio and the chamber area ratio respectively. It is important to note here that the author draws a comparison between the air restriction caused by the turbine's rotation into the opening: chamber area ratio and both fully closed and full open conditions is situations where the wave energy generator is not functioning. This prediction method, furthermore, ignores the sea-ward pressure that occurs on the rear part of the front wall because that will reduce the total landward horizontal force experience by the whole structure thus making the prediction conservative.

After determining the case of interest, then the pressure distribution on the in-chamber rear can be calculated using the following formula for fully closed ($(F_t)_{closed}$)

$$(F_t)_{closed} = K_t.F_{Goda} = p_c d$$

where p_c denotes the chamber pressure and d denotes the distance between the still water level and the OWC chamber's floor. For fully open condition the following formula can be implemented to predict the wave pressure acting at the still water level (p_{tswl}) and at the OWC chamber's floor elevation (p_{tbot})

$$H_t = H_{max}\,K_t$$

$$\alpha_{swl} = \left[0.6 + 0.5\left(\frac{4\pi h/L}{\sinh(4\pi h/L)}\right)^2\right] + \left[\min\left\{\frac{h_b - d}{3h_b}\left(\frac{H_{tr}}{d}\right)^2, \frac{2d}{H_{tr}}\right\}\right]$$

$$\alpha_{bot} = 1 - \frac{d}{h}\left[1 - \frac{1}{\cosh(2\pi h/L)}\right]$$

$$p_{tswl} = \alpha_{swl}\, \rho\, g\, H_t$$
$$p_{tbot} = \alpha_{bot}\, p_{tswl}$$
$$\eta_t = 1.5\, H_t$$

where H_{max} stands for the maximum wave height (H) measured in a series of the wave train, L denotes the deep-water wavelength, h denotes the water depth in front of the structure, and h_b denotes the water depth at the location of $5H$ in front of the structure. So that the integrated pressure $(F_t)_{open}$ can be predicted using

$$(F_t)_{open} = 0.5((\eta_t + d)p_{tswl} + dp_{tbot})$$

For the operating condition, the integrated pressure ($(F_t)_{operate}$) can be estimated using the proportion coefficient mentioned earlier and using the following formula

$$(F_t)_{operate} = P(F_t)_{closed} + (1 - P)(F_t)_{open}$$
$$(F_t)_{operate} = P(p_c(d + \eta_{tr})) + 0.5(1 - P)((\eta_t + d)p_{tswl} + dp_{tbot})$$

where the η_{tr} should follow

$$\eta_{tr} = \begin{cases} 0, & if\ P = 1 \\ \min(1.5\, H_t\, ,\ h_{ceil}), & if\ 0 \le P < 1 \end{cases}$$

With h_{ceil} indicates the distance between the OWC chamber's ceiling and the still water level. Recent findings also suggested that introducing an OWC caisson may lead to a reduction in wave reflection of a vertical breakwater [25].

The aforementioned front wall load (e.g. [18, 23]) and the in-chamber wall load studies (e.g. [23, 24]) are mostly focused on the non-impulsive wave loads, but one of the biggest concerns in a conventional vertical breakwater design is in impact loads. Unlike non-impulsive wave loads, impulsive wave loads (commonly as a result of wave breaking in front of or at the location of the breakwater) can results in a much higher peak local pressure (e.g. [26, 27]). Several examples were shown before where ignoring such pressure during design processes may lead to dire consequences for breakwaters by Oumeraci [28]. Since the water column of an OWC is "protected" behind the front wall and ideally is not exposed to the ocean surface, the water column inside the chamber is mostly assumed to be well-behaved during design stages, including when determining the device performances. Subsequent research [29], on the other hand, observed that an impact phenomenon might also occur inside the chamber. A Particle Image Velocimetry (PIV) exploration also suggested that such impact may occur due to a vortex that developed behind the front wall opening when the water column is moving upward [30]. Apart from the structural influence, non-ideal water column conditions can be expected to have a negative effect on the wave energy conversion efficiency of the device. One could imagine that the chaotic water column movement may increase the amount of larger-scale turbulent motions in the chamber, which could lead to a significant reduction in energy extraction capability. Additionally, due to the fact that if the moisture in the air chamber exceeds 65%, the power extraction capability may decrease by about 50–70% [31]. Focusing on the impact force acting on the structure, [32] shows the wave condition at risk at various power take-off setting conditions. The author further argues that an impact that occurs at the front of the wall may lead to an impact occurs inside the OWC chamber, with the highest impact pressure recorded on the ceiling of the chamber instead of the internal rear wall. This location increases the possibility of sliding and overturning of the entire coastal structure, thus increases the importance of impulsive wave loads consideration to be included during the design process.

4.4 Oscillating Water Column Type WEC Projects Developments History

Over the years, there have been several OWC pilot projects developed around the world, especially in Europe. Among many early examples, there were two shoreline OWC projects that stand out due to their early full-scale prototype testing called Land Installed Marine Power Energy Transmitter (LIMPET) in the Isle of Islay, UK, and Pico power plant in The Azores, Portugal. Soon after that, the world's first grid-connected OWC installed breakwater was constructed at the port of Mutriku, Baque Country, Spain between 2009–2012, and the U-OWC type WEC in the Port of Civitavecchia in Italy followed in 2011 [33], which mark the beginning of the integration between coastal protection structure with energy generating device which, hopefully, will become more mainstream in the future. This section will discuss the development of OWC, especially in Europe, through the aforementioned OWC devices. In addition to the bottom-stand OWC projects, several examples of early floating OWCs development around the world will be discussed briefly.

4.4.1 Land Installed Marine Power Energy Transmitter (LIMPET)

Named after a small marine mollusk, the Land Installed Marine Power Energy Transmitter or LIMPET was designed and developed by the Queen's University of Belfast and constructed on the Isle of Islay, Scotland. When it started its operation in 2001, LIMPET was the first grid connected WEC in the world. The wave energy converter is an onshore oscillating water column and was designed to generate 500 kW of electricity. Figure 4.6 shows the schematic of LIMPET power plant which comprises of three 6 m by 6 m concrete chambers with 40° inclined front wall.

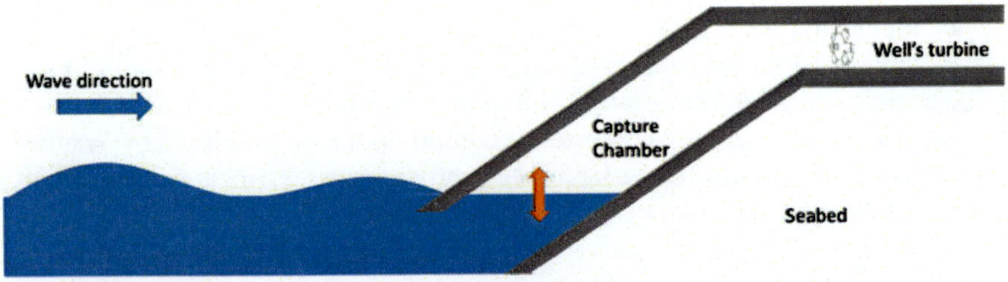

Fig. 4.6 Simplified schematics of LIMPET power plant in the Isle of Islay after [34]

The front wall inclination was selected after a small-scale test of 1:36 scale model experiments found that compared to a vertical structure, the 32.7° front wall inclination led to a 36% wave pressure reduction, and the −32.7° angle (inner chamber part of the front wall) reduced the wave loads experienced by the inner chamber by as much as 56% [17]. However, the experiment only used a single regular wave period of 1.49 s and the same small-scale test later showed large vortices appeared at the "lip" of the front wall during the water inflow and outflow and a breaking wave hitting the chamber's rear wall in a high incident wave cases [29] indicating possible high-efficiency losses during such cases.

A LIMPET performance report published after the conclusion of the project indicates breaking wave loads of up to 690 kN/m^2 experienced by the prototype which consistent with the small-scale test [34]. These results may show some of the reasons behind LIMPET's poor performance during operation, with only generating up to 100 kW of its peak, about 20% of the rated capacity. Anecdotal evidence points also to a lot of wave energy being dissipated by wave breaking in the shallower part of the water seaward of the plant, including the loud noise made by the turbine during operation. Despite these performance issues, the project demonstrated that the OWC wave energy converter's simplicity can offer excellent reliability by performing 98% of its lifetime [35]. This is shown by the past experienced that the plant didn't require a lot of routine maintenance check-ups and was quite robust against any weather condition. It is also important to remember that the power plant was located on the shoreline instead of the nearshore.

4.4.2 Pico Power Plant

In parallel with LIMPET design and production, another full-size fixed-structure OWC was initiated on the island of Pico, Azores, Portugal. The power plant was built under the coordination of Instituto Superior Tecnico (IST). The structure consists of a single 12 m by 12 m cross-section water column in a concrete chamber, standing on the sea bottom. The device was rated at 400 kW and was completed in 1999. The highest sustained energy output recorded was about 7.5 MWh during a near 400 hours of continuous production in November 2001. The development of the device included a wide range of scales physical model tests from 1:35th test in a wave flume to a 1:4th physical model test in actual seawater. After almost 20 years of operation, the wave power plant is finally closed due to some partial failure in the base of the wave energy converter, according to the press release by WavEc in 2018. Another site of the same island is currently under assessment for wave energy generation power plants following the decommissioning of the Pico power plant [36].

4.4.3 *Mutriku Wave Energy Plant*

Unlike coastline OWC structures such as LIMPET and Pico plants, the Mutriku wave energy plant is located in the deeper and more energetic location of the ocean where a breakwater is needed as harbor protection. The OWC type chamber is installed in an already designed approximately 440 m long breakwater as shown in Fig. 7a from above, courtesy of an open street map. Figure 7b shows the same structure taken from the ocean in front of the vertical wall, courtesy of [37]. As can be seen here, some of the front wall's outer layer "peels-off" due to extreme wave pressure during the storm. The construction was promoted by Basque utility Ente Vasco de la Energia (EVE). The main challenges faced by the WEC's designers at that time were making a WEC design that will not interfere with the harbor accessibility and will not cause many alterations to the breakwater's original design [38]. The team later decided that an OWC concept could best fulfill both requirements, with the final design of the OWC installed breakwater specified 16 OWC chambers, each connected to 18.5 kW self-rectifying Wells turbine with a rated power of 296 kW in total. The turbine used in this project is the same design as LIMPET. Unlike LIMPET, however, this project used separate power take-off (PTO) system for each chamber.

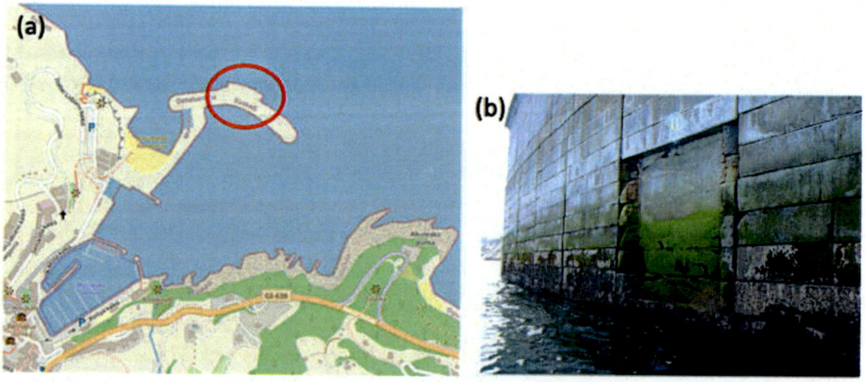

Fig. 4.7 The OWC type chamber is installed in an already designed approximately 440 m long breakwater **a** from above, courtesy of Open Street Map and **b** the same structure taken from the ocean in front of the vertical wall, courtesy of [37] with permission

The developer claims that the power generator will avoid 600 tonnes of CO_2 emission annually, although it is interesting to consider the embodied CO_2 in raw materials and the construction process. Nevertheless, the breakwater is much-needed harbor protection for the area, so included a wave energy harvesting mechanism should bring an advantageous addition to the existing design. Another intention of the project is to make the Mutriku a reference point in marine energy development which was successfully achieved upon the commissioning of the project and marked as the world's first grid connected OWC installed vertical breakwater in the world.

Although now it is operating, the construction was not a smooth process, suffering from severe storm damage in December 2007, March 2008, and January 2009. These storms caused 4 of the 16 chambers to be partially destroyed (Fig. 7b), including the loss of the entire section of the front wall [37]. A subsequent study suggests that the pressure experienced by the structure may have reached 6 times the operational limits of the front wall during these storms [39].

4.4.4 Resonant Wave Energy Converter (REWEC) or U-OWC

Europe's newest wave energy power plant project based on the OWC design as per writing is the Resonant Wave Energy Converter 3 (REWEC3) or as also known as the U-OWC type WEC credited to the additional wall extending in the front of the front wall forming a "U" shaped wave duct as shown in Fig. 4.8 [40]. By adding this vertical wall, the resonance of the device can be better tuned to a larger wave condition, which should lead to a higher absorption rate under-designed wave characteristics. Subsequent research [41] further estimated that for the identical size structure, the U-OWC may absorb 90.7% of the wave energy compared to only 15.6% of the simple conventional OWC configuration, or about 5.5 times more electricity generation. The results, however, was only done for irregular sea condition with

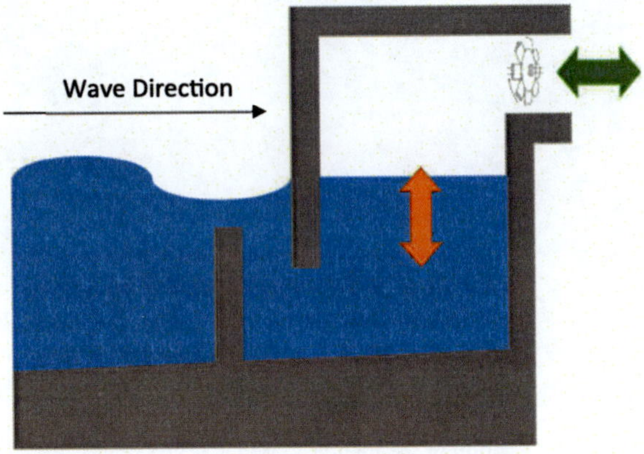

Fig. 4.8 The Resonance Wave Energy Converter (REWEC) or U-OWC schematics after Boccoti [40]

significant wave height (H_{m0}) of 2.5 m. So, it would be interesting to see the real sea operating performance once the construction is finished.

The construction of the U-OWC prototype was started at the Port of Civitavecchia in 2011 [42]. The 578 m wave energy power plant consists of 124 chambers integrated into caisson breakwaters. One caisson comprises 8 chambers. An early report of the first U-OWC plants can be seen in [35].

4.4.5 *Siadar Wave Power Project*

In 2009, N power proposed a 4 MW wave energy plant, that if finished, would be the largest in the world, located in Siadar Bay, Lewis, Scotland (Fig. 4.9). The proposal was submitted in 2010 as a result of Scotland's ambition to increase its renewable energy production from 50 to 80% of the overall electricity consumption by 2020. A Scottish-based wave energy company, WaveGen (LIMPET's and Mutriku's turbine provider) was then given the opportunity to build the wave power plant. Site measurements were used in a detailed study to establish the design wave condition and optimum orientation of the power plant [22].

As can be inferred from the remote island location, grid connection was one of the main challenges faced by this project. The project needed a subsea cable to connect the electricity produced to the main grid of the UK. The project was delayed due

Fig. 4.9 Location of the then proposed 4 MW Siadar wave energy converter project, indicated by the red triangle

to the delays in the interconnector and high transmission charges according to the spokesperson for Comhaile nan Eilean (Western Isle Council) in Stornoway Gazette before its ultimate abandonment in 2012. WaveGen subsequently re-designed the project into a 20 MW scheme. Sadly, it did not progress due to continued uncertainty on grid connection capacity and with the abandonment of the revised Siadar Project, WaveGen sadly folded.

4.4.6 Floating OWC Development

The European projects discussed so far have all been fixed structure, either on the shoreline or integrated within a fixed breakwater. Elsewhere, the floating OWC has been explored too. An early example was an open sea test done in Gokasho Bay, Nansei Town, Japan in September 1998. The structure, called "Mighty Whale", was a floating oscillating water column wave energy harvester device consisting of three air chambers. It was tested in an open sea and even endured a typhoon with recorded significant wave height (H_{m0}) up to 5.57 m and a significant wave period (T_s) of about 12.5 s [44]. Unlike a fixed structure OWC device, a floating body of a floating OWC structure enables the chamber to oscillate along with the water column oscillation and relative to the wave motion. This characteristic makes the mooring system of the whole device to be an integral part of the overall device energy generation efficiency. One significant advantage of a floating OWC, due to the multiple body oscillating motions, is the frequency range at which the device best operates. The structure will have two peak frequencies due to the movement of the OWC chamber and the water column inside the chamber. This allows the device to be tuned to a wider operational range [45].

Another example of a floating OWC was a 1:3 scaled floating OWC deployed in 2010 by Oceanlinx, the then Australian energy company, at Port Kembla, New South. Extreme weather, unfortunately, broke the mooring of the device in the same year of installation although fortunately was still able to be recovered soon after. The company continued to develop a 1 MW energy generator device using the same principle called greenWAVE in 2014 [46]. The completed device, unfortunately, met its end once again after damaged beyond repair during its transport from Port Adelaide to Port MacDonnell. This kind of failure, regrettably, was not unprecedented. In 1995, UK's first offshore wave power plant called Ocean Swell Powered Renewable EnergY or OSPREY 1, which intended to be installed near Dounreay, Scotland, sank to the bottom of the ocean after caught by a 3 m swell of Hurricane Felix. The unfortunate wave energy device was developed by Applied Research and Technology of Inverness. It was concluded that repairing OSPREY 1 was unlikely to be viable and the project was abandoned as reported by the Herald. The most recent floating OWC design took advantage of a cylindrical shape with a tapper-heavy bottom to stabilised the so-called Spar-buoy OWC device [47]. The 1:16 scale prototype of the device has been tested at National Renewable Energy Centre, NAREC, UK, in 2012 and is subject to further optimisation. One of the latest Floating OWC devices is the 5 m in diameter and 42 m in length floating spar buoy with OWC mechanism inside called the MARMOK-A-5. The 80 tonnes in weight floating OWC is designed to generate as much as 30 kW of power. Initially, the device was developed by Oceantec, but later inquired by IDOM inc. in September 2019. Under the initiatives of the Horizon 2020 European Union project called OPERA, the device was tested in Biscay Marine Energy Platform (BiMEP), near the village of Armintza-Lemoiz in Spain, where the device faces up to 14 m high wave [48].

Although the floating type OWC can be exposed to a more energetic offshore location, integration of an OWC device into a breakwater is the easiest and, so far, the most financially viable option.

4.5 Brief Summary of Wave Overtopping Devices' Development Over the years

4.5.1 General Introduction to Wave Overtopping Mechanism

Apart from the Oscillating Water Column (OWC) typed Wave Energy Converter (WEC), the Wave Overtopping mechanism is also another strong candidate for wave energy development into commercialisation. In general, the concept is equally, if not more, as simple as an OWC. As the wave passes, the wave overtopped and is collected into a water reservoir and then channeled back into the ocean (Fig. 4.10 with water flow indicated by the arrows). A turbine is then placed inside the channel between the water reservoir and the ocean to generate electricity using the water flow. However, unlike the OWCs, the turbine is constantly interacting with a unidirectional seawater flow under this mechanism. One advantage that the wave overtopping has over the OWC's mechanism is that the uplift force is at minimum when the horizontal force is at maximum [49]. In this case, one can imagine that wave overtopping is relatively more stable to higher wave loads against sliding and overturning when compared to

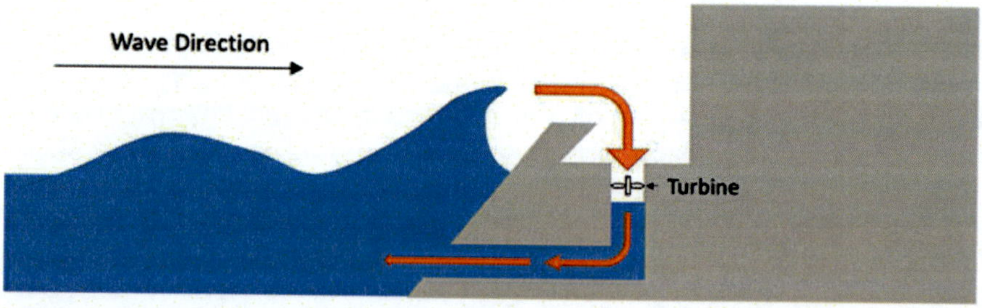

Fig. 4.10 Schematics of Wave Overtopping energy converters, with orange arrows indicating the flow of the seawater to be collected in the water reservoir and channeled back into the ocean through a turbine

an OWC device. This occurs due to higher wave pressure (commonly associated with higher wave height) resulted in more water being collected inside the water reservoir which in turn provides a larger downward hydrodynamic pressure to counter the uplift pressure. Here the uplift force is defined as the force acting in the upward direction resulted in structure being "lifted" and mostly caused by the combination of buoyancy and turning moment due to wave loads. Although it is less explored compared to an OWC-type device, there were several waves overtopping WEC projects both the floating type, like Wave Dragon and WaveCat, and the bottom standing type, such as the Sea-wave Slot-cone Generator (SSG) and the Overtopping BReakwater for Energy Conversion (OBREC), that gain a lot of attention and gathered a significant amount funding over the years. Here the development of the bottom standing type Wave Energy Converters, the two later projects mentioned, are presented as a brief introduction on the development of such structures.

4.5.2 Wave Loadings Analysis and Development of Sea-Wave Slot-Cone Generator (SSG)

In 2004 the Sea-wave Slot-cone Generator (SSG) wave overtopping wave generator concept was developed by WaveEnergy AS company based in Stavanger, Norway. The device, in general, is a multi-stage wave overtopping water reservoir with a multi-stage turbine installed in between the stages to generate electricity. The concept of the device first developed in the idea that using multiple reservoirs would lead to higher overall efficiency when compared to a single water reservoir [50]. This device, furthermore, is intended to be a standalone energy converter standing on a steep foreshore called "*concentrator*" [51] or "*focuser*" [52], which functioned as a way to increase the potential run-up height and also forced the incident wave to break, thus dissipating the incident wave load. The developer managed to secure 5million Euros in funding from both the private sector and European Union to build a prototype in the isle of Kvitsoy-Norway [53]. Shortly after that, another prototype was also planned to be built on the neighbouring West Norwegian Coast of Svaheia [52]. On the initial estimation, the 500 m long device should be able to generate electricity with the cost of about 0.12 EUR/kWh, which is already competitive when compared to the island main power generation source of diesel generators, with further forecast reduction to be as low as 0.04 EUR/kWh once deployed in a larger scale.

The early focus of the SSG's design was on the front wall wave loading. Since wave overtopping requires the incoming wave to overtopped into the water reservoir, the front wall is angled in such a way to increase the amount of overtopping. The SSG itself has 3 reservoirs with 35° inclined front wall on each reservoir stage as demonstrated in Fig. 4.11. To analyse the wave loads, [53] provides the wave forces acting on the front wall of the structure using a 1:60 physical scale model tested in a wave tank at Aalborg University.

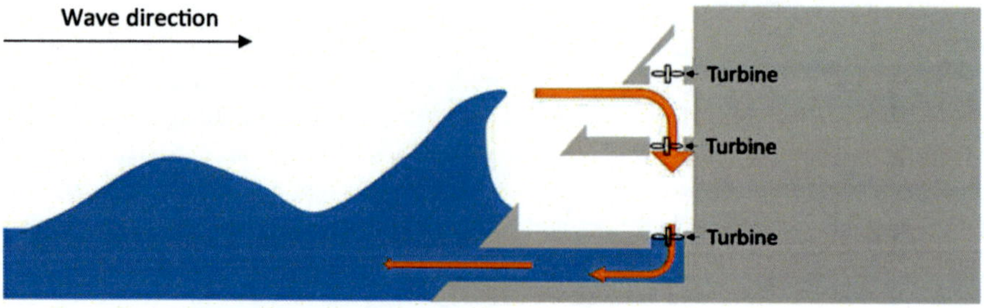

Fig. 4.11 A simplified schematics of Seawave Slit-cone Generator (SSG) wave energy device after [50]

Due to the shape of the bathymetry, the study found two types of wave endured by the structure: a surging wave attack to the front wall with no breaking wave (a rapid rise of the wave along the three stages of inclined front wall) and a partially damped plunging breaking which hits the sidewall and resulting in a high vertical jet. The study, furthermore, found that similar trapezoidal pressure distribution as to the one suggested by Goda [20] appeared in the pressure measurement results on each front plate for each stage.

In subsequent experiments, [50, 54] estimated the SSG has about 50% hydraulic efficiency by comparing the wave loads on the front wall of the device and the amount of hydrostatic pressure collected on the bottom of the water reservoir due to overtopping. This efficiency is extracted based on the 2-D study with a wave coming from the front of the device and no spreading occurs before reaching the device location. When exposed to wave spreading, the efficiency is quickly reduced to be between 30 and 25% depending on the angle of the incident wave with efficiency as low as 15% measured during severe unfavourable incident wave angle condition [50, 55]. Further exploration pointing to the water reservoir width: depth ratio to be one of the most determining factors for this efficiency number. It is also interesting to note that impact pressure was observed on both the front wall and the internal rear wall of the physical model. The 35° inclined front wall was designed to "force" the incident wave to pass the breaking wave height, thus dissipating the wave energy. The impact pressure measurement on the internal rear wall, on the other hand, showed that the angle is not optimal to break all of the incident waves.

4.5.3 Overtopping BReakwater for Energy Conversion (OBREC) Development

Based on the lesson gathered during the Sea-wave Slot-cone Generator (SSG) development, a wave overtopping device called Overtopping BReakwater for Energy Conversion (OBREC) is currently under development, led by The University of Campania Luigi Vanvitelli or formerly known as The Second University of Naples

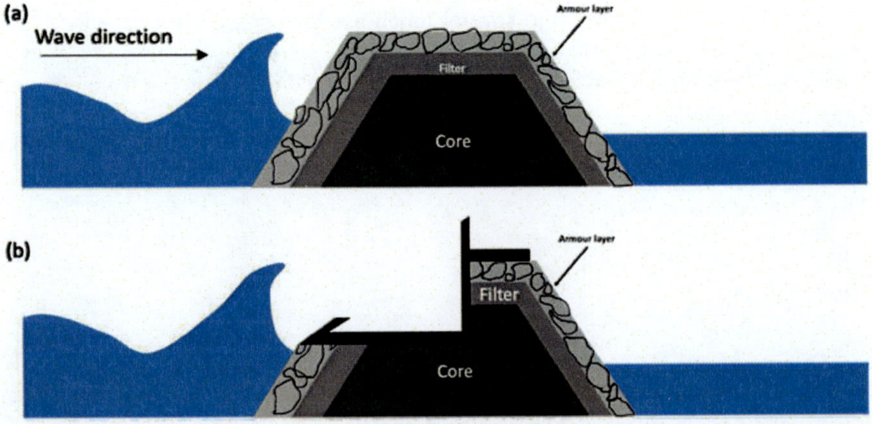

Fig. 4.12 A simple schematics of a traditional rubble mound breakwater (**a**) and the proposed Overtopping BReakwater for Energy Conversion (OBREC) (**b**) after Vicinanza et al. [56]

[56]. The device takes advantage of the traditional rubble mound breakwater configuration (Fig. 12a) and added a water reservoir sitting on top with a front opening facing the seaward (Fig. 12b). The water reservoir is then connected back into the ocean via a power take-off mechanism.

Early exploration of design suggested that the addition of the WEC increases the stability of the overall structure by providing a counter momentum to the wave loading direction due to the hydrostatic pressure provided by the water collected in the reservoir. The subsequent experiment, however, showed that the inclusion may slightly increase the amount of overtopping water discharged to the rear part of the breakwater, especially in the case where a large amount of water entered the reservoir and created an impacting water jet to the OBREC's rear wall. The same experiment suggested that this problem can be mitigated by adding a triangle on top of OBREC's upper crown wall pointing seaward [57].

When designing a breakwater, the height needs to be determined to be able to balance between the allowable overtopping rate, i.e., amount of water managed to go to the rear part of the breakwater, and the cost and the esthetic of the structure. In such a case, it is more likely that the height of the OBREC should be similar to the original breakwater design to stay in line with the esthetic consideration of the design. To gain more understanding, a prototype scaled version of OBREC was installed at San Vincenzo Breakwater in 2015 [58]. The prototype covered an area of about 75m^2 and consist of two different front wall heights. In order to get the incoming wave height, 2 buoys made out of 15 inches in diameter ABS spheres were deployed 100 m in front of the structures. The prototype was originally planned for a 3 year observation campaign and recently the effectiveness of a curved wall to be the front wall is also explored using a small-scale physical model [59]. It is interesting to see the results of the prototype testing once the project has been concluded. Even though the utilisation of wave energy as a renewable energy source has a rough start, with the inclusion of WECs into several much-needed coastal protection structure projects all around the world, especially when more extreme weather is expected in the future, the future of wave energy development could only increase from now on.

Acknowledgements The author would like to thank Prof. Tom Bruce from the University of Edinburgh for his guidance and suggestion during the writing of the OWC chapter.

References

1. Allen, M. R., Barros, V. R., & Broome, J., et al. (2014, November). In In P. Aldunce, T. Downing, S. Joussaume et al. (Eds.), *Intergovernmental Panel on Climate Change (IPCC) fifth assessment synthesis report—Climate Change 2014 synthesis report.*
2. Intergovernmental Panel on Climate Change (IPCC). (2018, October) *Special Report on Global Warming of 1.5 °C*. Incheon, Republic of Korea.
3. Smart, G., & Noonan, M. (2018). *Tidal stream and wave energy cost reduction and industrial benefit summary analysis*. Catapult Offshore Renewable Energy.
4. Renewables in Numbers 2019. https://www.scottishrenewables.com/forums/renewables-in-numbers/. Accessed May 07, 2019
5. Barstow, S., Mørk, G., Mollison, D., & Cruz, J. (2008). The wave energy resource. In *Ocean wave energy* (pp. 93–132). Berlin, Heidelberg: Springer.
6. World Energy Council. (2017). *World energy resources 2016*. Available at https://www.worldenergy.org/publications/2016/world-energy-resources-2016/. Accessed at May 14, 2018
7. IRENA (International Renewable Energy Agency). (2014). *Ocean energy technology: Innovation, patents, market status and trends*. International Renewable Energy Agency, June 2014.
8. IRENA (International Renewable Energy Agency). (2020). *Innovation outlook: Ocean energy technologies*. https://www.irena.org/publications/2020/Dec/Innovation-Outlook-Ocean-Energy-Technologies. International Renewable Energy Agency, June 2020.
9. Quirapas, M. A. J. R., Lin, H., Abundo, M. L. S., Brahim, S., & Santos, D. (2015). Ocean renewable energy in Southeast Asia: A review. *Renewable and Sustainable Energy Reviews, 41*, 799–817.
10. Ly, D. K., Aboobacker, V. M., Abundo, S. M. L., Srikanth, N., & Tralich, P. (2014, November). Wave energy resource assessment for Southeast Asia. In *Proceedings of the 5th International Conference on Sustainable Energy and Environment (SEE), Science, Technology, and Innovation for Association of Southeast Asian Nations (ASEAN) Green Growth, Bangkok, Thailand* (pp. 19–21).
11. Masuda, Y. (1986). An experience of wave power generator through tests and improvement. In *Hydrodynamics of ocean wave-energy utilization* (pp. 445–452). Berlin, Heidelberg: Springer.
12. Falcão, A. F., & Henriques, J. C. (2014). Model-prototype similarity of oscillating-water-column wave energy converters. *International Journal of Marine Energy, 6*, 18–34.
13. Falcão, A. F., & Henriques, J. C. (2016). Oscillating-water-column wave energy converters and air turbines: A review. *Renewable Energy, 85*, 1391–1424.
14. Jarlan, G. E. (1961). A perforated vertical wall breakwater. *The Dock and Harbour Authority, 486*, 394–398.
15. Takahashi, S. (1989). Hydrodynamic characteristics of wave-power-extracting caisson breakwater. In *Coastal engineering 1988* (pp. 2489–2503).
16. Evans, D. V., & Porter, R. (1995). Hydrodynamic characteristics of an oscillating water column device. *Applied Ocean Research, 17*(3), 155–164.
17. Müller, G. U., & Whittaker, T. J. T. (1993). An investigation of breaking wave pressures on inclined walls. *Ocean engineering, 20*(4), 349–358.

18. Morris-Thomas, M. T., Irvin, R. J., & Thiagarajan, K. P. (2007). An investigation into the hydrodynamic efficiency of an oscillating water column. *Journal of Offshore Mechanics and Arctic Engineering, 129*(4), 273–278.
19. Preen, S., & Robertshaw, G. (2010). Development of a generic caisson design for an oscillating water column power generator. In *Coasts, marine structures and breakwaters: Adapting to change: Proceedings of the 9th International Conference Organised by the Institution of Civil Engineers and Held in Edinburgh on 16 to 18 September 2009* (pp. 2–266). Thomas Telford Ltd.
20. Goda, Y. (1975). New wave pressure formulae for composite breakwaters. In *Coastal engineering 1974* (pp. 1702–1720).
21. Goda, Y. (2010). *Random seas and design of maritime structures* (Vol. 15). World Scientific Publishing Company.
22. Patterson, C., Dunsire, R., & Hillier, S. (2010). Development of wave energy breakwater at Siadar, Isle of Lewis. In *Coasts, marine structures and breakwaters: Adapting to change: Proceedings of the 9th International Conference Organised by the Institution of Civil Engineers and Held in Edinburgh on 16 to 18 September 2009* (pp. 1–738). Thomas Telford Ltd.
23. Kuo, Y. S., Lin, C. S., Chung, C. Y., & Wang, Y. K. (2015). Wave loading distribution of oscillating water column caisson breakwaters under non-breaking wave forces. *Journal of Marine Science and Technology, 23*(1), 78–87.
24. Pawitan, K. A., Dimakopoulos, A., Vicinanza, D., Allsop, W., & Bruce, T. (2019). Loading model for an OWC caisson based upon large-scale measurement. *Coastal Engineering*. https://doi.org/10.1016/j.coastaleng.2018.12.004
25. Viviano, A., Naty, S., Foti, E., Bruce, T., Allsop, W., & Vicinanza, D. (2016). Large-scale experiments on the behaviour of a generalised Oscillating Water Column under random waves. *Renewable Energy, 99*, 875–887.
26. Hattori, M., Arami, A., & Yui, T. (1994). Wave impact pressure on vertical walls under breaking waves of various types. *Coastal Engineering, 22*(1–2), 79–114.
27. Allsop, N. W. H., Vicinanza, D., & McKenna, J. E. (1996). *Wave forces and vertical and composite breakwater* (pp. 40–41). HR Wallingford: Report SR 443.
28. Oumeraci, H. (1994). Review and analysis of vertical breakwater failures—lessons learned. *Coastal Engineering, 22*(1–2), 3–29. https://doi.org/10.1016/0378-3839(94)90046-9
29. Müller, G., & Whittaker, T. J. (1995). Visualisation of flow conditions inside a shoreline wave power-station. *Ocean engineering, 22*(6), 629–641.
30. López, I., Castro, A., & Iglesias, G. (2015). Hydrodynamic performance of an oscillating water column wave energy converter by means of particle imaging velocimetry. *Energy, 83*, 89–103.
31. Medina-López, E., Ferrando, A. M., Gilabert, M. C., Del Pino, C., & Rodríguez, M. L. (2016). Note on a real gas model for OWC performance. *Renewable Energy, 85*, 588–597.
32. Pawitan, K. A., Vicinanza, D., Allsop, W., & Bruce, T. (2020). Front wall and in-chamber impact loads on a breakwater-integrated oscillating water column. *Journal of Waterways, Port, Coastal, and Ocean Engineering*. https://doi.org/10.1061/(ASCE)WW.1943-5460.0000595
33. Arena, F., Romolo, A., Malara, G., Fiamma, V., & Laface, V. (2017, June). The first full operative U-OWC plants in the Port of Civitavecchia. In *ASME 2017 36th International Conference on Ocean, Offshore and Arctic Engineering* (pp. V010T09A022–V010T09A022). American Society of Mechanical Engineers.
34. Whittaker, T. J. T., Beattie, W., Folley, M., Boake, C., Wright, A., Osterried, M., & Heath, T. (2004). The Limpet Wave Power Project—The first years of operation. *Renewable Energy*.
35. The Queen's University of Belfast. Islay Limpet Wave Power Plant [Publishable Report] 1 November 1998 to 30 April 2002
36. Henriques, J. C. C., Cândido, J. J., Pontes, M. T., & Falcão, A. D. O. (2013). Wave energy resource assessment for a breakwater-integrated oscillating water column plant at Porto, Portugal. *Energy, 63*, 52–60.
37. Horvath, E. (2009). *Wave loading at coastal wave energy converters* (MSc dissertation). University of Edinburgh.

38. Torre-Enciso, Y., Ortubia, I., López de Aguileta, L. I., Marqués, J. (2009, September). Mutriku wave power plant: from the thinking out to the reality. In *Proceedings of the 8th European Wave and Tidal Energy Conference, Uppsala, Sweden* (Vol. 710, p. 319329).
39. Medina-Lopez, E., Allsop, N. W. H., Dimakopoulos, A., Bruce, T. (2015). Conjectures on the failure of the OWC Breakwater at Mutriku. In *Proceedings of Coastal Structures and Solutions to Coastal Disasters Joint Conference, Boston, Massachusetts.*
40. Boccotti, P. (2003). On a new wave energy absorber. *Ocean Engineering, 30*(9), 1191–1200.
41. Boccotti, P. (2007). Comparison between a U-OWC and a conventional OWC. *Ocean Engineering, 34*(5–6), 799–805.
42. Malara, G., Gomes, R. P. F., Arena, F., Henriques, J. C. C., Gato L. M. C., Falão, A. F. O. (2015). Hydrodynamic characteristics of a U_OWC plant: comparison between analytical and numerical results. In *Proceedings of the 11st European Wave and Tidal Energy Conference, Nantes, France.*
43. Washio, Y., Osawa, H., Nagata, Y., Fujii, F., Furuyama, H., & Fujita, T. (2000, January). The offshore floating type wave power device" Mighty Whale": open sea tests. In *The Tenth International Offshore and Polar Engineering Conference*. International Society of Offshore and Polar Engineers.
44. Gomes, R. P. F., Henriques, J. C. C., Gato, L. M. C., Falcão, A. D. O. (2011). Design of a floating oscillating water column for wave energy conversion. In *Proceedings of the 9th European Wave and Tidal Energy Conference, Southampton, UK*.
45. Parkinson, G. (2013). "Oceanlinx launches world's first 1 MW wave energy machine in S.A." RenewEconomy, 25 Oct. 2013, reneweconomy.com.au/oceanlinx-launches-worlds-first-1mw-wave-energy-machine-s-88176/.
46. Falcao, A., Henriques, J., Gomes, R., Vicente, P., Fonseca, F., Varandas, J., & Trigo, L. (2015). Dynamics of oscillating water column spar-buoy wave energy converters deployed in array and its survivability in extreme conditions. *Marinet Infrastructure Access Report*.
47. OPERA (Open Sea Operating Experience to Reduce Wave Energy Cost). http://opera-h2020.eu/
48. Vicinanza, D., & Frigaard, P. (2008). Wave pressure acting on a seawave slot-cone generator. *Coastal Engineering, 55*(6), 553–568.
49. Kofoed, J. P., Vicinanza, D., & Osaland, E. (2006, January). Estimation of design wave loads on the SSG WEC pilot plant based on 3-D model tests. In *The Sixteenth International Offshore and Polar Engineering Conference*. International Society of Offshore and Polar Engineers.
50. Polinder, H., & Scuotto, M. (2005, November). Wave energy converters and their impact on power systems. In *2005 International Conference on Future Power Systems* (p. 9). IEEE.
51. Buccino, M., Vicinanza, D., Ciardulli, F., Calabrese, M., & Kofoed, J. P. (2011). Wave pressures and loads on a small scale model of the Svåheia SSG pilot project. In *Proceedings of 9th European Wave Tidal Energy Conference* (pp. 1–7).
52. Frigaard, P., Lykke Andersen, T., Margheritini, L., & Vicinanza, D. (2008, September). Design; construction; reliability and hydraulic performance of an innovative wave overtopping device. In *Proceedings of the 8th International Congress on Advances in Civil Engineering, Famagusta, North Cyprus* (pp. 15–17).
53. Vicinanza, D., Ciardulli, F., Buccino, M., Calabrese, M., & Koefed, J. P. (2011). Wave loadings acting on an innovative breakwater for energy production. *Journal of Coastal Research*, 608–612.
54. Margheritini, L., Vicinanza, D., & Frigaard, P. (2009). SSG wave energy converter: Design, reliability and hydraulic performance of an innovative overtopping device. *Renewable Energy, 34*(5), 1371–1380.
55. Vicinanza, D., Margheritini, L., Kofoed, J. P., & Buccino, M. (2012). The SSG wave energy converter: Performance, status and recent developments. *Energies, 5*(2), 193–226.
56. Vicinanza, D., Stagonas, D., Müller, G., Nørgaard, J. H., & Andersen, T. L. (2012b). Innovative breakwaters design for wave energy conversion. *Coast. Engineering Proceedings, 1*(1).
57. Vicinanza, D., Contestabile, P., Nørgaard, J. Q. H., & Andersen, T. L. (2014). Innovative rubble mound breakwaters for overtopping wave energy conversion. *Coastal Engineering, 88*, 154–170.

58. Contestabile, P., Ferrante, V., Di Lauro, E., & Vicinanza, D. (2016, June). Prototype overtopping breakwater for wave energy conversion at port of Naples. In *The 26th International Ocean and Polar Engineering Conference*. International Society of Offshore and Polar Engineers.
59. Iuppa, C., Contestabile, P., Cavallaro, L., Foti, E., & Vicinanza, D. (2016). Hydraulic performance of an innovative breakwater for overtopping wave energy conversion. *Sustainability, 8*(12), 1226.

Chapter 5
Performance Characteristics of an OWC in Regular and Random Waves

V. Sundar, S. A. Sannasiraj, and Daniel Raj David

Abstract Typically, the dimensions of OWC chamber have a single resonant frequency which captures efficiently wave energy over a narrow range of frequency of the waves. The high-frequency waves are reflected. Provision of a pair of additional harbour walls in front of the device creates an added resonance enabling it to capture substantial energy from a wider range of frequency of the waves. Although, there exists a substantial loss in energy due to wave deformation as the wave from the offshore approach shallow waters, in which case, the harbour walls prevent any further dissipation till it reaches the device. This chapter reports the influence of harbour walls on the hydrodynamic characteristics of a bottom - mounted OWC. The study was carried out with different configurations of harbour walls, i.e., the projecting length was varied in terms of the ratio of harbour wall length *(c)* to the breadth of OWC *(b)*, *(c/b* = 1, 1.5 and 2). The opening angle of harbour wall was varied in the range of [$\pi/2$, $7\pi/8$] in an interval of $\pi/8$ with respect to the front lip wall of OWC. The results reveal that the efficiency of the OWC with the harbour walls has been significantly improved. The details of such a behaviour are herein discussed.

5.1 Introduction

The concept of OWC was introduced by Yoshio Masuda in the 1940s and since then, several OWC plants have been deployed at several locations around the globe. The concept of harbour walls at the entrance of an OWC was later introduced by [10]. Theoretical studies revealed that the harbour wall concept could improve the

V. Sundar (✉) · S. A. Sannasiraj
Dept of Ocean Engineering, Indian Institute of Technology Madras, Chennai, India
e-mail: vsundar@iitm.ac.in

S. A. Sannasiraj
e-mail: sasraj@iitm.ac.in

D. R. David
University of Western Australia, Perth, Australia

A. Samad et al. (eds.), *Ocean Wave Energy Systems*, Ocean Engineering & Oceanography 14, https://doi.org/10.1007/978-3-030-78716-5_5

performance of the device and arguably may reduce the cost of electricity. This concept caught the attention later, and several researchers investigated the effect of harbour walls in enhancing the performance of OWC (e.g. [8], [17]).

The inclusion of harbour walls improved the performance of OWC based on creating an added resonance to the system, thus causing the incident wavefield to amplify between the harbour area (the area between the two parallel projecting side-walls) which is explained by the quarter-wavelength analogy in acoustics [10]. This amplification forces the incoming wave with apparent power having both active and reactive elements in it. Such wave amplification triggers the active elements present in the wave to rush into the device chamber, producing maximum oscillation of trapped water and hence the air column. The active elements define the degree of the kinetic energy of the system, whereas the potential energy of the system is controlled by the reactive elements. When the device experiences wave frequency differing from the resonance frequency of OWC, standing waves are likely to be formed in the harbour area, which results in the absorption of more power from the incoming wavefield over a wider range of frequencies. Whereas the conventional OWC chamber has a single resonant frequency that captures wave energy over a narrow band of wave frequencies.

It is noteworthy to mention, several existing studies focussed on different geometric properties of the conventional OWC device through both analytical or numerical and physical experiments e.g. [11, 19, 21, 24]. Furthermore, to absorb wave power over a wide range of wave frequencies and to increase the competitiveness of OWC with other devices, the multi-resonant oscillating water column was proposed and studied e.g. ([18], [22, 23, 25, 27]. For more comprehensive reviews on the history of OWCs, set-up and operation and theoretical optimal control and basis, readers are referred to [9, 13, 14].

Although a few full-scale prototypes OWCs have been tested till date, e.g., the Mutriku power plant in Spain [15], the LIMPET shore connected experimental plant installed at Islay, Scotland [3], U-OWC plants in the port of Civitavecchia, Italy [1], the PICO power plant in Portugal [20], there still remains a lot to be investigated on improving on the performance of OWCs. As harbour walls could play a vital role in improving the performance, for instance, by allowing larger energy of the incident wave to penetrate into the chamber the aim is to improve the performance by optimizing the geometries of harbour walls. In addition to improving the performance of the system, harbour walls can also be useful by preventing the rubbles from blocking the path of the incoming wave to the chamber (if it were to be integrated with rubble-mound breakwater structures) and could also be competent in an environment dominated by oblique waves. To the best knowledge of authors, there is no distinct efficient geometry of harbour walls (side wall length and opening angle) has been reported in the literature. This necessitates the goal of the present work, which is to optimally convert the incident wave energy into electrical power by analysing the influence of harbour walls and optimizing the projecting length and opening angle.

5.2 Experimental Investigations

5.2.1 *OWC Model*

A scale ratio of 1:20 as per Froude's law was considered for the present study considering the wave characteristics that could be achieved in the experimental test facility. The scale effects on air compressibility in scaled-down model experiments related to OWCs were emphasized by [26] as the fact that air compressibility cannot be accurately represented by the geometrical scaling of the device. In addition, the exact dynamic similarity of the air-compressibility effect in the chamber requires the atmospheric pressure (or room pressure) during the model testing to be much smaller than the atmospheric pressure at full scale as proposed by [12]. This may, however, not be practically feasible.

The geometric dimensions of the OWC model were adopted based on the various optimization studies carried out by researchers, e.g. [2, 11, 19, 21, 27]. The model was fabricated with 12 mm thick acrylic material. The interior length of the water column was 540 mm, and the width was 283 mm. The interior breadth was 180 mm with a circular orifice diameter of 21 mm modelled at the top of the device. The bottom rectangular opening was fixed as 180 mm by 283 mm with a circular curve bottom profile of radius 180 mm. A detailed plan and sectional view of the model [6] with the location of all sensors is shown in Fig. 5.1.

5.2.2 *Experimental Program*

Although contemporary theoretical models provide valuable information, a physical model study can also classify and isolate specific problems, often system configurations, and provide a significant contribution to the future generation of models. The experimental investigation was carried out in a 72.5 m long, 2 m wide, and 2.5 m deep wave flume at the Indian Institute of Technology Madras, India. The wave flume is well-equipped with a wavemaker at one end, which can function in the piston mode or hinged mode to generate shallow water or deep-water waves, respectively. At the other end, a perforated wave absorber is equipped with the capability of absorbing the incident waves effectively. The model rig was fixed rigidly at a distance of 36 m from the wavemaker. Throughout the experimental investigation, the water depth '*d*' and front lip wall submergence '*s*' were kept constant at 0.312 m and 0.120 m, respectively. The illustration of the experimental setup and the position of the model and wave gauges in the wave flume is shown in Fig. 5.2. All the instruments were calibrated at every change in the harbour wall configurations for accurate measurements.

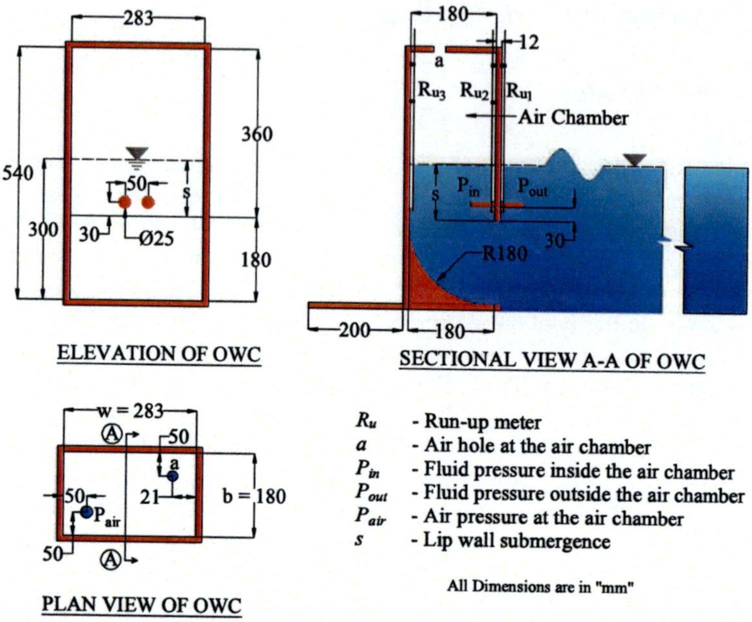

Fig. 5.1 Plan and sectional view of the OWC model

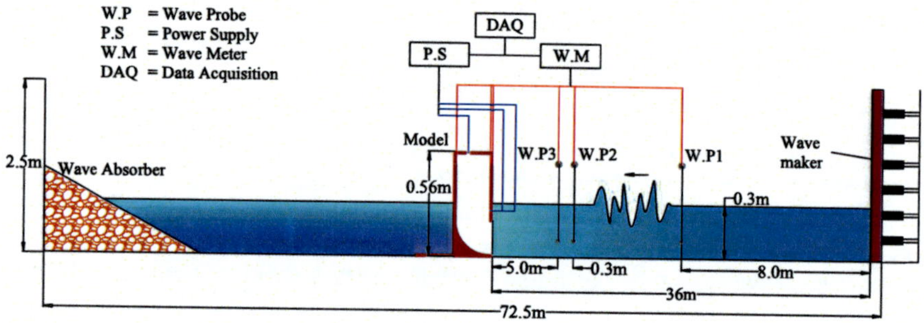

Fig. 5.2 The experimental setup

5.2.3 *Harbour Walls in OWC*

For illustrative purpose, experiments were carried out in two stages, with the first stage involving the identification of efficient resonant length and the second stage focussed on the opening angle of the harbour walls. The resonant length is defined in terms of *c/b*i.e. the ratio of harbour wall length (*c*) to the breadth of OWC (*b*). To understand the influence of harbour walls resonant length, experimental results for various *c/b* (1, 1.5, 2) were compared with the conventional OWC (without harbour walls). The second stage involved the identification of effective opening angle (θ) by varying its inclination in the range of [$\pi/2$, $7\pi/8$] at an interval of $\pi/8$ with respect to the front lip wall of OWC. The selection of the wall inclinations has been adopted based on an analytical study of [8] in which the efficiency was found to be maximum for the harbour wall inclination ranging between $\pi/2$ and $3\pi/4$. In total, 13 sets of experiments were carried out by changing the harbour wall configurations as stated above. The results from the physical model studies with models exposed to both regular and random waves were reported by the authors [4], [6]. The plan view representation of harbour wall configurations is shown in Fig. 5.3a and b, respectively.

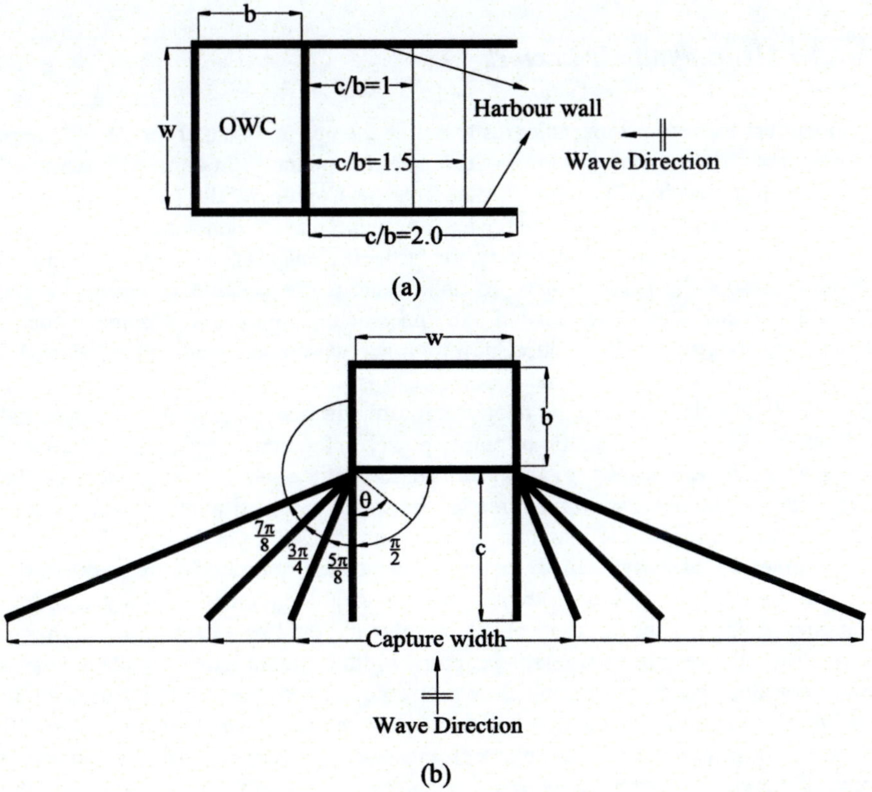

Fig. 5.3 **a** Plan view representation of resonant length formations. **b** Definition of the orientation of the harbour walls

5.2.4 Wave Characteristics for the Study

The waves of periods varying between 1 and 2.5 s in steps of 0.1 s were considered for experimental investigation with regular waves. For each of the wave periods, three-wave heights (0.03 m, 0.06 m, and 0.09 m) were employed. The said wave characteristics covered the relative water depth, *d/L* of range from 0.074 to 0.23 and wave steepness, *H/L* of range 0.0074 to 0.065, where '*H*' denotes the wave height and *L* denotes the wavelength. A similar range of wave conditions was adopted for the case of random waves defined by the JONSWAP spectrum.

5.2.5 Hydrodynamic Factors

As there are numerous terminologies to define the performance of the device, there is no consensus or standard term to designate this quantity. In a three-dimensional study, Non-dimensional Capture Width, Relative Capture Width (*RCW*), and non-dimensional absorption length are the terminologies used to define the performance. In the case of a two-dimensional study, the term efficiency is used in place of capture width to assess the performance. To understand a few definite variables, mostly two-dimensional studies are carried out, and efficiency played a prominent role in laboratory comparison. In addition, developing a two-dimensional numerical model is easier than three-dimensional models as it involves a number of variables. In general, both efficiency and the relative capture width are similar non-dimensional measures; efficiency is the ratio of output power to the input power for a unit width of wave crest with respect to the width of the device. As it is evaluated for the unit width of the device and unit crest of the wave, the maximum possible value it can reach is unity i.e., the maximum percentage of 100%. On the other hand, as a three-dimensional study involves total wave crest and not limited to the width of the device, the relative capture width may possess a value greater than unity. Both the measure of performance is applicable to regular and random waves. In wave energy conversion, the definition of relative capture width is unambiguous, while the same definition does not hold well in the case of efficiency. In this work, we used *RCW* as a measure to compare the hydrodynamic performances. For estimating the *RCW*, the incident wave power flux for plane progressive waves are calculated based on the relation defined according to the linear wave theory as given by,

$$incident\ power_{Reg} = \frac{\rho g H^2}{8} \frac{C}{2} \left(1 + \frac{2kd}{\sinh 2kd}\right) w \tag{5.1}$$

In which, celerity, $C = L/T$, wave number, $k = 2\pi/L$, ρ = mass density of fluid, g = acceleration due to gravity, and w = width of OWC. However, in the case of the random wave field, the time-average input power is calculated as by the expression (5.2),

$$incident\ power_{Ran} = w\rho g \sum_{i=1}^{N} (S_i (C_g)_i \nabla F) \tag{5.2}$$

whereas N = number of frequency bands having a width of Δf, S_i = spectral density at ‘i’th band and $(C_g)_i$ = group velocity at ‘i’th band, which is calculated using (5.3) and (5.4).

$$(C_g)_i = n_i C_i \tag{5.3}$$

$$n_i = \frac{1}{2}\left(1 + \frac{2k_i d}{\sinh 2k_i d}\right) \tag{5.4}$$

The pneumatic output power due to the volume flux of water inside the OWC chamber is calculated by the instantaneous integration of free surface oscillation inside the chamber fluctuating with a velocity *(V_w)*.

$$output\ power = \frac{1}{T}\int_{t}^{t+T} p_{air} A V_w dt \tag{5.5}$$

Here, *‘A’* is the area of the OWC chamber, V_w is discretised by the first derivative of third-order approximation as given below,

$$V_w = \frac{2Z_{i+1} + 3Z_i - 6Z_{i-1} + Z_{i-2}}{6\Delta t} \tag{5.6}$$

where, $Z_i$is the free surface elevation at time t_i and Δt = sampling interval.

The energy conversion performance of OWC is evaluated by the ratio of power incident on OWC to the output power.

$$Relative\ Capture\ Width(RCW) = \frac{input\ power}{output\ power} \tag{5.7}$$

A detailed methodology of the study to evaluate the performance of OWC is represented in the form of the flow chart in Fig. 5.4.

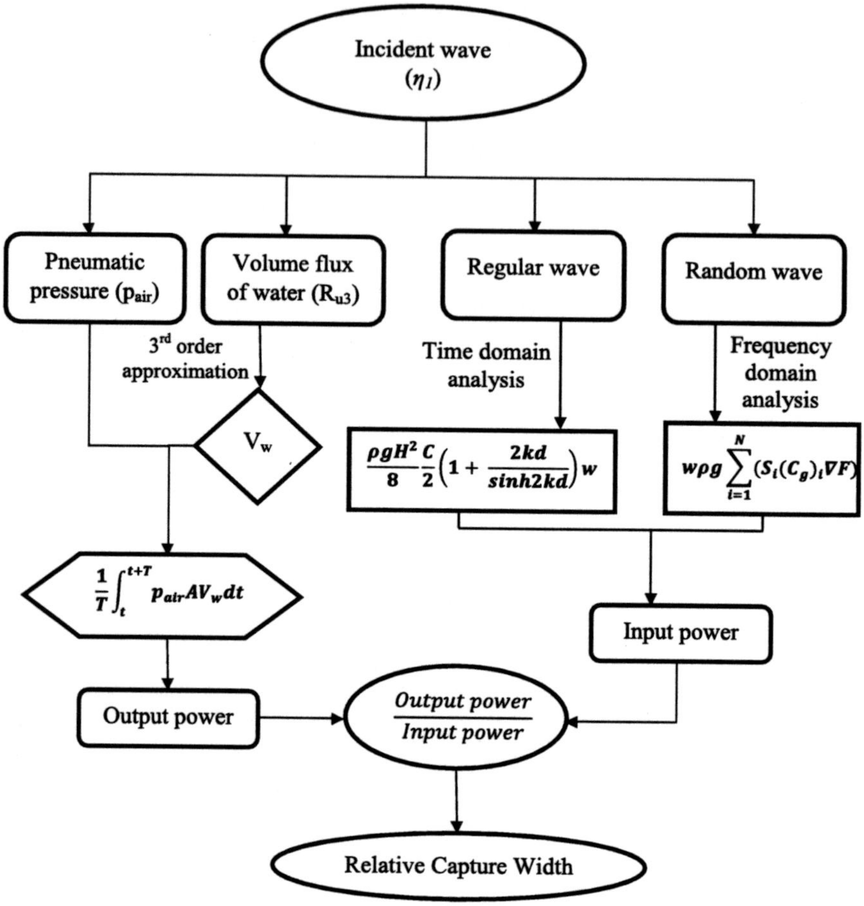

Fig. 5.4 Flow chart for a methodology for the evaluation of the performance of the OWC

5.3 Results and Discussion

5.3.1 Regular Wave Tests

5.3.1.1 Wave Amplification Factor

The performance of the OWC device mainly depends on the rate at which the compressed/decompressed air pressure exits/enters the vent, which in turn relies on the volume flux of water entering inside the chamber. The wave amplification factor (β_{out}) is one of the key parameters in defining the performance of the device. For illustrative purpose, hydrodynamic factors subjected to regular and random waves are differentiated by $(\ldots)_{reg}$ $(\ldots)_{ran}$ respectively. Wave amplification in between the areas of harbour walls (β_{out}) unswervingly enhances the volume flux of water inside the chamber (β_{in}). The wave amplification factor $(\beta_{out})_{reg}$ is established based on the ratio of measured wave run-up, R_{u1} to the wave height, H measured from the incident wave probe (W.P1, see Fig. 5.2). The comparison of $(\beta_{out})_{reg}$ for different configurations of harbour wall *(c/b)* for the different relative wave heights, *H/s* of 0.25, 0.50, and 0.75, is depicted in Fig. 5.5a, b and c, respectively. Herein, *'s'* is the front lip wall submergence.

The efficiency of the OWC increases when it functions closer to the resonant frequency, and in the case where wave frequencies are away, its performance is optimised through phase control mechanisms. In this study, it is observed that the influence of harbour walls on wave amplification (in-between the harbour area) is found to be consistent for all the wave frequencies tested, which implies that even for the out of resonance condition, the performance could be improved. The quarter-wavelength analogy in acoustics explains the large amplification of incident wave field inside the harbour area due to a frequency of disturbance having a wavelength corresponding to four times the length of a parallel-sided duct with one end closed.

In the absence of a harbour wall, the variation of $(\beta_{out})_{reg}$ is uniform and does not vary much over the entire frequency range tested. The trend in its variation is observed

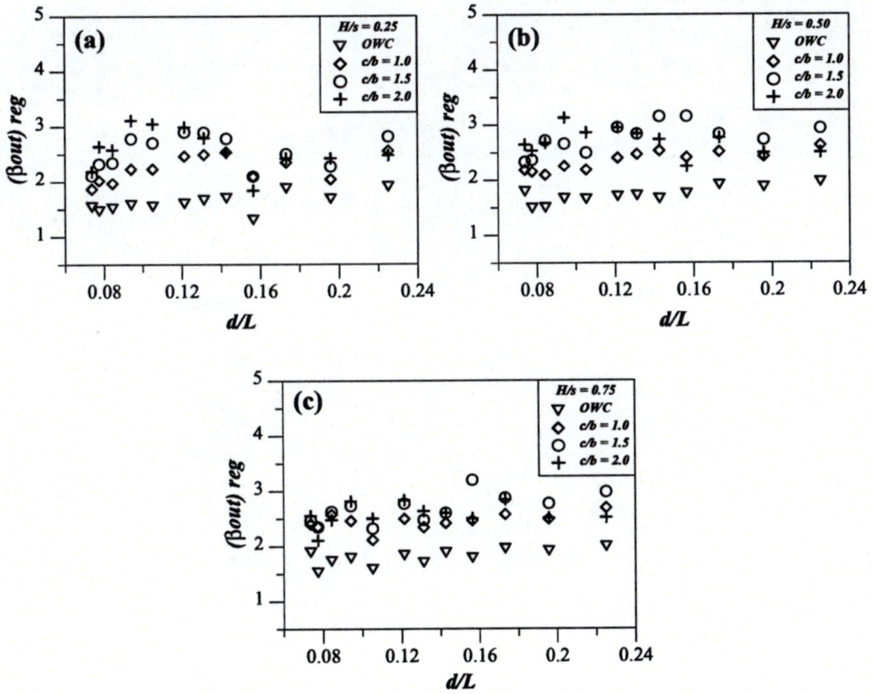

Fig. 5.5 Variation of $(\beta_{out})_{reg}$ with *d/L* for different *c/b* **a** *H/s* = 0.25 **b** *H/s* = 0.50 and **c** *H/s* = 0.75

to be similar for the different *H/s* as well. For all the *H/s* tested, the introduction of harbour walls is found to result in an increase in $(\beta_{out})_{reg}$ to a maximum extent of thrice the incident wave height. Although $(\beta_{out})_{reg}$ is found to be high for *c/b* of 2.0 for the least *H/s* tested, as *H/s* increases, the models with *c/b* = 1.5 and 2.0 exhibit almost similar results in their peak values. The increase in $(\beta_{out})_{reg}$ further leads to an increase in the volume flux of water inside the chamber, which eventually enhances the device efficiency. The observed trend also reveals that *H/s* has less influence on $(\beta_{out})_{reg}$ for different harbour wall configurations over the entire frequency range tested.

The amplification in front of the device is also due to the interaction of reflected waves from the front lip wall of OWC and diffracted waves from the harbour walls with the incident wave field. The wave amplification factor $(\beta_{in})_{reg}$ inside the water column for different *c/b* is shown in Fig. 5.6 and compared with different *H/s* i.e., (a) *H/s* = 0.25, (b) *H/s* = 0.50, and (c) *H/s* = 0.75. Though the wave amplification inside the water column is measured at both the rear wall, and behind the lip wall of OWC, due to linear variation, both the measured data showed a similar quantity and hence the results pertaining to the rear wall measurements are plotted in the above figure.

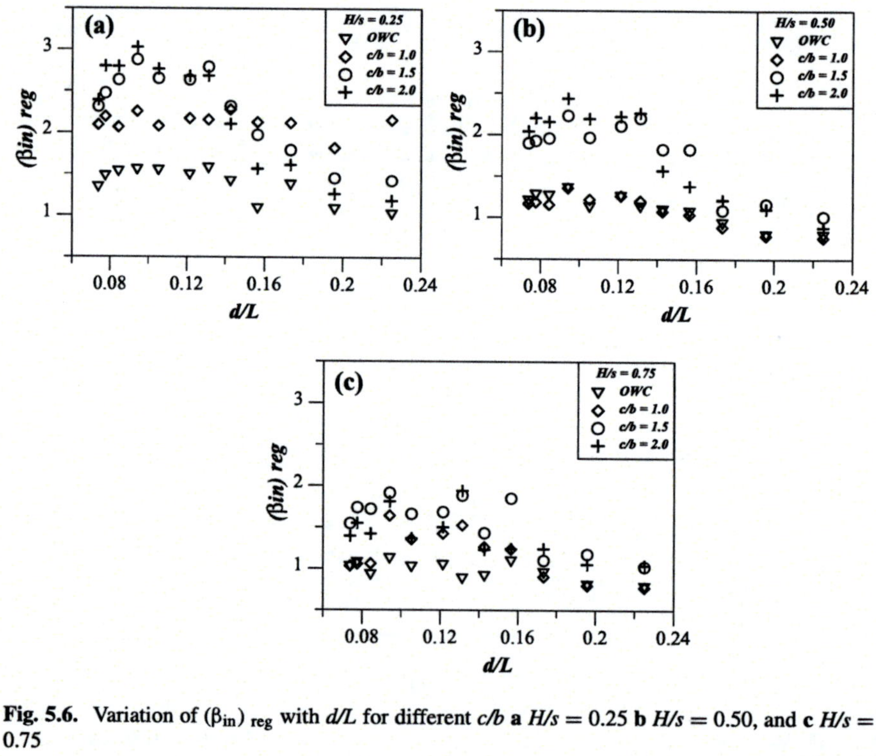

Fig. 5.6. Variation of $(\beta_{in})_{reg}$ with *d/L* for different *c/b* **a** *H/s* = 0.25 **b** *H/s* = 0.50, and **c** *H/s* = 0.75

When the incoming wave frequency entering the chamber is out of resonance with the natural frequency of the water column, phase variation between the rear and lip walls will be a maximum. This results in a resultant sub-harmonic component due to the sum of propagating waves in the opposite direction, which will lead to the circulation of pneumatic pressure inside the chamber, leading to stalling of the turbine. It results in an increase in the horizontal water particle motion, whereas the relative capture width (*RCW*) is directly proportional to the vertical motion inside the water column. Theoretical studies also specified that the dominant reflections from the walls of OWC could be partial standing waves, as reported by [11].

The above results clearly illustrate that the presence of harbour walls enhances the volume flux of water inside the chamber. On comparing the variation of $(\beta_{in})_{reg}$ as a function of *d/L* for different *H/s*, the wave amplification factor is higher for low-frequency waves and less for higher frequency waves due to reduced run-up inside the chamber similar to the variation observed for $(\beta_{out})_{reg.}$ However, as *H/s* increases, the $(\beta_{in})_{reg}$ for all the *c/b* decreases gradually, and for $d/L > 0.16$, it tends to be less than unity for OWC and $c/b = 1.0$. This shows that *H/s* has significant influence over $(\beta_{in})_{reg.}$ The least amplification has been observed for the short period waves incident on the OWC with the least *c/b* of 1.0. The wave amplification results paved the way to the understanding of the device configuration with optimum harbour wall projected length to have maximum volume flux of water.

Though the magnitude of $(\beta_{in})_{reg}$ is found to be a maximum for *c/b* of 2.0, the rate of change of volume flux of water plays a key role in the energy conversion process from hydrodynamic energy to pneumatic. The velocity with which the water oscillates inside the water column is established based on the relation, Eq. (5.6). Each of the frequency components contributes to the volume flux of water. However, when the wave frequency close to the natural frequency of the OWC, a larger flux of water with a maximum velocity is likely to happen.

5.3.1.2 Pneumatic Pressure

The analysis of the pneumatic pressure developed inside the chamber is another important factor in dictating the performance of the OWC. The rate at which pressure enters/exits from the air hole depends on the volume flux of water inside the water column. The top plate with an air hole of $a/A = 0.68\%$ acts as a damper which allows the oscillating water to compress air at the upper end of the chamber. This air is then forced out through a small orifice which drives the turbine to generate clean energy. Essentially, this is an air-flow exchange between the air in the pneumatic chamber and the atmosphere outside of the OWC chamber. The volume flux of water inside the chamber acts as a piston by the dynamic interaction and produces a bi-directional airflow through the orifice on the top of OWC. The issue of scale effect, particularly on-air compressibility, has been discussed briefly in Sect. 1.2, however, with careful attention, all the measurements were carried out [5].

A pressure transducer mounted on the top of the OWC model registered the pneumatic pressure developed inside the chamber. The velocity at which the water oscillates inside the column and pneumatic pressure measured is used to calculate the hydrodynamic power developed within the conversion chamber. The dimensionless air pressure $(P_{air})_{reg}$ due to the action of regular waves is based on the relation of measured air pressure $(p_{air})_{reg}$ inside the OWC chamber to the pressure head due to the incident wave as given in Eq. (5.8),

$$(P_{air})_{reg} = \frac{p_{air}}{\gamma H} \tag{5.8}$$

wherein, γ is the specific weight of the fluid. The effect of *c/b* on the variation of pneumatic pressure inside the air chamber as a function of *d/L* for different *H/s* is presented in Fig. 5.7.

The air pressure associated with wave amplification inside the chamber is greatly influenced by the harbour walls. In the absence of harbour walls, the $(P_{air})_{reg}$ is found to be the minimum for all the *H/s* tested. This is due to the fact that there exists dissipation of a large amount of incident wave energy in the system as the model does not extend over the entire width of the flume. The trend in the variation of normalized air pressure is influenced and varies as *H/s* increases. The introduction of the harbour

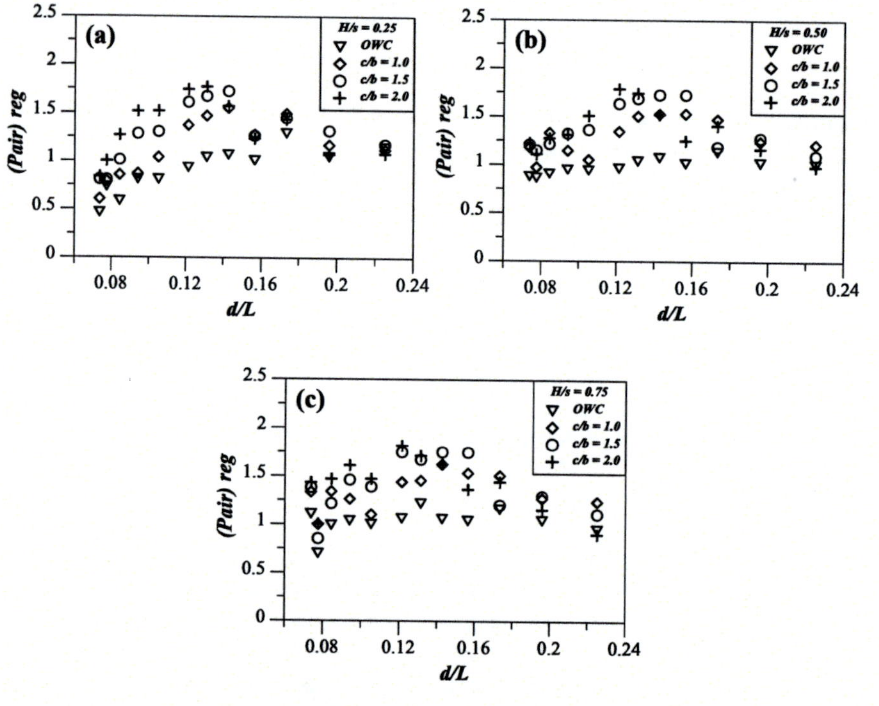

Fig. 5.7 Variation of $(P_{air})_{reg}$ with *d/L* for different *c/b* **a** *H/s* = 0.25 **b** *H/s* = 0.50 and **c** *H/s* = 0.75

walls creates an added resonance to the system and also confines the wave energy leak, which otherwise is expected to be more in the absence of the walls. Further, the results of the wave amplification show that the harbour wall increases the wave amplification in between them and thereby increases the volume flux of water inside the chamber, which influences the air pressure.

The $(P_{air})_{reg}$ increases with an increase in *c/b* up to 1.5, and a further increase in *c/b* to 2.0 does not have a significant influence on $(P_{air})_{reg}$ for high-frequency waves. However, $(P_{air})_{reg}$ is found to decrease gradually towards high-frequency waves. This trend in variation in $(P_{air})_{reg}$ is observed for all the *H/s* tested. The peak magnitude of $(P_{air})_{reg}$ for all *H/s* is found to follow a similar trend which substantiates that the device resonates closer to the natural frequency as designed. Further, for each of the configurations of harbour walls, the peak magnitude shifts towards low frequency as the natural frequency of the water column changes. This behaviour indicates the frequency dependence nature of the system and it is not influenced by the addition of harbour walls.

5.3.1.3 Pressure Outside OWC

The pressure acting on the front lip wall of OWC due to non-breaking regular waves and the influence of harbour walls on the front lip wall pressure are discussed in this section. The pressures on the lip wall induced by the incident waves can be a combination of the static pressure forces, the dynamic pressure forces, and the shock forces or impact pressures. The structure exposed to unbroken waves, a standing wave will be generated seaward of the wall, in which case only the static and dynamic forces exist, which can be determined from linear wave theory [7]. Though impact pressure forces are of short duration (of the order of tenths of a second) and consequently affect the dynamic response of the structure rather than its static equilibrium, the pressure due to non-breaking waves is equally important for the understanding of the cyclic loading experienced by the OWC which could result in fatigue failure. The results pertaining only to non-breaking waves are discussed here.

The non-dimensional pressure outside the lip wall $(P_{out})_{reg}$ is established based on the ratio as given in Eq. 5.9.

$$(P_{out})_{reg} = \frac{p_{out}}{\gamma H} \tag{5.9}$$

The influence of harbour walls on pressure variation outside the lip wall subjected to regular waves for different *H/s*is plotted and depicted in Fig. 5.8.

When a vertical impermeable wall with a bottom opening to the water column obstructs a major portion of the kinetic energy in the waves is reflected, and in addition, some spent in the wave run-up over the lip wall, and some are spent due to propagation into the water column. The upward component of the energy over the wall can result in the wave crests to rise to double the incident wave in deeper waters. However, the harbour walls create an added resonance, which results in further

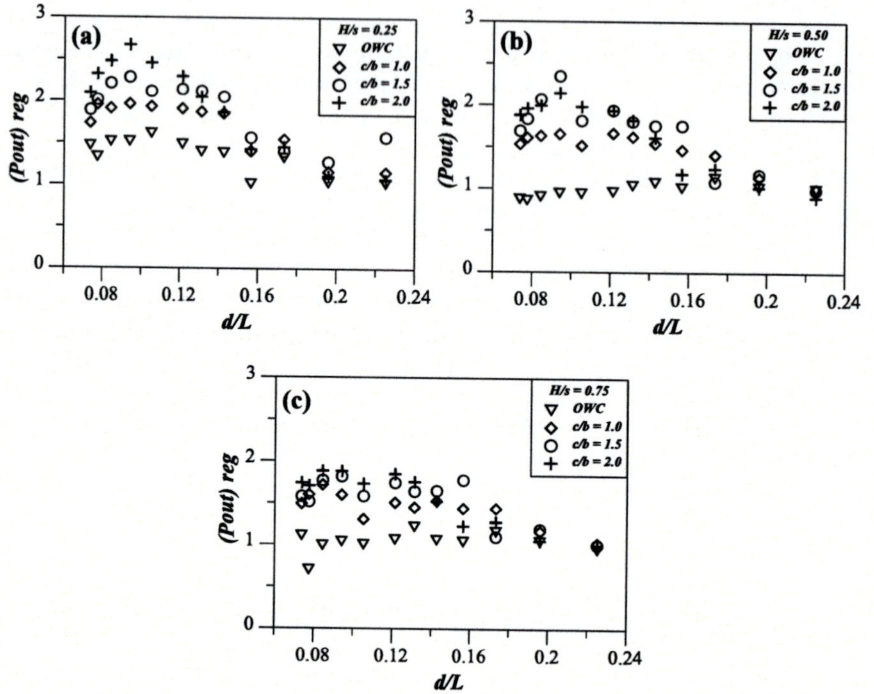

Fig. 5.8 Variation of $(P_{out})_{reg}$ with *d/L* for different *c/b* **a** *H/s* = 0.25 **b** *H/s* = 0.50 and **c** *H/s* = 0.75

amplification of the waves, thereby, increases the pressure acting on the lip wall. For *OWC*, the non-dimensional pressure is of the order of 1.5 in the low-frequency zone and slightly reduces towards the higher frequencies for *H/s* of 0.25. As *H/s* increases $(P_{out})_{reg}$ reduces to the order of unity and does not fluctuate throughout the entire frequency range. Introducing harbour walls is found to significantly influence $(P_{out})_{reg}$ due to the additional resonance. According to the fundamental principle, the pressure would be more for low-frequency waves. Similar trends in variation are observed, as can be seen from the results in the earlier figure. Increasing the resonant length of harbour walls from *c/b* = 1.0, 1.5, and 2.0 further increases $(P_{out})_{reg}$ towards the low-frequency zone. However, as *H/s* increases, the variation in magnitude for low-frequency waves decreases gradually, whereas, for the high-frequency waves, the variation is insignificant.

5.3.1.4 Pressure Inside OWC

The influence of harbour walls on the pressure variation inside the OWC subjected to regular waves for different *H/s* are superposed in Fig. 5.9. The non-dimensional

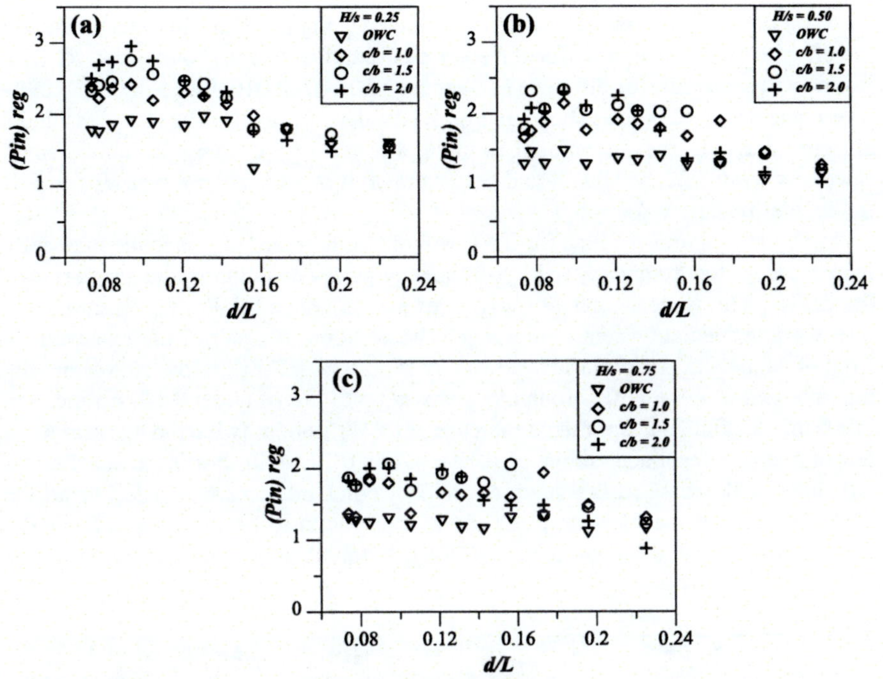

Fig. 5.9 Variation of $(P_{in})_{reg}$ with *d/L* for different *c/b* **a** *H/s* = 0.25 **b** *H/s* = 0.50 and **c** *H/s* = 0.75

pressure inside OWC $(P_{in})_{reg}$ is established based on the relation given in Eq. 5.10.

$$(P_{in})_{reg} = \frac{p_{in}}{\gamma H} \tag{5.10}$$

The variation of $(P_{in})_{reg}$ is directly related to $(P_{out})_{reg}$ as its variation is a function of $(P_{in})_{reg}$ Similar to $(P_{out})_{reg}$, the magnitude of variation due to the harbour wall's resonant length is larger at the low-frequency zone and marginal for high-frequency waves for *H/s* of 0.25. Further, as *H/s* increases, the variation of $(P_{in})_{reg}$ due to harbour walls gradually reduces for the entire frequency range tested. On comparing with the results of $(P_{out})_{reg}$, the overall magnitude of $(P_{in})_{reg}$ is slightly less than $(P_{out})_{reg}$. This is due to the fact that, as most of the energy is spent in reflection and run-up of the wave along the seaside of the lip wall, less energy flows into the OWC chamber.

5.3.1.5 Effect of Wave Steepness

The relative capture width due to regular waves, RCW_{reg} is established based on the ratio of output power in terms of air pressure to the incident wave power as per Eq. (5.7). The incident wave power flux for plane progressive waves is calculated based on Eq. (5.1), and output power is calculated using Eq. (5.5). In order to analyse the effect of wave steepness for different configurations of harbour walls, RCW_{reg} is plotted against *d/L* for different *H/L* ranges of 0.007–0.016, 0.018–0.027, 0.030–0.037, and 0.041–0.065, as shown in Fig. 5.10a–d, respectively. The steepness limits are considered in the operational limit of wave energy device, i.e., above small steepness wave (*H/L* > *0.0062*) and below the limit of very steep waves (*H/L* < *0.1*) in the intermediate water range.

The hydrodynamic RCW_{reg} for OWC without the harbour walls is found to steadily increase from low frequency to high-frequency waves up to the resonance period of the device. The wave steepness (*H/L*) does not influence the RCW_{reg} significantly over the intermediate water depth range. The addition of a pair of harbour walls in front of the device greatly influences the RCW_{reg}. The RCW_{reg} of the convertor with harbour wall of *c/b* = 1.0 is found to increase up to *d/L* of about 0.14. An increase in *c/b* to 1.5 exhibits a significant increase in RCW_{reg} about twice the incident wave power near the resonance range of 0.12 < *d/L* > 0.14. A further increase of *c/b* to 2.0 shows a decrease in the value of RCW_{reg} compared to *c/b* = 1.5. For all the configurations of the harbour walls, RCW_{reg} is found to be less than 0.6 for *d/L* < 0.05. Among the three *c/b* tested, RCW_{reg} is found to be a maximum for *c/b* =

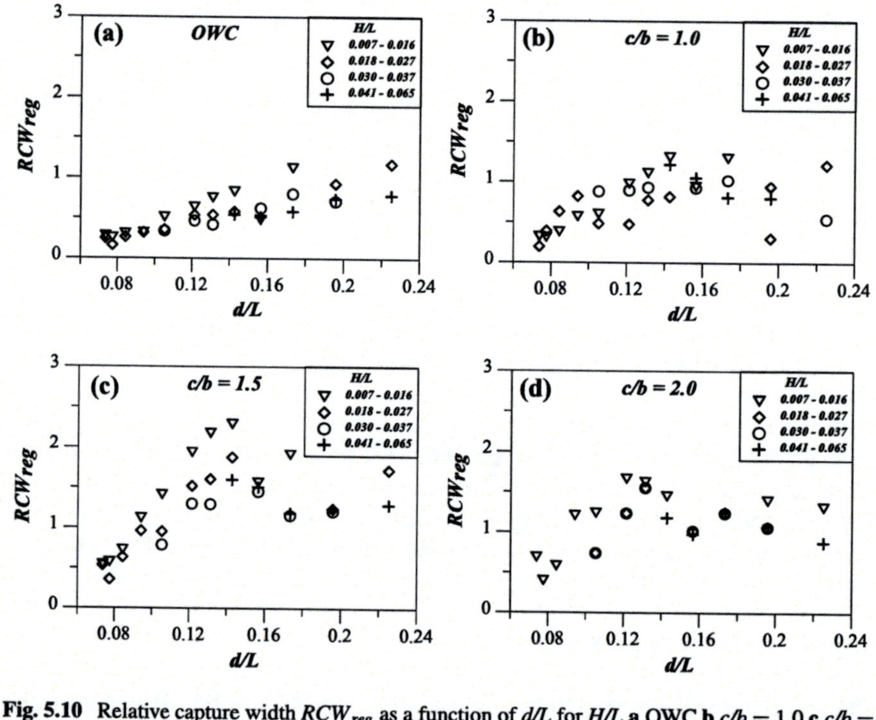

Fig. 5.10 Relative capture width RCW_{reg} as a function of *d/L* for *H/L* **a** OWC **b** *c/b* = 1.0 **c** *c/b* = 1.5 and **d** *c/b* = 2.0

1.5. Towards high-frequency waves, RCW_{reg} is found to reach an asymptotic value of more than 1.0.

5.3.1.6 Effect of Wave Frequency

The hydrodynamic RCW_{reg} as a function of *d/L* for different *c/b* including the case for without the harbour walls, *OWC* for *H/s* = 0.25, 0.50, and 0.75 are superposed in Fig. 5.11a, b, and c respectively. A comparison of RCW_{reg} for different configuration of harbour walls for a close range of wave conditions, as stated above, enables to identify the efficient system. The effect of wave frequency on each configuration also helps in locating the peak RCW_{reg}, and its variation with respect to *d/L* can be related to the natural frequency of the oscillating water column.

The trend in the variation of RCW_{reg} with *d/L* for different *c/b* is found to be similar to RCW_{reg} found to rapidly increase up to *d/L* = 0.142 for both *c/b* = 1.5 and 2.0. However, the difference between these two *c/b* being marginal. The said trend in variation is observed to be similar for the three *H/s* tested. The RCW_{reg} is found to be a maximum of 2.3 for the least *H/s* of 0.25, and the corresponding peak is found to be 2.0 and 1.6 for *H/s* = 0.50 and 0.75 respectively. An increase in *c/b* to

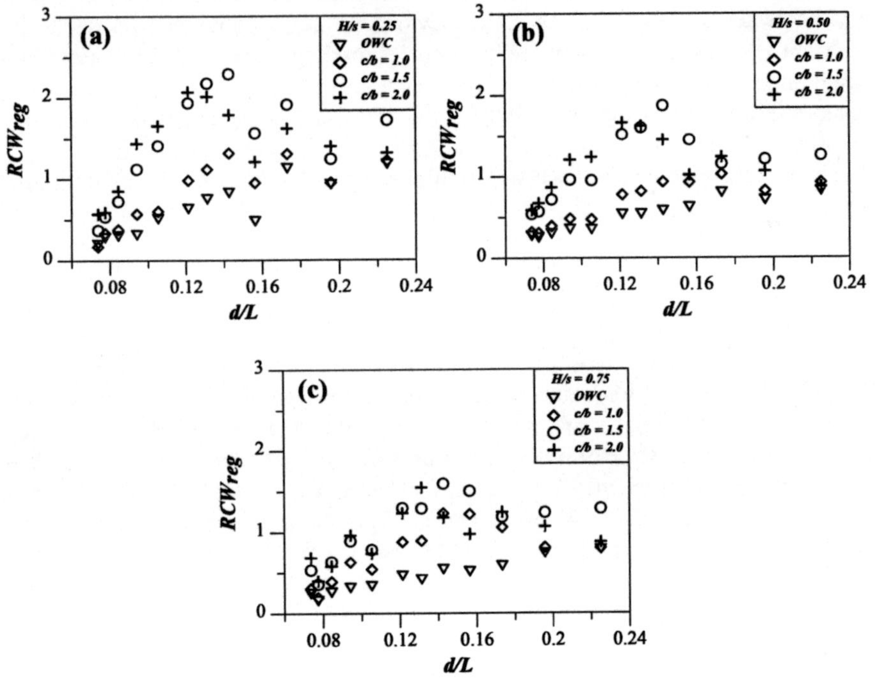

Fig. 5.11 Variation of RCW_{reg} with *d/L* for different *c/b* **a** *H/s* = 0.25 **b** *H/s* = 0.50 and **c** *H/s* = 0.75

Table 5.1 Theoretical natural frequency (T_n) for different harbour walls

Harbour wall configuration	T_n (sec)	d/L
OWC	0.98	0.232
$c/b = 1.0$	1.3	0.156
$c/b = 1.5$	1.43	0.139
$c/b = 2.0$	1.55	0.126

2.0 has shown a decrease in RCW_{reg}, which proves that there is an optimum c/b for maximum RCW_{reg} to be at its peak. For the long period waves, $d/L < 0.142$, although RCW_{reg} is found to be higher for $c/b = 2.0$, its rate of increase is found to be marginal for all H/s. Furthermore, for $H/s = 0.25$, the variation exhibits a secondary peak in RCW_{reg} for all c/b corresponding to $d/L = 0.173$ in the laboratory model scale. This secondary peak is not observed for higher H/s.

As long waves are incident on the OWC, a greater part of the energy is also likely to be reflected, and the less amount of energy entering the chamber contributes to the air pressure that leads to a reduction in RCW_{reg}. The chamber oscillation does not amplify from the incident wave surface profile in this case. Further, the harbour walls have facilitated the concentration of energy for higher frequency waves by focussing on the OWC. Similar to the trend observed from the results of $(P_{air})_{reg}$, the peak of RCW_{reg} is found shifting towards low frequency for all the H/s. This peak shift indicates that as the natural frequency of the water column changes due to the resonant length of the harbour walls its corresponding resonance also varies. This again implies that the device takes in energy closer to the natural period of the water column. Furthermore, the OWC system still relies on the wave frequency, and the addition of harbour walls alters the natural frequency of the water column.

The natural frequency of the water column is established based on the relation given as,

$$T_n = 2\pi\sqrt{\frac{l}{g}} \tag{5.11}$$

wherein, T_n = natural frequency of water column in secs and l = length of flow line (outside to inside the water column). According to this relation, the length of the flow line of water particles determines the natural frequency of the water column. The theoretical natural frequency established for the models are tabulated and shown in Table 5.1. The peak RCW_{reg} for each of the configurations of harbour walls and OWC compares well with the theoretical natural frequency of the water column estimated.

5.3.1.7 Percentage of Increase

For the purpose of quantifying the amount of increase in RCW_{reg} for OWC with harbour walls compared that for without the harbour walls, defined as compared for the three *c/b* for *H/s* = 0.25, 0.50, and 0.75 are projected in Fig. 5.12a–c. As the said percentage increase is found to be negligible for *d/L* < *0.09,* the results for the percentage increase is calculated for *d/L* > *0.09* alone are plotted. These results clearly demonstrate the effectiveness of harbour walls.

The above results imply an increase in RCW_{reg} thus, proving that harbour walls enhance the performance of the system. The harbour walls of *c/b* = 1.0 show a considerable increase in the RCW_{reg} for all the wave frequencies and for all the *H/s* tested. Increasing the harbour walls to *c/b*1.5 and 2.0 shows a tremendous increase in RCW_{reg} to an extent of about 70% towards low-frequency waves and about 40% for high-frequency waves. However, the increase for *c/b* = 1.5 is found to be a maximum and is consistent for the entire *d/L* range, particularly closer to the natural frequency region. This proves that there is an optimum resonant length of the harbour wall. Even though there is only a marginal variation in the RCW_{reg} for *c/b* = 1.5and 2.0, the optimization should also consider other parameters such as the cost of construction, maintenance, workability etc. In such a case, the cost of additional length with only a marginal variation of RCW_{reg} indicates that *c/b* = 1.5can be adopted as an efficient resonant length of the harbour wall.

5.3.1.8 Phase Difference

In the OWC system, generally, the phase variation exists between the incident wave and the volume flux of water as it is separated by the front lip wall with a bottom opening. It also leads to phase variation between the pneumatic pressure and the pressure at the front lip wall of the OWC. The instruments being fixed at a certain horizontal space in the direction of wave propagation, there exists a default phase between all measurements. The phase variation between outside and inside the OWC is important as it qualifies the energy conversion capacity of the system. As the quantity of power captured by an OWC due to the flow due to incident waves is controlled by the amplitude as well as the phase of the oscillation, the phase between the wave oscillations outside and inside the OWC should be as small as possible to achieve maximum performance.

The phase difference, ϕ between the wave amplification outside and inside the OWC is computed based on the relation,

$$\phi = 360 \times \frac{\Delta T}{T} \tag{5.12}$$

$$\Psi(\Delta T) = \langle R_{u1}(t)\ R_{u3}(t + \Delta T)\rangle \tag{5.13}$$

In which, $\Psi(\Delta T)$ is the absolute maximum phase lag is obtained from the cross-correlation between wave amplification outside (R_{u1}) at the front lip wall of OWC and the wave amplification inside (R_{u3}) at the rear wall of OWC. The phase difference between outside and inside OWC for a typical wave condition i.e., T = 1.5 s and H

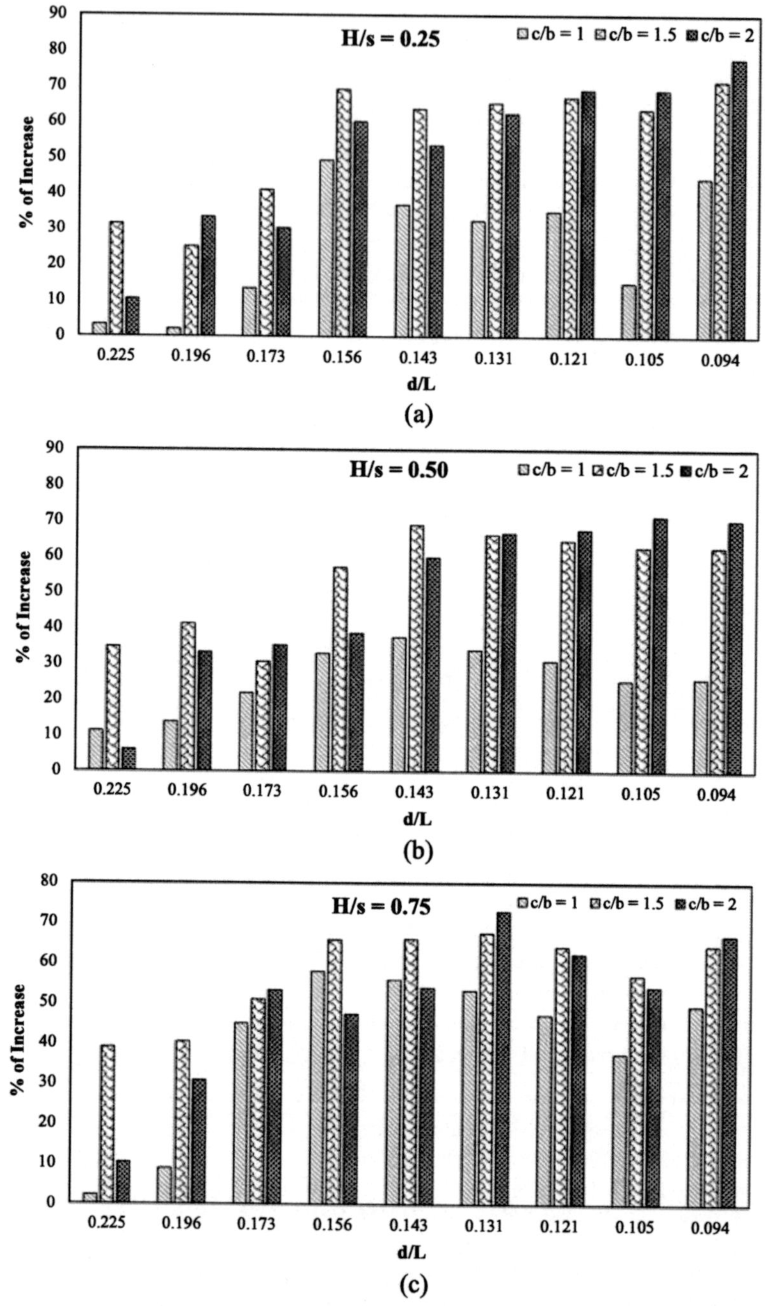

Fig. 5.12 **a** Percentage increase in RCW_{reg} as a function of *d/L* for $H/s = 0.25$. **b** Percentage increase in RCW_{reg} as a function of *d/L* for $H/s = 0.50$. **c** Percentage increase of RCW_{reg} as a function of *d/L* for $H/s = 0.75$

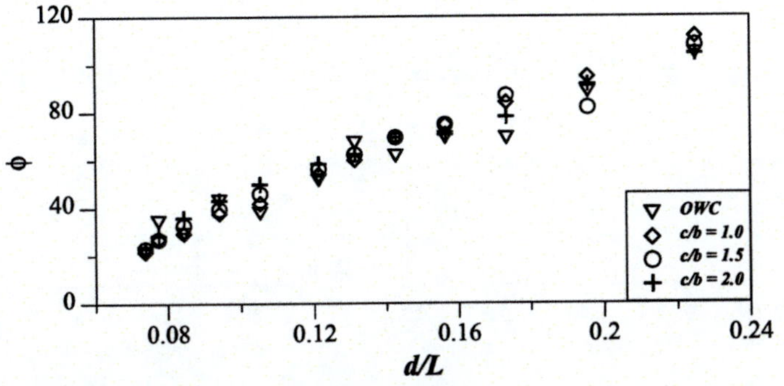

Fig. 5.13 Phase difference between outside and inside OWC

= 0.06 m for all the models tested in the first stage as a function of *d/L*is depicted in Fig. 5.13.

From the above result, it is explicit that the phase difference is found to vary linearly with respect to the wave frequency for the OWC models tested. Further, the natural frequency of the water column is observed to have the least influence over the phase variation even if the flow of streamline of water particles varies due to the variation in the harbour wall resonant length. On examining the phase variation between the R_{u2} and R_{u3}, both being measured inside the water column, ϕ showed negligible variation as the volume flux varies simultaneously inside the water column.

5.3.2 *Random Waves*

5.3.2.1 General

The study on the behaviour of models using random waves is crucial as it increases the reliability of the system performance under realistic sea states. In the laboratory, the random wavefield with pre-defined spectral characteristics is generated through a wavemaker. Though ocean waves are multidirectional, only unidirectional waves are considered for the present study. The response of the harbour walls integrated with OWC subjected to oblique waves will be considered in the future.

5.3.2.2 Wave Amplification Factor

Similar to regular waves, the non-dimensional wave amplification factor due to random waves $(\beta_{out})_{ran}$is established based on the ratio of significant R_{u1} to the significant wave height (H_s) as given in Eq. 5.14.

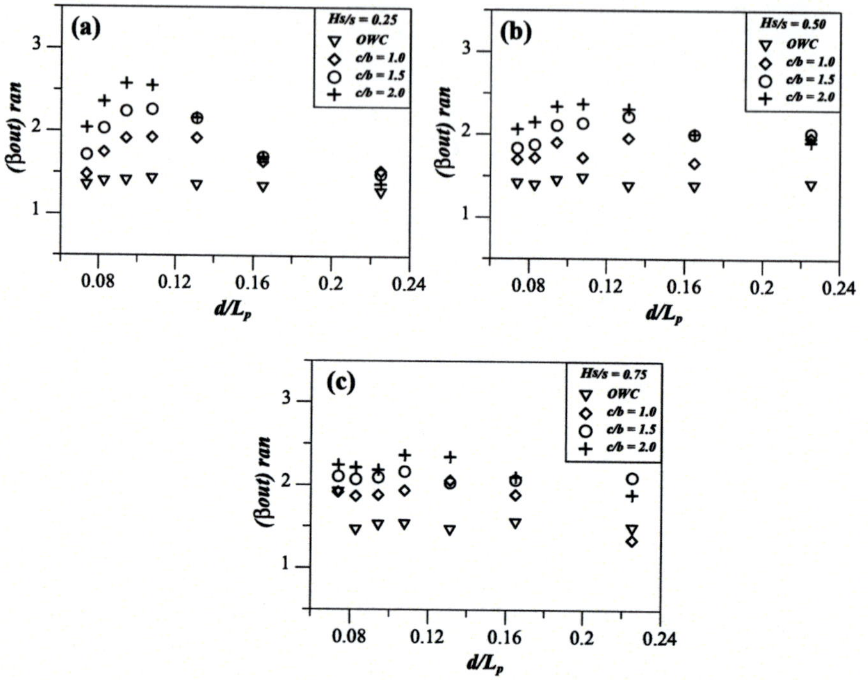

Fig. 5.14 Variation of (β_{out})$_{ran}$ with d/L_p **a** $H_s/s = 0.25$ **b** $H_s/s = 0.50$ and **c** $H_s/s = 0.75$

$$(\beta_{out})_{ran} = \frac{(R_{u1})_{m0}}{H_s} \tag{5.14}$$

The influence of harbour walls on $(\beta_{out})_{ran}$ due to different resonant lengths of harbour walls for different H_s/s are shown in Fig. 5.14a–c correspondingly.

From the results of$(\beta_{out})_{ran}$, it is observed that for *OWC* (without harbour walls), the variation of $(\beta_{out})_{ran}$ is not fluctuating throughout the entire d/L_p range examined. However, the introduction of harbour walls has amplified the incoming waves to a greater extent by increasing the resonant length which, the chamber oscillation is further amplifies the $(\beta_{out})_{ran}$. For $H_s/s = 0.25$, the amplification shows a drastic change in the low-frequency region for d/L_p between0.083 and 0.109, whereas, towards high frequency, the variation is observed to be marginal. In contrast, as H_s/s increases to 0.50 and 0.75, the amplification gradually reduced, and the variation is identical towards the entire d/L_p range.

The non-dimensional wave amplification factor inside the water column subjected to random waves $(\beta_{in})_{ran}$ is established based on the ratio of significant run-up measured from R_{u3} to the significant wave height (H_s) as given in Eq. 5.15.

$$(\beta_{in})_{ran} = \frac{(R_{u3})_{m0}}{H_s} \tag{5.15}$$

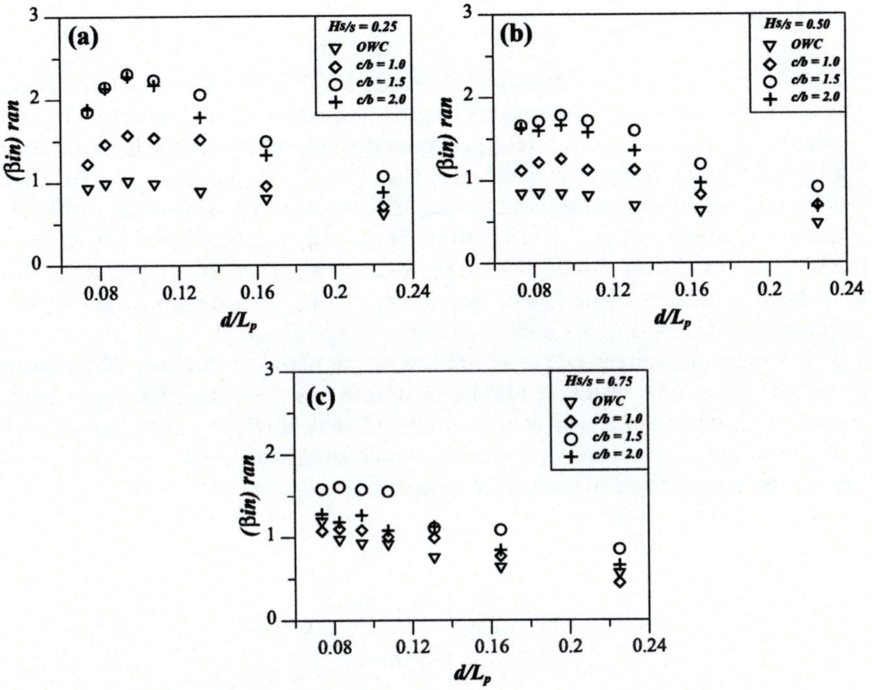

Fig. 5.15 Variation of $(\beta_{in})_{ran}$ with d/L_p **a** $H_s/s = 0.25$ **b** $H_s/s = 0.50$ and **c** $H_s/s = 0.75$

The influence of harbour walls on $(\beta_{in})_{ran}$ due to different resonant lengths of harbour walls for different H_s/s is brought out in Fig. 5.15a– c, respectively.

As the calculation of total power output from the OWC involves the volume flux of water and pneumatic pressure, the study of wave amplification due to real field condition is essential for the present study. On comparing the results of $(\beta_{in})_{ran}$ with that of $(\beta_{out})_{ran}$, both the overall magnitude inside the water column is less than the wave amplification outside the water column, especially towards the high-frequency zone which is similar to the results of $(\beta_{in})_{reg}$. Unlike the trend observed in $(\beta_{out})_{ran}$ that increasing the resonant length increases the amplification, $(\beta_{in})_{ran}$ increases up to $c/b = 1.5$ and beyond, which decreases the volume flux of water. This trend in variation of $(\beta_{in})_{ran}$ is found to follow for all the H_s/s. However, the influence of H_s/s on $(\beta_{in})_{ran}$ reveals that the magnitude of flux reduces as H_s/s increases, particularly for low-frequency waves which shows a drastic reduction in volume flux, whereas a considerable reduction is observed for high-frequency waves. Based on the above results, the overall wave amplification is significantly influenced due to the presence of the harbour walls.

5.3.2.3 Pneumatic Pressure

The variation of normalized pneumatic pressure $(P_{air})_{ran}$ due to random waves as a function of d/L_p for varying projecting length of harbour walls is shown in Fig. 5.16.

Unlike $(\beta_{in})_{ran}$ and $(\beta_{out})_{ran}$, $(P_{air})_{ran}$ resonates differently for varying significant wave height. From the results for different H_s/s, it is observed that $(P_{air})_{ran}$ for *OWC* (without harbour walls) is minimal compared to the presence of harbour walls. The trend in variation of $(P_{air})_{ran}$ appears to be similar for H_s/s 0.25 and 0.50, whereas, for $H_s/s = 0.75$, due to an increase in the nonlinearity of waves, a slight difference in its trend is observed. The rate of increase in $(P_{air})_{ran}$ by increasing the length of harbour walls from *c/b* of 1.5 and 2.0 is found marginal.

The pneumatic pressure resonates closer to the natural frequency of the water column, similar to the results of regular wave tests. As the natural frequency of the water column changes for each configuration, the peak of $(P_{air})_{ran}$ shifts correspondingly. In addition, the device configuration should aim for higher pneumatic pressure flux for the efficient performance of the system.

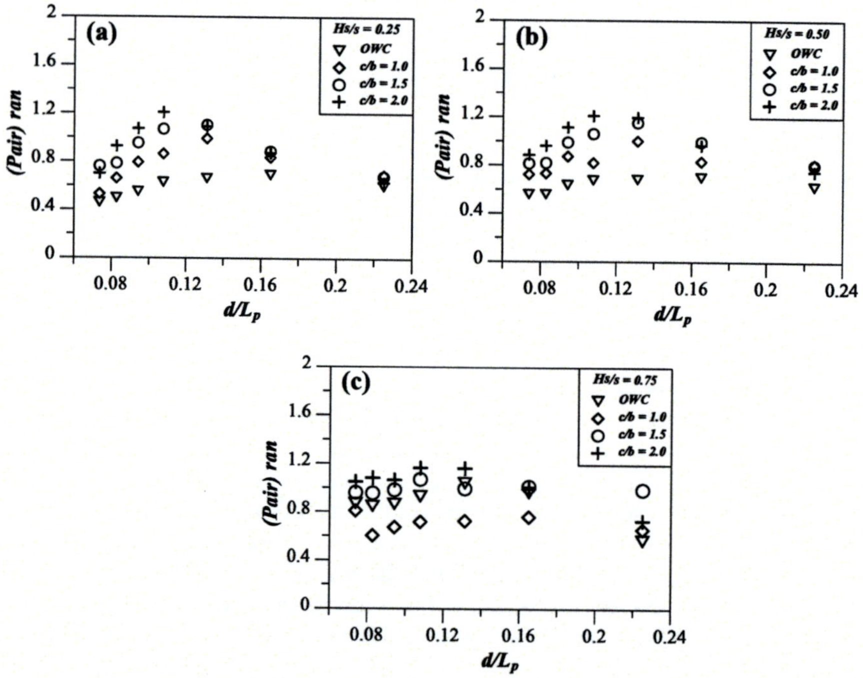

Fig. 5.16 Variation of $(P_{air})_{ran}$ with d/L_p. **a** $H_s/s = 0.25$ **b** $H_s/s = 0.50$ and **c** $H_s/s = 0.75$

5.3.2.4 Pressure Outside OWC

The dynamic pressures due to random waves acting on an OWC are lesser than the wave-induced pressure on a vertical wall. The analysis and applicability of dynamic wave pressure are still important for estimating the cyclic pressure and the momentum, which could result in horizontal sliding and structure overturning. The influence of harbour walls on the variation in the dynamic pressures along the front lip wall due to different peak frequencies for varying configurations of harbour walls is shown in Fig. 5.17a–c, respectively.

The measured $(P_{out})_{ran}$ due to varying resonant length of harbour walls show a peak of about 2.2 when d/L_p is around 0.09 for H_s/s of 0.25. For smaller H_s/s, a smooth variation of $(P_{out})_{ran}$ was observed, whereas, increasing the H_s/s, the variation in trend gradually reduces, which could be associated with the non-linearity of incoming waves.

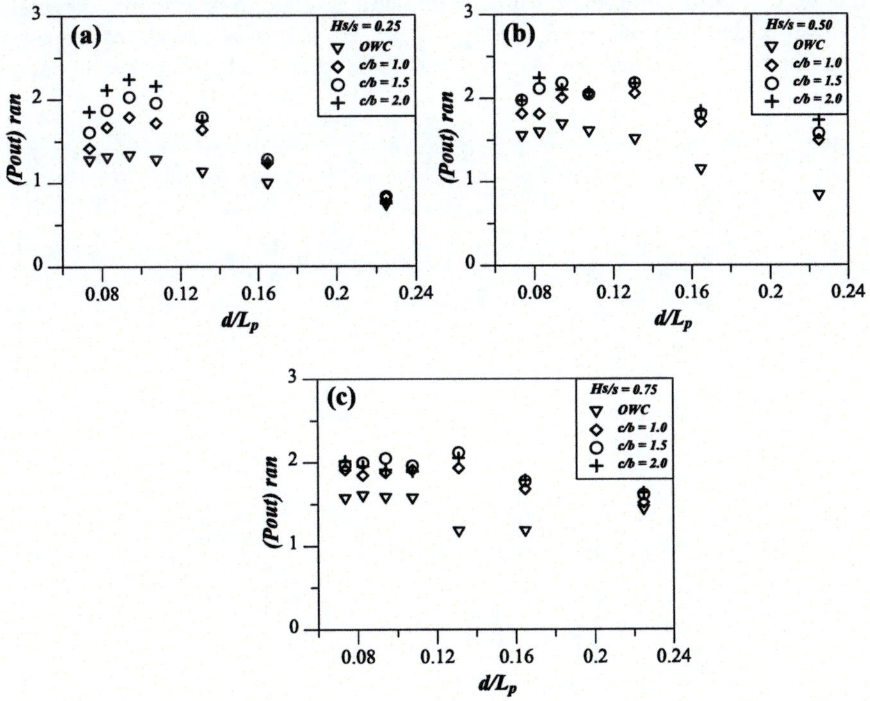

Fig. 5.17 Variation of $(P_{out})_{ran}$ with d/L_p. **a** $H_s/s = 0.25$ **b** $H_s/s = 0.50$ and **c** $H_s/s = 0.75$

5.3.2.5 Pressure Inside OWC

The influence of harbour walls on dynamic pressure variation inside the water column under different sea states due to the volume flux of water is shown in Fig. 5.18a–c for the different H_s/s.

Inside the rectangular chamber of OWC, the volume flux of water, dynamic pressure, and pneumatic pressure are inter related to each other as one influences the other. On comparing both the results, the variation in trend coincides with each other with a slight reduction in the magnitude of $(P_{in})_{ran,}$ as $(P_{out})_{ran}$ being exposed to the direct incoming waves. The reduction in magnitude is mostly due to energy loss by front lip wall reflection. Similar to $(P_{out})_{ran}$, due to the non-linearity of the waves for higher H_s/s (e.g., 0.75), the variation for all configurations of OWC models exhibits a marginal variation, especially for *c/b* 1.5 and 2.0.

Based on the comprehensive experimental study, the results show that the non-dimensional dynamic pressure outside the front lip wall is more than that of inside the OWC. Furthermore, the pneumatic pressure is found to be minimal compared to other pressure results i.e., $P_{out} > P_{in} > P_{air.}$ This is because, as waves propagate from the free surface into the water column, there exists a substantial loss of energy

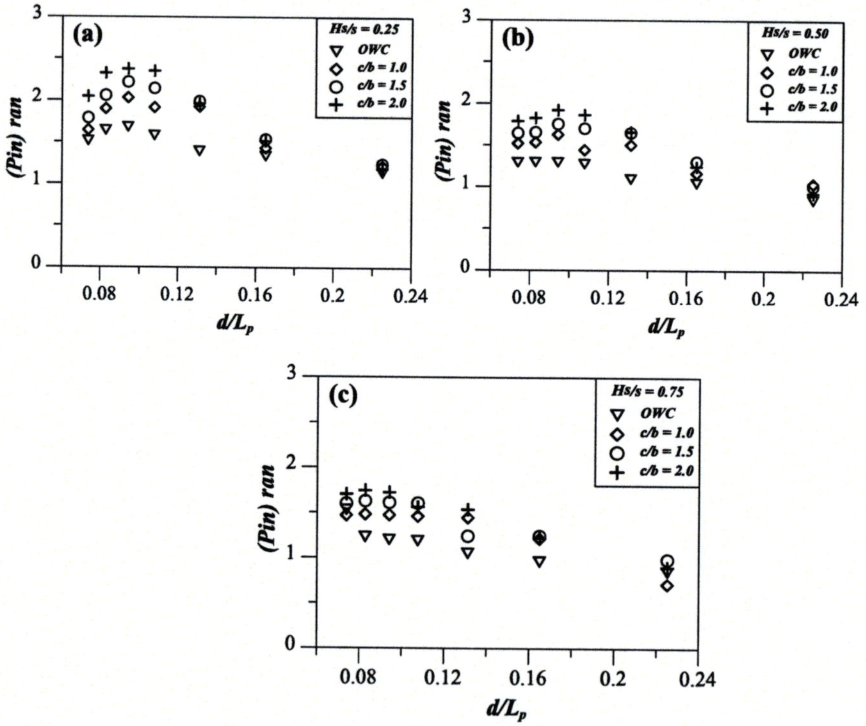

Fig. 5.18 The variation of $(P_{in})_{ran}$ with d/L_p **a** $H_s/s = 0.25$ **b** $H_s/s = 0.50$ and **c** $H_s/s = 0.75$

due to reflection from the front lip wall of OWC and the transfer of energy from hydrodynamics to aerodynamics. This variation of pressure is identical for both regular as well as random wave studies. The observed results are in agreement with the conclusions drawn by [16].

5.3.2.6 Relative Capture Width (RCW)

The Relative capture Width $(RCW)_{ran}$ subjected to random sea states is also calculated based on the ratio of output power to the input power similar to regular waves. The variation of $(RCW)_{ran}$ for different OWC models of varying resonant length of harbour walls *H/s* = 0.25, (b) *H/s* = 0.50, and (c) *H/s* = 0.75 are shown in Figs. 5.19a–c, respectively.

The results from random wave studies are found similar to those obtained through regular wave tests. For *OWC* without harbour walls, the RCW_{ran} is less than unity for all peak wave frequency except during the resonant frequency. Similar to regular wave studies, random wave studies also reveal that the harbour walls with *c/b* of 1.5 show an increase in RCW_{ran} with a maximum of about 2.5 corresponding to d/L_p of

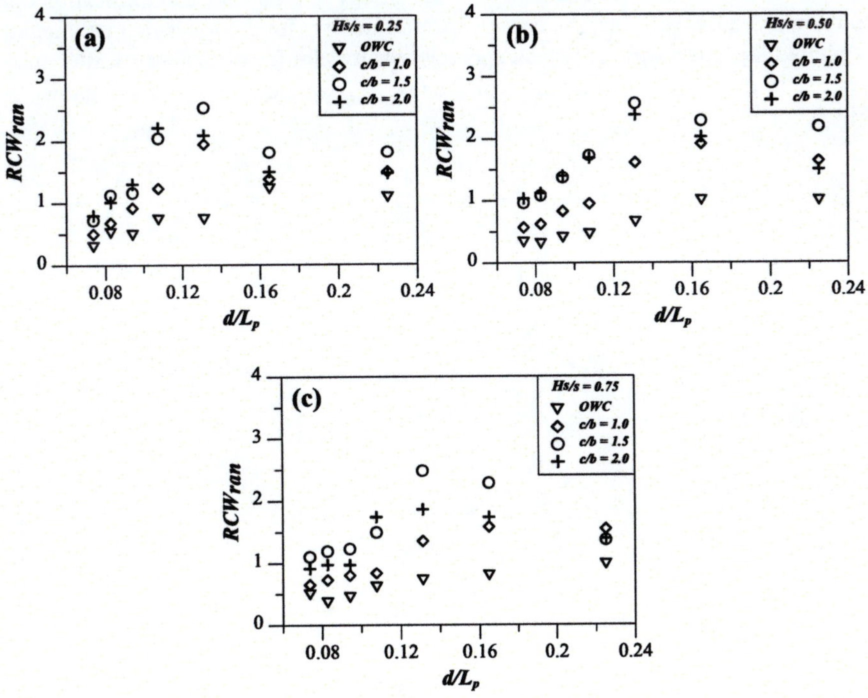

Fig. 5.19 Variation of RCW_{ran} with d/L_p for different *c/b* **a** $H_s/s = 0.25$ **b** $H_s/s = 0.50$ and **c** $H_s/s = 0.75$

0.1312. It is also evident that the energy due to low-frequency waves are not absorbed appreciably and beyond d/L_p of about 0.16. The variation in trend seems to be almost asymptotic. This observed variation in trend specifies that wave energy conversion due to random wave field mainly is governed by the system configuration and peak wave period similar to the findings of [27]. Proving that the energy conversion due to water surface oscillation inside the chamber can be maintained optimum if $b/L =$ 0.1, where '*b*' is the breadth of OWC normal to the wave crest.

Though the experimental runs with random waves are carried out with a peak period interval of 0.25 s, the observed peak of RCW_{ran} still can be observed closer towards the natural frequency of the water column, which shifts for each of the OWC models with and without harbour walls. Unlike regular wave study, the *c/b*1.5 shows a considerable increase in RCW_{ran} for all H_s/s tested under random sea states, which implies an optimum *c/b* could be adopted. The study of hydrodynamic RCW_{ran} enables to identify the efficient resonant length of harbour wall system under random sea states.

5.3.2.7 Spectral Density Comparison

In order to have a clear understanding of the energy loss and wave amplification due to harbour wall resonance, the energy spectral density for the measured parameters is plotted and compared. A typical spectral density plot for the harbour wall configuration of $c/b = 1.5$ during the resonance peak frequency of $T_p 1.5$ s is shown in Fig. 5.20.

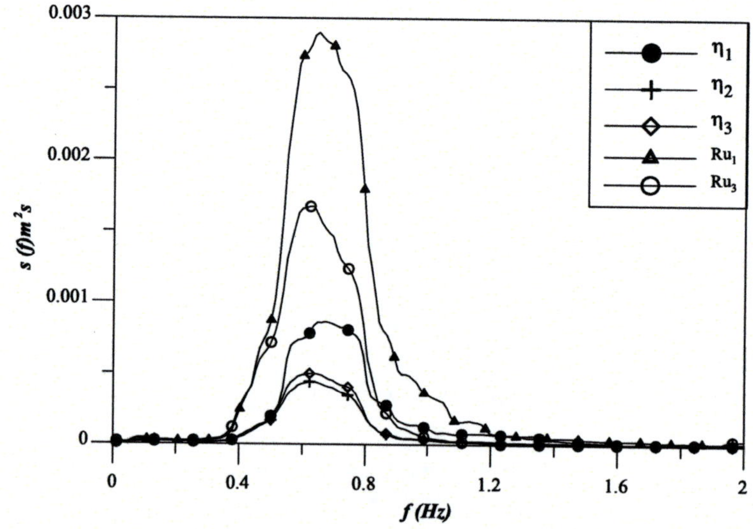

Fig. 5.20 Spectral density of η_1, η_2, η_3, R_{u1} and R_{u3} as a function of *f(Hz)*

From the above spectral density result, the variation of energy density measured at different locations is quantified and compared with the incident wave spectrum (η_1), η_2, and η_3 are the measurements from W.P2 and W.P3 (see Fig. 5.2) respectively. The energy density of R_{u1} measured in the harbour area shows the extent of amplification of wave energy which could be attributed to the harbour resonance. As the wave amplification outside the OWC leads to volume flux of water inside the chamber, and it also gets reflected due to the vertical front lip wall, the spectral density of R_{u3} shows a slight reduction of energy density compared to $R_{u1.}$ However, the energy density of R_{u3} is still more than that of incident energy density. The energy density of $\eta_2$and $\eta_3$showed a significant reduction compared to the energy density of $\eta_{1.}$ A slight shift of peak spectral density towards the high-frequency region for all the parameters is observed compared to that of incident peak wave frequency. The variations of spectral density for pressure outside, inside, and pneumatic pressure comparing with incident wave pressure are depicted in Fig. 5.21.

The spectral density of pressure outside the lip wall has been greatly amplified due to harbour walls which agree with the spectral density of wave amplification outside the lip wall, as shown in Fig. 5.20. Further, due to the volume flux of water inside the OWC, the spectral density of pressure inside the OWC shows a marginal reduction compared to $p_{out.}$ The reduction of spectral density is explained by the front lip wall reflection; however, the energy density of the p_{in} is more than the incident wave pressure. All these pressure variations contribute to the pneumatic pressure, of which the spectral density is slightly higher than the incident wave pressure. For *OWC,* the energy density of pneumatic pressure would fall well below the incident wave pressure. This increase of pneumatic pressure is solemnly due to the resonance in the harbour area, which amplified the incident wave energy to a considerable

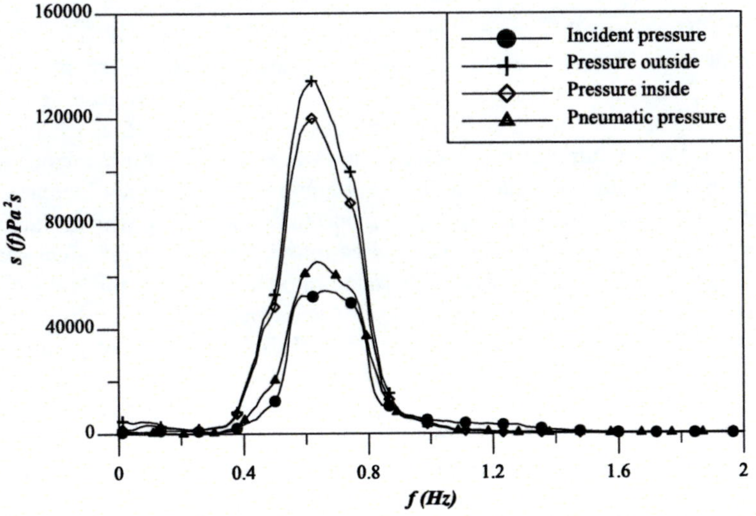

Fig. 5.21 Spectral density of $\rho g \eta_1$, p_{out}, p_{in}, and p_{air} as a function of f(Hz)

extent. The spectral density of pressure plots also showed a slight shift of peak pressure towards the high-frequency region. This illustrates how the harbour area resonance alters the incident wave frequency.

5.3.3 *OWC with Inclined Harbour Walls in Regular and Random Wave Fields*

In the second stage of the experiments, we investigate the influence of the opening angle of the harbour walls. The concept of providing the harbour walls in an inclined manner instead of straight is to trap a large amount of incident wave energy in front of the bottom opening. Providing harbour walls in such a manner increases the width trapping the incoming waves and thereby increases the volume flux of water inside the column. A system with a wider capture width also competent in an environment dominated by oblique waves. As the efficient resonant length identified in the present study as $c/b = 1.5$ (discussed in Sect. 1.4), we first focus on the opening angle variation for $c/b = 1.5$. However, inclination corresponding to other resonant lengths is summarized in Sect. 1.6. The harbour wall inclination,θ is studied by changing the outward inclination of each harbour wall in the range of [$\pi/2$ to $7\pi/8$] with respect to the front lip wall of OWC at an interval of $\pi/8$. The range of wall inclination has been adopted based on the analytical study by [8] in which the efficiency was found to be a maximum for θin the range of $\pi/2$ to $3\pi/4$.

5.3.3.1 Wave Amplification Factor

When the harbour wall is perpendicular to the front lip wall of OWC, i.e., $\theta = \pi/2$, the width traps incoming wave equal to the bottom opening width of OWC. As θincreases, the width trapping the incoming waves also increases, leading to a huge amplification of waves inside the harbour region (see Fig. 5.22 and 5.23). In particular, the presence of the harbour wall increases the wave run-up as the θ increases up to $3\pi/4$. The results also indicate that the wave amplification does not increase with a further increase in θ. Though $(\beta_{out})_{reg}$ is found to be increasing with an increase in θup to $3\pi/4$, a further increase in the inclination yielded only a marginal variation in $(\beta_{out})_{reg}$. For $H/s = 0.25$, $(\beta_{out})_{reg}$ is found to be a maximum for low-frequency waves, whereas it increases considerably for short period waves, in particular for the lower H/s. The variation of $(\beta_{out})_{reg}$ with d/L is observed to be similar for all the H/s. Though, the inclination of the harbour wall was varied, the resonant length being constant resulted in the peak of $(\beta_{out})_{reg}$ to occur at the same wave frequency. It may also be noted that the variation of $(\beta_{out})_{reg}$ for each of the inclination shows a considerable variation for the least H/s tested, whereas for higher H/s, the variation in magnitude for each of the inclination diminishes, and a marginal variation being noticed.

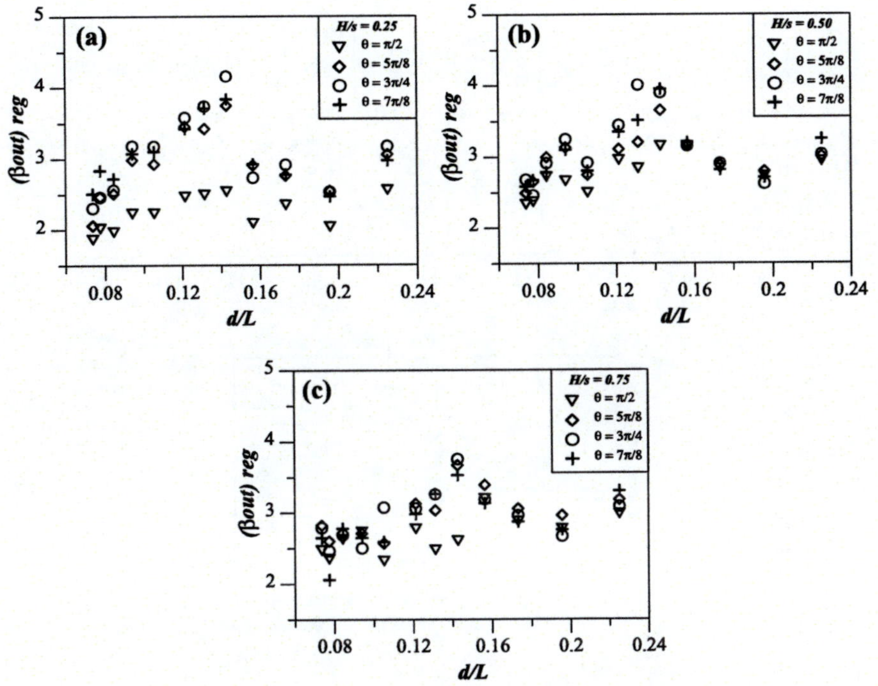

Fig. 5.22 Variation of $(\beta_{out})_{reg}$ with *d/L* for different θ. **a** *H/s* = 0.25 **b** *H/s* = 0.50 and **c** *H/s* = 0.75

The trend in the variation of $(\beta_{out})_{ran}$ is found to be different from that of $(\beta_{out})_{reg}$. Though the peak of $(\beta_{out})_{ran}$ occurs at the same wave frequency for all the H_s/s, the trend in variation vanishes and shows marginal variation for all the d/L_p. On comparing both $(\beta_{out})_{ran}$ and $(\beta_{out})_{reg}$, the maximum amplification occurs for $\theta = 3\pi/4$, however, the magnitude of $(\beta_{out})_{ran}$ is comparatively less than $(\beta_{out})_{reg}$. This can be illustrated by the presence of a number of frequency components in the random waves which give rise to a number of harmonics, and the resonance is found to decrease considerably in the harbour area. Overall, the above results confirm that the inclination of the harbour walls has a significant effect on wave amplification.

The effect of θ on the variation in the volume flux of water inside OWC subjected to regular and random waves are shown in Figs. 5.24 and 5.25, respectively.

Due to the wave amplification inside the harbour area, the active and reactive elements in the wave particles influence the volume flux of water inside the OWC. As the resonance inside the OWC and harbour resonance are two different consequences, the trend in the variation of $(\beta_{out})_{reg}$ and $(\beta_{in})_{reg}$ is different from each other. On comparing the results of $(\beta_{out})_{reg}$ with $(\beta_{in})_{reg}$, it is observed that the wave amplification outside is more than that inside the OWC chamber. Unlike $(\beta_{out})_{reg}$, the wave amplification inside the chamber is a maximum for low-frequency waves and a minimum for high-frequency waves. Furthermore, as *H/s* increases, the rate

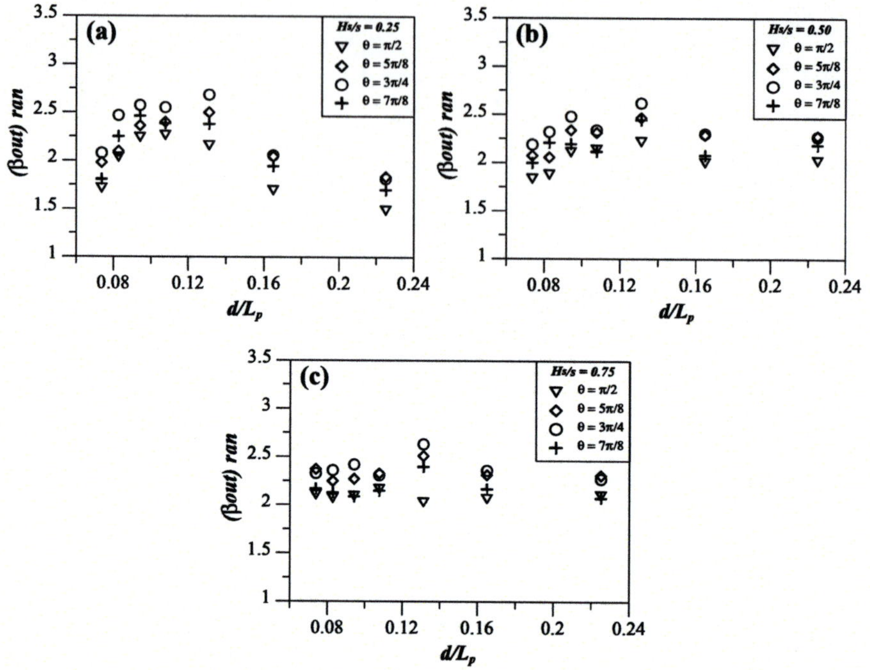

Fig. 5.23 Variation of $(\beta_{out})_{ran}$ with d/L_p for different θ **a** $H_s/s = 0.25$ **b** $H_s/s = 0.50$ and **c** $H_s/s = 0.75$

of growth of $(\beta_{in})_{reg}$ decreases gradually, and when $H/s = 0.75$, the harbour wall inclination has a marginal influence on $(\beta_{in})_{reg}$.

In the case of random waves, smooth variation with higher volume flux for low-frequency waves and slightly lesser flux for higher frequency waves were observed. The observed trend for each of the θ is quite similar to the trend observed for $(\beta_{in})_{reg}$. The presence of a harbour wall increases the wave run-up from 1.5 times to a maximum of 2.5 times for low-frequency waves, and it increases to a considerable extent for high-frequency waves. This volume flux of water further compresses the air column above and leads to an increase in the pneumatic pressure which eventually enhances the performance of the device.

5.3.3.2 Pneumatic Pressure

The change in the pneumatic pressure inside the air chamber subjected to the action of regular and random waves are shown in Figs. 5.26 and 5.27, respectively.

Similar to the variation found with $(\beta_{in})_{ran}$ and $(\beta_{in})_{reg}$, the $(P_{air})_{reg}$ and $(P_{air})_{ran}$. Also, follow a similar trend for the different θ. The peak of $(P_{air})_{reg}$ and $(P_{air})_{ran}$ observed to fall on the same wave frequency (d/L_p of 0.131), which again implies that

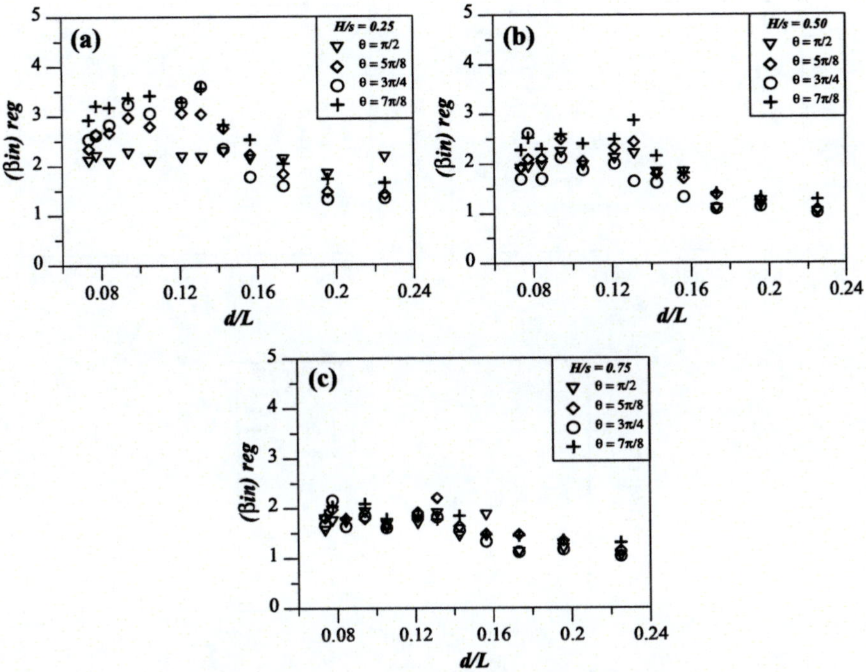

Fig. 5.24 Variation of $(\beta_{in})_{reg}$ with *d/L* for different θ **a** *H/s* = 0.25 **b** *H/s* = 0.50 and **c** *H/s* = 0.75

the device takes in energy from the frequency components closer to its natural period. The results indicate that widening of harbour walls increased the pneumatic pressure, especially near the resonant frequency of the structure. The contributions from other frequency components remain marginal. This proves that the natural frequency of the system remains unaltered with the variation of harbour wall inclination. It is also observed that the overall peak pneumatic pressure is a maximum for greater θ and minimum for $\theta = \pi/2$. Unlike $(P_{air})_{reg,}$ the results pertaining to $(P_{air})_{ran}$ show smooth variation for all the inclinations. Furthermore, similar to the results observed for varying resonant length (see Fig. 5.7 and 5.16), the inclination of harbour walls has also yielded a similar conclusion that$(P_{air})_{reg} > (P_{air})_{ran.}$

5.3.3.3 Pressure Outside OWC

The influence of harbour walls inclinations on the non-dimensional pressure variation (on the lip wall)subjected to the action of regular and random waves for different *H/s* are plotted in Figs. 5.28 and 5.29, respectively.

Similar to the wave amplification factor and pneumatic pressure, the non-dimensional pressure outside the front lip wall subjected to the action of regular waves

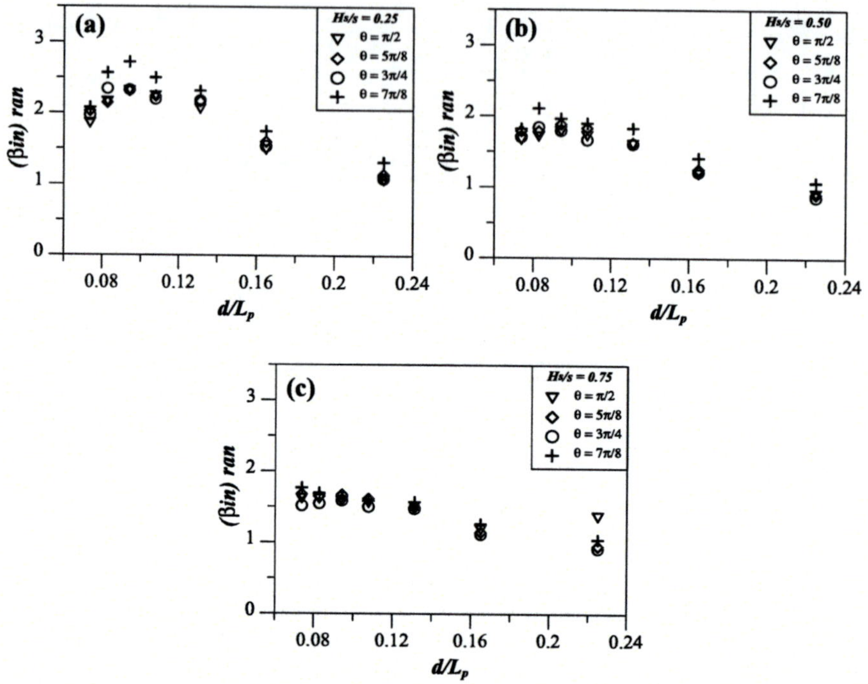

Fig. 5.25 Variation of $(\beta_{in})_{ran}$ with d/L_p for different θ **a** $H_s/s = 0.25$ **b** $H_s/s = 0.50$ and **c** $H_s/s = 0.75$

expressed a higher magnitude compared to random waves. Besides, the observed trend follows the same for both regular and random waves.

5.3.3.4 Pressure Inside OWC

The non-dimensional pressure in the lee of the lip wall (inside OWC) due to regular and random waves are shown in Figs. 5.30 and 5.31, respectively.

In general, due to the convergence of direct incoming waves to the OWC system, the internal pressure, and the volume flux of water are expected to be increased and the results indicate a similar pattern as reported in the above figures. Furthermore, the variations of $(P_{in})_{reg}$ and $(P_{in})_{ran}$ are found to follow similar trends with the other hydrodynamic parameters discussed so far.

5.3.3.5 Relative Capture Width

In the calculation of hydrodynamic *RCW*, the width traps the incoming wave is taken into account for establishing incident power. It may also be noted that as

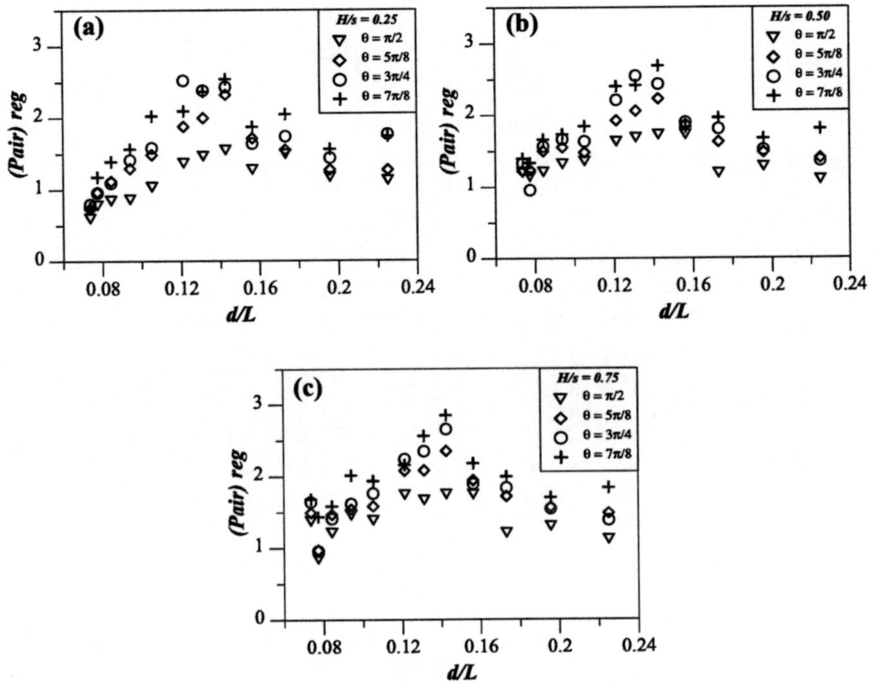

Fig. 5.26 Variation of for different θ **a** *H/s* = 0.25 **b** *H/s* = 0.50 and **c** *H/s* = 0.75

the inclination of harbour walls increases, the width capturing the incoming wave energy increases as well. However, the bottom opening width of the water column remains constant thus it makes two different entities. The arguments of calculating the incident wave power either by considering the width of the bottom opening or the width capturing the wave energy appear to be reasonable, and hence it was calculated in both ways for a better comparison. The variation of RCW_{reg} with *d/L* by considering the constant bottom opening width of the water column is plotted as a function of relative water depth is shown in Fig. 5.32.

The RCW_{reg} increases up to *d/L* of about 0.142, beyond which it decreases. Further, RCW_{reg} gets increased as θ increases. The peak of RCW_{reg} appears to be about 4.6 for θ = 7π/8for the least *H/s* of 0.25, which reduces by about 22% for the highest *H/s* of 0.75. On the contrary, as the capture width is at its maximum, the incoming waves converge towards the OWC, leading to a large run-up on the front of the lip wall of the device, and splashing was noticed for steeper waves of *H/L > 0.041* during the experimental investigation. RCW_{reg} was found to be less for long waves, as a major part of wave energy is liable to get reflected back, and only a minor part of the energy is driven into the system to contribute to the air pressure. Even the widest θ failed to counterpart the natural frequency of the device resulting in a less RCW_{reg} for low-frequency waves. However, the provision of the flared opening of the harbour

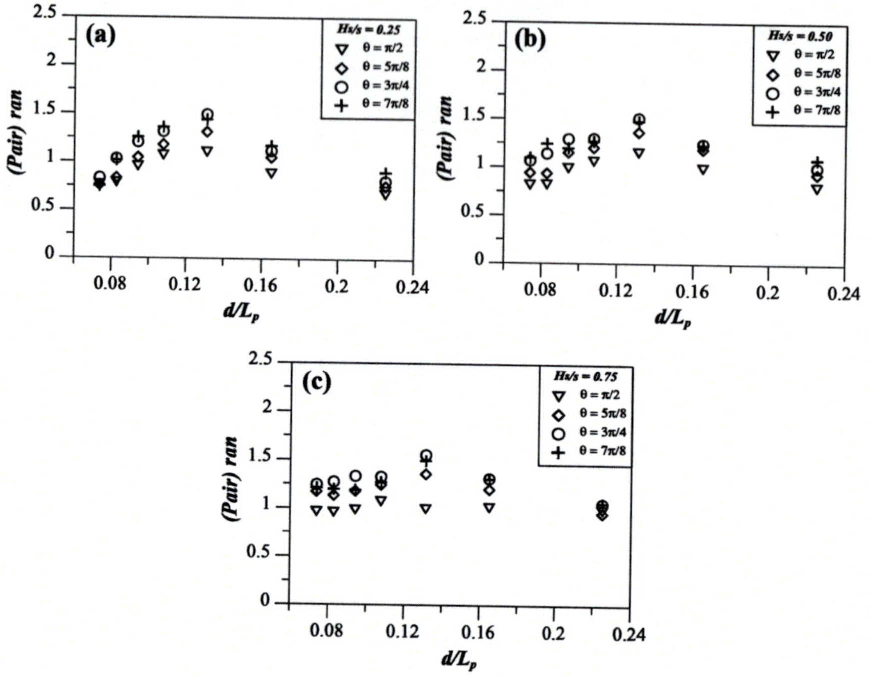

Fig. 5.27 Variation of $(P_{air})_{ran}$ with d/L_p for different θ **a** $H_s/s = 0.25$ **b** $H_s/s = 0.50$ and **c** $H_s/s = 0.75$

wall has resulted in an increase of output power from that of the model with a straight harbour wall. On the other hand, the harbour walls have facilitated the concentration of energy due to higher frequency waves.

The foregoing discussion of the results again reveals that the wave energy conversion due to the incoming wave field is mainly governed by the system configuration and the wave period. However, the inclination of the harbour wall has no influence on the frequency dependence nature of the system. The resonance condition of the device remains unaltered if the projecting length is fixed same. The output power of the device, however, is enhanced by the inclination of harbour walls, and so does RCW_{reg}.

5.3.3.6 Percentage of Increase in RCW$_{reg}$

In order to have a clear understanding of the effect of θ on RCW_{reg}, the percentage of increase in RCW_{reg} from that for $c/b = 1.5$ is plotted as a function of d/L for different H/s in Fig. 5.33. As the change in RCW_{reg} for d/L, less than 0.09 is merely negligible, the percentage increase is depicted for d/L greater than 0.09.

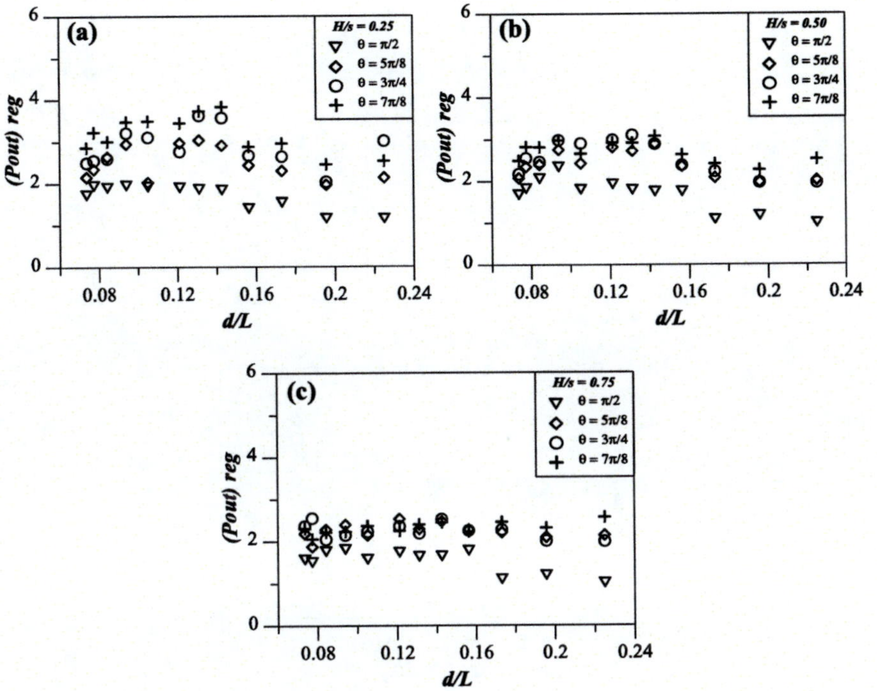

Fig. 5.28 Variation of $(P_{out})_{reg}$ with *d/L* for different θ and wave conditions **a** *H/s* = 0.25 **b** *H/s* = 0.50 and **c** *H/s* = 0.75

It is observed that although, the width trapping the incoming waves (capture width) is increased to about 80% of '*w*' for $\theta = 7\pi/8$, the corresponding average increase in RCW_{reg} is found to be about 40%. Similarly, an increase of capture width by 65% $(\theta = 3\pi/4)$ and 45% $(\theta = 5\pi/8)$ yields an average increase in RCW_{reg} to an extent of about 25% and 20% respectively. Hence, a trade-off to be adopted while deciding on the inclination of the harbour walls. If the cost of additional lengths of the walls could be justified for the benefit in the efficiency that is attained with higher θ, it could be considered.

However, if the capture width of the harbour wall is taken into account instead of the OWC chamber width of the OWC, the magnitude of RCW_{reg} for different θ follows a different pattern. The value of RCW_{reg}, considering the varying capture width of the harbour wall is depicted in Fig. 5.34.

In contrast to the results projected in Fig. 5.33, an increase in θ results in a reduction of RCW_{reg} for all the *H/s* tested. This is due to the fact that, as RCW_{reg} is the ratio of output power to the input power, and even though, the output power is a maximum for the widest θ, its capture width for trapping the incoming wave is also maximum and thereby, the corresponding ratio appears to be decreasing. The results calculated based on varying the capture width is contrary to the results on the

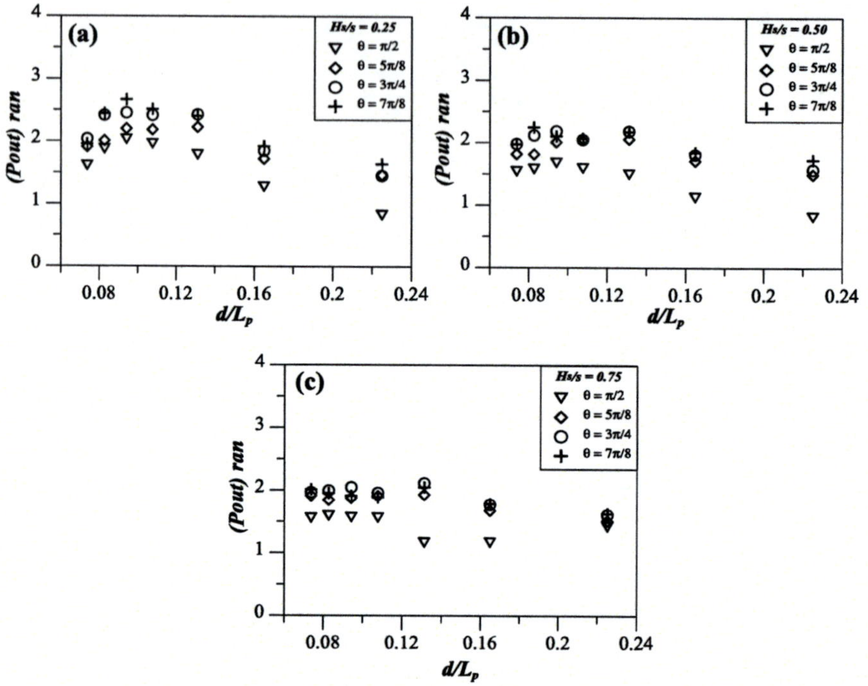

Fig. 5.29 Variation of $(P_{out})_{ran}$ with d/L_p for different θ **a** $H_s/s = 0.25$ **b** $H_s/s = 0.50$ and **c** $H_s/s = 0.75$

variation in the air pressure and wave amplification which showed maximum values for the widest θ.

The influence of harbour walls on *RCW* due to the action of random waves was analysed for further information. Therefore, RCW_{ran} by considering the constant bottom opening of the water column is studied, and the results for different H_s/s is shown in Fig. 5.35.

Unlike RCW_{reg} the variation of RCW_{ran} shows a different trend for each inclination of harbour walls. The trend of increase in *RCW* for each inclination of harbour walls is not observed in the case of random waves, though RCW_{ran} has been enhanced due to the inclination of harbour walls. Only marginal variations were observed with the increment of H_s/s, and the results are slightly contradicting the regular wave studies. We noted that RCW_{ran} is very minimum for the widest inclination, both H_s/s of 0.50 and 0.75, which could be associated with the non-linear dynamics inside the harbour area for higher significant wave heights. Furthermore, this could also be illustrated by the reflection of incident waves, and the interaction of a number of frequency components with the reflected waves from harbour walls disturbs the resonance in the harbour area and thereby the resonance inside the water column is greatly affected. Furthermore, RCW_{ran}, by considering the capture width of the harbour walls, makes it more complex as it reveals the similarity with the regular

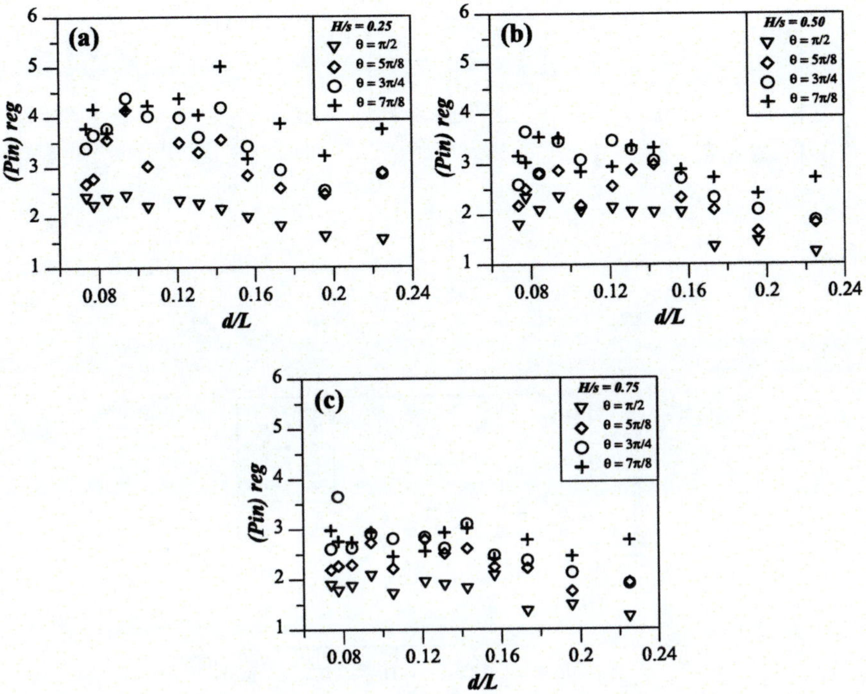

Fig. 5.30 Variation of $(P_{in})_{reg}$ with d/L for different θ **a** $H/s = 0.25$ **b** $H/s = 0.50$ and **c** $H/s = 0.75$

waves results of varying capture width. The influence of harbour walls on RCW_{ran} and calculation considering the varying capture width is depicted in Fig. 5.36.

The above results show similar kinds of variations observed in the case of regular waves. The RCW_{ran} decreases with an increase in θ with peaks occurring at the same wave frequency for all θ. The widest θappears to fall less than unity for all the H_s/s tested. We also carried out the analysis for all inclinations with varying resonating lengths of harbour walls e.g., c/b1 and 2. As we noted, similar variations as discussed with c/b1.5 it therefore not presented here. However, for a better comparison, all the resonant lengths and various inclinations subjected to regular and random waves are compared in Fig. 5.37a and b, respectively.

It is evident from the results that the hydrodynamic characteristics of inclined harbour walls for all the resonant length have shown appreciable results compared to that of straight harbour walls. This proves the concept of inclined harbour walls is worth considering. The above results unveil that θ of $3\pi/4$ shows enhanced *RCW* during the regular wave study and for random wave, maximum *RCW* is observed for θ of $7\pi/8$ when the resonant length is at its maximum i.e., $c/b = 2.0$. The above statement conveys that, increasing the width trapping incoming wave increases the performance. The percentage increase of width for the widest inclination is 85% compared to straight harbour walls, and the corresponding increase of *RCW* is not

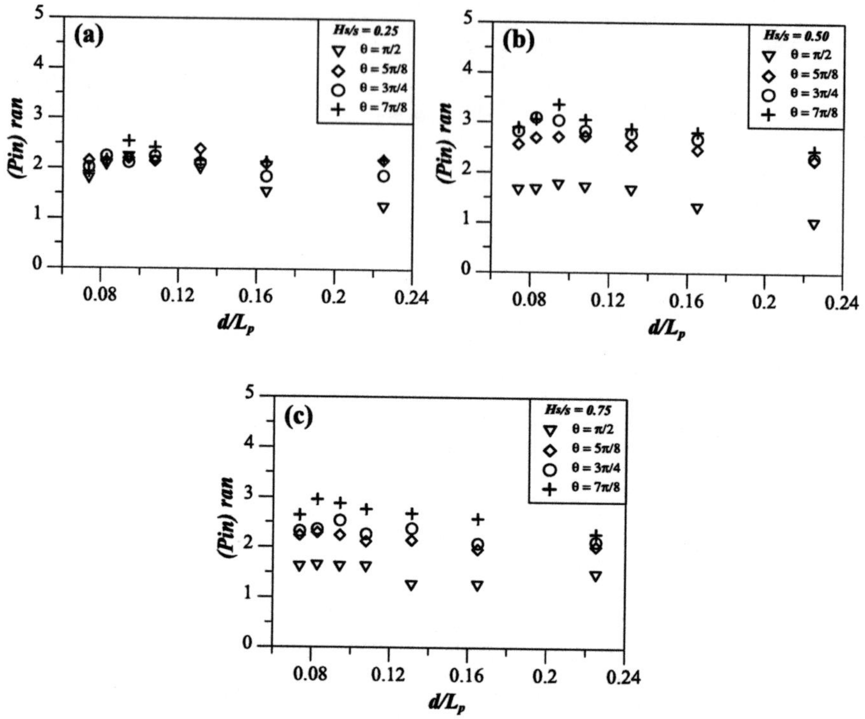

Fig. 5.31 Variation of $(P_{in})_{ran}$ with d/L_p for different θ **a** $H_s/s = 0.25$ **b** $H_s/s = 0.50$ and **c** $H_s/s = 0.75$

appreciable. The trend in variation of regular and random waves shows minimum *RCW* towards the low-frequency region, and maximum *RCW* occurs closer towards the natural frequency of water column region and moderate *RCW* towards the high-frequency region.

5.4 Summary and Conclusions

The hydrodynamic characteristics of the terminator category, OWC type wave energy device, was studied by integrating the harbour walls of different configurations. As the sustainable generating potential from ocean waves is of the order of 1700 TWh per year which is about 10% of the global energy needs, researchers from the multi-disciplinary field have been claiming that ocean energy is promising and fulfilling the deficiency at least partially. Due to the nascent scientific development in ocean science and technology, the capability to improvise the performance of wave energy devices are promising. Numerous works on the conversion of wave energy into clean

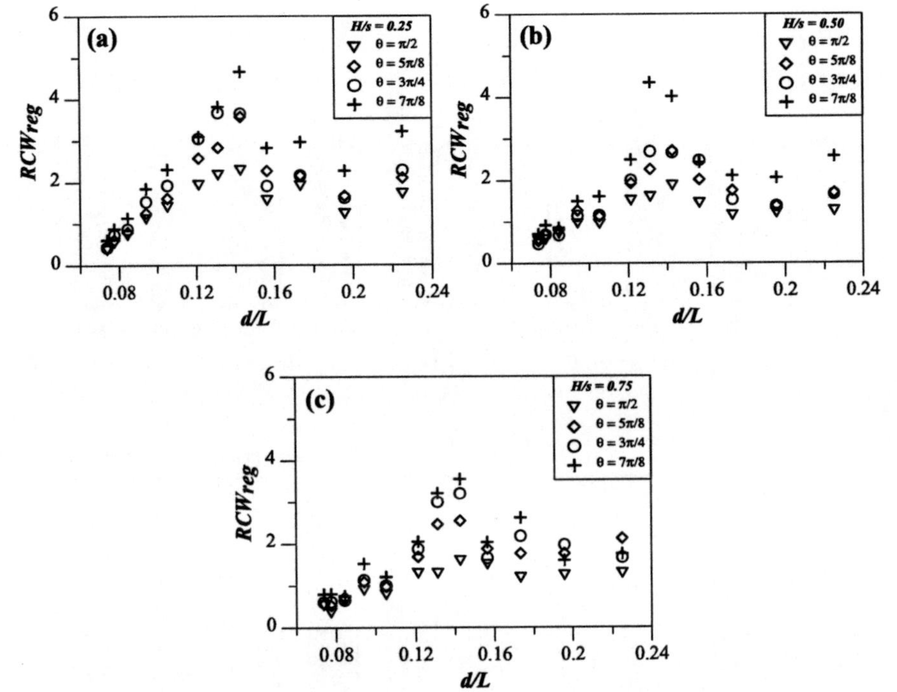

Fig. 5.32 Variation of RCW_{reg} with *d/L* for different θ **a** *H/s* = 0.25 **b** *H/s* = 0.50 and **c** *H/s* = 0.75

electricity have evinced intense interest amongst researchers worldwide over the last few decades.

As the working and performance of the OWC system rely on the geometric parameters (which influences the energy conversion based on the natural period of the system), the influence of harbour walls on the performance of OWC is addressed in this work.

The intent of the harbour walls is to create an added resonance such that it enables the device to resonate for a wider range of frequencies. The extended study on the energy output from the OWC system for different configurations of harbour walls indicates that there exists an optimal geometry of harbour walls for the device to extract energy from the incident waves. Moreover, no major disagreement has been observed on the study of the natural frequency of water column till this point between theory and experiment. The addition of harbour walls increases the natural period, and thereby, energy from longer period waves can also be trapped.

The well-controlled experimental program paved the way to understand the hydrodynamic characteristics due to the influence of the harbour walls when subjected to the action of regular as well as random waves. The energy conversion capacity of conventional OWC is compared with the OWC integrated with harbour walls which revealed that the concept of harbour walls is earnest of going over. The quantitative

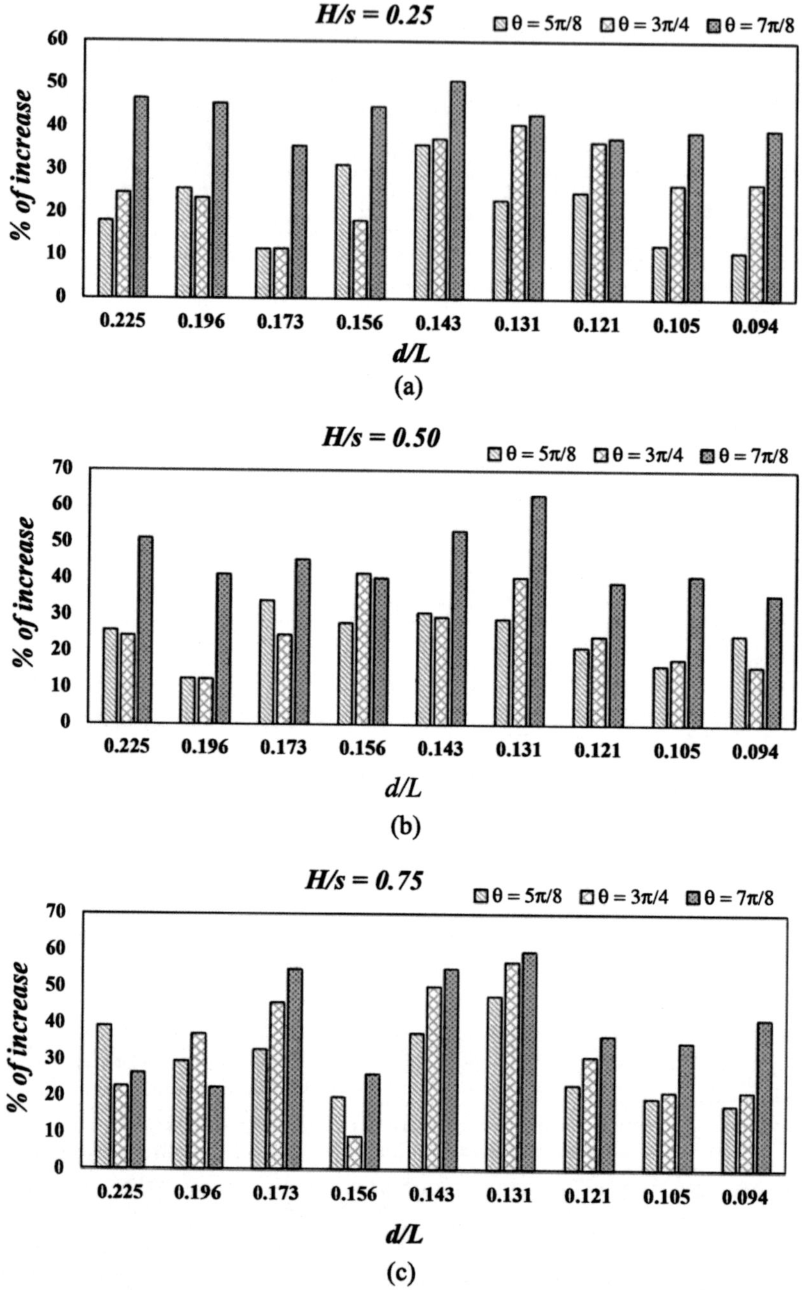

Fig. 5.33 **a** Percentage of increase with *d/L* for *H/s* = 0.25. **b** Percentage of increase with *d/L* for *H/s* = 0.50. **c** Percentage of increase with *d/L* for *H/s* = 0.75

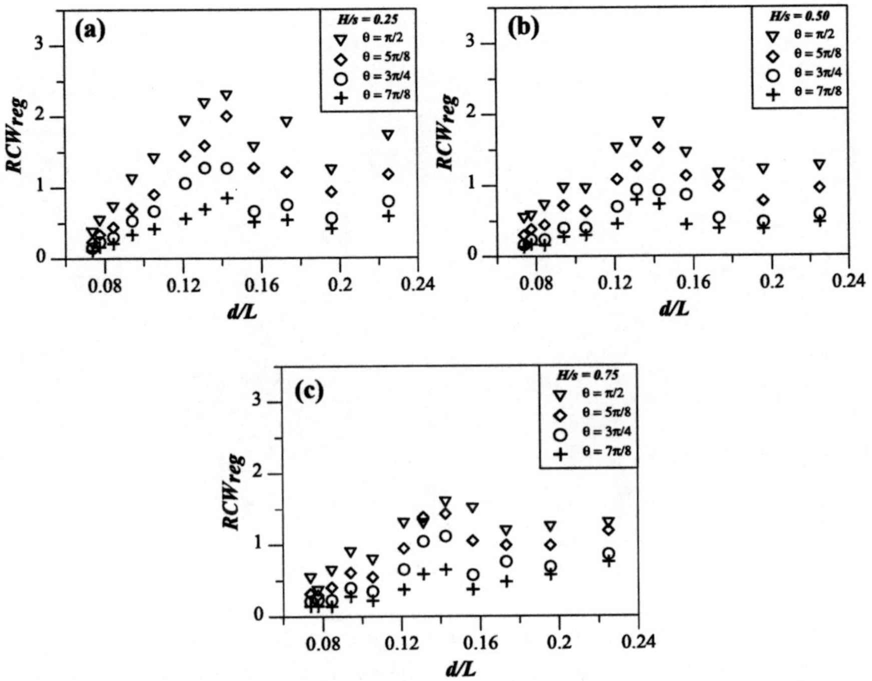

Fig. 5.34 Variation of RCW_{reg} with *d/L* for varying capture width **a** *H/s* = 0.25 **b** *H/s* = 0.50 and **c** *H/s* = 0.7

analysis of the phase difference between inside and outside the water column and the influence of harbour walls are reported. The pneumatic compressibility and hydrodynamics necessitate different model scales for the complete understanding, and the influence of vortex shedding, flow separation, and viscous effects which are difficult to infer from small-scale experiments. The hydrodynamic *RCW* due to the harbour wall's resonant length and its opening angle is studied distinctly.

Even though the magnitude of air pressure and wave amplification is found to be a maximum for harbour walls of length *c/b = 2.0*, the overall *RCW* seems to be maximum and consistent for *c/b = 1.5*. The results are also in agreement with the analytical studies of [10] indicate that the presence of the harbour halls enhances the performance of OWC in terms of its efficiency.

The influence of the harbour wall's inclination on air pressure and volume flux of water inside OWC indicate that for θ of $7\pi/8$the*RCW*is found to be a maximum. As *RCW* increases with an increment of θ, it makes it more difficult to report an optimum inclination. Furthermore, an optimum θ cannot be adopted based on the *RCW* alone as it involves various other factors such as the overall output power (if deployed in arrays), the location where space is a constraint, and the cost investment, including the maintenance cost over the design life of the OWC and grid connection, etc. In addition, as the behaviour of the device is not known for OWC providing

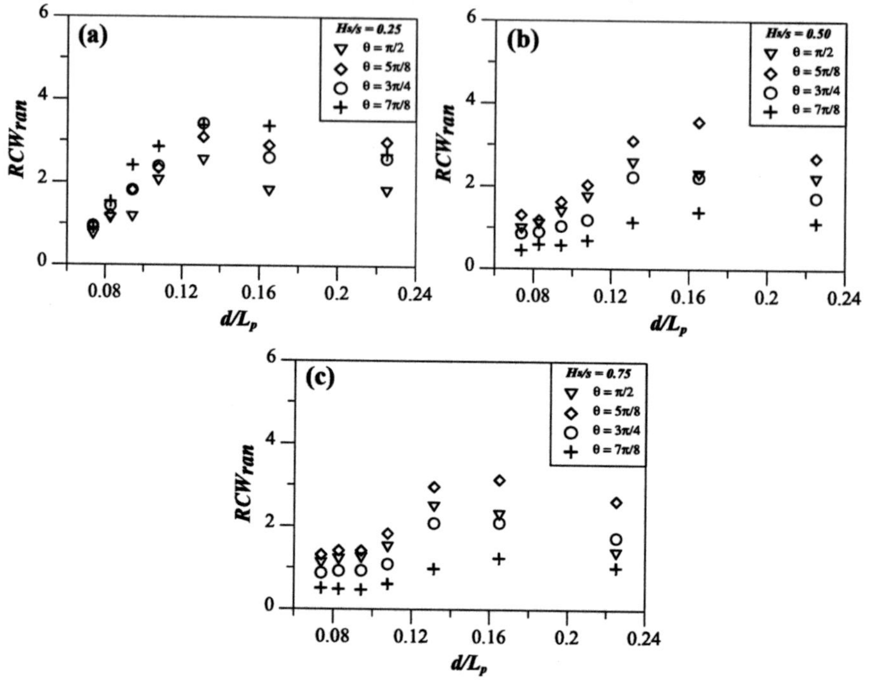

Fig. 5.35 Variation of RCW_{ran} with d/L_p for uniform capture width **a** $H_s/s = 0.25$ **b** $H_s/s = 0.50$ and **c** $H_s/s = 0.75$

with harbour walls in a field dominated by oblique waves, where the principal axis of the OWC is not exactly parallel to the wavefronts, the results could be different for different θ.

Considering a model scale of 1:20, the experiments tested here may be applied to a shoreline of 10 m exposed to a uniform wave climate, which can accommodate 32 devices with a harbour wall angle $\theta = \pi/2$ without any spacing. While increasing the harbour wall angle $\theta = 5\pi/8$ would reduce the number of devices to 18. Similarly, when accommodating OWCs with harbour wall angles of $\theta = 3\pi/4$ and $\theta = 7\pi/8$, the number of devices reduces to 12 and 6 respectively. In such a case, the overall *RCW* is at a maximum for $\theta = \pi/2$, which also comes with a huge capital cost as it involves more number of devices. This overall *RCW* for a given number of devices decreases as θ increases. Practically, the cost involved for the widest opening angle would be a maximum, and hence a compromise between conversion efficiency and construction cost needs to be adopted while deciding on the opening angle of harbour walls.

However, while integrating OWC with breakwater, considering the fact of space constraint, overall *RCW* based on the number of devices that can be accommodated for a given length of the breakwater and the outcome based on the cost investment, the shortcomings of "θ" of $7\pi/8$ is more compared to that of other θ This leaves to

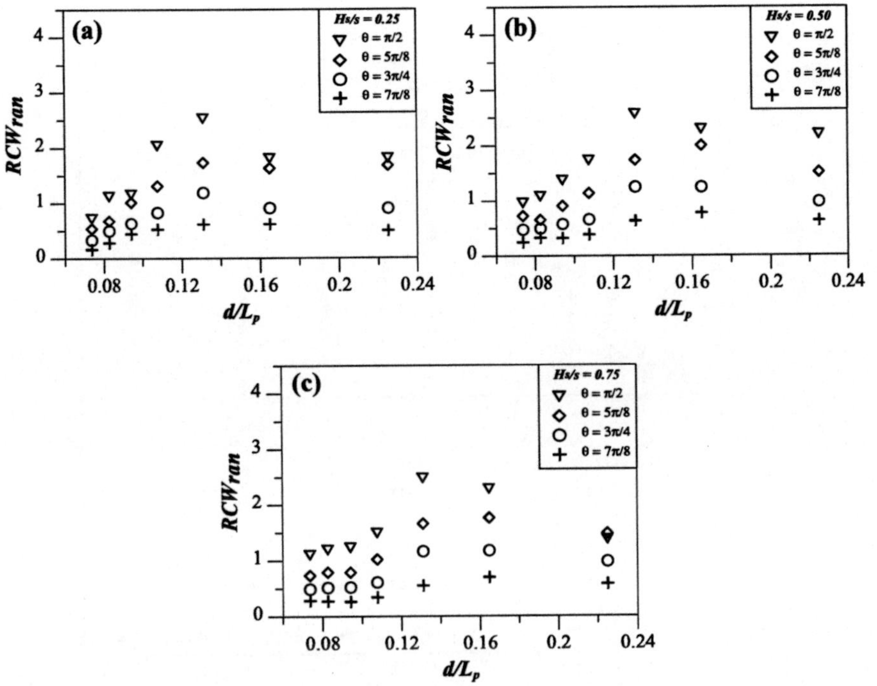

Fig. 5.36 Variation of RCW_{ran} with d/L_p for varying capture width **a** $H_s/s = 0.25$ **b** $H_s/s = 0.50$ and **c** $H_s/s = 0.75$

a conclusion that the harbour wall of θ between $5\pi/8$ and $3\pi/4$ could be consistent. The results also flagged a way to understand the natural frequency of the system, which is independent of the harbour wall inclination, whereas the *RCW* of the system relies on the wave frequency. Furthermore, the performance of the device in wave power absorption is appreciable when the relative water depth is $0.12 < d/L < 0.16$, and it shows poor performance for $d/L < 0.09$.

The salient conclusions drawn from the present study are given below,

1. The addition of harbour walls substantially enhances the performance of the device compared to that of an OWC without harbour walls.
2. The Harbour wall's resonant length, *c/b* of 1.5, has been identified to be optimum and efficient as the hydrodynamic characteristics and *RCW* appears to be consistent for both regular and random waves. The efficiency of the configuration is recognized to be in the vicinity of the resonant region of the device. In addition, the optimum *c/b*is close to the range suggested by [8] based on his studies with circular OWC. Hence, it is suggested to consider a system with a harbour wall configuration of *c/b*1.5.
3. The opening angle, θ of the harbour wall, can be adopted between $5\pi/8$ and $3\pi/4$ by maintaining the resonant length constant as *c/b*1.5.

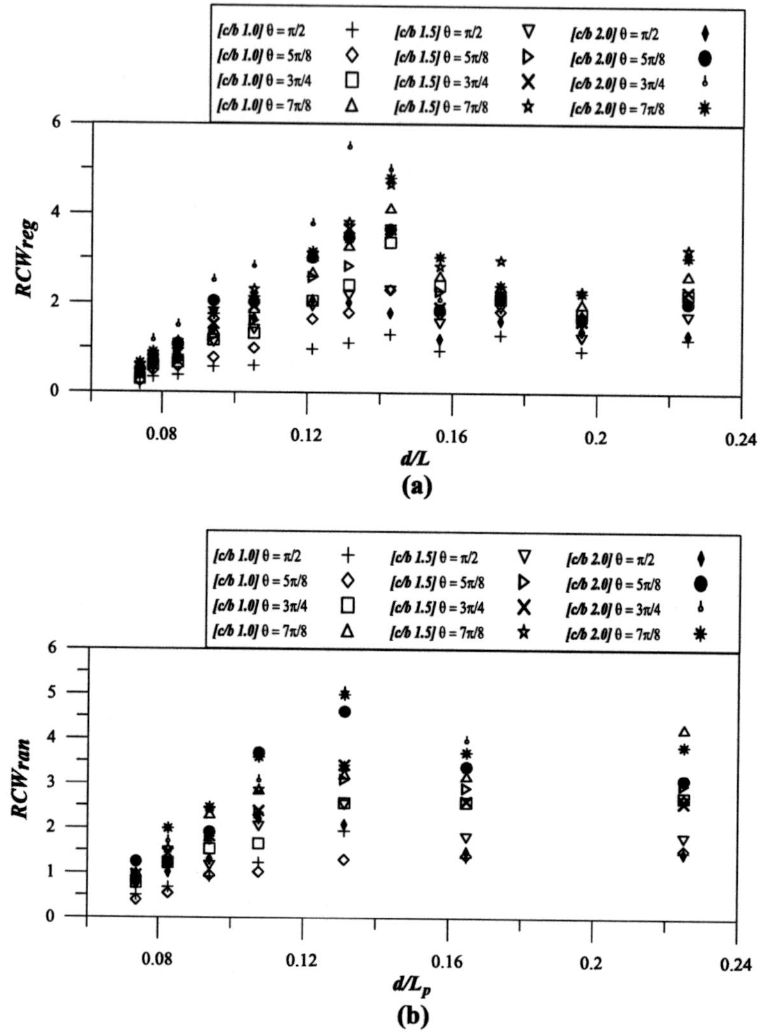

Fig. 5.37 **a** Comparison of RCW_{reg} for different configurations of harbour walls. **b** Comparison of RCW_{ran} for different configuration of harbour walls

4. The resonating length of the harbour wall determines the natural frequency of the water column, whereas the inclination of harbour walls does not influence natural frequency.
5. The pneumatic pressure resonates with respect to the natural frequency of the water column, and the peak of P_{air} shifts towards low frequency as the resonant length of the harbour wall increases. However, the inclination of the harbour has no significance towards pneumatic pressure as the resonant length was

maintained constant for every increment of θ. This agrees with the fact that pneumatic pressure also resonates close to the natural frequency of the water column.

6. The phase difference of water level oscillation between inside and outside the OWC chamber varies linearly with the wave frequency.
7. The variation of dynamic pressure inside and outside the OWC chamber due to the incident wave field is mainly governed by the system configuration and peak wave period.
8. The P_{out} is greater than P_{in}, and P_{air} is found to be minimum.
9. The energy conversion capacity of the device can be maintained optimum if the natural frequency of the OWC device is closer to the predominant wave frequency, which is in agreement with the claims made by [27].
10. The results of RCW indicate that the performance of the device relies on the wave frequency, and this confirms the frequency dependence nature of the device. The performance results indicate that substantial changes occur on adding harbour walls.

References

1. Arena, F., Romolo, A., Malara, G., Fiamma, V., & Laface, V. (2017). The first full operative U-OWC plants in the port of Civitavecchia. In: Proceedings of the international conference on offshore mechanics and arctic engineering—OMAE. https://doi.org/10.1115/OMAE2017-62036
2. Ashlin, S. J., Sundar, V., & Sannasiraj, S. A. (2016). Effects of bottom profile of an oscillating water column device on its hydrodynamic characteristics, *Renewable Energy*, *96*(Part A), 341–353, https://doi.org/10.1016/j.renene.2016.04.091.
3. Boake, C. B., Whittaker, T. J. T., Folley, M., Ellen, H. (2002). Overview and initial operational experience of the LIMPET wave energy plant. In: Proceedings of the international offshore and polar engineering conference.
4. Daniel Raj, D., Sundar, V., Sannasiraj, S. A. (2019a). Enhancement of hydrodynamic performance of an oscillating water column with harbour walls. Renew. Energy. https://doi.org/10.1016/j.renene.2018.07.089
5. Daniel Raj, D., Sundar, V., Sannasiraj, S. A. (2019b). Influence of harbour wall on pressure variation in an oscillating water column. In: Proceedings of the fourth international conference in ocean engineering (ICOE2018). Lecture Notes in Civil Engineering, Vol. 23. Springer. https://doi.org/10.1007/978-981-13-3134-3_56
6. David, D. R., Vallam, S., Sannasi Annamalaisamy, S. (2018). Effect of harbor walls on the efficiency of an oscillating water column, *144*, 1–16. https://doi.org/10.1061/(ASCE)WW.1943-5460.0000429.
7. Dean, R. G., & Dalrymple, R. A. (1991). *Advanced series on ocean engineering: water wave mechanics for engineers and scientists.* World Scientific.
8. Deng, Z., Huang, Z., Law, A. W. K. (2014). Wave power extraction from a bottom-mounted oscillating water column converter with a V-shaped channel. Proc R Soc A Math Phys Eng Sci. https://doi.org/10.1098/rspa.2014.0074
9. Drew, B., Plummer, A. R., Sahinkaya, M. N. (2009). A review of wave energy converter technology. Proc Inst Mech Eng Part A J Power Energy. https://doi.org/10.1243/09576509JPE782

10. Evans, D. V. (1984). The influence of projecting sidewalls on the hydrodynamic performance of wave-energy devices. *Journal of Fluid Mechanics*. https://doi.org/10.1017/S0022112084002962
11. Evans, D. V., & Porter, R. (1996). Hydrodynamic characteristics of an oscillating water column device, *1187*, 155–164.
12. Falcão, A. F. O., Henriques, J. C. C. (2014). Model-prototype similarity of oscillating-water-column wave energy converters. *Int J Mar Energy*. https://doi.org/10.1016/j.ijome.2014.05.002
13. Falnes, J. (1993). Optimum control of oscillation of wave-energy converters.
14. Heath, T. V. (2012). A review of oscillating water columns. *Philos Trans R Soc A Math Phys Eng Sci, 370*, 235–245. https://doi.org/10.1098/rsta.2011.0164
15. Ibarra-Berastegi, G., Sáenz, J., Ulazia, A., Serras, P., Esnaola, G., & Garcia-Soto, C. (2018). Electricity production, capacity factor, and plant efficiency index at the Mutriku wave farm (2014–2016). *Ocean Engineering*. https://doi.org/10.1016/j.oceaneng.2017.10.018
16. John Ashlin, S., Sundar, V., Sannasiraj, S. A. A., Ashlin, S. J., Sundar, V., & Sannasiraj, S. A. A. (2016). Effects of bottom profile of an oscillating water column device on its hydrodynamic characteristics. *Renewable Energy, 96*, 341–353. https://doi.org/10.1016/j.renene.2016.04.091
17. Malmo, O., & Reitan, A. (1985). Wave-power absorption by an oscillating water column in a channel. *Journal of Fluid Mechanics*. https://doi.org/10.1017/S0022112085002592
18. Malmo, O., & Reitan, A. (1985b) Development of the Kvaerner Multiresonant OWC. https://doi.org/10.1007/978-3-642-82666-5_4
19. Morris-Thomas, M. T., Irvin, R. J., & Thiagarajan, K. P. (2007). An Investigation Into the Hydrodynamic Efficiency of an Oscillating Water Column. *Journal of Offshore Mechanics and Arctic Engineering, 129*, 273. https://doi.org/10.1115/1.2426992
20. Sarmento, A. J. N. A., Brito-Melo, A., Neumann, F. (2006). Results from sea trials in the OWC european wave energy plant at Pico, Azores. Invit. Pap. WREC-IX.
21. Thiruvenkatasamy, K., & Neelamani, S. (1997). On the efficiency of wave energy caissons in array. *Applied Ocean Research, 19*, 61–72. https://doi.org/10.1016/S0141-1187(97)00008-4
22. Thiruvenkatasamy, K., Neelamani, S., & Sato, M. (2005). Nonbreaking wave forces on multiresonant oscillating water column wave power caisson breakwater. J Waterw Port Coastal Ocean Eng. https://doi.org/10.1061/(asce)0733-950x(2005)131:2(77)
23. Tseng, R. S., Wu, R. H., & Huang, C. C. (2000). Model study of a shoreline wave-power system. *Ocean Engineering, 27*, 801–821. https://doi.org/10.1016/S0029-8018(99)00028-1
24. Vyzikas, T., Deshoulières, S., Barton, M., Giroux, O., Greaves, D., Simmonds, D., 2017. Experimental investigation of different geometries of fixed oscillating water column devices. Renew. Energy. https://doi.org/10.1016/j.renene.2016.11.061
25. Wang, C., Deng, Z., Wang, P., & Yao, Y. (2019). Wave power extraction from a dual oscillating-water-column system composed of heave-only and onshore units. *Energies 12*. https://doi.org/10.3390/en12091742
26. Weber, J. (2007). Representation of non-linear aero-thermodynamic effects during small scale physical modeling of OWC wave energy converters. Proc 7th Eur Wave Tidal Energy Conf.
27. Wilbert, R., Sundar, V., & Sannasiraj, S. A. (2013). Wave Interaction with a Double Chamber Oscillating Water Column Device. *Int. J. Ocean Clim. Syst., 4*, 21–39. https://doi.org/10.1260/1759-3131.4.1.21

Chapter 6
Wave Induced Pressures and Forces on an OWC Device

V. Sundar and S. A. Sannasiraj

Abstract The research over several decades on the Oscillating Water Column has been on the enhancement of its efficiency by optimizing its configuration and its wave power absorption. However, the stability of such a device is of great concern, which requires a detailed evaluation of wave induced loadings on such structures. This chapter details a comprehensive experimental investigation to understand the dynamic pressures on the different components of the OWC including the total force on the structure induced by both regular and random waves. The design of the lip wall of OWC is dictated by the seaward exerted pressure it experiences, which is a critical parameter, as it governs the stability of the structure towards the seaside. The total horizontal wave force is noted to be 2.5 to 3 times the total vertical wave force. The pressures on the rear wall of the OWC chamber and the air pressure over the roof of the caisson exhibit a reverse trend in their variation compared to the pressures on the front of the lip wall. A comparison of the measured force with the formula of (Goda Y Random seas and design of maritime structures. University of Tokyo press (1985)) has revealed that the formula overestimates the shoreward force and underestimates the seaward force in the low frequency zone.

6.1 Introduction

Several researchers in the past have contributed significantly to the enhancement of the efficiency of the OWC. A detailed review on the effects of power take-off, the structural dimensions that are vital in governing the efficiency are discussed by Wilbert et al. [26] and Ashlin et al. [12]. Numerical modeling of the power take-off damping in an OWC was covered in detail by Kamath et al. [13]. The review of literature on the other important aspect of the bottom configuration of the OWC that is

V. Sundar (✉) · S. A. Sannasiraj
Department of Ocean Engineering, Indian Institute of Technology Madras, Chennai, India
e-mail: vsundar@iitm.ac.in

S. A. Sannasiraj
e-mail: sasraj@iitm.ac.in

A. Samad et al. (eds.), *Ocean Wave Energy Systems*, Ocean Engineering & Oceanography 14, https://doi.org/10.1007/978-3-030-78716-5_6

responsible for increasing its efficiency is reported by Ashlin et al. [1] and Sundar et al. [21]. For an efficient ocean wave energy conversion system, apart from the criteria as mentioned, the aspects that are essential to consider are reliability, survivability, and maintainability. Wave-power-extracting OWC-type caisson breakwaters have been widely scrutinized for their effective conversion of wave energy under normal conditions, however, it is most important to ascertain the loads resulting from waves acting on them to ensure their survivability. The high impact breaking force acts on the structure over a small fraction of time, although it is a crucial factor in the structural design of caissons. However, the nonbreaking wave force, which has a longer action, is important for the design against horizontal sliding and overturning [24].

Apart from wave energy absorption, an OWC caisson has the advantage of the wave force reduction because of wave phase dispersion, which means that the different components of the OWC device experience the maximum force at different instants of time during the propagation of a wave, which leads to a reduction in the overall construction cost to be economically competitive with the other renewable energy systems. As a result, the nature and magnitude of peak wave forces on an OWC device would be different from the forces on vertical-walled-type structures. Hence, a rational estimation of the nature and extent of the dynamic pressures and forces exerted on the individual components of the OWC subjected to both periodic and random wave incidence is vital for its safe and optimal design. This is also important for the design of the structure against horizontal sliding and overturning. Therefore, a comprehensive understanding of the pressures and forces induced by nonbreaking waves that are the inputs for the design of the device is essential.

6.2 Literature Review

The wave loading on an OWC caisson is complicated to be investigated through analytical or numerical methods. This is mainly because of the unconventional geometry of the structure, with an opening in front, a PTO device, and a dual-phase chamber. Hence, this calls for the measurement of pressures and forces through physical modeling. Prior to detailing the present experimental investigation, a brief literature review is reported herein. Malmo and Reitan [16] have investigated the wave power absorption by an OWC device and reported that the natural period of an OWC device predominantly depends on the submerged depth of its front lip wall. A design method for wave-power-extracting caissons by considering impact air pressure in the air chamber of the OWC device was proposed by Takahashi [22]. The theoretically predicted impact air pressure inside the air chamber for the prototype was found to be less than that estimated through model tests, which was claimed to be due to scale effects. The pressure distribution for the design of the air chamber walls on an OWC device was proposed, and the wave reflection from the breakwater was relatively smaller because of the high wave power absorption. McIver and Evans [17] showed that the response of the OWC system depends on the magnitude of the dynamic pressure and its excitation period. Goda et al. [8] and Nakada et al. [19] reported the field study conducted on the wave-power-extracting caisson breakwater at Sakata Port, Japan. Muller and Whittaker [18] explained the development and installation of the wave load monitoring system for the Islay LIMPET wave power plant with a capacity of 75 kW. The field measured return pressures of 50 years were extrapolated and compared with the available formulae in literature such as those from the USACE [24], Standard BS 6349 [4], and Goda [7]. The comparison showed that none of the suggested formulae resulted in satisfactory agreement.The pressures measured inside the OWC chamber were reported to be maximum to an extent of 1.5–2. Jayakumar [9] reported that the wave force on the multi-resonant oscillating water column (MOWC) caisson model was less than that exerted on a traditional rectangular caisson when there is a decrease in the air damping inside the OWC. Muller and Whittaker [18] through flow visualization experiments on an OWC suggested that due to internal breakers, the load on the rear wall of the chamber was more predominant than the pressure distribution on the front side of the lip wall. Thiruvenkatasamy et al. [23] observed that the magnitude of the shoreward force on the structure was always greater than that of the seaward. Kuo et al. [15] reported that the method of Sainflou [20] overestimated the wave load on the OWC caisson breakwater for shallow water wave conditions. Ashlin et al. [2] through their experimental investigations reported that the wave-induced pressures and forces on the OWC device and inferred that the total horizontal wave force is 2.5–3 times larger than the total vertical force. Ashlin et al. [2, 10] have also reported the literature referred by them in addition to their own test results, part of which are discussed in this chapter.

Krisna [14] carried out tests of a small-scale physical model of 1:79 on an OWC installed vertical breakwater model in two different wave flumes and concluded that the pressure recorded from the smaller-scale model was over-estimated by 2.45 times than that of the larger scale model. It was also pointed out that the existing methods of evaluation of wave forces on conventional vertical breakwaters can be used for OWC with certain modifications. A non-linear numerical model was developed by Wang et al. [25] based on the potential flow and time-domain higher-order boundary element method, the results of which were compared with experiments. It was reported that the seaside curtain wall experienced higher loads induced by waves compared to that on the inner curtain wall.

Although the magnitude of wave loads dictates the optimal design criteria, the foregoing review of the literature reveals that well-defined and generalized empirical methods to estimate wave forces on an OWC device are rather scanty. The wave forces could be obtained through experimental studies. Hence, through a well-controlled experimental program, the results on the evaluation of wave loads on an OWC device are presented in this chapter.

The dynamic pressure distribution on different structural components of the OWC device and the total horizontal and vertical wave forces acting on the different components of an OWC model due to both periodic and random wave incidences are measured by a well-controlled experimental program. The measured wave forces and pressure on the OWC system are compared with theoretical and standard empirical formulae for evaluating the design wave forces and pressures on the vertical wall to have a clear understanding of the difference in behaviour between an OWC device and a conventional vertical wall.

6.3 Experimental Investigation

6.3.1 Test Facility

The present study was carried out in a wave flume, 72.5 m long, 2 m wide, and 2.5 m deep. The wavemaker can be operated with a piston or hinge mode, thus capable of generating both shallow and deep-water waves. An artificial beach (wave absorber) on the other end is provided to absorb the wave energy, which consists of a perforated sheet shaped like a parabola with a rubble mound underneath the sheet.

6.3.2 Test Model and Experimental Set-Up

The model setup in the wave flume consisted of two parts: the scaled-down OWC model and the top steel frame supporting arrangement. The OWC model was made up of acrylic sheets of 20 mm thickness. The width of the chamber parallel to the wave crest was 1.91 m (w), and the length along the wave direction was 0.3 m (b). The height of the model, h, was 0.9 m with a front lip wall of 0.6 m deep. The bottom opening at the chamber was 0.3 m. The bottom profile inside the OWC chamber was a circular curve with a radius, R of 300 mm, made by rolling an aluminium sheet of 2 mm thickness, and stiffeners were added for its rigidity. The OWC model was positioned 45 m away from the wavemaker, with firm support provided by steel frame arrangements connected to the immovable trolley situated at the top of the flume. The test duration and the model position in the wave flume were chosen to avoid re-reflection from the wavemaker. The details of the OWC device with a circular curve bottom profile model in plan and sectional elevation are shown in Fig. 6.1.

Two force transducers were mounted on the OWC model at 1/3 and 2/3 width of the model. Thus, the OWC model was rigidly suspended from a steel frame arrangement through the force transducers. Clearances of 25 mm between the model and flume sidewalls and 70 mm between the model and the flume bottom surface were maintained. A false bottom surface of 90 mm thickness was provided in front of the model for a length of 30 m toward the wave maker to avoid direct wave penetration below the OWC device. The experiments were carried out by Ashlin et al. [11] with an air vent of 0.68% of the plan area of the air chamber. The longitudinal section and the plan view of the flume depicting the position of the wave gauges and the

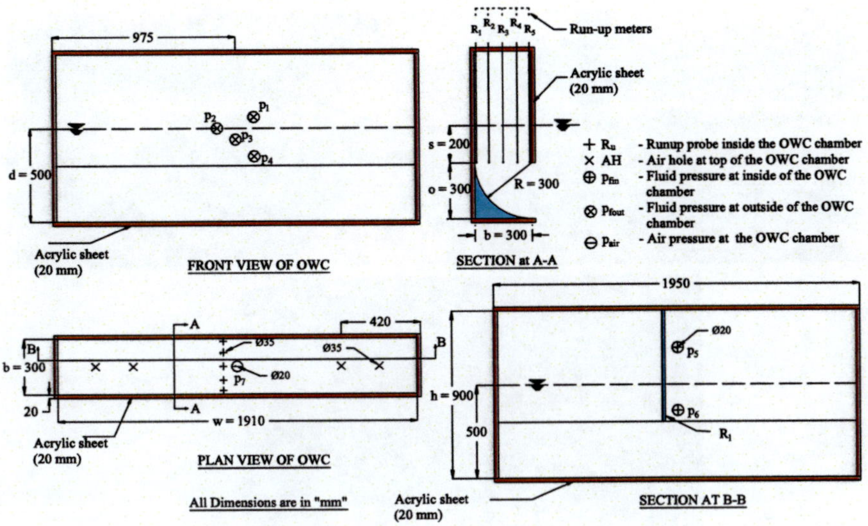

Fig. 6.1 Details of OWC model

test model are shown in Fig. 6.2. The force transducers and steel frame arrangement fixed to the OWC model inside the wave flume are shown in Fig. 6.3.

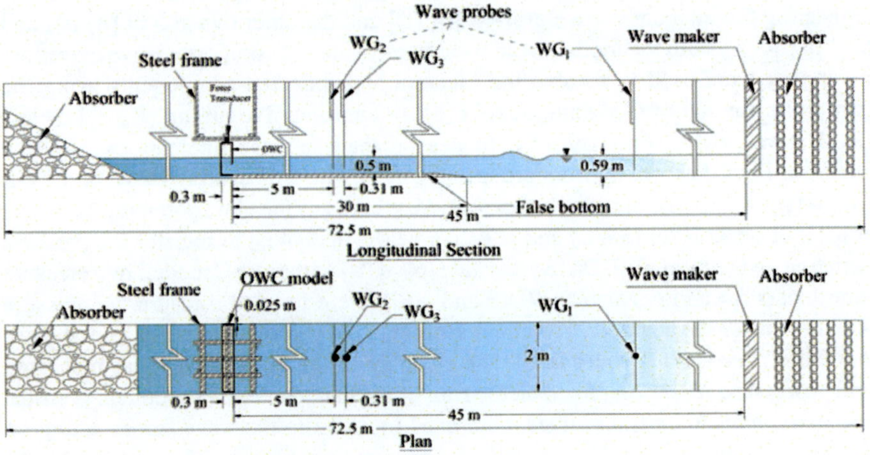

Fig. 6.2 Schematic diagram of the experimental setup

Fig. 6.3 Model set-up inside the wave flume

6.3.3 Instrumentation

Three resistance-type wave probes (WG_1–WG_3), one at 7 m from the wavemaker for registering the incident wave elevation [η_1 (t)] and the other two at 5 m [η_2 (t)] and 5.31 m [η_3 (t)] from the model, were used to measure the composite wave elevation. η_2 and η_3 were used to resolve the incident and reflected components using the two-probe method of Goda and Suzuki [6]. In addition, five run-up [R_u (t)] meters (R_1–R_5) were rigidly fixed vertically to measure the water level fluctuations inside the chamber. Four pressure transducers (p_1–p_4), each with a measuring range of 0.2 bar [$p_{fout,\,i}$ (t)], were fixed at different levels [z/d, where z is the elevation with the origin at the still water level (SWL) and negative below the SWL] to register the pressure variation on the front wall. On the rear side [$p_{fin,\,i}$ (t)] of the OWC model, two pressure transducers (p_5 and p_6) at z/d = 0.40 and –0.28 were fixed to measure the pressure variation inside the chamber. The air pressure [p_{air} (t)] variation resulting from the rise and fall of free surface water inside the chamber of the OWC model was measured with a pressure sensor (p_7) fixed on the roof of the caisson. Two-component (vertical and horizontal) force transducers were used to measure the wave forces acting on the structure. In essence, 19 channels of signatures consisting of 7 pressure signals, 5 run-up signals, incident water wave elevation, 2 wave probes to measure reflection from the model, and 4 strain gauge readings were simultaneously acquired with a sampling interval of 0.02 s.

6.3.4 Wave Characteristics

The tests were performed at two water depths, d/h = 0.56 and 0.62, where d is the water depth in front of the model. The regular wave tests were conducted with wave steepness (H/L) ranging from 0.01 to 0.05 and a relative water depth (d/L) ranging from 0.1 to 0.56. Regular waves of period (T) varying between 0.8 and 2.4 s in steps of 0.1 s were employed for the tests. For each of the wave periods, four-wave heights (H) in the range of 0.03–0.12 m were considered. Similarly, the random wave tests covered wave steepness, (H_{m0}/L_p, defined as the ratio of the spectral estimate of significant wave height, H_{m0}, to the wavelength corresponding to peak wave period, L_p) varying from 0.01 to 0.05 with d/L_p ranging between 0.08 and 0.37. For the random wave generation, the two-parameter Pierson Moskowitz (PM) spectrum was employed. The details of the PM spectrum were discussed in detail under Sect. 2.1.4.

6.4 Hydrodynamic Parameters

The different hydrodynamic pressures that are considered herein are the shoreward peak $[(p_{fout,i})_{max}]$ and seaward peak $[(p_{fout,i})_{min}]$ pressures exerted on the front face of the lip wall and are normalized by dividing by the incident wave pressure head (0.5 ρgH). The dimensionless pressures are indicated as $\{(PR_{fout,i})^n{}_{reg}\}$, where 'n' may represent either the crest $[(PR_{fout,i})^c{}_{reg}]$ or the trough $[(PR_{fout,i})^t{}_{reg}]$; and 'i' represents the number of the pressure sensor. Here, 'ρ' is the mass density of the fluid; and 'g' is the gravitational acceleration. A similar definition will hold for the shoreward $[(p_{fin,i})_{max}]$ and seaward $[(p_{fin,i})_{min}]$ pressures induced on the rear wall of the chamber, which, on normalising, are defined as $\{(PR_{fin,i})^n{}_{reg}\}$, and also for air pressure as $\{(PR_{air})^n{}_{reg}\}$. For random waves, the zeroth spectral moment estimate of wave elevation (H_{m0}) and pressure are considered. The analysis based on the induced pressures on the lip wall, rear wall, and roof provide an insight into the nature and magnitude of the wave load distributions corresponding to the different frequencies and amplitudes of the waves.

Further, the total horizontal and vertical wave forces acting on the system can be measured using the load cells. The shoreward $(F_{HZL/VER})_{max}$ and seaward $(F_{HZL/VER})_{min}$ [HZL/VER = horizontal/vertical] wave forces are normalized by dividing by 0.5ρgwdH and are represented by $\{(F_{HZL})^n{}_{reg}\}$ or $\{(F_{VER})^n{}_{reg}\}$, where 'n' represents either the shoreward or seaward peak. The zeroth spectral moment estimate of the force is used for random waves.

The measured wave forces were compared with the results from the standard theoretical wave force estimation method of Goda [7]. Even though Goda's theoretical method pertains to wave force estimation on vertical walls, comparisons were made to determine the difference between the nature of wave force acting on the vertical wall and the OWC device that has a bottom opening. This attempt has been made in the absence of any similar wave force measurements or predictions. The expression for estimating the shoreward horizontal wave force acting on the vertical wall based on the theoretical method of Goda [7], $(GM_{HZL})^c$, is given by

$$(GM_{HZL})^c = \frac{(F_{HZL})_{\max}}{0.5\rho g w d H} = \alpha_1\left[\left(1 + \frac{1}{\cosh kd}\right) + \left(\frac{1.5H}{d}\right)\right] \quad (6.1)$$

$$\alpha_1 = 0.6 + 0.5\left[\frac{4\pi\frac{d}{L}}{\sinh\left(4\pi\frac{d}{L}\right)}\right]^2 \quad (6.2)$$

where $k = 2\pi/L$. For comparative purposes, the seaward horizontal wave force acting on the vertical wall based on the theoretical method of Goda [7], $(GM_{HZL})^t$, has been taken.

The water surface oscillations $[(R_i)_{max}$ and $(R_i)_{min}]$ inside the OWC chamber are normalized with the amplitude of incident wave height, 0.5H, to yield the wave amplification factors $[(\beta_i)^c_{reg}$ and $(\beta_i)^t_{reg}]$ that represent the normalized shoreward and seaward oscillations, respectively. Herein, 'i' denotes the run-up meter number. In this chapter, the results of the investigation of the effect of wave steepness, H/L, and relative water depth, d/L, on the dimensionless pressures and wave forces acting on the OWC model are reported.

6.5 Results and Discussion

6.5.1 Time Histories of Measured Signatures

Typical time-series plots of wave surface elevation (η_1, η_2, and η_3), pressure in front of the lip wall ($p_{fout,\,i}$), rear wall pressure ($p_{fin,\,i}$), air pressure (p_{air}), vertical wave force (F_{VER}), and horizontal wave force (F_{HZL}) measured on the OWC device subjected to a regular wave field characterized by $H/L = 0.0159$, and $d/L = 0.131$ are depicted in Fig. 6.4. It was found that p_{fout} was greater than p_{fin} and p_{air} because of reflection, vortices formation in the vicinity of the lip wall, and the radial return flow of waves from the chamber. The magnitude of wave force measured over the OWC in the vertical direction is approximately half of the wave force acting along the horizontal direction.

6.5.2 Pressure Distribution in Front of the Lip Wall

6.5.2.1 At the SWL (Z/d = 0)

The variation of the non-dimensional shoreward pressure on the seaward side of the lip wall, $(PR_{fout,2})^c_{reg}$, as a function of d/L is shown in Fig. 6.5. As this pressure

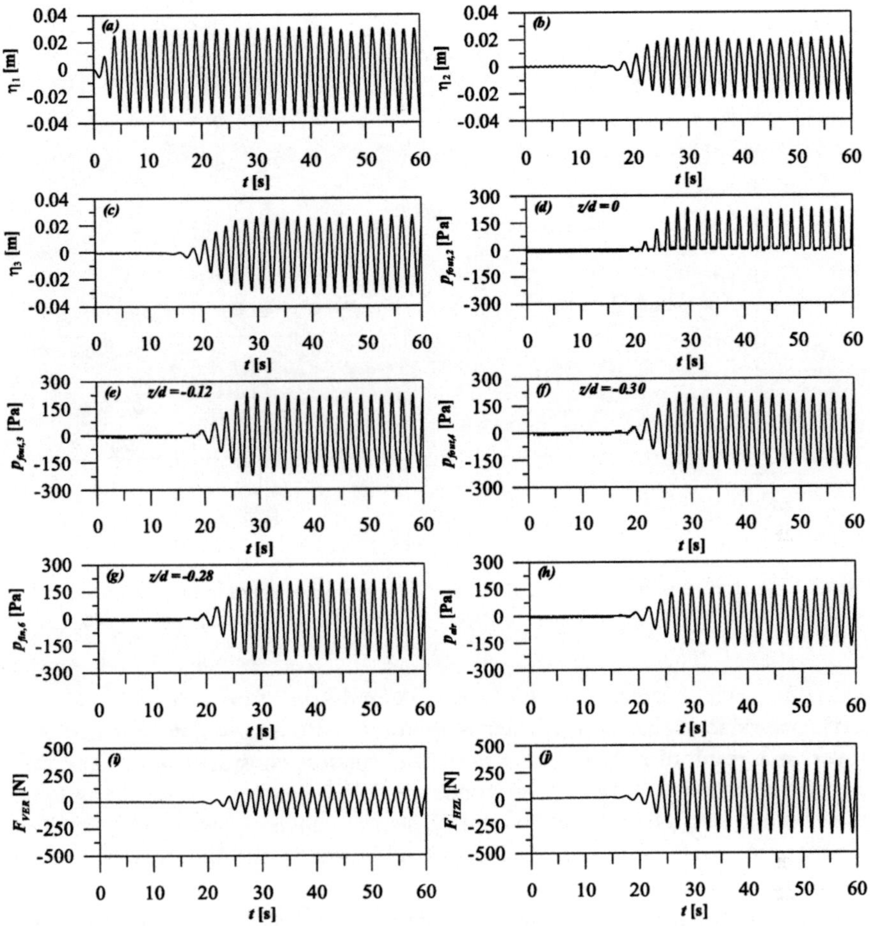

Fig. 6.4 Time series of the OWC device measurements: **a** η_1, **b** η_2, **c** η_3, **d** $p_{fout,2}$, **e** $p_{fout,3}$, **f** $p_{fout,4}$, **g** $p_{fin,6}$, **h** p_{air}, **i** F_{VER}, **j** F_{HZL} (*Note* H/L = 0.0159; d/L = 0.131)

measurement is taken at the SWL, only the intermittent pressure resulting from the crest is recorded. In general, $(PR_{fout,2})^c{}_{reg}$ increases with wave steepness. Even though the pressure is normalized with wave height, the dynamic pressure increases up to 25% for a higher wave steepness of 0.05, which is because the OWC model absorbs a larger part of the wave energy from waves of relatively small steepness compared with that from steeper waves. Due to this effect, a substantial change in the direction of the orbital motion of the water particles from horizontal to vertical reduces the pressure build-up in front of the lip wall caused by small-steepness waves. The reflection attributable to the front lip wall increases with wave steepness and hence, increases the induced pressure. For a particular wave steepness, $(PR_{fout,2})^c{}_{reg}$ is lower in the low-frequency region than in the high-frequency region. With an increase in

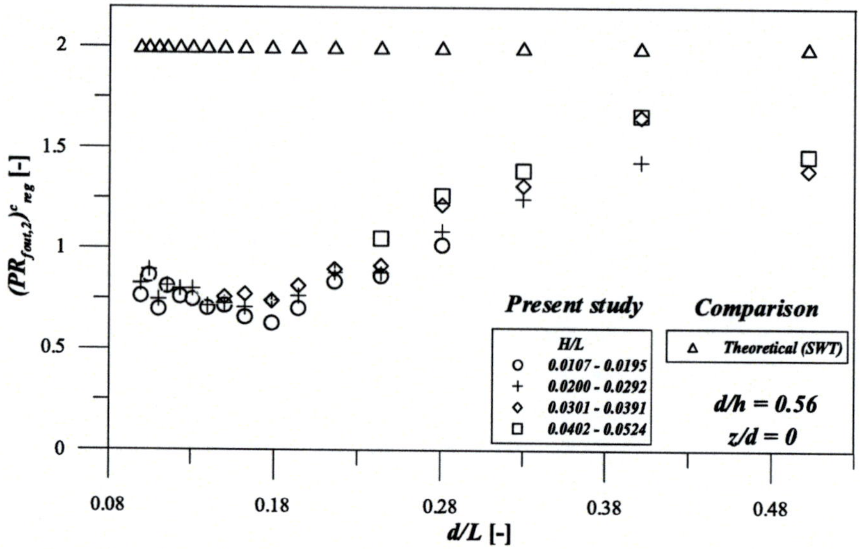

Fig. 6.5 Effect of relative water depth (d/L) on the shoreward pressure in front of the lip wall of the OWC device in comparison with theoretical SWT for different H/L ranges at z/d = 0

d/L, $(PR_{fout,2})^c{}_{reg}$ decreases up to d/L = 0.18, and then, it increases with d/L. The maximum value of $(PR_{fout,2})^c{}_{reg}$ reaches up to 1.7 at d/L = 0.40. For high-frequency waves, as a result of reduced water particle excursion, considerable incident wave energy incident on the lip wall cannot enter the OWC chamber, causing the system to behave like a vertical-wall-type structure. Therefore, there is a tremendous build-up of water mass in front of the OWC chamber caused by standing wave formation, which provokes large shoreward pressure.

The dynamic pressure acting on the surface of the structure with closed geometries is attributable to wave run-up around them, whereas for vertical-walled structures, in particular, when exposed directly to steep waves, the impact pressure is because of the action of the water mass progressing almost perpendicular to the wall. However, in the case of an OWC caisson, the forces exerted are attributable to a combination of both vertical oscillation of the water mass and a horizontal wave impact. Furthermore, it was found that the pressure on the lip wall was on the order of approximately 75% of the pressure head resulting from wave height for d/L less than approximately 0.20. A drastic increase in pressures up to approximately 1.5 times the pressure head resulting from wave height was seen for d/L greater than 0.20. This might be due to the free entry of the long-period waves into the chamber, allowing more energy to be trapped and resulting in less energy available to induce pressures on the lip wall because of relatively less build-up of water mass on the seaward side of the lip wall. For comparison, the theoretical pressure measurement on the vertical wall based on the standing wave theory (SWT) [5] is also shown in Fig. 6.5. Even though the OWC model looks like a vertical-wall-type structure, it absorbs wave power from incident waves and behaves like a perforated caisson breakwater. This is evident because the measurements never attained values closer to those of a standing wave, although the pressure exerted by short-period waves tested with $d/L > 0.40$ tended toward it.

6.5.2.2 Below the SWL (Z/d = -0.12)

A similar variation of $(PR_{fout,3})^{c}{}_{reg}$ and $(PR_{fout,3})^{t}{}_{reg}$ as a function of d/L along with the theoretical SWT values has been shown in Fig. 6.6a and b, respectively. With an increase in d/L, $(PR_{fout,3})^{c}{}_{reg}$ and $(PR_{fout,3})^{t}{}_{reg}$ decrease up to $d/L = 0.18$, and, beyond which, they increase. At the low-frequency region ($d/L < 0.20$), the present experimental measurements were found to be less than half those of the theoretical SWT values because of higher wave power absorption and asymptotic convergence toward the high-frequency region ($d/L > 0.40$). The values for the theoretical SWT and the observed physical phenomenon for the OWC device were found to be inverse in nature; with SWT, the pressure was found to decrease with wavelength, whereas the present measurements increase with a decrease in wavelength. At $d/L = 0.18$, $(PR_{fout,3})^{c}{}_{reg}$ and $(PR_{fout,3})^{t}{}_{reg}$ were found to be the least. Conventional coastal structures (e.g., breakwater, seawall) are inefficient in dissipating wave energy for long-period waves. Because the OWC device showed better performance for long-period waves, the combination of the OWC device and breakwater will enhance both structural stability and wave energy conversion, thus serving a dual purpose.

The variation of $(PR_{fout,3})^{c}{}_{reg}$ and $(PR_{fout,3})^{t}{}_{reg}$ as a function of H/L for different d/L values is shown in Fig. 6.7. The trend in variation of the pressure at $z/d = -0.12$ was found to be similar to that of pressure at the SWL; that is, the effect of H/L on $(PR_{fout,3})^{n}{}_{reg}$ was found to be marginal. It is important to note that the seaward pressure was greater than the shoreward pressure for $d/L > 0.20$, like the observations of [3], and it attained a maximum value of 1.6. This would mean that when the trough of an incident wave acted on the lip wall, the exerted dynamic pressure was less than the incident static wave pressure head under the SWL, that is, the net pressure experienced by the lip wall was directed toward the seaside. Hence, the seaward pressure also should be considered as a critical parameter in the design of the front wall because the induced moment alternated from sagging to hogging while the shoreward pressure reversed to seaward pressure.

The importance of seaward pressure is attributable to the presence of second harmonic components of considerable amplitude resulting from wave nonlinearity. As the waves propagate from deep to shallow water, the non-linear effects made the wave crest sharper and the trough flatter. This resulted in higher dynamic wave pressures under the crest compared with under the trough. In contrast to this phenomenon, for the OWC device at the high-frequency region, because of the appreciable standing wave formation, the pressure under the crest of the antinode was lower than that under the trough as a result of the presence of the second harmonic. Under such a condition, the lip wall may pull toward the offshore during heavy wave action. A similar variation in $(PR_{fout,4})^{n}{}_{reg}$ was observed for z/d of -0.30.

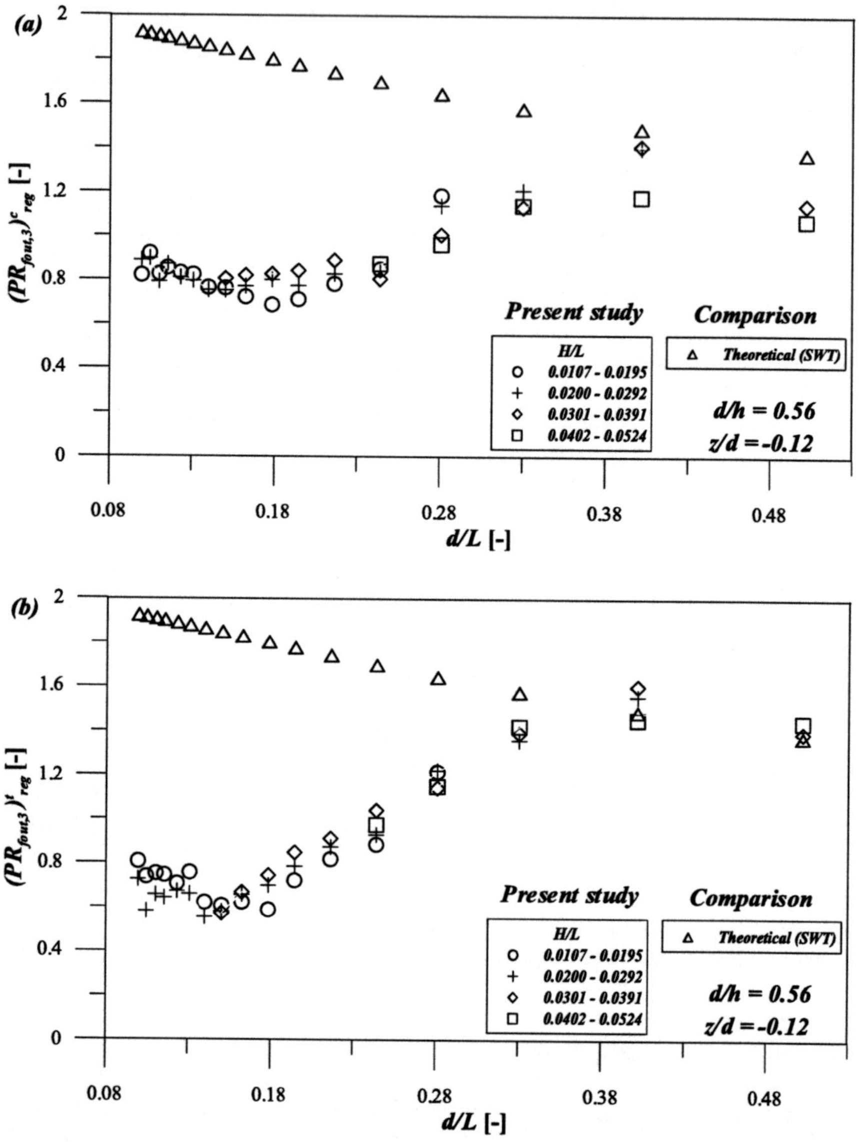

Fig. 6.6 Effect of relative water depth (d/L) on the pressure in front of the lip wall of the OWC device in comparison with theoretical SWT for different H/L ranges at z/d = –0.12 **a** shoreward; **b** seaward

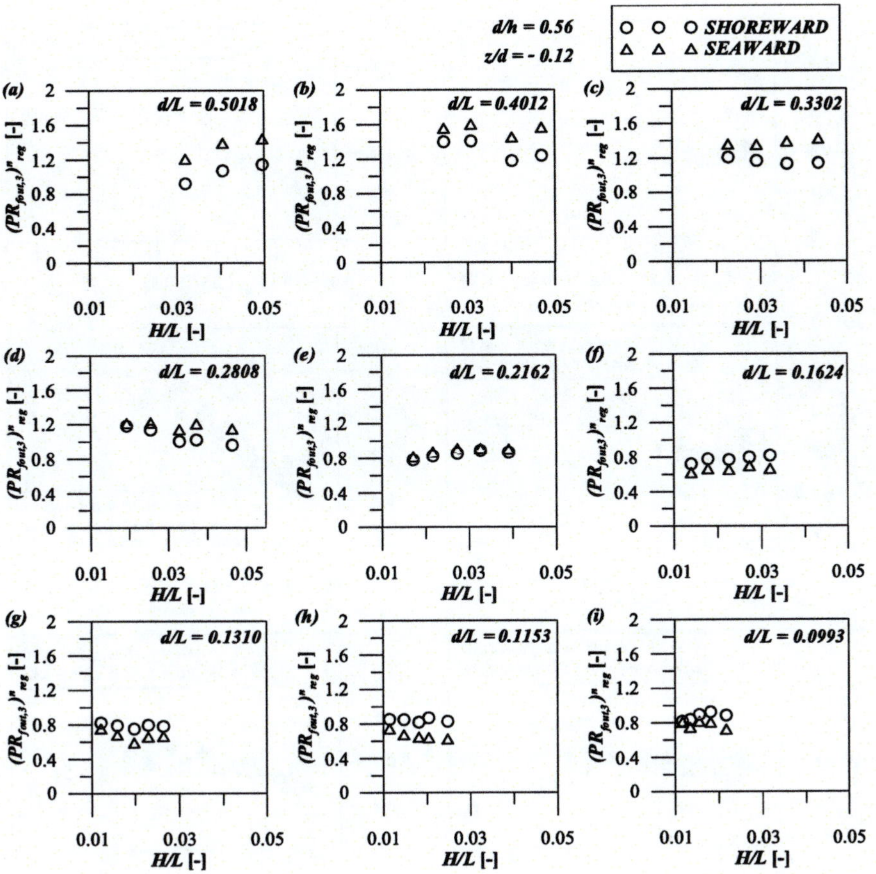

Fig. 6.7 Effect of wave steepness (H/L) on the pressure in front of the lip wall for different relative water depths (d/L) at z/d = −0.12

6.5.2.3 Effect of Water Depth on the Pressure at the Front Lip Wall

The effect of water depth d/h on the variation of $(PR_{fout,i})^n_{reg}$ with d/L for all H/L values tested is shown in Fig. 6.8. Since the pressure sensor locations remained the same, the results are superposed for closer values of z/d, as indicated in the plots in Fig. 6.8. It is observed that $(PR_{fout,i})^n_{reg}$ decreases with an increase in the absolute value of z/d. The outside pressure at the front wall near the bottom opening (e.g., z/d = −0.3) for d/h = 0.56 was 5–10% higher than that for d/h = 0.62. This confirms the selection of d/h = 0.56 as a design water depth based on the ratio between the bottom opening and the water depth (o/d). Further, the pressure variation near the bottom opening was found to increase for longer waves. This observation is contrary to that measured near the SWL (at z/d = 0 and z/d = −0.12). The high wave power absorption at the low-frequency region induced higher pressure when waves penetrated through

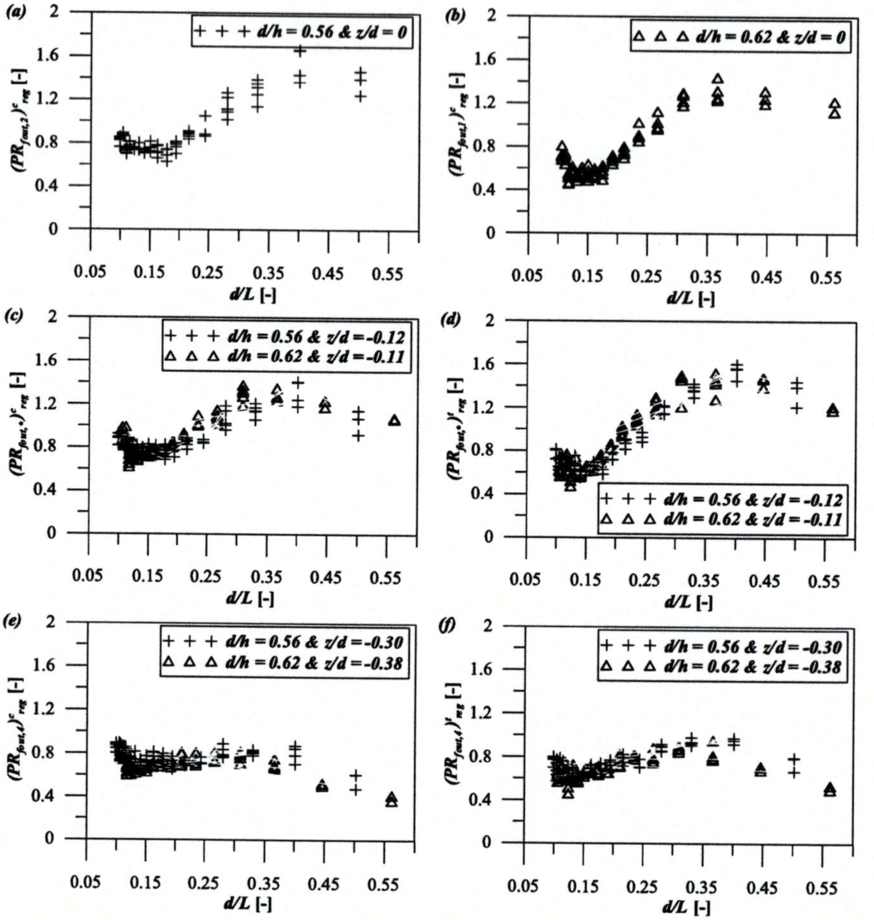

Fig. 6.8 Effect of relative water depth (d/L) on the pressure in front of the lip wall of the OWC device for different water depths at different z/d values

the bottom opening into the chamber. Based on the position of the pressure sensor on the lip wall, the maximum $(PR_{fout})_{reg}$ value measured was about 1.7 in the shoreward direction and about 1.6 in the seaward direction. These parameters dictate the design criteria for the lip wall.

The variation of $(PR_{fout,i})^{n}{}_{ran}$ with relative water depth for different water depth conditions is shown in Fig. 6.9. The OWC system was excited using a random wave packet.

The trend in the variation of $(PR_{fout,i})^{n}{}_{ran}$ was found to be similar to that of $(PR_{fout,i})^{n}{}_{reg}$ with the magnetic reduction of 5–10%. This was due to the part consisting of the number of wave components involved in the description of the wave climate. Based on the position of the pressure sensor on the lip wall, the maximum

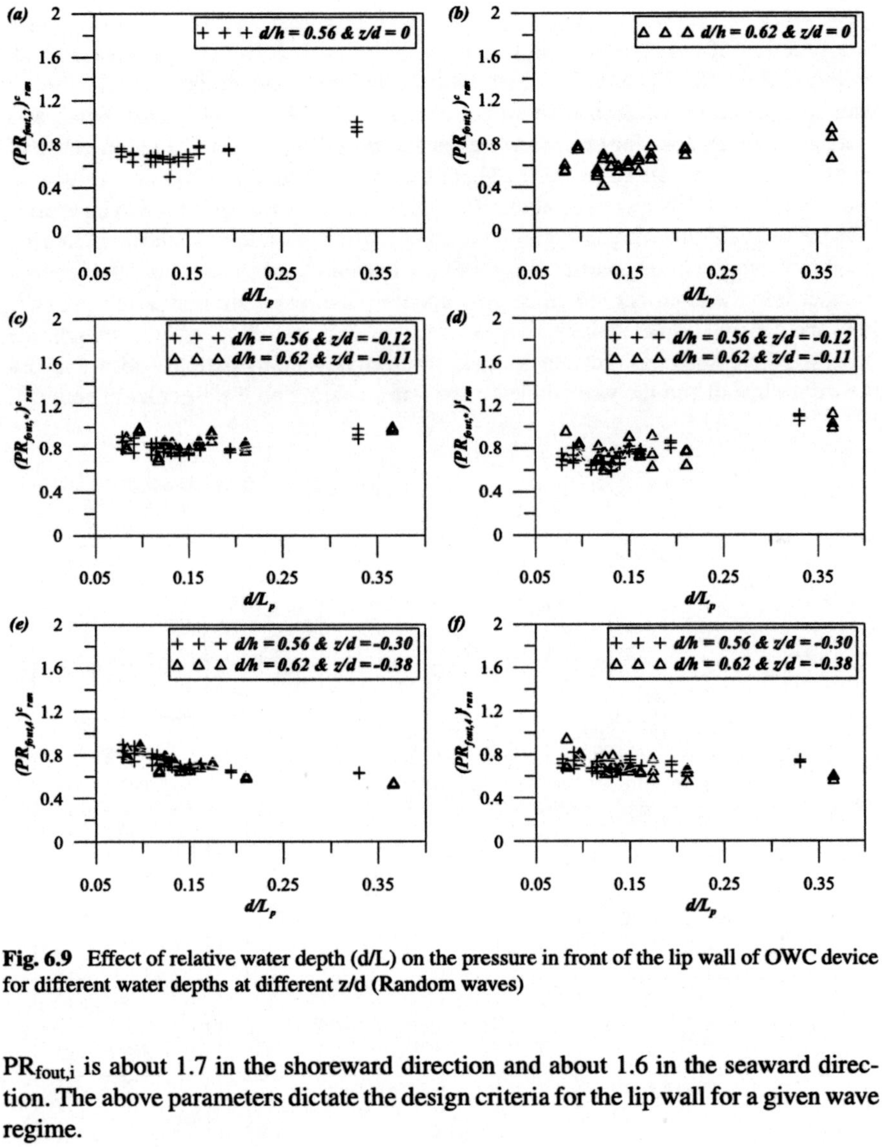

Fig. 6.9 Effect of relative water depth (d/L) on the pressure in front of the lip wall of OWC device for different water depths at different z/d (Random waves)

$PR_{fout,i}$ is about 1.7 in the shoreward direction and about 1.6 in the seaward direction. The above parameters dictate the design criteria for the lip wall for a given wave regime.

6.5.3 *Pressure Distribution at the Rear Wall of OWC Device*

The rise and fall of the water column inside the OWC chamber induce rear wall pressures, $(PR_{fin,6})^n{}_{reg}$. A higher $(PR_{fin,6})^n{}_{reg}$ indicates a larger amount of wave power absorption, resulting in a more efficient system. The variation of the shoreward

$(PR_{fin,6})^{c}_{reg}$ and seaward $(PR_{fin,6})^{t}_{reg}$ pressures on the rear wall of the OWC device as a function of H/L is shown in Fig. 6.10. In general, $(PR_{fin,6})^{n}_{reg}$ decreases with an increase in H/L. It is to be noted that the induced outside pressure on the lip wall was found to increase with an increase in H/L. This might be attributable to increased energy loss for steeper waves at the tip of the lip wall as an outcome of energy dissipation caused by vortex shedding, turbulence, the radial flow of waves, and other factors. The pressure at the shoreward rear wall was observed to be greater than the seaward pressure throughout the wave regime considered in the tests. During the flow from the high-pressure region to the low-pressure region, the wave energy transfer into the chamber would be obstructed by the rear wall, thus amplifying the flow and causing higher shoreward pressure at the rear wall. However, the magnitude of seaward pressure was reduced because of return flow being partially obstructed by the front lip wall and the work done by the water column on the air column above it.

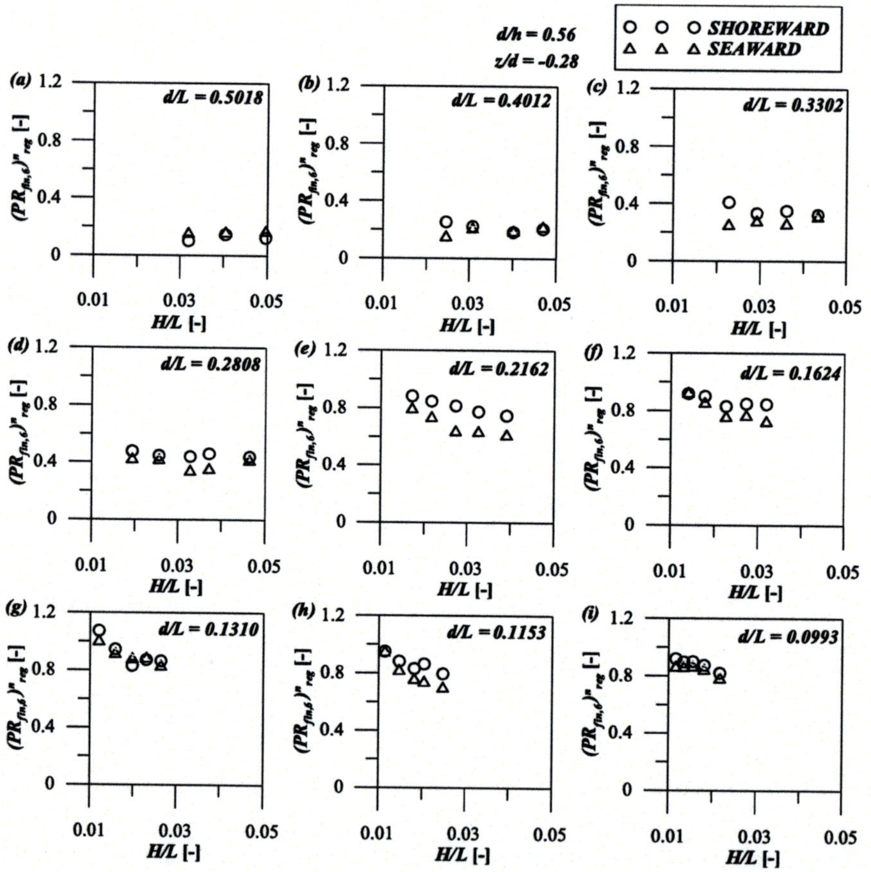

Fig. 6.10 Effect of wave steepness (H/L) on rear wall pressure for different relative water depths (d/L)

For the constant water depth condition, $(PR_{fin,6})^{n}{}_{reg}$ increases with wavelength and thus, proves that the system is more efficient when exposed to long waves. In the low-frequency region, the pressure at the front lip wall and that at the rear wall balance each other because of optimum wave power absorption. The maximum $(PR_{fin,6})^{c}{}_{reg}$ is approximately 1.1 times the incident static wave pressure head at $d/L = 0.131$. Towards the high-frequency region, the pressure at the rear wall is the least, even though the pressure available in front of the chamber was the largest near the SWL.

The physical processes that cause the variation between $(PR_{fout,2})^{c}{}_{reg}$ and $(PR_{fin,6})^{c}{}_{reg}$ can be explained in terms of the time series of the phase difference between the hydrodynamic parameters $(PR_{fout,4})^{c}{}_{reg}$ and $(\beta_1)^{c}{}_{reg}$ for a few combinations of wave characteristics that are superposed in Fig. 6.11. For $d/L = 0.401$, while the flow passed into the chamber, $(PR_{fout,4})^{c}{}_{reg}$, is found to be nearly out of phase ($\theta = 160^\circ$) with the flow oscillation inside the chamber, $(\beta_1)^{c}{}_{reg}$. Thus, the wave inflow into the chamber was flow suppressed, resulting in lesser energy available inside the OWC device. The quantitative phase difference between the hydrodynamic parameters $(PR_{fout,4})^{c}{}_{reg}$ and $(\beta_1)^{c}{}_{reg}$ from these results as a function of t/T is reported in Table 6.1. The phase difference between $(\beta_1)^{c}{}_{reg}$ and $(PR_{fout,4})^{c}{}_{reg}$ is large for high-frequency waves. Further, the phase lag decreases with an increase in wavelength, and it indeed induced significant flow into the chamber and thus $(PR_{fin,6})^{c}{}_{reg}$ increases. The phase lag between $(\beta_1)^{c}{}_{reg}$ and $(PR_{fout,4})^{c}{}_{reg}$ is observed to be the least at $d/L = 0.131$ when the pressure at the rear wall is maximum. Along with the phase difference, the orbital motion of waves exciting the system is found to play a major role in the efficiency of the system. At the high-frequency region, the orbital motion of waves was smaller and was subjected to a significant amount of interference by the lip wall, which induced vortex formation on its lee side. This vortex formation reduced the power available inside the system. Both the phase difference and the vortex formation are the reasons for the poor performance of the OWC device at the high-frequency region.

The effect of water depth on shoreward and seaward pressures at the rear wall subjected to both regular and random waves as a function of d/L is shown in Fig. 6.12. The trend in the variation of $(PR_{fin,6})^{n}{}_{reg}$ for both water depths is found to be similar. However, for $d/L < 0.25$, the depth effect is found to have significance with up to 40% higher pressure magnitude at the rear wall for lesser d/h. The trend in variation of $(PR_{fin,6})^{n}{}_{ran}$ corresponds to d/L_p is found to be similar to that observed for regular wave tests. The shoreward rear wall pressure is found to be higher than the seaward for both periodic and random waves. A comprehensive analysis of the pressures inside the OWC chamber shows that the rear wall should be designed for a value of approximately 1.1 times the incident static wave pressure head.

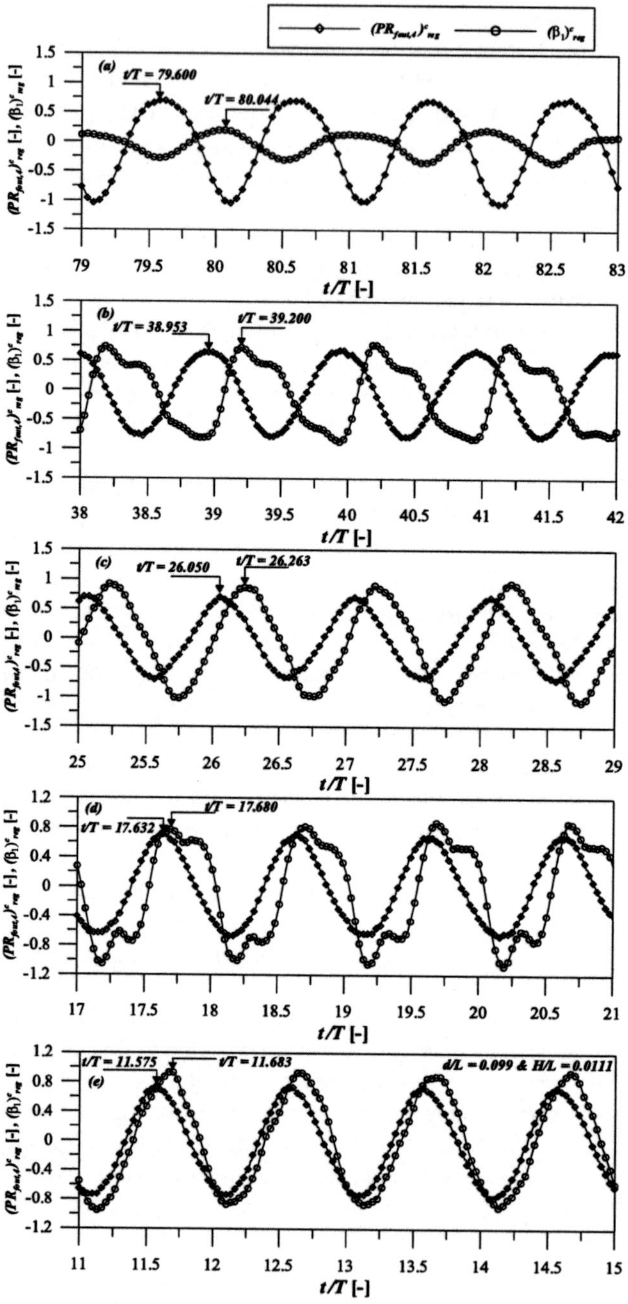

Fig. 6.11 Time series in quantifying phase difference between $(PR_{fout,4})^c_{reg}$ and $(\beta_1)^c_{reg}$ against t/T for different combinations of d/L and H/L at constant H of 0.06 m

Table 6.1 Quantitative phase difference between $(PR_{fout,4})^c_{reg}$ and $(\beta_1)^c_{reg}$ in terms of t/T for different relative water depths at constant H = 0.06 m

d/L	$(PR_{fout,4})^c_{reg}$ (t/T)	$(\beta_1)^c_{reg}$ (t/T)	$(\beta_1)^c_{reg}$ (t/T) – $(PR_{fout,4})^c_{reg}$ (t/T)	$(\beta_1)^c_{reg}$ – $(PR_{fout,4})^c_{reg}$ (°)
0.401	79.600	80.044	0.444	160
0.216	38.953	39.200	0.247	89
0.162	26.050	26.263	0.213	77
0.131	17.632	17.68	0.048	17
0.099	11.575	11.683	0.108	39

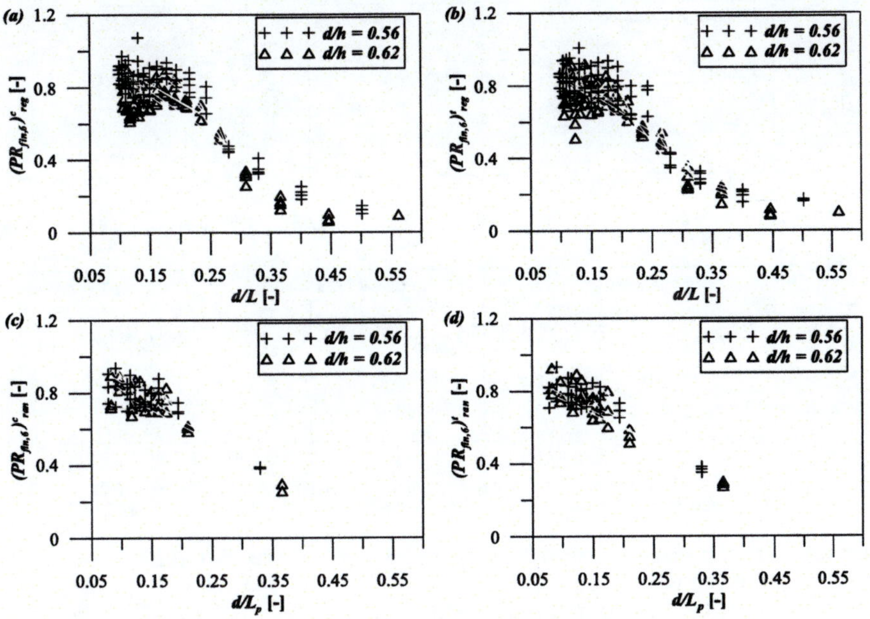

Fig. 6.12 Effect of relative water depth (d/L) on rear wall pressure inside the OWC device for different water depths at z/d = –0.28, **a**, **b** Regular wave, **c**, **d** Random waves

6.5.4 Air Pressure Inside the OWC Caisson

The magnitude of air pressure inside the chamber is essential for the design of the roof of the OWC caisson. The values for normalized upward and downward air pressure $[(PR_{air})^n_{reg}]$ developed inside the chamber of the caisson as a function of H/L are presented in Fig. 6.13. The values for upward and downward air pressure correspond to compression and expansion of the air column, respectively. The variation in air pressure inside the chamber correlates with the pressure at the rear wall. From high frequency to d/L = 0.131, the induced air pressure $[(PR_{air})^n_{reg}]$ in the chamber

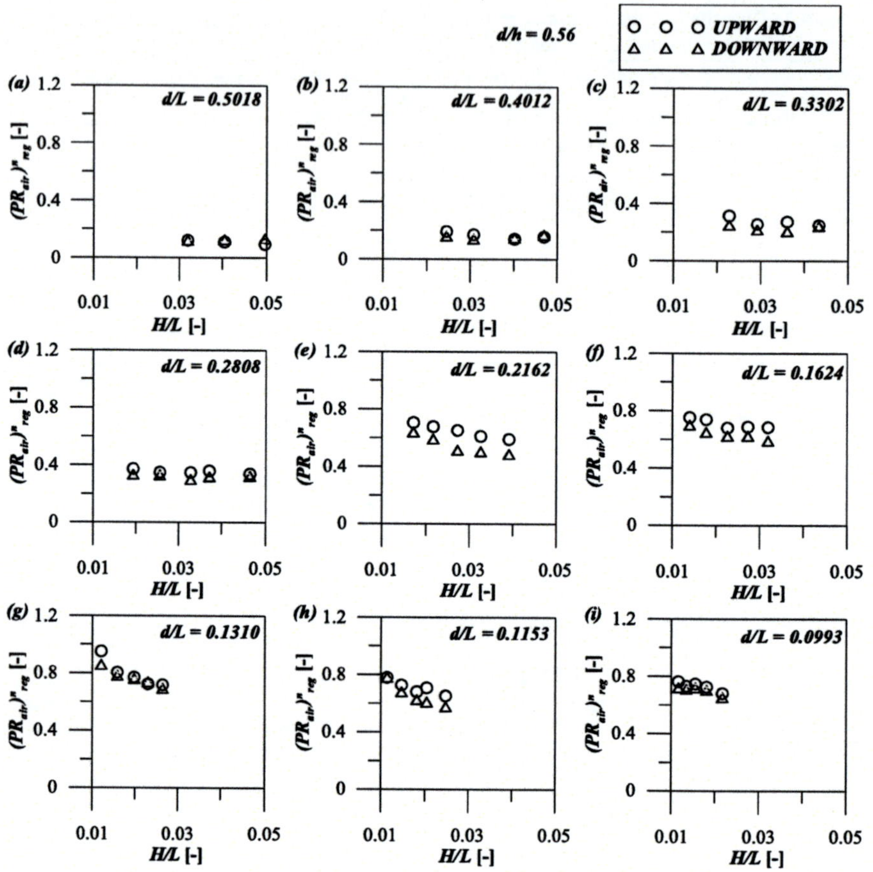

Fig. 6.13 Effect of wave steepness (H/L) on the air pressure inside the OWC chamber for different relative water depths (d/L)

increases and slightly reduces for the further longer wave incidences. The reason for the reduction in the air pressure beyond d/L = 0.131 is an increase in phase lag between $(PR_{fout,4})^{c}{}_{reg}$ and $(\beta_1)^{c}{}_{reg}$ that enhances the suppression of inflow into the chamber. At d/L = 0.131, the normalized upward and downward air pressures reached a maximum value close to that of the incident static wave pressure head. The upward air pressure was always noticed to be slightly greater throughout the considered wave characteristics.

The effects of water depth on the upward and downward normalised air pressures, subjected to both regular and random waves as a function of relative water depth, are shown in Fig. 6.14. At d/h of 0.62, 10% reduction in air pressure is observed compared to the design water depth (d/h = 0.56). The air pressure measured inside the chamber subjected to random waves is found to be 20% lesser than the regular

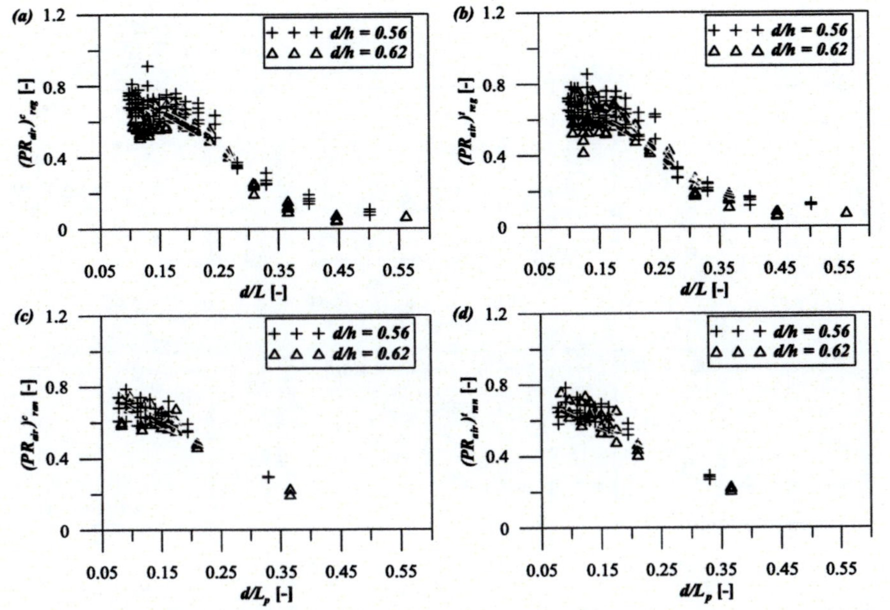

Fig. 6.14 Effect of relative water depth (d/L) on the air pressure inside the OWC device for different water depths, **a**, **b** Regular waves **c**, **d** Random waves

wave tests. The trend in the variation was observed to be similar for both regular and random waves.

6.5.5 Horizontal Wave Force

The variation of normalised shoreward and seaward total horizontal wave force, $(F_{HZL})^n{}_{reg}$, as a function of H/L, is shown in Fig. 6.15. $(F_{HZL})^n{}_{reg}$ is found to be the least at the high-frequency region and increases towards the low-frequency region. It should be mentioned that the nature and magnitude of the pressures distributed on the individual components of the OWC device differ from the total wave load on the OWC device. For instance, the pressure distribution on the front lip wall, $(PR_{fout,2})^c{}_{reg}$, at the SWL increases with a decrease in wavelength; conversely, the pressure inside the chamber, $(PR_{fin,6})^c{}_{reg}$, at the rear wall decreases with the wavelength. The addition of pressures $(PR_{fout,i})^n{}_{reg}$ and $(PR_{fin,i})^n{}_{reg}$ induces the total horizontal wave force on the OWC caisson model and to the wave, the force increases with the wavelength. In general, $(F_{HZL})^n{}_{reg}$ is observed to increase with H/L. The shoreward and seaward total horizontal wave force $(F_{HZL})^n{}_{reg}$ attained the maximum values of 1.3 and 1.1, respectively. On average, the magnitude of shoreward total horizontal wave force was about 10% more than that of the seaward force.

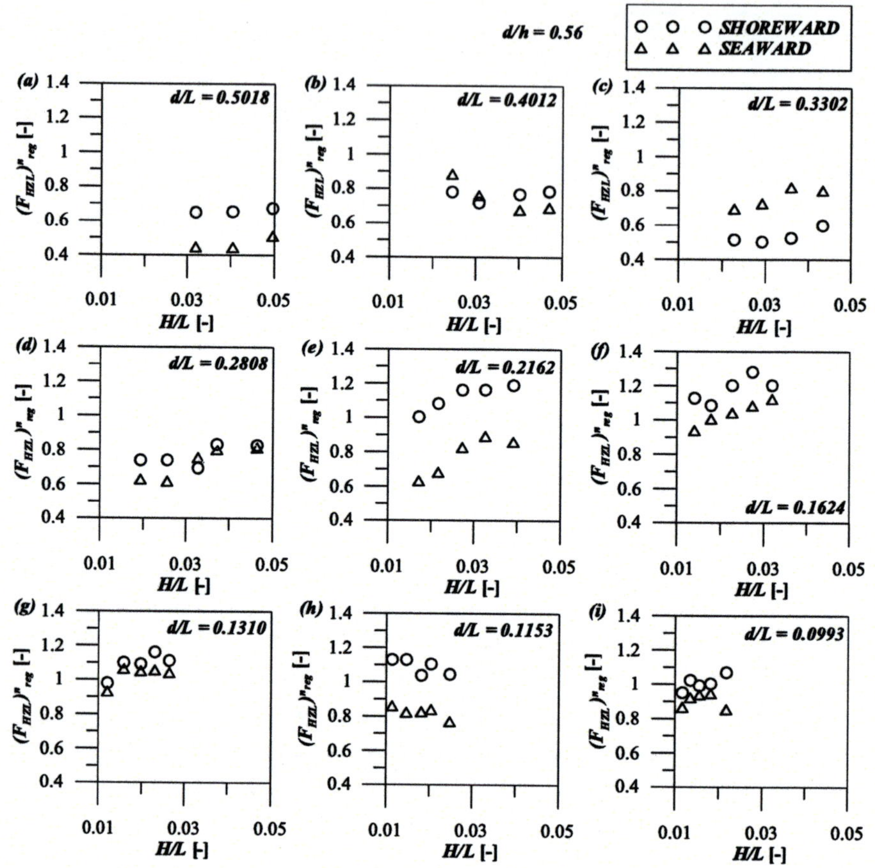

Fig. 6.15 Effect of wave steepness (H/L) on the total horizontal wave force acting on the OWC device for different relative water depths

6.5.6 Vertical Wave Force

The variation of normalised upward and downward total vertical wave forces, $(F_{VER})^n{}_{reg}$, acting on the OWC caisson model as a function of H/L is presented in Fig. 6.16. Because of the presence of the partially submerged lip wall, the flow enters the chamber causing significant water oscillation. This additional feature is responsible for the differences in the nature of the total vertical wave force on the OWC caisson compared with a conventional rectangular caisson of similar dimensions. The trend in the variation of the total vertical wave force, $(F_{VER})^n{}_{reg}$, is found to be like that of horizontal wave force, $(F_{HZL})^n{}_{reg}$. The total vertical wave force acting on the model has contributions from two different sources: the oscillation inside the chamber and the vertical component of wave impact on the structure. The

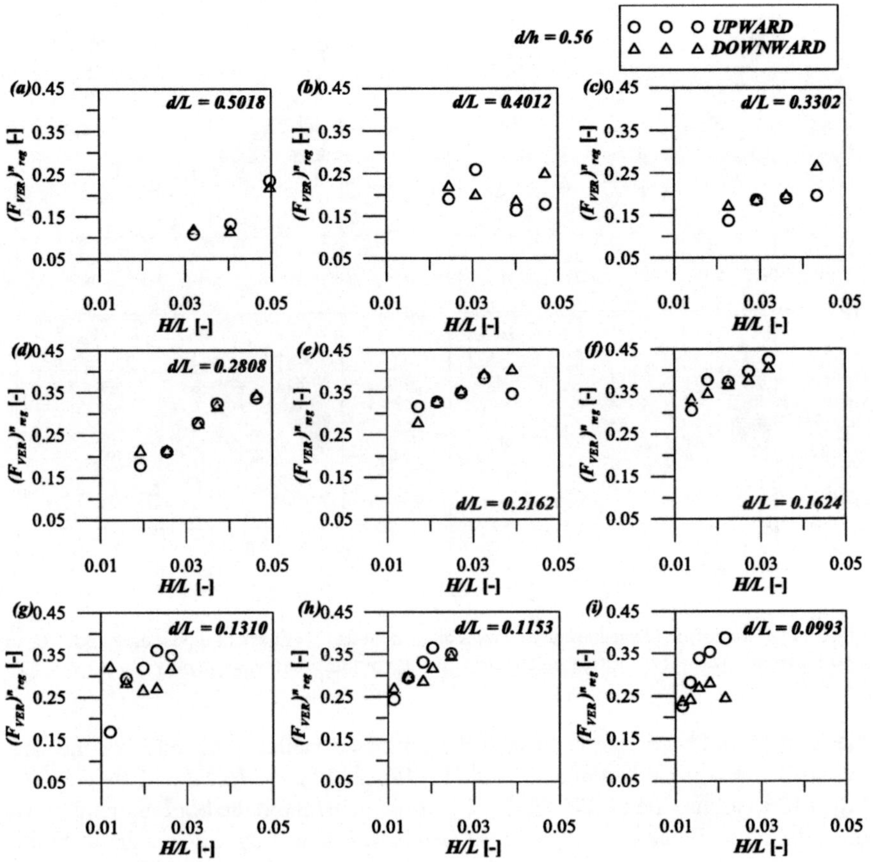

Fig. 6.16 Effect of wave steepness (H/L) on the total vertical wave force acting on the OWC device for different relative water depths

vertical wave force is found to be the least at d/L = 0.131 due to higher wave energy conversion. Further, with an increase in H/L, the reflection from the model increases, which contributed to the higher vertical wave force component from wave impact so that it induces higher $(F_{VER})^n{}_{reg}$ on the model. The upward and downward total vertical wave forces, $(F_{VER})^n{}_{reg}$, reached a maximum value of approximately 0.45.

6.5.7 Comparison of Measured and Estimated Horizontal Wave Force

To understand the difference in the nature and magnitude of the force between the OWC caisson model and a vertical wall, the horizontal wave forces measured in

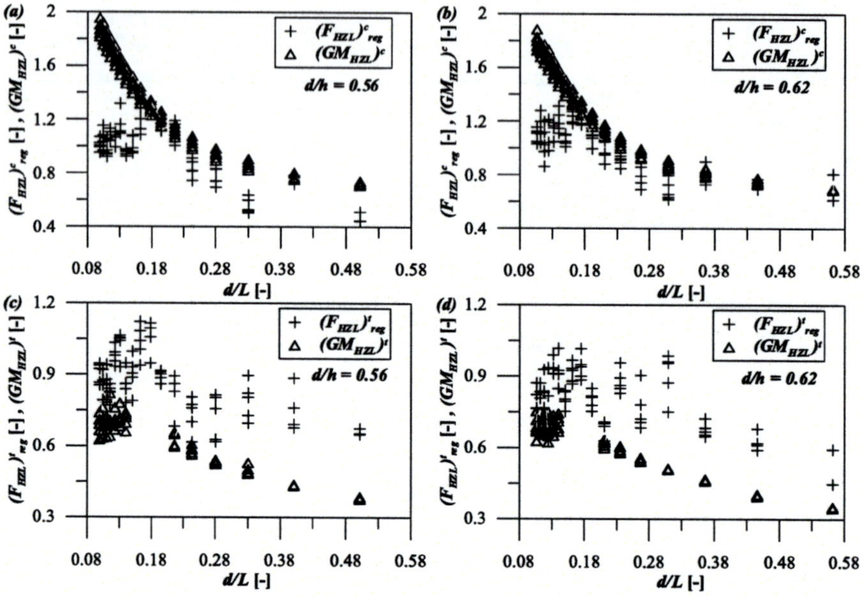

Fig. 6.17 Comparison of the measured values and the Goda [7] empirical model in quantifying the total horizontal wave force acting on the OWC device for different water depths

the present study are compared with the estimated values obtained from the well-established empirical formula of Goda [7] [Eqs. (6.1) and (6.2)] and are shown in Fig. 6.17 as a function of d/L. Goda [7] has not considered the bottom opening, and thus, the hydrodynamic behaviour between OWC and the caisson without bottom opening as considered by Goda [7] would be quite different. Attempts are made to quantify the difference in wave loads acting on them at the two different water depths tested. The results exhibit a consistent trend in their variation. In general, the method of Goda [7] over-predicts the shoreward peak forces, particularly at lower d/L, whereas it under-predicts the seaward peak forces. Because of the high wave power absorption at the low-frequency region, the partial standing wave formation in front of the OWC chamber led to a reduction in the shoreward total horizontal wave force. Further, it should be mentioned that the efficiency of the OWC system is found to be a maximum of 71% for $d/L < 0.15$. This could be the reason for the Goda method to over-estimate the total horizontal wave force. However, a closer examination of the results shows that for $d/L > 0.15$, the agreement between the measurement and estimation is reasonable. The shoreward total horizontal wave force for the two different water depths compared with $(GM_{HZL})^c$ shows that the difference between them is in the range of 46–90% for the lesser d/h of 0.56, and 27–80% for the larger d/h of 0.62. The seaward total horizontal wave force was observed to be always greater than $(GM_{HZL})^t$ by about 5–50%. Under the action of the wave trough, flow takes place from inside the chamber to the seaward, which causes an increased force on the rear side of the lip wall. This phenomenon could not be observed in the vertical wall because there is no obstruction for a wave trough receding from the structure.

To determine the effect of wave steepness on the wave forces on the OWC device, the variation of the normalized shoreward and seaward total horizontal wave forces, $(F_{HZL})^n{}_{reg}$, is compared with Goda [7] empirical formula as a function of H/L for a constant water depth of d/h = 0.56, as shown in Fig. 6.18. A comparison of the variation of $(GM_{HZL})^n$ and $(F_{HZL})^n{}_{reg}$ for different d/L shows that the former is larger than the latter, with the difference being higher for longer wavelengths (i.e., lesser d/L). In terms of the effect of wave steepness for each of the d/L values, in general, it is found that $(F_{HZL})^n{}_{reg}$ increases with an increase in H/L for d/L > 0.15, which is probably attributable to the nonlinear effect. For d/L < 0.15, the effect of H/L on

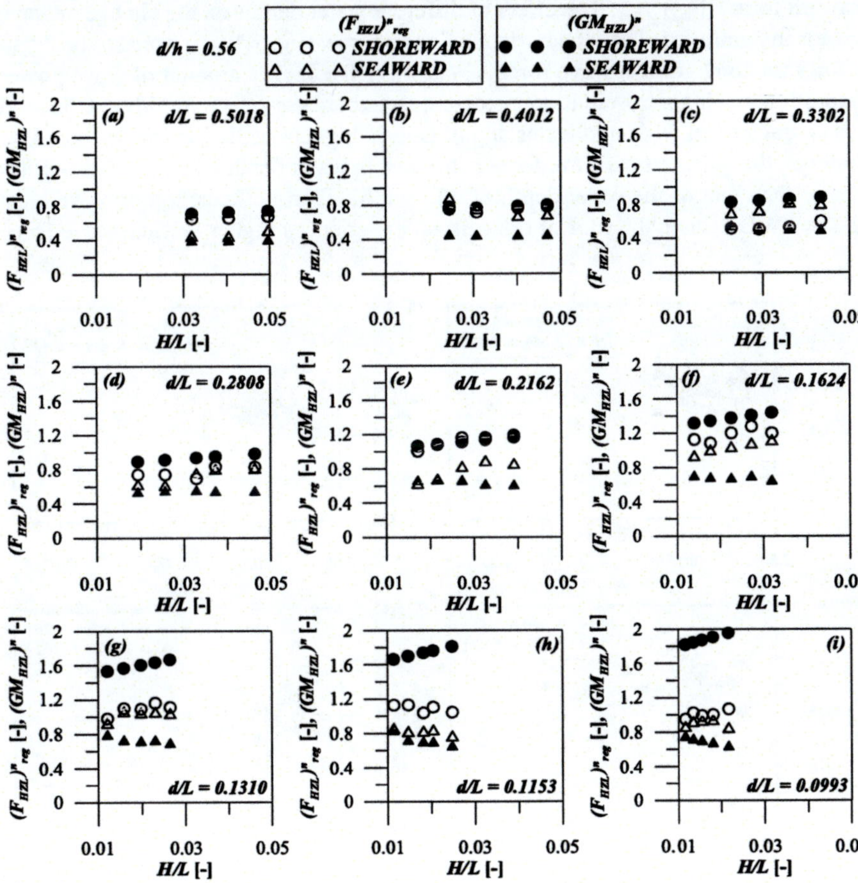

Fig. 6.18 Effect of wave steepness (H/L) on the total horizontal wave force acting on the OWC device compared with the Goda [7] empirical model

$(F_{HZL})^{n}_{reg}$ is found to be negligible in the low-frequency region because of the high wave power absorption.

6.5.8 Total Horizontal and Vertical Wave Forces Due to Random Waves

The changes in total horizontal and vertical wave forces acting on the OWC model subjected to random waves for different water depths are shown in Fig. 6.19. The trend in variation of total wave force under random waves is observed to be similar to the regular wave incidences along with a reduction in the magnitude of about 10%. The shoreward horizontal wave force, $(F_{HZL})^{c}_{ran}$, is observed to be greater than the seaward force, $(F_{HZL})^{t}_{ran}$. The effect of different water depths on the change in wave force is insignificant. The total horizontal wave force, $(F_{HZL})^{n}_{ran}$, is noted to be 2.5 to 3 times the total vertical wave force, $(F_{VER})^{n}_{ran}$. The higher amount of wave power absorption and higher oscillation inside the chamber takes place for d/h = 0.56, and it is characterized as design water depth. Due to higher oscillation inside the OWC chamber, the total vertical wave force at the design water depth is found to be slightly greater than that of the water depth of d/h = 0.62. During the action of wave crest on the OWC device, the roof of the caisson experiences higher vertical wave force

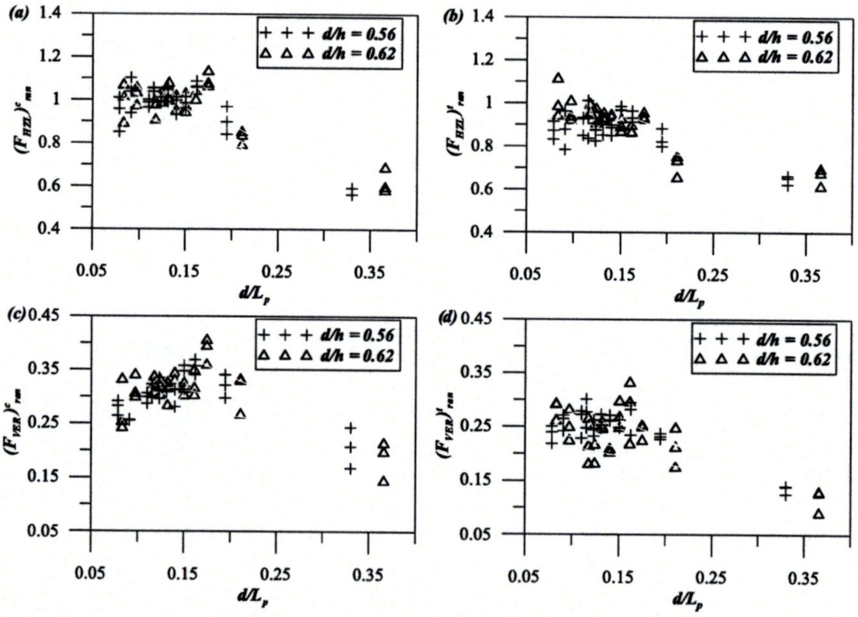

Fig. 6.19 The total horizontal and vertical wave forces acting on the OWC model subjected to random waves

due to compressed air than the air expansion during the trough of wave action. The shoreward and seaward horizontal wave forces, $(F_{HZL})^n{}_{ran}$, reached a maximum value of 1.1 and 1.0, respectively. At the same time, the total vertical wave force in the upward and downward direction, $(F_{VER})^n{}_{ran}$, reached the maximum value of 0.4 and 0.35, respectively.

6.6 Summary and Conclusions

The results from measurements of the dynamic pressures on the different components of the OWC device as well as the total force induced by regular and random waves on the structure are presented and discussed in this chapter. The following are conclusions drawn from the comprehensive study.

- The forces and the pressure outside the front wall of the OWC increase with wave steepness.
- The pressures on the rear wall of the chamber and the air pressure on the roof of the caisson have a reverse trend in their variation compared to the pressures on the front of the lip wall.
- The seaward pressure on the Lip wall should also be considered as a critical parameter in the design as it governs the stability toward the seaside. The design water depth [based on the given ratio between the bottom opening and the water depth (*o/d*)] is proposed as d/h = 0.56, as higher pressure on the rear wall is exerted and hence resulting in the availability of higher wave energy conversion.
- The average shoreward total horizontal wave force is 10% more than that of the seaward force.
- The magnitude of total horizontal wave forces is about 2.5 to 3 times greater than the total vertical wave forces.
- The wave load distribution on the OWC device exposed to random waves is 0% lesser than that due to regular waves.

References

1. Ashlin, S. J., Sannasiraj, S. A., & Sundar, V. (2014). Performance of an oscillating water column device with different bottom profiles subjected to random waves. In *Proceedings of 11th International conference in hydro science and engineering* (pp. 729–736). ICHE, Hamburg, 28 Sep–2 Oct 2014. ISBN-978-3-939230-32-8.
2. Ashlin, S. J., Sannasiraj, S. A., & Sundar, V. (2015). Wave forces on an oscillating water column device. In *Proceedings of the 8th international conference on Asian and Pacific coasts* (pp. 1019–1026). Chennai, India, 7–10 Sept (2015), Procedia Engineering. https://doi.org/10.1016/j.proeng.2015.08.336.
3. Boccotti, P., Arena, F., Fiamma, V., Romolo, A., & Barbaro, G. (2012). Small-scale field experiment on wave forces on upright breakwaters. *Journal of Waterway, Port, Coastal, and Ocean Engineering, 138*, 97–114.

4. BS 6349 (British Standards). (1984). *British standard code of practice for marine structures, Part 1* (4th ed.). London: British Standard Institution.
5. Dean, R. G., & Dalrymple, R. A. (1992). *Water wave mechanics for engineers and scientists*. Singapore: World Scientific.
6. Goda, Y., & Suzuki, Y. (1976). Estimation of incident and reflected waves in random wave experiments. In *Proceedings of the 15th coastal engineering conference*. Hawaii.
7. Goda, Y. (1985). *Random seas and design of maritime structures*. University of Tokyo Press.
8. Goda, Y., Kanda, K., Ohneda, H., Odani, H., Suzuki, H., Hirano, M., & Takahashi, S. (1990). Construction of a wave power extracting caisson breakwater for field experiments and electric power generation. In *Proceedings of the Pacific congress on marine science and technology PACON90*. PACON International, Honolulu.
9. Jayakumar, D. (1994). *Wave forces on oscillating water column type wave energy caisson—An experimental study* (Ph.D. Thesis). Ocean Engineering Centre, Indian Institute of Technology Madras, India.
10. John Ashlin, S., Sundar, V., & Sannasiraj, S. A. (2017). Pressures and forces on an oscillating water column—type wave energy caisson breakwater. *Journal of Waterway, Port, Coastal and Ocean Engineering, 143*(5), 1–18. https://doi.org/10.1061/(ASCE)WW.1943-5460.0000405.
11. John Ashlin, S., Sannasiraj, S. A., & Sundar, V. (2018). Performance of an array of oscillating water column devices integrated with an offshore detached breakwater. *Ocean Engineering, 163*, 518–532. ISSN 0029-8018. https://doi.org/10.1016/j.oceaneng.2018.05.043.
12. John Ashlin, S., Sannasiraj, S. A., Sundar, V., Kamath, A., & Bihs, H. (2019). Effects of power take-off damping and model scaling on the hydrodynamic performance of oscillating water column device. *Lecture Notes in Civil Engineering, 23*, 807–821. https://doi.org/10.1007/978-981-13-3134-3_60
13. Kamath, A., Bihs, H., & Arntsen, Ø. A. (2015). Numerical modeling of power take-off damping in an oscillating water column device. *International Journal of Marine energy., 10*, 1–16. https://doi.org/10.1016/j.ijome.2015.01.001
14. Krisna, K. A. P. (2019). *Wave loadings and scaling effects on and within an Oscillating Water Column (OWC) caisson breakwater*. Ph.D. thesis in Energy Systems. The University of Edinburgh.
15. Kuo, Y., Lin, C., Chung, C., & Wang, Y. (2015). Wave loading distribution of oscillating water column caisson breakwaters under non-breaking wave forces. *Journal of Marine Science and Technology, 23*, 78–87.
16. Malmo & Reitan (1985). Wave-power absorption by an oscillating water column in a channel. *Journal of Fluid Mechanics, 158*, 153–175.
17. McIver, P., & Evans, D. V. (1988). An approximate theory for the performance of a number of wave-energy devices set into a reflecting wall. *Applied Ocean Research, 10*, 58–65.
18. Muller, G. U., & Whittaker, T. J. T. (1995). *Wave load measurement on a prototype wave power device, Coastal structures and breakwaters* (pp. 365–372). London: Thomas Telford.
19. Nakada, H., Suzuki, M., Takahashi, S., & Shikamori, M. (1992). *Development test of wave power extraction caisson breakwater* (pp. 146–164). Civil Engineering in Japan.
20. Sainflou, G. (1928). Essai sur les digues maritimes, verticals. *Annales Ponts et Chausses, 98*(4).
21. Sundar, V., Sannasiraj, S. A., John Ashlin, S., Jegatheeswaran, B. (2014). An experimental study of an oscillating water column with different bottom configuration. In *Paper No: 798, Presented in 3rd IAHR Europe congress* (p. 96). University of Porto, 14–16 April 2014. ISBN-978-989-96479-2-3.
22. Takahashi, S. (1988). Hydrodynamic characteristics of wave-power-extracting caisson breakwater. In *Proceedings of the 21st coastal engineering conference* (pp. 2489–2501). ASCE, Spain.
23. Thiruvenkatasamy, K., Neelamani, S., & Sato, M. (2005). Nonbreaking wave forces on multi-resonant oscillating water column wave power caisson breakwater. *Journal of Waterway, Port, Coastal and Ocean Engineering, ASCE, 131*(2), 77–84.
24. USACE (U.S. Army Corps of Engineers). (1984). *Shore protection manual* (4th ed.). Washington, DC: U.S. Government Printing Office.

25. Wang, R. Q., Ning, D. Z., & Qing-ping Zou, Q. P. (2020). Wave loads on a land-based dual-chamber Oscillating Water Column wave energy device. *Coastal Engineering, 160*(103744), 1–15.
26. Wilbert, R., Sundar, V., & Sannasiraj, S. A. (2013). Wave interaction with double chamber oscillating water column device. *International Journal of Ocean and Climate Systems, 4*(1), 21–39.

Chapter 7
Hydrodynamic Performance Characteristics of U-OWC Devices

V. Sundar and S. A. Sannasiraj

Abstract OWC is reported to perform well when exposed to low frequency waves in narrow frequency band. On several occasions, the trough of the potential long wave might reach below the front lip wall of OWC, inducting air into the air chamber. To solve this problem, Boccotti (Boccotti, P. Ocean Engineering 34(5–6), 806–819 (2007a)) modified the conventional OWC concept by incorporating a duct in front of the opening with an intention was to capture the maximum magnitude of dynamic pressure available for oscillation inside the air chamber. This is widely referred to as U-OWC. The performance of U-OWC is explored in a laboratory wave flume. A Froude model scale of 1:20 was adopted to carry out a comprehensive experimental study to investigate the effect of water plane area on pneumatic damping with two aspect ratios (W/B, ratio of chamber width parallel to the wave crest to the width of the chamber in the wave direction) of 0.50 and 1.67 in the model geometry. To investigate the effect of dynamic pressure intensity variation with depth, three h/di of 0.15, 0.30, and 0.45 were considered in the study, where di is the water depth in the chamber. At each of the h/di, the effect of relative bottom opening (O/di) for better phase control and energy conversion was studied. The maximum impact of pressure excitation had occurred at h/di of 0.30 and b/B of 0.50. The performance of the U-OWC is found to be better over a relatively wider frequency bandwidth near the natural frequency of the system.

7.1 Introduction

The Oscillating Water Column (OWC), among the several wave energy convertors, is well-established with several proven field trails and hence stands a front to become the most sought device. The gap between the lip wall and the bottom of the

V. Sundar (✉) · S. A. Sannasiraj
Department of Ocean Engineering, Indian Institute of Technology Madras, Chennai, India
e-mail: vsundar@iitm.ac.in

S. A. Sannasiraj
e-mail: sasraj@iitm.ac.in

A. Samad et al. (eds.), *Ocean Wave Energy Systems*, Ocean Engineering & Oceanography 14, https://doi.org/10.1007/978-3-030-78716-5_7

OWC allows the incident waves to propagate with the water mass inside the caisson, responsible for the generation of pneumatic power. An increase in the front duct opening increases the dynamic pressure thus, enhancing the capture rate of energy leading to a reduction in the natural period of the system due to the lesser streamline length for the flow. Hence, the conventional OWC captures the energy efficiently from the high-frequency waves. A reduced bottom opening results in a reduction in the magnitude of dynamic pressure for oscillation and leads to an increase in the natural period of the OWC. Thus, the system performs well only during the propagation of low-frequency waves. Furthermore, in the event the trough of a wave passes the lip wall without being interrupted, the air gets sucked into the air chamber. To combat this problem, modified the conventional OWC concept by incorporating a duct in front of the opening was proposed by [8–9]. The intention was to capture the maximum magnitude of dynamic pressure available during the oscillation inside the air chamber. Later through a field study was carried out under limited wave conditions. Boccotti et al. [11, 12], Arena and Filianoti [1], Filianoti and Camporeale [15], Filianoti and Piscopo [16] have reported that the Double Chamber Oscillating Water Column (U-OWC), which is also referred to as U-OWC, has a better prospect of rectifying the limitations of Single Chamber OWC (SCOWC) which is widely referred to as OWC. In 2012, the construction details of the first full-scale U-OWC (REWEC3) incorporated in a breakwater of 19 caissons, each including 8 cells, 34 m long for the harbor of Civitavecchia, Italy, the details of which are discussed by Arena et al. [2]. Filianoti and Gurnari [17] measured the horizontal wave force on a small-scale U-OWC breakwater erected off the eastern coast of the Strait of Messina (Southern Italy) and claimed that the pressure formula of Boccotti [7] and Goda [14] tend to overestimate by up to 25% the actual horizontal loads on the breakwater.

Malara and Arena [19] applied a semi-analytical approach for calculating both the infinite frequency added mass and the retardation function for understanding the performance characteristics of U-OWC arrays. The approach provided the hydrodynamic parameters of an isolated U-OWC of finite length, as well as the added mass matrix and the retardation function matrix associated with a U-OWC array. A reduction in the power output was claimed limited to about 9% with respect to the case of isolated U-OWC. Ashlin et al. [5] attempted to validate a numerical model using REEF3D with an analytical model and field measurement of the oscillation of the water surface and air pressure inside a U-OWC device chamber from a series of field experiments carried out in the natural wave basin of Natural Ocean Engineering Laboratory (NOEL). REEF3D of Bihs et al. [6] is an open-source numerical model that solves the fluid flow problem using the Reynolds-averaged Navier–Stokes (RANS) equations along with the k–ω model for turbulence and the level set method for the free surface is utilized. The comparison between the numerical and the field measurements reported to be good. It was also reported that a larger bottom opening allows incrementing the fraction of extracted energy, while limited variation is observed in the case of different lip wall submergences.

The U-OWC plants, as detailed by Boccotti [8–10], Arena and Filianoti [1], Filianoti and Camporeale [15], Filianoti and Piscopo [16], as stated earlier, are breakwaters incorporated in an OWC with an additional vertical duct on the front wall. There are two types of U-OWC that have been widely studied: REWEC1 and REWEC3. The REWEC1 is fully beneath the sea surface, with water-driven turbines in the vertical duct, the details of which are reported by Boccotti et al. [13], Boccotti [8, 9], Filianoti and Camporeale [15], Filianoti and Piscopo [16]. The REWEC3 is in part beneath the sea level and in part above the sea level. The air pocket of the REWEC3 relates to the atmosphere through a tube where a Wells turbine is installed. The details of REWEC3 are reported by Boccotti [10], Filianoti and Camporeale [15].

Filianoti and Camporeale [15] has reported the performance characteristics of a U-OWC with a Wells turbine through small-scale field experiments.

To bring out the factors influencing the energy conversion process, an experimental investigation has been carried out on a scaled-down model. The definition sketches of OWC and U-OWC are projected in Fig. 7.1. In its physical form, this device consists of two chambers for the water flow in the energy conversion process. Hence it is termed as the Double Chamber Oscillating Water Column (U-OWC). Since the U-OWC device operates as a dynamic system and is under the influence of radiation damping and added mass in wave structure interaction, it is essential to know the influence of the front duct on its hydrodynamic performance [26]. A detailed experimental investigation was carried out on U-OWC in random waves to examine its energy conversion efficiency, the variation of dynamic pressures on the front and rear walls of the U-OWC, air pressure, spectral characteristics of the pressures, and the phase difference between the excitation pressure and air pressure as a function relative water depth.

(a) OWC device (b) U-OWC device.

Fig. 7.1 Definition sketch

7.2 Experimental Set-Up

7.2.1 General

As the gravity and inertia force influence the hydrodynamic behavior, Froude's similitude criteria were adopted. The model scale was selected based on the capacity of the wavemaker to generate waves of required characteristics. Based on the physics of wave structure interaction, Kaldenhoff [18] recommended a model scale within the range of 1:20 to1:60 or larger if laboratory facility permits. In line with the above requirement, also as per the guidelines of Sarmento and Thomas [21], a model scale of 1:20 in a water depth of 1 m was considered.

7.2.2 Details of the Models and Test Set-Up

The U-OWC unit consists of two chambers, as mentioned earlier, the front one is defined as the energy absorption chamber, while the rear one is defined as an energy conversion chamber. Four units of U-OWC, of which three smaller units each of size 0.30 × 0.60 × 1.45 m and one bigger size unit of dimensions of 1.0 × 0.60 × 1.45 m, were considered for simultaneous testing in the wave flume. The width of the energy absorption chamber for both smaller and bigger units was in conformity with the width of the energy conversion chamber. Each of the smaller units had the provision for varying its front energy absorption chamber breadth from 0.15 to 0.45 m in steps of 0.15 m and its height from 0.55 to 0.85 m in steps of 0.15 m. The parameters of interest considered in the investigation are the bottom opening (O) of the energy conversion chamber width (b) and the mouth clearance (h) of the energy absorption chamber. B and h were varied as 0.15, 0.30, and 0.45 m. The adjacent units were separated by plywood partitions to avoid the interference effect between them while performing the simultaneous testing. The plan and section of the model units are shown in Fig. 7.2. The positions of pressure transducers to measure vertical wall pressure, rear wall pressure, and air pressure in each of the U-OWC are marked in the plan and sectional view of the figure. To simulate the pneumatic damping introduced by a turbine in the prototype, an air hole having 0.65% of waterplane area was provided in the model units as per the suggestions of several investigators [4, 20, 22].

7.2.3 Test Facility

The experimental study was carried out in a wave flume of 72.5 m long, 2.0 m wide, and 2.7 m deep in the Department of Ocean Engineering, Indian Institute of Technology Madras, India. A wavemaker at one end of the flume can generate

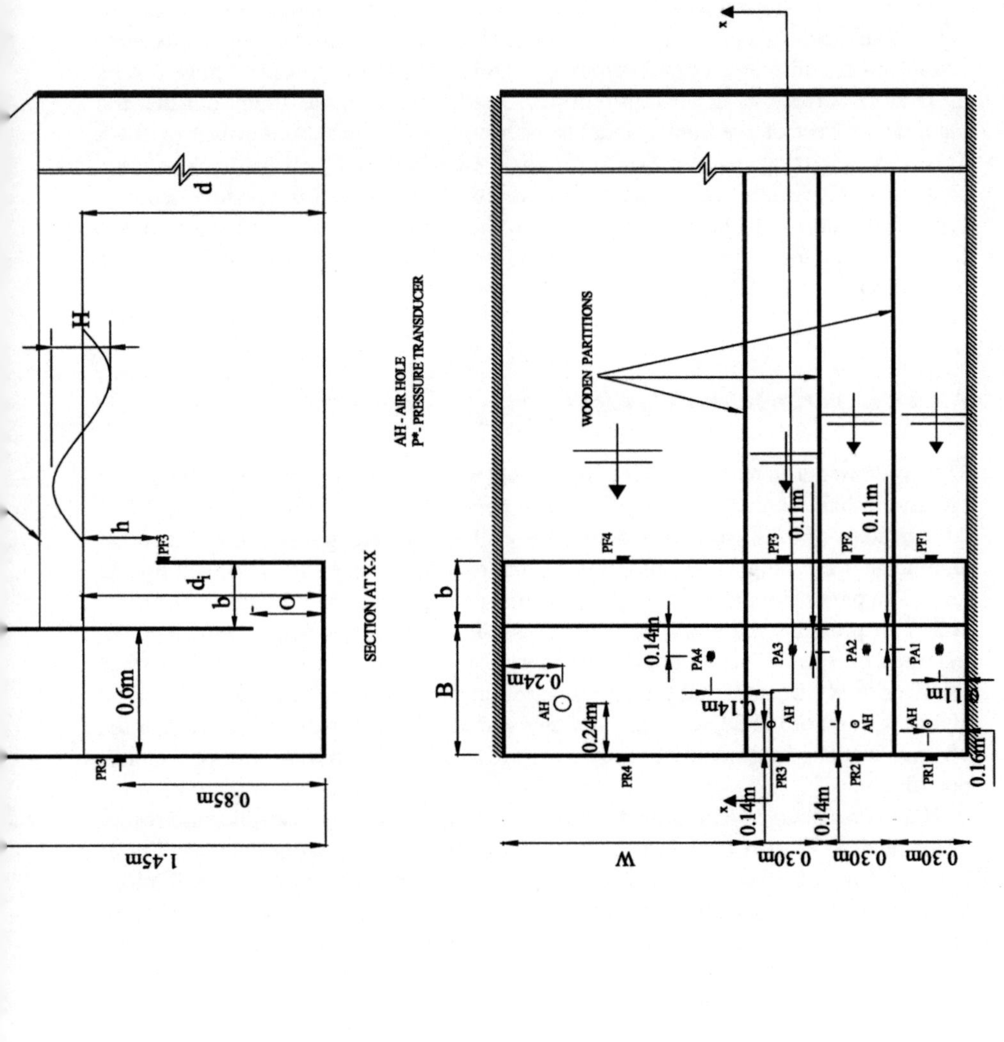

Fig. 7.2 Plan and sectional elevation of the model

random waves of predefined spectral characteristics, while the other end of the flume is provided with an absorber to absorb the incident water waves. The model units were rigidly fixed to the flume floor in the four compartments, as detailed in Fig. 7.2. To study the effect of waterplane area on pneumatic damping, two aspect ratios (W/B, a ratio of chamber width parallel to the wave crest to the width of the chamber in the wave direction) of 0.50 and 1.67 were considered in the model geometry. To investigate the effect of dynamic pressure intensity variation with depth, three h/di of 0.15, 0.30, and 0.45 were considered in the study, where di is the water depth in the chamber. At each of the h/di, the effect of relative bottom opening (O/di) for better phase control and energy conversion were investigated. Although tests were conducted for different combinations of regular and random waves, only results on random wave tests are herein reported. Furthermore, it was found out that the maximum effect of pressure excitation occurred at h/di of 0.30 and b/B of 0.50. To assess the performance of SCOWC (single chamber OWC) critically, the energy conversion efficiencies at the maximum possible relative bottom opening depth of 0.80 were studied. To have more clarity on the residual effect of reflection by the front wall, at O/di = 0.80, tests were conducted by keeping the front wall at h/di of 0.30. The results on the tests with regular waves are reported by [23–25].

7.2.4 Experimental Procedure

The experiments were carried out with random waves defined by Pierson Moskowitz spectrum characterized by the peak wave periods (T_P) varying between 1.2 and 2.0 s in steps of 0.2 s were considered. For each of the peak periods, three different significant wave heights, H_S of 0.055, 0.065, and 0.095 m, were employed for the tests. The parameters that were simultaneously measured from the four model units are (i) air pressure and rear wall water pressure in the energy conversion chamber, (ii) water pressure at the mouth of the energy absorption chamber, (iii) water surface oscillations at the centre of the energy absorption mouth, (iv) water surface oscillation at 4 m in front of the device and, (v) water surface oscillation at 9 m in front of the wavemaker. The time adopted for data collection was 160 s irrespective of the peak period.

Based on a detailed experimental investigation on the power absorption capability of the device, Wilbert et al. [23] concluded that the power absorption capability of the device increases with an O/d value of 0.30 and above, the dynamic oscillations inside the OWC are independent of the volume of the device, and the power absorption efficiency reaches closer to 100 at higher wave periods exhibiting higher wave amplification near the mouth of the device. The efficiency of the system, that is, the ratio of hydrodynamic power to wave energy flux is reported as a function of relative opening depth, relative wave heights, and frequency parameter has been discussed by Wilbert et al. [25].

This chapter focuses on the effect of the elevation of the mouth of the U-OWC device on its efficiency through the tests conducted without the front duct to simulate

SCOWC. For a relative bottom opening, O/di of 0.45, the energy conversion capacity of U-OWC is nearly 10% higher than that for a SCOWC. The maximum energy conversion capacity reaching nearly 78% established the possibility of U-OWC also being a technological solution in wave energy conversion programs. The energy conversion efficiency significantly depends on h/di, and the maximum performance has been observed for h/di = 0.30. The effect of O/di in combination with h/di was found to have a significant influence on the efficiency of the device. The phase variation between the excitation pressure and the air pressure for varying O/di and h/di demonstrated that by suitably selecting the depth of water di within the device, the tuning effect could be brought forth. The wave amplification in front of U-OWC reaches about 1.2, and it indicates the potential stability of the device to absorb the severe wave impact in critical environmental conditions.

7.3 Results and Discussion

7.3.1 *Spectral Width Parameter*

The frequency distribution of the variation of front wall pressure, rear wall pressure, and air pressure relative to the incident wave asserts the necessity of understanding the influence on the relationship between O/d_i and h/d_i. A characteristic feature of the spectrum, i.e., the spectral width parameter of the wave elevation, η, (ε_η), has been analysed to understand the energy capturing capacity of the device from the various frequency bands. The variation of spectral width parameter of the incident wave, ε_η for h/di = 0.15, 0.30 and 0.45 is shown in Fig. 7.3a–c. The ε_η is found to be close to 0.50 for all the tests.

7.3.2 *Dynamic Pressures*

7.3.2.1 General

Typical plots showing the variation of the spectral density of η, p_f, p_r, and p_a for random waves defined by $H_s/L_P = 0.017$ for the three different O/d_i and $h/d_i = 0.15$, 0.30, and 0.45 are shown in Fig. 7.4a, b, and c, respectively. The peak front wall pressure is higher by about 40% of the peak of incident energy for O/d_i and h/d_i of 0.15. The above-said difference is found to decrease with an increase in O/d_i for a constant h/d_i. This may be attributed to the fact that the larger depth of submergence of the front wall induces more reflection of incident wave energy. The peak rear wall pressure reaching around 50% of peak incident wave energy for h/d_i of 0.30 and O/d_i of 0.45 indicates that the system performs well at these system parameters for the wave frequencies up to resonant wave frequencies. In general, the peak air

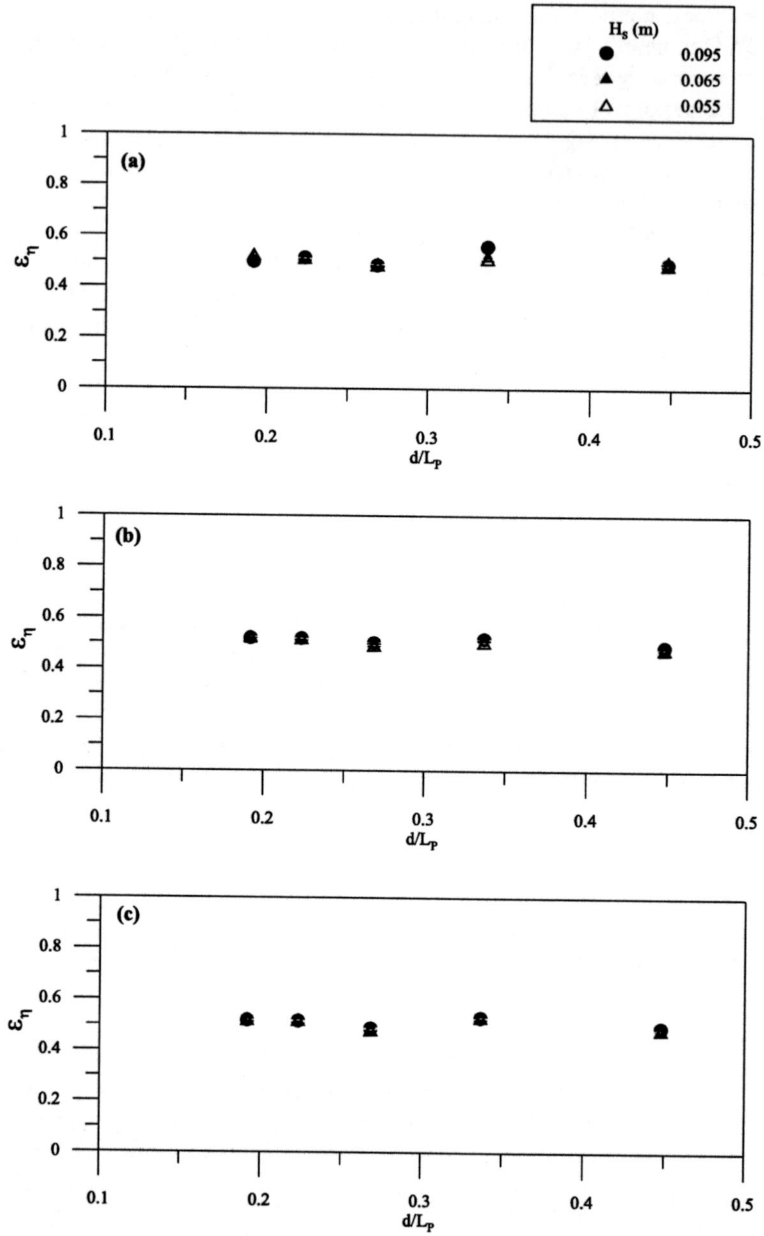

Fig. 7.3 Spectral width of incident random waves at (**a**) $h/d_i = 0.15$, (**b**) $h/d_i = 0.30$ and (**c**) $h/d_i = 0.45$

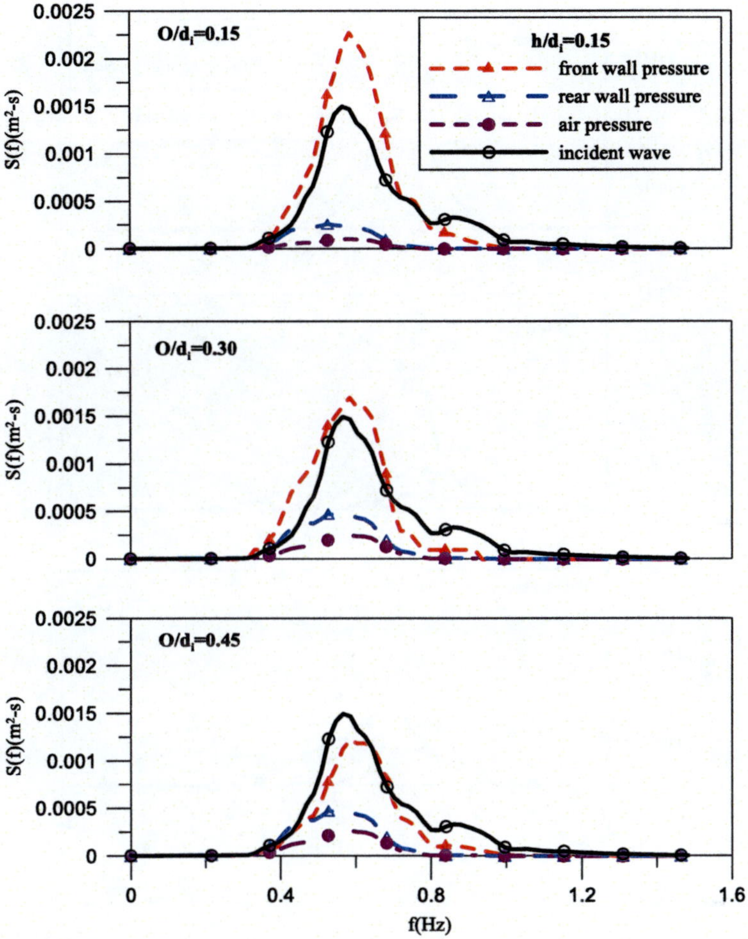

Fig. 7.4 **a** Variations of incident wave and pressure spectra for $h/d_i = 0.15$. ($H_s = 0.065$ m and $T_p = 1.6$ s). **b** Variations of incident wave and pressure spectra for $h/d_i = 0.30$. ($H_s = 0.065$ m and $T_p = 1.6$ s). **c** Variations of incident wave and pressure spectra for $h/d_i = 0.45$. ($H_s = 0.065$ m and $T_p = 1.6$ s)

pressure is found to be about 50% less than the rear wall peak pressure. Further, as h/d_i increases for a constant O/d_i, the front wall pressure keeps decreasing compared to the incident wave energy. As the depth of the pressure port is away from the free surface, the pressure decay is evident. From the above discussion, it is confirmed that hydrodynamic efficiency is subjected to the system configuration, and to bring more clarity, further analysis has been done over the system parameter variables, which are discussed below.

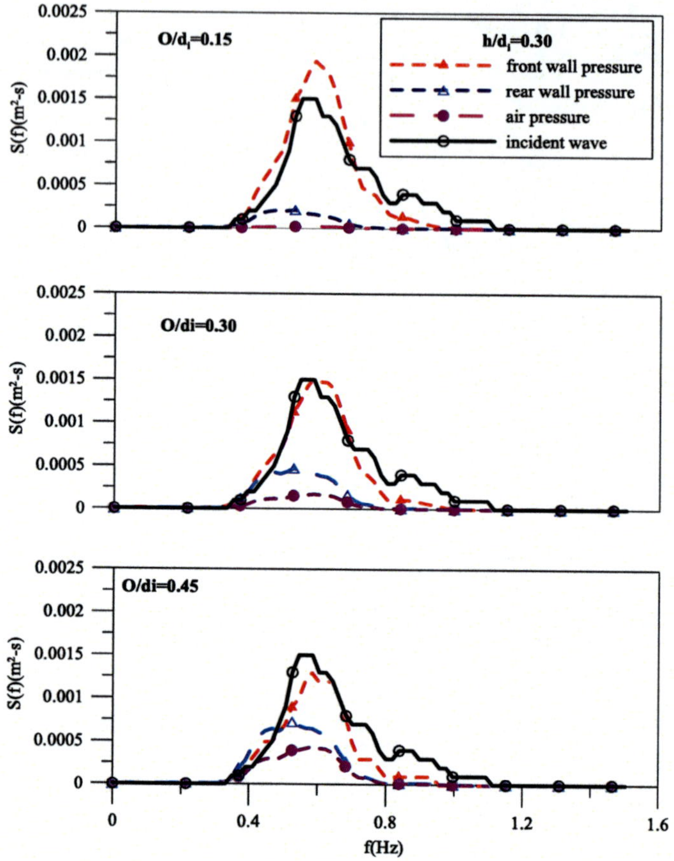

Fig. 7.4 (continued)

7.3.2.2 Front Wall Pressures

To understand the interaction effects between frequency components on energy conversion, the spectral width parameter for the front wall pressure spectrum ε_{p_f} is presented as a function of the relative water depth (d/L$_P$) for O/d$_i$ = 0.15, 0.30, and 0.45, and three different h/d$_i$ are shown in Fig. 7.5a–c, respectively. ε_η adopted for all the tests was about 0.5. However, only for h/d$_i$ of 0.3 ε_{p_f} is found to be close to 0.5 (least probably due to wave amplification at the mouth), and for the other h/d$_i$, a significant increase ε_{p_f} is seen in particular for d/L$_p$ greater than about 0.35.

The efficiency of the system is found to be good if ε_{p_f} it is closer to ε_η, which is seen for h/d$_i$ = 0.30 for the entire range of d/L$_P$. Further, higher ε_{p_f} attributed to the lack of energy concentration around the peak frequency is seen to be associated with short period waves, i.e., higher d/L$_p$, which is expected to decay fast. The hydrodynamic source causing the oscillation inside the chamber is being attenuated due to the

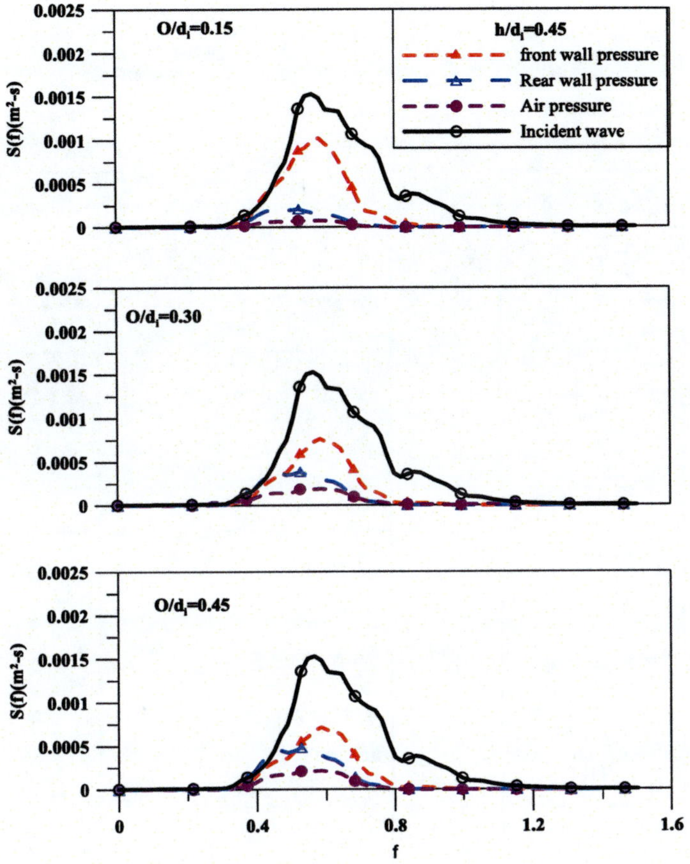

Fig. 7.4 (continued)

interaction of higher frequency components. Further, for $h/d_i = 0.30$, the mouth elevation at this level would not cause an interference effect on pressure excitation due to low-frequency waves. To bring in more clarity over the wave structure interaction, the ratio of zeroth spectral moment of front wall pressure, $(m_0)_{pf}$ to that of the incident wave, $(m_0)_\eta$, hereinafter termed as '$[(m_0)_r]_{front}$' is presented as a function of the relative water depth d/L_P for $O/d_i = 0.15$, 0.30 and 0.45 and three different h/di as shown in Fig. 7.6a–c, respectively. The datasets were separated into two different ranges of H_s/L_p. This ratio varies between 0.2 and 1.8 and is found to be large for less d/L_p since low-frequency waves exert more pressure on an obstruction. The propagation of waves felt closer to the surface where the dynamic pressures are maximum, and hence lesser relative depth of submergence of the pressure port location results in a higher $[(m_0)_r]_{front}$. Evidently, as d/Lp increases, a drastic decrease in $[(m_0)_r]_{front}$ is observed due to the pressure decay of high-frequency components. The pressure values of $[(m_0)_r]_{front}$ below one may be due to the obstruction effect

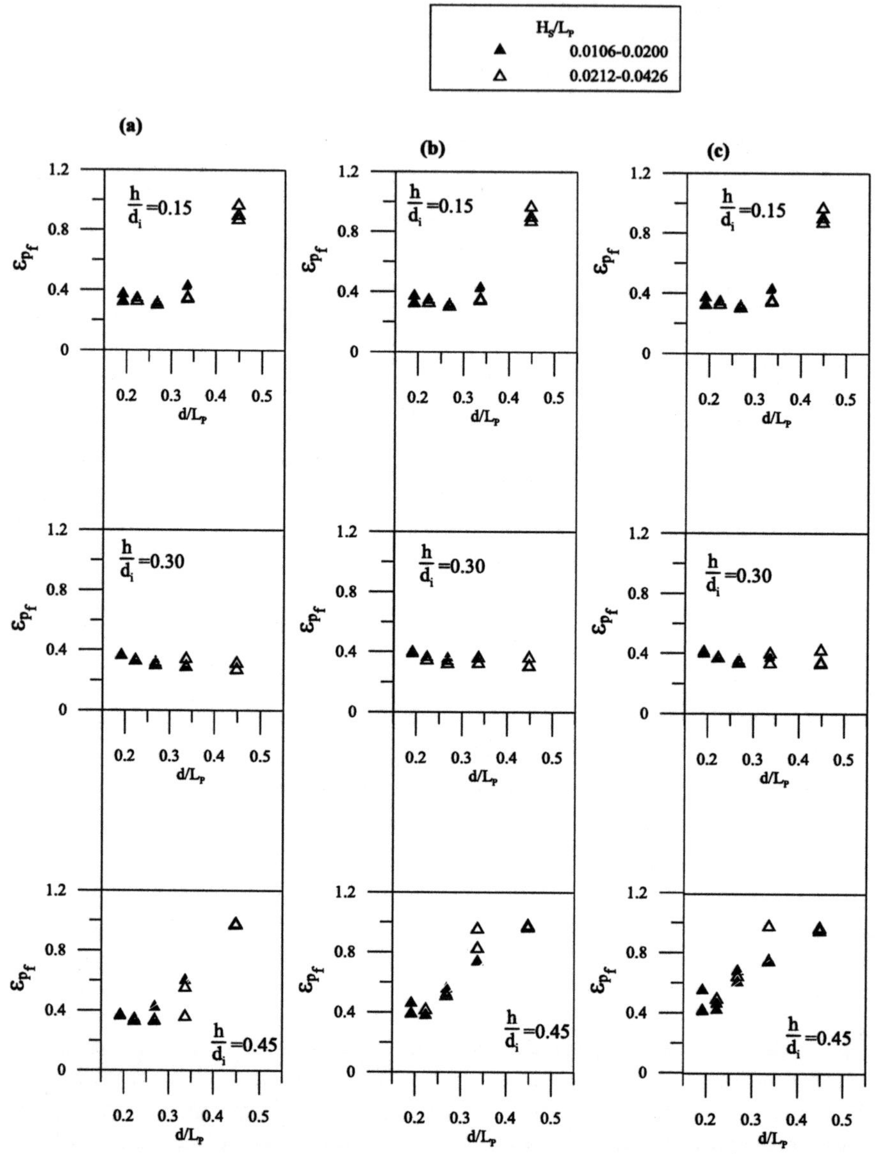

Fig. 7.5 Effect of relative opening depth and relative mouth elevation on spectral width of front wall pressure, **a** $O/d_i = 0.15$, **b** $O/d_i = 0.30$, **c** $O/d_i = 0.45$

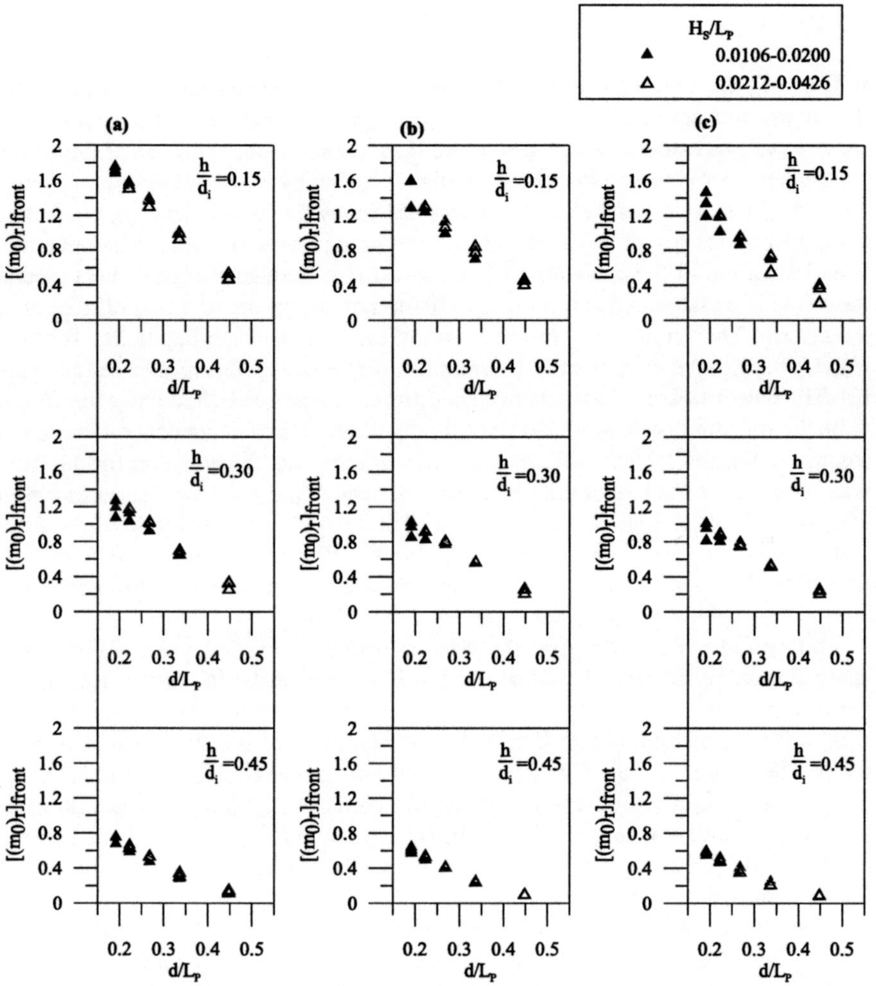

Fig. 7.6 Variation of zeroth spectral moment of front wall pressures with respect to the zeroth spectral moment of incident wave, (**a**) $O/d_i = 0.15$, (**b**) $O/d_i = 0.30$, (**c**) $O/d_i = 0.45$

and the corresponding reflection effect by the front wall induces on the path of the wave, whereas values above one may be attributed to the effect of the dynamic wave set up that would occur near the mouth when the power absorption capacity of the device is low. It can be concluded that irrespective of mouth elevation, a U-OWC system with an O/d_i of 0.15 is expected to be inefficient.

7.3.2.3 Rear Wall Pressure

In U-OWC, the magnitude of free surface oscillation inside the chamber determines the energy conversion capacity of the device. Here the spectral width parameter of the rear wall pressure spectrum provides information about the effects of frequency components that can enter into the oscillation chamber by virtue of the system parameters. An increase ε_{p_r} would indicate less efficiency due to the decay of frequency components closer to the resonance. On the contrary, the narrowness of the spectrum is an indication of the concentration of most of the frequency components around the resonance that would result in amplification of energy and thus the efficiency of the system. The variation ε_{p_r} as a function of the relative water depth d/L_P for $O/d_i = 0.15, 0.30$, and 0.45 and three different h/d_i are shown in Fig. 7.7a–c, respectively [3]. The data sets are separated into two different ranges of H_s/L_p. Only for $d/L_P < 0.30$, the spectral distribution of rear wall pressure is mainly dependent on excitation frequency for all O/d_i and h/d_i. Although the spectral width parameter for the front wall pressure for the h/d_i of 0.30 is found to around 0.5, ε_{p_r} it keeps increasing with d/L_p for h/d_i of 0.15 and 0.45. For $h/d_i = 0.30$, a systematic variation of the front wall pressure is observed except for O/d_i of 0.45. While ε_{p_r} and ε_{p_f} for h/d_i of 0.30 and O/d_i of 0.45 are close to ε_η, one can anticipate the maximum efficiency for a combination of the system parameters.

To bring in more clarity over the wave structure interaction, the ratio of zeroth spectral moment of rear wall pressure, $(m_0)_{pr}$ to that of the incident wave, $(m_0)_\eta$, hereinafter termed as '$[(m_0)_r]_{rear}$' as a function of the relative water depth (d/L_P) for $O/d_i = 0.15, 0.30$ and 0.45 and three different h/di is shown in Fig. 7.8a–c, respectively. The datasets were separated into two different ranges of H_s/L_p. A similar trend in its variation has been observed with the variation of $[(m_0)_r]_{front}$. The magnitude of the rear wall pressures for h/d_i of 0.15 and 0.45 are found to be drastically decreasing and is found to be 2–4 times less than the front wall pressures. This only indicates the loss of energy during the wave propagation into the chamber. It is interesting to note that for $O/d_i = 0.45$ and $h/d_i = 0.30$, the pressures on both the walls are of the same order, indicating that the said dimensions are likely to result in higher efficiency.

7.3.2.4 Air Chamber Pressure

In the primary energy conversion process, the physics connected with the air chamber is complex due to the interaction effect between the water column within the air chamber and atmospheric air above the water column inside the chamber. It influences the performance characteristics in two ways: by applying pneumatic damping in addition to the secondary flow that can cause effects like the radiation damping of floating structures. The variation of ε_{p_a} as a function of the relative water depth d/L_P for $O/d_i = 0.15, 0.30$, and 0.45 for three different h/d_i is shown in Fig. 7.9a–c, respectively. Two different ranges of wave steepness (H_s/L_p) are grouped to evaluate the importance of larger wave height. In general, for all h/d_i and O/d_i, ε_{p_a} it is found to be higher than 0.5. However, for $h/d_i = 0.3$ and $O/d_i = 0.45$, ε_{p_a} is found to be closer

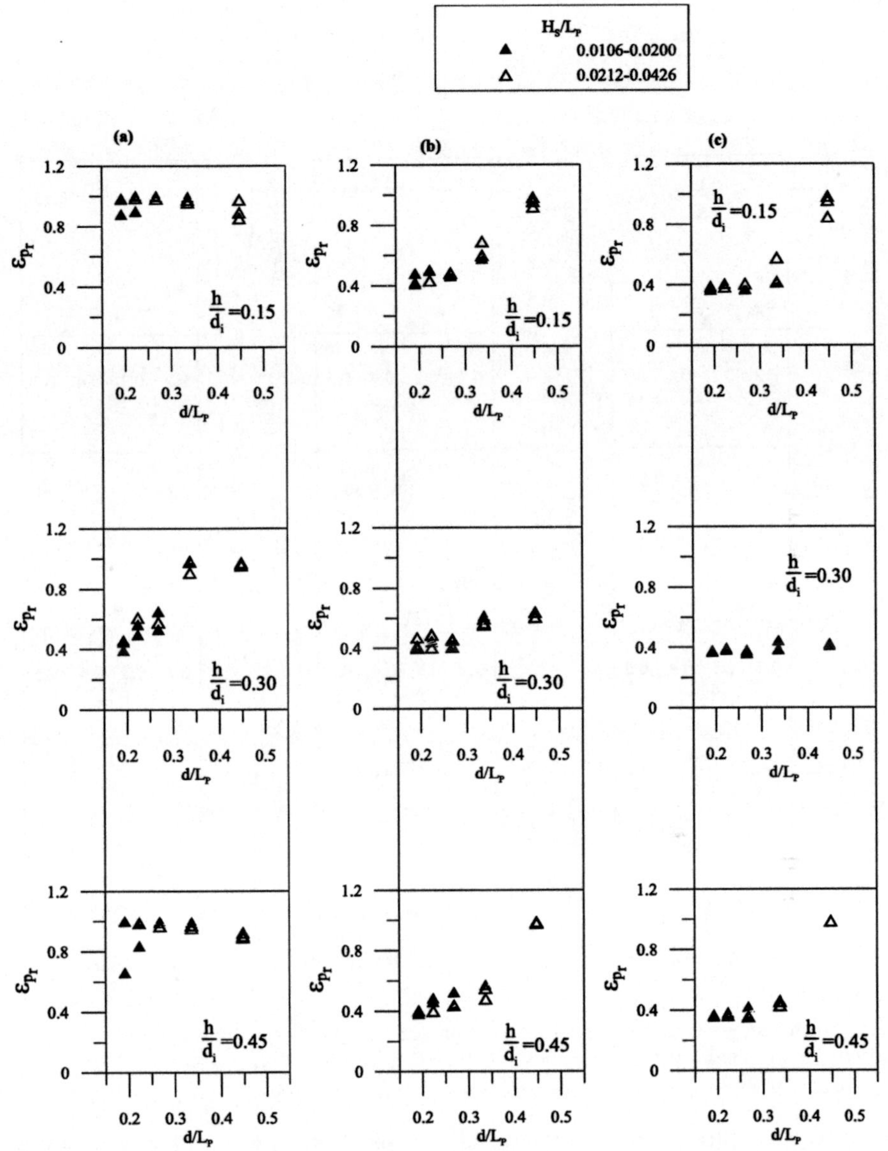

Fig. 7.7 Effect of relative opening depth and relative mouth elevation on spectral width of rear wall pressure, **a** $O/d_i = 0.15$, **b** $O/d_i = 0.30$, **c** $O/d_i = 0.45$

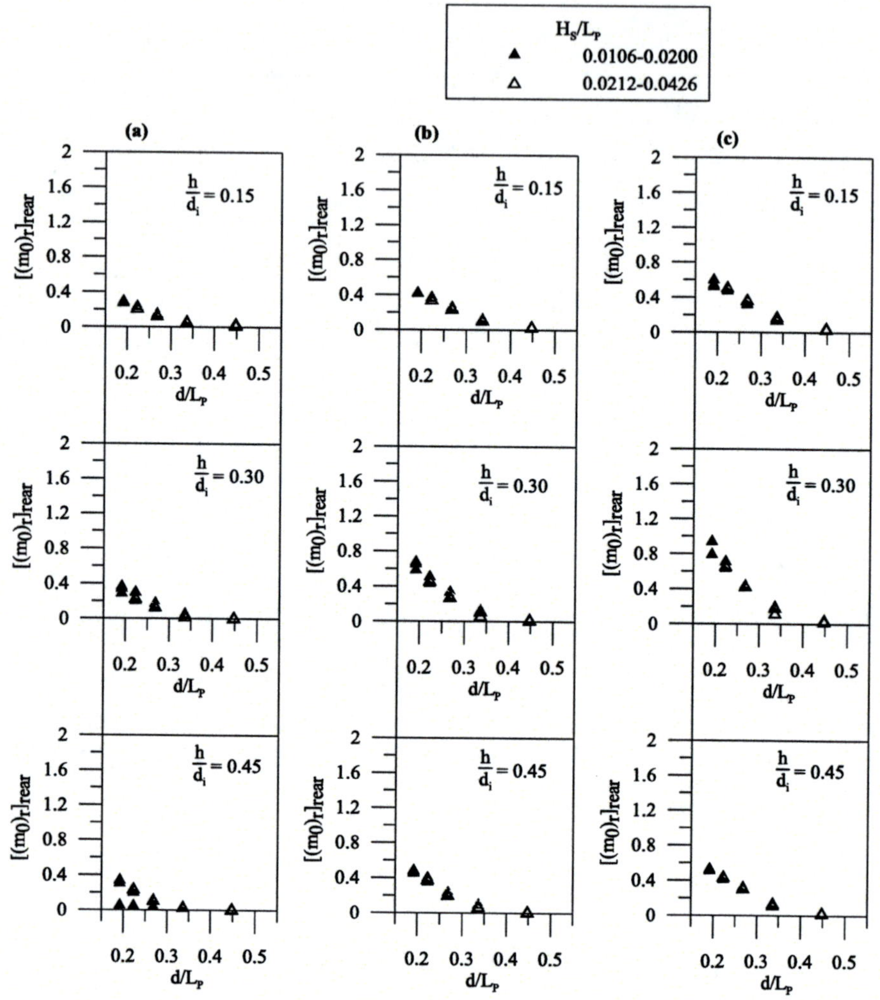

Fig. 7.8 Variation of zeroth spectral moment of rear wall pressures with respect to the zeroth spectral moment of incident wave, **a** $O/d_i = 0.15$, **b** $O/d_i = 0.30$, **c** $O/d_i = 0.45$

to ε_η of 0.5, which could be the probable combination for better efficiency of the system. The variation of the ratio of zeroth spectral moment of the air pressure, $(m_0)_{pa}$ to that of the incident wave, $(m_0)_\eta$, hereinafter termed as '$[(m_0)_r]_{air}$' as a function of the relative water depth d/L_P for $O/d_i = 0.15$, 0.30 and 0.45 for three different h/d_i is shown Fig. 7.10a–c, respectively. The U-OWC having $h/d_i = 0.30$ and $O/d_i = 0.45$ gives better performance for the range of wave conditions. For justifying the above-mentioned aspects of U-OWC, a quantitative assessment is carried out in terms of average power development under varying wave conditions.

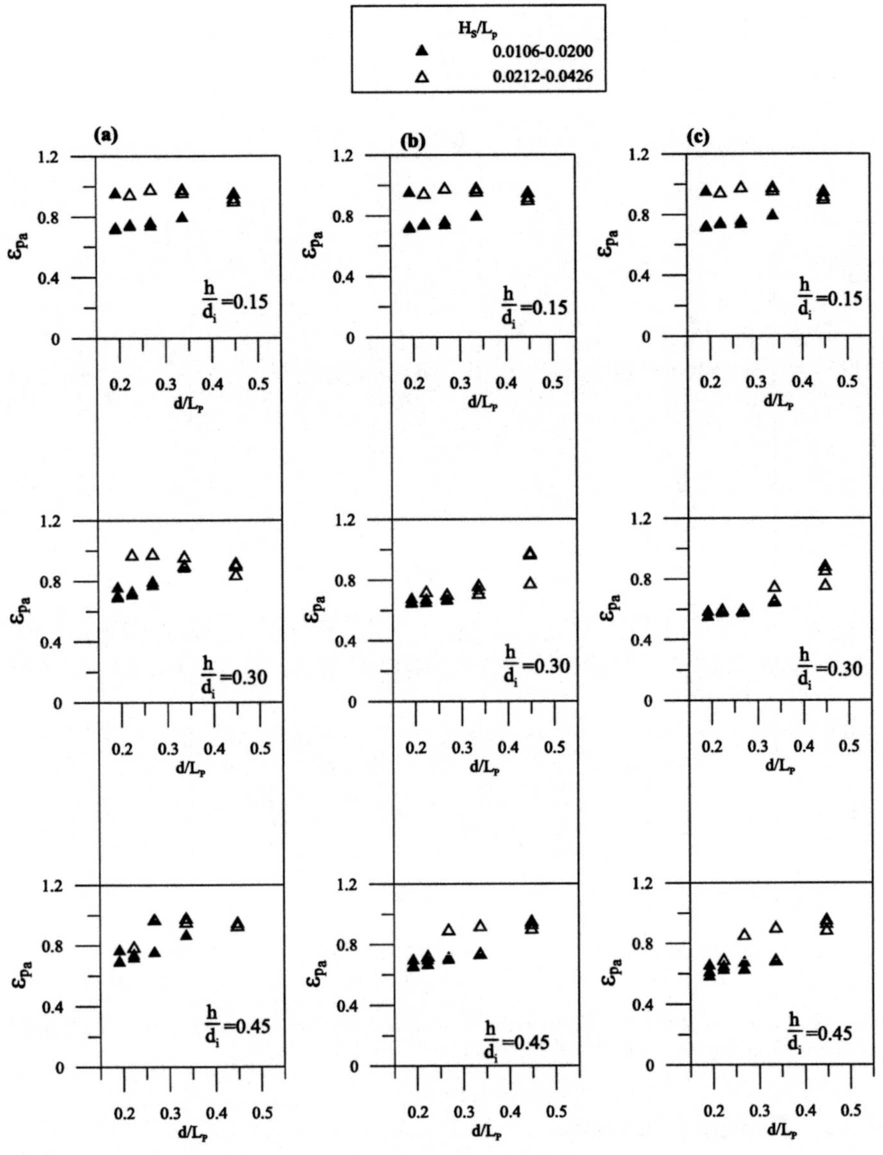

Fig. 7.9 Effect of relative opening depth and relative mouth elevation on spectral width of air pressure, **a** $O/d_i = 0.15$, **b** $O/d_i = 0.30$, **c** $O/d_i = 0.45$

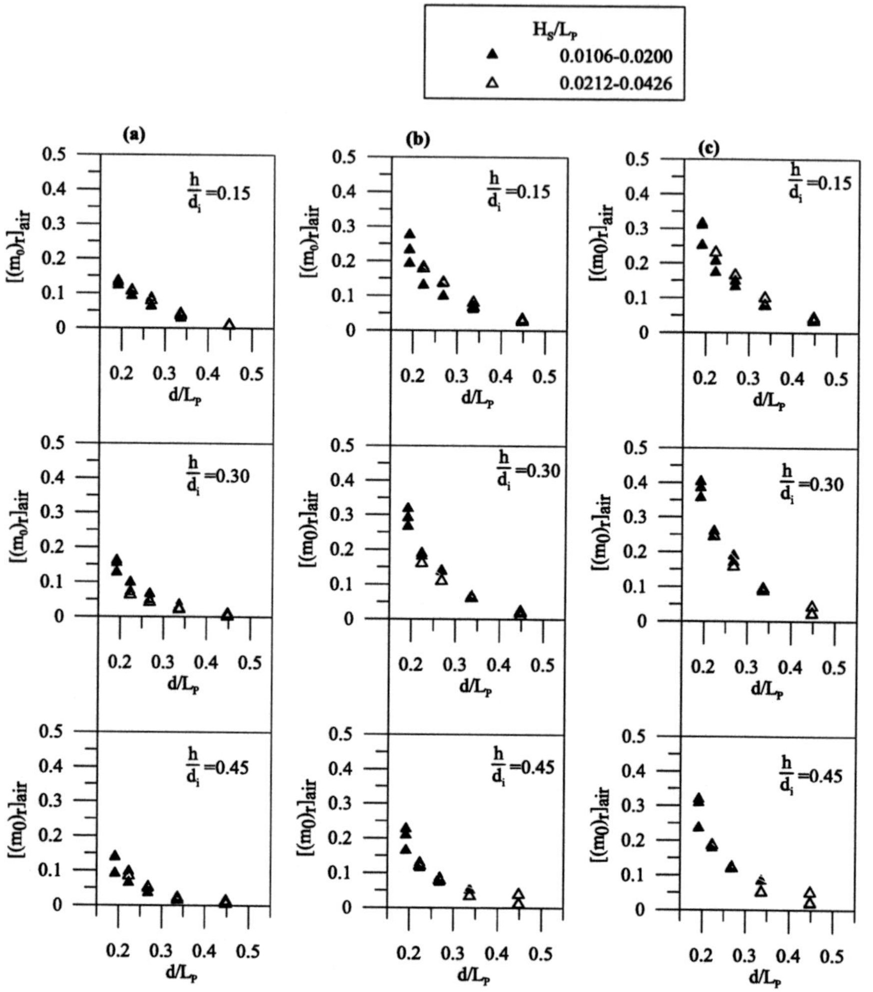

Fig. 7.10 Variation of zeroth spectral moment of air pressures with respect to the zeroth spectral moment of incident wave **a** $O/d_i = 0.15$, **b** $O/d_i = 0.30$, **c** $O/d_i = 0.45$

7.3.3 Energy Efficiency

Under the random wave incidence, the air pressure and the velocity of the free surface are also random in nature. Thus, the device must be designed for average power development. The average power (P_{ave}) is calculated from the significant velocity of free water surface oscillation (v_O) and significant air pressure (p_O) following the general methodology as,

$$P_{ave} = \frac{1}{2} p_0 A v_0 \tag{7.1}$$

wherein, A is the cross-sectional area of the oscillation chamber. The incident energy flux (P_{in}) is calculated using significant wave height and group celerity (C_g) corresponding to peak wave period as,

$$P_{in} = \frac{1}{8} \rho g H_S^2 C_g. \tag{7.2}$$

The energy conversion capacity (λ) is determined as

$$\lambda = \frac{P_{ave}}{P_{in}} \times 100\% \tag{7.3}$$

The variation of energy conversion capacity (λ) with the relative water depth (d/L_P) for three h/d_i and $O/d_i = 0.15, 0.30$, and 0.45 are presented in Figs. 7.11, 7.12 and 7.13 under a, b and c, respectively. λ achieves relatively higher values for the highest O/d_i under each h/d_i. This would mean a larger bottom opening allows more penetration of energy into the chamber. However, a further increase in the bottom gap allows a deeper trough to pass through without resulting in wave energy conversion. U-OWC is found to be more effective in absorbing the long waves. The effect of O/d_i is found to reduce with an increase in d/L_P. The variation in λ can be explained by the linear circuit theory of electrical engineering as it has been explained for regular waves. The lowest performance for U-OWC with $O/d_i = 0.15$ and for all h/d_i deserves special attention compared to $O/d_i = 0.30$ and 0.45. The trend can be attributed mainly to the constraint in the flow due to the low area of the bottom opening. Here, the frictional losses are not considered.

Even though the function of the front duct is to capture the high magnitude of incident wave-induced dynamic pressure. It also determines the natural frequency of the system. Since the variation of dynamic pressure under the wave is hyperbolic, the mouth elevation has to be relatively at a higher elevation in order to absorb maximum pressure. The study, however, yields a better efficiency for $h/d_i = 0.30$ than that for 0.15 and 0.45. The possible reason for the low performance for $h/d_i = 0.15$, apart from the variation in the phase difference, is the additional reflection due to the duct wall. The interaction of the reflected wave with the incident wave might further reduce the energy absorption, even though the dynamic pressure might be high due to the likely formation of a standing wave. Boccotti [11, 12], Boccotti et al. [13] also proposed a critical limit for the mouth elevation for optimum capture of pressure excitation.

The performance characteristics detailed in Fig. 7.13 for $h/d_i = 0.45$ agree with the hydrodynamic theory of pressure decay with increasing depth of submergence. For the lowest wave frequency parameter (d/L_P), λ takes about 40% for $O/d_i = 0.45$, while λ for $O/d_i = 0.30$ is about 25%. As far as the geometry of the device is concerned, this configuration with $h/d_i = 0.45$ produced a lesser phase difference between pressure

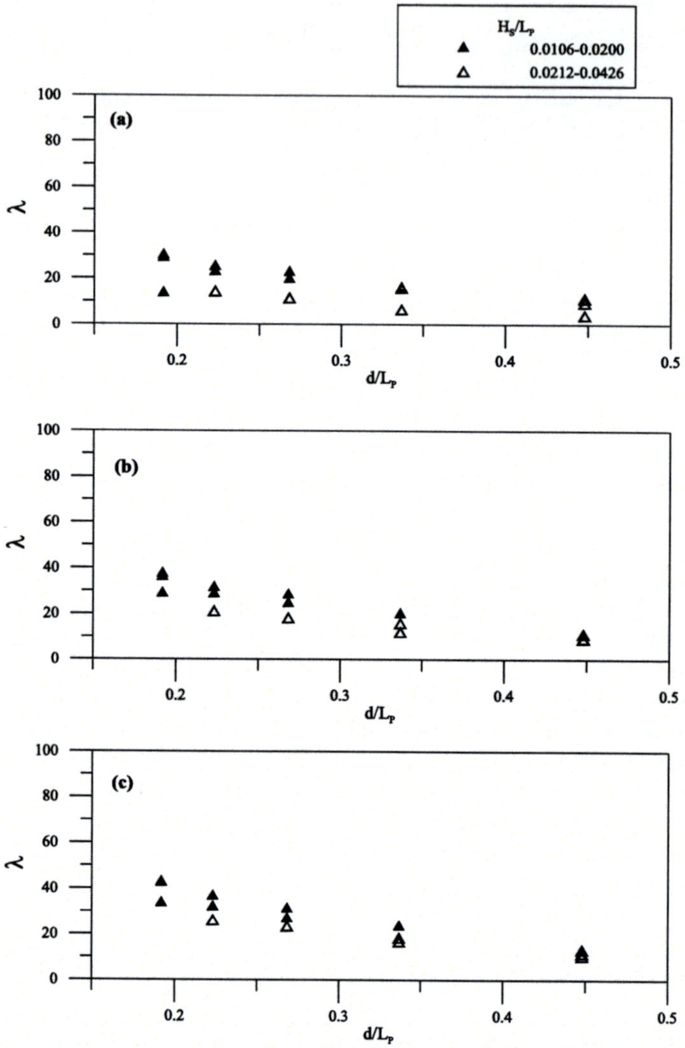

Fig. 7.11 Energy conversion against frequency parameter at $h/d_i = 0.15$, **a** $O/d_i = 0.15$, **b** $O/d_i = 0.30$, **c** $O/d_i = 0.45$

excitation and air pressure, but the advantage is offset by the decrease in dynamic pressure intensity.

To evaluate the effectiveness of U-OWC in energy conversion under random waves, the efficiency computed under specific device configuration is compared with its counterpart values for regular waves in Fig. 7.14a and b. In view of the easiness of understanding, the efficiency under the maximum wave height is considered here. The results prove conclusively that U-OWC can convert energy from random waves.

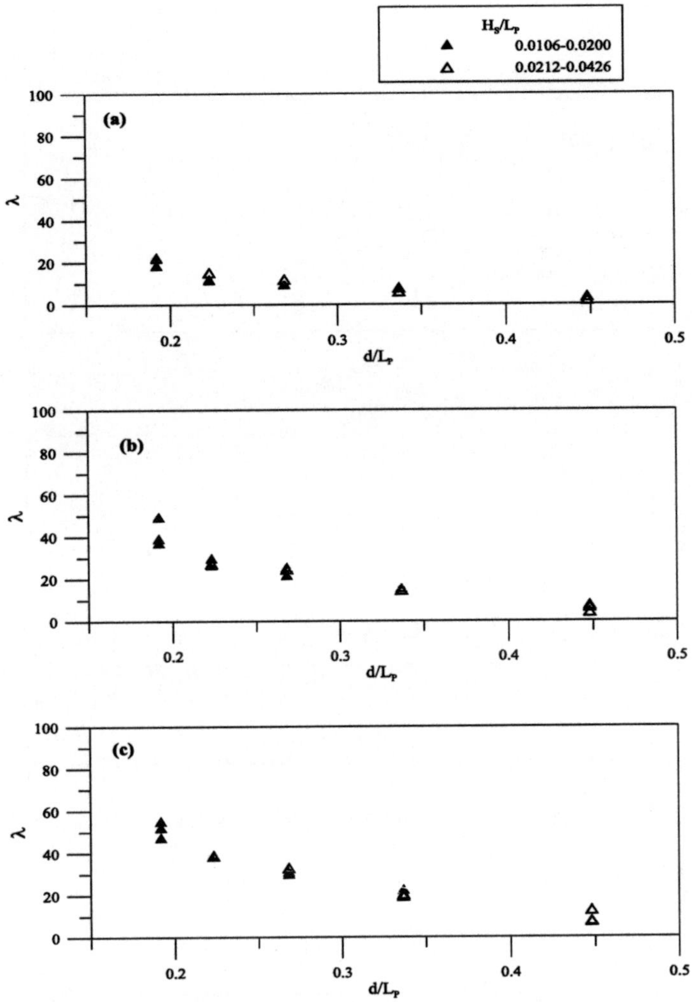

Fig. 7.12 Energy conversion against frequency parameter at $h/d_i = 0.30$, **a** $O/d_i = 0.15$, **b** $O/d_i = 0.30$, **c** $O/d_i = 0.45$

7.3.4 Air Pressure Variation

The pneumatic pressure developed inside the oscillation chamber rotates the turbine, which remains coupled with the electric generator. In random sea conditions, the oscillation inside the air chamber assumes a range of values. Thus, generally, the turbine is designed to work under average conditions that, in general, can be the significant air pressure. When the air pressure reaches the maximum level, there might be adverse effects of turbine stalling and subsequent functional shutdown of power

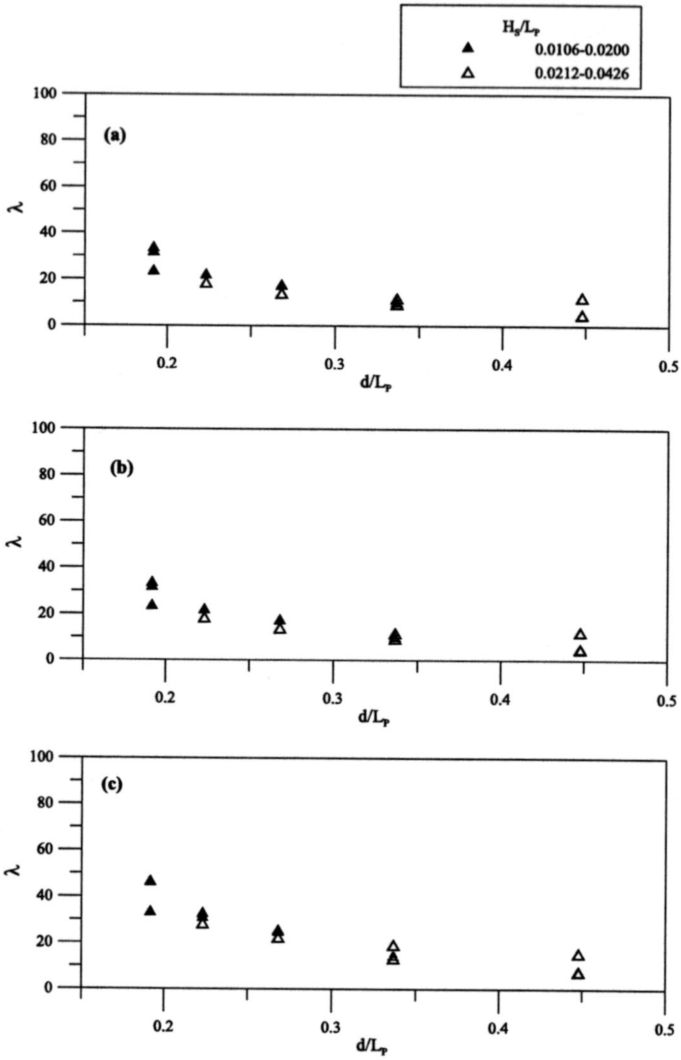

Fig. 7.13 Energy conversion against frequency parameter at $h/d_i = 0.45$, **a** $O/d_i = 0.15$, **b** $O/d_i = 0.30$, **c** $O/d_i = 0.45$

generation. Hence, for the smooth functioning of the turbine, constant air pressure must be assured. A simple method to release the excessive air pressure is with the provision of relief valves, which work in synchronization with the development of excessive air pressure through a control system. For the proper design of the control system, the aerodynamic aspect of the air chamber is to be clearly understood. In this work, it is designated as air pressure ratio (A_R).

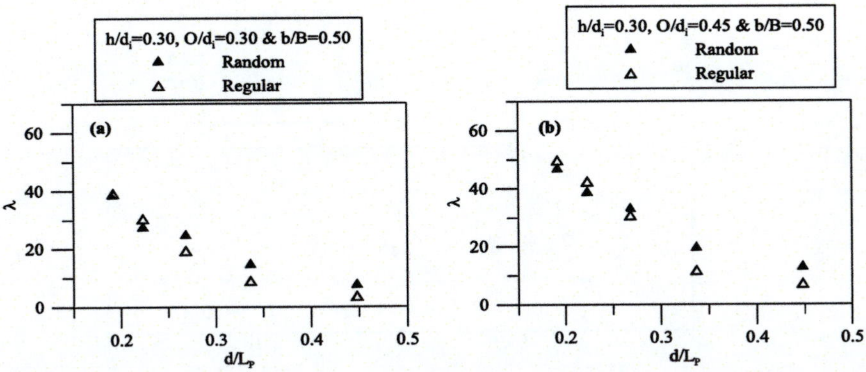

Fig. 7.14 **a, b** Comparison on energy conversion between random waves and regular waves, (H_S = 0.095 m)

$$A_R = \frac{\text{Maximum air pressure}}{\text{Average air pressure}} \tag{7.5}$$

The effect of geometrical parameters of U-OWC on the variation of A_R with d/L_p is presented in Fig. 7.15a–c. For a fine-tuned system, there should be a fixed relationship between average and maximum air pressures. The variation of A_R with d/L_P for h/d_i = 0.30 and 0.45 shows that the oscillation depends mainly on the incident wave spectral variation.

7.3.5 Phase Difference

The variation of phase difference ϕ_d between excitation pressure and air pressure with d/L_P for the maximum wave height is presented in Fig. 7.16a–c. These variations provide explicit information about the degree of control that the system parameters can have over the phase control in oscillation. The energy conversion under random waves follows a similar process in regular waves. The phase difference here is defined as the time delay measured in degrees between air pressure waveform and excitation pressure waveform. The phase difference plays a major role in the energy conversion capacity of the device. When the phase difference is zero, the oscillation inside the chamber is in tune with the incident wave, and the energy conversion reaches a maximum. Due to the existence of phase difference, the air pressure variation within the chamber causes radiated oscillatory flow that can interfere with the incoming waves to reduce its magnitude of pressure intensity in excitation and subsequent lower performance in energy conversion capacity of the device. This interference effect increases with the phase difference.

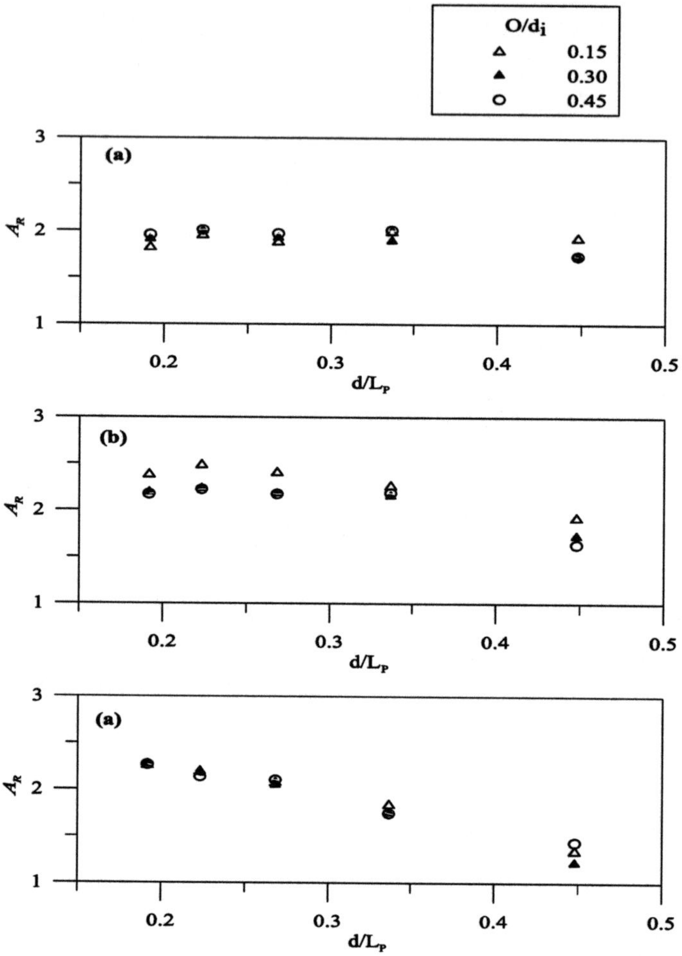

Fig. 7.15 Effect of relative mouth opening on air pressure ratios at **a** $h/d_i = 0.15$, **b** $h/d_i = 0.30$, **c** $h/d_i = 0.45$, $H_S = 0.095$ m

7.3.6 *Wave Amplification*

The knowledge about wave energy in front of coastal structures is one important aspect in free board design. To form a notion of its significance, the variation of the amplification factor β with d/L_P for three O/d_i and $h/d_i = 0.15$, 0.30, and 0.45 is presented in Fig. 7.17a, b and c, respectively. The amplification is calculated for the maximum wave height. β increases drastically with d/L_P up to 0.275 and, beyond which, although the same trend exists, the variation is gradual. It is seen that the lowest values of β for $O/d_i = 0.45$ and $h/d_i = 0.30$ show an improvement in the hydrodynamic efficiency of U-OWC. The explanation given in regular waves holds

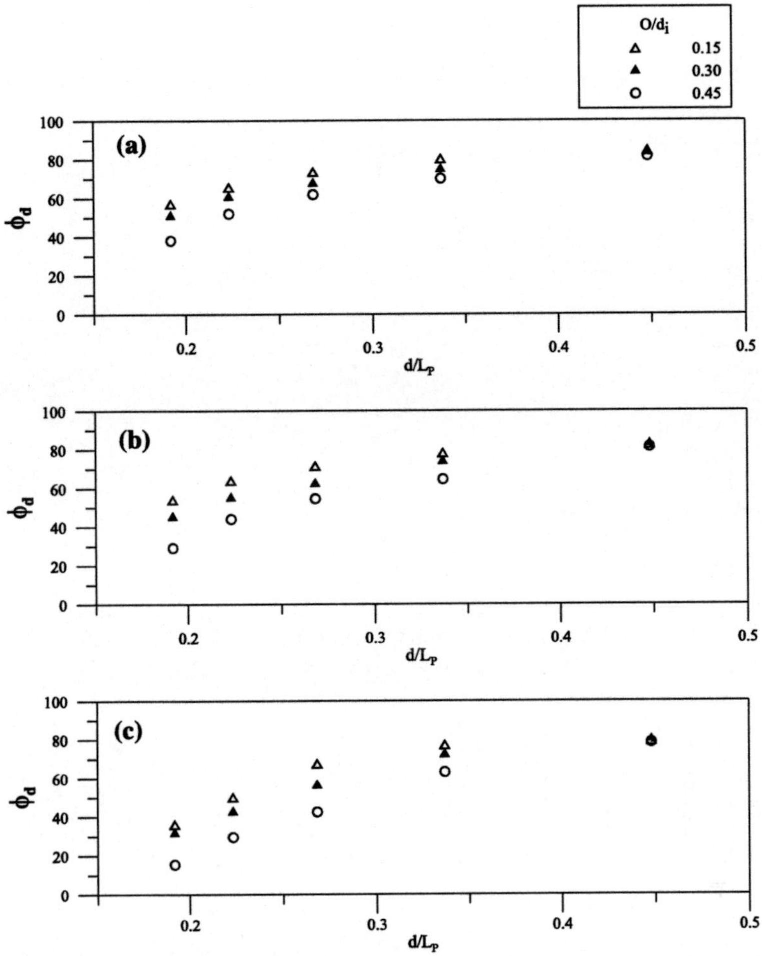

Fig. 7.16 Effect of relative mouth opening on phase difference at **a** $O/di_i = 0.15$, **b** $O/di_i = 0.30$, **c** $O/di_i = 0.45$, $H_S = 0.095$ m

good here also. The slight flatness in the trend line over β at $h/d_i = 0.45$ shows the influence of reflection that could occur from the more exposed area above the mouth level. Thus, for a tuned system, wave amplification is not an important parameter in design consideration.

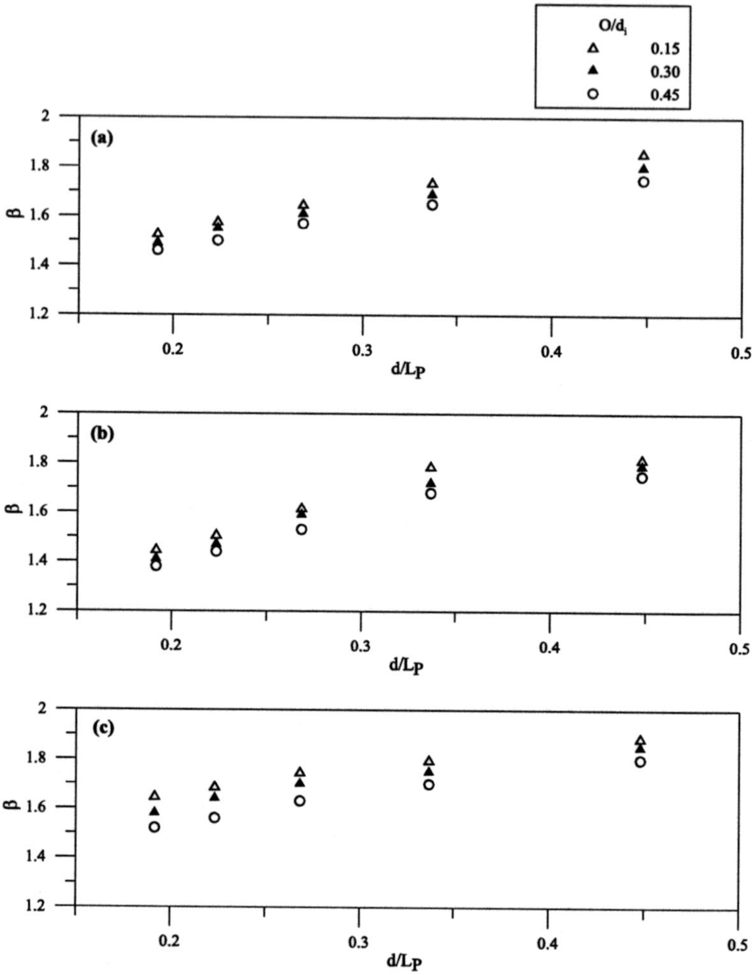

Fig. 7.17 Effect of relative mouth opening on wave amplification at **a** $h/d_i = 0.15$, **b** $h/d_i = 0.30$, **c** $h/d_i = 0.45$, $H_S = 0.095$ m

7.4 Conclusions

U-OWC is an improved OWC system with an outer chamber to achieve a better performance in two ways. The flow path line gets increased, and hence, the frequency of the system drifts towards low frequency. This would mean a wider low-frequency band from which the energy from waves can be extracted, including that from swells and long wind waves. Next, the intensity of the pressure fluctuations on the front chamber is greater than for the conventional OWC. This would result in capturing short wind waves. The possible air entry through the lip wall is prevented with the front chamber. Additional advantages of sediment entry into the chamber due to bedload transport could be greatly avoided. However, structurally, the system receives the environmental forces, and hence, the capital cost would increase. The balance of hydrodynamic performance and economics will decide between OWC and U-OWC.

References

1. Arena, F., & Filianoti, P. (2007). A small-scale field experiment on a submerged breakwater for absorbing wave energy. *ASCE Journal of Waterway, Port, Coastal, and Ocean Engineering, 133*(2), 161–167. https://doi.org/10.1061/(ASCE)0733-950X(2007)133:2(161)
2. Arena, F., Romolo, A., Malara, G., & Ascenelli, A. (2013). On design and building of a u-owc wave energy converter in the Mediterranean Sea: a case study. In: Proceedings of ASME 2013 32nd International conference on ocean, offshore and arctic engineering. Paper No: OMAE2013-11593, Nantes, France, 9–14 June 2013.
3. Ashlin, S.J., S.A. Sannasiraj, and V. Sundar (2014). Performance of an Oscillating Water Column Device with Different Bottom Profiles Subjected to Random Waves. Proceedings of *11th International conference on hydro science & engineering*. Universität Hamburg, Germany. https://doi.org/10.13140/RG.2.1.1258.5844.
4. Ashlin, S.J., Sannasiraj, S.A., Sundar, V. (2017). Influence of power take-off damping on the hydrodynamic performance of an oscillating water column. In *Proceedings of the 12th European wave and tidal energy conference*. University College Cork, Ireland, Sept 27–02 Oct 2017.
5. Ashlin, S.J., Malara, G., Sannasiraj, S. A., Sundar. V., Arena, F., and Romolo, A. (2018) Numerical validation of hydrodynamic characteristics of open-sea U-type oscillating water column wave energy converter. In *3rd International conference on renewable energies offshore* (pp. 569–577). Instituto Superior Técnico, Lisbon, Portugal, 8–10 Oct 2018.
6. Bihs, H., Wang, W., Pakozdi, C., & Kamath, A. (2020) REEF3D::FNPF—A Flexible Fully Nonlinear Potential Flow Solver. *Journal of Offshore Mechanics and Arctic Engineering, 1142*(4), 1–16. https://doi.org/10.1115/1.4045915
7. Boccotti, P. (2000). *Wave mechanics for ocean engineering*. Oxford, UK: Elsevier Science.
8. Boccotti, P. (2003). On a new wave energy absorber. *Ocean Engineering, 30*, 1191–1200.
9. Boccotti, P., (2003). European Patent EP 1133602 B1.
10. Boccotti, P., (2003). PCT N. PCT/IT03/00310.
11. Boccotti, P. (2007). Caisson breakwaters embodying an OWC with a small opening—Part I: Theory. *Ocean Engineering, 34*(5–6), 806–819.
12. Boccotti, P. (2007). Comparison between a U-OWC and a conventional OWC. *Ocean Engineering, 34*, 799–805.
13. Boccotti, P., Filianoti, P., Fiamma, V., & Arena, F. (2007). Caisson breakwaters embodying an OWC with a small opening-Part II: A small -scale field experiment. *Ocean Engineering, 34*, 820–841.
14. Goda, Y. (1992). The design of upright breakwater. In Proceedings of the 23rd international conference on coastal engineering (pp. 547–568). Venice, Italy, 4–9 Oct 1992.
15. Filianoti, P., & Camporeale, S. (2008). A linearized model for estimating the performance of submerged resonant wave energy converters. *Renewable Energy, 33*(4), 631–641.
16. Filianoti, P., & Piscopo, R. (2015). Sea wave energy transmission behind submerged absorber caissons. *Ocean Engineering, 93*, 107–117.
17. Filianoti, P. G. F., & Gurnari, L. (2020). A field experiment on wave forces on an energy-absorbing breakwater. *Energies, 13*, 1563. https://doi.org/10.3390/en13071563

18. Kaldenhoff, H. (1990). Breakwaters-layout and design. Short term course at the Department of Ocean Engineering, Indian Institute of Technology, Madras, India.
19. Malara, G., & Arena, F. (2019). Response of U-oscillating water column arrays: Semi-analytical approach and numerical results. *Renewable Energy, 138*(2019), 1152–1165.
20. Rapaka, E. V. (2007). Experimental investigation on a moored oscillating water column (MOWC) wave energy device. Ph.D. thesis, Ocean Engineering Department, Indian Institute of Technology Madras, Chennai, India.
21. Sarmento, A., & Thomas, G., (1993). Laboratory testing of wave energy devices. Wave energy converters generic technical evaluation study, Annex report B1, Device fundamentals/hydrodynamics. C.E.C., Brussels.
22. Wang, D. J., Katory, M., & Li, Y. S. (2002). Analytical and experimental investigation on the hydrodynamic performance of onshore wave-power devices. *Ocean Engineering, 29*, 871–885.
23. Wilbert, R., Sundar, V., Sannasiraj, S. A. (2012). An experimental investigation on the wave energy conversion capability of double chamber oscillating water column device. *Proceedings of 10th International conference on hydro science and engineering (ICHE 2012)*. Orlando, Florida, 4–7 Nov 2012.
24. Wilbert, R., Sundar, V., & Sannasiraj, S. A. (2013). Wave interaction with a double chamber oscillating water column device. *International Journal of Ocean and Climate Systems, 4*(1), 21–40.
25. Wilbert, R., Sundar, V., & Sannasiraj, S. A. (2013). Hydrodynamic performance characteristics of double chamber oscillating water column device. In *Proceedings of 35th IAHR World Congress*. Chengdu, China, 8–13 Sept 2013.
26. Wilbert, R., Sundar, V., & Sannasiraj, S. A. (2014). Asymmetry effect on hydrodynamic characteristics of double chamber oscillating water column device. *Journal of Ocean Mechanical and Aerospace*.

Chapter 8
CFD Modelling of OWC Devices for Wave Energy Harnessing

Marco Torresi, Pasquale G. F. Filianoti, and Sergio M. Camporeale

Abstract This book chapter is focused on how to study the fluid dynamic behavior of a Wave Energy Converter (WEC) of the Oscillating Water Column (OWC) type by means of Computational Fluid Dynamics (CFD). The U-OWC plant installed in the harbor of Civitavecchia (Italy) is chosen as the case study. The CFD approach is described referring to ANSYS Fluent, release 17.2. The model is mainly oriented to investigate the interaction of single harmonic waves with the OWC device and allows the user to perform a preliminary wave-to-wire energy conversion process analysis. To limit the computational cost, a 2D model is developed. The presence of the Power Takeoff (PTO) system generally breaks the 2D configuration of this kind of device, and here an approach to overcome this problem is described in detail. The solution is the introduction of a porous medium region in the 2D computational domain, which emulates the effect of the PTO system on the fluid dynamic behavior of the flow inside the WEC.

8.1 Introduction

Marine energy has great potential, even if exploiting this resource is more difficult than it may seem. Among the most performing WEC technologies, surely there are the Oscillating Water Column (OWC) devices. OWCs are devices with a partially submerged collector chamber open to the sea. Inside the chamber, due to the wave motion, the water column oscillates and cyclically pressurizes and expands

M. Torresi (✉) · S. M. Camporeale
Department of Mechanics, Mathematics and Management, Polytechnic University of Bari, Via G. Re David, 200, 70125 Bari, BA, Italy
e-mail: marco.torresi@poliba.it

S. M. Camporeale
e-mail: sergio.camporeale@poliba.it

P. G. F. Filianoti
Department of Civil, Energy, Environment and Materials Engineering, Mediterranean University of Reggio Calabria, Via dell'Università, 25, 89124 Reggio Calabria, RC, Italy
e-mail: filianoti@unirc.it

A. Samad et al. (eds.), *Ocean Wave Energy Systems*, Ocean Engineering & Oceanography 14, https://doi.org/10.1007/978-3-030-78716-5_8

the air trapped inside it, which is forced to flow in alternate directions through a self-rectifying turbine used as a Power Take-Off (PTO) system [1].

During the last 25 years, several OWC plants have been developed [2]. The first wave energy plant based on the OWC technology was installed in Vizhinjam, a region located in the city of Thiruvananthapuram, capital of the Kerala state (India), to run a desalination plant [3]. In the meantime, studies on a floating OWC device were carried out in Gokasho Bay, Nansei Town, Mie Prefecture (Japan), giving birth to the "Mighty Whale" [4]. In Europe, two full-sized fixed-structure OWC plants were designed and constructed: one on the island of Islay (Scotland), [5]; the other one on the island of Pico, Azores (Portugal) [6]. All these installations embed Wells turbines as Power Take-Off units. However, Wells turbines are not the only choice. Several are examples of OWC plants operating with impulse turbines. For instance, in the Zhelang district of Shanwei City, Guangdong Province, (China), there is a 100 kW OWC plant [7], and in Niigata Port (Japan), a small size OWC, both with impulse turbines as PTO systems [8]. OWC devices can be integrated within port breakwaters (nearshore plants), as in the case of the harbors of Civitavecchia (Italy) [9] and Mutriku (Spain) [10], with a significant capital cost reduction.

The first OWC devices used Wells turbines for mechanical power generation. The Wells turbine was invented by Dr. Alan A. Wells in 1976 when he was at Queen's University of Belfast, UK [11]. It is an axial-flow turbine composed of a runner with untwisted symmetrical airfoil blades, usually belonging to the double zero NACA four-digit series (for instance, NACA0018), with a 90° stagger angle. Several works can be found in the literature, either numerical or experimental, since the early 80s [12–18].

Alternative options to Wells turbines are the impulse turbines, patented by I. A. Babintsev in 1975 [19]. Impulse turbine rotors are very similar to the ones adopted in the first stage of conventional high-pressure Curtis wheels [20].

Recently, a novel concept of a self-rectifying turbine was introduced, the biradial impulse turbine, which aims to overcome several limitations of the Wells turbine, such as the sharp drop in efficiency above a critical flow rate. Its rotor is surrounded by radial-flow guide vanes [21].

A particular OWC device is the U-OWC system, developed by the University of Reggio Calabria [22]. In this WEC, the waves do not enter the structure directly but act as an external forcing (see Fig. 8.1).

In order to analytically investigate the fluid dynamic behavior of the free water surface motion inside an OWC device subject to the wave action, a simplified one-dimensional mathematical model can be derived from the generalized Bernoulli's theorem applied along a streamline from the free water surface inside the OWC (A) to the sea wave surface at the OWC opening (B) (see Fig. 8.1):

$$\frac{1}{g}\int_A^B \frac{\partial \bar{v}}{\partial t} \cdot d\bar{s} = H_A - H_B - \sum Y,$$

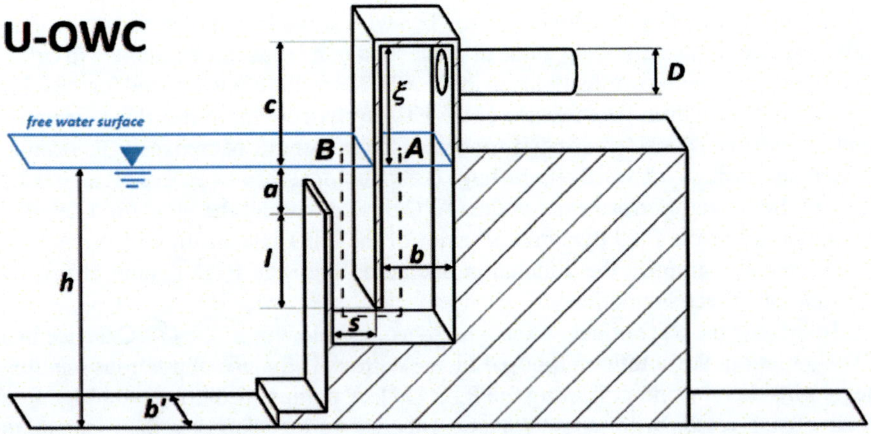

Fig. 8.1 Vertical cross-section of the U-OWC embedded in the breakwater at Civitavecchia's harbor. Schematic adapted from [9]

where $\sum Y$ is the sum of the head losses along the streamline. The total head H_A depends, among others, on the absolute pressure inside the OWC chamber, which is a function of both the pressure drop across the turbine and the air compressibility. In the hypothesis that the air inside the chamber experiences polytropic (for instance, isentropic) transformations, the pressure is a function of the plenum chamber volume, hence of the inner free water surface displacement, ξ. Regarding the pressure drop across the turbine, in the case of a Wells turbine, this is approximately linearly proportional to the volume flow rate. Finally, the terms on the left-hand side of the generalized Bernoulli's theorem are related to the inertia of the water column inside the OWC device. This means that the fluid dynamic behavior of the inner free water surface displacement, ξ, can be represented by a non-linear second-order equation. Under the hypothesis of small oscillations, this governing equation can be linearized becoming an ordinary second-order differential equation equivalent to the one of a mechanical mass-spring-damper system, with the inertial term related to the mass of the water column inside the OWC, the elastic term related to the air compressibility, and the viscous term related to the Wells turbine characteristic. This means that an OWC device can be characterized by its own natural damped frequency. More details can be found in Filianoti and Camporeale [23] and Malara et al. [24].

Since waves behave like a harmonic forcing for the OWC device, if the wave frequency equals the OWC natural damped frequency, the inner free water surface displacement, ξ, will experience amplified oscillations operating under resonant conditions. In such a case, the energy absorbed by the damper will be maximized, and consequently, the energy converted by the PTO of mechanical energy. For this reason, it is of great importance that the OWC device and its PTO are correctly coupled.

When investigating the fluid dynamic behavior of both on-shore and near-shore OWC devices, either standing alone or embedded into breakwaters, subject to cylindrical waves traveling orthogonally to them, a 2D approach could be used for the CFD simulations. However, the presence of the PTO system compromises this symmetry, and is often placed inside a conduit connected to the plenum. Moreover, a 3D simulation of the entire system will require huge computational resources, even if interested only in the overall performance of the OWC device without the need for a detailed investigation of the fluid dynamic behavior of the PTO system. In such a case, we can extremely simplify the simulation, including only the PTO system effect, by avoiding its direct simulation.

To include the PTO effect on the fluid dynamic behavior of the OWC device in a 2D simulation, the solution proposed by the authors is the use of a porous medium region inside the conduit hosting the PTO [22]. A porous medium is nothing more than a mesh region with whom source terms can be introduced in the momentum balance equations. Actually, a porous medium is designed for determining a pressure drop as a function of the velocity field inside it. In addition, it can account for both viscous (i.e., linearly proportional to the velocity) and inertial (i.e., quadratically proportional to the velocity) pressure drops. A typical PTO system for OWC can be based on a Wells turbine which determines a pressure drop proportional to the flow rate. This means that the porous medium is a better solution to the problem with respect to an orifice, which can only return a parabolic relationship between pressure drop and flow rate.

The chapter is organized in paragraphs. Initially, the reader is guided through a description of the numerical experiment in terms of the computational domain, mesh, and boundary conditions. Then, the chapter continues with the explanation of the governing equations in the CFD model. Afterward, the third paragraph describes the approach to deal with multiphase flows. The fourth concerns the setup of the porous medium region to account for the effect of the PTO system on the momentum balance equations. The fifth looks at the simulation setup, data-saving, and post-processing. Finally, the chapter ends with some concluding remarks.

8.2 The Numerical Experiment

To describe the CFD approach for modeling an OWC device, we have considered the case of the U-OWC system developed by the University of Reggio Calabria and installed in the harbor of Civitavecchia (Italy) [9].

8.2.1 Computational Domain

The U-OWC is characterized by a vertical duct with its opening exposed to the pressure oscillations due to the wave motion. This vertical duct communicates with

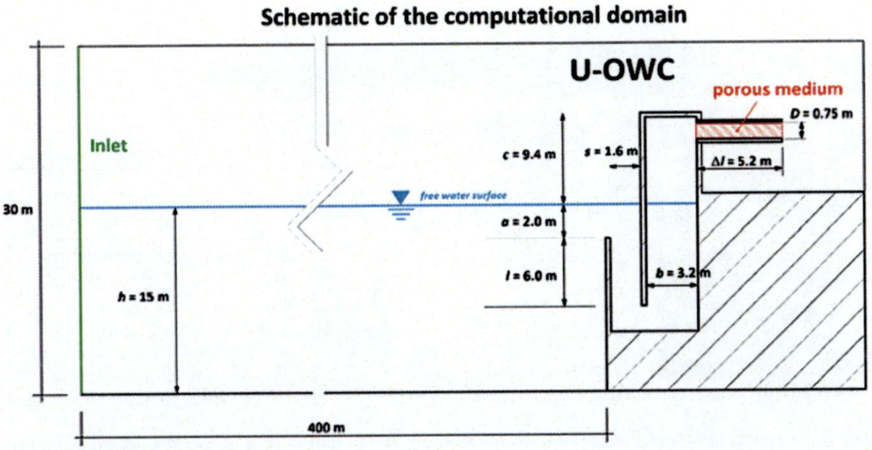

Fig. 8.2 Schematic of the computational domain designed for the U-OWC CFD simulation. On the left end side, the inlet boundary where the waves are generated by means of the Open Channel Velocity boundary condition. The porous medium is the zone where source terms are introduced in order to take into account the effect of the PTO on the flow field

the collector chamber forming the U-shaped duct. The widths of the vertical duct and the chamber are $s = 1.6$ m and, $b = 3.2$ m respectively; the length of the duct is $l = 6.0$ m, the height of the chamber with respect to the undisturbed free water surface is $c = 9.4$ m. The depth of a single U-OWC unit in the third dimension is $b' = 3.87$ m. The level of the undisturbed free water surface has been set at $h = 15.0$ m, while the duct opening depth is $a = 2.0$ m. Figure 8.1 also shows the conduit that hosts the PTO system.

In order to perform the 2D CFD simulations, the computational domain is extended 400 m upstream and is 30 m high and includes the U-OWC device, which consists of the collector chamber and the porous medium, substituting the conduit that houses the PTO (see Fig. 8.2).

8.2.2 Mesh

The spatial discretization of the computational domain is performed using a multi-block approach, meshing the different blocks with either structured or unstructured grids (see Fig. 8.3). The basic idea is to use a structured mesh both in the region where the sea wave motion evolves and inside the U-OWC, whereas an unstructured mesh around the U-OWC guarantees a smooth transition between the coarser mesh of the overall domain and the finer mesh size inside the WEC. In particular, the mesh density has been increased around the free water surface to obtain a better reconstruction of the wave elevation profile. The mesh consists of about 657,000 cells.

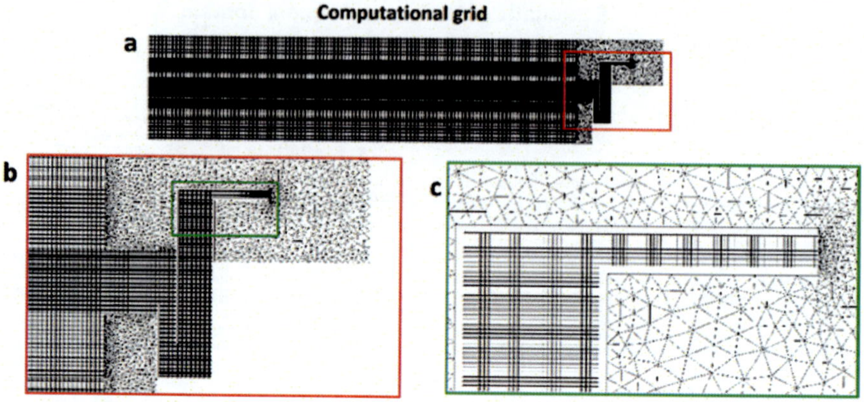

Fig. 8.3 Visualization of the entire computational grid (**a**) and details of the structured computational grid in the U-OWC (**b**) and the conduit where the porous zone accounts for the effect of the PTO (**c**)

Before starting the CFD simulation, the mesh quality must be evaluated by analyzing the mesh metric information. High quality of the mesh will significantly increase the stability and the accuracy of the simulation. Independently of the type of mesh used, the main parameters to be checked are the minimum orthogonal quality, the maximum ortho skewness, and the maximum aspect ratio [25].

The orthogonal quality considers the area vectors of the cell faces,$\bar{A}_i$, the vectors from the centroid of the cell to the centroids of those faces, $\bar{f}_i$, the vectors from the centroid of the cell to the centroids of the adjacent cells that share the corresponding faces,$\bar{c}_i$. The orthogonal quality, q_0, is defined as $q_o = min\left[\frac{\bar{A}_i \cdot \bar{f}_i}{|\bar{A}_i||\bar{f}_i|}; \frac{\bar{A}_i \cdot \bar{c}_i}{|\bar{A}_i||\bar{c}_i|}\right]$. The orthogonal quality ranges from 0 (the worst) to 1 (the best). The minimum orthogonal quality for all types of cells should be $q_{o,min} > 0.01$.

The skewness refers to the difference between the shape of a cell and the shape of the equilateral cell of equivalent volume. For a 2D mesh, the ortho skewness,S_o, is defined as $S_o = max[(\alpha_{max} - \alpha_e)/(180° - \alpha_e); (\alpha_e - \alpha_{min})/\alpha_e]$ with α_{max} being the largest cell angle, α_{min} the smallest cell angle and α_e the angle of the corresponding equilateral cell. Ortho skewness equal to one implies degenerated cells. However, maximum values $S_{o,max} = 0.95$ could lead to convergence difficulties.

The aspect ratio (AR) measures the cell stretching. The AR is computed as the ratio of the maximum value to the minimum value of any of the following distances: the distances between the cell centroid and face centroids and the distances between the cell centroid and nodes. Values of the $AR > 10$ must be avoided inside the computational domain but can be accepted close to the walls.

In the mesh taken into account, the minimum orthogonal quality is $q_{o,min} = 0.21$. The maximum ortho skewness is $S_{o,max} \cong 0.52$. The maximum Aspect Ratio is $AR = 7.39$.

8.2.3 *Boundary Conditions*

Regarding the boundary conditions (see Fig. 8.2), all the walls are considered smooth no-slip walls. The edge on the left of the computational domain is the velocity inlet where the open channel condition is implemented, enabling the user to define the characteristics of the waves entering the domain in terms of free surface level, wave height, wavelength, and phase difference. To simulate gravity waves, it is necessary to set the gravitational acceleration (here, $g = 9.81\,\mathrm{m/s^2}$). Furthermore, the solver needs to define the operating pressure ($p_0 = 101,325\,\mathrm{Pa}$) and its location. According to the ANSYS Fluent Tutorial Guide [26], the reference pressure location needs to be in a domain region where the fluid phase is just one and preferably the phase with the lowest density (i.e., 100% air). This condition is essential for a smooth and rapid convergence. The reference pressure location has been chosen in the upper left side of the computational domain where only air exists, and precisely at $x = 0.1\,\mathrm{m}$ and $y = 29\,\mathrm{m}$. Regarding the Variable-Density Parameters, after having turned on the "Specified Operating Density" option, this has been set to zero, so excluding the buildup of hydrostatic pressure within the lighter phase, hence improving the accuracy of the momentum balance solution. Air is assumed to be a compressible ideal-gas to take into account the compressibility of the air pocket inside the plenum, whereas the water is a fluid of constant density ($\rho_w = 1025\,\mathrm{kg/m^3}$, considering sea water at 25 °C).

8.3 Governing Equations

The ANSYS Fluent CFD code provides the numerical solution of the differential equations governing the flow by means of a finite volume discretization approach. Considering a multiphase flow (water and air), the governing equations to be solved are the continuity equation, the momentum balance equations, and the conservation equations of volume fractions and energy. Even if the heat exchanges are negligible, the solution of the energy equation is still necessary to determine the thermodynamic variables of the air, which will be assumed as an ideal compressible gas.

In ANSYS Fluent, there are two different methodologies for solving the system of governing equations:

- Pressure-based solver;
- Density-based solver.

In both solvers, the velocity field is obtained from the solution of the momentum balance equations. In the pressure-based solver, the pressure field is obtained directly from an appropriate formulation of the continuity equation, while in the density-based solver, the continuity equation is used to obtain the density field, whilst the energy equation returns the enthalpies and hence temperatures. Then, the pressure is obtained from the equation of state. The reason for this different approach derives from the fact that originally the pressure-based solver was developed for the resolution of incompressible flows, whilst the density-based solver for the resolution of compressible flows (high Mach numbers). Successively, both models have been extended and reformulated so that, currently, they can be adopted for both types of flow.

After having imported the mesh into ANSYS Fluent, the process of solving the governing equations, which leads to the update of the values of the dependent variables at each timestep, is essentially divided into two stages:

- Integration of the equations in each cell (i.e., the single control volume of the mesh) to build the system of discretized equations;
- Linearization of the discretized equations and solution of the resulting linear system through an iterative cycle to update the values of the dependent variables at each timestep of the simulation.

The integration phase leads to a spatial discretization of the governing equations. In this phase, the solver adopts numerical interpolation techniques to obtain the values along with the faces by starting from the values of the properties in the cell centroid. In the case of transient simulations, a further time discretization is required, which concerns the time derivatives of the variables.

8.3.1 Pressure-Based Solver

Considering that the pressure-based solver is the only model compatible with a multiphase problem, this is the one adopted and described here.

In the pressure-based solver, the constraint of mass conservation of the velocity field is actually satisfied by a pressure correction equation.

Two different pressure-based algorithms are available:

- Sequential calculation method (segregated solver);
- Coupled calculation method (coupled solver).

The solution procedure of both methods is analyzed in the flow charts of Fig. 8.4.

The two algorithms differ substantially in that in the segregated solver, all the equations, and therefore also those of continuity and momentum balance, are solved separately and sequentially (i.e., in a segregated way), while in the coupled solver, the equations of continuity and momentum balance are solved simultaneously.

In both cases, all the other governing equations (e.g., turbulence, energy, etc.) are separately and sequentially solved. Being the equations nonlinear and coupled to one another, the solution procedure is necessarily iterative.

The segregated solver is efficient from the point of view of computer memory usage since the individual equations are loaded one at a time. However, the convergence of the solution is relatively slow since the equations are solved in a decoupled manner, introducing a segregation error.

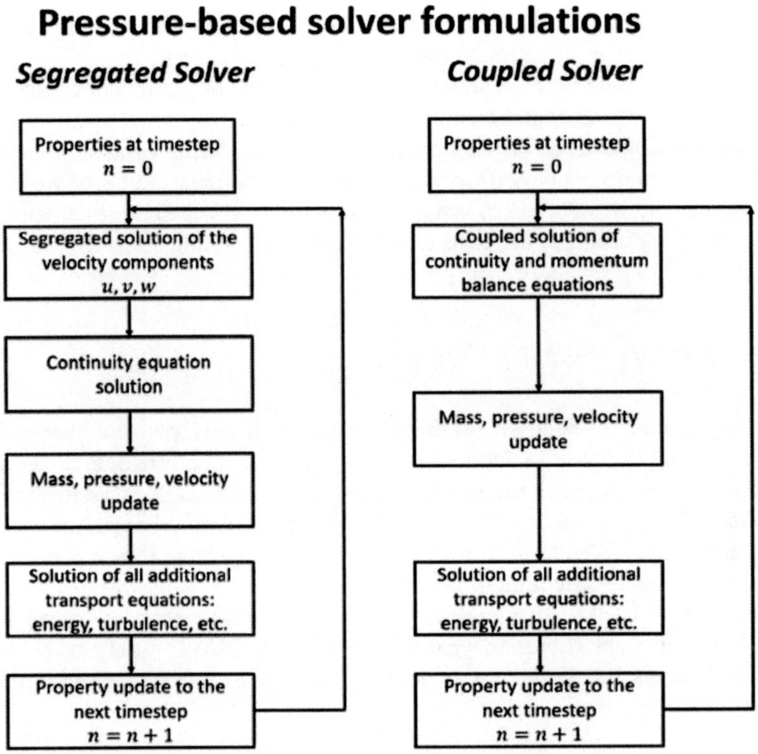

Fig. 8.4 Algorithms of the segregated (left) and coupled (right) solvers

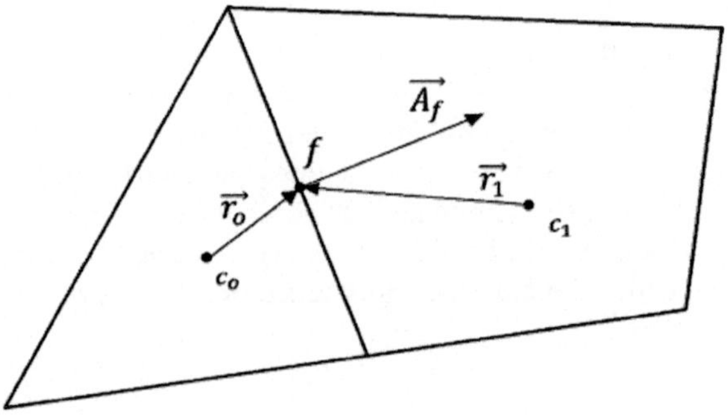

Fig. 8.5 Scheme for the reconstruction of the gradients on the face of a cell starting from the values of the quantities in the centers of the adjacent cells

On the contrary, since in the coupled method, the equations of continuity and momentum balance are solved simultaneously, the tendency of the solution to converge improves since the segregation error is not introduced. However, the computer memory usage significantly increases since, to solve simultaneously the pressure and velocity field, the three/four equations (2D/3D) of continuity and momentum balance must be loaded into memory at the same time.

The coupled solver must be preferred in the presence of significant compressible effects or subject to high rotational or buoyancy forces.

In the simulations of sea wave propagation, even considering the air as a compressible ideal gas, compressible effects are limited, and the pressure-based Segregated solver model can be adopted in order to increase the computational efficiency, in terms of memory usage, at the costs of an increase in solution convergence time.

8.3.2 Pressure–Velocity Coupling

In the segregated formulation of the pressure-based solver, a pressure–velocity coupling is used to derive an additional condition for the pressure field by enforcing mass conservation. Among others, the SIMPLEC (SIMPLE-Consistent) algorithm is applied. This is a variant of the original Semi-Implicit Method for Pressure-Linked Equations (SIMPLE), which can accelerate convergence in problems where the pressure–velocity coupling is significant.

In the original SIMPLE algorithm, the momentum equation is solved with a guessed pressure field. However, the resulting face fluxes do not satisfy the continuity equation. Hence a flux correction is needed. Writing the flux correction terms as a function of a pressure correction, the pressure field can be computed.

8.3.3 Solution Control Parameters

During the segregated solver setup, together with the definition of the pressure–velocity coupling scheme and the spatial discretization of all the equations involved in the problem, the solution control parameters, such as the under-relaxation factors, must be defined. The latter defines the duration of the iteration cycle.

In a transient simulation, the iteration cycle duration is governed by the maximum number of iterations per timestep. Hence, the control parameters of the solver coincide with only the under-relaxation factors of the equations.

8.4 Under-Relaxation Factors

For a single equation, an under-relaxation factor defines the correction fraction to apply on the previous value of a variable following the advancement of the solution to the next iteration. If we denote by ϕ the current value of the variable under consideration, this is equal to:

$$\phi = \phi_{old} + \alpha \Delta\phi$$

being α the value of the under-relaxation factor and $\Delta\phi$ the value of the correction calculated starting from the previous variable value, ϕ_{old}.

Higher under-relaxation factors speed-up convergence but could lower the stability with the risk that the solution diverges. On the contrary, lower under-relaxation factors slow down convergence but increase stability.

In our case, the under-relaxation factors for each individual equation were kept equal to their default values (i.e., 0.3 for pressure, 0.7 for the momentum, and 1 for density, body forces, and energy).

8.4.1 Spatial Discretization of Equations

The finite volume method (FVM) consists of the integration of the partial differential equations in each control volume, thus providing discrete equations that express the conservation law of the scalar quantity, which they refer to, in the specific control volume. Considering the transport equation of a generic scalar quantity, ϕ, the integral form of the equation, discretized on an arbitrary control volume, V, is the following:

$$\int_V \frac{\partial \rho\phi}{\partial t} dV + \oint \rho\phi\vec{v} d\vec{A} = \oint \Gamma_\phi \nabla\phi \cdot d\vec{A} + \int_V S_\phi dV$$

where we indicate with ρ the density, with $\vec{v}$ the velocity vector, with $d\vec{A}$ the surface area vector, with Γ_ϕ the diffusion coefficient for ϕ and with S_ϕ the source term of ϕ per unit of volume. The discretized equation is integrated on each control volume, i.e., on each cell of the computational domain. The result of the integration is the following:

$$\frac{\partial \rho\phi}{\partial t} V + \sum_f^{N_{faces}} \rho_f \vec{v}_f \phi_f \cdot \vec{A_f} = \sum_f^{N_{faces}} \Gamma_\phi \nabla\phi_f \cdot \vec{A_f} + S_\phi V$$

The subscript f indicates that the corresponding quantity value refers to one of the N_{faces} faces that surround the control volume.

Regarding the obtained discretized equation, it is necessary to consider that ANSYS Fluent actually stores the values of the scalar quantity, ϕ, at each cell centroid. However, the convective terms are calculated starting from the values on the cell faces, ϕ_f, which must therefore be obtained starting from the cell centroid values by means of suitable interpolation schemes (for instance, an upwind scheme). The upwind schemes allow to derive the value of ϕ_f from the quantities present in the upstream cells relative to the direction of the velocity, v_n, normal to the face considered.

ANSYS Fluent allows one to choose among different schemes: first-order upwind, second-order upwind, power-law, and QUICK.

When the first order accurate upwind scheme is considered, the value of the generic quantity, ϕ, on a face, ϕ_f, is calculated assuming that the value in the cell centroid is the average value of the quantity over the entire cell. Therefore, using the first-order upwind scheme, the value of the quantity on the face, ϕ_f, is considered equal to the value of the quantity, ϕ, in the centroid of the cell immediately upstream.

On the other hand, if the upwind scheme is second-order accurate, the value of the quantity on a face, ϕ_f, is equal to the sum of the value of the quantity in the centroid of the cell immediately upstream, ϕ, plus the dot product between the gradient, $\nabla\phi$, calculated in the upstream cell and the distance vector, $\overrightarrow{r}$, of the face with respect to the upstream cell centroid:

$$\phi_f = \phi + \nabla\phi \cdot \vec{r}$$

In general, the first-order upwind discretization scheme is more stable than the second-order one, compatibly with the use of the explicit multiphase VOF model.

8.4.2 *Reconstruction of Gradients*

The gradient of a generic scalar quantity, ϕ, needs to be evaluated to compute the convective and diffusive terms. To obtain the gradient value of a scalar quantity in a cell, different methods can be used. Two of them are based on the Green-Gauss theorem. This theorem relates the volume integral of the gradient of a generic scalar quantity, ϕ, to the surface integral of the scalar, ϕ. In particular, if you have a volume, C, formed by a surface, S, the theorem suggests for a generic scalar quantity, ϕ, that:

$$\iint\limits_S \phi \, d\bar{S} = \iiint\limits_C \nabla\phi \, dC.$$

Discretizing the Green-Gauss theorem over a cell, the gradient of a scalar quantity, ϕ, at the cell centroid can be calculated according to the equation:

$$(\nabla\phi)_{c_o} = \frac{1}{v}\sum_f \bar{\phi}_f \overrightarrow{A_f}$$

where $\bar{\phi}_f$ is the value of ϕ at the cell face, and $\overrightarrow{A_f}$ is the face area vector (i.e. a vector orthogonal to the cell face, as shown in Fig. 8.5, and magnitude equal to the face surface area). In the cell-based Green-Gauss method, the value on the face of a cell c_O is extrapolated from the arithmetic mean of the values in the centroids of the neighboring cells:

$$\bar{\phi}_f = \frac{\phi_{c_o} + \phi_{c_{neighboring}}}{2}$$

In the Green-Gauss method based on nodes, however, the value of $\bar{\phi}_f$ is calculated as the arithmetic mean of the values, $\bar{\phi}_n$, present at the N_f nodes of the faces, therefore:

$$\bar{\phi}_f = \frac{1}{N_f}\sum_{n=1}^{N_f}\bar{\phi}_n.$$

In turn, the values $\bar{\phi}_n$ are calculated from the weighted average of the values of the cells surrounding the node. This second method is mainly used for triangular (2D) or tetrahedral (3D) cell grids.

When structured meshes are involved, the cell-based Green-Gauss method is applied for the gradients, which is also the default method for the Pressure Based solver.

The third method for gradient reconstruction is the Least Squares Cell-Based method. This method is based on the hypothesis that the solution, hence each generic scalar, ϕ, varies linearly. This means that the change in cell values between cell c_0 and the generic neighboring cell, c_i, along the vector $\delta\bar{r}_i$ from the centroid of the cell c_0 to the centroid of cell c_i, can be formulated as follows:

$$(\nabla\phi)_{c_0}\Delta r_i = \left(\phi_{c_o} - \phi_{c_i}\right).$$

Writing similar equations for each neighboring cell, we obtain a system that can be written in a compact form:

$$[J](\nabla\phi)_{c_0} = \Delta\phi.$$

$[J]$ is the coefficient matrix, a pure function of geometry.

This is an over-determined linear system of equations, and the cell gradient $(\nabla\phi)_{c_0}$ can be determined by solving a minimization problem in a least-square sense.

The least-squares gradient method is as accurate as of the Green-Gauss node-based gradient on irregular, skew, and distorted meshes and both are superior compared to the Green-Gauss cell-based gradient. Moreover, the least-squares gradient method is less computationally expensive with respect to the Green-Gauss node-based.

8.4.3 Time Discretization

Considering the discretized equation for the transport of a generic scalar quantity, ϕ, when the flow is unsteady (either intrinsically or due to transient boundary conditions), the term $\partial\phi/\partial t$ is not zero and must be discretized in time over the timestep Δt.

The temporal discretization method is based on the finite-difference approach, which can be accurate to the first or second order. We have that according to a first-order accurate scheme, the discretized form of the time derivative of ϕ will be:

$$\frac{\partial \phi}{\partial t} = \frac{\phi^{n+1} - \phi^{n}}{\Delta t}$$

while according to a second-order accurate scheme, we get:

$$\frac{\partial \phi}{\partial t} = \frac{3\phi^{n+1} - 4\phi^{n} + \phi^{n-1}}{2\Delta t}$$

It must be pointed out that, in the presence of a deforming mesh, it is possible to adopt the only first-order accurate scheme.

Finally, the discretized time derivative is equated, in the transport equation of a generic quantity, ϕ, to other terms which are functions of the quantity, ϕ; for instance, in case of first-order accurate time discretization, it gives:

$$\frac{\phi^{n+1} - \phi^{n}}{\Delta t} = F(\phi).$$

According to whether the value of the variable ϕ considered in $F(\phi)$ is that calculated at the current time step $n + 1$, ϕ^{n+1}, or at the previous time step n, ϕ^{n}, the temporal discretization model will be implicit or explicit, respectively.

When adopting the Pressure Based solver, only the implicit model can be used. The value of ϕ^{n+1} is obtained by solving the following implicit equation:

$$\phi^{n+1} = \phi^{n} + \Delta t F\left(\phi^{n+1}\right)$$

Being this an implicit formulation, the corresponding equation must be solved iteratively at each time level before moving on to the next time step. The implicit discretization method turns out to be more onerous from a computational point of view, but it has the advantage of being stable with respect to any timestep size, unlike the explicit model for which the use of a timestep of limited size is recommended.

8.5 Multiphase Flow

The ANSYS Fluent CFD code offers two different approaches to study multiphase flows: either the Eulerian–Lagrangian approach or the Eulerian-Eulerian approach [25]. The former approach is indicated if one of the two phases (the secondary one) is dispersed in the form of particles in the other (the primary phase), occupying only a much smaller volume fraction than the primary phase (maximum acceptable volume fraction lower than 12%). The latter approach must be used in all those cases where the volume fractions of all the phases are comparable.

The Eulerian-Eulerian approach considers the different phases present in the computational domain as interpenetrating continua. It is supposed that the volume of a phase cannot be occupied by another phase hence the concept of phasic volume fraction is introduced. Several Eulerian-Eulerian approaches exist the volume of fluid (VOF) model, the Eulerian model, and the mixture model. The VOF model can be used in the presence of immiscible fluids. In this model, a single set of momentum equations is solved, and the different species volume fractions are tracked in the computational domain. The Eulerian multiphase model deals with multiple interacting but separate phases, which are treated according to an Eulerian approach. The mixture model is a simplified version of the previous model that can be used when single phases move at different velocities, but still, there is a strong coupling among them.

In the case of sea waves, we are in the presence of two immiscible fluids (water and air) whose volumes fraction are of the same order of magnitude. Therefore, an Eulerian-Eulerian approach should be adopted, and, among the three possible multiphase models, the Volume of Fluid (VOF) model is considered the most appropriate, being able to obtain the best definition of the air–liquid interface.

8.5.1 The Volume of Fluid (VOF)

When using the VOF model, the definition of the interface between any two phases among the n phases present in the mixture starts from the solution of the continuity equation for the volume fraction, α_q, of each qth phase, with the exception of the primary phase. Indeed, the primary phase volume fraction value, α_1, is obtained starting from the values calculated for the other phases to guaranty respect of the non-penetrability condition, i.e.:

$$\sum_{q=1}^{n} \alpha_q = 1.$$

The continuity equations for each volume fraction, α_q,

$$\frac{1}{\rho_q}\left[\frac{\partial}{\partial t}(\alpha_q \rho_q) + \nabla \cdot \alpha_q \rho_q \vec{v}_q = S_{\alpha_q} + \sum_{p=1}^{n}(\dot{m}_{pq} - \dot{m}_{qp})\right]$$

are solved together with the flow balance equations in each cell of the mesh.

In the continuity equations, ρ_q indicates the density of the qth phase, $\vec{v}_q$ the velocity vector and the terms $\dot{m}_{pq}$ and $\dot{m}_{qp}$ the mass exchange between phases p and

q and between phases q and p, respectively. Finally, S_{α_q} is an eventual source term of the volume fraction of phase q.

In our case, neither source terms or mass exchange among the two phases (namely, water and air) are considered. Hence the continuity equation for the secondary phase (water) volume fraction reads:

$$\frac{\partial}{\partial t}(\alpha_w \rho_w) + \nabla \cdot \alpha_w \rho_w \vec{v}_w = 0$$

and $\alpha_a = 1 - \alpha_w$.

The volume fraction continuity equation is a partial differential equation that is solved in the CFD code by replacing the differential operators with their discretized form, developed using the finite volume approach. Once this equation is discretized, it can be solved according to either an explicit or an implicit scheme.

In order to determine which discretization scheme must be chosen, it is important to consider the aim of the simulation and the main characteristics of both. An implicit scheme allows the use of larger timestep sizes with respect to an explicit scheme. However, an implicit scheme suffers from a diffusion of the interphase more than an explicit scheme. Another advantage of the explicit scheme in ANSYS Fluent is that it enables the use of the Geo Reconstruction method for the volume fraction, which is the best method available for interface reconstruction. Regarding the use of larger time steps, it is noteworthy that the advection of a volume fraction of more than one cell per timestep is not reasonable, and the solution will not be time accurate. This means that if the interest is the transient description of the free-water surface, it is recommendable the use of the VOF explicit scheme. Finally, the explicit scheme requires less memory and time to solve the equations.

8.6 Explicit Scheme

Using the explicit scheme, the volume fraction and the other properties of the fluid at the current time step $n + 1$ are calculated starting from the values they had at the previous time step n, therefore the discretized form of the water-liquid phase volume fraction continuity equation will be as follows:

$$\frac{\alpha_w^{n+1}\rho_w^{n+1} - \alpha_q^n \rho_w^n}{\Delta\tau} V + \sum_f \rho_w^n U_f^n \alpha_{w,f}^n = 0$$

where the subscript f indicates that we refer to the fth face of the cell in which the equation is integrated. Moreover, the term U_f^n is the volumetric flow rate in the direction normal to face f at the instant n, and the term V is the volume of the cell.

The time step $\Delta\tau$ (with respect to which the volume fraction equation is discretized) is generally different from the time step Δt, which defines the time advancement of the flow balance equations and is automatically refined by the solver based on the value of the Courant Number (*Co*) in the proximity of the interface between the two fluids.

When using the explicit scheme, the maximum value of the Courant Number must be set among the inputs of the Multiphase model panel, being the default value equal to 0.25.

The Courant number is a dimensionless parameter defined as the ratio between the time step size, Δt, used in the solution of the volume fraction equation, and the minimum time, $\Delta\tau$, that the fluid takes to cross a control volume located at the interface between the two fluids:

$$Co = \frac{\Delta t}{\Delta \tau} = \frac{\Delta t}{\Delta x_{cell}/v_{fluid}}$$

If the value of the ratio between Δt and $\Delta\tau$ is high (it should not exceed 1000), there is an increase in the entity of the computational effort required and lower stability of the solution [27]. Therefore, when using the explicit VOF model, the simulation time step Δt must be chosen consistently with the parameters used to calculate the solution time step $\Delta\tau$ of the volume fraction equation, i.e., the maximum value of Courant Number, the minimum cell size, $\Delta x_{cell,min}$, and maximum flow velocity, $v_{fluid,max}$, present at the interface between the two fluids. In fact, if the dimension of Δt is much larger than that of $\Delta\tau$, the value of $\Delta\tau$ is automatically increased in order to limit the relationship between the two timesteps. However, this involves an increase in the value of the Courant Number, which leads to the stopping of the iterative loop if it becomes greater than 250.

Therefore, a relatively small simulation time step must be adopted when using the explicit VOF multiphase model.

8.7 Implicit Scheme

On the other hand, in the implicit scheme, the volume fraction and the properties of the fluids at the current time step n + 1 are calculated simultaneously. The discretized form of the volume fraction equation will be of the following type:

$$\frac{\alpha_w^{n+1}\rho_w^{n+1} - \alpha_w^n \rho_w^n}{\Delta\tau} V + \sum_f \rho_w^{n+1} U_f^{n+1} \alpha_{w,f}^{n+1} = 0$$

The value of the volume fraction at the current time step n + 1 is obtained through an iterative solution process. Therefore, when using the implicit method, in addition to momentum balance equations and the conservation equation, the transport equation of the volume fraction will also be solved iteratively at each time step. For this reason, the implicit scheme requires a computational effort greater than that required by the explicit scheme. Furthermore, when using the implicit scheme, the time step used in the calculation of the volume fraction equation is not defined starting from the Courant Number value, which, in this case, is not an input to be entered in the definition of the multiphase model. Therefore, there are no limitations related to the maximum value of the Courant Number, and the solution trend is stable regardless of the size of the time step adopted.

The implicit scheme allows an optimal solution of flows that have a stationary trend over time.

The sea wave motion is intrinsically a time-dependent problem; therefore, the explicit VOF model should be adopted.

8.7.1 Interpolation Near the Water–air Interface

When the simulated flow is biphasic, the solver also provides for the discretization and subsequent solution of the continuity equation of the volume fraction.

Both the explicit and implicit VOF models imply that for the cells that are completely occupied by one phase or the other, the interpolation of the quantities is carried out with the upwind schemes indicated for the other governing equations, whilst for the cells containing the interface (in which both phases are present), two specific interpolation schemes (see Fig. 8.6) can be used:

- Geometric-Reconstruction scheme;
- Donor–Acceptor scheme.

For wave motion simulations, the Geometric-Reconstruction interpolation scheme is used, which allows to obtain a better definition of the interface between the two fluids compared to the Donor–Acceptor scheme. Furthermore, the Geometric-Reconstruction interpolation scheme can also be used in the presence of unstructured meshes.

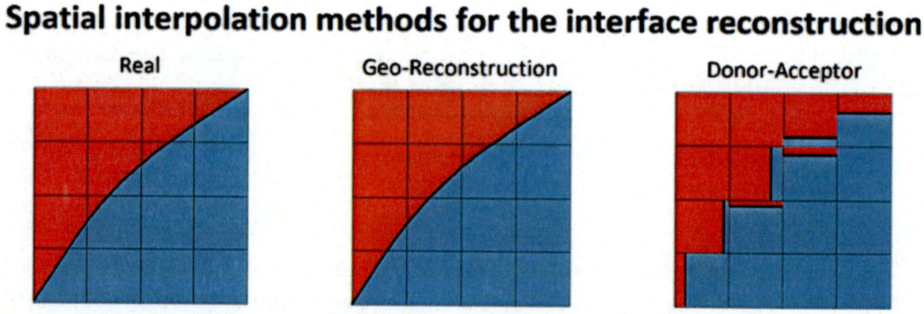

Fig. 8.6 Definition of the interface position with the different spatial interpolation methods for the VOF equation: (left) Real shape of the interface; (center) Shape of the interface according to the Geometric-Reconstruction scheme; (right) Shape of the interface according to the Donor–Acceptor scheme

Using the Geometric-Reconstruction scheme, the resulting separation surface between the two fluids is a piecewise linear curve in which each stroke derives from the tracing of the interface in a cell.

The process of calculating the position of the interface is divided into three steps. Initially, for each cell occupied by both fluids, the position of the linear interface with respect to the cell centroid is calculated based on information of the volume fractions of the two phases. Then, the amount of fluid flow through each face is evaluated based on the calculation of the normal and tangential velocities with respect to the surface on each face. Finally, on each face, the volume fractions are recalculated for the two phases based on the flow balance calculated during the previous step.

8.7.2 Wave Generation

In order to generate waves in a numerical flume, two approaches can be chosen:

- Using dynamic mesh to simulate the motion of a piston-type wavemaker;
- Using the open channel inlet boundary condition.

To simulate the wave propagation, it is fundamental to enable the gravity contribution by setting the gravitational acceleration fields in the operating condition dialog box. In the same dialog box, it is possible to set the operating pressure value (by default, the atmospheric pressure equals to 101,325 Pa) and its position. Here, having both air and water, it is preferred to define the operating pressure position in the region filled with air. This allows a reduction of the round-off error. Regarding the operating density, this needs to be set equal to zero due to the presence of the air; otherwise, an incorrect hydrostatic pressure will be built up also in the region which is filled with air.

8.7.3 Dynamic Mesh

The dynamic mesh model can be used in ANSYS Fluent when the shape of the computational domain changes over time due to the movement of a boundary. The mesh motion can be defined a priori when the velocity and the position of the center of gravity of the moving solid body are specified, or it can be the consequence of an event that occurs during the simulation, as in the case in which the displacements are calculated through the balance of the forces acting on the moving solid body.

In the simulations of sea wave generation, we impose the motion of the wall boundary, which acts as a piston. The piston motion is defined by the velocity field returned by a User Defined Function (UDF), i.e., a script file in C++ language.

```
/*********************************************************
* 1-degree of freedom equation of motion (x-direction)
* compiled UDF
*********************************************************/
#include "udf.h"
real pi = 4. * atan (1.);
real period = 6;
real stroke = 0.794;
DEFINE_CG_MOTION(piston, dt, vel, omega, time, dtime)
{
Thread *t;
face_t f;
/*get time*/
real flow_time = RP_Get_Real("flow-time");
/*get the thread pointer for which this motion is defined */
t = DT_THREAD(dt);
/* set x-component of velocity */
vel[1] = (stroke*pi/period)*sin(flow_time*2.*pi/period);
}
```

The UDF tells the boundary wall on the left side of the computational domain to move sinusoidally, starting from its farthest position from the U-OWC. It is worth noticing that the UDF defines the wall velocity rather than its displacement.

Since the computational domain contains both moving and fixed wall boundaries, the mesh must be divided into different zones, some of which should be subject to deformation with consequent cell volume variation.

When setting up a dynamic mesh, the moving boundaries must be specified, defining the motion parameters, the mesh zones that are subject to deformation, and the fixed mesh zone.

Using a dynamic mesh, the integral form of the conservation equation for a scalar quantity, ϕ, in an arbitrary control volume, V, with moving edges can be written as follows:

$$\frac{d}{dt}\int_V \rho\phi dV + \int_{\partial V} \rho\phi(\vec{u} - \vec{u}_g)\cdot\overrightarrow{dA} = \int_{\partial V} \Gamma_\phi \nabla\phi \cdot \overrightarrow{dA} + \int_V S_\phi dV.$$

Unlike the equations which are integrated over a cell that does not deform, here the velocity vector in the convective term is the vector difference between the velocity vector of the flow, $\overrightarrow{u}$, and the velocity vector of the cell center of gravity, $\overrightarrow{u}_g$, due to the dynamic mesh motion. The time discretization of the first term of the equation involves the calculation of the volume V^{n+1} of the cell at instant $n + 1$:

$$\frac{d}{\partial t}\int_V \rho\phi dV = \frac{(\rho\phi V)^{n+1} - (\rho\phi V)^n}{\Delta t}$$

in which:

$$V^{n+1} = V^n + \frac{dV}{dt}\Delta t$$

The time derivative of V is calculated from the continuity equation on the cell:

$$\frac{dV}{dt} = \int_{\partial V} \vec{u}_g \cdot \overrightarrow{dA} = \sum_j^{N_f} \vec{u}_{g,j} \cdot \overrightarrow{A_j}$$

where N_f is the number of faces delimiting the control volume while $\overrightarrow{A_j}$ is the area vector corresponding to the jth face. The dot product $\vec{u}_{g,j} \cdot \overrightarrow{A_j}$ is calculated as follows:

$$\vec{u}_{g,j} \cdot \overrightarrow{A_j} = \frac{\delta V_j}{\Delta t}$$

where $\delta V_j/\Delta t$ is the volume part moved during the timestep Δt.

8.8 Dynamic Mesh Update

The mesh volume update is automatically managed by ANSYS Fluent in each timestep, based on the position of the moving boundaries, using one of the three possible methods available:

- Elastic smoothing method;
- Layering method;
- Remeshing method.

The elastic smoothing method is then analyzed, being the one suggested for performing the CFD simulations of sea waves generation.

8.9 Elastic Smoothing Method

The elastic smoothing method considers the cell edges inside the deforming mesh region as a network of interconnected springs. The initial distance between the edges before each boundary movement represents the equilibrium condition of the mesh.

When the nodes belonging to the moving boundary move, elastic forces will be generated and transmitted to all the interconnected springs. In accordance with Hook's law, the overall force, $\bar{F}_i$, that the ith moving node globally transmits to the n_i nodes located at the other end of the springs connected to the ith node is the following:

$$\vec{F_i} = \sum_{j}^{n_i} k_{ij}(\Delta\vec{x_j} - \Delta\vec{x_i}),$$

where $\Delta\vec{x_i}$ is the displacement of the ith node, $\Delta\vec{x_j}$ is the displacement of the jth node, n_i is the total number of nodes connected to node i with a spring of elastic constant k_{ij} defined as follows:

$$k_{ij} = \frac{1}{\sqrt{x_i - x_j}}$$

.

For the balance of forces, the force on a node due to all the connected springs must be zero. Starting from this condition, the following iteration algorithm is built:

$$\Delta x_i^{m+1} = \frac{\sum_j^{n_i} k_{ij}\Delta x_j^m}{\sum_j^{n_i} k_{ij}}$$

Since the displacements of the nodes on the moving boundary are known, it is possible to find, by solving the iterative process with the Jacobi method, the displacement of all the internal nodes. Called $\Delta x_i^{m,conv}$ the displacement value for which the iterative process converges, the position of the ith node at time step $n+1$ will be calculated as a function of the position at the previous timestep n according to the formula:

$$x_i^{n+1} = x_i^n + \Delta x_i^{m,conv}$$

The elastic smoothing method can be applied to meshes with quadrangular cells (in the two-dimensional case) if the motion of the moving contour occurs mainly in one direction (see Fig. 8.7). Therefore, considering the occurrence of this condition, this is the suggested method to be applied for updating the mesh during the mesh motion.

However, it should be noted that the elastic smoothing method does not support any systems for replacing or reconstructing excessively deformed cells. For this reason, it could be occurred the presence of low-quality cells, which can lead to convergence problems, or even worse, the presence of cells with negative volume, which would lead to the immediate stop of the simulation. To have an optimal quality of the deformed cells, it is necessary that, in the preprocessing phase, the size of the mesh region that will be subject to deformation is appropriately defined. For

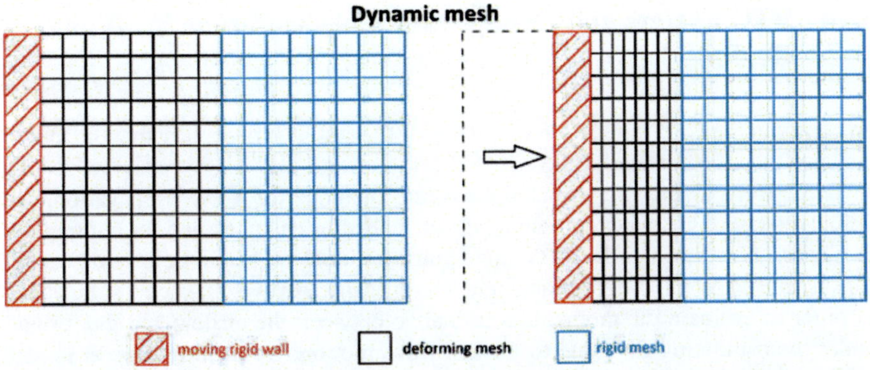

Fig. 8.7 Example of elastic smoothing applied to a 2D cylinder in the compression phase

instance, increasing this dimension of the deforming region will increase the number of deforming cells, but each one will be subjected to a minor deformation.

8.9.1 Open Channel Boundary Condition

The ANSYS Fluent CFD also allows the definition of a specific inlet boundary condition to simulate the propagation of waves, the Open Channel Wave Boundary Condition. This is a sub-model of the Volume of Fluid Multiphase model. After the selection of this sub-model, the corresponding dialog box becomes available in the velocity inlet boundary condition.

First of all, it must be specified the wave theory to be used and the Wave Boundary Condition Options. The available wave theories are First order Airy wave theory (which is a linear theory and the default one); Higher order Stokes Wave theory; Higher order Cnoidal/Solitary wave theory. The available options are shallow/intermediate waves or short gravity waves.

The right theory must be chosen based on the wave steepness, H/L, and the relative depth, H/h, where H is the wave height (the difference between the crest and the trough), L is the wavelength, and h is the water depth. In order to be able to apply the linear theory, the Ursell number, $Ur = HL^2/h^3$, should be lower than $32\pi^2/3 \cong 100$.

Finally, there is the need of specifying: the wave amplitude, the wavelength, the free surface level, the phase difference.

For the simulation under consideration, the wave amplitude is $H = 1.41$ m, the wavelength is $L = 53.07$ m, and the free surface level is $h = 15$ m.

8.10 PTO System and Configuration of the Porous Medium Region

To carry out the CFD simulation of the U-OWC plant, it is necessary that the model includes the effect of the PTO.

To perform a fully 3D simulation, which includes the wave propagation, the fluid dynamic of the water column inside the collector chamber, and the consequent oscillating flow through the PTO, huge computational costs and times are needed. This is caused by the different time and length scales involved in such a simulation. In order to evaluate the momentum exchange between the airflow and the turbine under the oscillating flow due to the sea wave motion, we need to have at least a mesh size of the order of the main turbulent structures close to the turbine blade ($\mathcal{O}(10^{-5}\,m)$) and a corresponding time step size ($\mathcal{O}(10^{-5}s)$); on the other hand, the wave motion has length scales of order $\mathcal{O}(1m)$ and time scale of the order $\mathcal{O}(1s)$. Furthermore, the physical time duration of a simulation should be in the range of tens of seconds. To resolve these scale problems and to be able to perform preliminary CFD simulations enabling us to estimate the performance of the WEC, we propose to perform a 2D simulation with a mesh refinement adequate for the investigation of the fluid dynamics inside the U-OWC but avoiding directly simulate the PTO and instead introducing only its effect. To do so, the air conduit which houses the PTO is substituted by a porous medium. A porous medium is a region of the computational domain where source terms can be introduced in the momentum balance equations, giving the same effect of the PTO in terms of pressure drop. In fact, the PTO simply acts as a damping term in the airflow motion of the OWC excited by the incident waves (e.g., [23]).

The use of a porous medium was firstly used by Torresi et al. in other applications (e.g., see [28, 29]) and opportunely adapted here. Several examples of similar approaches can be found in the literature. For instance, Didier et al. [30] explored the use of porous media to develop a numerical model of the PTO damping effect considering a linear relationship between the air pressure in the chamber and the airflow velocity across the turbine. The numerical results were validated by comparing them with the experimental results collected from a physical model in which the turbine damping effect was actually generated by a porous membrane (textile) placed on the top of the conduit.

The improvement of using the porous medium is that we correctly take into account the characteristic curve of the PTO, rather than simply approximating it as an orifice, as it is usually performed by several authors [31–33].

The porous medium characterization starts from the individuation of the PTO. In our CFD simulation, we suppose that the U-OWC is equipped with a monoplane Wells turbine (without guide vanes) geometrically similar to the one studied by Curran and Gato [34]. In Fig. 8.8, the characteristic curves are reported in terms of non-dimensional parameters (torque coefficient, T^*, pressure drop coefficient, Δp^*, and efficiency, η, vs. flow coefficient, U^*).

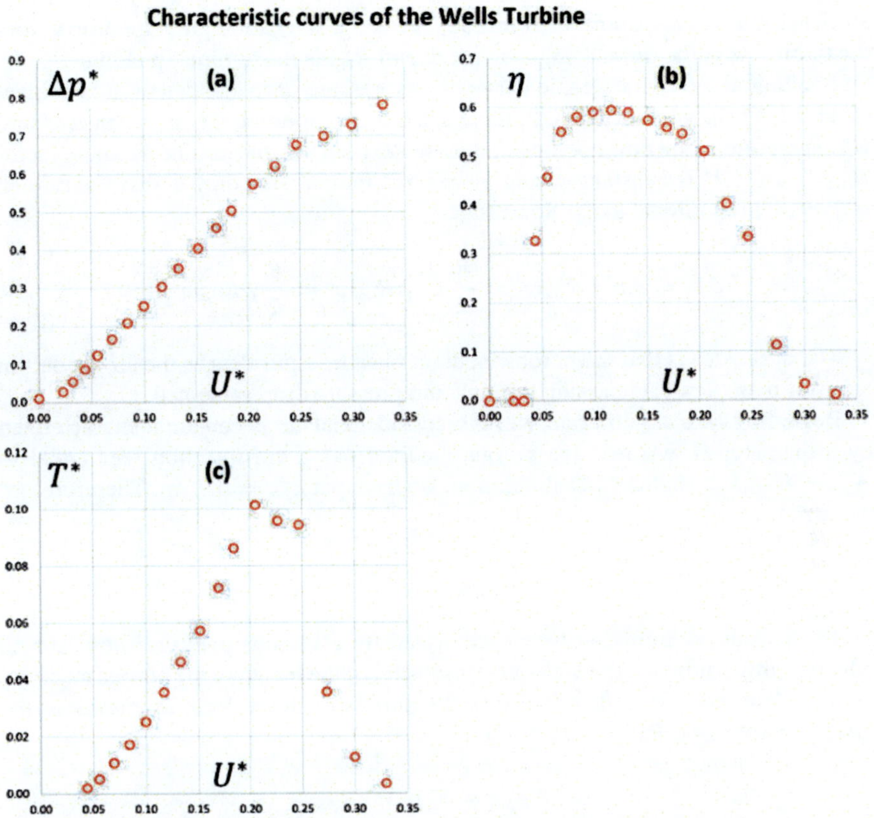

Fig. 8.8 Characteristic curves of the Wells turbine used as a PTO, similar to the one referenced by Curran and Gato [34]: a) non-dimensional pressure drop, Δp^*, vs. flow coefficient, U^*; b) turbine efficiency, η, vs. flow coefficient, U^*; c) non-dimensional shaft torque coefficient, T^*, vs. flow coefficient, U^*

$$U^* = \frac{V_a}{\omega R_{tip}}$$

$$T^* = \frac{T}{\rho R_{tip}^5 \omega^2}$$

$$\Delta p^* = \frac{\Delta p}{\rho R_{tip}^2 \omega^2}$$

$$\eta = \frac{T\omega}{Q\Delta p}$$

In the non-dimensional parameter definitions, $V_a = Q/A_a$ is the bulk velocity, Q the volumetric flow rate, A_a the annular area swept by the turbine blades, ρ is the air density, T is the turbine shaft torque, Δp is the stagnation pressure drop across the turbine, ω is the turbine angular speed, and R_{tip} is the turbine tip radius.

Looking at the characteristic curve of the pressure drop coefficient, it is evident that for U^* values lower than 0.2, there is a direct proportionality between Δp^* and U^*. Therefore, if the flow coefficient does not exceed 0.2, the turbine damping factor $B_t = 2\Delta p^*/U^*$ is constant, which means that the pressure drop across the turbine is directly proportional to the flow rate:

$$\Delta p_t = \Delta p^* \rho R_{tip}^2 \omega^2 = \frac{B_t}{2} U^* \rho R_{tip}^2 \omega^2 = \frac{B_t}{2} \rho R_{tip} \omega \frac{Q}{A_a}$$

The basic idea is that the porous medium must be equivalent to the air conduit in terms of mass flow rate passing through and pressure drop generated.

Regarding the mass flow rate, we must consider that the air conduit is characterized by a diameter D, whereas the porous medium has a cross-section area equal to $A_{pm} = b' \times h_{pm}$. In the CFD simulation, we have put $h_{pm} = 0.75$ m. Therefore, we can write:

$$\rho A_{pm} V_{pm} = \rho A_a V_a$$

The porous medium can be characterized by a viscous pressure drop, Δp_{pm}, which is proportional to the molecular viscosity, μ, the length of the porous medium, $\Delta l = 5.2$ m, and the bulk velocity in the porous medium, V_{pm}, by means of the viscous coefficient, R_v:

$$\Delta p_{pm} = R_v \mu \Delta l V_{pm}.$$

Therefore, equating the two pressure drop, which must be equal,

$$\Delta p_{pm} = \Delta p_t$$

we can determine the viscous coefficient of the porous medium, i.e., the inverse of the absolute porous medium permeability:

$$R_v = \frac{\Delta p_{pm}}{\mu \Delta l V_{pm}} = \frac{\Delta p_t}{\mu \Delta l V_{pm}} = \frac{\frac{B_t}{2} \rho R_{tip} \omega \frac{Q}{A_a}}{\mu \Delta l V_{pm}} = \frac{A_{pm}}{A_a} \frac{B_t}{2} \frac{\rho}{\mu} \frac{R_{tip} \omega}{\Delta l}.$$

The turbine damping factor of the Wells turbine proposed by Curran and Gato [34] is approximately equal to $B_t = 4.834$. Considering a geometrically similar Wells turbine ($h_{cg} = R_{hub}/R_{tip} = 0.65$) with a tip radius $R_{tip} = 0.375$ m and rotating at $n = 3200\ rpm$, we get a value of the inverse of the absolute porous medium permeability $R_v = 45490390\ \mathrm{m}^{-2}$.

8.11 Simulation, Data Saving, and Post-Processing

A rapid summary of the simulation setup is needed before the result analysis. The CFD code was ANSYS Fluent, release 17.2.

The first step is the reading of the mesh generated by means of any available software. Then, in the General Task Page (see Fig. 8.9), the user has to select the solver type (*Pressure-Based*), the Velocity Formulation (*Absolute*), the type of 2D Space (*Planar*). Being an unsteady simulation, the time formulation is *Transient*. For gravitational waves, the gravitational acceleration must be defined ($\bar{g} = -9.81\ \bar{j}$ [m/s^2]).

The next step (see Fig. 8.10) is the definition of the pressure velocity scheme (*SIMPLEC*), the technique to evaluate gradients at the cell faces (*Least-Square Cell-Based*), the pressure scheme (*PRESTO!*), the density scheme (*First Order Upwind*), the momentum scheme (*First Order Upwind*), the way to define the interface among different phases (*Geo-Reconstruction*), the energy conservation equation (*First Order Upwind*). Furthermore, the corresponding under-relaxation factor are assigned for each equation (see Fig. 8.11).

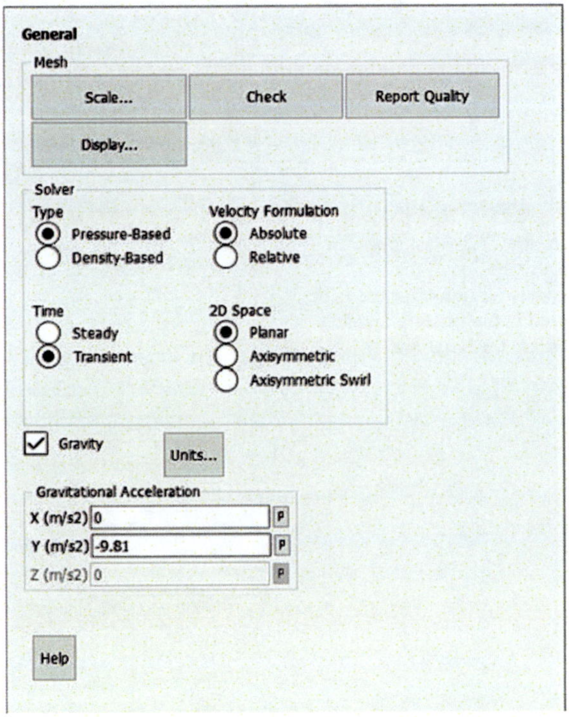

Fig. 8.9 ANSYS fluent, release 17.2—task page for the general configuration of the CFD model. It enables the user to set the solver type, the velocity formulation, the time formulation, the type of 2D space. Finally, the user sets the gravitational acceleration vector

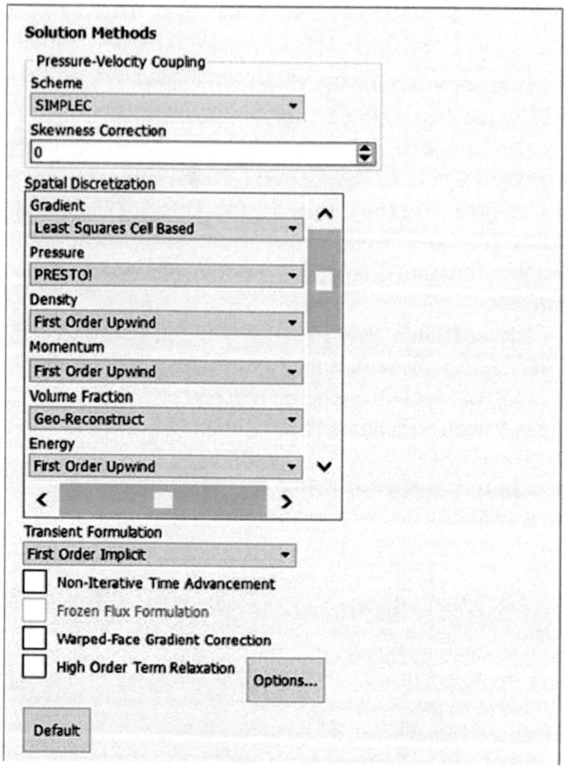

Fig. 8.10 ANSYS fluent, release 17.2—task page for the solution methods. It enables the user to set the pressure–velocity coupling scheme and the different spatial discretization schemes for the evaluation of the variable gradients, and the solution of the pressure field, the density field, the momentum balance equation, the volume fraction, and the energy equation

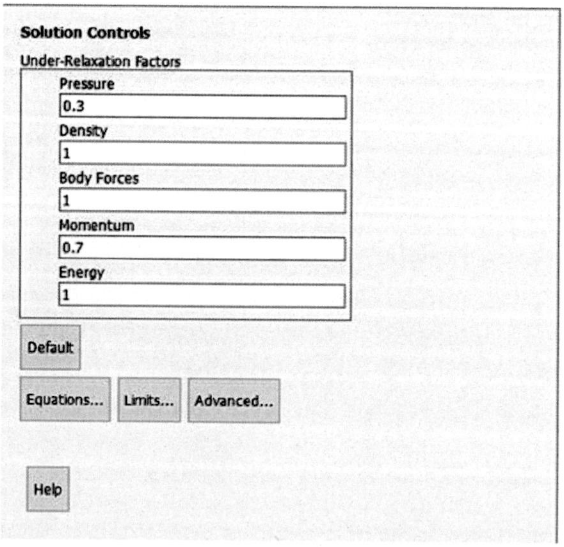

Fig. 8.11 ANSYS fluent, release 17.2—task page for the definition of the under-relaxation factors to be used in the variable update during the iterations

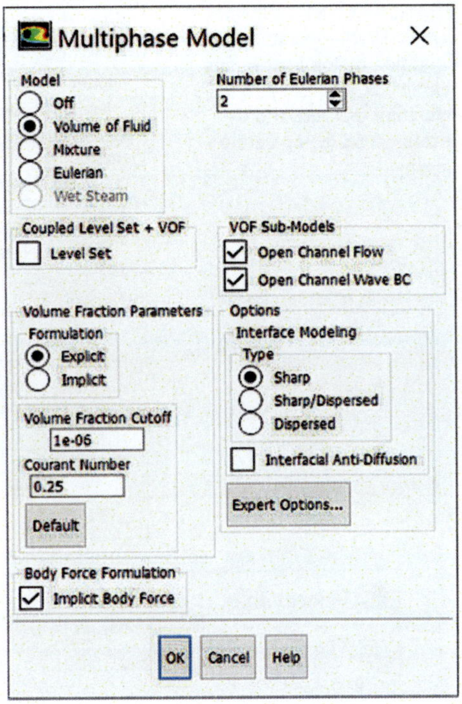

Fig. 8.12 ANSYS fluent, release 17.2—multiphase model dialog box. The user selects the multiphase model, its formulation and can activate the open channels VOF sub-model

In the Multiphase Model dialog box (see Fig. 8.12), the user sets the model (*Volume of Fluid*), the formulation (*Explicit*), and activates the *Open channel Flow* and the *Open Channel Wave Boundary Condition* VOF sub-models.

For the correct estimation of the pressure field in the presence of gravitational waves, it is suggested to set the operating conditions (see Fig. 8.13) as follows:

1. The operating pressure equal to the standard atmospheric pressure ($p_{atm} = 101,325\,\text{Pa}$) and its location in a point filled by 100% of the lightest phase (in our case, the air);
2. The operating density equal to zero (being the air compressible) in order to exclude the buildup of hydrostatic pressure within the lightest phase and improving the round-off accuracy for the momentum balance).

In this simulation, for the sake of simplicity, a single harmonic wave is considered, with a wave height equal to $H = 1.41\,\text{m}$, and a period $T = 6\,\text{s}$.

In order to set the CFD simulation, the Open Channel Dialog Box under the velocity inlet boundary condition must be selected (see Fig. 8.14). Here, after the activation of the Open Channel Wave BC, the user can set the parameter for the wave generation. The user sets the free surface level ($h = 15\,\text{m}$). Under the hypothesis of first-order Airy waves, the parameter to be defined are the wave height ($H = 1.41\,\text{m}$), the wave length ($L = 53.07\,\text{m}$), the phase difference (90° in order to start with a zero elevation), wave heading angle (0° being this a 2D simulation).

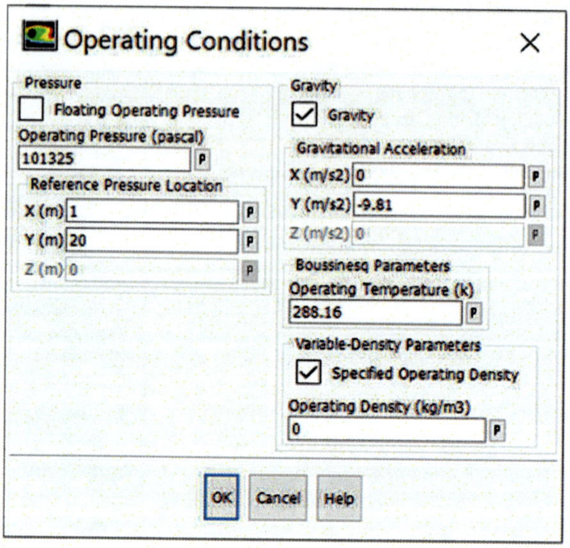

Fig. 8.13 ANSYS Fluent, release 17.2—operating condition dialog box. In this dialog box, the user sets the operating pressure and its location, and the operating density

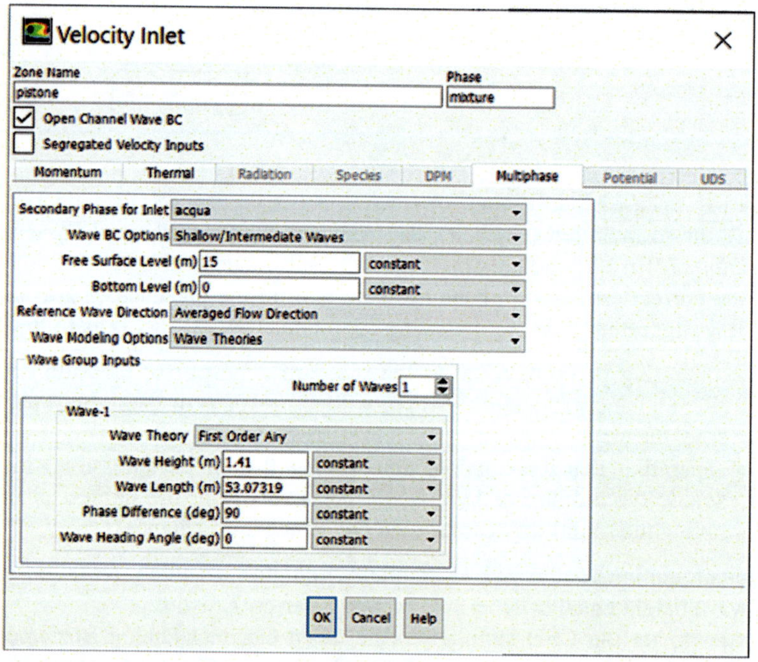

Fig. 8.14 ANSYS fluent, release 17.2—open channel dialog box. Under the Multiphase panel of the velocity Inlet dialog box, after the activation of the open channel wave BC, the user can set the parameter for the wave generation

The user needs to compute the correct wavelength to obtain the required wave period, T. According to the first order Airy wave theory, in the case of transitional water ($0.04 < h/L = 15/53.07 = 0.283 < 0.5$), the wavelength and the wave period are implicitly related:

$$L = c \cdot T = \frac{gT^2}{2\pi} \tanh\left(\frac{2\pi}{L} h\right)$$

Once the CFD model setup is completed, the simulation can start (see Fig. 8.15).

During the simulations, numerical data concerning the main properties of the analyzed flow need to be saved. The ANSYS Fluent software coordinates the data saving of the main properties of the flow through the Surface Monitors option. Data can be recorded on files at each timestep. The CFD code allows one to record either the time history of a property in a specific position or integral values defined in specific regions (computational domain zones, iso-surfaces, etc.). Among others, it is possible to define a particular iso-surface, which allows to visualize the water-free surface elevation. To do so, one needs to define an iso-surface over the entire computational domain of the water volume fraction, α_w, equal to 0.5. During the

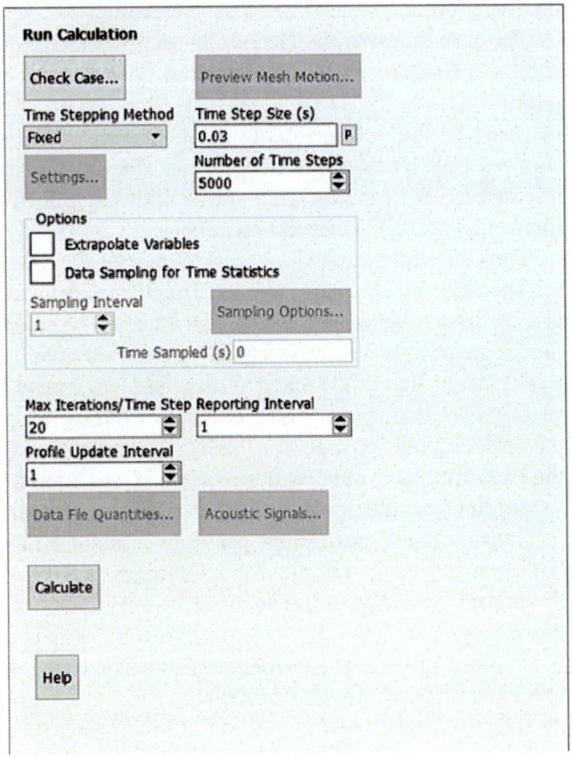

Fig. 8.15 ANSYS fluent, release 17.2—task page for running the calculation

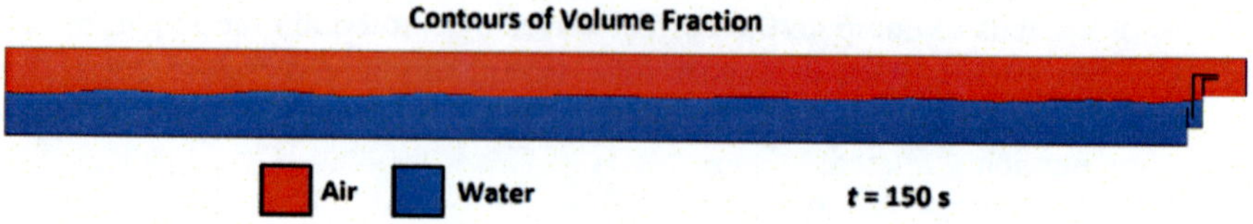

Fig. 8.16 Flow field visualization in terms of contours of volume fraction at $t = 150$ s

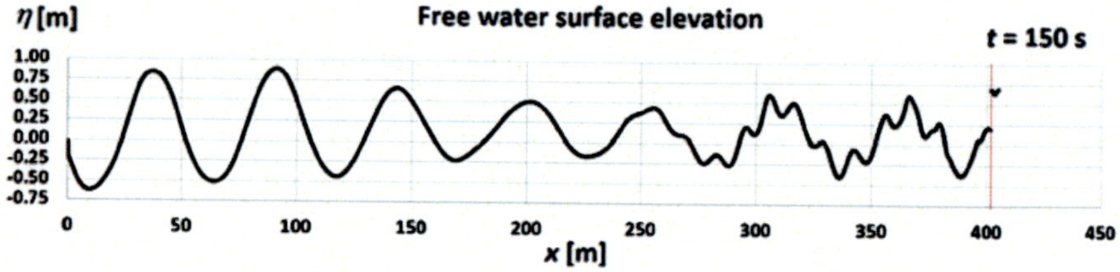

Fig. 8.17 Free water surface elevation at $t = 150$ s. The red line evidences the position of the U-OWC

simulation, the code is able to recompute the iso-surface following the dynamic movement of the water-free surface elevation.

The simulation of the U-OWC device was carried out for 5000 timesteps, equivalent to 150 s, since each timestep has been defined to be equal to 1/200 of a wave period ($\Delta t = T_{wp}/200 = 0.03$ s). Figure 8.16 shows the flow field in terms of contours of the volume fractions at the end of the simulation, whereas Fig. 8.17 shows the elevation of the free water surface at the same instant. It is possible to see the interaction of the incident waves with the reflected ones and the water elevation inside the U-OWC collector chamber.

The CFD simulation allows one to register the time history of the surface elevation at a specific distance from the inlet. The first thing to do is the creation of an iso-surface at a constant x value on the free water surface. Actually, the iso-surface corresponds to a single point being the intersection of two curves, i.e., the free water surface that is a curvilinear line in 2D and a vertical straight line. The point definition is dynamic, so when the free water surface changes during the simulation, the coordinates of the defined point change too. Finally, to register the surface elevation at a specific distance from the inlet, it is sufficient to monitor the average vertex value of the y-coordinate of the point.

Considering two different positions, distant 50 and 200 m from the inlet at the left boundary of the computational domain, it is possible to evidence how at 200 m the perturbation due to the generated waves at 200 m arrives later than at 50 m (see Fig. 8.18).

Another important parameter to monitor is the wave energy flux, Φ, across a vertical section, defined as follows:

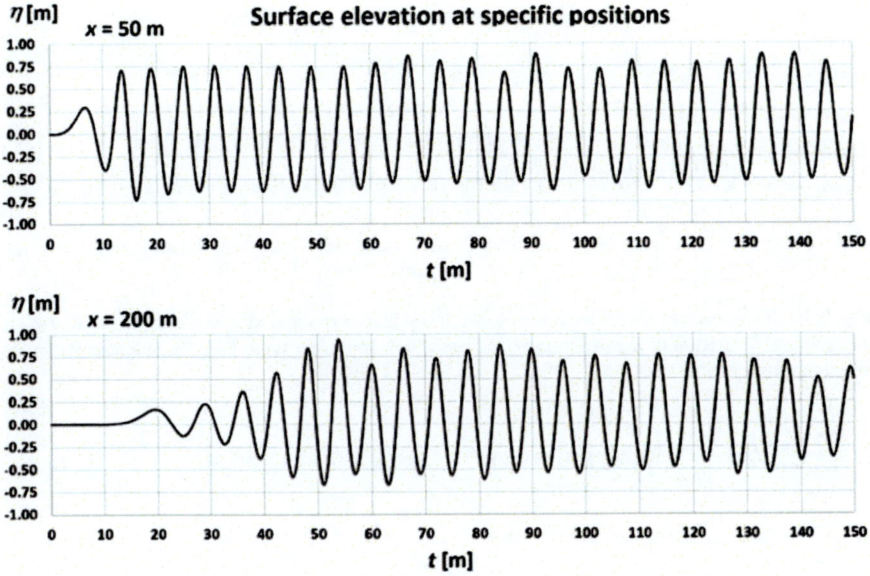

Fig. 8.18 Time history of the water surface elevation registered in two different points, 50 m and 200 m away from the inlet, at the left boundary of the computational domain

$$\Phi(t) = \int_{0}^{h+\eta} \Delta p(x, y, t)\, v_x(x, y, t)\, dy,$$

where $\Delta p(x, y, t)$ and $v_x(x, y, t)$ are the pressure perturbation from the hydrostatic pressure field and the horizontal velocity, respectively, at position (x, y) and time t. Starting from $\Phi(t)$, one can also compute the mean energy flux, $\bar{\Phi}(t)$:

$$\bar{\Phi}(t) = \frac{1}{T}\int_{t}^{t+T} \Phi(t)dt = \frac{1}{T}\int_{t}^{t+T}\int_{0}^{h+\eta} \Delta p(x, y, t)\, v_x(x, y, t)dydt,$$

In ANSYS Fluent, to compute the instantaneous energy flux, $\Phi(t)$, it is sufficient to define a custom field function that is the product of the water volume fraction (to limit the integral just to the water phase) multiplied by the difference of the gauge static pressure minus the hydrostatic pressure and the x-velocity. This custom field function is then integrated along a vertical line.

In Fig. 8.19, the time history of the wave energy flux across the vertical section at $x = 50$ m, whereas Fig. 8.20 reports the evolution of the moving mean of the energy flux, computed at the same vertical section.

According to the first order Airy wave theory, in the case of a wave height $H = 1.41$ m and a wave period $T = 6$ s, the mean energy flux, $\bar{\Phi}_{Airy}$, can be computed as

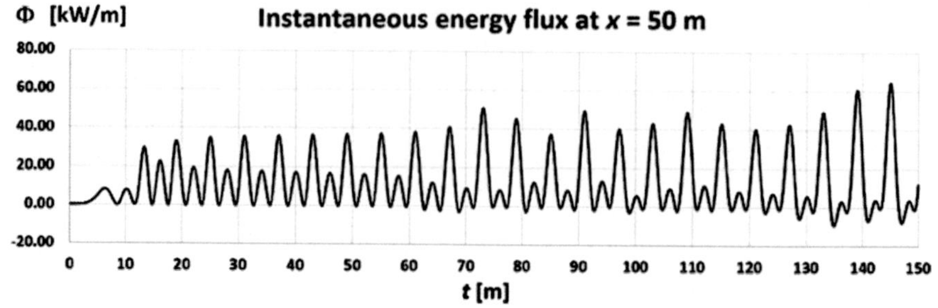

Fig. 8.19 Time history of the instantaneous energy flux computed at $x = 50$ m. In the first 20 s, we have the formation of the progressive waves, which last unperturbed till the reflected waves do not reach this section

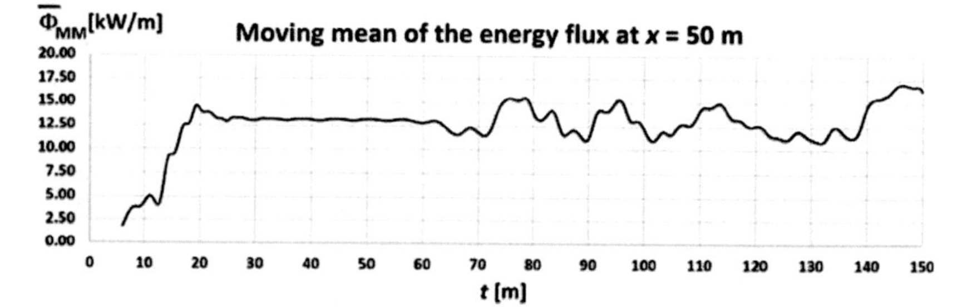

Fig. 8.20 Evolution of the moving mean (computed over a period T) of the energy flux at $x = 50$ m

follows:

$$\bar{\Phi}_{Airy} = \frac{1}{8}\rho g H^2 c_g = \frac{1}{16}\rho g H^2 \sqrt{\frac{\mathrm{g}}{\mathrm{k}}\tanh(kh)}\left(1 + \frac{2kh}{\sinh(2kh)}\right),$$

being $k = 2\pi/L$ the wavenumber.

In this simulation $\bar{\Phi}_{Airy} = 12.98\,\mathrm{kW/m}$. If we compute the mean energy flux starting from $t = 4T = 24\,\mathrm{s}$ over the next 6 periods, we obtain $\bar{\Phi}_{4T-10T} = 13.00\,\mathrm{kW/m}$, a very accurate result.

In order to evaluate the amount of the energy flux entering the U-OWC device, we have considered a horizontal section, 1 m below the U-OWC inlet.

In this section, the definition of the wave energy flux, Φ, is different since it refers to a horizontal section:

$$\Phi(t) = \int_0^l \Delta p(x, y, t)\, v_y(x, y, t)\, dx,$$

where $\Delta p(x, y, t)$ and $v_y(x, y, t)$ are the pressure perturbation from the hydrostatic pressure field and the vertical velocity, respectively, at position (x, y) and time t. In Fig. 8.21, the time history of the instantaneous energy flux computed 1 m below the U-OWC inlet is reported, and its moving means in Fig. 8.22.

Monitoring the volume flow rate (see Fig. 8.23) and the total pressure drop (Fig. 8.24) across the porous medium, we can evaluate the amount of energy subtracted to the U-OWC and made available to the PTO system. Actually, the flow rate is computed per unit depth since we are performing a 2D simulation. The volume flow rate is positive when the flow goes outside from the plenum. Regarding the total pressure drop, this is positive when the pressure in the plenum is higher than the atmospheric pressure outside the U-OWC.

The energy flux in the porous medium is computed by multiplying the volume flow rate and the total pressure drop in the porous medium (see Fig. 8.25). Furthermore, the moving mean of the energy flux is reported in Fig. 8.26.

From the air volume flow rate through the porous medium and the total pressure drop across it, it was possible to compute both Δp^* and U^* values. Their absolute values where reported on the Δp^* versus U^* diagram and compared with the experimental data of Curran and Gato [34] (see Fig. 8.27). The good agreement confirms that the porous medium has been correctly configured and is able to simulate the effect of the Wells turbine on the U-OWC fluid dynamic behavior.

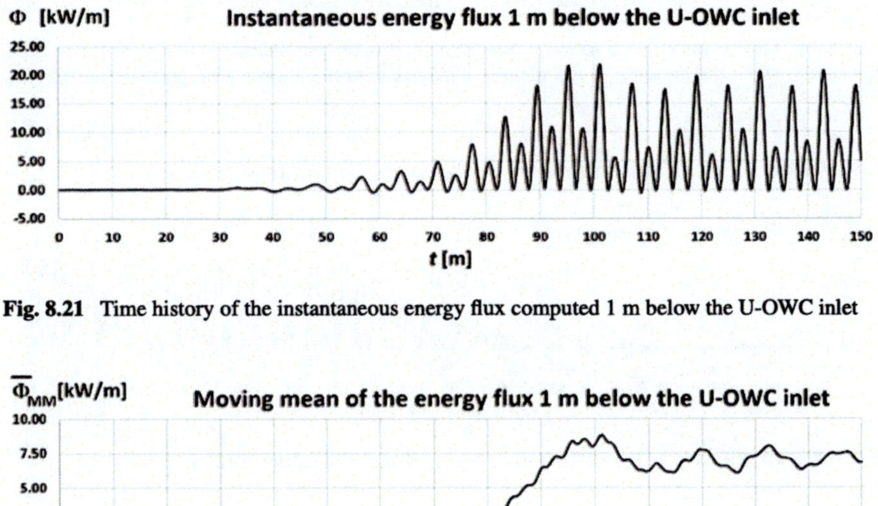

Fig. 8.21 Time history of the instantaneous energy flux computed 1 m below the U-OWC inlet

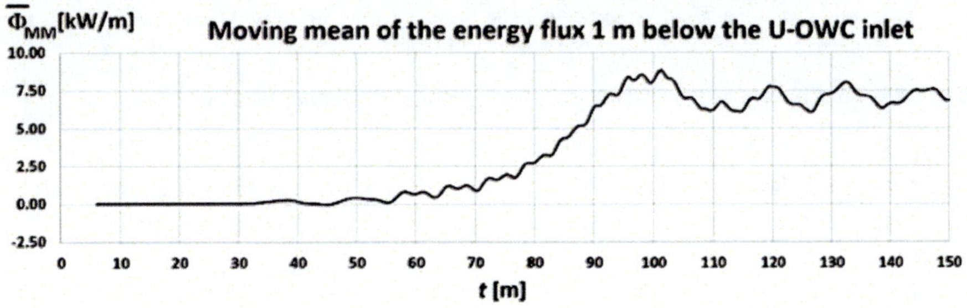

Fig. 8.22 Evolution of the moving mean (computed over a period T) of the energy flux 1 m below the U-OWC inlet

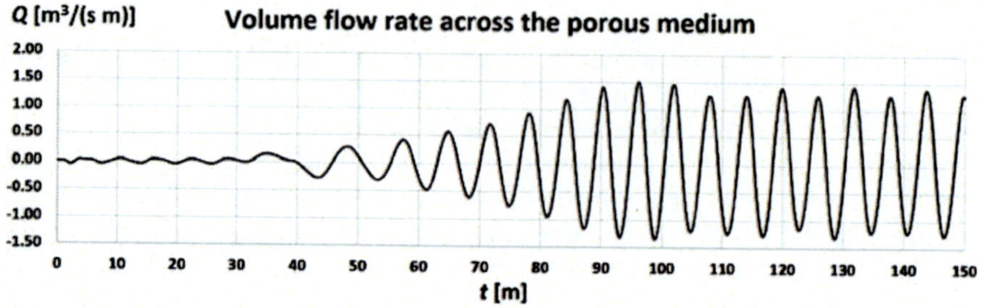

Fig. 8.23 Time history of the volume flow rate across the porous medium. The flow rate is computed per unit depth since we are performing a 2D simulation. The volume flow rate is positive when the flow goes outside from the plenum

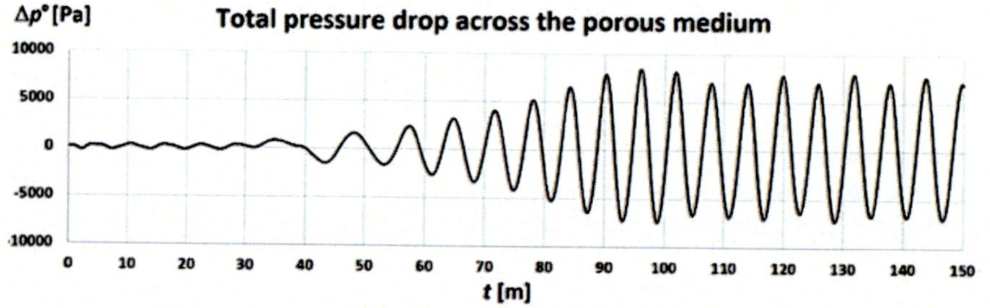

Fig. 8.24 Time history of the total pressure drop across the porous medium. The pressure drop is positive when the pressure in the plenum is higher than the atmospheric pressure outside the U-OWC

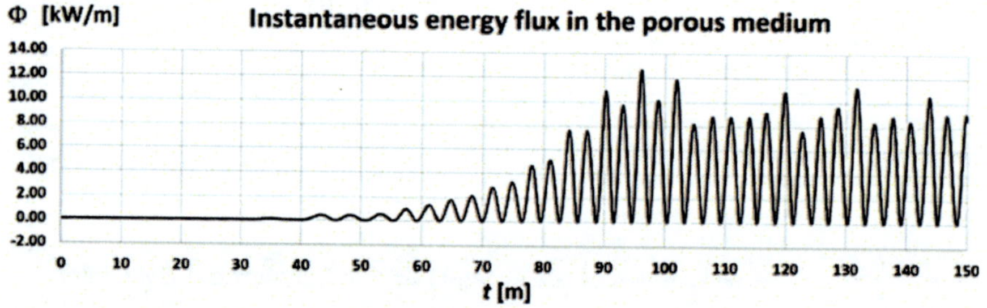

Fig. 8.25 Time history of the instantaneous energy flux in the porous medium

From those data, it was possible to determine both the turbine efficiency values (see Fig. 8.28) and the non-dimensional turbine shaft torque values (see Fig. 8.29) at each flow coefficient by interpolation of the experimental curves.

With these data, it is possible to evaluate the instantaneous Wells turbine power output (see Fig. 8.30) and to report the moving mean diagram (see Fig. 8.31). Clearly, these data are reported in terms of power per unit depth of the U-OWC device.

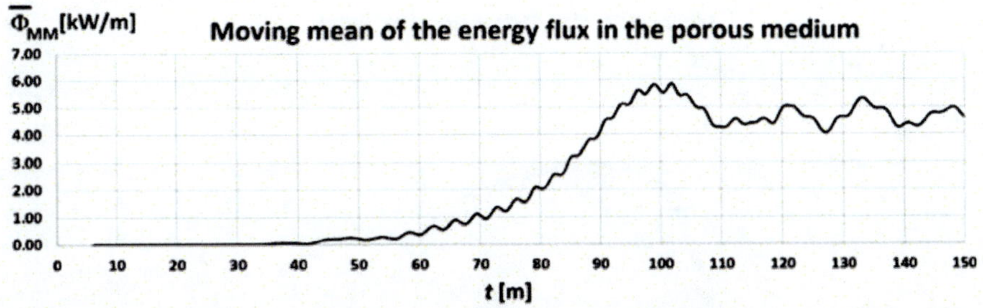

Fig. 8.26 Evolution of the moving mean (computed over a period T) of the energy flux in the porous medium

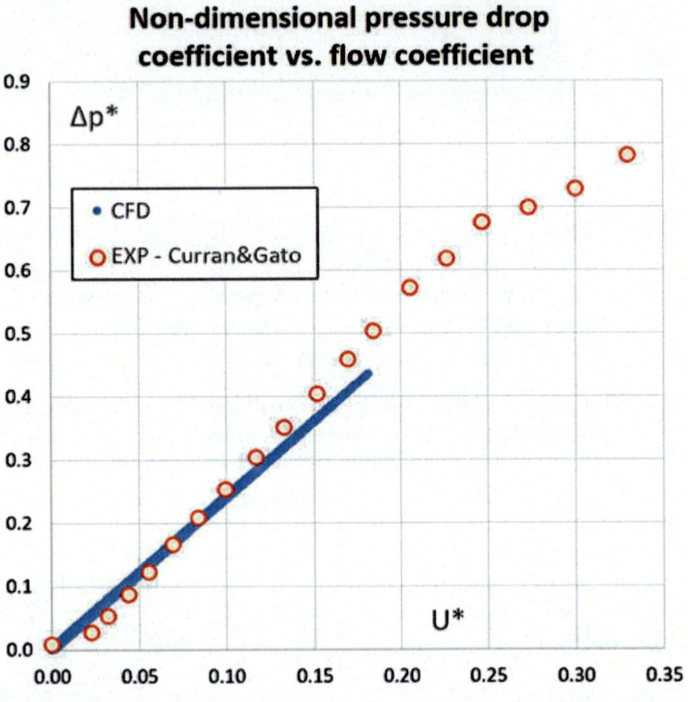

Fig. 8.27 Non-dimensional total pressure drop coefficient vs. flow coefficient: comparison of the computed characteristic curve against experimental data of the Wells turbine by Curran and Gato [34]

Averaging the results on the last 7 time periods (from $t = 108$ s to $t = 150$ s), it is possible to analyze the energy conversion process from the inlet wave energy flux to the Wells turbine shaft power.

The inlet wave power is equal to $P_{wave} = 13.00\,\text{kW/m}$. Actually, only 53.4% of this enters the U-OWC ($P_{abs} = 6.94\,\text{kW/m}$), whereas the remaining share is reflected

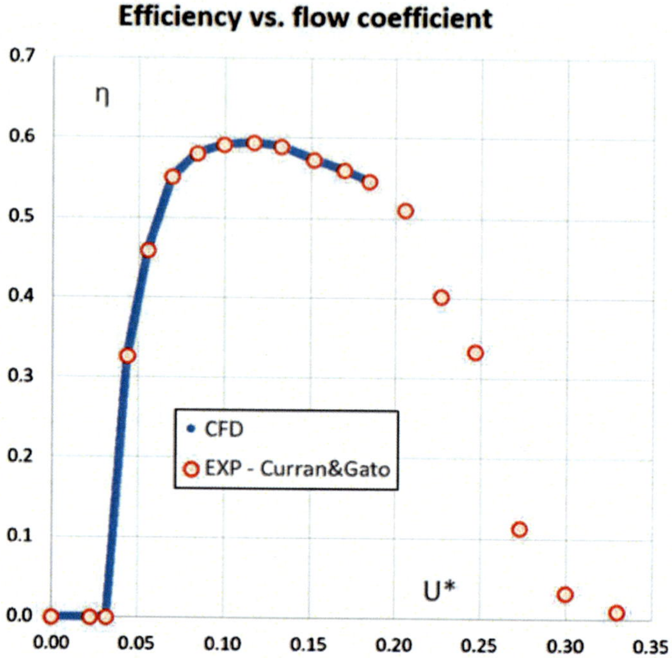

Fig. 8.28 Turbine efficiency versus flow coefficient: interpolation of the computed characteristic curve from experimental data of the Wells turbine by Curran and Gato [34]

and dissipated at the U-OWC inlet. Due to the fluid dynamic losses inside the U-OWC, only 66.3% of the absorbed power becomes pneumatic power ($P_\pi = 4.60\,\text{kW/m}$) available for the Wells turbine, which is able to convert only a 57.6% of it into shaft power ($P_{Wells} = 2.65\,\text{kW/m}$). This means that the wave energy conversion efficiency of the U-OWC is equal to 20.4%, as summarized by Table 8.1.

8.12 Conclusions

This book chapter gives the basis for the CFD simulation of an Oscillating Water Column (OWC) device. The interest is to investigate the interaction of the OWC device with the waves. For the sake of simplicity, single harmonic waves are considered. Referring to the ANSYS Fluent code, release 17.2, the user is guided in the model setup. Moreover, in order to limit the computational cost, the performed simulation is 2D. To correctly account for the presence of the PTO, the proposed solution is the introduction of specific source terms in the momentum balance equations by means of a porous medium in the region where the PTO in installed. The user is guided in the way to configure the porous medium and, in particular, the evaluation of the inverse of the porous medium permeability.

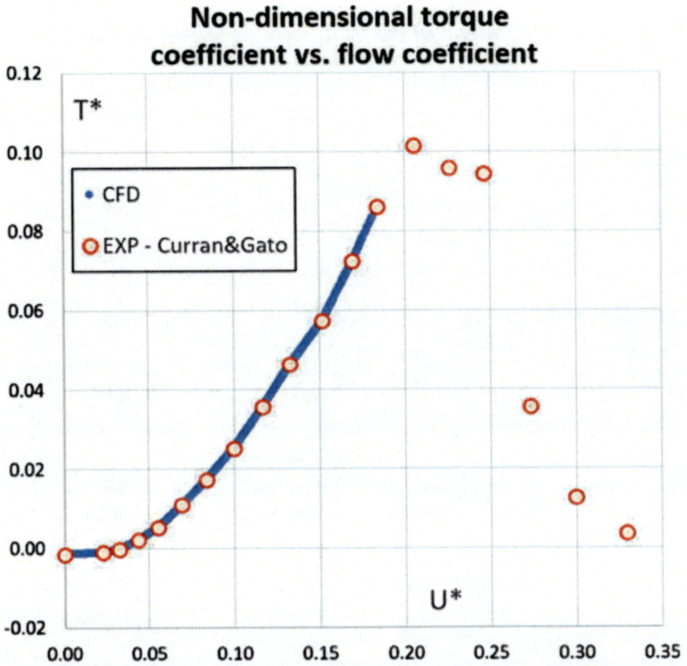

Fig. 8.29 Non-dimensional shaft torque coefficient versus flow coefficient: interpolation of the computed characteristic curve from experimental data of the Wells turbine by Curran and Gato [34]

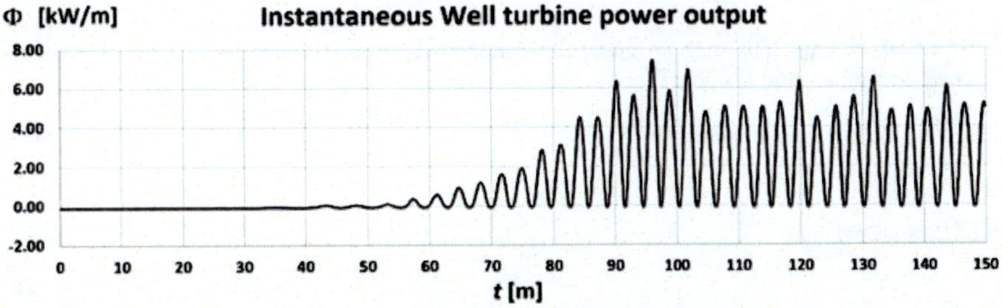

Fig. 8.30 Time history of the instantaneous Wells turbine power output

With the proposed approach, the user can perform a preliminary wave-to-wire energy conversion process investigation. In the described example, the incident wave power is equal to $P_{wave} = 13.00\,\text{kW/m}$. The U-OWC captures only a fraction of this power ($P_{abs} = 6.94\,\text{kW/m}$). Furthermore, due to the fluid dynamic losses inside the U-OWC, the pneumatic power available for the Wells turbine reduces to $P_{\pi} = 4.60\,\text{kW/m}$. At the end, the shaft turbine power is $P_{Wells} = 2.65\,\text{kW/m}$, with a wave energy conversion efficiency of the U-OWC equal to 20.4%.

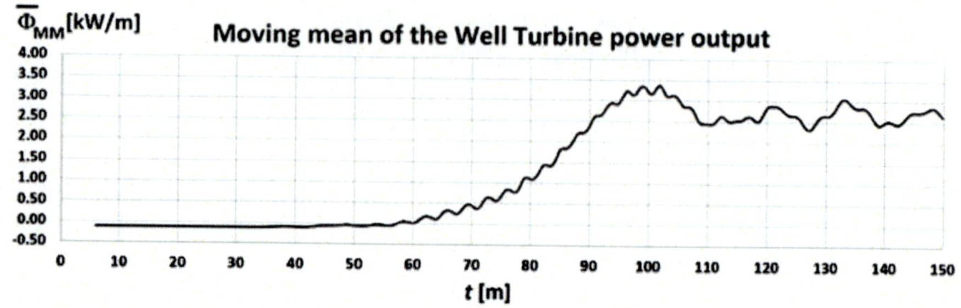

Fig. 8.31 Evolution of the moving mean (computed over a period *T*) of the Wells turbine power output

Table 8.1 The energy conversion process from the inlet wave power to the Wells turbine shaft power

Wave power (kW/m)	Power entering the U-OWC (kW/m)	Pneumatic power (kW/m)	Wells turbine shaft power (kW/m)
13.00	6.94	4.60	2.65
	Absorption efficiency	Pneumatic efficiency	Wells turbine shaft power (kW/m)
	53.4%	66.3%	57.6%
			Overall energy conversion efficiency
			20.4%

Acknowledgements The authors gratefully acknowledge the collaboration of Luana Gurnari, Michele Stefanizzi, and Daniele Pice.

References

1. Heath, T. V. (2012). A review of oscillating water columns. *Philosophical Transactions of the Royal Society A, 370*, 235–245. https://doi.org/10.1098/rsta.2011.0164
2. Cui, Y., Liu, Z., Zhang, X., & Xu, C. (2019). Review of CFD studies on axial-flow self-rectifying turbines for OWC wave energy conversion. *Ocean Engineering, 175*, 80–102. https://doi.org/10.1016/j.oceaneng.2019.01.040
3. Sharmila, N., Jalihal, P., Swamy, A. K., & Ravindran, M. (2004). Wave powered desalination system. *Energy, 29*(11), 1659–1672. https://doi.org/10.1016/j.energy.2004.03.099
4. Hotta, H., Washio, Y., Yokozawa, H., & Miyazaki, T. (1996). R&D on wave power device "Mighty Whale." *Renewable Energy, 9*(1–4), 1223–1226. https://doi.org/10.1016/0960-1481(96)88497-7
5. Boake, C. B., Whittaker, T. J. T., Folley, M., Hamish, E. (2002). Overview and initial operational experience of the LIMPET wave energy plant. In *Proceedings of the twelfth International Offshore and Polar Engineering Conference*, Kitakyushu, Japan.

6. Falcão, A. F. O., & Henriques, J. C. C. (2016). Oscillating-water-column wave energy converters and air turbines: A review. *Renewable Energy, 85*, 1391–1424. https://doi.org/10.1016/j.renene.2015.07.086
7. Zhang, D., Li, W., & Lin, Y. (2009). Wave energy in China: Current status and perspectives. *Renewable Energy, 34*(10), 2089–2092. https://doi.org/10.1016/j.renene.2009.03.014
8. Suzuki, M., Takao, M., Sato, E., Nagata, S., Toyota, K., & Setoguchi, T. (2008). Performance prediction of OWC type small size wave power device with impulse turbine. *Journal of Fluid Science and Technology, 3*(3), 466–475. https://doi.org/10.1299/jfst.3.466
9. Arena, F., Fiamma, V., Laface, V., Malara, G., Romolo, A., Strati, F. M. (2015). Monitoring of the U-OWC under construction in Civitavecchia (Rome, Italy). In *Proceedings of the 11th European Wave and Tidal Energy Conference (EWTEC)*, Nantes, France, 6–11 September 2015.
10. Torre-Enciso, Y., Ortubia, I., de Aguileta, L. L., Marqués, J. (2009). Mutriku wave power plant: From the thinking out to the reality. In *Proceedings of the 8th European Wave and Tidal Energy Conference* (Vol. 710, p. 319329), Uppsala, Sweden, 7–10 September 2009.
11. Wells, A. A. (1976). Fluid driven rotary transducer. Br. Patent 1595700.
12. Raghunathan, S., Tan, C.P., Wells, N. A. J. (1981). *AIAA Journal, 19*(11), 1490–1492. https://doi.org/10.2514/3.60084.
13. Curran, R., Stewart, T. P., & Whittaker, T. J. T. (1997). Design synthesis of oscillating water column wave energy converters: Performance matching. *Proceedings of Institution Mechanical Engineering Part A Journal Power Energy, 211*, 489–505. https://doi.org/10.1243/0957650981537375
14. Kinoue, Y., Kim, T. H., Setoguchi, T., Mohammad, M., Kaneko, K., & Inoue, M. (2004). Hysteretic characteristics of monoplane and biplane wells turbine for wave power conversion. *Energy Conversion and Management, 45*, 1617–1629. https://doi.org/10.1016/J.ENCONMAN.2003.08.021
15. Falcão, A. F. O., & Gato, L. M. C. (2012). Air turbines. Comprehensive. *Renewable Energy, 8*, 111–149. https://doi.org/10.1016/B978-0-08-087872-0.00805-2
16. Ghisu, T., Puddu, P., & Cambuli, F. (2015). Numerical analysis of a Wells turbine at different non-dimensional piston frequencies. *Journal of Thermal Science, 24*(6), 535–543. https://doi.org/10.1007/s11630-015-0819-6
17. Halder, P., Samad, A., & Thévenin, D. (2017). Improved design of a Wells turbine for higher operating range. *Renewable Energy, 106*, 122–134. https://doi.org/10.1016/j.renene.2017.01.012
18. Kumar, P. M., & Samad, A. (2018). Effect of blade profiles on the performance of bidirectional wave energy turbine. In *MATEC Web of Conferences*, 172, art. no. 06002. https://doi.org/10.1051/matecconf/201817206002.
19. Babintsev, I. A. (1975). Apparatus for converting sea wave energy into electrical energy. US Patent 3922739.
20. Setoguchi, T., Santhakumar, S., Maeda, H., & Takao, M. (2001). Kaneko K. *A review of impulse turbines for wave energy conversion, 23*(2), 261–292. https://doi.org/10.1016/S0960-1481(00)00175-0
21. Henriques, J. C. C., Sheng, W., Falcão, A. F. O., & Gato, L. M. C. (2017). A comparison of biradial and wells air turbines on the mutriku breakwater OWC wave power plant. In *Proceedings of the ASME 2017 36th International Conference on Ocean, Offshore and Arctic Engineering* (Vol. 10) Ocean Renewable Energy. Trondheim, Norway. June 25–30, 2017. V010T09A037. https://doi.org/10.1115/OMAE2017-62651.
22. Gurnari, L., Filianoti, P. G. F., Torresi, M., & Camporeale, S. M. (2020). The wave-to-wire energy conversion process for a fixed U-OWC Device. *Energies, 13*(1), 283. https://doi.org/10.3390/en13010283
23. Filianoti, P. G. F., & Camporeale, S. M. (2008). A linearized model for estimating the performance of submerged resonant wave energy converters. *Renewable Energy, 33*(4), 631–641. https://doi.org/10.1016/j.renene.2007.03.018

24. Malara, G., Romolo, A., Fiamma, V., & Arena, F. (2017). On the modelling of water column oscillations in U-OWC energy harvesters. *Renewable Energy, 101*, 964–972. https://doi.org/10.1016/j.renene.2016.09.051
25. ANSYS, Inc. (2016). ANSYS fluent theory guide, release 17.2. ANSYS, Inc. Southpointe, 275, Technology Drive Canonsburg, PA 15317.
26. ANSYS, Inc. (2016). ANSYS fluent tutorial guide, release 17.2 ANSYS, Inc., 275 Technology Drive Canonsburg, PA 15317.
27. Bakker, A. (2002). Free surface flow. Applied computational fluid dynamics.
28. Torresi, M., Saponaro, A., Camporeale, S. M., Fortunato, B. (2008). CFD analysis of the flow through tube banks of HRSG. In *Proceedings of the ASME Turbo Expo 2008: Power for Land, Sea and Air, GT2008-51300*, Berlin, Germany, 9–13 June 2008. https://doi.org/10.1115/GT2008-51300
29. Torresi, M., Postiglione, N., Filianoti, P. G. F., Fortunato, B., & Camporeale, S. M. (2016). Design of a ducted wind turbine for offshore floating platforms. *Wind Engineering, 40*(5), 468–474. https://doi.org/10.1177/0309524X16660226
30. Didier, E., Conde, J. M. P., Teixeira, P. R. F. (2011). Numerical simulation of an oscillation water column wave energy converter with and without damping. In *Proceedings of the fourth International Conference on Computational Methods in Marine Engineering* (pp. 206–217), Lisbon, Portugal, 28–30 September 2011.
31. Lopez, I., Pereiras, B., Castro, F., & Iglesias, G. (2014). Optimisation of turbine-induced damping for an OWC wave energy converter using a RANS-VOF numerical model. *Applied Energy, 127*, 105–114.
32. He, F., & Huang, Z. (2017). Characteristics of orifices for modeling nonlinear power take-off in wave-flume tests of oscillating water column devices. *Journal of Zhejiang University. Science, 18*, 329–345.
33. He, F., Li, M., & Huang, Z. (2016). An experimental study of pile-supported OWC-type breakwaters: Energy extraction and vortex-induced energy loss. *Energies, 2016*(9), 540.
34. Curran, R., & Gato, L. M. C. (1997). The energy conversion performances of several types of wells turbine designs. *Proceedings of Institution Mechanical Engineering, 1997*(211), 133.

Chapter 9
Numerical Modelling Techniques for Wave Energy Converters in Arrays

Vengatesan Venugopal, Zhi Yung Tay, and Tirumaleswara Reddy Nemalidinne

Abstract Wave and tidal renewable energy systems have received a great deal of attention in recent years worldwide. A number of Wave Energy Converter (WEC) technologies to harvest wave energy have been proposed, developed, tested and in operation in ocean at different parts of the globe with varying maturity levels. Over the past two decades, a broad range of academic research output has emerged, covering topics that include numerical and physical modelling of WEC arrays systems. However, to date, there have only been limited examples of WEC array installations, and these have been small in terms of the number of devices. Many developments in the future will focus on arrays of devices, as such an improved understanding of how arrays can be modelled is essential. This article attempts to describe the numerical modelling techniques used to study the hydrodynamic interaction of WEC arrays. An understanding of these techniques is necessary for not only device developers and researchers, but also for authorities, investors, insurers and other stakeholders. Such work will provide evidence for the expected energy output, control requirements, array configurations and to evaluate any environmental impact the array deployments may have on the ocean environment.

9.1 Introduction

Wave and tidal renewable energy systems have received a great deal of attention in recent years worldwide. Although several estimates for global wave energy potential are given in the literature, the Intergovernmental Panel on Climate Change (IPCC) in 2012 reported a theoretical potential of around 29,500 terawatt-hours per year

V. Venugopal (✉)
Institute for Energy Systems, School of Engineering, The University of Edinburgh, Edinburgh, UK
e-mail: V.Venugopal@ed.ac.uk

Z. Y. Tay
Engineering Cluster, Singapore Institute of Technology, Singapore, Singapore

T. R. Nemalidinne
DHI (India) Water and Environment Pvt Ltd, Delhi, India

A. Samad et al. (eds.), *Ocean Wave Energy Systems*, Ocean Engineering & Oceanography 14, https://doi.org/10.1007/978-3-030-78716-5_9

(TWh/yr) considering all areas with wave energy densities higher than 5 kW/m [1]. A number of device developers have been endorsed by large utility companies hoping to exploit these resources by developing various Wave Energy Converter (WEC) technologies. Over the past two decades, a broad range of academic research output has emerged, covering topics that include numerical and physical modelling of WEC arrays systems. The foundation for all research, however, remains in the 1970s and '80s with pioneering work by Salter, Budal, Falnes, Evans, Newman, Mei, Jefferys, Count and others. There are a number of fascinating accounts on the historical developments of wave energy extraction included in Salter [2], Clement et al. [3], Falnes [4], Drew et al. [5], Falcão [6], Langhamer [7], Lindroth [8], Moriarty [9] and Falnes [10]. These articles cover a range of subjects in the context of wave energy conversion. The European Marine Energy Centre (EMEC) in the UK has outlined device type categories [11]. The technology maturity levels of the various devices vary considerably. A degree of early commercial interest in some of the devices has helped identify the relatively advanced proposals. To date, there have only been limited examples of WEC array installations, and these have been small in terms of the number of devices. Many developments in the future will focus on arrays of devices, as such an improved understanding of how arrays can be modelled is essential. This article attempts to describe the numerical modelling techniques used to study the hydrodynamic interaction of WEC arrays. An understanding of these techniques is necessary for not only device developers and researchers, but also for authorities, investors, insurers and other stakeholders alike, to support backing for large scale developments. Such work will provide evidence for the expected energy output, control requirements, array configurations, environmental impact, etc.

9.2 Review of Hydrodynamic Modelling of WEC Arrays

Modelling of hydrodynamic interactions between floating, fixed and constrained rigid bodies with ocean waves and currents has been the focus of many studies in the field of marine systems. This has traditionally been carried out for the design of ships and military vessels, but also applies to oil storage and production platforms, harbours and coastal defence systems, and more recently, wave energy converters (WECs). Folley et al. [12] provide a review of hydrodynamic modelling methods for WEC arrays, giving benefits of each method by evaluating three characteristics: fundamental modelling ability, computational processing requirements and usability. This highlights issues such as code availability, stability and processing time. Folley et al. [12] also score the key numerical methods against their suitability to various types of study, showing relative merits. Another important review is that of Li and Yu [13], who also consider the numerical methods available, specifically for modelling point absorber arrays.

Two distinct groups of modelling approaches are apparent, the first including potential flow based codes (semi-analytical, boundary element method) and CFD solvers, both of which are well suited to addressing the device interaction problem.

Diffraction and radiation can be accurately modelled, giving realistic device responses within the scattered and radiated wave fields. What is more, there is a broad range of flexibility in terms of efficiency or accuracy. The potential flow methods include the point absorber (PA) [14–19], plane wave (PW) [19–22], multiple scattering (MS) [19, 23, 24], direct matrix (DM) [25–29] and boundary element methods (BEMs) [30, 31]. These are generally considered semi-analytical methods (with the exception of BEM) in that various assumptions are used to provide expressions for the velocity potential of the incident waves, the diffracted waves and radiated waves. Some of these assumptions that are made can remove or simplify certain components, for example, scattering effect omitted from the PA method.

9.2.1 Point Absorber Method

The point absorber (PA) method has seen a widespread application since it was introduced by Budal in 1977 [14]. The individual body dimensions are assumed to be small compared to incident wavelength allowing the scattering effect on the incoming wave to be ignored. The q-factor can therefore be taken from the solution of the radiation problem alone, which is determined using the excitation forces and damping coefficients. As with the other semi-analytical methods, the linear potential theory is used to formulate expressions to evaluate the velocity potential. It should be noted the term 'point absorber' is often used to describe floating buoy type WECs in a more general sense and does not always imply that the point absorber method is used.

The PA approximation has been used by Thomas and Evans [17] to assess the power capture of five and ten-body arrays of spheres, making use of the fact that with the PA approximation, no knowledge of the precise device geometry is required in order to compute the interaction factor q. The work carried out is essentially an array optimisation study, considering variations in wavenumber, device spacing and wave heading. For the range of spacings studied the maximum and minimum q-factors for unconstrained motions of the five-body array are found to be $q = 2.25$ and 0.57, respectively. These limits occur within a relatively narrow range of conditions ($ka =$ 5.1 and 6.75, respectively, where k is the wavenumber and a is the device radius), confirming a strong variance of q with device spacing. Constrained motions are also studied, as the device displacements are considerably larger than the incident wave amplitude and are in violation of linear wave theory.

Mavrakos and McIver [19] carried out detailed studies on arrays of five equally spaced truncated cylinders using the PA, PW and MS methods. As the 'exact' MS approach is known to accurately represent the total wave field around each body, this has been used to benchmark the comparative study. Two values are chosen for the ratio of device spacing over radius, i.e. $d/a = 5$ and 8, respectively. Mavrakos and McIver found that the results for the PA method show reasonable agreement for longer wavelengths, up to around $ka < 1.0$ and $ka < 1.4$ for $d/a = 5$ and 8 respectively, and beyond this large discrepancies are found. The better performance of the PA method at wider device separations is to be expected, as the no-scattering approximation applies.

9.2.2 Plane-Wave Method

The plane-wave (PW) method assumes wide device spacing relative to the incident wavelength, it ignores evanescent waves and approximates non-planar outgoing waves as plane waves. This produces a set of simultaneous equations for the plane wave amplitude, allowing straightforward assessment of the hydrodynamic interactions after the scattering and radiation problems are solved for a single device. In this respect, it can be regarded as a simplified variant of the direct method described in Sect. 9.2.4.

Initial work on the PW method emerged from Simon [20], in which simplifications were sought for representations of the scattered component of the wave field. The assumptions that allowed these simplifications were that the bodies were axisymmetric, widely spaced and body motions were in heave only, and the result was a method that allows a compromise between the efficiency of the PA method and accuracy of the MS method.

McIver and Evans [21] modified Simon's solution by including a correction term and consequently making a marked improvement to the accuracy. The PW method is applicable to a wide range of problems and has demonstrated good performance in evaluations of the wave amplitude down to a surprisingly close device spacing. McIver and Evans confirmed a good correspondence between body forces obtained using their method and that of the "exact" Spring and Monkmeyer method [25]. Linton and Evans [27] then extended the validation of the PW method by comparing the more onerous free-surface elevation with their rework of [22], also finding good agreement down to a relatively close device spacing (unlike the forces, the wave amplitudes are not integrated quantities, making the potential for errors much more significant).

In the comparative study by Mavrakos and McIver [22], the authors showed that the hydrodynamic forces can be accurately calculated using the PW method. Mavrakos and McIver found that the modulus of the complex exciting forces agrees well with the MS method, with small deviations of $<5\%$ around the peak forces. The phase of these forces also agrees well with the MS method, with the biggest discrepancies occur around the 'edge-of-array' bodies. There may however be errors in the forces measured within the array, with these errors increasing with array size. Erratic behaviour was found for very long wavelengths, due to an inversion procedure applied to the damping matrix. The onset of this instability of the PW method (shown in Fig. 9.7, in [19]) occurs at $ka \approx 0.45$ and $ka \approx 0.275$ for $d/a =$ 5 and 8, respectively. By normalising wave number to account for these two device spacings, the value of ka for which instability occurs appears to fall in the range 2.20–2.25. Pending additional work, this may provide a useful working limit for generic PW studies.

9.2.3 Multiple Scattering

The multiple scattering (MS) technique takes the superposition of the various propagating and evanescent waves scattered and radiated by the array. The computations are simplified due to the fact that there is no need to simultaneously retain the spectra of partial wave amplitudes around all the individual bodies, as the boundary conditions for each are satisfied successively. Ohkusu [24] applied it to floating bodies. The effect on a wavefield caused by any number of fixed obstructing objects is not just the sum of the scattering components of this incident wave at each obstruction, but the sum of the scattered incident wave plus the scattering of the scattered waves (first-order scattering), and scattering of these waves (second-order scattering), and so on. Twersky [23] asserted that these effects are not necessarily as insignificant as had been previously believed because phase alignments may cause small scattered interference to superpose into much larger spikes (as confirmed by existing experimental anomalies in the field of acoustics and electromagnetics). This lends itself well to the problem of interactions between closely spaced WECs in water waves where multiple scattering interference is generally significant. Ohkusu adopted this MS approach, adding radiating components for the case of floating (and hence oscillating) bodies. Mavrakos and Koumoutsakos [22] extended this methodology to also include evanescent waves with solutions for arbitrary numbers of vertically axisymmetric bodies (in any array configuration with any individual body geometries). Any number of orders of interactions may be obtained to give the total wave field.

9.2.4 Direct Matrix Method

The term *direct matrix* (DM) is associated with methods that perform an *exact* analytical assessment of a problem in which the boundary conditions are applied directly to all bodies simultaneously, allowing all unknowns to be evaluated in a single matrix inversion procedure. The velocity potential for any field point is calculated by taking the superposition of the incident wave and all scattered waves for each of the N bodies in the problem. To evaluate the wave amplitudes at each body, the set of equations that result are split into their real and imaginary parts. The force components on the bodies are finally evaluated after integrating the pressure on the wetted surface of each body.

Spring and Monkmeyer [25] first carried out this technique for a two-cylinder, bottom-mounted problem. They found that, when compared to a single cylinder, forces increase by as much as 60–65%. Spring and Monkmeyer also noted that the force ratios obtained from their studies are periodic with the variation of the device spacing and resemble the Bessel functions that are used to translate local coordinate frames. As the application of this particular method is limited to simple bottom-mounted geometries, it can only contribute to evaluations of certain types of WECs, such as OWCs.

Kagemoto and Yue [26] carried out an exact method (as far as is possible using linearized theory) to give the wave excitation forces, hydrodynamic coefficients and second-order drift forces in a selection of array problems, including that of floating bodies (and is in fact applicable to arbitrary configurations and numbers of arbitrarily shaped objects). This combines the work of Spring and Monkmeyer [25], Simon [20] and Ohkusu [24], becoming applicable to a far wider range of WEC problems due to the inclusion of evanescent waves. It is executed using only the diffraction interactions from a single body and hence removes the need to evaluate all of the n-order scattering components for the other bodies interfering with the incident wave. Kagemoto and Yue found excellent agreement of their solution with the *exact* hybrid element method (HEM) of Yue et al. [32], in which an entire array is modelled in one assessment.

Other more recent work on the DM method include Linton and Evans' [27] rework of Spring and Monkmeyer's method [25] in which a major simplification is identified. This again applies only to bottom mounted cylinders and may find more applications in studies of fixed offshore platforms. Linton and Evans ultimately found very efficient expressions for the free surface amplitude, for both the near-field and far-field regions.

Child and Venugopal [29] also used the direct method and optimised layouts of buoy type WECs by applying a *genetic algorithm* (GA) and alternative *parabolic intersection* (PI) approach. Using a partial wave notation, based on the assumption that the components can be combined linearly, a set of discrete expressions are used to describe the scattered and radiated wave fields. As GAs are simply optimisation routines, their application is not limited the DM method. Child and Venugopal demonstrated that the GAs are particularly well suited to problems to which an absolute optimum solution is not deemed essential. It would usually be assumed that this type of problem would take a long time to solve using conventional techniques and in many cases, there would be limited knowledge of the rational optimisation functions. Given the complex nature of interactions within an array of WECs, a GA routine can be used to rapidly generate acceptable solutions. The alternative PI method involves a more analytical approach, positioning devices on the peaks of waves that are diffracted by neighbouring devices. Each WEC is positioned in a way that maximises the constructive interference with its neighbours, so that the scattered waves are in phase with incident waves as they reach the neighbouring device. PI identifies very regular shaped array layouts, which tend to be symmetrical and highly sensitive to changes in incident wave direction. The highest q-factors obtained by Child and Venugopal in [29] come from the application of their GA method, rather than PI. In the GA case, it has been accepted that an element of randomness exists with the generation of the array shapes.

The second group of software codes (see Sect. 9.2.5) include those used for evaluating the far-field effects due to the presence of an array of WECs. These codes model wave propagation through a larger geographical scale domain, approximating the effects of energy extraction for a real ocean site with capacity to model ecological and morphological changes around a WEC array. These can be phase resolving or phase averaging techniques and can capture the various coastal processes that are observed on a medium to large scale.

9.2.5 Geographical Scale Studies

Global and regional scale numerical wave modelling are undertaken at various research and commercial institutions for wave forecasting and hindcasting, using models like WAM 4.5, WAVEWATCH III, SWAN, TOMAWAC, MIKE 21 and Delft 3D. More recent applications include the modelling of wave energy extraction processes and the resulting impact on wave climate. The scarcity of field data for this specific purpose means however that validation is problematic. Literature indicates that this methodology might serve as a preliminary assessment tool for energy extraction at real sites and also as a key technique for environmental impact assessments (EIAs). In general wave models can be classified as 'phase resolving' or 'phase averaged'. This section gives an explanation of the various methods and applications with this distinction in mind.

9.2.5.1 Mild-Slope Models

The mild-slope method treats wave diffraction and refraction over a mildly varying water depth, where it is known that wavelengths and amplitudes will also vary. The dependence of wave number k on water depth h can be inferred from the dispersion relation. A new expression is therefore required for the free surface elevation $\eta(x, y)$, on which a number of publications (including an early derivation by Berkhoff [33]) have reported variations of the same result [18]. The form used in the MILDwave solver [34] for example, which emerged from the work of Troch [35], takes the Radder and Dingemans [36] depth-integrated form of the mild-slope equations:

$$\frac{\partial \eta}{\partial t} = B\phi - \nabla \cdot (A\nabla\phi)\,, \quad \frac{\partial \phi}{\partial t} = -g\eta \tag{9.1}$$

where,

$$A = \frac{\bar{C}\bar{C}_g}{g}, \; B = \frac{\omega^2 - k^2\bar{C}\bar{C}_g}{g}.$$

Equation (9.1) describes the evolution of the free surface elevation with time, where either the velocity potential ϕ or free surface elevation η can be eliminated to give a single time-dependent mild-slope expression. Here, $\bar{C}$ is the phase velocity and $\bar{C}_g$ is the group velocity.

The provision for the spatial dependency makes the mild-slope equations well suited to near coast problems with traditional application to coastal defence, breakwaters, harbour design and environmental impact assessments. This has been readily extended to applications in wave energy conversion processes, particularly for fixed WECs, e.g. overtopping devices and OWCs. Power extraction is not implicitly solved, however, sponge layers have been shown to give effective representations of WEC absorption characteristics.

Several studies have also applied the mild-slope method to problems with floating bodies. Mendes et al. [37] carried out an assessment of the Pelamis device installed at the offshore pilot zone in Portugal using REF/DIF mild-slope software with some auxiliary modifications. Beels [38] went a step further and carried out an evaluation of a motion-based device (FO3) in MILDwave and included radiated wave inputs from an independent WAMIT solution. This approach takes the radiation outputs from WAMIT and reproduces a corresponding and similar wave train in the MILDwave model from a circular generation line that surrounds the WEC. These results are compared to the full WAMIT diffraction and radiation results for a single FO3 device and show reasonable agreement. Application of the mild-slope solver in this way avoids the unmanageable burden on computational resources that would otherwise be necessary to model a large scale array of multiple-float devices.

A common problem with phase-resolving models is a suitable treatment of the domain boundaries. Conditions must be specified at domain boundaries so that waves are not reflected towards the area of interest. This includes the boundaries that lie behind any wave generation lines (absorption boundaries and generation lines are discrete).

For irregular and omnidirectional long-crested waves, Beels [38] gave a good comparative study of generation line configurations. A U-shaped configuration (two parallel lines with connecting arc) is found to be the most effective at generating wave components in the range $0° < \theta < 90°$. Short crested waves are also considered, with the directionality of the separate components accounted for through application of a spreading factor s. The parameter s_{max} is used as a limit for the cases of wind-generated seas ($s_{max} = 10$), swells with short decay distances ($s_{max} = 25$) and swells with long decay distances ($s_{max} = 75$).

Finally, Beels includes a single obstruction to represent an array of devices. The power captured by the simplified single interference is shown to be $\pm 40\%$ that of more detailed evaluations of the array. On the basis of this result, Beels suggests that any economic evaluations of proposed schemes should take a more onerous route, modelling discrete WECs in a solver that can account for interactions. Ultimately, these types of mild-slope studies remove the difficult coupling of the reflection and transmission which is present in Boussinesq methods but also allows detailed geometry modelling within the array as a result of the small grid size, something which is not possible with the spectral wave models like SWAN.

9.2.5.2 Boussinesq Models

The linear theory discussed above requires the ratio of wave amplitude and water depth to be small, i.e. the nonlinearity $A/h << 1$. When studying a wave as it approaches a coastal area, the combined effects of reduced water depth and shoaling means $A/h > 1$, thus the hydrodynamics can become highly nonlinear. The Boussinesq equations are well suited to this type of problem as they omit the vertical components of the wave equations. This happens to be a reasonable assumption that addresses the fact that the circular particle motions associated with deep water waves become distorted in shallow water. Another important parameter in the derivation of the Boussinesq equations is the dispersion $\mu^2 \equiv (kh)^2$, wherein the limiting cases $(A/h) \sim 0$ and $\mu^2 \sim 0$, the solution becomes linear. Numerical codes that implement variations of the classical Boussinesq equations include the MIKE21 Boussinesq Wave software, first developed by Madsen and Sørensen [87].

Boussinesq models have been used in the context of wave energy conversion to model interactions around fixed structures such as OWCs or overtopping devices. Venugopal and Smith [39] used a Boussinesq model (MIKE21) to investigate changes to the wavefield around an array of five hypothetical overtopping devices. Sponge layers are configured around the boundary of a domain (measuring 5×4.5 km) representing a coastal region to the west of Orkney. Waves are generated along lines within the domain, with input conditions provided by a much larger spectral wave model (130×110 km, also validated against buoy data). Venugopal and Smith evaluated the wave disturbance coefficients h_{wdc} (local significant wave height, normalised by the input significant wave height) in the direction of wave propagation at each WEC location (waves are irregular and unidirectional). Results showed largely consistent profiles, with discrepancies in part caused by differing bathymetries at each specific location. Venugopal and Smith found that the wave amplitude is between 13 and 69% lower in the lee of the WECs depending on the porosity. The wave amplitudes recover briefly at a distance of around 500–600 m in the lee of the array regardless of porosity, which might suggest a suitable location for a second row of WECs.

Venugopal et al. [40] used the Boussinesq modelling code MIKE 21 to investigate h_{wdc} around an array of bottom fixed OWCs. A significant challenge when studying OWCs is the quantification of radiation characteristics. This includes representation of the PTO mechanisms, as it is not possible to model the airside processes of the OWC as an integrated part of the Boussinesq solver. An iterative investigation is carried out at each wave period in order to resolve sponge layer. The authors concluded that the array spacing and peak wave periods are both factors that affect wave disturbance coefficients. The maximum disturbances experienced under the incident JONSWAP wave condition peaked at +39% upstream and −41% downstream of the array. The benefit of using the Boussinesq codes in this type of application is that they are capable of resolving a comprehensive set of hydrodynamic mechanisms. This includes the combined effects of diffraction, refraction, shoaling, wave breaking, nonlinear wave-wave interactions, bottom dissipation, partial wave reflection and transmission from structures, directional wave spreading and internal wave generation. Solid and permeable structures can be modelled, however, objects in the flow domain must be fixed to the ocean floor (e.g. OWCs), as there are no means to account for the dynamics of a moving object.

9.2.5.3 Phase Averaged Spectral Wave Model

Spectral wave models are based on the principle of wave action conservation, where wave action is a ratio of spectral density to intrinsic frequency. In contrast to other methods that focus on predictions of the surface elevation, the spectral wave methods are phase-averaging, predicting how the characteristics of a wave evolve with time. This method originated from the idea that the ocean waves could be decomposed into components of various frequency, amplitude and direction. The JONSWAP [41] project was an important step in the development, with first, and then second-generation models emerging from such work, with the latter including parameterised representations of nonlinear waves generated by the wind.

A later collaboration titled WAM provided the first of the third generation models that explicitly included these nonlinear waves. WAM is still commonly used, along with other global wave prediction models such as WAVEWATCH III, to provide input boundary conditions to other solvers intended for local or coastal interaction studies, such as SWAN (Delft University of Technology) and TOMAWAC (Electricité de France). Global wave models (WAM, WAVEWATCH III, etc.) and local models (SWAN, TOMAWAC, MIKE21 SW, etc.) work on similar principles, with the latter including additional descriptions for shallow water behaviour. Local models can theoretically execute oceanic scale problems however this inevitably proves unnecessary and inefficient.

It is common for both model scales to use a spherical coordinate system however it can be more convenient to work with a Cartesian system at the smaller scale. Discretization of the local domain can be achieved using structured or unstructured grids. Structured grids can feature more refined nested grid(s) in regions of greatest interest, whereas with an unstructured grid it is possible to have a continuously variable cell density. Care must be taken to ensure that the grid captures all relevant seabed features and avoids oversimplify step changes, as model bathymetry is interpolated between grid points. At Wave Hub, Millar et al. [42] and Smith et al. [43] used SWAN to examine the impact on the surrounding wave climate following a reduction in spectral energy density at the array location. Various configurations of barrier and transmission coefficients are used. Millar et al. [42] impose this energy reduction equally across all wave frequencies, while in the subsequent study, Smith et al. [43] employ a frequency-dependent power transfer function (PTF). The latter study is understood to be a significant improvement on [42] in terms of accurately estimating the wave height reduction, but both studies present useful applications of spectral wave models.

9.3 Boundary Element Methods

The boundary element method (BEM) is a well-established technique used to study floating bodies for which again potential flow theory is assumed. Wetted surfaces are panelled, either into planar surface elements curved patches (e.g. Higher Order in WAMIT) using splines. A normal force is then applied at the centre of each element, inducing a body motion as a result of fluid pressure fluctuations around the body. BEMs can be simplified by formulating in the frequency domain, however, use of the time domain enables the inclusion of non-linear external loads and non-linear hydrodynamics.

An outline requirement of BEMs is that there must be an appropriate Green's function solution to translate the volume problem into a surface problem and hence evaluate the velocity potential in the whole fluid domain. Added complexity when dealing with nonlinear codes includes discretisation of the free surface which changes upon every time step, allowing any floating bodies to reorientate prior to the application of the Green's function.

Codes such as WAMIT, ANSYS AQWA, Aquaplus and Nemoh operate in the frequency domain, and are generally used to provide matrices for the added mass A_{ij}^{added}, added damping B_{ij}^{added} and excitation force F_i. These terms are required to evaluate the complex excursion amplitudes ξ_j using the 6×6 linear system based on Newton's law (2).

$$\left[-\omega^2\left(M_{ij} + M_{ij}^E + A_{ij}^{added}(\omega)\right) + \mathrm{i}\omega\left(B_{ij}^{added}(\omega) + B_{ij}^E\right) + \left(C_{ij} + C_{ij}^E\right)\right]\xi_j = F_i, \quad \text{for } i, j = 1, 2, \ldots, 6 \tag{9.2}$$

where M_{ij} is the inertia matrix and C_{ij} is the matrix of hydrostatic and gravitational coefficients. The externally applied inertia M_{ij}^E, damping B_{ij}^E and stiffness C_{ij}^E matrices in Eq. (9.2) can be tuned to control the motion of the WECs in a desirable way, such that the power output is maximised. It is common to either tune the external damping alone (real tuning) or the external damping and stiffness (reactive tuning). Applicability of the linear BEM codes requires that the hydrodynamic problem is suitably linear, and that the motion of any floating bodies remains small compared to the wave height. Larger excursions that can alter for example the hydrostatics of a WEC, must be captured in the time domain.

The WAMIT is used to determine A_{ij}^{added}, B_{ij}^{added}, F_i and response amplitude operators $\bar{\xi}_j$ (non-dimensional excursion amplitudes). Equation (9.3) highlights the principle steps required in the procedure that follows, in which the excursion amplitudes are normalised, the auxiliary power P_{aux} is evaluated (expressed in W/m^2), the average absorbed power per farm element $\bar{P}_n$ is found and summated to give the average absorbed power by the array.

$$P_{aux} = \frac{1}{2}\omega^2 B_{ij}^E |\bar{\xi}_j|^2 \Rightarrow \bar{P}_n = \int 2P_{aux} S(\omega) d\omega \Rightarrow \bar{P} = \sum_n^N \bar{P}_n \tag{9.3}$$

As discussed above, a plethora of methods have been developed and applied in WEC arrays modelling. This article provides the application of two methods, namely the Boundary Element Method and Phase Averaged Spectral Wave Model for evaluating the wave-device array interactions, where the BEM is used in assessing the performance and wave climate of the WEC array in nearfield (Sect. 9.4), whereas the Phase Averaged Spectral Wave Model is used in the ocean scale (Sect. 9.5).

9.3.1 Problem Definition

This paper considers arrays of two different types of WECs, i.e. the terminators (e.g. Oyster WEC) and attenuators (e.g. Pelamis WEC). The terminator type WEC is shown in Fig. 9.1(a), which comprises a flap-type floating body hinged at the bottom to a foundation, hence only allowing for the rotational motion about the hinge (pitch, Θ_y). Note that the local-coordinate (x, y, z) is located at the hinge. The flap has a width a_f, immersion depth d_f, thickness h_f and the hinge is located at a height c_f from the seafloor. The attenuator type WEC is shown in Fig. 9.1(b), which comprises four modules connected by three hinges. The modules are allowed to pitch (Θ_y,

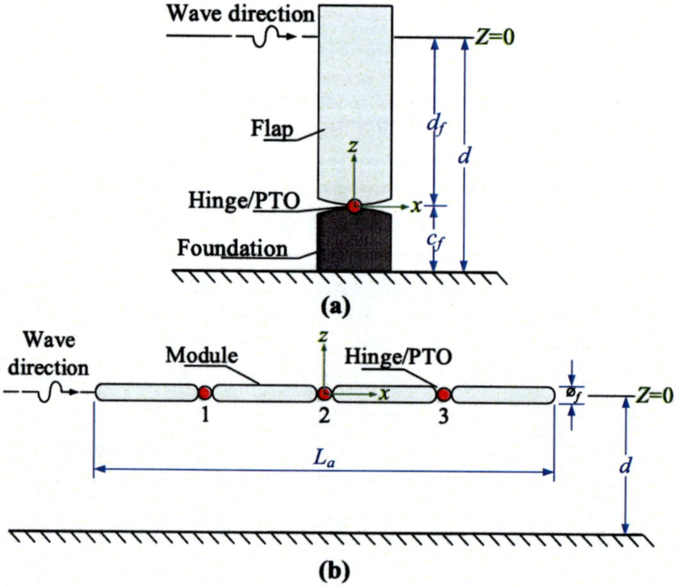

Fig. 9.1 Dimensions for: a Terminator WEC, b Attenuator WEC

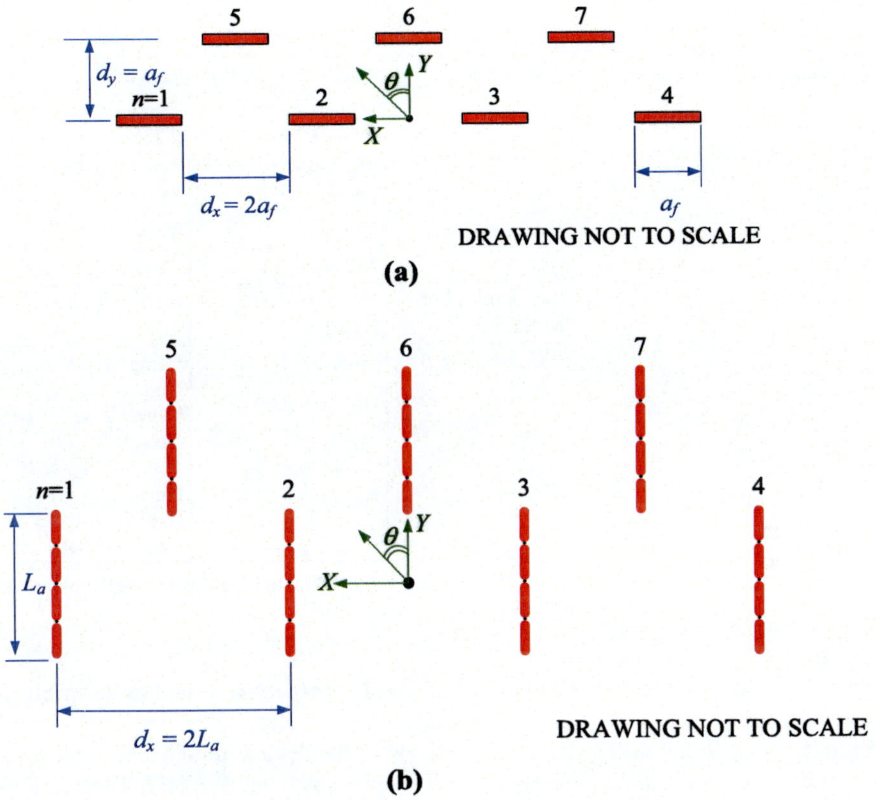

Fig. 9.2 Array layout for: **a** Terminator WEC, **b** Attenuator WEC

rotation about y-axis) and yaw (Θ_z, rotation about z-axis) at their connected hinges. Note that the local-coordinate (x, y, z) is located at the hinge 2 (see Fig. 9.2b). The attenuator has a total length L_a, a diameter of $\O_a$ and an immersed depth $d_a = \O_a/2$, where the length of the hinges is assumed to be negligibly small so that each module has an equal length of $L_a/4$. The PTO system is modelled by a force represented by the damping coefficient B^E to convert the kinetic energy into electricity. The seabed is considered to be flat with a constant water depth of d. The principal dimensions for the terminator and attenuator are presented in Table 9.1.

The array layouts for the WECs are presented in Fig. 9.2(a)-(b), in which the waves approach the arrays at an angle θ from the Y-axis. Each terminator and attenuator array has seven WECs and is arranged in a two-row array configuration as shown in Fig. 9.2(a) and (b), with four devices at the front row and three devices at the back row. For the terminator array, the devices are spaced at an equal horizontal distance d_x of $2a_f$ and vertical distance d_y of a_f. Note that these are the optimal spacings obtained via an optimisation scheme [44] to produce the maximum q-factor at the operating wave period T. On the other hand, the horizontal spacing d_x and vertical

Table 9.1 Principal dimensions and properties for terminator and attenuator WECs

	Terminator WEC	Attenuator WEC
Dimensions	Width $a_f = 26.0$ m	Length $L_a = 120$ m
	Immersion depth $d_f = 9.0$ m	Diameter $Ø_a =$ 3.75 m
	Thickness $h_f =$ 4.0 m	
Water depth d	12.5 m	42 m
Mass m	Not required	750 Metric-Tonne
Mass moment of inertia I	9.1455×10^6 kg.m^2	0.73×10^6 kg.m^2
PTO Damping B^E	16×10^6 N m s	[a]Hinge 1: 1.9×10^6 Nms [a]Hinge 2: 1.2×10^6 Nms [a]Hinge 3: 1.9×10^6 Nms (see Fig. 9.1b for hinge location)

[a]The PTO damping B^E is taken to be the same for the pitch and the yaw of the attenuator WEC

spacing d_y for the attenuator array are $2L_a$ and L_a, respectively, and these spacings are determined to ensure that the devices do not interfere with each other under oblique wave direction. The rigid body motion of the devices together with the total wave elevation surrounding the arrays are generated by using WAMIT. These values are expressed in terms of the response amplitude operator RAO. The stochastic post-processing is then performed in MATLAB to obtain the average q-factor of the array and the significant wave height of the wave climate under the multi-directional sea. The details of the post-processing will be discussed in Sect. 9.3.2.

9.3.2 *Mathematical Formulation*

9.3.2.1 WEC Under Regular Wave

Consider a generic WEC operating in a constant water depth d as shown in Fig. 9.3. The WEC is subjected to an incoming wave of wave period T and a wave height $2A$, where A is the wave amplitude, which impacts the structure at a wave angle θ measured from the Y-axis (see Fig. 9.2). The motion of the WEC is assumed to be $W = \left(W_x, W_y, W_z, \Theta_x, \Theta_y, \Theta_z\right)$, where W_x, W_y and W_z are the respective translations about the x-, y- and z-axes and Θ_x, Θ_y and Θ_z the respective rotations about the x-, y- and z-axes. The water domain is denoted by Ω whereas the symbols S_F, S_B, S_s and S_∞ denote the boundary for the free surface, the seabed, the wetted

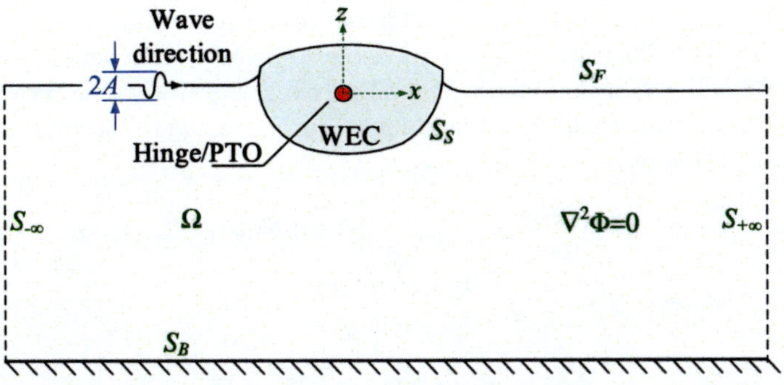

Fig. 9.3 Computational domain for a generic wave energy convertor

surface of the WEC and the artificial boundary at infinity, respectively, as shown in Fig. 9.3.

Governing Equation for Water Motion

The water is assumed to be an ideal fluid with no viscosity, incompressible and the fluid motion is irrotational. Based on these assumptions, the fluid motion may be represented by a velocity potential $\Phi(x, y, z, t)$. We consider the water to oscillate in a steady-state harmonic motion with the circular frequency ω. The velocity potential $\Phi(x, y, z, t)$ could be expressed into the following form

$$\Phi(x, y, z, t) = \mathrm{Re}\{\phi(x, y, z)e^{-i\omega t}\} \tag{9.4}$$

The single frequency velocity potential $\phi(x, y, z)$ must satisfy the Laplace equation [45] and the boundary conditions on the surfaces as shown in Fig. 9.3. These boundary conditions are given in Faltinsen [45]. The Laplace equation together with the boundary conditions on the surface S are transformed into a Boundary Integral Element (BIE) by using the Green's 2nd Theorem via a free surface Green's function given in the WAMIT manual [46] that satisfies the surface boundary condition at the free water surface S_F, the seabed S_B and at the infinity S_∞. Hence, only the wetted surface of the body S_s needs to be discretised into panels so that the boundary element method could be used to solve the diffracted and radiated velocity potentials. For details on the Green's function used in solving the BIE, refer to the WAMIT manual [46].

Governing Equation for WEC Motion

The WEC with a moment of inertia I, Mass m, PTO damping B^E and hydrostatic stiffness C is assumed to be a rigid body oscillating with six degrees of freedom rigid body motion $\mathbf{W}(x, y, z, t)$ at a frequency ω and is subjected to wave forces F. $\mathbf{W}(x, y, z, t)$ could then be written as

$$\mathbf{W} = Re\{W(x, y, z)e^{-i\omega t}\} \tag{9.5}$$

where $W = (w_1, w_2, w_3, w_4, w_5, w_6)$. It is noted that w_1, w_2 and w_3 are the WEC translation motion along the x, y and z-axes, respectively, whereas w_4, w_5 and w_6 are the rotational motion about the x-, y- and z-axes, respectively. The corresponding equation of motion for the terminator or attenuator expressed in Einstein summation convention is given by

$$\left[\omega^2\left(M_{ij} + A_{ij}^{added}\right) - i\omega\left(B_{ij}^{added} + B_{ii}^{E}\right) + C_{ij}\right]w_j = F_i,$$
$$\text{where}\begin{cases} i, j = 5 & \text{for terminator} \\ i, j = 3, 5, 6 & \text{for attenuator} \end{cases} \tag{9.6}$$

where, M_{ij} is the inertia matrix for the pitch component as given in the WAMIT manual [46]. For the case of a terminator, the external damping B_{55}^E in (9.6) is the optimum PTO damping for the rotational (pitch) motion and is obtained from Eq. (9.7) [10]. On the other hand, B_{55}^E and B_{66}^E for the attenuator WEC are obtained via a trial and error calibration process with hydrodynamic properties found in the literature,

$$B_{55}^E = \sqrt{\frac{\left[C_{55} - \omega^2\left(M_{55} + A_{55}^{added}\right)\right]^2}{\omega^2} + \left(B_{55}^{added}\right)^2} \tag{9.7}$$

where, A_{55}^{added} and B_{55}^{added} are the added inertia and radiated damping for the pitch component, respectively whereas C_{55} is the rotational stiffness. It is to note that M_{55} in (9.7) is the mass moment of inertia I for the terminator given in Table 9.1. Note that B_{55}^E varies with respect to the wave frequency ω; however, the PTO damping is taken as a constant in WAMIT by taking the minimum value of the B_{55}^E generated from (9.7). This value is found from calibration with known results published in the open literature as shown later in Sect. 9.4.

The excitation force F_i in Eq. (9.6) comprises the wave force components which can be derived from the velocity potential $\phi(x, y, z)$ as,

$$F_i = i\rho\omega \int_S \phi \cdot n_i \cdot dS, \quad \text{where}\begin{cases} i = 5 & \text{for terminator} \\ i = 3, 5, 6 & \text{for attenuator} \end{cases} \tag{9.8}$$

where, ρ is the mass density of the seawater. n_5 and n_6 are, respectively, the pitch and yaw components of the unit normal vector to the K-th panel in local coordinate system defined as $(n_4, n_5, n_6) = \boldsymbol{r} \times \boldsymbol{n}$, where $\boldsymbol{n} = (n_1, n_2, n_3)$ is the unit normal vector which is defined to point out of the fluid domain and $\boldsymbol{r}$ is the vector of coordinates of the WEC panel. It is to note that the subscripts 1, 2 and 3 denote the translation motion about the x, y and z-axes, respectively, whereas 4, 5 and 6 denote the rotational motion about the x, y and z-axes, respectively. As the velocity potential ϕ in Eq. (9.4) could be further decomposed into the diffracted ϕ_D and radiated ϕ_R part, this gives us the exciting moment F_i which is derived from the diffracted velocity potential, and the added inertia A_{ij}^{added} and radiated damping B_{ij}^{added} which are derived from the radiated velocity potential.

For N numbers of WECs, the equation of motion of the terminator body n due to the body p is written as

$$\left\{\omega^2\left[M_{ij}+\left(A_{ij}^{added}\right)_{nn}\right]-i\omega\left[B^E+\left(B_{ij}^{added}\right)_{nn}\right]+C_{ij}\right\}(w_j)_n - \sum_{\substack{p=1\\p\neq n}}^{p}\left[\omega^2\left(A_{ij}^{added}\right)_{pn}+i\omega\left(B_{ij}^{added}\right)_{pn}\right](w_j)_p=(F_i)_n \tag{9.9}$$

where, $i, j = 5$ for terminator WEC and $i, j = 3,\ 5,6$ for attenuator WEC. It is to note that W in (9.5) can be expanded as a series of product of the complex excursion amplitudes ξ_j and rigid body modes, thus, the equation of motion can also be represented by (9.2).

9.3.2.2 WEC Array Under Multi-Directional Sea

The JONSWAP (JS) wave spectrum is developed for limited fetch North Sea by the offshore industry and is given by Goda [47]

$$S_{JS}(\omega)=\beta_\gamma\cdot 8.1\times 10^{-3}\frac{g^2}{\omega^5}\exp\left[-0.032\left(\frac{g}{H_s\omega^2}\right)^2\right]\gamma^b \tag{9.10}$$

where, the normalising function $\beta_\gamma = 1-0.287\ln\gamma$ [48], $b=\exp\left[-\frac{(\omega-\omega_p)^2}{2\sigma^2\omega_p^2}\right]$ and bandwidth parameter $\sigma=\begin{cases}0.07 \text{ for } \omega<\omega_p\\ 0.09 \text{ for } \omega\geq\omega_p\end{cases}$. ω and ω_p are the wave frequency and peak wave frequency, respectively, g the gravitational acceleration and H_s is the significant wave height. It is noted here that the shape parameter γ is taken as a mean value of 3.3.

The multi-directional wave spectrum $S_{JS}^{MD}(\omega,\theta)$ is then obtained by multiplying the wave spectrum $S_{JS}(\omega)$ with a spreading function $D(\theta)$ as given in Eq. (9.11),

$$S_{JS}^{MD}(\omega,\theta)=\sum_{i=1}^{mf}S_{JS}(\omega_i)\cdot D(\theta) \tag{9.11}$$

where, mf is the number of wave frequency ω_j considered in the JONSWAP wave spectrum $S_{JS}(\omega)$. The spreading function $D(\theta)$ is given by,

$$D(\theta) = \sum_{j=1}^{m} \left\{ \frac{\sqrt{\pi}\,\Gamma(\hat{s}+1)}{2\theta_{\max}\Gamma\left(\hat{s}+\frac{1}{2}\right)} \cdot \left| \cos\left(\frac{\pi\left(\theta - \overline{\theta_j}\right)}{2\theta_{\max}} \right) \right|^{2\hat{s}} \right\}, \tag{9.12}$$
$$for - \theta_{\max} < \left(\theta - \overline{\theta_j}\right) < \theta_{\max}$$

where, m is the number of wave direction θ_i considered in the spreading function $D(\theta)$, $\bar{\theta}_j$ is the mean wave direction, $\hat{s}$ the wave spreading parameter and $\theta_{\max} = \pi/2$ in (9.12).

9.3.3 Generated Power and Interaction Factor

By solving the equation of motion (9.9), w_j of the WEC can be obtained. This value can then be used to derive the average power generated by the nth WEC over the range of wave frequency ω considered under the multi-directional sea by using the following expression [30],

$$\left(\overline{P_n}\right) = \int_{-\pi/2}^{\pi/2} \left(\omega^2 \cdot B^E \cdot |w_i|_n^2\right) \cdot S_{JS}^{MD}(\omega,\theta) \cdot d\omega, \quad \text{where} \begin{cases} i = 5 & \text{for terminator} \\ i = 5,6 & \text{for attenuator} \end{cases} \tag{9.13}$$

Noted also that (9.13) provides the average power generation $(\overline{P_n})$ of the nth WEC in the wave energy farm when subjected to the multi-directional sea in the long term statistical sea state. In order to quantify the interaction between devices, Budal [14] defines the q-factor which is modified here in Eq. (9.14) for the multi-directional sea to facilitate the discussion on the performance of the array.

$$q = \frac{\sum_{n=1}^{N} \left(\overline{P_n}\right)}{N \times \left(\overline{P_0}\right)} \tag{9.14}$$

where, $\overline{P_0}$ is the average generated power of an isolated WEC and the total average generated power in the array $\bar{P} = \sum_{n=1}^{N} \left(\overline{P_n}\right)$. For simplicity, the total average generated power will be referred as P and the hat of the wave spreading parameter $\hat{s}$ is dropped and represented by s. Equation (9.14) is used as a performance evaluator for the array where a constructive interaction is denoted by a value greater than 1.0 and a destructive interaction when smaller than 1.0.

9.3.4 Wave Disturbance Under Multi-Directional Sea

The diffracted and radiated wave elevations surrounding the WEC arrays under regular wave could be obtained directly from the WAMIT software. The response amplitude operator for the wave elevation $\bar{\xi}_\eta$ are expressed by normalising the wave elevation η with the incident wave amplitude A, i.e.

$$\bar{\xi}_\eta = \frac{\eta}{A} \tag{9.15}$$

In order to obtain the significant wave height of the wave disturbance under the multi-directional sea, the square of the amplitude for $\bar{\xi}_\eta$ is multiplied with the multi-directional wave spectrum to obtain the response spectrum $S_{res}(\omega, \theta)$ as follows

$$S_{res}(\omega, \theta) = \int \left|\bar{\xi}_j\right|^2 \cdot S_{JS}(\omega, \theta) \cdot d\omega \tag{9.16}$$

The significant wave height of the wave field is obtained as:

$$(H_s)_{wf} = 4\sqrt{m_0} \tag{9.17}$$

where, m_0 is the moment of order 0 of spectrum $S_{res}(\omega, \theta)$.

9.4 Verification of the Numerical Model

The numerical models for the terminator and attenuator were verified with the hydrodynamic properties found in the literature review. The verification of the numerical model for the terminator WEC was undertaken by Tay and Venugopal [44, 49, 50]. The results obtained from our numerical model were found to be in very good agreement as with those published by Renzi et al. [51] and Retzler [52]. With the numerical model verified with the existing data found in the literature, the hydrodynamic interaction analysis is then performed on the array layouts presented in Fig. 9.2 with the particulars given in Table 9.1. The number of terminator and attenuator WECs formed in the array is the same as those presented in Fig. 9.2. Note that the terminator and attenuator WEC arrays shown in Fig. 9.2 generate an approximately 5 MW of power, where each terminator and attenuator WECs has a rated power of 800 kW and 750 kW, respectively.

The arrays are subjected to multi-directional sea generated using the JONSWAP wave spectra. Three different wave spreading parameters s as given in Eq. (9.12) are considered, i.e. $s = 4$, 15 and 100. It is noted here that the larger s value, i.e. $s = 100$ corresponds to the uni-directional sea whereas the smaller s value, i.e. $s = 4$ corresponds to a combination of wave components approaching from different directions with the most wave coming from $\bar{\theta}$.

9.4.1 *Performance of Arrays*

9.4.1.1 Response of WECs Under Multi-Directional Sea

The responses of the individual and arrays of terminator and attenuator WECs when subjected to waves generated by the JONSWAP spectrum with a significant wave height of 3 m are plotted in Fig. 9.4 to study the motion behaviour of the various WECs under the multi-directional sea. It is noted that the response θ_s of the attenuator, WEC is presented as the total summation of the magnitude response (pitch w_5 and yaw w_6) at the three joints of the WEC. The results shown here are only for the spreading parameter = 4. It is to be noted that the responses of the WECs for $s = 15$ and 100 have similar patterns as their counterparts of $s = 4$, but with higher magnitudes when the wave spreading parameter s increases. This is because of the more concentrated wave energy distribution when waves approach close to the uni-directional sea.

The response of the terminator WEC is found to increase with the increase of wave period, i.e. wavelength and when the wave approaches from the headsea. Hence, this suggests that the response of the terminator WEC is driven by the exciting torque and could be more effective when operating at harsh environment. On the other hand, the

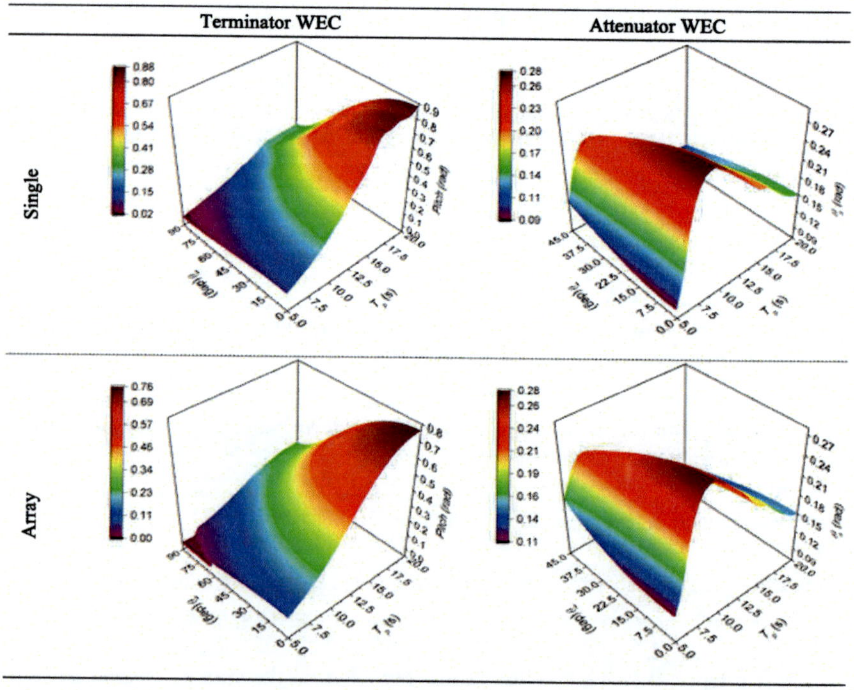

Fig. 9.4 Performance of WECs for JONSWAP spectrum $H_S = 3$ m, $s = 4$. Top: single terminator and attenuator, and, bottom: an array of terminators and attenuators

attenuator WEC is effective when the sea state is mild with its optimal performance occurs when the wave period is around 7–12 s. The motion is also found to be the highest when wave approaches from the headsea with decreasing response when wave approaches from the oblique direction. This is because of the higher pitching motion at the connection joints between the connected modules when subjected to the headsea condition, thus making the turret mooring system more preferable for the attenuator WEC as the device is able to weathervane in accordance to the different wave directions.

The average responses of the WECs when deployed in arrays are also presented in Fig. 9.4. It is noted that the average response is obtained by normalizing the sum of the response of the WECs in the array with the numbers of WEC considered in the array, i.e. $\left(\sum_{n=1}^{N} \bar{\xi}_n\right)/N$. It is important to note also that the average displacement for each WEC in the array may not be a good measure to quantify the performance of the array as the individual displacement may vary considerably about the average displacement value. Nevertheless the former is able to provide an overall insight to the performance of the WEC array under the multi-directional sea as the average displacement has similar patterns as their individual counterpart presented in Fig. 9.4. Due to the constructive and destructive interferences as a result of the hydrodynamic interaction between the devices, the effectiveness of the array configurations in generating energy could be quantified by using the q-factor (9.14), which is a quantity defined as the ratio of the average total power produced in an array to the power produced by an individual WEC. It is noted that the q-factor is related to the WECs' response via Eq. (9.13) and is used to quantify the efficiency of the array in producing energy (Fig. 9.5) as compared to its individual counterpart, where constructive interference is denoted by q-factor value larger than 1.0. This is to say that q-factor greater than 1.0 is desirable as it implies that more energy could be generated by each WEC in the array as compared to a single isolated WEC.

The performances of the WEC arrays expressed as the q-factor subjected to multi-directional sea are presented in Fig. 9.5. The q-factor is plotted against the mean wave direction $\bar{\theta}$ and the peak wave period T_p, where $\bar{\theta}$ ranges from 0 ° (headsea) to 90° (beam sea) and T_p from 5 to 20 s. The maximum $\bar{\theta}$ for the attenuator is however taken only up to 45° as the attenuator is usually weathervane against the wave direction. Focusing on the multi-directional sea with $s = 4$, it is observed that the terminator WEC has a higher q-factor with its value greater than 1.0 (denoting constructive interference) when the array is subjected to smaller T_p, i.e. smaller wavelength at headsea condition. At headsea direction, the q-factor is found to reduce with the increase of T_p. This is because when the terminator WEC is placed across the incoming wave, they behave like a breakwater, and a greater disturbance is generated as waves are diffracted and radiated, which contributes to the constructive wave interference between the wave energy devices.

On the other hand, the q-factor for the attenuator arrays are found to converge to 1.0 with the increase of T_p indicating that the hydrodynamic interactions of these types of WEC arrays when operating at a longer wavelength are similar to their counterparts of a single device. This is because the attenuator WECs oscillate at the

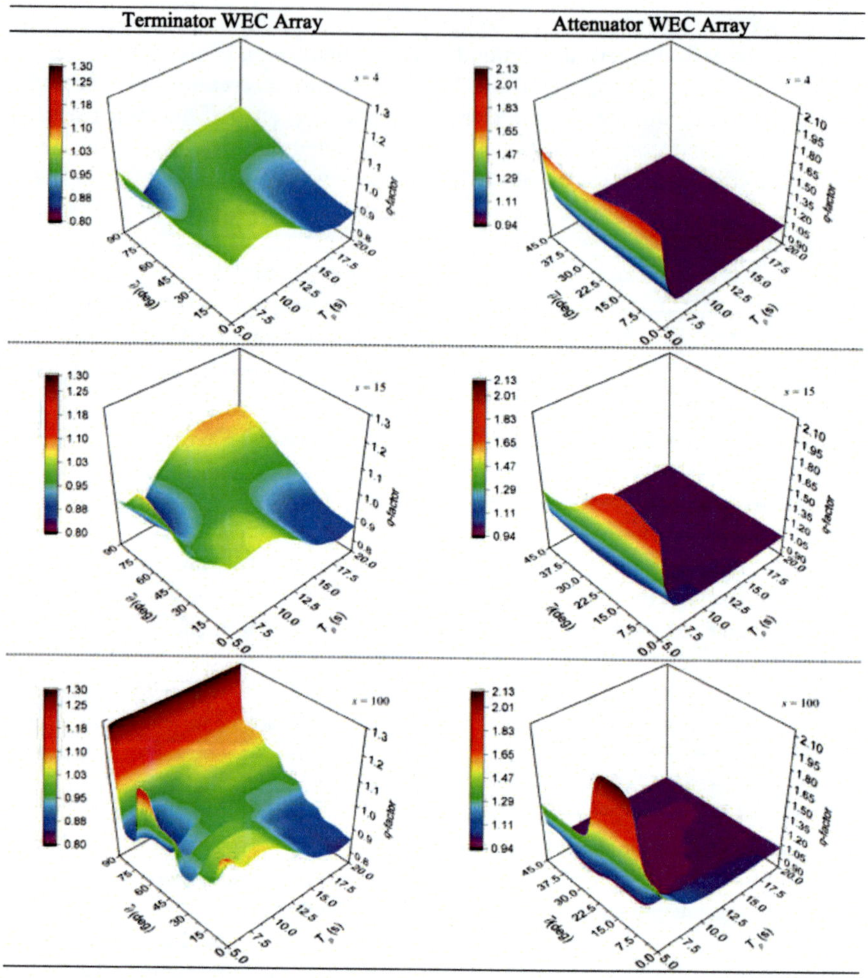

Fig. 9.5 Comparison of q-factor for terminator and attenuator WEC arrays for JONSWAP Spectrum for different s and $H_s = 3$ m

same phase and amplitude as the incoming wave when the wavelength is large, hence resulting in a smaller relative motion. It is found that the q-factor for the attenuator WEC array is greater than 1.0 when the T_p is smaller indicating the occurrence of constructive interference. This is attributed to the larger relative movement between the modules in the attenuator WEC and hence generating greater power when the wavelength is small. It is observed from the time series simulation that the radiated waves generated by the attenuator devices could be as small as 1.0 m (as a result of energy being extracted from the wave) which contribute significantly to the constructive interference between the devices. This shows that the diffracted and radiated waves that arise from the presence of the devices aid in increasing the power absorption of the wave farm. For all the WEC arrays, the maximum q-factor occurs when wave approaches from the headsea direction ($\bar{\theta} = 0°$).

Similar observation on the performance of the WEC arrays is also found when s increases to 15. The q-factors for the terminator WEC array are found to be affected significantly by the changes in $\bar{\theta}$ with the increase of s for the whole range of T_p considered whereas the counterpart for the attenuator WEC array is only influenced by the changes in $\bar{\theta}$ when T_p is small. This finding is affected by the absorption bandwidth of the WECs where the terminator is known to have a wider bandwidth [10], hence enabling it to generate energy at a wider range of wave frequency, i.e. wave period.

It can be seen from Fig. 9.5 that the terminator and attenuator WEC arrays have a higher q-factor when subjected to uni-directional sea, i.e. $s = 100$. This is because the encountered waves have a very small spreading and travel in the headsea direction. This also emphasizes the importance of taking into account the realistic sea by using the multi-directional sea and suggests that the q-factor could sometimes be overestimated if the uni-directional sea is assumed instead of the multi-directional sea. The q-factor for the attenuator WEC array has the same trend for varying s value but in general, it is observed that the q-factor varies significantly with the changes in wave directions at smaller T_p. However, for the attenuator type WECs such as the Pelamis WEC, the device is always weathervane with respect to the incoming waves, thus allowing the WEC to generate maximum wave power at headsea direction.

9.4.1.2 Wave Climate Surrounding Arrays

The wave disturbance is represented by the significant wave height $(H_s)_{wf}$, which is obtained via post-processing of the regular wave elevations approaching from different directions as given in Eqs. (9.16) and (9.17). Three different spreading parameters s = 4, 15 and 100 are considered. The array layout as presented in Fig. 9.2(a) and (b) are used for the terminator and attenuator WEC arrays, respectively, where the terminator WEC array has a computational domain of 600 × 1200 m whereas the attenuator WEC array has a computation domain of 1400 × 3000 m. The computational domain of the arrays for these two types of WEC is different as it depends on the attenuated wave downstream the array which will be presented in the subsequent sections.

The significant wave heights of the wave field $(H_s)_{wf}$ surrounding the terminator and attenuator WEC arrays subjected to the multi-directional sea are presented in Fig. 9.6. The mean wave propagation direction considered is $\bar{\theta} = 0°$ and the significant wave height is $H_s = 3$ m. The wave disturbances upstream and downstream of the array depend significantly on the mechanism of wave generation by the WECs. In general, it can be seen from Fig. 9.6 that the maximum wave height of the wave disturbance due to the presence of the WEC arrays is the highest for the attenuator WEC array, followed by the terminator array. It can also be seen that the maximum $(H_s)_{wf}$ occurs at the wave field near the WECs due to the greater interaction of

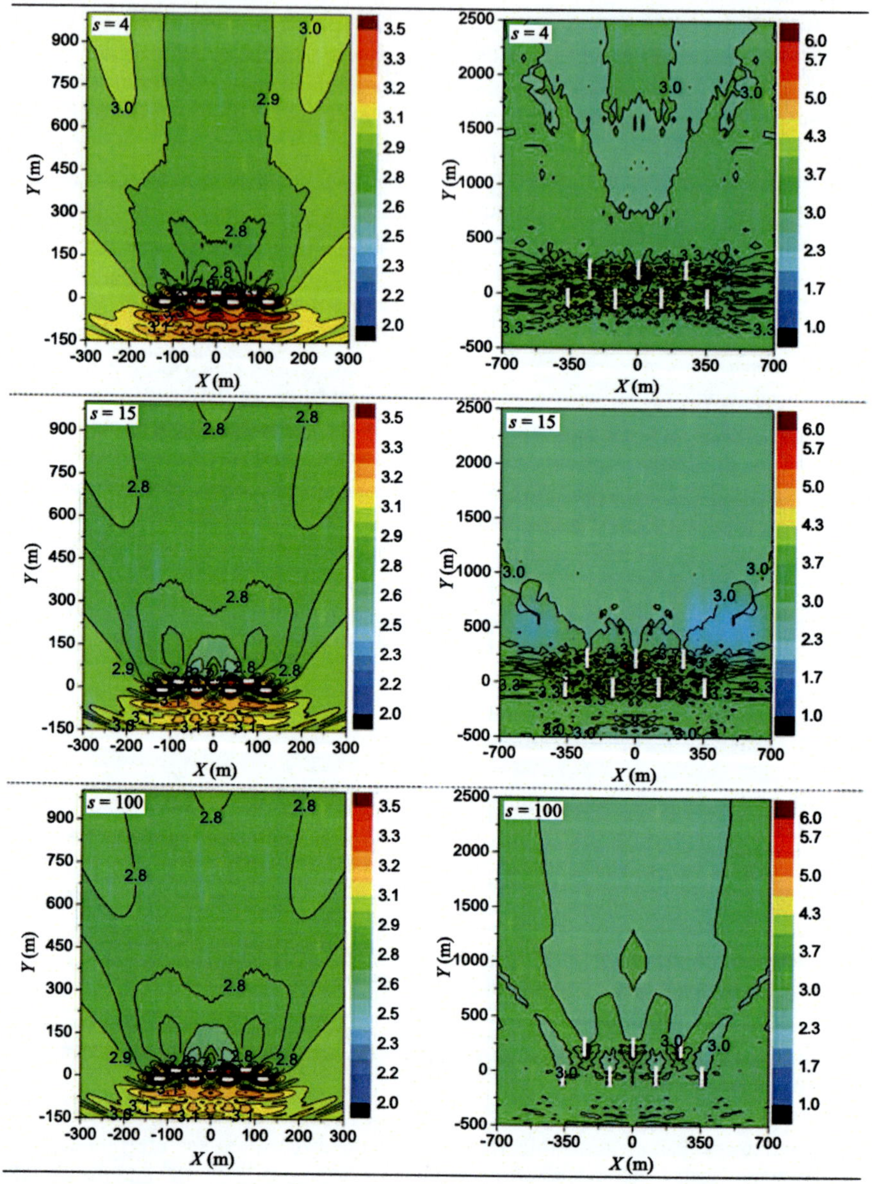

Fig. 9.6 Comparison of $(H_s)_{wf}$ for terminator and attenuator WEC arrays for JONSWAP spectrum for different s. $\bar{\theta} = 0°$, $T_p = 8s$

the diffracted and radiated waves, and these wave elevations are found to be attenuated when they are propagating away from the arrays. The presence of the highest $(H_s)_{wf}$ in the wave field surrounding the attenuator array is attributed to the greater intensity of waves being radiated by the attenuator WECs. The intensity of the interaction between the diffracted and radiated waves will affect the wave attenuation at the downstream of the array. For the case when $s = 4$, the wave interaction is the greatest for the attenuator WEC array, hence explains the occurrence of the slowly attenuated waves downstream of the array as compared to their counterparts of the terminator WEC arrays. Note that this phenomenon can be checked by observing the $(H_s)_{wf}$ of the wave disturbance, where the wave is fully attenuated when it approaches $H_s = 3$ m, i.e. the significant wave height of the wave spectrum. This implies that the radiated and scattered waves have diminished and only the incident waves are present.

When comparing the effect of the spreading parameters s, it can be seen that the wave disturbances surrounding the terminator WEC arrays are greater when $s = 100$, i.e. when the arrays are subjected to an almost uni-directional sea. However, the opposite is observed for the attenuator WEC array where greater disturbance is noticed when $s = 4$. The difference is attributed to the means of power generation by different types of WECs. As the attenuator generates power via articulation in two degrees of freedom, there are greater interactions with the waves approaching from different directions, which is represented by the multi-directional sea when $s = 4$. On the other hand, the terminator WEC pitches to generate energy with respect to the wave oscillation when the wave approaches in a single direction, i.e. $s = 100$, and thus contributes to a lesser disturbance in the wave field. As the terminator WECs generate energy via a single degree of freedom, their presence in the uni-directional sea i.e. $s = 100$ will produce greater wave disturbance because of the higher wave energy in the governing direction. It can also be seen that due to the geometry of the terminator WEC that is oriented perpendicular to the wave propagation direction, greater waves are being reflected upstream as compared to their counterparts of the attenuator WECs.

In Fig. 9.6, the wave disturbances surrounding the terminator WEC arrays are found to behave in the similar manner with respect to the changes in s. However, the opposite trend is found for the WEC array under the oblique sea (see Fig. 9.16) as compared to those presented when $\bar{\theta} = 0°$ (in Fig. 9.15), where the wave disturbance increases with greater s value, i.e. $s = 100$. For the terminator WEC array under oblique sea, larger wave separation occurs when the wave hits on the terminator WEC array, which in a group behaves like a single floating breakwater. This wave separation phenomena result in a larger wave disturbance downstream as observed in Fig. 9.7. It is also found from Fig. 9.7 that the wave attenuates at a slower pace downstream the WEC arrays as compared to their counterparts of the headsea presented in Fig. 9.6. This shows a greater interference of diffracted and radiated waves occur between the devices in the array when subjected to an oblique sea condition.

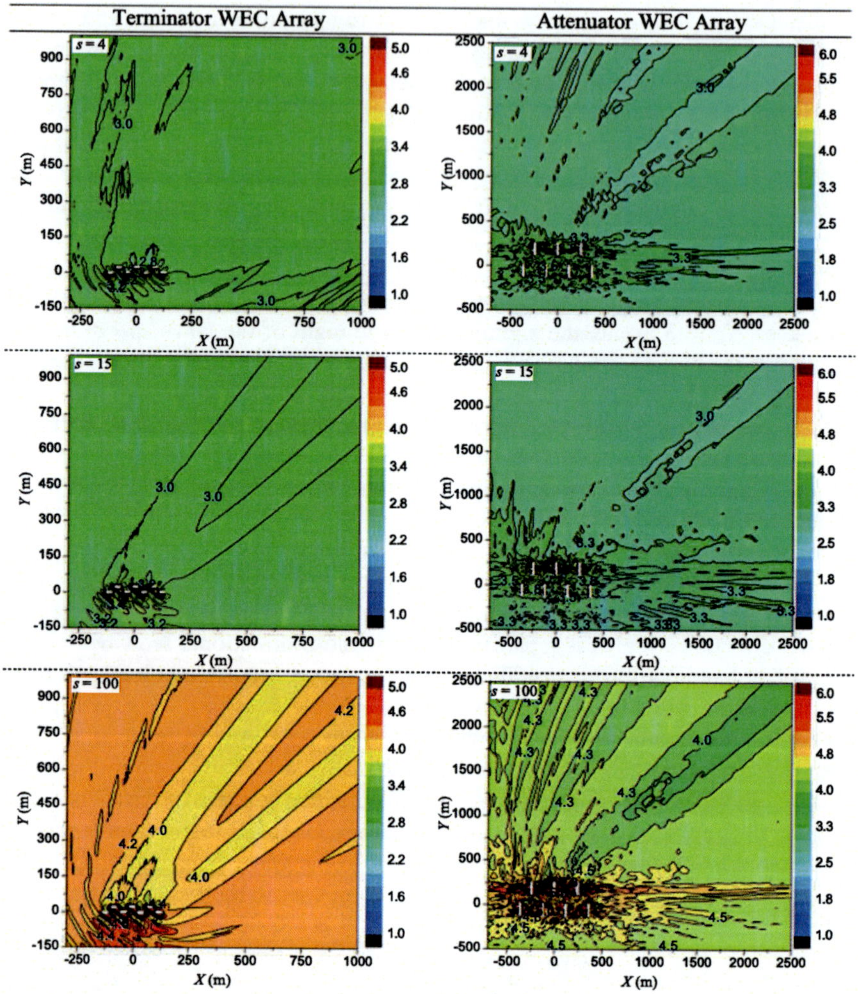

Fig. 9.7 Comparison of $(H_s)_{wf}$ for terminator and attenuator WEC arrays under JONSWAP wave spectrum for different *s*. $\bar{\theta} = 45°$, $T_p = 8s$

9.5 WEC Array Modelling by Ocean Scale Numerical Models

The extraction of wave energy from a wave farm produces a wave energy deficit or shadow down region which in turn may affect the downstream sediment transport and may thus result in beach erosion/deposition. The quantification of the wave energy reduction and identification of induced wave height gradients are desired for an accurate assessment of the environmental impact of a Wave Energy Converter (WEC) array on the nearby marine environment. Here we investigate the consequences wave energy extraction by large scale wave arrays in the areas of the Crown Estate Round 1 lease sites in the Pentland Firth and Orkney Waters (PFOW) in the United Kingdom (Fig. 9.8). MIKE 21 Spectral Wave, in association with the wave-structure interaction software tool WAMIT, has been employed to study the impact of energy extraction by large arrays of Wave Energy Converters (WECs) on the wave height alteration in the neighbourhood of WEC arrays. As with Sect. 9.4, two generic types of WEC, one representing surface attenuators (deployed in deep water) and other representing terminators (deployed in shallow waters) are used for numerical modelling. The power extraction performance of the WECs are initially modelled using WAMIT (Sect. 9.4) and validated with data from the literature. As MIKE 21 SW has limited ability in modelling complete dynamics of a moving structure, each WEC has been modelled as a generic structure but with appropriate reflection, transmission and energy absorption properties derived from WAMIT, and this methodology is found to have worked well.

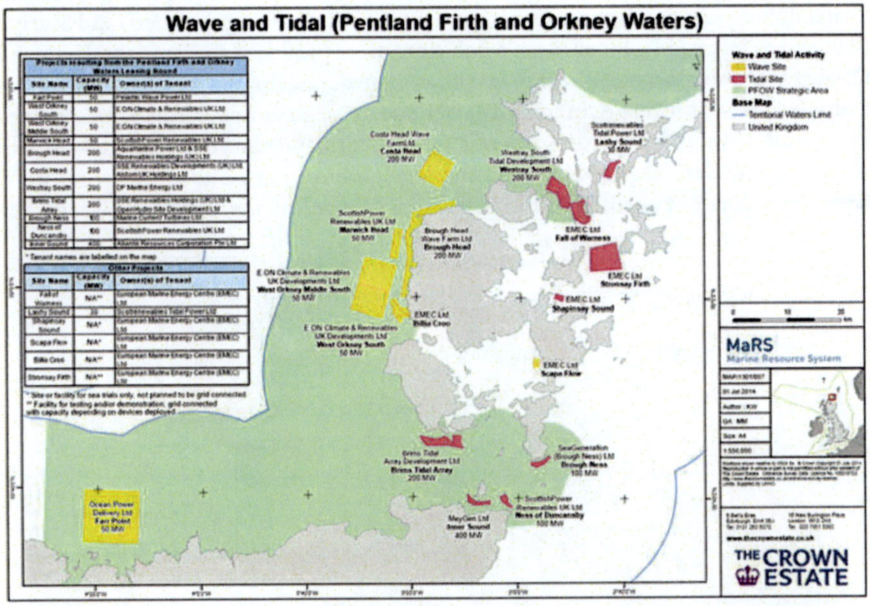

Fig. 9.8 Location of Pentland Firth showing wave and tidal energy leasing sites (http://www.thecrownestate.co.uk/energy-and-infrastructure/wave-and-tidal/the-resources-and-technologies/3)

9.5.1 Numerical Model Set-Up

MIKE 21 SW is a third-generation spectral wind-wave model utilising unstructured meshes [53] to simulate the growth, decay and transformation of the wind-generated sea and ocean swells in offshore and coastal areas. The wind waves are expressed by the wave action density spectrum. The model accounts for the physical phenomenon of wave growth from wind, energy transfer due to non-linear quadruplet or triad wave-wave interaction, and includes energy dissipation terms for white-capping, bottom friction and depth-induced breaking. A cell-centred finite volume method is applied in the discretization of the governing equations in geographical and spectral space and a multi-sequence explicit method is applied for the wave propagation with the time integration carried out using a fractional step approach. The model generates phase averaged wave parameters as output for either the total computational domain or selected parts thereof. Further details can be found in the MIKE 21 wave modelling user guide [53].

An unstructured computational mesh (see Fig. 9.9) was constructed for the North Atlantic region (10°E–75°W and 10°N–70°N) using bathymetry data compiled from General Bathymetric Chart for Oceans (GEBCO, http://www.gebco.net/) and Marine Scotland Science [54]. Finer mesh resolutions were produced for Pentland Firth and Orkney Waters with a mesh area of 0.0005 square degrees (approx. 1700 m^2), for the Hebrides and northwest Scotland of 0.001 square degrees and 0.75 square degrees (approx. 2.5 km^2) for the North Atlantic Ocean. Further details on the model domain and setup are available in Venugopal and Nemalidinne [55]. The model was forced with wind data obtained from the operational model of the European Centre for

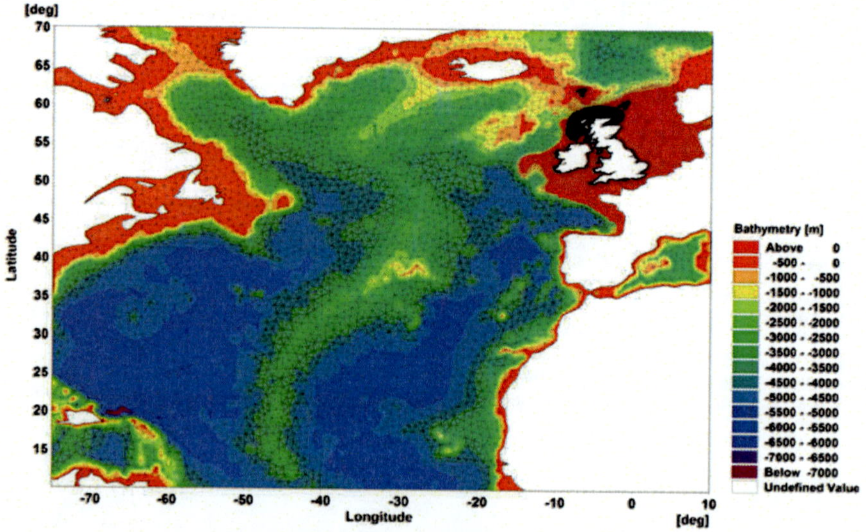

Fig. 9.9 Computational domain for North Atlantic wave model

Medium-Range Weather Forecasts (ECMWF, http://www.ecmwf.int/) at 6 hourly intervals with a spatial resolution of 0.125° × 0.125°. To ensure fetch unlimited wave growth, decay and transformation of wind sea and swells, the model was run in 'fully spectral' mode with 'Instationary formulation'. The number of frequencies used for the model was 25 with $f_{min} = 0.04$ Hz and a logarithmic frequency distribution with a frequency factor of 1.1. The directional discretisation had 24 directional bins, each with 15° resolution. A low order fast algorithm has been chosen as the solution technique with the 'maximum number of levels in the transport calculation' set as 32. A quadruplet-wave interaction has been applied. No current, ice coverage and diffraction were included into the model. Dissipation due to whitecapping, bottom friction and depth-induced wave breaking were considered in the simulations and the energy transfer was activated. For a detailed description of the above source terms and the basis for selection of the appropriate values, the reader is referred to the MIKE 21 SW user guide [53] and [55]. Measured wave data from scientific buoys deployed around Scotland was used to calibrate and validate the model. The successful calibration and validation of the model are described in detail in [55].

9.5.2 Predictions Without Energy Extraction

The efficiency of WECs depends on the characteristics of the site-specific resource, and key parameters are the wave height and period of the different sea states. Power matrices are a standard way of expressing WEC performance against particular sea states and the selection of the most appropriate site and technology for wave farm operation in a specific region should be based on a thorough analysis of the energy production of different combinations of different types of WECs at a range of sites. To be able to undertake such an assessment an accurate and detailed resource characterisation at each location of interest has to be conducted. The approach taken here is to run the MIKE 21 Spectral Wave model for the year 2010, with and without WECs, and to subtract the results from each other to produce maps of the differences in wave parameters following the inclusion of WECs. The significant wave height and conditions simulated for the year 2010 are shown in Figs. 9.10 and 9.11. The evolution of mean significant wave height for the pre-device model, for the study area around Orkney, is shown in Fig. 9.12 for the period January to December 2010, which is extracted from the validated model described in Sect. 9.3.1. Note that this is

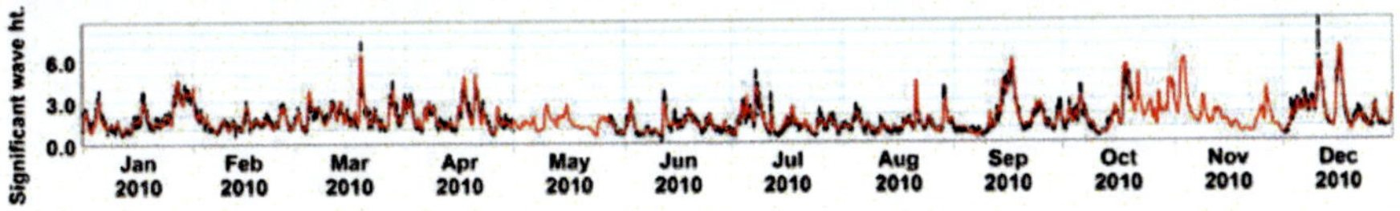

Fig. 9.10 Comparison of significant wave height, between measurements (black line) and model (red line) for Orkney Islands wave deployment site (58.970200°N–3.390900°W)

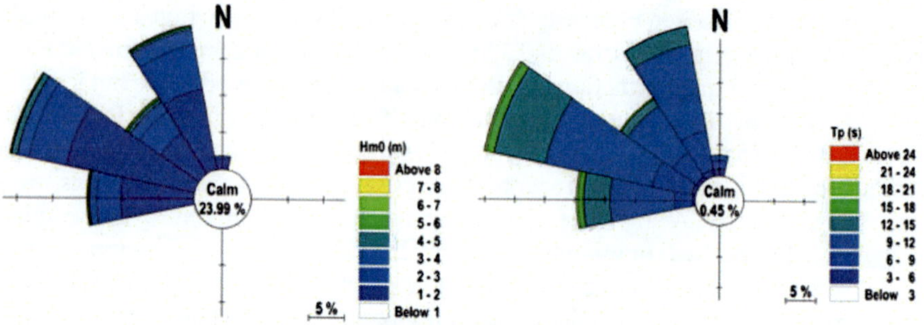

Fig. 9.11 Wave rose diagram of the model data at Orkney location: left hand—significant wave height (Hm0) and right hand—peak wave period Tp

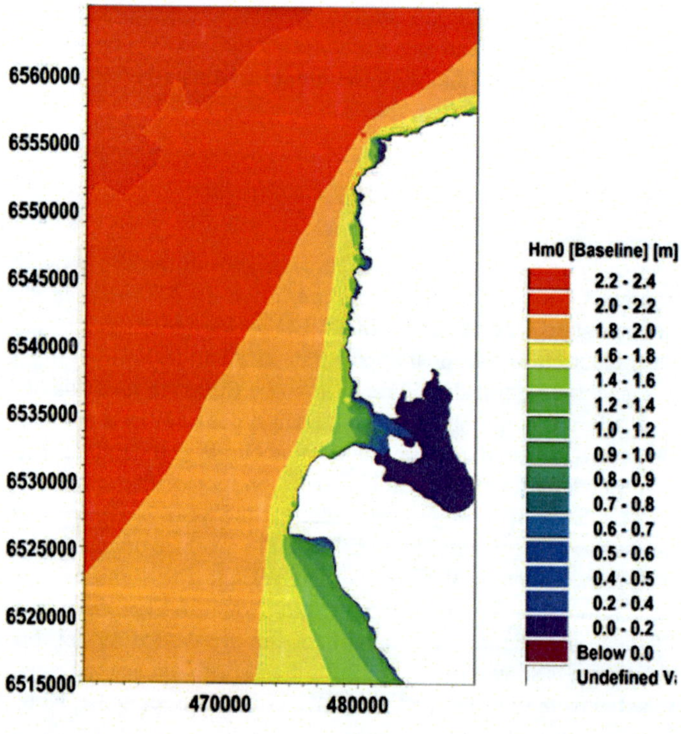

Fig. 9.12 Mean significant wave height for the period January to December–2010 without energy extraction

one of the strategic deployment zones identified by the Crown Estate in Fig. 9.8. As expected, the wave height decreases as the wave progresses from a westerly direction across the domain into reduced water depths. The annual average wave height varies from about 1.6 m to about 2.5 m at the sites where nearshore arrays are proposed.

9.5.3 Implementation of Energy Extraction

Marine Scotland Science developed possible array layout scenarios from Environmental Statements submitted by developers. Two generic types of WEC are used, a surface-based line attenuator and terminators (Oscillating Wave Surge Converters, OWSC). The first device is based on the Pelamis WEC. The second device, an OWSC, used in this study is based on the Oyster 800 by Aquamarine Power. The WEC arrays proposed by the developers are:

(i) ***Marwick Head***: An array of 66 terminator type devices with a 350 x 400 m staggered spacing across 4 rows was considered.
(ii) ***West Orkney***: Arrays of 132 terminator devices with a 400 x 400 m distance between the WECs in two staggered rows, and 10 times the device length between the arrays (1800 m) were considered.
(iii) ***Brough Head***: Single row of 120 attenuator devices (26 m wide and 45 m spacing) was distributed as 4 arrays along the 12.5 m depth contour.

For the detailed methodology used for the array layouts see O'Hara Murray and Gallego [56]. These array configurations are modelled using MIKE 21 (see Fig. 9.13). This study explores a new method of removing energy from the model domain as the MIKE 21 SW model has no built-in algorithm for simulating WECs. By including individual WECs or WEC arrays as additional source terms representing the energy extracted and redistributed by the WECs in spectral wave models as used in this study, it is possible to assess the hydrodynamic behaviour and power performance of WEC array. Such an approach in the representation of WECs in the most commonly used coastal modelling packages has significant practical implications on the development of software tools for the planning of wave farms. Since the horizontal dimensions of WECs are usually smaller than the mesh resolution used in the computational grid, the generic wave energy device in the MIKE 21 SW are modelled using a sub-grid scaling technique. Further details may be found in [57]. The location of a WEC is given by a number of geo-referenced points which together make up a polyline. The location and geometry of these polyline structures are included when the mesh is created. The values of the wave energy transmission factors are still not completely understood (due to a lack of installed systems for validation) or openly disclosed by the WEC developers. Moreover, these values depend not only on the farm geometry, but on a wider range of factors, and they often appear to have a dynamic behaviour relationship with the impacting wave conditions.

Given that this work is concerned with surface attenuator and terminator type technology as represented by Pelamis and Oyster devices respectively, the values

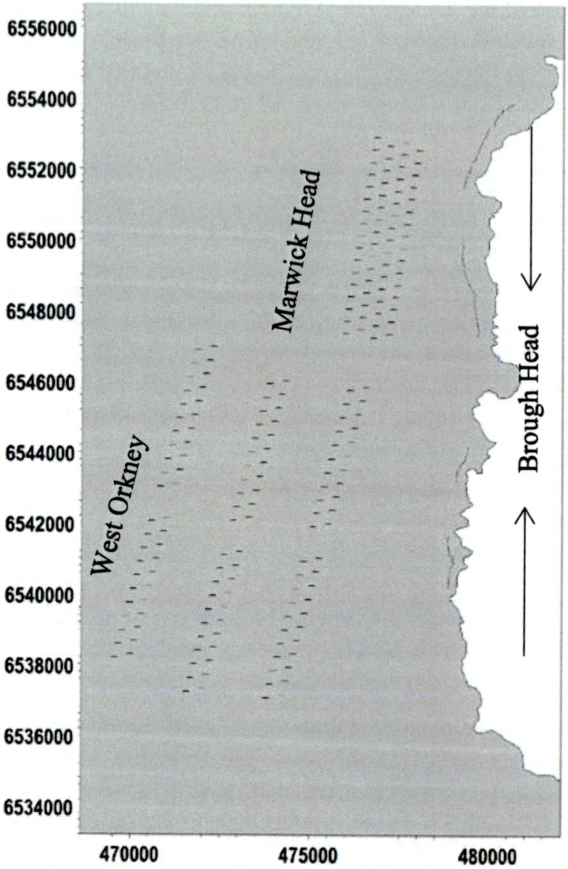

Fig. 9.13 Map of the Orkney showing the layout of 300 WEC devices

of the reflection, transmission and energy loss coefficients were obtained from the software tool WAMIT as presented in Sect. 9.3.2, to represent various scenarios.

Although the MIKE21 SW model does not have the ability to model the entire dynamics of a floating wave energy device, it can model wave propagation accurately over varying complex bathymetry in coastal to ocean scale regions, thus it makes a highly suitable tool to study wave forecasting or hindcasting. On the other hand, the wave structure interaction tool WAMIT, as demonstrated through various studies is powerful and accurate in modelling any type of fixed and floating offshore structures, but its downside is that it cannot be applied to varying bathymetry or wave forecasting or hindcasting and therefore its use in modelling realistic oceanic scale array to assess environmental impact is less proved. Hence the method to use WAMIT to determine the energy absorption, reflection and transmission characteristics of WECs and then transfer these into MIKE 21 for large scale array modelling and at the same time for studying WECs interaction with the marine environment on a wider scale is chosen. The facility available with MIKE 21 allows one to model the WEC as a line or point structure, and the former was chosen for this study to model the WECs. MIKE 21 also allows the structure to be modelled as a submerged, emergent, or sub-aerial structure.

The following methodology was adopted in modelling the devices. The energy transmission through any WEC structure can be represented by the energy balance equation,

$$C_R^2 + C_T^2 + C_L^2 = 1 \tag{9.18}$$

where, C_T is the transmission coefficient, C_R is the reflection coefficient and C_L is the energy loss coefficient. The reflection and transmission coefficients are calculated in normalised form as ratios of reflected wave height to incident wave height, and, transmitted wave height to incident wave height respectively.

The information on energy loss or energy absorbed by a device could be obtained from the power matrix produced by developers. The power matrices for Pelamis and Oyster devices are extracted using the data from [58] and a metric known as Power Capture Ratio (*PCR*) is produced. The *PCR* which is a measure of wave power absorbed by a device can be calculated for the different range of significant wave heights (*Hm0*) and energy periods (T_e) as,

$$PCR = \frac{P_m}{P_{flux}} \tag{9.19}$$

where, P_m is the power produced by a device based on its power matrix for a chosen pair of wave height and energy period and P_{flux} is theoretical energy flux calculated for the same pair of wave parameters, using Eqs. (9.20) and (9.21),

$$P_{flux} \approx \rho g \frac{H_{m0}^2}{16} C_g(T_e, d) \tag{9.20}$$

$$C_g(T_e, d) = \frac{1}{2}\left[1 + \frac{2kd}{\sinh 2kd}\right]\sqrt{\frac{g\lambda}{2\pi}\tanh(kd)} \tag{9.21}$$

where, $C_g(T_e, d)$ is the group velocity corresponding to the energy period T_e, ρ is the seawater density taken as 1025 kg/m^3, g is the gravitational constant, k and λ are the wavenumber and wavelength respectively, computed with T_e for the depth d using the linear wave 'dispersion relationship'. Now the energy conservation Eq. (9.18) may be rewritten as,

$$C_R^2 + C_T^2 + (PCR)^2 = 1 \tag{9.22}$$

While C_T, C_R and C_L (or PCR) are wave period dependent which means they change with sea states, determining the energy coefficients for every wave frequency, while it is possible to do, is laborious, and considering the time and resources available, the wave periods that correspond to the maximum power output from both the attenuator and terminator have been selected and corresponding energy coefficients have been chosen to model the WECs in MIKE 21 SW model. This approach may be justified as the main interest is to evaluate the impacts on the environment when the devices operate at its best.

However, before using these coefficients in MIKE 21, the WAMIT results are to be validated which were done by comparing the results produced from WAMIT with published literature and from Sect. 9.3. To verify the validity of the above coefficients and method, wave climate modification for an array of devices from both WAMIT and MIKE 21 have been compared in Fig. 9.14, for significant wave height $H_{m0} = 3$ m and peak wave period $T_p = 7.5$ s. In total, three attenuator type devices and five terminators were simulated using JONSWAP wave spectrum with a peak enhancement factor of 3.3 and a directional standard deviation = 5°. Despite some differences in wave height distribution, the general wave propagation pattern in MIKE 21 agreed well with WAMIT. It is also clear that MIKE 21 shows a large reduction in wave height immediately downstream of the device and no alteration to the upstream wave field; this is expected as the wave diffraction which was included in WAMIT was not considered in MIKE 21. Nevertheless, as the far-field wave conditions being less affected in MIKE 21, and no other single software tool that can model both WEC and wave environments, it was decided to accept this solution for further modelling. A maximum difference of up to 30% in wave heights between WAMIT and MIKE 21 results are seen, particularly very close to downstream of the WECs, however, this difference is found to be minimum further downstream. Thus the WAMIT model has been verified, further simulations were carried out for the wave conditions corresponding to the maximum power production for each device and the values of transmission and reflection coefficients have been obtained. The values of the wave transmission/reflection coefficients for attenuator type devices selected were $C_T = 0.99$ and $C_R = 0.01$, and for terminator OWSC, $C_T = 0.75$ and $C_R = 0.25$.

The WEC array models were simulated for the same time period as the baseline model (as in Fig. 9.12), to be able to evaluate the wave farm impact by comparing model outputs with and without WECs. Following the implementation of the WEC arrays (both attenuator and terminators) in the model for the year 2010, it is observed that the mean significant wave height is decreased downwave of the arrays and this is visible in Fig. 9.15. Further, the absolute mean difference in the significant wave height as a result of the inclusion of the WEC arrays is shown in Fig. 9.16. A clear reduction of wave height is observed downwave following the inclusion of WEC arrays with the largest differences being visible in the region immediately behind the wave array. At the point of maximum impact, i.e., downstream of array close to the coast, a large decrease in wave height based on the annual mean wave conditions is noticed and this constitutes a reduction of a maximum of up to 1 m or just above (i.e. difference in mean wave height computed with WEC minus no WEC) of the incident wave height. The wave height is decreased because of the energy extraction by the WEC array, which is indicated by a negative value in the downstream of the arrays (Fig. 9.16). Furthermore, the results show that as the downstream distance increases the individual effects of each device reduces and an evenly distributed reduction in

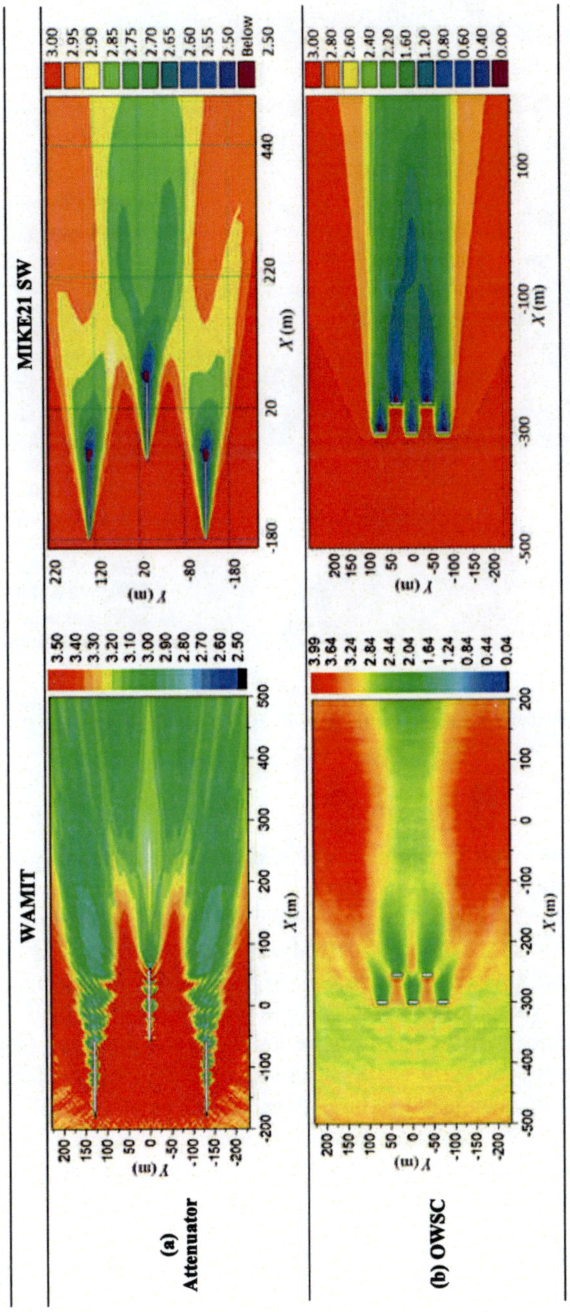

Fig. 9.14 Comparison of WEC performance between WAMIT and MIKE21 SW models for **a** attenuator and **b** terminator. Wave conditions: $H_{m0} = 3$ m, $T_p = 7.5$ s and $\bar{\theta} = 0°$

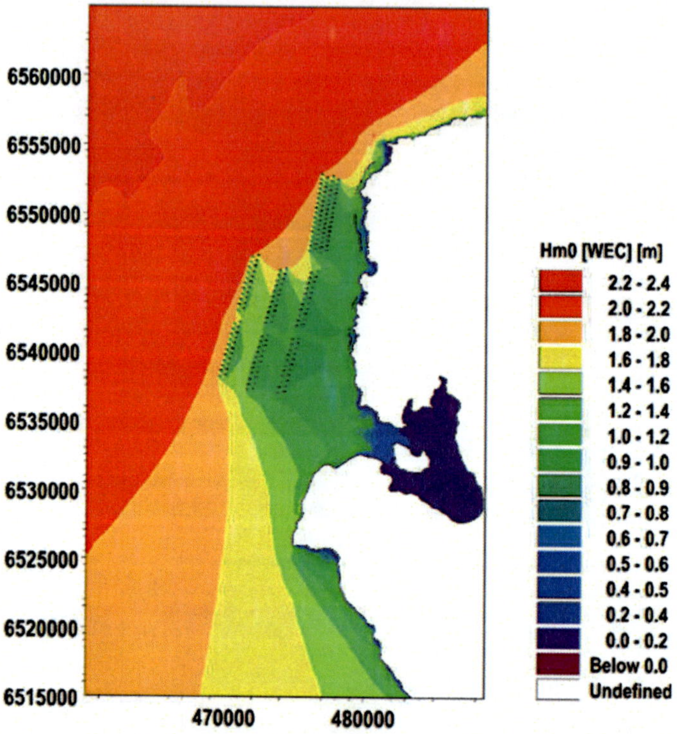

Fig. 9.15 Mean significant wave height for the period January to December 2010 with energy extraction

wave height is observed. The impact of the wave farms decreases with increasing distance, and this is due to restoring processes e.g. diffracted wave energy penetrating into the lee of the wave farm from both sides.

A comparison of the baseline data time series against the WEC array scenario is shown in Fig. 9.17 for a location indicated by the red coloured point to the west of the Bay of Skaill in a water depth of 8 m. Significant wave height, peak wave period and mean wave direction of both cases are presented for the year 2010, and, it is clear that a significant reduction in wave height is obvious. The fact that the wave heights are reduced and the wave directions and wave periods are less affected, which demonstrates that changes to sediments and beach erosion at this site might be less significant, however, this needs further research.

To provide a detailed analysis of energy extraction, the percentage change in significant wave height for each node is calculated and shown in Fig. 9.18. The location of each device is indicated by a large reduction in wave height leeward of the device as indicated in the theoretical test in Fig. 9.14. The maximum percentage height difference results from a situation where a very small original wave height undergoes a substantial change due to wave farm even though the absolute difference

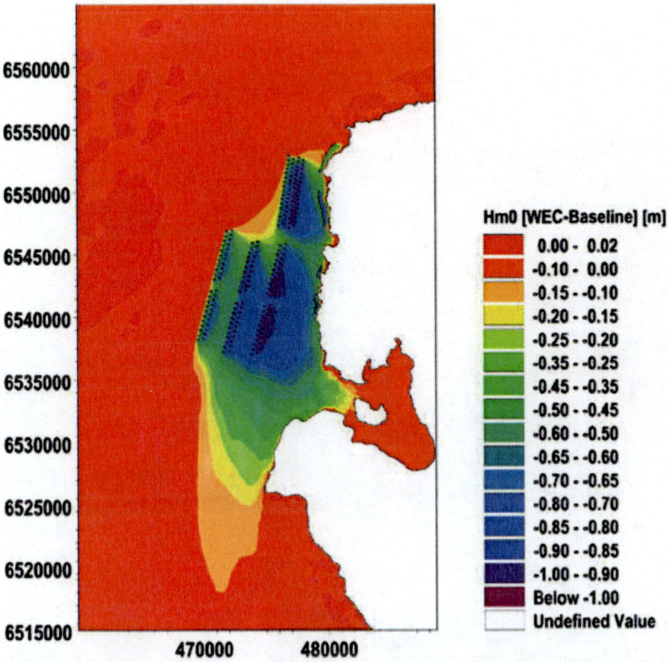

Fig. 9.16 Absolute difference of mean significant wave height with and without energy extraction for the period January to December 2010

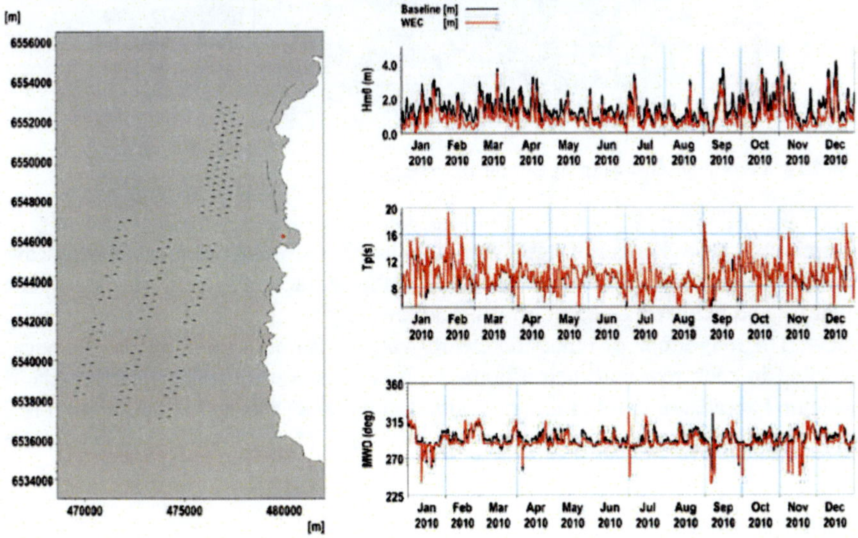

Fig. 9.17 Comparison of time series of wave height, wave period and wave direction at Bay of Skaill between Baseline and WEC scenarios

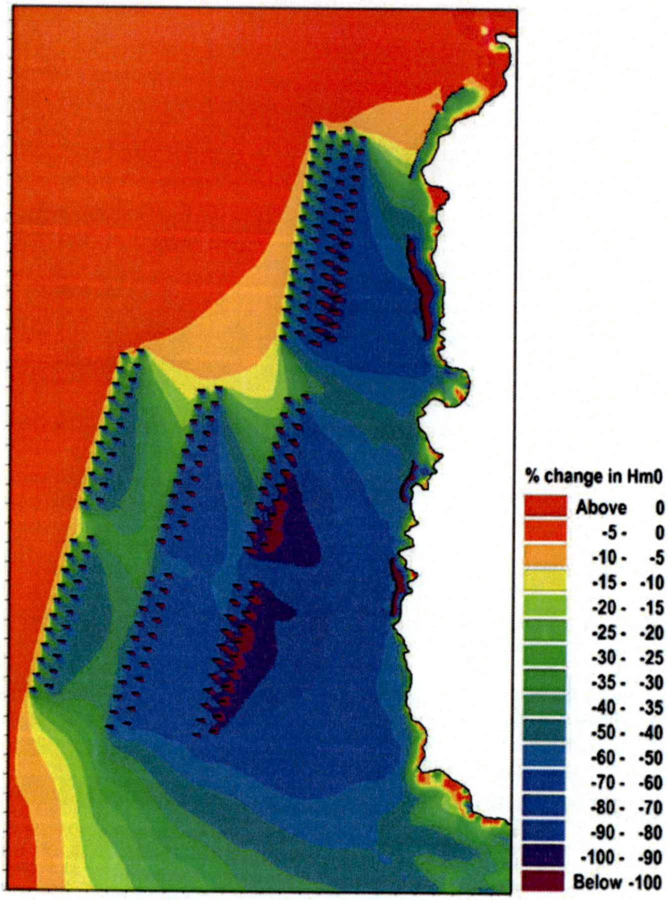

Fig. 9.18 Percentage change in significant wave height

is minimal. These changes of significant wave height represent the net change of wave conditions due to energy absorption by WECs, the interaction between devices within an array and between adjacent arrays, scattering by the array and local bathymetry effect. The magnitude of reduction clearly varies with both the number of devices installed in each row and their alignment. It is to be kept in mind that the results presented above are purely numerical simulation-based and no validation with real scale deployments has been done, hence the usage of this data to be treated with caution.

9.6 Concluding Remarks

The possibilities of modelling wave energy converter arrays by using two different numerical models were illustrated in this article. Hydrodynamic performance of two types of energy converter arrays, i.e. a terminator and an attenuator was successfully investigated. These arrays were subjected to multi-directional realistic seas.

The arrays' performance determined by the Boundary Element Method, presented in terms of the q-factor, was obtained by using stochastic analysis. The seas close to uni-directional wave conditions ($s = 100$) tend to produce a much larger q-factor because of the greater energy density in the single wave component approaching from one direction. The q-factor differs in a relatively greater magnitude with respect to the peak wave period as compared to the wave direction, when subjected to multi-directional sea. Also, the q-factor of the terminator and attenuator WEC arrays are greater when the peak wave period is small. Comparison of the significant wave height of the wave field showed that the wave patterns upstream and downstream of the arrays depends significantly on the type of WECs where the WEC that generates greater diffracted and radiated waves tend to produce greater wave disturbance. Thus, the attenuator WEC generates a greater wave disturbance followed by the terminator; and these phenomena are even more obvious when waves approach from the oblique wave direction. The intensity of the wave disturbance was very much affected by the changes in the wave spreading parameters used in defining the spreading function and encountering wavelength (peak wave period) but not much by the variation in wave direction. In general, this depends on the diffracted and radiated waves generated by the WECs where greater wave disturbances are created when the WECs are subjected to smaller wavelength or when two uni-directional waves interfere with each other.

The array's performance simulated using ocean scale numerical model demonstrated that the WEC arrays altered the wave climate within an array and also the neighbouring array. Cumulative wave height reduction downstream of the WEC array and farther down the coastline appears to be significant, and, its magnitude depends on the array layout and number of co-located arrays. The average yearly reduction in wave height immediately to the lee side of the array was observed to be very high, and its magnitude varies from arrays to arrays. With increasing distance from the arrays towards the shoreline, a recovery in wave heights or energy restore is evident, and energy progresses into the downwave shadow region of the arrays from the sides. The devices modelled for this article used frequency-independent transmission coefficients, rather than using a more detailed approach, due to unavailability of precise device-specific data. Further work is required in investigating frequency-dependent behaviour and dynamic response characteristics of coefficients for absorption, transmission and reflection.

The hydrodynamic interaction of the WEC arrays presented here has added values for WEC designers working on optimising the performance of the array. Also, the wave disturbance upstream and downstream of the arrays can be used as a guideline to coastal engineers in investigating the effect of the presence of the WEC arrays on the environment.

Acknowledgements Work presented in this paper is part of the TeraWatt (EP/J010170/1) and Ecowatt2050 (EP/K012851/1) projects and the authors wish to acknowledge support from the UK Engineering and Physical Sciences Research Council (EPSRC). The authors are also grateful to Cefas (UK) for wave buoys data, European Centre for Medium-Range Weather Forecasts (ECMWF) for providing wind data, European Marine Energy Centre (EMEC) for providing wave buoy data for Orkney, The Crown Estate, UKHO and Marine Scotland Sciences for providing bathymetry data.

References

1. Edenhofer, O., Pichs-Madruga, R., Sokona, Y., Seyboth, K., Kadner, S., Zwickel, T., Eickemeier, P., Hansen, G., Schlömer, S., & von Stechow, C. (2011) *Renewable energy sources and climate change mitigation: Special report of the intergovernmental panel on climate change.* Cambridge University Press.
2. Salter, S. H. (1989). World progress in wave energy—1988. *International Journal of Ambient Energy, 10*(1), 3–24.
3. Clément, A., McCullen, P., Falcão, A., Fiorentino, A., Gardner, F., Hammarlund, K., Lemonis, G., Lewis, T., Nielsen, K., & Petroncini, S. (2002). Wave energy in Europe: Current status and perspectives. *Renewable and Sustainable Energy Reviews, 6*(5), 405–431.
4. Falnes, J. (2007). A review of wave-energy extraction. *Marine Structures, 20*(4), 185–201.
5. Drew, B., Plummer, A. R., & Sahinkaya, M.N. (2009). *A review of wave energy converter technology.* Sage Publications Sage UK.
6. Falcão, A. F. O. (2010). Wave energy utilization: A review of the technologies. *Renewable and Sustainable Energy Reviews, 14*(3), 899–918.
7. Langhamer, O., Haikonen, K., & Sundberg, J. (2010). Wave power—Sustainable energy or environmentally costly? A review with special emphasis on linear wave energy converters. *Renewable and Sustainable Energy Reviews, 14*(4), 1329–1335.
8. Lindroth, S., & Leijon, M. (2011). Offshore wave power measurements—A review. *Renewable and Sustainable Energy Reviews, 15*(9), 4274–4285.
9. Moriarty, P., & Honnery, D. (2012). What is the global potential for renewable energy? *Renewable and Sustainable Energy Reviews, 16*(1), 244–252.
10. Falnes, J. (2002). *Ocean waves and oscillating systems.* Cambridge University Press.
11. EMEC. *Wave devices.* [25 Oct 2020]; Available from: http://www.emec.org.uk/marine-energy/wave-devices/.
12. Folley, M., Babarit, A., Child, B., Forehand, D., O'Boyle, L., Silverthorne, K., Spinneken, J., Stratigaki, V., & Troch, P. (2012). A review of numerical modelling of wave energy converter arrays. In *13th International conference on offshore mechanics and arctic engineering.* American Society of Mechanical Engineers.
13. Li, Y., & Yu, Y.-H. (2012). A synthesis of numerical methods for modeling wave energy converter-point absorbers. *Renewable and Sustainable Energy Reviews, 16*(6), 4352–4364.
14. Budal, K. (1977). Theory for absorption of wave power by a system of interacting bodies. *Journal of Ship Research, 21*(04), 248–254.
15. Evans, D. (1979). Some theoretical aspects of three-dimensional wave-energy absorbers. In *Proceedings of the 1st symposium on wave energy utilization.*
16. Falnes, J. (1980). Radiation impedance matrix and optimum power absorption for interacting oscillators in surface waves. *Applied Ocean Research, 2*(2), 75–80.
17. Thomas, G. P., & Evans, D. V. (1981). Arrays of three-dimensional wave-energy absorbers. *Journal of Fluid Mechanics, 108*, 67–88.
18. Mei, C.C. (1989). *The applied dynamics of ocean surface waves* (vol. 1). World Scientific.

19. Mavrakos, S. A., & McIver, P. (1997). Comparison of methods for computing hydrodynamic characteristics of arrays of wave power devices. *Applied Ocean Research, 19*(5–6), 283–291.
20. Simon, M.J. (1982). Multiple scattering in arrays of axisymmetric wave-energy devices. Part 1. A matrix method using a plane-wave approximation. *Journal of Fluid Mechanics,* **120**, 1–25.
21. McIver, P., & Evans, D. V. (1984). Approximation of wave forces on cylinder arrays. *Applied Ocean Research, 6*(2), 101–107.
22. Mavrakos, S. A., & Koumoutsakos, P. (1987). Hydrodynamic interaction among vertical axisymmetric bodies restrained in waves. *Applied Ocean Research, 9*(3), 128–140.
23. Twersky, V. (1952). Multiple scattering of radiation by an arbitrary configuration of parallel cylinders. *The Journal of the Acoustical Society of America, 24*(1), 42–46.
24. Ohkusu, M. (1974) Hydrodynamic forces on multiple cylinders in waves. In *Proceedings of international symposium on the dynamics of marine vehicles and structures in waves, 1974*. Institute of Mechanical Engineers.
25. Spring, B.H., & Monkmeyer, P.L. (1975). Interaction of plane waves with vertical cylinders. In *Coastal Engineering 1974* (pp. 1828–1847).
26. Kagemoto, H., & Yue, D. K. P. (1986). Interactions among multiple three-dimensional bodies in water waves: An exact algebraic method. *Journal of Fluid Mechanics, 166*, 189–209.
27. Linton, C. M., & Evans, D. V. (1990). The interaction of waves with arrays of vertical circular cylinders. *Journal of Fluid Mechanics, 215*, 549–569.
28. Yılmaz, O., & Incecik, A. (1998). Analytical solutions of the diffraction problem of a group of truncated vertical cylinders. *Ocean Engineering, 25*(6), 385–394.
29. Child, B. F. M., & Venugopal, V. (2010). Optimal configurations of wave energy device arrays. *Ocean Engineering, 37*(16), 1402–1417.
30. Cruz, J., Sykes, R., Siddorn, P., & Taylor, R. E. (2010). Estimating the loads and energy yield of arrays of wave energy converters under realistic seas. *IET Renewable Power Generation, 4*(6), 488–497.
31. Matsui, T., & Tamaki, T. (1981). Hydrodynamic interaction between groups of vertical axisymmetric bodies floating in waves. In *International symposium on hydrodynamics in ocean engineering*.
32. Yue, D. K. P., Chen, H. S., & Mei, C. C. (1978). A hybrid element method for diffraction of water waves by three-dimensional bodies. *International Journal for Numerical Methods in Engineering, 12*(2), 245–266.
33. Berkhoff, J.C.W. (1973) Computation of combined refraction—diffraction. In *Coastal Engineering 1972* (pp. 471–490).
34. University Gent. *Infrastructure and Services—MILDwave*. [25 Oct 2013]. Available from: http://www.ugent.be/ea/civil-engineering/en/research/coastal-bridges-roads/coastal-engineering/infrastructure-services/mildwave.
35. Troch, P. (1998). *A numerical model for propagation and transformation of linear water waves*. Department of Civil Engineering, Ghent University.
36. Radder, A. C., & Dingemans, M. W. (1985). Canonical equations for almost periodic, weakly nonlinear gravity waves. *Wave Motion, 7*(5), 473–485.
37. Mendes, L., Palha, A., Fortes, C. J., Sarmento, A. (2008). Analysis of the impact of a pilot zone for wave energy conversion offshore Portugal. In *The Eighteenth International offshore and polar engineering conference*. International Society of Offshore and Polar Engineers.
38. Beels, C. (2009). *Optimization of the lay-out of a farm of wave energy converters in the North Sea: Analysis of wave power resources, wake effects, production and cost*. Ghent University.
39. Venugopal, V., & Smith, G. H. (2007). Wave climate investigation for an array of wave power devices. In *Proceedings of the 7th European wave and tidal energy conference*.
40. Venugopal, V., Bryden, I. G., & Wallace, A. R. (2010). On the interaction of waves with an array of open chambered structures: application to wave energy converters. In *International conference on offshore mechanics and arctic engineering*.
41. Hasselmann, K., Barnett, T. P., Bouws, E, Carlson, H, Cartwright, DE, Enke, K, Ewing, JA, Gienapp, H, Hasselmann, DE, and Kruseman, P, *Measurements of wind-wave growth and swell decay during the Joint North Sea Wave Project (JONSWAP)*. Ergänzungsheft 8–12, 1973.

42. Millar, D. L., Smith, H. C. M., & Reeve, D. E. (2007). Modelling analysis of the sensitivity of shoreline change to a wave farm. *Ocean Engineering, 34*(5–6), 884–901.
43. Smith, H. C. M., Pearce, C., & Millar, D. L. (2012). Further analysis of change in nearshore wave climate due to an offshore wave farm: An enhanced case study for the Wave Hub site. *Renewable Energy, 40*(1), 51–64.
44. Tay, Z. Y., & Venugopal, V. (2017). Optimization of spacing for oscillating wave surge converter arrays using genetic algorithm. *Journal of Waterway, Port, Coastal, and Ocean Engineering, 143*(2), 04016019.
45. Faltinsen, O.M. (1993). *Sea loads on ships and offshore structures* (vol. 1). Cambridge University Press.
46. Lee, C.-H., & Newman, J.N. (2006). *WAMIT User manual* (p. 42). WAMIT Inc.
47. Goda, Y. (2010). *Random seas and design of maritime structures.* World Scientific.
48. Det Norske Veritas. (2010). *Recommended practice DNV-RP-C205: Environmental conditions and environmental loads.* DNV.
49. Tay, Z. Y., & Venugopal, V. (2017). Hydrodynamic interactions of oscillating wave surge converters in an array under random sea state. *Ocean Engineering, 145*, 382–394.
50. Tay, Z. Y., & Venugopal, V. (2019). The impact of energy extraction of wave energy converter arrays on wave climate under multi-directional seas. *Journal of Ocean Engineering and Marine Energy, 5*(1), 51–72.
51. Renzi, E., Abdolali, A., Bellotti, G., & Dias, F. (2014). Wave-power absorption from a finite array of oscillating wave surge converters. *Renewable Energy, 63*, 55–68.
52. Retzler, C. (2006). Measurements of the slow drift dynamics of a model Pelamis wave energy converter. *Renewable Energy, 31*(2), 257–269.
53. Danish Hydraullic Institute. (2020). *MIKE21: Wave modelling user guide.* DHI.
54. Marine Scotland. Available from: http://www.scotland.gov.uk/Topics/marine/science/MSInteractive/datatype/Bathymetry/data.
55. Venugopal, V., & Nemalidinne, R. (2015). Wave resource assessment for Scottish waters using a large scale North Atlantic spectral wave model. *Renewable Energy, 76*, 503–525.
56. O'Hara Murray, R., & Gallego, A. (2017). Data review and the development of realistic tidal and wave energy scenarios for numerical modelling of Orkney Islands waters, Scotland. *Ocean & Coastal Management, 147*, 6–20.
57. Venugopal, V., Nemalidinne, R., & Vögler, A. (2017). Numerical modelling of wave energy resources and assessment of wave energy extraction by large scale wave farms. *Ocean & Coastal Management, 147*, 37–48.
58. Babarit, A., Hals, J., Muliawan, M. J., Kurniawan, A., Moan, T., & Krokstad, J. (2012). Numerical benchmarking study of a selection of wave energy converters. *Renewable Energy, 41*, 44–63.

Chapter 10
Hydrodynamic Performance of an Array of OWC Devices Integrated with Breakwater

V. Sundar, S. A. Sannasiraj, and S. J. Ashlin

Abstract Marine renewable energy has potential benefits like the enhancement of the livelihood and can solve the energy-security. For cost-effective solution, a multi-purpose wave energy system with an integration into coastal protection has been explored in detail. Having evaluated the performance of an OWC in isolation in the previous chapters, an effort is made to understand the performance of the combination of wave energy technology with detached offshore breakwaters. The said multi-purpose option, termed as OWCBW system is investigated for optimizing the spacing between an array of OWC devices subjected to regular and random waves in normal and oblique wave incidences. It is observed that the OWCBW system with thrice the width of the model spacing exhibited a better performance. The OWC system is found to absorb more power when being placed in an array of devices than compared to an isolated device. The extent of decrease in relative capture width at the resonant frequency of the system for oblique wave incidence compared to that of normal wave incidence is found to be about 30%.

10.1 Introduction

An overview and the importance of integrating wave energy convertors, WEC's to structures for the multi-purpose option was stressed in Sect. 2.8. Studies on the performance characteristics of integration of WEC, i.e., OWC with caisson breakwaters through experimental, numerical, and field investigations have been carried out by several researchers, the option of integration of an array of OWC devices with a rubble mound offshore detached breakwater, (OWCBW), is yet to be explored.

V. Sundar (✉) · S. A. Sannasiraj
Department of Ocean Engineering, Indian Institute of Technology Madras, Chennai, India
e-mail: vsundar@iitm.ac.in

S. A. Sannasiraj
e-mail: sasraj@iitm.ac.in

S. J. Ashlin
COAST Engineering Research Group of University of Plymouth, Plymouth, UK

A. Samad et al. (eds.), *Ocean Wave Energy Systems*, Ocean Engineering & Oceanography 14, https://doi.org/10.1007/978-3-030-78716-5_10

Hence, this chapter focuses on the hydrodynamic characteristics of OWCBW, for the purpose of harnessing wave energy and serving as a coastal protection measure. Prior to evaluating the performance of the OWCBW system, the influence of different structural components of the OWC device for improving its hydrodynamic performance and the evaluation of the design wave loads on the OWC system for deriving its design guidelines should be performed. In this chapter, the functional ability of the OWCBW is evaluated by investigating the hydrodynamic performance of an array of OWC devices embedded in the breakwater.

The most dynamic environment around the globe is the coastal zone, the inhabitants along which is several times more than in the inner region. The factors are responsible for the stability of the coastal zone broadly classified as geomorphological, hydrodynamic, biological, climatic, and man-made factors. Due to the ever increase in the developments along the coast for the purpose of fishing, harbors, power plants, etc., the above instability of several stretches of coast has become a common feature calling for coastal protection measures. The key physical parameters that need to be understood to identify coastal erosion as a problem in the coastal zone are coastal geomorphology, wind, waves, tides. The coastal protection is broadly classified as soft and hard Engineering. The hard structural engineering options that are constructed along the coast as coastal erosion countermeasures are seawalls, groins, or offshore detached breakwaters, DBW. The effect of the DBW to serve not just as coastal protection but also win the lost beach is discussed in Sect. 2.8. Among the different coastal protection structures, detached breakwaters (DBW), although difficult to construct compared to seawalls and groins, have been proved to be an efficient coastal protection structure. The first step towards this would examine its feasibility. Realizing this importance, Sundar et al. [32] have presented and discussed the conceptual design of the integration of OWC devices with DBW for both electricity production and coastal protection or for harbor formation.

Although OWC device technology is a proven technology, the initial investment is the one that does not make it competitive in comparison to the alternative renewable technologies. Again, protecting an eroding coast with DBW can be expensive compared to the other protection measures. Hence, it would be wise to the possibility of applying a multi-utilitarian system. The performance characteristics in terms of its efficiency of an array of OWC devices placed in a row and that in isolation was investigated by Malmo and Reitan [21, 22] and reported that the former, i.e. when in an array, is more effective. It was inferred by McIver and Evans [24] that the PTO mechanism for an array of units exhibited a better efficiency over a range of wave frequencies. It was also reported that the efficiency of the system reduces when exposed to oblique waves due to the reduction in the capture width. Whittaker and Stewart [34] reported that the maximum power output could increase for OWC devices in a row by a scale of two compared to the one in isolation. Through an experimental investigation on OWC wave pumps, Magagna et al. [19] reported that a limited spacing between the devices exhibits higher efficiency.

It is to be mentioned that offshore detached breakwaters (ODBW) have proved to be an efficient coastal protection structure. In a recent study, the influence of the chamber bottom profile configuration (i.e., Flat, Circular curve, Slope 1 in 1, and

Slope 1 in 5 bottom profiles) has been reported by Ashlin et al. [1]. It was resolved that the circular curve bottom profile OWC device is more effective in terms of its hydrodynamic performance, which has been adopted in this paper. Malara and Spanos [20] confirmed the reliability and the efficiency of an array of OWC through approximate solutions by relying on the concept of equivalent linearization. Zheng et al. [35], through a theoretical model based on the linear potential flow theory and eigenfunction matching method, evaluated the hydrodynamic performance of an array of OWCs. It was reported that the wave reflection at the coast mostly plays a constructive role in wave power absorption, and the hydrodynamic interaction between the coast-integrated OWCs significantly increases the efficiency of the OWCs. Through an experimental investigation, Simeon and George [29] proved that with multi OWC's, the performance in terms of its efficiency improves as the spacing is increased. It was stated that an OWC array is less efficient by having individual power take-off systems that operate in isolation compared to that of a modular M-OWC. Howe et al. [11] have proved the spacing between OWC is the most important factor if it needs to be considered as multiple systems, in which case the interaction between the devices also play a key role as it can increase or decrease the efficiency. The development of floating offshore multi-purpose structures and their feasibility was also emphasized. Recently, Nguyen et al. [25] have presented a comprehensive review on the hydrodynamic analysis of integrating wave energy converters with large floating platforms.

The available literature widely pertains to the evaluation of the hydrodynamic performance of an array of OWCs setting into a reflecting wall. There arises a need to examine its hydrodynamic performance while being constructed along with porous structures such as rubble mound breakwaters. The concepts explained by Sundar et al. [32] are believed to minimize the cost of its construction with multi-purpose: energy extracting devices and protecting the coast from erosion or harbor formation. The mutual benefits of avoiding erosion and harnessing energy could be achieved through a detailed study on the performance of OWC devices integrated with DBW could raise the confidence level in working towards deriving the combined benefits of protecting the coast and extraction of energy from ocean waves.

An OWCBW system is investigated through a well-controlled experimental program to evaluate the optimum spacing between an array of OWC devices exposed to the action of the normal and oblique angle of incidence of both regular and random waves. The details of the testing facility, test model, model setup and instrumentation, wave characteristics, uncertainty analysis, hydrodynamic parameters, results, and discussion of the study are reported in this chapter.

10.2 Experimental Investigation

10.2.1 Test Facility

The present comprehensive experimental study on the (OWCBW) to assess the hydrodynamic performance of OWC devices built in the DBW was carried out in a shallow wave basin of dimensions 17 m long, 15 m wide, and 1 m deep in the Department of Ocean Engineering, Indian Institute of Technology Madras, India.

One end of the wave basin is fixed with a piston-type wavemaker for the generation of waves.

The desired wavemaker control signals are generated through a personal computer using an efficient and user-friendly interface, the WaveGen Software of Sriram et al. [30], which is compatible with different operating systems. The five independent wave paddles of the wavemaker are driven by linear electric actuators, which in turn are controlled by the Parker ED controller (Compax3S). The complete details of the wave generation and analysis are discussed in detail by John Ashlin [15].

The maximum stroke length of the wavemaker is 0.2 m with a frequency ranging between 0.3 and 4 Hz. An artificial beach (wave absorber) on the other end is provided with a combination of a parabolic perforated fiber-reinforced polymer (FRP) sheet and rubble mound to absorb the incident waves. The water depth in the shallow wave basin can be varied from 0.3 to 0.6 m.

10.2.2 Data Acquisition Sensors

In the present experimental program, the wave elevation, wave run-up, water pressure, air pressure, and wave forces on the OWC devices were measured using a set of instruments, the details of which are discussed in this section. The host computer, which is used for controlling the wavemaker, is also used for acquiring the signals from the different instruments used in this study. The present physical model studies were designed based on the guidance provided by Hughes [14], Holmes [9], and EMEC [5]. The hydrodynamic performance due to the variation of space between the OWC units when integrated with a breakwater and exposed to oblique wave attacks is studied. The functional ability of the breakwater is also studied in terms of reflection and transmission characteristics.

10.2.3 Test Model and Experimental Setup

The schematic plan view of the shallow wave basin, which depicts the position of wave gauges and the test model, is shown in Fig. 10.1. The locations of the wave

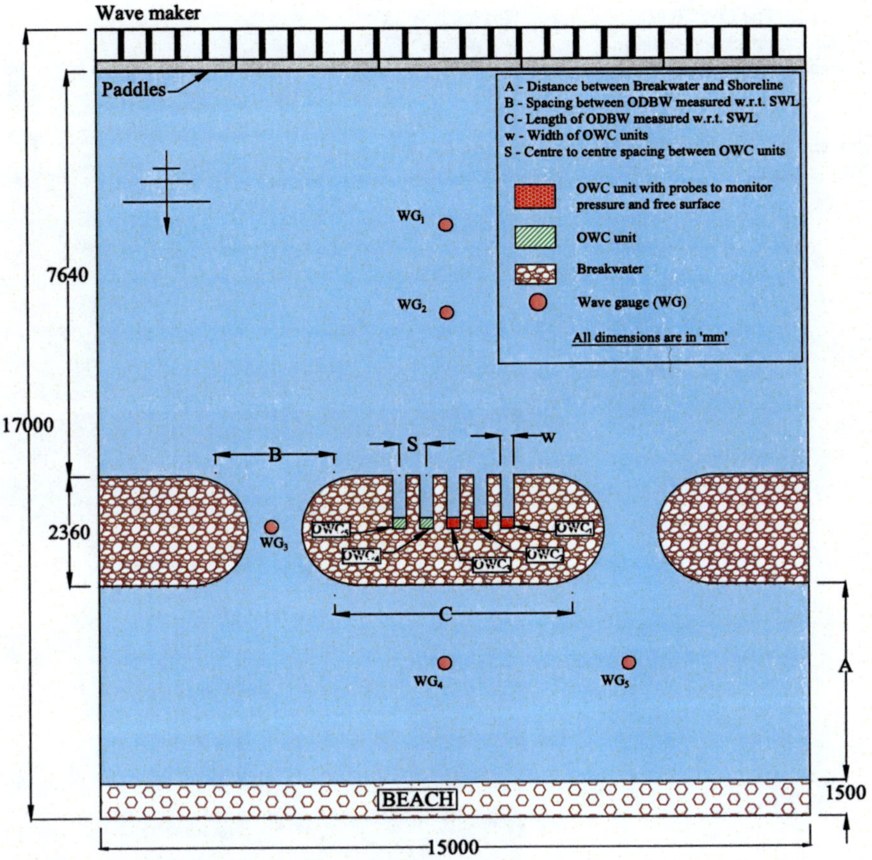

Fig. 10.1 Schematic plan view of the experimental set-up in the shallow wave basin

gauges WG_1 to WG_5 as discussed in the next section, are indicated in the above figure.

The test model OWCBW consists of the OWC models and their integration with ODBW. Views of the model setup with the wave absorber and the wave makers in the background are projected in Fig. 10.2a and b, respectively. A closer view of the OWC models placed inside the center of the three breakwaters representing the detached breakwaters is projected in Fig. 10.2c. As the performance of an OWC device primarily depends on its geometrical dimensions as well as the configuration was finalized through a comprehensive literature survey as detailed by Ashlin et al. [1, 2]. Each of the OWC chambers with dimensions of 283 mm wide (*w*) (parallel to the wave crest) and 180 mm length (*b*) (along the wave direction) was made of a 12 mm thick acrylic sheet. The total height (*h*) of the OWC chamber was 540 mm, and the water depth (*d*) in front of the OWC chamber was 300 mm. The bottom opening (*o*) in the front lip wall was 180 mm. It is found that the natural frequency

Fig. 10.2 The OWCBW system experimental setup in the shallow wave basin: **a** View from wavemaker end; **b** View from absorber end; and **c** Closer view of OWC models

(a)
(b)
(c)

of the system was around $d/L = 0.131$ (L is wavelength). The PTO damping on the OWC chamber was represented by an air hole of optimized a/A, i.e., 0.68% ('a' is an air vent area and 'A' is the waterplane area of the OWC chamber). The dimensional details of the OWC model through front and plan views as well as cross-sectional elevation are shown in Fig. 10.3.

The OWC models were positioned at 10.5 m from the wavemaker and rigidly fixed to the basin floor. A plate on the lee side of the OWC device was surrounded by the rubble mound and formed as an integral of the rubble mound breakwater,

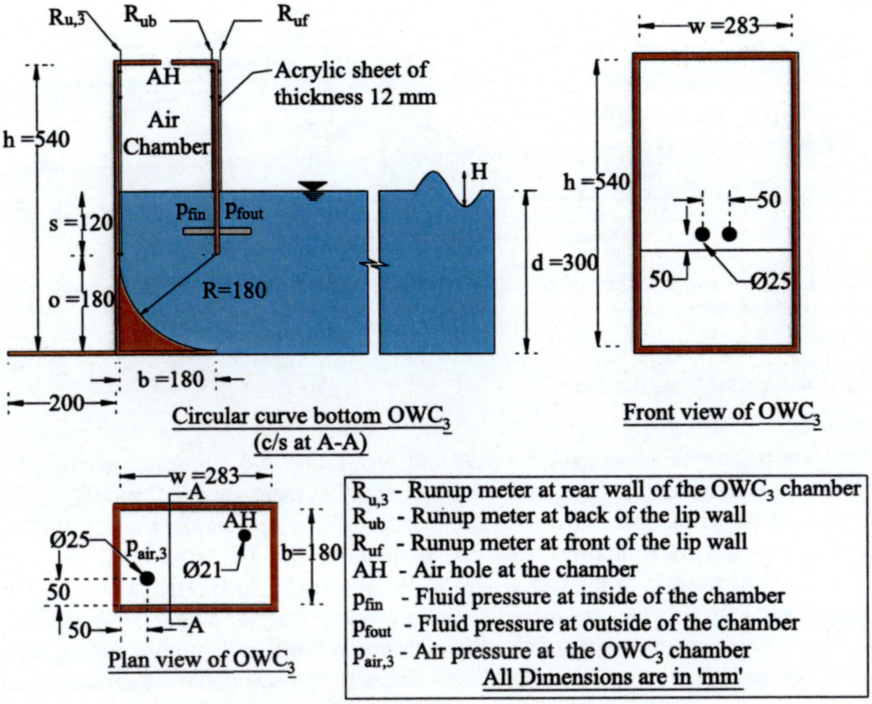

Fig. 10.3 Cross-sectional details of OWC device

which was designed as per the formula of Hudson [12]. The distance between the shoreline and ODBW, A, of 3.5 m was to allow propagation of at least one wave-length corresponding to the longest period of the wave that could be generated in the basin. The length of the central breakwater, C of 5 m, and the spacing between ODBW, B of 2.5 m with reference to still water level (SWL) was finalized as per the recommendations of Dally and Pope [4], Pope and Dean [26], Seiji et al. [28] and Suh and Dalrymple [31].

To avoid sliding of rubble materials from the integrated breakwater on either side of the OWC, harbor walls of length, l, same as its length, b allowed direct propagation into the OWC. This was done as per the recommendations of Malmo and Reitan [21, 22] to achieve maximum efficiency. A Froude scale of 1:20 was employed for integrating five similar OWC devices to the ODBW (OWCBW) for the experimental investigation. A cross-sectional view of the details of the OWCBW system is projected, shown in Fig. 10.4.

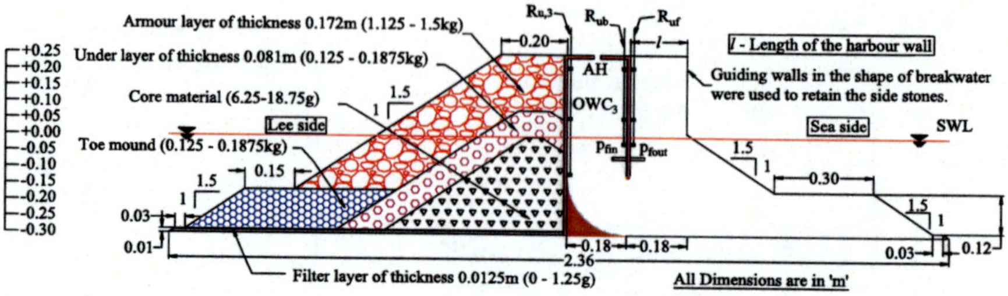

Fig. 10.4 Cross-sectional view of OWCBW system with guiding walls on its adjacent side with measurement details

10.2.4 Instrumentation

Two resistance type wave gauges (WG_1 and WG_2), one at 3.7 m (one wavelength) from the wavemaker, $\eta_1(t)$, and then another at 6.5 m from the OWC model, $\eta_2(t)$, registered the composite wave elevation, from which the incident and reflected components were decomposed as per the method of Goda and Suzuki [8].

Five OWC models of the same sizes (OWC_1–OWC_5), integrated with a rubble mound breakwater, were considered for laboratory testing. Among the five models, the first three models were fixed with pressure sensors and run-up meters. The basic assumption, as per the conclusion of Evans [6, 7], was that the hydrodynamic behaviour of the OWCBW system is symmetric about the mid-point of the array of OWCs. Pressure sensors with a measuring range of 0.2 bar in the central OWC device (OWC_3) in the OWCBW system, one in the front wall at the mouth of the OWC chamber facing the waves ($p_{fout}(t)$) and the other towards the inner side ($p_{fin}(t)$) registered the respective pressure time histories. The air pressure variation due to the oscillation of free surface water inside the chamber of the first three OWC models was measured with a pressure sensor fixed on its top ($p_{air,\,i}(t)$). Each of the three instrumented models was fixed with a run-up gauge inside its chamber over the rear wall to register the run-up ($R_{u,\,i}(t)$). Herein, '*i*' denotes the respective OWC devices' number in which the measurements were registered.

The measured differential air pressure and chamber free surface elevation were adopted to calculate the efficiency of the system. The central unit alone was fixed with an additional run-up gauge on the lee side ($R_{ub}(t)$) and seaside ($R_{uf}(t)$) of the lip wall. This was essential to understand the sloshing inside the chamber as well as the phase difference between the wave elevation inside and outside of the chamber. To understand the diffraction pattern around the OWCBW system, three-wave gauges (WG_3–WG_5) were placed in between the beach and the structure. To evaluate the phase difference between incident waves and the water oscillation inside the chamber, videos were taken close to the OWC system from the front and side view. In total, 15 signatures from five pressure sensors, five run-up meters, two wave gauges in front of the model, and three-wave gauges around the breakwater, were acquired simultaneously with a sampling interval of 0.01 s. The analysis for the measurements was carefully executed in the initial smaller window of the steady-state region.

10.2.5 Wave Characteristics and Hydrodynamic Parameters

The range of frequency, wave steepness, *H/L*, and relative water depth, *d/L* covered under the tests with regular waves were 0.47–1.11 Hz, 0.0065–0.0638, and 0.08–0.26, respectively. Similarly, the H_{m0}/L_p (the ratio of the spectral estimate of significant wave height, H_{m0} to the wavelength corresponding peak wave period, L_p) and d/L_p covered under random waves defined by the Texel MARSEN ARSLOE (TMA) spectrum [13] covered wave steepness, varied from 0.0154 to 0.0364 and 0.0921 and 0.2186, respectively.

The different hydrodynamic parameters that were considered to evaluate the hydrodynamic performance of the structure are discussed herein. For the three-dimensional model study on an array of OWC devices, the fundamental quantity employed to evaluate its performance is the capture width (*CW*), which considers its overall power absorption. For regular waves, this is defined to be the ratio between the total mean pneumatic output power produced by an OWC device to the mean power offered by a unit crest wave width of the incident wave [3], wherein 'mean' refers to the average power per wave period as follows,

$$(\text{Capture width})_{\text{reg}},\ (\text{CW}_{\text{i}})_{\text{reg}} = \frac{\text{P}_{\text{out, i}}}{\text{P}_{\text{w}}} \tag{10.1}$$

Herein, '*i*' denotes the respective OWC devices' number in which the capture width measurement was taken up. The time average incident wave power (P_w) as a function of wave height (*H*), wave period (*T*), and water depth (*d*) is given as,

$$P_w = \frac{\rho g H^2}{8}\frac{C}{2}\left(1 + \frac{2kd}{\sinh 2kd}\right) \tag{10.2}$$

where ρ is the mass density of the fluid, g is the gravitational acceleration, $C = L/T$. Due to the rise and fall of water inside the OWC chamber, the pneumatic output power ($P_{out,\,i}$) is developed. It is calculated by integration of instantaneous free surface oscillation inside the chamber moving with a velocity ($V_{w,\,i}$) under the air pressure ($p_{air,\,i}$) as follows [18],

$$\text{P}_{out,i} = \frac{1}{T}\int_{t}^{t+T} P_{air,i}\, A\, V_{w,i}\, dt \tag{10.3}$$

herein, $V_{w,\,i}$ is the first derivative of the third-order approximation to the free surface elevation inside the respective OWC chambers calculated by the expression of López et al. [18],

$$V_{w,i} = \frac{2z_{j+1} + 3z_j - 6z_{j-1} + z_{j-2}}{6\Delta t} \tag{10.4}$$

where z_j is the elevation of the free surface at time t_j, and Δt is the sampling interval. The capture width for the OWC in random wave field $(CW_i)_{ran}$, is defined as,

$$(Capture\,width_i)_{ran},\ (CW_i)_{ran} = \frac{P_{out,i}}{P_{ave(in)}} \tag{10.5}$$

The average input power is calculated through the expression,

$$P_{ave(in)} = w\rho g \sum_{i=1}^{N} S_j (C_g)_j\, df \tag{10.6}$$

where N is the number of frequency bands (each with a width of df), S_j is the spectral density at the jth band, and $(C_g)_j$ is the group velocity at the jth band. The group velocity is given by,

$$(C_g)_j = n_j C_j \tag{10.7}$$

herein,

$$n_j = \frac{1}{2}\left(1 + \frac{2k_j d}{\sinh(2k_j d)}\right) \tag{10.8}$$

An alternative non-dimensional parameter, viz., relative capture width denoting the ratio of the capture width to the width of the device is proposed for assessing the performance of the system.

$$(\mathrm{Rel}ative\,capture\,width_i)_{reg},\ (RCW_i)_{reg} = \frac{(CW_i)_{reg}}{w} \tag{10.9}$$

$$(\mathrm{Rel}ative\,capture\,width_i)_{ran},\ (RCW_i)_{ran} = \frac{(CW_i)_{ran}}{w} \tag{10.10}$$

The pressure in front of the lip wall $(p_{fout})_{max}$ (from crest to trough), and that inside the air chamber $(p_{air,\,i})_{max}$ (from crest to trough), are normalized by 'ρgH' to represent $(PR_{fout})_{reg}$ and $(PR_{air,\,i})_{reg}$, respectively. Herein, 'H' is measured at the model location in the absence of a model. The wave amplification factor $(\beta_i)_{reg}$, is defined as the ratio of the maximum free surface oscillation inside the OWC chamber, $(R_{u,i})_{max}$ (from crest to trough), to the incident wave height. In the case of random waves, zeroth spectral moment of (p_{fout}), $(p_{air,i})$, and $(R_{u,i})$ are considered. In this chapter, the effect of wave steepness, H/L, relative water depth, d/L, and non-dimensional spacing, kS

($k = 2\pi/L$), on the non-dimensional front wall outside pressure, wave amplification factor, air pressure, capture width, and relative capture width measured in the OWC model subjected to normal and oblique wave incidence has been investigated.

10.3 Results and Discussion

10.3.1 Time Histories

Typical time histories of wave elevation (η_1 and η_2), run-up measurement at the seaside of the lip wall (R_{uf}), run-up measurement at the lee side of the lip wall (R_{ub}), run-up measurement at the back wall of first three respective OWC devices ($R_{u,1}, R_{u,2}, and\ R_{u,3}$), pressure measurement at the front side of the lip wall (p_{fout}), air pressure measurement at first three respective OWC devices ($p_{air,1}, p_{air,2}, and\ p_{air,3}$), and wave transmission behind the ODBW (η_4) subjected to regular wave field characterized by $H = 0.03$ m and $T = 0.9$ s are depicted in Fig. 10.5a. The time series for the above parameters for the model subjected to random waves for $H_{m0} = 0.05$ m and $T_p =$ 1.0 s are projected in Fig. 10.5b. In general, R_{ub} is as low as about half of the R_{uf} and $p_{air,3}$ reduces for low period waves (away from the natural frequency of the system), as a major portion of the energy incident on the lip wall gets reflected, leading to lesser energy into the chamber.

10.3.2 Effect of Wave Characteristics

10.3.2.1 General

An array of OWC devices with five devices exhibited better performance than the infinite number of devices. In the present experimental study, OWC devices were integrated with the layout of detached breakwaters, and the functional requirement of the length of the breakwater limited the maximum number of OWC devices up to five for the purpose of its integration with ODBW. For systems of more than a single device, an appropriate spacing must be chosen based on the average performance of an array of devices, as pointed out by Cruz [3]. To understand the effect of spacing between OWC units on the average hydrodynamic performance of an array of OWC devices, three different spacing, S, i.e., one, two, and three times the width of the OWC model, were considered. When the spacing equals the width of the OWC device, the array of OWC units positioned together in an OWCBW system will act as a partially absorbing vertical wall type structure when exposed to incident waves. For other cases, the OWCBW system will act as a porous structure in dissipating diffracted waves around the OWC devices. This phenomenon has been confirmed based on the reflection coefficient, as discussed later under Sect. 10.3.7. The influence of wave

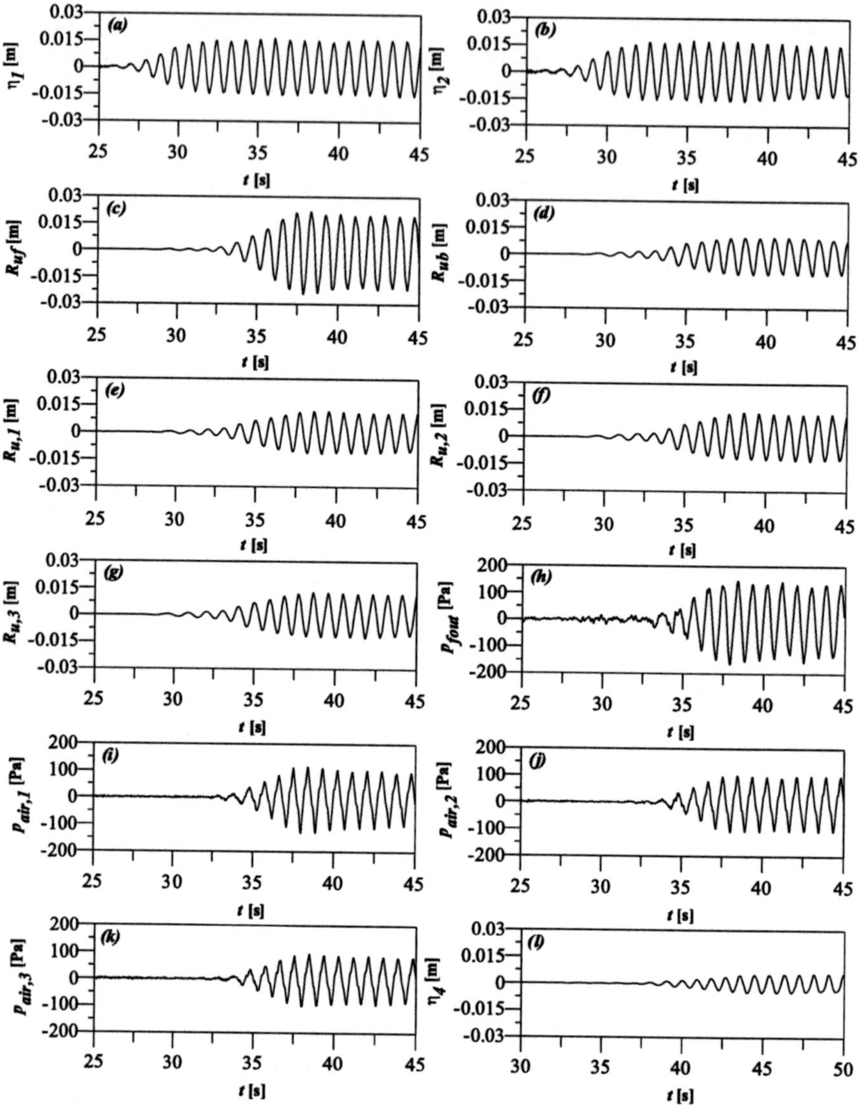

Fig. 10.5 **a** Typical time histories of (a) η_1, (b) η_2, (c) R_{uf}, (d) R_{ub}, (e) $R_{u,1}$, (f) $R_{u,2}$, (g) $R_{u,3}$, (h) p_{fout}, (i) $p_{air,1}$, (j) $p_{air,2}$, (k) $p_{air,3}$, and (l) η_4 for thrice the width of the OWC model spacing in an OWCBW system. ($H = 0.03$ m, $T = 0.9$ s).**b** Typical time histories of (a) η_1, (b) η_2, (c) R_{uf}, (d) R_{ub}, (e) $R_{u,1}$, (f) $R_{u,2}$, (g) $R_{u,3}$, (h) p_{fout}, (i) $p_{air,1}$, (j) $p_{air,2}$, (k) $p_{air,3}$, and (l) η_4 for thrice the width of the OWC model spacing in an OWCBW system. (H_{mo}= 0.05 m, T_p= 1.0 s)

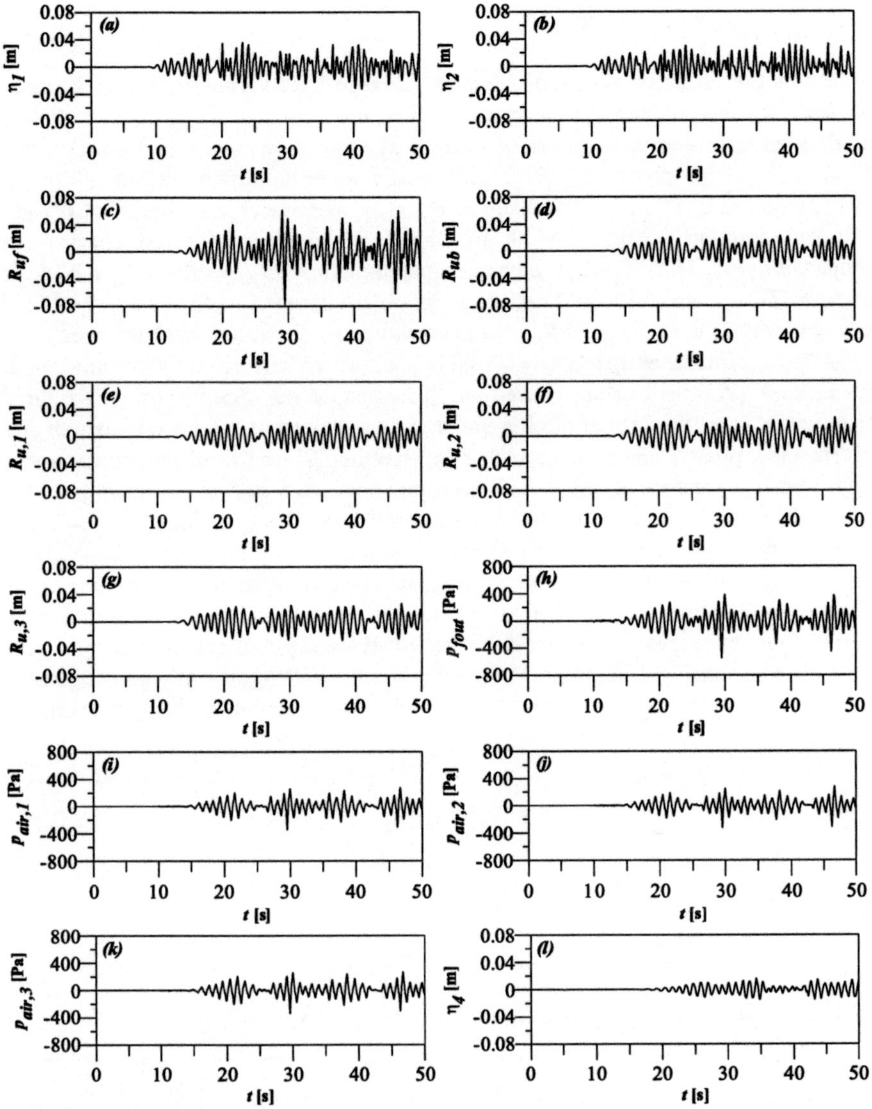

Fig. 10.5 (continued)

characteristics on the average performance of the OWCBW system for different spacing has been evaluated by exposing to tests of wave heights from 0.03 m to 0.09 m with different wave steepness.

10.3.2.2 Effect of Wavelength

To assess the effect of wavelength on the average hydrodynamic performance of an array of OWC devices for different spacing, the regular wave experiments were carried out with a wave height *H* of 0.03 m. The variations in the non-dimensional front wall outside pressure $(PR_{fout})_{reg}$, average wave amplification factor $<\beta>_{reg}$, average non-dimensional air pressure $<PR_{air}>_{reg}$, average capture width $<CW>_{reg}$, and average relative capture width $<RCW>_{reg}$ with *d/L* for different spacing are projected in Fig. 10.6. The hydrodynamic parameters, $<\beta>_{reg}$, $<PR_{air}>_{reg}$, $<CW>_{reg}$, *and* $<RCW>_{reg}$ are discussed here in terms of the average of the first three OWC devices (OWC_1 to OWC_3). $(PR_{fout})_{reg}$ is presented for the mid-OWC chamber.

$(PR_{fout})_{reg}$ increases up to $d/L = 0.104$, beyond which it decreases with wavelength (increase in *d/L*; the tests are carried out in a constant water depth) for all the three spacings, *S* since the orbital motion under the wave decreases with wavelength and exerts lesser pressure near the mouth of the chamber. Since the orbital motion under high-frequency waves is less, the lip wall prevents the wave energy from entering the OWC device chamber and increases the reflected energy. Around $d/L = 0.128$ (natural frequency of the system), there was observed to be a large $(PR_{fout})_{reg}$ due to lesser reflection coefficient, which increases with a decrease in wavelength, as this can be seen in the later Sect. 10.3.7. Due to the higher reflection coefficient at high-frequency waves that increase the reflected energy and exerts lesser pressure near the mouth of the chamber. The higher value of $(PR_{fout})_{reg}$ around the natural frequency of the system indicates the possibility of more energy being transmitted

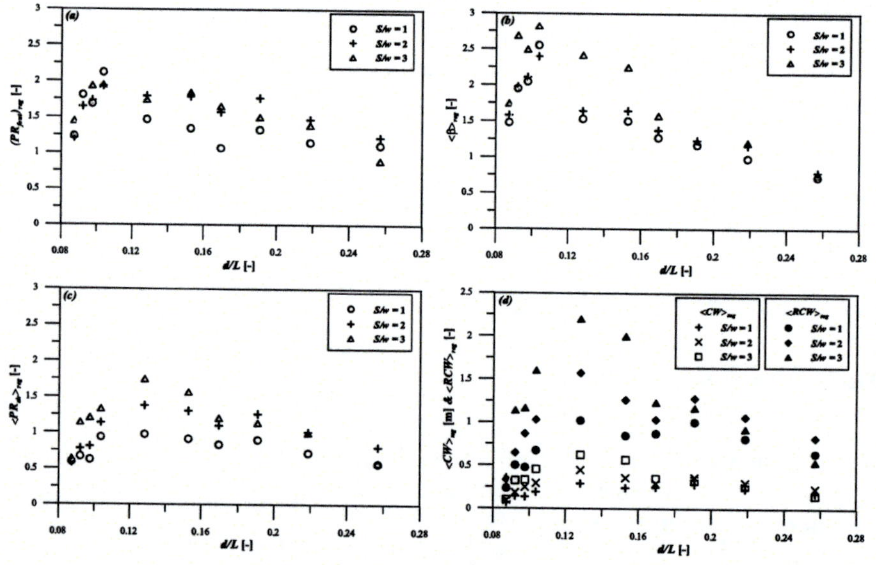

Fig. 10.6 Variation of **a** $(PR_{fout})_{reg}$, **b** $<\beta>_{reg}$, **c** $<PR_{air}>_{reg}$, **d** $<CW>_{reg}$, and $<RCW>_{reg}$ corresponds to d/L for different spacing (H = 0.03 m)

past the lip wall into the chamber. Among the three spacings tested, the maximum $(PR_{fout})_{reg}$ was found to be 2.12 for $S/w = 1$. It is higher than the value observed from the similar two-dimensional (2D) model tests [Ashlin et al., [1] in which the maximum PR_{fout} was observed to be 1.3 under regular and random wave incidences. It is believed that due to higher reflection from the OWCBW system as well as the convergence of water waves, the energy concentration in front of the device is high and absorbs more power from the total incident wavefront greater than the width of the device, which in fact tends to increase $(PR_{fout})_{reg}$ up to a maximum of about 2.12, i.e., more than the value for a fully reflected wave condition in front of a structure.

The influence of *d/L* and *S* on the free surface amplification, $<\beta>_{reg}$, inside the OWC chamber is shown in Fig. 10.6b. The energy conversion in the device occurs due to the wave oscillation inside the chamber and is determined by the hydrodynamics of the system. However, the effect of radiated flow around the device also influences the energy conversion. The maximum $<\beta>_{reg}$ is about 2.8 times the incident wave height for $S/w = 3$ for *d/L* up to about 0.16, beyond which the factor reduces to about 0.75.

$(PR_{fout})_{reg}$ and $<\beta>_{reg}$ for $S/w = 1$ are 2.12 and 0.8, respectively. Thus, the wave amplification factor is high around the natural frequency of the system and increased spacing. The dynamic air pressure inside the chamber at the natural frequency of the system ($d/L = 0.128$) is almost equal to that of the dynamic pressure available outside the chamber, $(PR_{fout})_{reg}$, for $S/w = 3$. Further, $<PR_{air}>_{reg}$ is found to be maximum at $d/L = 0.128$ (natural frequency of the system) and reduces on either side.

The variation of $<CW>_{reg}$ and $<RCW>_{reg}$ with *d/L* for different spacing is depicted in Fig. 10.6d. The efficiency in the case of 2D tests is like $<RCW>_{reg}$, wherein the energy incident over the entire width of the OWC device is considered. However, in the case of 3D tests, to accommodate the diffraction phenomena, the term "capture width (*CW*)" is introduced. The capture width is not dimensionless, and thereby, it is scale-dependent. However, the concept of *CW* is meaningful [27] in the sense that it represents the length of the unit wavefront that contains the absorbed power. For spacing equals to thrice the width of the OWC model, the $<CW>_{reg}$ is observed to be a maximum to the extent of about 0.6 near the natural period of the system ($d/L = 0.128$) and decreases on either side.

The physical processes causing the reduction in capture width on either side of the natural period of the device are explained by quantitative analysis of the phase difference between the variations in the hydrodynamic parameters $(\beta_3)_{reg}$ and $(PR_{air,3})_{reg}$. The phase difference between these parameters for the different *d/L* as a function of *t/T* is provided in Table 10.1. The phase difference between $(\beta_3)_{reg}$ and $(PR_{air,3})_{reg}$ is found to be almost 270^0 with the least at $d/L = 0.128$ correspondings to the natural frequency of the system. On either side of the above-stated *d/L*, since the phase difference is found to be higher, the $<CW>_{reg}$ is found less despite $<\beta>_{reg}$ being higher. A similar observation has been discussed earlier by Kamath et al. [16]. The percentage of decrease in $<CW>_{reg}$ for $S/w = 1$ and $S/w = 2$ compared to that for $S/w = 3$ is found to be in the range of 11–59% and 8–43%, respectively.

The non-dimensional parameter $<RCW>_{reg}$ can possess values greater than unity. When OWC devices are placed in an array, the devices capture maximum power to

Table 10.1 Quantitative phase difference between $(\beta_3)_{reg}$ and $(PR_{air,3})_{reg}$ in terms of t/T in radians for different relative water depth ($S/w = 3$)

d/L	$(\beta_3)_{reg}$ (t/T)	$(PR_{air,3})_{reg}$ (t/T)	$(\beta_3)_{reg}$ (t/T)–$(PR_{air,3})_{reg}$ (t/T)	Phase lag (rad)
0.257	46.01	46.79	−0.78	4.90
0.170	24.77	25.53	−0.76	4.78
0.128	15.79	16.52	−0.73	4.59
0.104	12.53	13.30	−0.77	4.84
0.087	13.08	13.85	−0.77	4.84

the extent of about 2.2 times the given input power at the natural frequency of the system. This confirms that the OWC device could absorb more power when placed in an array than in isolation due to energy concentration. $\langle RCW\rangle_{reg}$ is found to decrease with spacing between the OWC units. In general, the key hydrodynamic parameters $(PR_{fout})_{reg}$, $\langle\beta\rangle_{reg}$, $\langle PR_{air}\rangle_{reg}$, $\langle CW\rangle_{reg}$, and $\langle RCW\rangle_{reg}$ increases with wavelength until the natural frequency of the system and the spacing between the OWC units.

10.3.2.3 Effect of Wave Height

The performance of the OWC device is highly influenced by wave period followed by wave height [17]. The effect of wave height on the average performance of the OWCBW system has been examined. Figures 10.7a–d show the variation of $(PR_{fout})_{reg}$, $\langle\beta\rangle_{reg}$, $\langle PR_{air}\rangle_{reg}$, $\langle CW\rangle_{reg}$, and $\langle RCW\rangle_{reg}$ for $H = 0.09$ m, respectively. It is observed that $(PR_{fout})_{reg}$ is higher than that measured for $H = 0.03$ m. For any *d/L*, an increase in *H* leads to an increase in the wave steepness. It would induce higher pressures on the lip wall. The trend in the variation of $\langle\beta\rangle_{reg}$, $\langle PR_{air}\rangle_{reg}$, $\langle CW\rangle_{reg}$, and $\langle RCW\rangle_{reg}$ with *d/L* for the three '*S*' tested is found to exhibit a similar trend as that seen for $H = 0.03$ m. However, the values of these four parameters for $H = 0.09$ m are found to be lesser than that obtained for $H = 0.03$ m. When the steeper waves pass into the chamber, the energy dissipation at the tip of the lip wall due to turbulence, vortex shedding, the radial flow of wave, etc., is higher than the low steepness waves and leads to higher energy loss [16]. This phenomenon decreases the water surface oscillation and, thus, the air pressure inside the chamber for a higher wave.

10.3.2.4 Effect of Wave Steepness

The effect of wave steepness, *H/L*, on the average performance characteristics through the parameters $(PR_{fout})_{reg}$, $\langle\beta\rangle_{reg}$, $\langle PR_{air}\rangle_{reg}$, $\langle CW\rangle_{reg}$, and $\langle RCW\rangle_{reg}$ for finer ranges of *H/L* (0.0065–0.0155, 0.0201–0.0278, 0.0302–0.0434 and 0.0597–0.0638) is reported in Fig. 10.8. *S/w* of 3 was considered. The increase in wave steepness

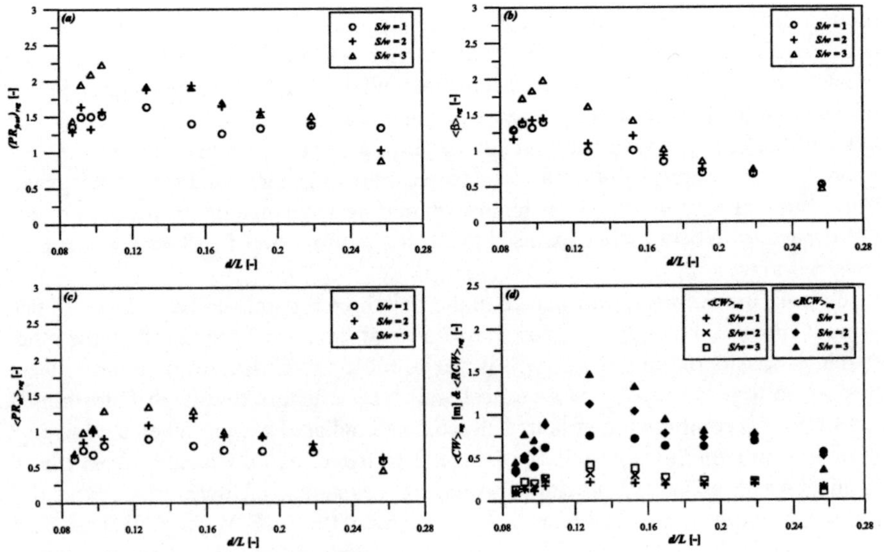

Fig. 10.7 Variation of **a** $(PR_{fout})_{reg}$, **b** $<\beta>_{reg}$, **c** $<PR_{air}>_{reg}$, **d** $<CW>_{reg}$, and $<RCW>_{reg}$ with *d/L* for different spacing ($H = 0.09$ m)

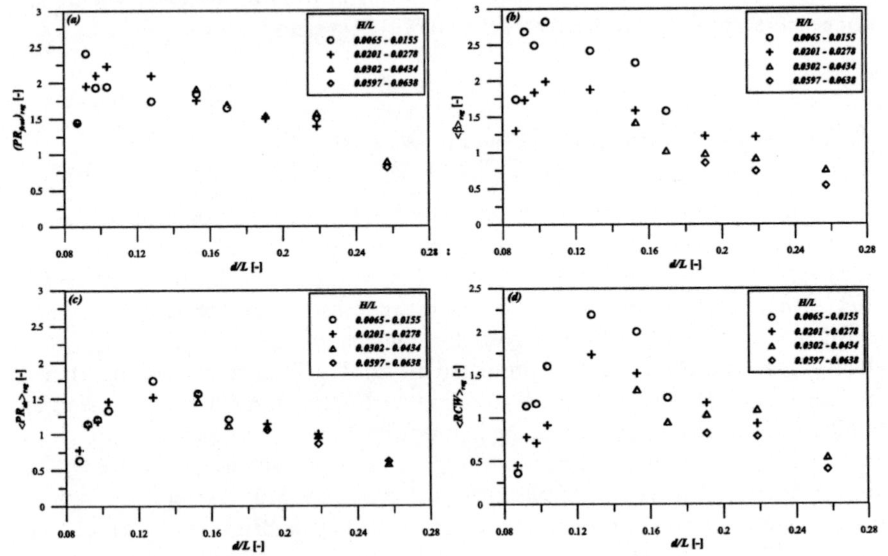

Fig. 10.8 Variation of **a** $(PR_{fout})_{reg}$, **b** $<\beta>_{reg}$, **c** $<PR_{air}>_{reg}$, and **d** $<RCW>_{reg}$ corresponds to *d/L* for different wave steepness range ($S/w = 3$)

induces more pressure on the front side of the lip wall. However, $<\beta>_{reg}$, $<PR_{air}>_{reg}$, *and* $<RCW>_{reg}$ are found to decrease with an increase in *H/L* due to energy loss when the wave passes into the chamber past the lip wall. Further, it is to be mentioned that the same model, when it was subjected to two-dimensional waves [1], exhibited a similar kind of hydrodynamic variation for different wave steepness regime, i.e., the hydrodynamic efficiency of the system decreases with an increase in wave steepness. The results clearly indicate that hydrodynamic performance characteristics of the OWC device are better when exposed to the action of waves of lesser steepness than steeper waves.

Overall, the average performance of the first three respective OWC devices in an array is found to be higher for thrice the width of the model spacing than the other spacings considered in this study. It is due to those OWC devices in an array were positioned together, acting as a vertical wall type structure that contributes higher reflection and combination of both diffracted and radiated waves. When the incident wave is acting on the structure, the reflected, radiated, and diffracted waves cancel incident waves and reduce the input power to the system [3]. However, for twice and thrice the width of the model spacing integrated with ODBW, the reflection from OWC devices decreases, and the combination of both diffracted and radiated waves are dissipated through the rubble materials in between the individual OWC units. This certainly reduces the cancellation of waves, and so increases the concentration of energy due to the convergence of water waves in front of the OWC model and further increases the performance of the OWCBW system.

10.3.3 Wave Interaction Between Devices

The average hydrodynamic performance of the OWCBW system for different spacing between devices has been presented in the previous section. It is further important to understand how the individual OWC units in an array perform due to the interaction effect when subjected to wave action. In order to assess the variability in energy conversion within the system, the hydrodynamic parameters $(\beta_i)_{reg}$, $(PR_{air,\,i})_{reg}$, and $(RCW_i)_{reg}$ as a function of *d/L* are presented in Fig. 10.9 for different spacing. Herein, '*i*' denotes the respective OWC devices in which the measurement was taken up. Similar to the earlier observation, the hydrodynamic parameters $(\beta_i)_{reg}$, $(PR_{air,i})_{reg}$, and $(RCW_i)_{reg}$ are found to increase with the spacing as well as the wavelength for all the three models. It is also observed that the array with $S/w = 3$ exhibits the maximum efficiency. For $S/w = 1$, the hydrodynamic parameters $(\beta_i)_{reg}$, $(PR_{air,\,i})_{reg}$, and $(RCW_i)_{reg}$ of the first three respective OWC models are found to decrease from the leading edge (OWC_1) towards the central OWC chamber (OWC_3). The reduction in the wave amplification factor at the natural frequency of the system ($d/L = 0.128$) for OWC_2 and OWC_3 devices with the base reference of OWC_1 device is 5% and 21%, respectively.

It is due to the OWC model placed on the leading edge subjected to higher wave amplification inside of its chamber than the other models, as observed from one of the

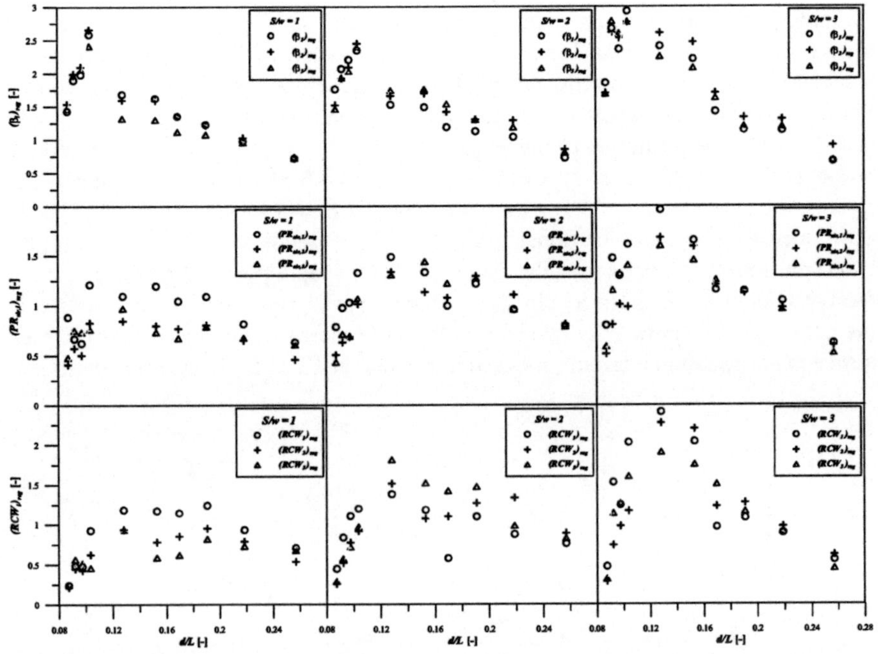

Fig. 10.9 The interaction between OWC models in an array influence on the first three respective OWC devices $(\beta_i)_{reg}$, $(PR_{air,i})_{reg}$, and $(RCW_i)_{reg}$ corresponds to *d/L* for different spacing (H = 0.03 m)

videos taken from the front view of the OWCBW system. Further, the leading-edge model in the spacing configuration of $S/w = 1$ is surrounded on one side by rubble materials, and that decreases the reflection from the system. This certainly increases the wave power input to the system and highly pushes and pulls the water mass inside the OWC model in the leading edge of the OWCBW system than the other models. For the OWCBW system with $S/w = 2$, the efficiency of the individual OWC device for d/L <0.12 is found to be a maximum for OWC_1 device and the least for OWC_2 device. For d/L> 0.12, OWC_1 device is found to exhibit the least efficiency. The OWCBW system with $S/w = 3$, for d/L <0.12, the performance characteristics is found to be similar to that seen for $S/w = 2$. However, for d/L> 0.12 a clear trend on the effect of position of the OWC device is not observed.

10.3.4 Effect of Spacing

Further, the effect of non-dimensional spacing (kS *or* k_pS in the case of random waves) on the average hydrodynamic performance of the OWCBW system subjected to both regular and random waves for their direct field applications is examined.

To understand the variation of the $\langle RCW\rangle_{reg}$ and $\langle RCW\rangle_{ran}$ as a function of their respective non-dimensional spacings is projected in Figs. 10.10 and 10.11, respectively. To avoid the influence of wave frequency on $\langle RCW\rangle$, the above-said figures are depicted in terms of different discrete representations of *d/L*. As discussed earlier, due to the limitation in the length of ODBW to serve as a coastal protection purpose, the $\langle RCW\rangle$ is given for *kS* in the range $0.5 \leq kS \leq 4.56$. In the above figures, the lower limit of *kS* corresponds to the case, wherein the OWC models are positioned together, and the upper limit of *kS* corresponds to the case, wherein the OWC units are positioned with the center to center spacing being thrice the width of the OWC device. For both regular and random waves, the trend in the variation of the hydrodynamic parameter $\langle RCW\rangle$ is found to increase with an increase in non-dimensional spacing, attaining a maximum value at $kS = 2.27$, which corresponds

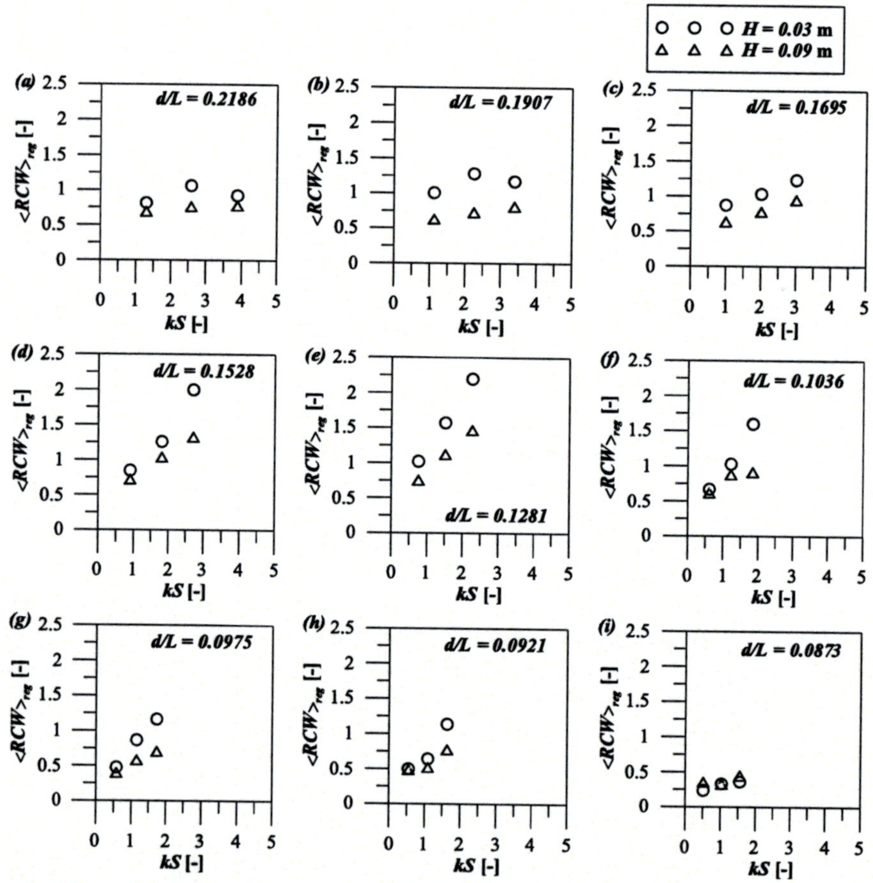

Fig. 10.10 Effect of spacing (*kS*) on average relative capture width ($\langle RCW\rangle_{reg}$) (*H/L* = 0.0065–0.0638 and *d/L* = 0.0873–0.2186)

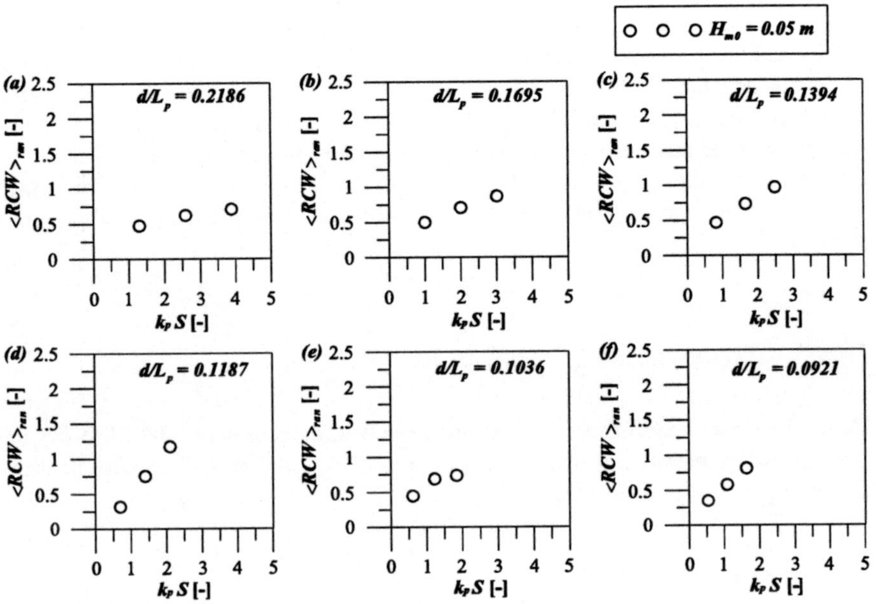

Fig. 10.11 Effect of spacing (k_pS) on average relative capture width (<RCW> $_{ran}$) (H_{mo}/L_p= 0.0154–0.0364 & d/L_p= 0.0921–0.2186)

to the natural frequency of the system ($d/L = 0.128$). For $d/L = 0.128$, the minimum value of <RCW> occurs at $kS = 0.75$. For the range of kS used amid the constraints in spacing, the maximum and minimum values of hydrodynamic parameters correspond to the resonant frequency of the system that occurs at 1.52 times of kS apart, which is similar to the results observed in the theoretical investigation of Thomas and Evans [33].

Even though the trend in the variation of results observed to be similar for both regular and random waves, the magnitude of upper bound random wave measurements is 50% lesser than the upper bound regular wave measurements. On the other hand, the magnitude of lower bound random measurements is 39% higher than the lower bound regular wave measurements. The regular wave measurements were conducted for the wave height ranges from 0.03 m to 0.09 m, and the random wave measurements adopted a significant wave height of 0.05 m. The increase in wave height leads to an increase in wave steepness, consequently reduces the performance of the system. So, when we are trying to compare the different ranges of wave steepness being tested for regular and random waves, the measurements are contradicting each other. By comparing the similar wave tests between regular and random wave measurements, the OWCBW system exhibits only about 17–22% of reduction in average hydrodynamic performance for random wave measurements over regular wave measurements. It is due to the number of wave components involved in the description of the random wave climate.

From the above discussions, we can conclude that the optimal spacing between the OWC units for the present model setup is $kS = 2.27$. Herein, the term wavelength in 'k' corresponds to the predominant wavelength prevail in the coast considered for integrating OWC models with ODBW. The optimal spacing may vary for the long breakwater than the one considered in this study and can be studied separately. Consequently, the key geometric parameters of the OWC device can be fixed based on the same predominant wavelength, as discussed in detail in the authors' previous work [1].

10.3.5 Performance of OWC in an Array

The hydrodynamic performance of an array of OWC devices in the OWCBW system is investigated, hereby normalizing the sum of CW_{reg} of five devices with its array occupying space as given below:

$$(Array\ RCW)_{reg} = \frac{(CW_3)_{reg} + 2\big((CW_1)_{reg} + (CW_2)_{reg}\big)}{5w + 4S} \tag{10.11}$$

The variations in $(Array\ RCW)_{reg}$ as a function of d/L for three different spacing configurations are shown in Fig. 10.12. From the results, it is observed that, unlike, as seen in the earlier results of $\langle RCW \rangle_{reg}$, which is always found to be high for S/w =3 throughout the d/L regime, the $(Array\ RCW)_{reg}$ is measured to be higher for S/w =1 corresponds to d/L greater than 0.16. The change in trend is mainly due to the uneven normalization because of different end-to-end array occupancy widths

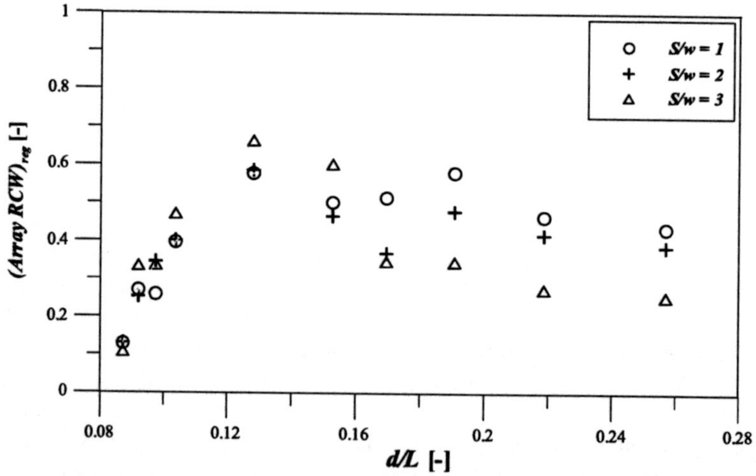

Fig. 10.12 Variation of $(Array\ RCW)_{reg}$ corresponds to d/L for different spacing ($H = 0.03$ m)

by OWC devices. The *(Array RCW)*$_{reg}$ for *S/w* =3, despite being normalized by the value of 90% more than that of *S/w* = *1*, the quantity difference between them is only about 50–70%. But with the increase in *d/L*, the trend in variation of *(Array RCW)*$_{reg}$ is reversed and found to be high for the spacing configuration of *S/w* = 3 of about 15%–30% more in comparison with other spacing configurations. This explains overall, the array configuration with *S/w* = 3 exhibits better performance throughout the *d/L* regime considered in this study. Further, in the next section, the hypothetical situation of OWC models occupying constant end to end width in the OWCBW system for all the three spacing configurations and its performance has been studied.

10.3.6 Total Performance vs. Average Performance

Till now, the discussion focussed on keeping five number of OWC devices while integrating with ODBW for the different center to center spacing. The average performance of the OWCBW system, the model setup with thrice the width of the model spacing, was found to exhibit a better efficiency than the other spacings considered in this study. The model occupying space between extreme ends is not the same for all three model setups. Herein, an attempt is made to analyze the relationship between the total efficiency of the system and the average efficiency of the system for a constant model occupying space. By doing so, the model setup with *S/w* = 1, *S/w* = 2, and *S/w* = 3 will occupy 13, 7, and 5 number of power absorbing devices, respectively, to fill up the excess spaces beyond the ends of extreme OWC device. The schematic plan view of the non-equal and equal end-to-end space occupancy by OWC devices in the central OWCBW system for different spacing configurations is shown in Figs. 10.13a and b, respectively. It is assumed that the average *RCW* calculated from five devices integrated with ODBW is multiplied by the total number of devices to obtain the *(Total RCW)*$_{reg}$ for the respective spacings. The relation between *(Total RCW)*$_{reg}$ and *<RCW>* $_{reg}$ corresponds to a number of devices for a constant *H* of 0.03 m is shown in Fig. 10.14. Herein, the *(Total RCW)*$_{reg}$ and *<RCW>* $_{reg}$ values are represented by symbols of circles and triangles, respectively.

In general, the parameters *(Total RCW)*$_{reg}$ and *<RCW>* $_{reg}$ are directly and inversely proportional to the number of devices, respectively. The *<RCW>* $_{reg}$ is seen to decrease with an increase in the number of devices throughout the *d/L* considered in this study. Similar to the above observations, the *(Total RCW)*$_{reg}$ is found to be increasing with an increase in the number of devices. For *d/L* $\leq$ 0.15, the difference between the *(Total RCW)*$_{reg}$ while using five and 13 number of power absorbing devices is only about 20% and for *d/L* $\geq$ 0.16, the difference between them is found to be large of about 100%. Though the increase in the number of devices shows a better performance in harnessing *(Total RCW)*$_{reg}$ compared to the lesser number of devices in a constantly occupied spacing, it indirectly concludes the ineffectiveness of individual devices in harnessing energy from waves. Consequently, the wave power absorption by the individual unit is fewer leads to higher wave force and scour in

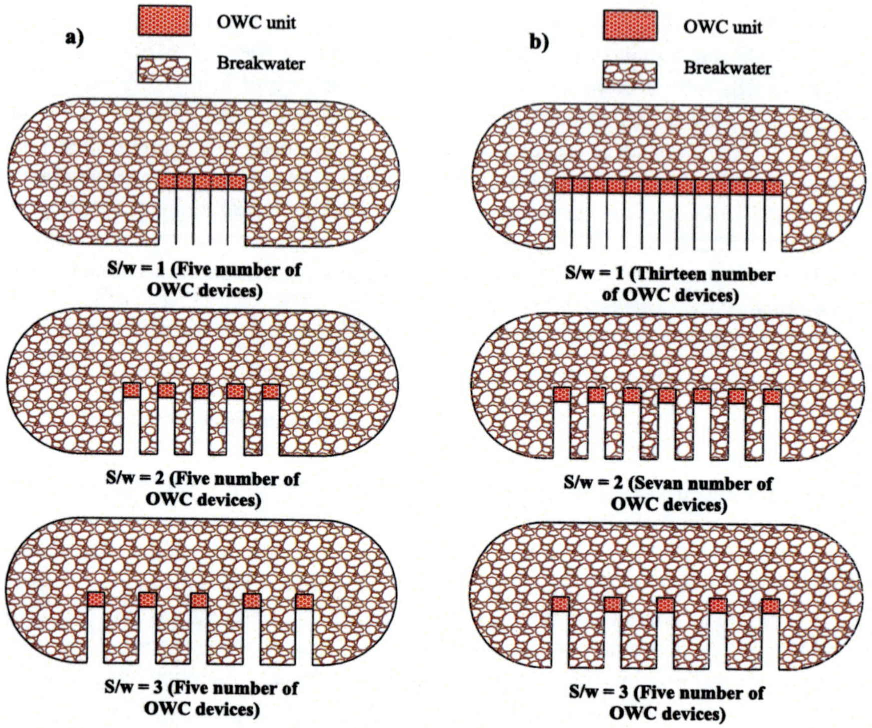

Fig. 10.13 The schematic plan view of the end-to-end space occupancy by OWC devices in the central OWCBW system for different spacing configurations: **a** non-equal; **b** equal

and around the structure, respectively, with increased maintenance. By considering the total cost of building many devices and less difference only in harnessing *(Total RCW)$_{reg}$* for a lesser number of devices compared to a large number of devices, lead us to prefer the model with thrice the width of the model spacing is better for the considered dual-purpose system.

10.3.7 Reflection Nature of OWCBW System

The performance of the coastal protection system is characterized by its reflective property [(Howe and Nader [10], Martins-Rivas and Mei [23] and is herein presented. The percentage of wave reflection from the OWCBW system is determined by the wave reflection coefficient (k_r). The k_r is defined as the ratio between reflected wave height (H_r) and incident wave height (H_i). The composite wave elevations from $\eta_1(t)$ and $\eta_2(t)$ have been decomposed into reflected and incident components as per the

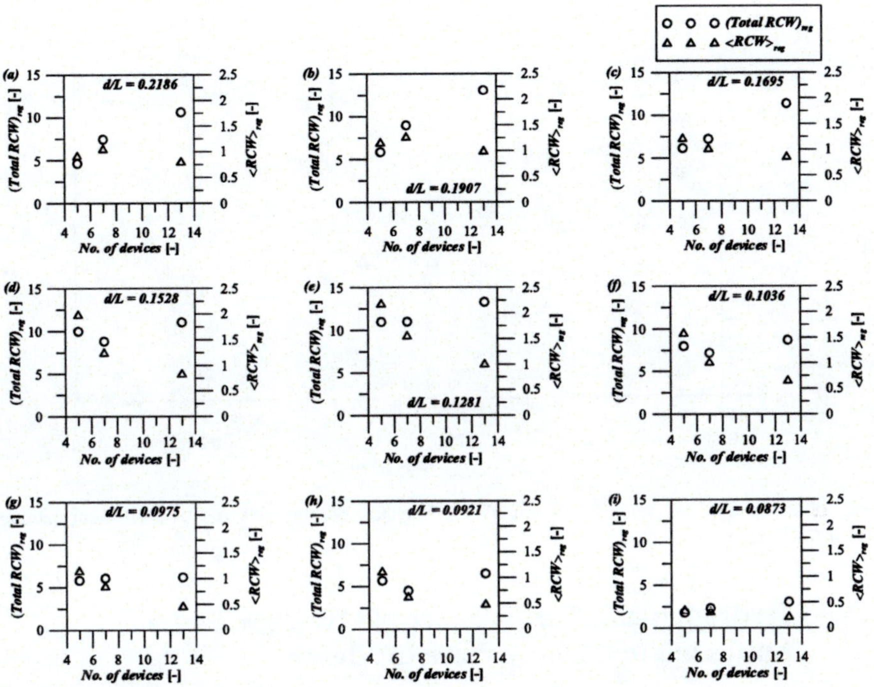

Fig. 10.14 The relation between $(Total\ RCW)_{reg}$ and $<RCW>_{reg}$ corresponds to the number of devices ($H = 0.03$ m)

method of Goda and Suzuki [8]. The variation of k_r as a function of *d/L* for different spacing is depicted in Fig. 10.15.

From the results, it is seen that k_r is maximum for the OWCBW system with a spacing, $S/w = 1$, followed by $S/w = 2$, and $S/w = 3$. The k_r is found to decrease up to $d/L = 0.128$, beyond which it increases with a decrease in the wavelength for all the three spacings stated above. The reason for the lowest k_r at $d/L = 0.128$ (resonant frequency of the system) is high wave power absorption, which maximizes the efficiency. For long-period waves, the k_r slightly increases compared to short period waves since, in general, the reflection due to low-frequency waves is higher than that of the high-frequency waves. The higher order of k_r under $S/w = 1$ is due to the reflective nature of the OWCBW system. Since at the OWCBW system with $S/w = 1$, the OWC units are positioned together and acting as a partially absorbing vertical wall type structure that induces more reflection. With an increase in *S/w* the OWCBW system act as a porous structure dissipating the incident wave energy by rubble materials. Consequently, the reflection due to the OWCBW system decreases and increases the performance of the individual OWC device. A lower k_r indicates a large amount of wave power absorption.

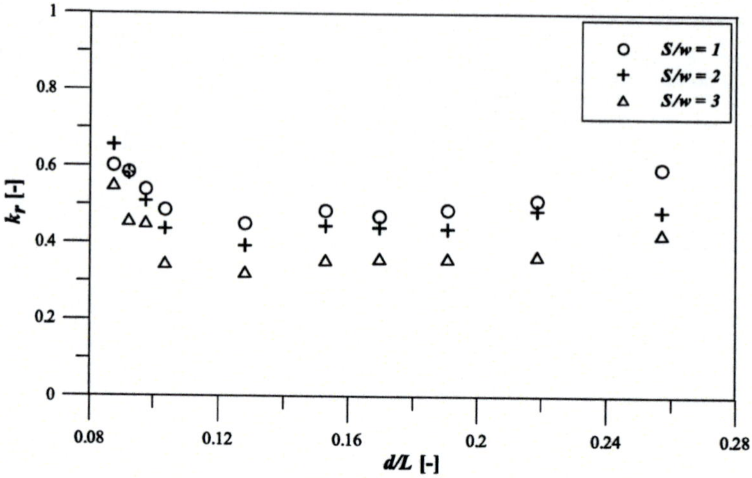

Fig. 10.15 Reflection coefficient (k_r) of OWCBW system corresponds to *d/L* for different spacing ($H = 0.03$ m)

10.4 Hydrodynamic Performance of OWCBW System Subjected to Oblique Wave Incidence

The waves in the deep ocean are multi-directional. But when it is approaching the shallow water, the propagating wave aligned itself to the orientation of the beach due to refraction. Since the present model study considers the seabed fixed OWC device situated in shallower water, the obliqueness of waves approaching the model can't be ignored. For that purpose, in this section, the effect of the different angle of wave incidences, θ, i.e., 90°, 80°, and 70°, with reference to the front face of the structure on the hydrodynamics of the OWC device are studied and presented for both regular and random waves. The variations of PR_{fout}, $<\beta>$, $<PR_{air}>$, $<CW>$, and $<RCW>$ due to both regular and random waves as a function of their respective relative water depth are projected in Figs. 10.16 and 10.17, respectively. It is clearly seen that $(PR_{fout})_{reg}$ increases with wavelength for all the three-wave angle incidence, θ. The effect of θ on the variation of the pressure is found to be significant for *d/L* beyond 0.16. The pressure increases initially with an increase in *d/L* up to 0.128, which corresponds to the natural frequency of the system, beyond which a gradual reduction in the pressure is seen. Compared to the normal angle of wave incidence, $(PR_{fout})_{reg}$ is found to decrease by 5–15% for waves with $\theta = 80^0$ and by 10–30% for waves with $\theta = 70^0$. This is mainly because of the leaking or lateral spread of energy that increases with a decrease in θ.

$<\beta>_{reg}$ is found to be higher for waves with $\theta = 90^0$, which reaches a maximum value of 2.8 at low-frequency region, thereafter, found to decrease. Compared to the normal wave angle of incidence, for waves with $\theta = 80^0$, the $<\beta>_{reg}$ decreases to the extent of about 15%, whereas the rate of decrease is found to be about 30% for waves

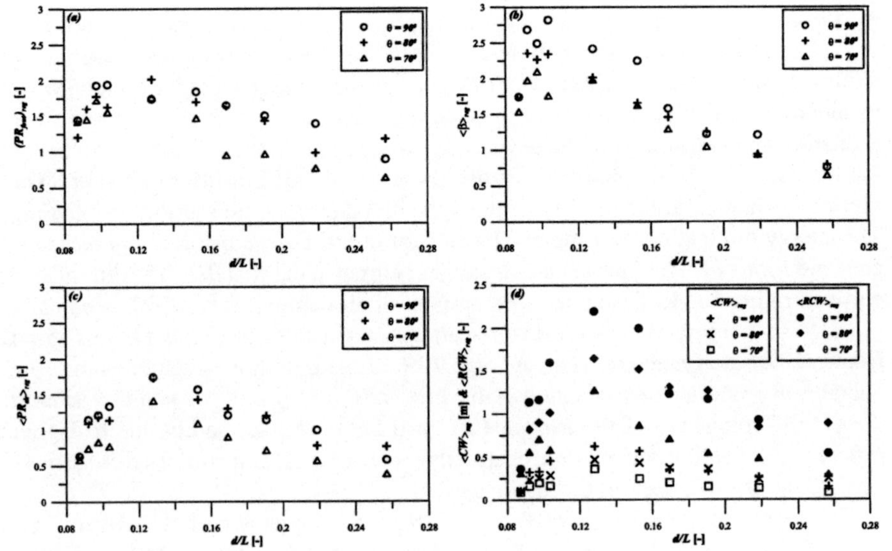

Fig. 10.16 Variation of **a** $(PR_{fout})_{reg}$, **b** $<\beta>_{reg}$, **c** $<PR_{air}>_{reg}$, **d** $<CW>_{reg}$, and $<RCW>_{reg}$ corresponds to *d/L* for different angle of wave incidence ($H = 0.03$ m & $S/w = 3$)

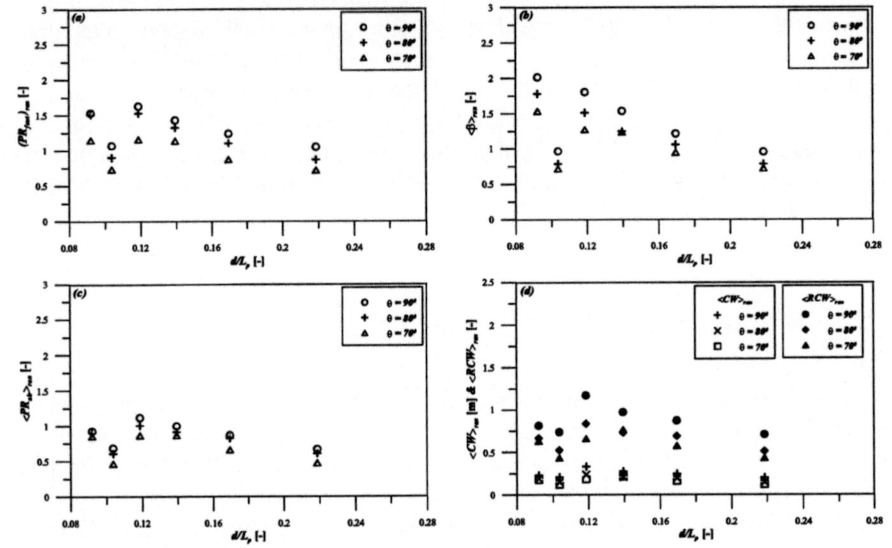

Fig. 10.17 Variation of **a** $(PR_{fout})_{ran}$, **b** $<\beta>_{ran}$, **c** $<PR_{air}>_{ran}$, **d** $<CW>_{ran}$, and $<RCW>_{ran}$ corresponds to *d/L* for different angle of wave incidence ($H_{m0} = 0.05$ m & $S/w = 3$)

with $\theta = 70^0$. It is mainly due to waves approaching normal to the structure able to excite the system directly without any obstruction and energy dissipation due to deformation of waves. The reduced $(PR_{fout})_{reg}$ as discussed earlier for oblique wave incidence agrees with reduced $<\beta>_{reg}$. Due to oblique wave incidence, non-uniform excitation on the frontline of the device causes uneven free surface oscillation inside the chamber. This further induces a radiated flow inside and outside the chamber. The vortices formation in the vicinity of the lip wall reduces the free surface oscillation. The energy dissipation over the breakwater for oblique wave incidence reduces the concentration of energy that leads to less energy input to the chamber. The diminished wave amplification leads to a reduced amount of air compression, which eventually yields lesser output power. The measured air pressure inside the chamber at the natural frequency of the system is on an average 70% of the pressure available outside the chamber for normal wave incidence. Further, $<PR_{air}>_{reg}$ reduces at either side of the natural frequency of the system ($d/L = 0.128$). This is mainly due to higher reflection, radial flow, diffraction, destructive interference, and vortices formation in and around the chamber, as discussed earlier.

The extent of decrease in $<CW>_{reg}$ at resonant frequency of the system ($d/L = 0.128$) for $\theta = 80^0$ and $\theta = 70^0$ compared to that $\theta = 90^0$ is 24% and 42%, respectively. But a substantial amount of power still available to the device. When OWC devices are placed in an array, the devices can capture the maximum power of 2.2 times the given input power at the natural frequency of the system. This confirms that the OWC device could absorb more power when placing in an array of devices than the isolated device due to energy concentration. The RCW decreases with an increase in oblique wave direction. The change in hydrodynamic behaviour of $(PR_{fout})_{ran}$, $<\beta>_{ran}$, $<PR_{air}>_{ran}$, $<CW>_{ran}$, and $<RCW>_{ran}$ subjected to random waves as seen in Fig. 10.17 is like that observed earlier for regular waves that the system exposed to the normal angle of wave incidence found to be better than the oblique wave incidence, θ, i.e., 80° and 70°.

10.5 Summary and Conclusions

A comprehensive experimental study to evaluate hydrodynamic performance characteristics of OWCs in an array integrated with a rubble mound breakwater for multi-purpose operation: i.e., to act as an energy-extracting device while protecting the coast from erosion, was taken up.

This chapter covers a detailed review on the above topic, which reveals that almost the entire research to date has been focussed mainly on the integration of OWC with vertical or caisson type breakwaters and none on other types, viz., rubble mound types.

Hence, the OWC type WEC was chosen for integrating with a rubble mound offshore detached breakwater to harness wave energy while simultaneously protecting the coastal zone from erosion. It is believed that such a combination will be highly useful for the Islands.

As per an exhaustive literature review, to the best of the knowledge of the authors, this is the first kind of study taken up.

This chapter reports only one phase of the comprehensive study on the performance evaluation of different spacings between an array of OWC devices integrated with offshore detached breakwaters subjected to normal and oblique wave incidence in both regular and random wave fields. The effect of wave steepness, relative water depth, and non-dimensional spacing on the performance of the OWC device has been investigated in terms of hydrodynamic parameters, i.e., a non-dimensional front wall outside pressure, wave amplification factor, air pressure, capture width, and relative capture width.

10.5.1 The Salient Conclusions Drawn from the Studies Are

- The induced pressures on the seaside of the lip wall, average wave amplification factor, average air pressure inside the chamber, average capture width, and average relative capture width are found to increase with an increase in wavelength up to the natural frequency of the system ($d/L = 0.128$), and then, it decreases with wavelength.
- The hydrodynamic performance of the system decreases with an increase in wave height. The OWC device shows a better performance for low steepness waves than the high steepness waves.
- The average performance is found to be higher for thrice the width of the model spacing in between successive OWC units.
- With an increase in the number of devices, the total relative capture width increases, and the average relative capture width decreases.
- The OWCBW system exhibits about 17–22% of reduction in average hydrodynamic performance for random waves over that due to regular waves.
- The convergence of water waves due to three-dimensional effects leads to the concentration of wave energy in front of the array of OWC chambers and increases the absorbed power to a maximum of 2.2 times the incident wave power. This confirms that the OWC system could absorb more power when being placed in an array of devices than compared to an isolated device due to energy concentration.
- The natural frequency of the system is observed to be unaltered with the change in spacing between individual OWC models and oblique wave incidence.
- The extent of decrease in average capture width at resonant frequency of the system ($d/L = 0.128$) for oblique wave incidences, i.e., $\theta = 80°$ and $\theta = 700$ compared to that of normal wave incidence, $\theta = 90°$, is 24% and 42%, respectively. But a substantial amount of power still available to the device.

References

1. Ashlin, S. J., Sundar, V., & Sannasiraj, S. A. (2016). Effects of bottom profile of an oscillating water column device on its hydrodynamic characteristics. *Renewable Energy, 96*(Part A), 341–353. https://doi.org/10.1016/j.renene.2016.04.091
2. Ashlin, S. J., Sannasiraj, S. A., & Sundar, V. (2018). Performance of an array of oscillating water column devices integrated with an offshore detached breakwater. *Ocean Engineering, 163*, 518–532. https://doi.org/10.1016/j.oceaneng.2018.05.043
3. Cruz, J. (2008). *Ocean wave energy: Current status and future perspectives*. Springer
4. Dally, W. R., & Pope, J. (1986). *Detached breakwaters for shore protection* (Technical Report CERC-86-1). Waterways Experiment Station, Corps of Engineers, Coastal Engineering Research Center, Vicksburg, Mississippi.
5. EMEC. (2009). *Tank testing of wave energy conversion systems* (Technical report).
6. Evans, D. V. (1979). *Some theoretical aspects of three-dimensional wave-energy absorbers*. Proceedings of the 1st Symposium on Ocean Wave Energy Utilization, Chalmers Institute of Technology, Sweden, pp. 77–113.
7. Evans, D. V. (1982). *Wave power absorption within a resonant harbor*. Proceedings of the 2nd International Symposium on Wave Energy Utilization, Trondheim, Norway, Papir, pp. 371–378.
8. Goda, Y., & Suzuki, Y. (1976). *Estimation of incident and reflected waves in random wave experiments*. Proceedings of the 15th Coastal Engineering Conference, Hawaii.
9. Holmes, B. (2003). Ocean wave energy: Development and evaluation protocol, Part 1: Wave power (Technical Report, HMRC)
10. Howe, D., & Nader, J. R. (2017). OWC WEC integrated within a breakwater versus isolated: Experimental and numerical theoretical study. *International Journal of Marine Energy, 20*, 165–182.
11. Howe, D., Nader, J. R., & Macfarlane, G. (2020). Experimental investigation of multiple oscillating water column wave energy converters integrated in a floating breakwater: Energy extraction performance. *Applied Ocean Research, 97*(102086), 1–18.
12. Hudson, R. Y. (1959). Laboratory investigation of rubble-mound breakwaters. *Journal of Waterways, Harbors Division, ASCE, 85*, 93–121.
13. Hughes, S. A. (1984). *The TMA shallow-water spectrum, description and applications* (Technical Report CERC-84-7). US Army Engineer Research Station, Vicksburg, Mississippi.
14. Hughes, S. A. (1993). Physical models and laboratory techniques in coastal engineering. Vol. 7 of Advanced Series on Coastal Engineering, World Scientific.
15. John Ashlin, S. (2019). *Hydrodynamics of an oscillating water column device integrated with breakwaters*. Ph.D. thesis, Indian Institute of Technology Madras, India.
16. Kamath, A., Bihs, H., & Arntsen, Ø. A. (2015). Numerical modeling of power take-off damping in an Oscillating Water Column device. *International Journal of Marine Energy, 10*, 1–16.
17. López, I., Pereiras, B., Castro, F., & Iglesias, G. (2014). Optimisation of turbine-induced damping for an OWC wave energy converter using a RANS-VOF numerical model. *Applied Energy, 127*, 105–114.
18. López, I., Pereiras, B., Castro, F., & Iglesias, G. (2015). Performance of OWC wave energy converters: influence of turbine damping and tidal variability. *International Journal of Energy Research, 39*, 472–483.
19. Magagna, D., Stagonas, D., Warbick, D., & Muller, G. (2010, October 6). *Physical investigations into the capture width of an array of OWC Wave Pumps for maximum efficiency*. Proceedings of the 3rd International Conference on Ocean Energy, Bilbao.
20. Malara, G., & Spanos, P. D. (2020, August 3–7). *Efficient nonlinear response determination of an array of OWC energy harvesters*. ASME 2020 39th International Conference on Ocean, Offshore and Arctic Engineering, (virtual) OMAE2020-19306, V009T09A040, 10 p. https://doi.org/10.1115/OMAE2020-19306
21. Malmo, O., & Reitan, A. (1986). Wave power absorption by an oscillating water column in a reflecting wall. *Applied Ocean Research, 8*(1), 42–48.

22. Malmo, O., & Reitan, A. (1986). Wave-power absorption by a finite row of oscillating water columns in a reflecting wall. *Applied Ocean Research, 8*(2), 105–109.
23. Martins-Rivas, H., & Mei, C. (2009). Wave power extraction from an oscillating water column at the tip of a breakwater. *Journal of Fluid Mechanics, 626*, 395–414. https://doi.org/10.1017/s0022112009005990mciver
24. McIver, P., & Evans, D. V. (1988). An approximate theory for the performance of a number of wave energy devices set into a reflecting wall. *Applied Ocean Research, 10*(2), 58–65.
25. Nguyen, C. M., Wang, C. M., Tay, Z. Y., & Luong, V. H. (2020). Wave energy converter and large floating platform integration: A review. *Ocean Engineering, 213*(107768), 1–17.
26. Pope, J., & Dean, J. L. (1986, November 9–14). *Development of design criteria for segmented breakwaters.* Proceedings of the 20th International Conference on Coastal Engineering. Taipei, Taiwan, pp. 2144–2158.
27. Price, A. A. E., Dent, C. J., & Wallace, A. R. (2009). On the capture width of wave energy converters. *Applied Ocean Research, 31*(4), 251–259.
28. Seiji, M., Uda, T., & Tanaka, S. (1987). Statistical study on the effect and stability of detached breakwaters. *Coastal Engineering in Japan, 30*(1), 121–131.
29. Simeon, D., & George, A. (2020). Experimental investigation and performance comparison of a single OWC, array, and M-OWC. *Renewable Energy.*
30. Sriram, V., Schlurmann, T., & Schimmels, S. (2015). Focused wave evolution using linear and second order wave maker theory. *Applied Ocean Research, 53*, 279–296.
31. Suh, K., & Dalrymple, R. A. (1987). Offshore breakwaters in laboratory and field, Journal of Waterway, Port. *Coastal, and Ocean Engineering, 113*(2), 105–121.
32. Sundar, V., Torgeir, M., & Jorgen, H. (2010). *Conceptual designs on integration of oscillating water column devices in breakwaters.* Proceedings of the ASME 29th International Conference on Ocean, Offshore and Arctic Engineering, pp. 479–489.
33. Thomas, G. P., & Evans, D. V. (1981). Arrays of three-dimensional wave-energy absorbers. *Journal of Fluid Mechanics, 108*, 67–88.
34. Whittaker, T. J. T., & Stewart, T. P. (1993, July). *An experimental study of nearshore and shoreline oscillating water columns with harbours.* Proceedings of the European Wave Energy Symposium, Edinburgh, Scotland, pp. 151–156.
35. Zheng, S., Antonini, A., Zhang, Y., Greaves, D., Miles, J., & Iglesias, G. (2019). Wave power extraction from multiple oscillating water columns along a straight coast. *Journal of Fluid mechanics, 878*, 445–480.

Chapter 11
Power Take-Off Devices for Wave Energy Converters

Vishnu Vijayasankar, Tapas K. Das, Paresh Halder, and Abdus Samad

Abstract In a Wave Energy Converter (WEC), the Power Take-off (PTO) mechanism does the energy conversion through a series of mechanical and electrical systems. The PTO system plays a major role in defining the WEC system's structural dynamics. Since the PTO system directly affects the efficiency of power conversion, it has a significant role to play in the annual energy production of the WEC system. This emphasizes the importance of selecting an appropriate PTO system for any WEC based on the working conditions and other related parameters. In this chapter, different types of PTO systems developed over the past few decades are discussed along with their corresponding advantages and shortcomings. On reading this chapter, the reader will get acquainted with different types of PTO mechanisms in existence and also gain an insight on when to use what type of PTO.

11.1 Introduction to Wave Energy

Despite being discussed in patents since the late eighteenth century, the emerging oil crisis of the 1970s stimulated modern research into harnessing wave energy. With global attention now being focused on climate change and rising CO_2 levels, once again generating electricity from renewable sources has emerged into an important research area. India has a 1.2 billion population, and a split in India's predominant energy demand reveals that fossil fuels supply more than 80% of this [1]. Given the scarce reserves of fossil fuel, energy security, and climate change, renewable energy is expected to play a major role in the long-term energy mix of India. Compared to other sources of renewable energy, wave energy is available continuously throughout the year. Primary estimates show that the annual potential of wave energy along the Indian

V. Vijayasankar · T. K. Das · A. Samad (✉)
Wave Energy and Fluids Engineering Laboratory (WEFEL), Department of Ocean Engineering, Indian Institute of Technology Madras, Chennai 600036, Tamil Nadu, India
e-mail: samad@iitm.ac.in

P. Halder
Institute of Ocean Energy, Saga University, 1-Honjo-machi, Saga-shi, Saga 840-8502, Japan

A. Samad et al. (eds.), *Ocean Wave Energy Systems*, Ocean Engineering & Oceanography 14, https://doi.org/10.1007/978-3-030-78716-5_11

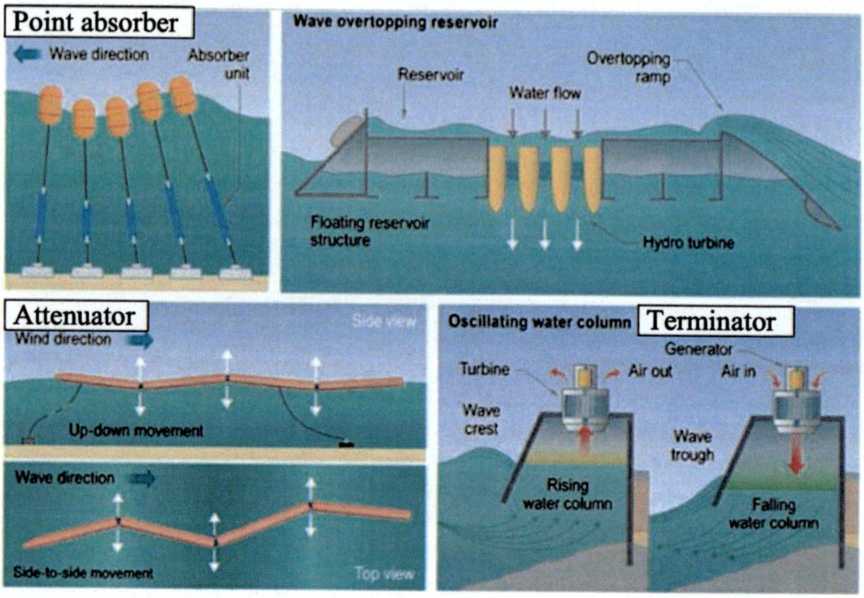

Fig. 11.1 Types of WECs (*Source* www.boem.gov/NREL-Ocean-Wave-Energy-Technology)

coast is between 5 and 15 MW/m. The theoretical energy potential for a coastline of approximately 6000 km and therefore, amounts to nearly 40,000–60,000 MW [2].

WEC devices can be categorized based on their dimension and alignment to the direction of the incoming wave; their distance from the coast; or their operating principle. Based on the dimension, a WEC is termed as a "point absorber" if its width is very small compared to the wavelength [3]. If the device width is equal to or larger than the wavelength, the WEC will be called a "terminator" or an "attenuator" (Fig. 11.1). If a WEC is positioned parallel or normal with the predominant wave direction, it can be called a terminator or attenuator, respectively.

11.2 Types of Power Take-Off Mechanisms Used in Point Absorbers

In a WEC, the PTO mechanism does the energy conversion through a series of mechanical and electrical systems. The PTO system plays a major role in defining the WEC system's structural dynamics. The different flow paths involved in the conversion of wave energy to electricity conversion are schematically shown in Fig. 11.2. Most of the existing WECs use a high-speed turbine that works on either pneumatics or hydraulics to extract wave energy. Even so, the implementation of these systems becomes complicated in the case of wave energy extraction, where the velocity of

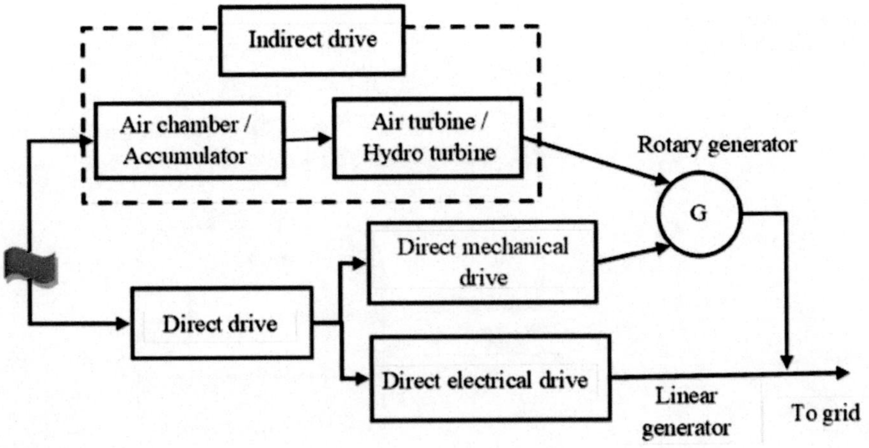

Fig. 11.2 Flow paths for different PTO mechanisms

oscillation is very low. Even though such systems are more reliable, their conversion efficiencies are very low.

A direct wave energy converter converts the energy associated with the waves directly into electrical energy, with the help of a linear or a rotary generator. The direct-drive systems have fewer stages of power conversions, which makes them more efficient. Among direct drive systems, direct electric drive system that uses a linear generator, careful placement of the mounting structure is important to ensure fine air gaps between the translator and the stator. Hence, in terms of simplicity and efficiency, the direct mechanical system has an advantage over other types of PTO.

Due to limited data, it is difficult to compare the different types of PTOs that are in existence. Despite that, PTO systems can be broadly classified into five main categories, and these categories are explained in the following sections.

11.2.1 Air Turbines

Air turbines are widely used in Oscillating Water Columns to convert wave power from sea waves into mechanical power. In an OWC, the oscillations in water level are used to compress air inside an enclosed chamber; this compressed air is in turn used to rotate an air turbine (Fig. 11.3). However, the bidirectional flow of compressed air poses a great challenge in the case of OWCs. One solution to this problem is to use a non-returning valve to rectify the airflow so that a conventional turbine can be used to extract mechanical power. However, this solution proves to be cumbersome due to the use of large valves and cost-ineffective due to higher maintenance costs. Another viable solution is a self-rectifying turbine which will rotate in the same

Fig. 11.3 Air turbine used as PTO mechanisms

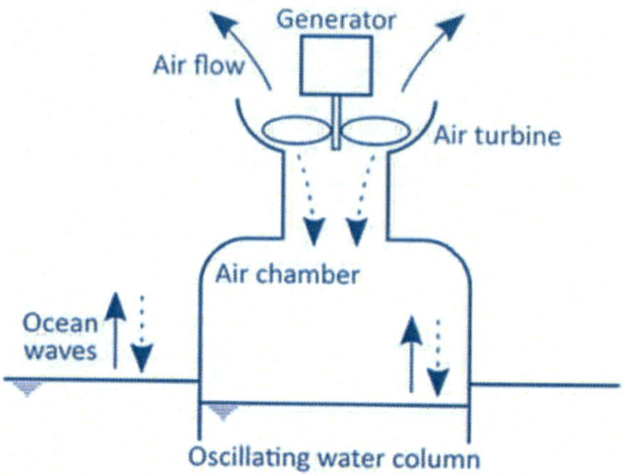

direction irrespective of the airflow direction. This ensures that alternating airflow will be converted to a unidirectional rotation.

Over the past four decades, several varieties of self-rectifying turbines were proposed by different researchers around the globe. Out of which, Wells turbine [2] was the first one to be implemented. It consists of a symmetrical rotor composed of many aerofoil blades positioned around a hub with zero stagger angle (Fig. 11.4).

The Wells turbine is considered to be the simplest and probably the most cost-effective option for wave energy conversion using air turbines. However, one major disadvantage of a Wells Turbine is that, it is not self-starting and hence it requires an external source to start its rotor initially. A brief comparison between two of the most widely used turbines in the field of Wave Energy Conversion is shown in Table 11.1.

Fig. 11.4 Wells turbine [2]

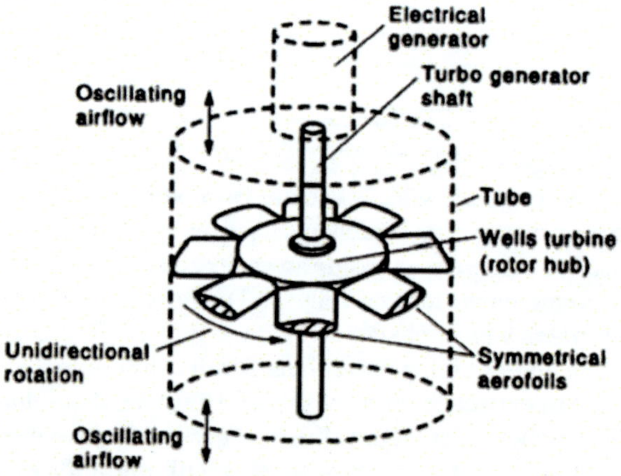

Table 11.1 Comparison of different self-rectifying turbines used in OWC based WECs

Sl. No	Turbine Type	Advantages	Disadvantages
1	Wells turbine	Simple technology	• The flow range for which the turbine operates with satisfactory efficiencies is very narrow • Starting characteristics of the turbine is very poor • High noise due to high operational speed • High axial thrust
2	Impulse turbine	• Better starting characteristics • Low noise since the operational speed is less • Range of flow coefficients for which the turbine works efficiently is much more wider	Self-pitching configuration requires a lot of additional moving parts

11.2.2 Hydraulic Converters

Conventional turbine-based power conversion technology is not useful in some energy conversion systems where the oscillation of water is transferred to the movement of a body, as in the case of WECs like attenuators and point absorbers.

Since the wave energy conversion deals with large forces with low frequencies, often a hydraulic converter is used as an interface between the wave energy converter and the electric generator due to its ability to absorb energy in these conditions. In the case of hydraulic cylinders used for wave energy conversion, the energy flow path is reversed in comparison to that in a conventional hydraulic system. The oscillation of water is captured by a moving body, which in turn energizes a hydraulic motor that translates this rotational energy to an electrical generator. A schematic of a WEC setup that uses a hydraulic PTO is shown in Fig. 11.5. In this setup, a point absorber moves up and down along with the incoming waves, and this movement is passed on to a hydraulic cylinder which then forces the working fluid to flow through controlled manifolds before reaching a hydraulic motor. This motor is coupled with an electric generator. Smoothening of the flow of high-pressure fluid is done with the help of accumulators that either provide or accumulate hydraulic energy whenever required. Radial piston motor is preferred in wave energy extraction since it is well suited for low velocity, high loads applications.

Choosing a hydraulic PTO system for WECs can lead to many issues that should be taken care of with great care for satisfactory performance. The fluid used as working fluid may undergo contamination and this may affect the performance of the system and may also pose a threat to the environment. Environmental impacts can be addressed by using biodegradable transformer oil. However, the efficiency of the WEC system is also of great importance. Considering the variable nature of the ocean waves, hydraulic PTO systems usually come with several accumulators

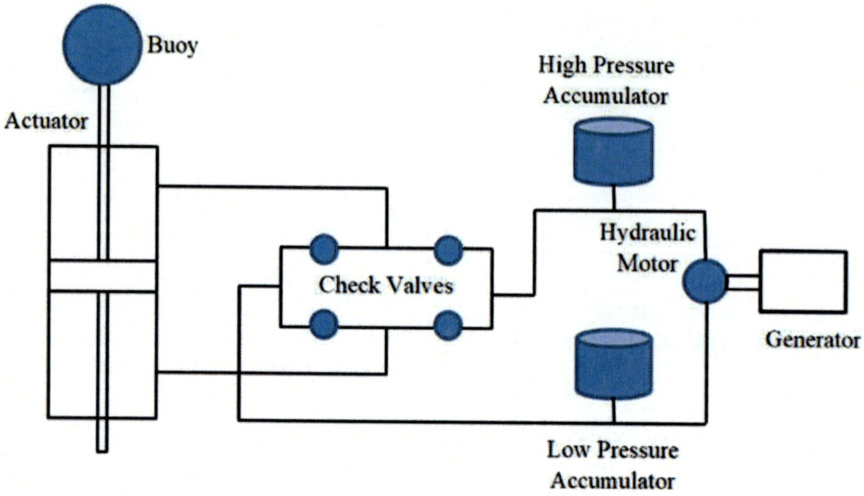

Fig. 11.5 Hydraulic PTO for wave energy extraction [4]

that can absorb the peak loads at times and ensure a smooth wave energy conversion from the motor end. Thus, it is common for hydraulic PTO-based WEC systems to have a lot of moving parts, and maintaining all of these parts will become costly and difficult. For example, the seals around the piston inside the hydraulic cylinder are expected to undergo wear over time and need to be replaced periodically. All these factors have to be kept in mind while adopting and designing a hydraulic converter.

Another major issue that needs attention is the protection of the PTO and WEC assembly in the event of extreme load conditions, where the wave loads are high and the piston may travel more than the designed displacement value. If proper damage control measures are not considered during the design phase, these extreme conditions may damage the PTO system or even the whole WEC assembly.

11.2.3 Hydro Turbines

Hydro turbines are employed in overtopping devices or hydraulic pump systems using seawater as fluid. Hydro turbines are used in wave energy converters (usually overtopping devices) where the working fluid is seawater. In overtopping devices, the potential energy of seawater is converted to rotational energy using a low-head axial flow turbine, like Kaplan Turbine (Fig. 11.6). Hydro turbines are known to operate at efficiencies exceeding 90% with low maintenance. From the point of view of wave energy conversion, the bottleneck in using hydro turbine-based PTO is the necessity to deliver the required flow and head to the Kaplan turbine to make the conversion process economical.

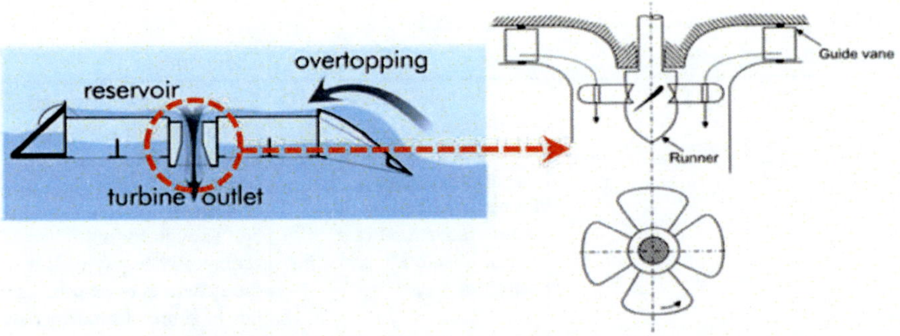

Fig. 11.6 Hydro turbine inside an overtopping WEC (*Source* http://www.wavedragon.net/)

11.2.4 Direct Mechanical Drive Systems

In a direct mechanical drive system, an extra mechanical attachment is used to translate the mechanical energy extracted by the actuator from ocean waves into rotational energy which in turn drives an electric generator. Such a PTO system is depicted in Fig. 11.7. In the illustration, the mentioned extra mechanical system comprises pulleys, gearbox, and cables.

Similar to the function of an accumulator in the case of a hydraulic converter, in a direct mechanical drive system, a flywheel is often used to smoothen power variations. Smoothening is achieved by accumulating or releasing energy according to requirements.

Direct mechanical drive PTO-based WEC systems have slightly higher efficiency compared to other WECs since it involves only up to three stages of energy conversion. However, the reliability of the direct mechanical drive systems is highly susceptible since these are subjected to uncountable and variable load cycles [6]. Table 11.2 includes some key features of a few of the existing direct mechanical drive PTO designs.

As derived from the literature survey shown in Table 11.2, the following observations have to be considered to propose an improved direct drive WEC.

Fig. 11.7 Direct mechanical drive PTO system [5]

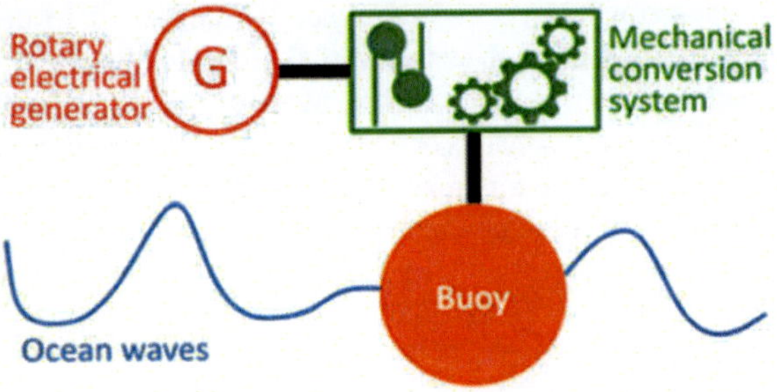

Table 11.2 Existing direct mechanical drive WECs

Sl. No	Existing design	Features	Shortcomings
1	CECO WEC [7]	Rack and pinion based PTO mechanism Rotary generator coupled to a gear box is used Power absorbed by the lateral movement of a streamlined body under the action of waves	Separate support systems are required to adjust the orientation of the device according to wave directions Not enough room to accommodate payload; if the system is to be used as a power source for ocean data acquisition
2	Rotational WEC [8]	Basic components match that of an IC engine Rotary generator is used	Complex system System may fail in the presence of breaking waves
3	Life saver [9]	PTO system is realized as a winch and rope system Rotary generator is used	The generator can produce power during upward motion only
4	Mechanical motion rectifier (MMR) based WEC [10]	MMR based PTO is used Rotary generator is used Power is generated in both upward and downward motions	The weights of the gears and the generator are supported by the buoy, hence increasing its size and the overall cost
5	A Mechanical WEC [11]	Rack and pinion based PTO Rotary generator is used Power is generated in both upward and downward motions	It is cost-inefficient to construct a separate structure to hold the generator and PTO in the middle of the ocean

1. A rotary generator is easy to maintain.
2. It is easier to draw power lines from the sea surface rather than the sea bed.
3. Compact design is advised for cost-effectiveness.
4. The generator should produce power in both upward and downward strokes.
5. Dependence on an additional offshore support structure is not ideal.

One of the most popular practical applications of wave energy, which has been in use for decades, is the wave-powered navigational buoy. In 1965, Masuda [12] proposed a unique and compact WEC system to deliver power to a navigational buoy. Since then, thousands of these navigational buoys, working on this concept made their way to the market. Most of these compact devices had a power rating of 60–100 W and consists of a buoy as big as 3 m in diameter, whose oscillations are used to pressurize the air in an OWC chamber, which in turn rotates a turbine-generator assembly to produce electricity.

Fig. 11.8 Direct electrical drive PTO system [4]

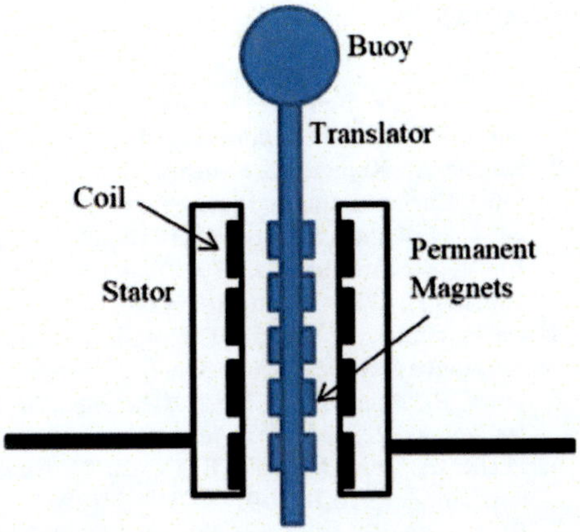

11.2.5 Direct Electrical Drive Systems

In direct electrical drive PTO systems, the mechanical energy absorbed by the actuator is directly connected to the translator of a linear generator. A simple direct electrical drive PTO system is illustrated in Fig. 11.8. In the shown setup, the buoy is directly coupled to a translator which has permanent magnets with alternating polarity mounted on its surface. The relative motion between the translator that oscillates with ocean waves and the stationary stator induces an electric current flow in the stator.

Since the output grid requires a sinusoidal waveform with fixed voltage and frequency, the output of the linear generator needs to be rectified before connecting it to the grid. In the case of linear generators, the mounting structure should be carefully designed to maintain fine gaps between the stator and the translator.

11.3 Conclusion

This chapter introduced the concept of the PTO system in a wave energy converter to the reader, explained different categories of PTO systems, and also showcased the importance of PTO systems in any wave energy converter system. Adopting a suitable PTO system is of prime importance for any wave energy converter since the efficiency of the WEC is highly dependent on the efficiency of the PTO system. Also, the choice of different components associated with a PTO system directly influences the initial cost and maintenance cost of the whole WEC system. A reliable and efficient PTO system is the key to reach the goal of a cost-effective and efficient wave energy conversion.

References

1. Pillai, I. R., & Banerjee, R. (2009). Renewable energy in India: Status and potential. *Energy, 34*(8), 970–980. https://doi.org/10.1016/j.energy.2008.10.016
2. Kumar, A., Kumar, K., Kaushik, N., Sharma, S., & Mishra, S. (2010). Renewable energy in India: Current status and future potentials. *Renewable and Sustainable Energy Reviews, 14*(8), 2434–2442. https://doi.org/10.1016/j.rser.2010.04.003
3. Budar, K., & Falnes, J. (1975). A resonant point absorber of ocean-wave power. *Nature, 256*(5517), 478–479. https://doi.org/10.1038/256478a0
4. Raghunathan, S. (1995). The wells air turbine for wave energy conversion. *Progress in Aerospace Sciences, 31*(4), 335–386. https://doi.org/10.1016/0376-0421(95)00001-F
5. Drew, B., Plummer, A. R., & Sahinkaya, M. N. (2009). A review of wave energy converter technology. *223*, 887–902. https://doi.org/10.1243/09576509JPE782.
6. Pecher, A., Jens Peter, K. (Eds.). (2017). *Handbook of ocean wave energy* (Vol. 7). Springer. https://doi.org/10.1007/978-3-319-39889-1.
7. Rosa-santos, P., & Taveira-pinto, F. (2015). CECO wave energy converter : Experimental proof of concept. *Journal of Renewable Sustainable Energy, 061704*. https://doi.org/10.1063/1.4938179.
8. Karayaka, H. B., Mahlke, H., Bogucki, D., Mehrubeoglu, M., Texas, A., Christi, M. U. (2011). A rotational wave energy conversion system development and validation with real ocean wave data. In *IEEE Power Energy Society General Meeting*, (pp. 5–11). IEEE. https://doi.org/10.1109/PES.2011.6038932.
9. Sjolte, J., Tjensvoll, G., & Molinas, M. (2013). Power collection from wave energy farms. *Applied Sciences, 3*, 420–436. https://doi.org/10.3390/app3020420
10. Liang, C., Ai, J., & Zuo, L. (2015). Design, fabrication, simulation and testing of an ocean wave energy converter with mechanical motion rectifier. *Ocean Engineering, 136*, 190–200. https://doi.org/10.1016/j.oceaneng.2017.03.024.
11. Chandrasekaran, S., & Harender. (2012). Power generation using mechanical wave energy converter. *The International Journal of Ocean and Climate Systems, 3*(1), 57–70. https://doi.org/10.1260/1759-3131.3.1.57.
12. Masuda, Y., Yamazaki, T., Outa, Y., & McCormick, M. E. (1988). The backward bend duct buoy-an improved floating type wave power device. In *Proceedings of OCEANS '88 "A Partnership of Marine Interests* (pp. 1067–1072). https://doi.org/10.1109/oceans.1988.23663.

Chapter 12
Wells Turbine as a Power Take-Off Mechanism for Wave Energy Converters

Tapas K. Das and Abdus Samad

Abstract Wells turbine is an axial flow bi-directional turbine used for wave energy conversion by Oscillating Water Column (OWC) system. Different variations of Wells turbine have been studied and tested to find out the best rotor geometry. A comprehensive review of Wells turbine is presented here to familiarize the reader with the state of the art of Wells turbine. Turbines are broadly classified into two types: monoplane and biplane, with further classifications including attachment of guide vanes and variation of blade pitch angles. Numerical optimization works carried out on air turbines for wave energy system are also presented in this chapter. Turbines with optimized geometry give a better performance compared to standard turbines in terms of overall system performance, efficiency, and energy absorption.

12.1 Introduction

With the ever-increasing threat of exhausting conventional energy sources in the near future, researchers are turning more and more towards unconventional sources of energy. Research in the area of ocean energy has gained popularity in the last few decades. Among the various forms of ocean energy, wave energy is one of the most promising renewable energy sources with huge potential as well as inherent challenges. The advantages of wave energy are numerous; it is abundant, and the resource is well characterized, as the wave data is readily available from several data collecting buoys. The energy density of wave is much higher than wind energy, and it is the highest among all available renewable energy sources [1]. The wave energy potential of the entire ocean is approximated to be 8000–80,000 TW/year, while each wave crest transmits 10–50 kW per meter [2]. However, several challenges exist to convert wave energy into useful electrical energy. The main difficulty is

T. K. Das (✉) · A. Samad
Wave Energy and Fluids Engineering Laboratory (WEFEL), Department of Ocean Engineering, Indian Institute of Technology Madras, Chennai, Tamil Nadu 600036, India

A. Samad
e-mail: samad@iitm.ac.in

A. Samad et al. (eds.), *Ocean Wave Energy Systems*, Ocean Engineering & Oceanography 14, https://doi.org/10.1007/978-3-030-78716-5_12

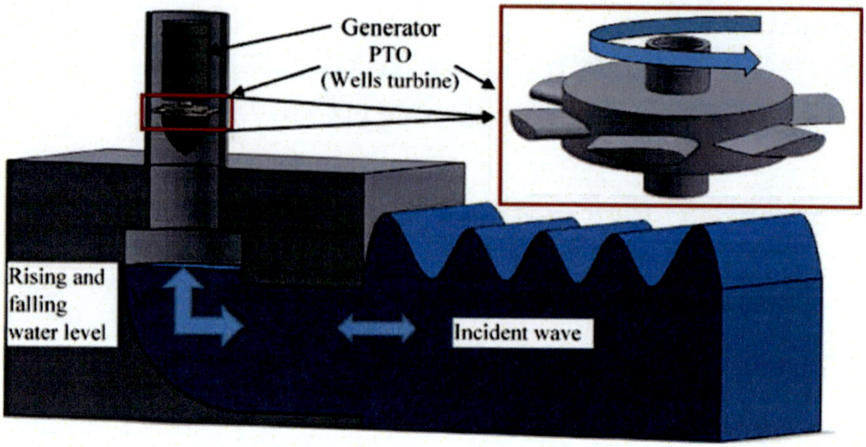

Fig. 12.1 Schematic representation of OWC [8]

the fluctuating nature of wave energy, which leads to power quality issues such as fluctuation of frequency or voltage. Any wave-energy conversion device must withstand the harsh marine environment. Therefore, the system needs to be robust. In addition, many moving parts lead to a problem in the maintenance of the device. Wave energy devices are characterized either by the depth at which they operate or by the working principle [3]. Several different types of wave energy conversion devices are reported in articles [4–7]. Among the various wave energy extraction devices available, the oscillating water column (OWC) is the most popular one and the maximum number of prototypes available all over the world.

OWC is a wave-energy conversion technology that uses an air turbine to convert the kinetic energy of air into useful mechanical energy. The energy conversion from wave energy to electrical energy occurs in three phases inside OWC (Fig. 12.1). Firstly, the oscillation of the sea surface has been converted to pressure energy inside the air chamber. The turbine converts this pressure energy into mechanical shaft power. Finally, a generator connected with a turbine shaft converts the mechanical power into electrical power.

The specialty of air turbines used in OWC is that it has to rotate in one direction for the bidirectional airflow. Both reaction and impulse turbine can be used in OWCs. Wells turbine is one type of reaction turbine that is used in most OWC prototypes. While the turbine creates mechanical power, a generator converts this mechanical power into electrical energy. The selection of a generator is an important exercise for the efficient operation of the OWC plant. Due to the unpredictability of the ocean waves, a control technique is required for the stable operation of any wave energy plant. Past researchers have given some detailed reviews on the Wells turbine [8–12]. This chapter presents a comparative analysis of different Wells turbine designs along with the optimization works carried out to improve the performance of the turbine.

12.2 Historical Overview

The history of wave energy conversion is dated back to the 1940s. Yoshio Masuda (1925–2009), a Japanese navy officer, is considered the pioneer of wave energy technology. The concept of OWC came from the navigation buoys created by Masuda. These buoys produced energy by using a conventional uni-directional turbine with rectifying valves [13]. Such systems were in commercial use in Japan since 1965. During 1978–80, Mashuda built a large (80 m × 12 m) barge named Kaimei that had thirteen OWC chambers. However, all these OWC chambers used a uni-directional turbine with the rectifying valve for power generation. In 1976, Alan Arthur Wells (1924–2005) invented the self-rectifying turbine. Since then, most of the OWC plants have used Wells turbines as power take-off devices. Table 12.1 gives a list of OWC plants that were set up or presently being used. It can be seen that most plants used Wells turbine for testing and running purposes. Since the invention of the Wells turbine, there have been many types of research to improve its performance. Some of the main patents related to Wells turbine are listed in Table 12.2. A literature search using the professional, scientific search tool Scopus (Fig. 12.2) shows that during

Table 12.1 OWC plants with turbines

Wave energy plant	Year of establishment	Type of turbine used	Rated power (kW)	Remarks
Sea of Japan [14]	1983	Wells turbine	40	Used as wave power buoy
Toftestallen, Bergen [11]	1985	Vertical Axis Wells turbine	500	Destroyed in storm
Port of Sakata, Japan [15]	1990	Wells turbine	60	Converter integrated with breakwater
Vizhinjam, India [16]	1990	Wells turbine	125	Later Impulse turbine installed and tested
Pico Island, Portugal [17]	1999	Wells monoplane turbine with guide vanes	500	Fixed structure OWC
LIMPET, Scotland [18]	2000	Wells turbine-in line contra-rotating	500	Fixed structure OWC
China, 2001 [19]	2001	Wells turbine	100	Shoreline OWC
Mighty Whale, Japan [1]	1998	Wells turbine-tandem configuration	110	Floating OWC
Energetech, Australia [1]	2006	Variable pitch Wells turbine	500	Fixed structure OWC
Mutriku, Spain [20]	2008	Wells turbine (16 turbines X18.5 kW)	296	Breakwater integrated OWC

Table 12.2 Important patents on Wells turbine

Patent Name and No.	Description
Fluid Driven Rotary Transducer, British Patent Spec. No. 1595700 [21]	Patent of the Wells turbine
Wells turbine for wave activated power generation, Patent No. JPS6385201 (A) [22]	New design to improve starting characteristics
A blade-pitch control mechanism, Patent No. US 2010/0290908 A1[23]	Blade pitch control mechanism for Wells turbine
Bi-directional axial flow turbine with self-pivoting blades for use in wave energy converter, Patent No. US2012/0124986 A1[24]	A self-pitch control mechanism for Wells turbine blades
Rotor blade design for a Wells turbine, Patent No. US8678745 B2 [25]	Design of blade sweep
Air turbine for a wave power station, Patent No. WO2009089902 [26]	Two-stage Wells turbine with stator blades in between

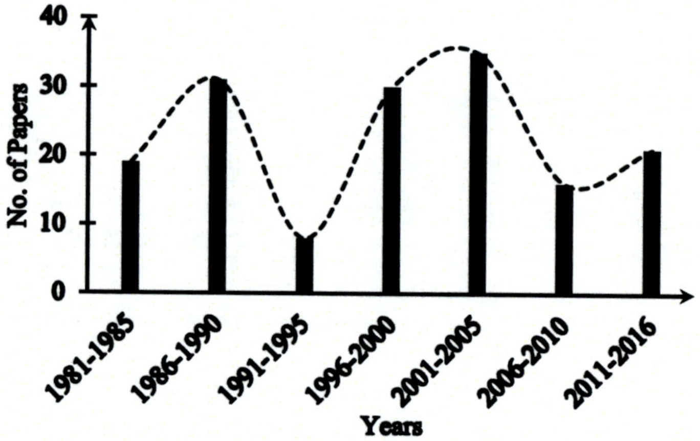

Fig. 12.2 Number of publications regarding Wells turbine based on search tool Scopus

mid 80s there were a lot of publications regarding Well turbines. The surge in wave energy research occurred due to the global oil crisis in the late 70s. These were mainly research works based on experimental techniques. As the oil price reduced during the first half of the 90s, the research on renewable energy was limited during that time. With the advent of numerical analysis techniques, there was renewed interest in Wells turbine since 2000. The recent trend in Wells turbine research shows the use of optimization techniques to create better geometry for performance improvement. A breakup of research works using Computational Fluid Dynamics (CFD) and other techniques is given in detail by Shehata et al. [8].

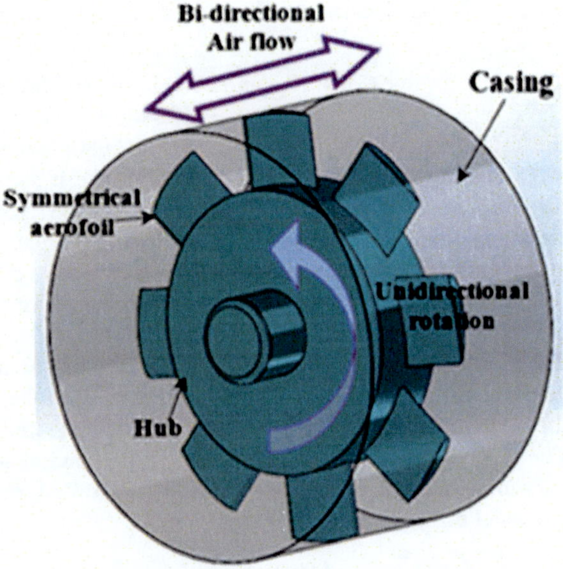

Fig. 12.3 Schematic of a Wells turbine

Fig. 12.4 The Wells turbine used in OWC plant at Vizinjham, India

12.3 Wells Turbine: Principle of Operation

The Wells turbine (Figs. 12.3 and 12.4) is an axial flow air turbine used in most of the OWC type wave-energy devices. Professor Alan Wells of the Queen's University of Belfast developed this novel turbine in the 1980s. Since then, it has been extensively researched to improve performance and make it suitable for use in OWCs. Wells turbine is made up of a symmetric aerofoil profile fixed around a central hub. Due to the symmetrical blade profile, the Wells turbine does not require any rectifying air valve, and it can rotate in one direction irrespective of bi-directional airflow.

Wells turbine works under the general principle of aerodynamics [27, 28]. Blades are set at 90°-stagger angle (the angle between the chord line and the turbine axial direction). The absolute velocity of air hits the blade axially, and the tangential velocity of the blade acts in a direction parallel to the plane of rotation. As a result, the relative velocity V_w acts at angle α, known as the angle of attack, to the blade. The relative velocity causes a lift force L perpendicular to the direction of V_w and a drag force in the direction of V_w. This lift and drag force can be resolved in tangential and axial direction:

$$F_T = L\sin\alpha - D\cos\alpha \tag{1a}$$

$$F_A = L\cos\alpha + D\sin\alpha \tag{1b}$$

where F_T and F_A are the tangential and axial forces, respectively. It can be observed in Fig. 12.5 that the direction of the tangential force will always remain the same direction irrespective of the direction of flow. Based on this principle, the turbine always rotates in the same direction with the bidirectional flow of air.

It is important to observe that the lift and drag forces for any aerofoil depend on the angle of attack (α). The forces increase for a certain value of α beyond which the

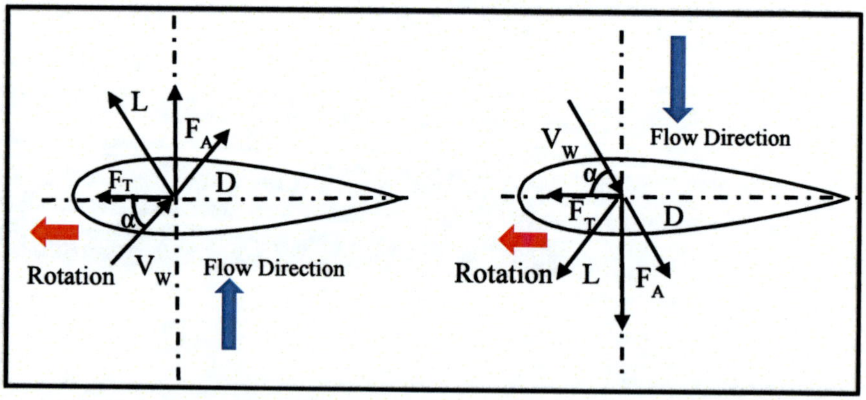

Fig. 12.5 Forces acting on Wells turbine blade

flow around the aerofoil separates [29]. The angle at which flow separation occurs is known as the stall angle. Increasing the angle of attack beyond the stall limit decreases the lift force and increases the drag significantly. As a result, tangential force decreases, and the performance of Wells turbine decreases drastically beyond stall point.

12.4 Variations of Wells Turbine

Over the years, the Wells turbine has been researched extensively to improve its performance. As a result, a different variation of Wells turbine came up, and these have been well studied both experimentally and numerically. The Wells turbines presently in use can be divided into two major groups:

- Monoplane Wells turbine
- Biplane Wells turbine.

Monoplane Wells turbine is the conventional turbine that was originally proposed by Professor Alan Wells. To utilize the total available pressure drop across the turbine, a biplane turbine was developed and tested [30]. Biplane turbines can be of two types: one where turbines in both the planes rotate in the same direction and the other where turbines in both the planes rotate in the opposite direction (contra-rotating). Both monoplane and biplane turbines can be used with or without guide vanes. Conventional turbines used fixed-pitch blades with zero pitch angles. However, fixed-pitch blades with non-zero pitch angles have also been studied. Turbines with variable pitch blades were introduced to increase the range of operation with respect to flow coefficient. However, these turbines have their own complexity due to the variable rotor geometry. This section gives a present state of the art of various types of Wells turbines (Fig. 12.6) followed by optimization techniques in the following section.

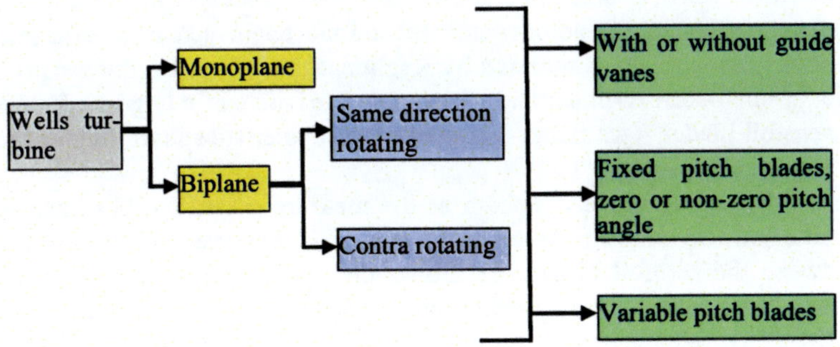

Fig. 12.6 Different modifications of Wells turbine

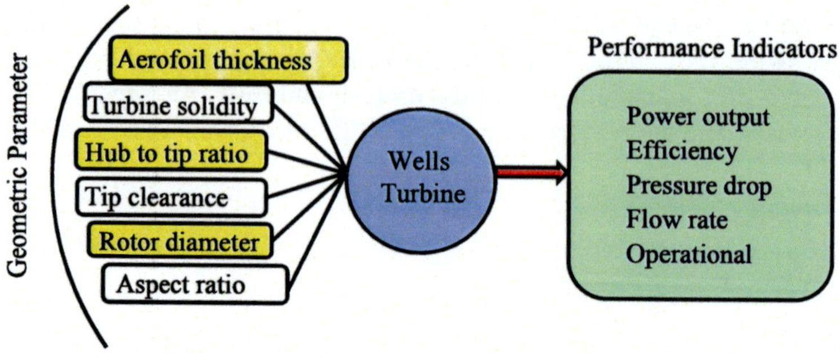

Fig. 12.7 Different parameters of Wells turbine (Reproduced from [31])

12.4.1 Monoplane Wells Turbine

The Monoplane Wells turbine is the basic turbine that has been studied and tested so far. In the majority of works, the Wells turbine performance is described by the following non-dimensional coefficients:

Flow coefficient $(\varphi) = \frac{Va}{Vt}$, where V_a is the axial air velocity, V_t is the tangential velocity of the turbine blade.

Torque coefficient $(C_T) = \frac{T}{\rho\omega^2 R^5}$, where T is the total torque.

Pressure coefficient $(\Delta P_0^*) = \frac{\Delta P_0}{\rho\omega^2 R^2}$, where ΔP_0 is the pressure drop across the turbine.

Efficiency $(\eta) = \frac{T\omega}{\Delta P_0 Q}$, where Q is the mass flow rate across the turbine.

In the above dimensionless coefficients, R, ρ, and ω are tip radius, the density of air, and rotational speed in rpm, respectively. Various parameters affecting the performance of the turbine are given in Fig. 12.7.

Based on the analytical description, Raghunathan [32] has given a detailed analysis of different parameters affecting the performance of monoplane turbines. Aerofoil thickness is an important parameter affecting the aerodynamic force coefficients. Isolated aerofoil test in a wind tunnel shows the effect of thickness ratio and Reynold's number are interrelated. The aerodynamic lift and stall angle increase with an increase in Reynold's number. It is suggested by Raghunathan [33] that a thicker profile is suitable for small-scale turbines, and a thinner profile is better for large-scale turbines. Thick aerofoil blades have better self-starting characteristics than thinner profiles [34].

The solidity of a turbine is known as the measure of airflow blockage within the turbine [32]. It is an important parameter as it measures the mutual interaction between the aerofoil blades. As the solidity increases, the blockage of flow increases. This creates a radial distribution of flow which reduces the axial component of velocity, reducing the angle of incidence at which stall occurs [35]. Low solidity turbines experience less radial distribution of flow hence the stall angle is comparatively greater (Fig. 12.8a). The increase in solidity increases the pressure drop and torque coefficient across the turbine blades. However, efficiency decreases

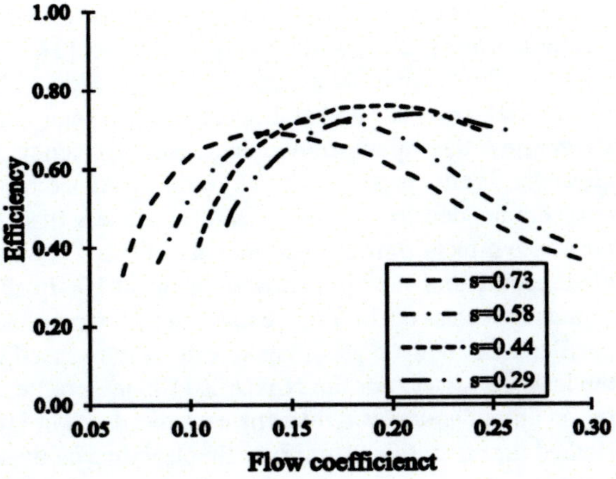

(a)

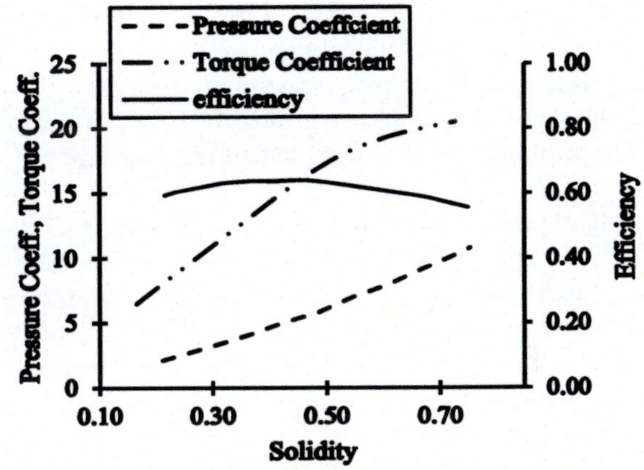

(b)

Fig. 12.8 **a** Effect of solidity on efficiency [28], **b** variation of pressure coefficient, torque coefficient, and efficiency with solidity [28]

for a solidity greater than 0.55–0.6 (Fig. 12.8b). The reduction in efficiency at high solidity is mainly due to the large kinetic energy loss at the blade exit. One option to use a high solidity turbine is to use a biplane turbine, where the losses to turbine exit are less. However, a high solidity turbine has good starting characteristics; it takes fewer number cycles to reach operational speed [36].

According to Raghunathan [32], the hub-to-tip ratio (h) is an important parameter as it controls the local airflow incidence angle near the hub, the tip leakage losses, and the relative interference effects at the hub. The effect of the hub-to-tip ratio is complex. For a given diameter, the hub-to-tip ratio controls the volume of air going inside the turbine. It is directly linked with the tip clearance ratio. For a fixed hub diameter and outer casing diameter, a decrease in h reduces the tip clearance. As a result, tip leakage losses decrease, increasing the turbine efficiency. However, a decrease in h advances stalls [32]. Results from experiments [28, 32, 33, 36] recommend a hub to tip ratio <0.6 for optimal design.

Tip clearance is another important parameter for Wells turbine performance. Decreasing the tip clearance increases efficiency by reducing leakage losses. However, Wells turbines with large tip clearance have a wider operating range. It is recommended to use a tip clearance of less than 2% of chord length [32]. Tip clearance effects were studied numerically in [37]. It is seen that for a NACA0015 blade profile, the maximum power occurs at 1% tip clearance. Whereas for other tip clearance such as 0.5 or 3%, the maximum power is less, but it occurs at a higher flow coefficient. A new tip grooving scheme is introduced by Halder et al. [38] to reduce the leakage losses. For the NACA0015 blade profile, 26% relative improvement in power output is achieved by using a groove depth of 3% chord length. Taha et al. [39] studied the effect of non-uniform tip clearance on the performance of Wells turbine. The tip clearance is increased gradually from the leading edge to the trailing edge. It is seen that a turbine with non-uniform tip clearance has similar performance compared to the turbine with uniform tip clearance in terms of torque coefficient, pressure coefficient, and efficiency. However, the turbine with non-uniform tip clearance seems to have a wider operating range.

Aspect ratio is defined as the ratio of blade span to the chord length. Aspect ratio influences turbine stall as it is directly related to the tip clearance and hub to tip ratio. As recommended by Raghunathan [32], an aspect ratio of 0.5 is good for turbine design. A numerical study by Kim et al. [40] gave the optimum blade geometry for NACA0020 blade profile as blade sweep ratio 0.35, solidity 0.67, the hub to tip ratio 0.7, and aspect ratio 0.5.

Apart from the geometric parameters, the Mach number at the blade tip is an important parameter affecting the performance of the Wells turbine. If the free stream Mach number at blade tip exceeds the value of unity, the flow over aerofoil becomes supersonic. This leads to a significant increase in drag and a decrease in efficiency. According to Raghunathan [32], for a monoplane Wells turbine of solidity 0.3, the tip Mach number should be less than 0.4 to avoid transonic effects.

12.4.1.1 Blade Shape

Several authors have used different blade shape modifications to improve the performance of Wells turbine. Blade shape modifications include the use of different blade profiles, blade sweep, and blade skew. Although NACA aerofoil profiles are most commonly used in Wells turbines, other types of profiles are also studied to compare the performances. Gato and Webster [41] carried out an experimental investigation to compare performance between NACA0015 and optimized HSIM blade. Results show that HSIM blades are able to produce power at higher flow rates. Although HSIM blades cannot delay the onset of stall, it is less abrupt, and it can continue to produce power until the complete stall occurs. This makes HSIM blade advantageous over NACA blades. Many authors have compared the performance of NACA0015,

Table 12.3 Comparison of different blade profiles

Authors	Blade profiles	Remarks	Conclusion
Setoguchi et al. [42]	NACA0020, NACA0015, CA9, HSIM 15-262123-1576 (two different solidity 0.48 and 0.64 are used)	Sweep ratio is 0.35 for all the blades	NACA0021blade with a sweep ratio of 0.35 is a suitable choice
Takao et al. [43]		An experiment carried out for large Reynolds no	For large Reynold no. of 4×10^5, NACA0015 is the suitable blade profile
Thakker and Abdulhadi [44]		Hysteretic characteristics studied experimentally	Maximum mean efficiency is achieved for the NACA0015 blade. However, CA9 blades have a wider operating range
Setoguchi et al. [45]		Sweep ratio is 0.35 for all the blades	NACA0021 blade with a sweep ratio of 0.35 is suggested

NACA0020, CA9, and optimized HSIM blades. The comparison is listed in Table 12.3.

Blade sweep is an important modification for the performance improvement of Wells turbine. Suzuki and Arakawa [46] studied the flow visualization around normal and swept blades. Results indicated that the swept forward blade is inferior to other blades as the flow separation occurs around the hub for a small angle of attack. Experimental comparison between swept and unswept blades [47] shows that backward swept design produces efficiency and positive torque for a wider operating range. However, the peak efficiency is lower for swept blades compared to standard unswept blades. Using numerical analysis, Kim et al. [48] suggested the optimum blade sweep ratio to be 0.35 for best performance.

Blade skew is another modification carried out to improve the performance of the Wells turbine. Experimental and numerical result [49] with a skewed blade (skewness limited to maximum 15°) indicates 106% relative improvement of stall-free operation from baseline. Blade-skew also decreases the flow-generated sound in the unstalled region by up to 4 dB as compared to the unskewed standard blades.

Rotor blade with variable chord is another important modification of blade shape that has been studied numerically. Unlike the standard constant chord rotor blades, variable chord blades have different chord lengths from hub to tip. Two different numerical studies of variable chord rotor and the comparison with standard rotor blade performance are listed in Table 12.4.

All the design modifications of blade shape mentioned before used 2-D blades, i.e., blade thickness remains constant in the radial direction. A three-dimensional blade is an important design modification where the blade has a different thickness at the hub and tip section. Experimental studies by Takao et al. [52, 53], and Takasaki et al. [54] used NACA0015 and NACA0025 at hub and tip, respectively. Results indicated

Table 12.4 Case study of variable chord rotor

Specification	Case I [50]	Case II [51]
Tip diameter	263 mm	298 mm
Hub to tip ratio	0.62	0.7
No. of blade	4	6
Solidity	0.58	0.67
Chord length from hub to tip	84–135.5 mm	72.92–115.15 mm
Blade profile from hub to tip	NACA0020 to NACA0010	NACA0022 to NACA0012
Remarks	The peak efficiency of 60% with variable chord rotor for flow coefficient 0.1–0.28 Peak efficiency 77% with guide vane	Static pressure difference increases around the blade; total entropy generation decreased by 26.02% throughout the operating region

that the 3-D blade has better performance in terms of peak efficiency. However, it fails to improve the onset of stall. Numerical studies carried out by Halder et al. [55] used a combination of NACA0015 and NACA0024 to create a 3-D blade profile. Four different 3-D blade profiles are numerically studied. It is recommended to use a lower thickness at mid-span and higher thickness at tip section. Improvement in operating range is noticed in this type of blades. Different blade shape modifications are shown in Fig. 12.9.

Apart from the blade shape modifications discussed above, one more important modification is to use an endplate in the blade tip. Takao et al. [56–58] carried out an experiment to study the effect of endplate on Wells turbine performance. Although experimental results indicated qualitative improvement in the performance by using endplate, quantitatively, the improvement is not so significant. Out of the three different types of end plates (Fig. 12.10) used for investigation, the forward type case showed the highest tangential force coefficient and highest efficiency. Cui and Hyun [59] proposed a ring-type endplate as a penetrating blade tip treatment. A numerical study showed that the stall occurs at a higher flow coefficient compared to blades without endplate. For the ring-type endplate, the peak efficiency is 16% higher than that of an unmodified turbine and 9% higher compared to the turbine with the standard endplate.

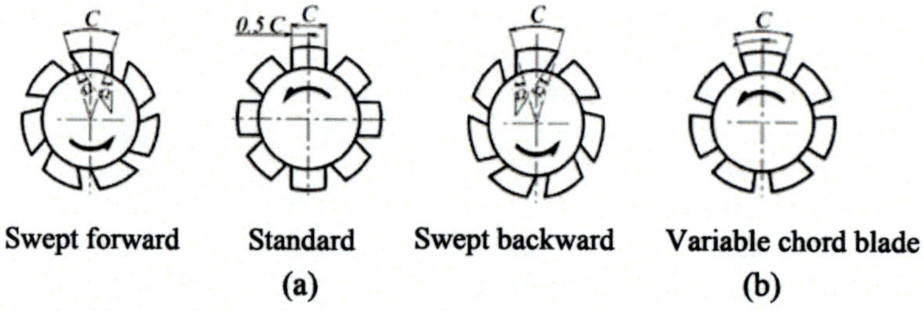

Fig. 12.9 Different blade shapes: **a** swept blade [46], **b** variable chord blade

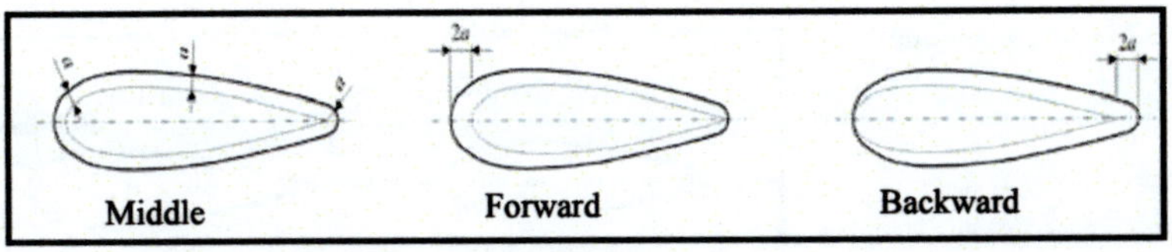

Fig. 12.10 Rotor blade with different endplates [56]

12.4.1.2 Biplane Wells Turbine

The pressure drop across the turbine blade is a function of the tip radius and the rotational speed of the turbine. From the expression of the pressure coefficient, it can be seen that the pressure drop across the turbine is proportional to the square of the rotational speed. If the pressure drop is too high, the turbine attains a high rotational speed subjecting it to transonic effects. Also, the experimental results indicated that a large amount of kinetic energy is lost in a monoplane turbine without the use of a guide vane. The use of a multiplane turbine (Fig. 12.11) can alleviate the transonic effect and reduces the exit kinetic energy loss. Gato and Curran [60] carried out an experiment of the biplane Wells turbine using a symmetrical blade profile. Results indicated that the biplane turbine has better peak efficiency and a wider operating range compared to an equal solidity monoplane turbine.

One important aspect of the biplane turbine is the mutual interaction between the planes. The gap between the plane and the stagger angle of blades plays an important role in biplane turbine performance. Raghunathan [32] recommended using a gap to chord ratio greater than 2 for avoiding mutual interaction between planes. From the experimental result, Gato and Curran [60] suggested using a gap to chord ratio of 1.4 and a stagger angle of 45°. Kaneko et al. [61] carried out an experiment to study the running and starting characteristics of a biplane turbine. Results indicated that the biplane turbine has better starting characteristics compared to standard Wells turbines. Based on this experiment, a gap to chord ratio of 0.5 and pitch angle between 2° and 4° is suggested for blade profile NACA0020. A numerical analysis of the biplane turbine by Shaaban [62] showed that the downstream rotor torque provides only 10–30% of the total torque produced by the turbine. The

Fig. 12.11 Biplane Wells turbine

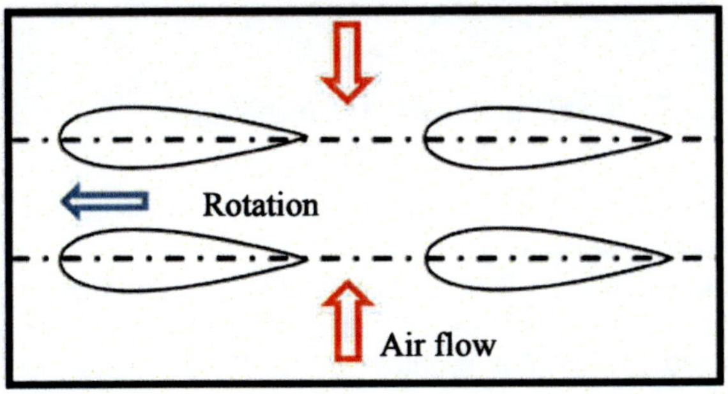

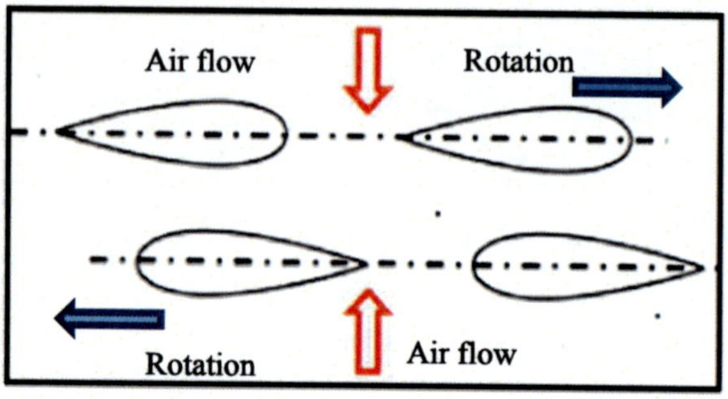

Fig. 12.12 Contra-rotating Wells turbine

upstream rotor efficiency is 1.5–5 times the downstream rotor efficiency. Exergy analysis results showed that the downstream rotor causes the reduction in turbine second law efficiency (defined as the ratio of exergy output to the exergy input of the rotor).

12.4.1.3 Contra-Rotating Biplane Turbine

One interesting variation of the biplane turbine is to use a contra-rotating turbine where the turbines in both the planes rotate in different directions (Fig. 12.12). Performance prediction of contra-rotating Wells turbine based on analytical method showed improvement in performance compared to standard biplane turbine. In order to achieve such improvement, it is suggested to have a high speed of the downstream rotor compared to the upstream rotor. Experimental results [63, 64] supported the fact that a significant increase in operating range is achieved when the downstream rotor operates at a higher speed than the upstream rotor. An increase in peak efficiency from 50 to 66% was achieved in a contra-rotating turbine with an overall solidity of 1.28. The first commercial wave power plant (500 kW capacity) LIMPET (Land Installed Marine Pneumatic Energy Transmitter) in the UK used a two-stage contra-rotating Wells turbine. However, the real-time results of this wave energy plant [65] indicated a poor efficiency compared to standard monoplane and biplane Wells turbine with guide vanes. Contra-rotating turbines also require an additional gearbox arrangement, which makes the system complex and expensive to implement for a practical purpose.

12.5 Turbines with Guide Vane

Experimental and numerical results of monoplane Wells turbine showed that a considerable amount of swirl kinetic energy is lost at the exit of the turbine. The use of a biplane turbine helps to reduce this loss. Another option is to use a guide vane in the upstream and downstream sections. Inoue et al. [66] studied 14 different combinations of NACA profiles with different solidity and blade numbers to compare the performance with and without guide vanes. All combinations with guide vane showed self-starting characteristics. The experimental result from other authors [67, 68] showed improvement in efficiency and range of operation by using guide vanes. The effect of the gap between rotor blade and guide vanes is studied in [69]. It can be seen from the experiment that an optimum gap exists between the rotor blade and guide vane for best performance. A smaller gap gives favorable starting characteristics. Govardhan et al. [70] carried out an experiment to study the flow phenomenon around the guide vanes. It is seen that the presence of inlet guide vane creates a pre-whirl, which increases the tangential velocity coefficient. The efficiency of the turbine with a guide vane is observed to be 8% higher than the turbine without a guide vane. As the flow separation occurs near the hub of downstream guide vanes, kinetic energy losses are experienced in this region. Setoguchi et al. [71] studied the effect of a gap to chord ratio and solidity of guide vane on the turbine performance. An optimum gap to chord ratio of 0.433 and guide vane solidity of 1.25 was recommended for the experimental results. Other authors [72–75] who studied the effect of guide vane on Wells turbine performance are listed in Table 12.5. The effect of 3-D guide vanes was studied in [76, 77]. Results showed better performance than 2-D guide vanes. A numerical study by Halder and Samad [78] showed that the highest efficiency was achieved for guide vane angle 11.8°.

A comparison of four turbines (a monoplane turbine without guide vane, a biplane turbine without guide vane, a contra-rotating turbine without guide vane, and a monoplane turbine with guide vane) with the same blade profile (NACA0015) and solidity (0.64) is carried out by Curran and Gato [79]. The variation of efficiency with flow coefficient for different turbines is reproduced in Fig. 12.13. It can be seen that a contra-rotating turbine has the best performance; however, the curve is not smooth. A biplane turbine has similar peak efficiency, but the stall is abrupt. Efficiency is less for monoplane and biplane turbines; however, the range of operation is wider compared to the turbine with guide vane. It can be concluded from an aerodynamic point of view that the contra-rotating turbine has the best performance among the four Wells turbines. However, considering the mechanical complicacy associated with contra-rotating turbine (it needs extra gearbox), Wells turbine with guide vanes is the best suitable choice.

12.6 Turbines with Non-zero Pitch Angles

All the turbine modifications discussed until now used blades fixed at zero pitch angles. In this section, Wells turbine with fixed non-zero pitch angle (Fig. 12.14a, b) is briefly discussed. Gato and Falcao [80] studied the effect of various blade pitch angle sets between ±15°. It is seen that significant performance improvement is possible if blades are fixed at a non-zero pitch angle. Kaneko et al. [61] carried out an experiment on a biplane turbine with a non-zero pitch angle (γ). Results showed that a positive pitch angle is favorable for self-starting the turbine. Recommended pitch angle for the biplane turbine, as suggested in [61], is $2° < \gamma < 4°$. Real sea tests from wave energy

Table 12.5 Summary of research work for a turbine with guide vane

Author	Blade Profile	Remarks
Inoue et al. [66]	NACA0015, NACA0018, NACA0020,	Rotor with guide vane shows better starting and running characteristic
Gato and Falcao [67]	NACA0012	Overall improvement in performance and range of operation
Gato et al. [68]	NACA0015, Variable thickness (NACA0021 at hub and NACA0009 at the tip)	Improvement in efficiency, considerable reduction in exit kinetic energy loss
Takao et al. [69]	NACA0015, NACA0020	An experiment was carried out at the various gap to chord ratio; maximum efficiency occurs at the gap to chord ratio 0.433; smaller gap gives favorable starting characteristics
Govardhan et al. [70]	NACA0021	Studied flow phenomenon due to guide vanes; the presence of inlet guide vane increases the tangential velocity as well as the efficiency; peak efficiency of 76% is observed for a turbine with guide vane
Setoguchi et al. [71]	NACA0020 (5 different solidities)	A Smaller gap between blade and guide vane is favorable. An optimum gap to chord ratio of 0.433 and guide vane solidity of 1.25 was recommended
Govardhan and Dhanasekaran [72, 73]	Variable chord blade (NACA0021 at the hub, NACA0010 at the tip)	Starting and running characteristics improved; turbine with guide vanes shows self-starting characteristics
Suzuki and Arakawa [74], Suzuki [75]	NACA0021	Explained the effect of guide vane based on momentum theory; results show upstream guide vanes are more effective in improving the efficiency; upstream guide vane increases the rotational speed of the rotor, downstream guide vane absorbs kinetic energy
Setoguchi et al. [76],Takao et al. [77]	NACA0020	3-D guide vane has better performance than 2-D guide vanes

(continued)

Table 12.5 (continued)

Author	Blade Profile	Remarks
Halder and Samad [78]	NACA0021	Different guide vane angles studied numerically, and the Highest efficiency is achieved at a guide vane angle of 11.8°

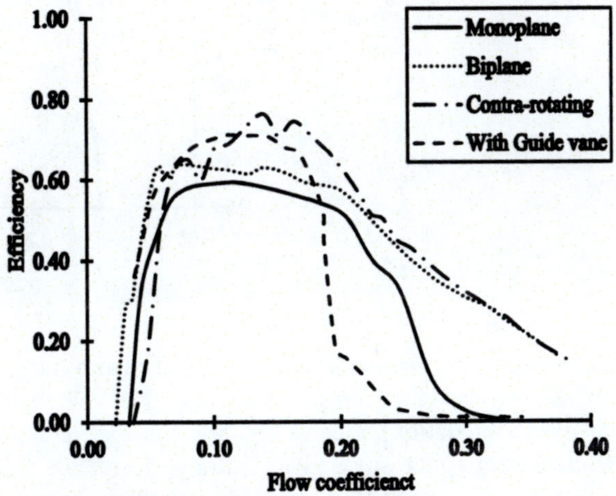

Fig. 12.13 Comparison of different turbine performance (Reproduced from [79])

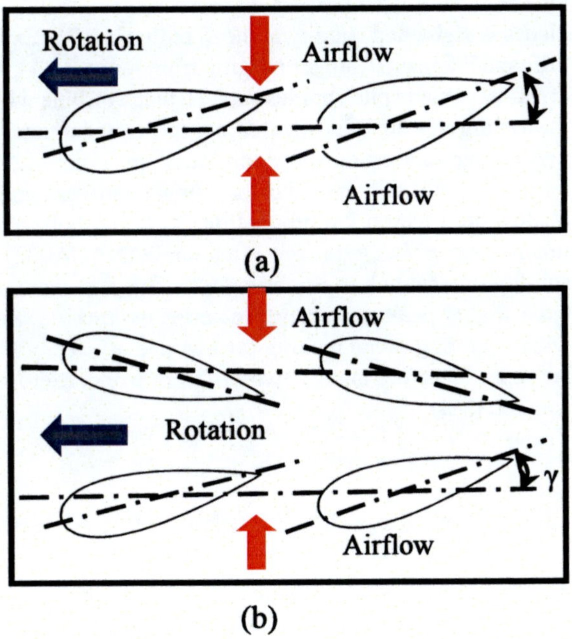

Fig. 12.14 Turbine with non-zero pitch angle **a** monoplane, **b** biplane

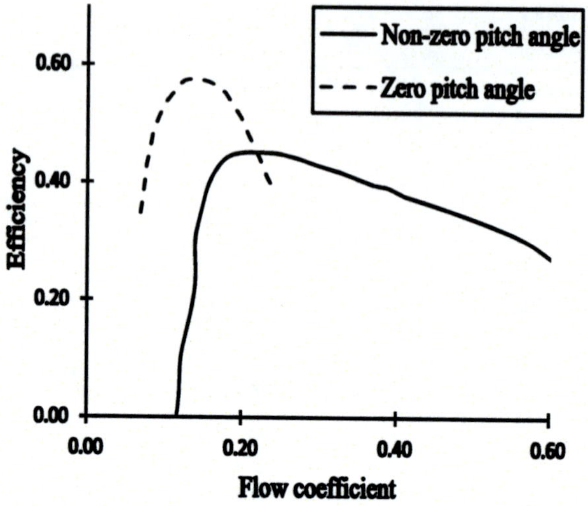

Fig. 12.15 Comparison of performance between zero and non-zero pitch angle turbines ([81, 82])

plants show that the axial airflow velocity from the air chamber to the atmosphere (exhalation) is higher compared to the airflow velocity from the atmosphere to the air chamber (inhalation). Due to this non-symmetric nature of airflow velocity, blades with non-zero pitch angle gives better performance. Setoguchi et al. [68, 69] carried out an experiment to study the effect of blade pitch angle on the monoplane Wells turbine. The test was carried out in a turbine with NACA0021 blades, the number of blades is eight, and solidity at mean radius is 0.75. Experiments were carried out for different blade pitch angles ranging from −4° to + 4°. A comparison of performance at different blade pitch angles showed that a turbine with +4° (positive sign indicates the trailing edge to the downstream side, i.e., reduction in angle of attack) gave the best performance among the other combinations. Figure 12.15 shows a comparison of two monoplane turbine; the solid line shows non-zero pitch angle turbine (NACA0021 blade, chord length 75 mm, solidity 0.7, and hub to tip ratio 0.7), and the dashed line shows turbine with zero pitch angle (NACA0021 blade, chord length 65 mm, solidity 0.5 and hub to tip ratio 0.62). The plot shows turbine with zero pitch angle has a higher peak efficiency. However, the turbine with non-zero pitch angle has a wide operating range making the average efficiency higher. The plots in Fig. 12.15 are reproduced combining two different works (Setoguchi et al. [81], Raghunathan and Tan [82]).

12.7 Turbines with Variable Pitch Angles

Another modification to the non-zero pitch blade angle is to use a turbine where pitch angle can be changed with the change in the airflow direction. The variable pitch angle turbines can be of two types: self-pitch-controlled blades and variable pitch control blades. In the case of self-pitch-controlled blades, the blades are pivoted near the leading edge and free to rotate up to a certain degree in both directions with the change of airflow (Fig. 12.16). For variable-pitch angle blades, a separate control mechanism is used to vary the pitch angle with the change of airflow direction. The concept of self-pitch-controlled blades was first introduced by Inoue et al. [83]. Turbines with self-pitch-controlled blades were studied by model testing and numerical analysis with the sinusoidal flow in [84–86]. Model test results showed that the turbine with a self-pitch control-blade had a better overall performance. The numerical analysis confirmed that a turbine with the self-pitch-controlled blade has better starting characteristics compared to fixed pitch blades. Kim et al. [87] studied the effect of air chamber efficiency on the performance of the self-pitch controlled blade. Through numerical simulation, it is found that the optimal pitch angle setting is 10° for the particular air chamber in concern. This is in contrast to the earlier studies [83], which suggested the optimum pitch angle of about 6°, neglecting the air chamber efficiency. The starting and running characteristics of this turbine with a 10° pitch angle are found to be much improved than other turbines. On a different note, the model test result of Kinoue et al. [88] suggests that the optimum setting angle during inhalation is 6° and 10° during exhalation. Setoguchi et al. [10] carried out a numerical simulation to compare the performance of the different types of turbines. It can be seen (Fig. 12.17) that among monoplane, biplane, and self-pitch-controlled turbine, the self-pitch-controlled turbine has the highest efficiency and wider operating range.

The OWC plant LIMPET [89] and the one at PICO Island [90] used variable pitch blades for the real sea operation. It is observed that the stall at a high flow coefficient can be avoided by pitching the blades at high angles. One of the main observations of the variable pitch turbine is that the maximum blade pitch angles can be extended to ±30° without any drop in performance [89].

Fig. 12.16
Self-pitch-controlled blade

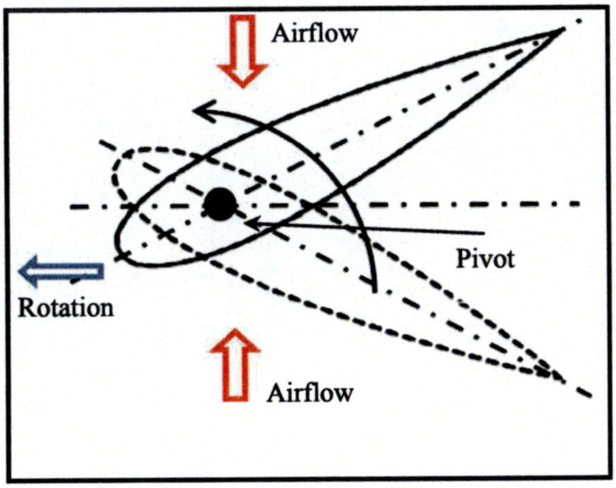

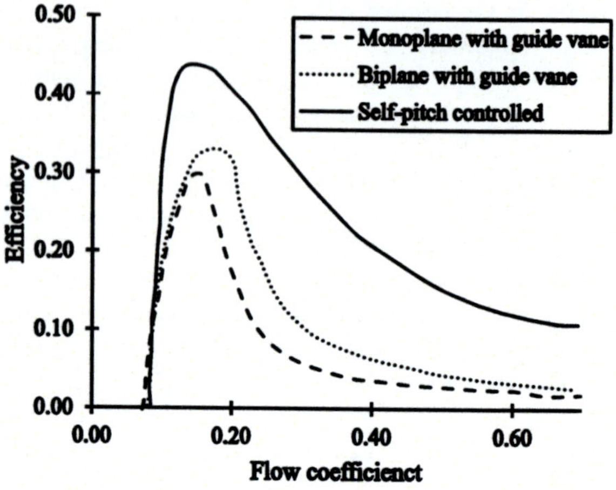

Fig. 12.17 Comparison of different turbines [10]

12.8 Unsteady Flow Analysis

All the works mentioned so far have used a steady analysis, where the inlet air velocity does not change with time, for evaluating the performance of the turbine. However, for real field applications, the inlet air velocity changes with time. Considering the airflow as sinusoidal, the inlet air velocity is divided into two regions; the accelerating flow and the decelerating flow (Fig. 12.18). It can be seen that the value of the torque coefficient is lower in accelerating flow compared to the decelerating flow (Fig. 12.19). This hysteretic behavior is associated with the change in the vortical flow near the blade suction surface. In the case of accelerating flow, the vortex is enlarged; as a result, the flow separation area increases. Whereas for the decelerating flow, the

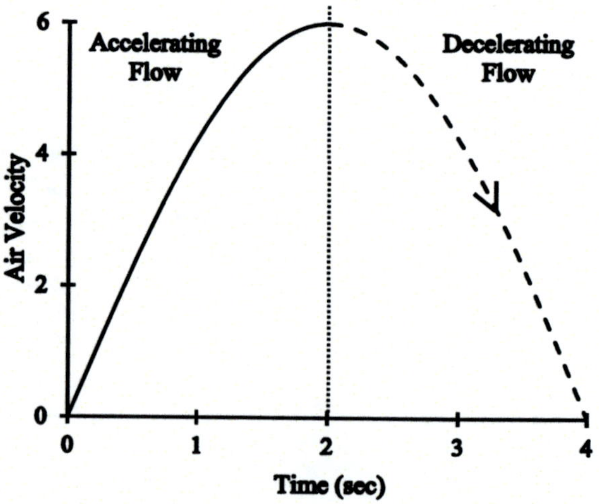

Fig. 12.18 One half-cycle of sinusoidal flow

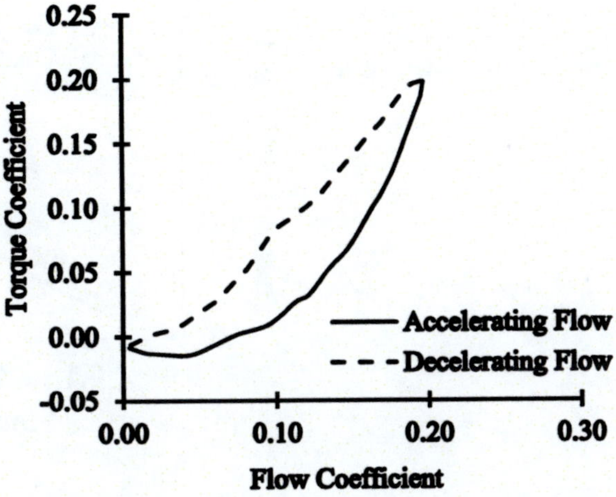

Fig. 12.19 Variation of torque coefficient for unsteady flow in unstalled condition [99]

strength of the vortex was less, and the flow separation area is also reduced [91]. Researchers have used experimental and numerical techniques to study the effect of different parameters on the hysteretic behavior of the Wells turbine. Thakker and Abdulhadi [44, 92] studied the effect of blade profile and solidity on the performance of Wells turbine under unsteady flow conditions. Four blade profiles, NACA0015, NACA0020, CA9, and HSIM 15-262123-1576 with two solidities, 0.48 and 0.64, were tested experimentally and compared with numerical results. CA9 blade with solidity 0.64 showed the best performance among these blades. Setoguchi et al. [93] numerically studied the effect of blade thickness and setting the angle on the hysteretic characteristic of Wells turbine. Results indicated that blade thickness had less influence on hysteretic characteristics compared to the solidity or the setting angle. Kim et al. [94] analyzed the effect of the hub-to-tip ratio and tip clearance on the hysteretic behavior of Wells turbine. While the increase in tip-clearance reduces the size of the hysteretic curve, the effect of the hub-to-tip ratio is very minimal on a hysteretic curve. Paderi and Puddu [95] and Puddu et al. [96] carried out an experiment to analyze the phenomenon of the hysteretic curve in Wells turbine. The experiment results revealed that hysteresis is prominent during outflow, and it is negligible during the inflow phase. It should be noted that all the works mentioned here discussed only the hysteresis in unstalled condition. The hysteretic curve in the deep stall condition is different from the unstalled condition. In comparison to the counter-clockwise hysteresis loop in unstalled conditions, the deep stall condition has a clockwise hysteresis loop (Fig. 12.20). The experimental investigation by Setoguchi et al. [97] and Kinoue et al. [98] showed that clockwise hysteresis in deep stall conditions was due to the flow separation near the tip of the suction surface.

Ghisu et al. [100] carried out a numerical simulation of the Wells turbine with similar geometry as in [91, 93, 101]. The authors stated that the origin of hysteresis was linked to the compressibility effect within the OWC system. This contradicts the explanations given by the authors in the previous research work [91, 93].

Fig. 12.20 Comparison of a hysteresis loop in unstalled and deep stall condition [99]

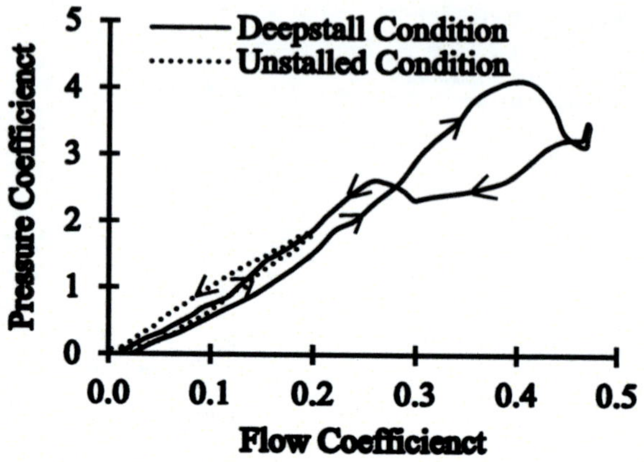

12.9 Starting Characteristics of Wells Turbine

One major problem with Wells turbine is its poor starting characteristics. When starting from rest, it is difficult for the turbine to reach its operational speed. The turbine can accelerate only up to a certain speed which is much below its operational speed. This phenomenon is known as crawling [34]. Experimental [102] and analytical [103] study by Inoue et al. showed that solidity is an important parameter for the starting characteristics of Wells turbine. Kim et al. [104] carried out an experiment to compare the starting characteristics of different types of Wells turbines. Monoplane turbines with guide vane and self-pitch-controlled blades have similar types of starting characteristics, whereas a biplane turbine takes more time to accelerate to its operational speed (Fig. 12.21).

Fig. 12.21 Comparison of starting characteristics for different Wells turbines [104]

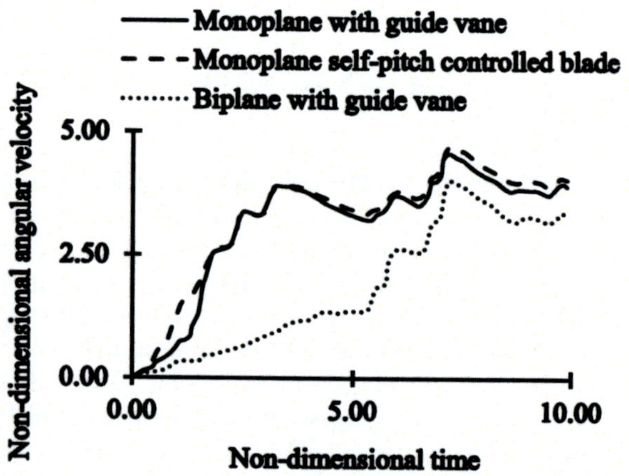

12.10 Optimization of Air Turbines

Traditionally optimization is known as finding the input of a function to maximize or minimize it, which can be subjected to different constraints. Optimization algorithms can be broadly divided into two categories: classical methods (direct or gradient-based methods) and heuristic or evolutionary methods. In the last few decades, there have been several improvements regarding optimization techniques. Unlike the traditional gradient-based techniques, the heuristic or evolutionary algorithms do not get trapped in the local optima and always find out the global optima. Because of this advantage, heuristic algorithms are used more in engineering optimization problems. Banos et al. [105] reviewed the optimization techniques applied to renewable energy systems. However, it does not have any mention of optimization methods applied to wave energy systems. Historically optimization in engineering is regarded as a trial-and-error method for finding the best performance. Few researchers have incorporated mathematical optimization in turbine blade design [106–110]. An extensive review of optimization techniques applied in wind turbine design can be found in [111]. However, literature concerning optimization in the wave energy system is limited. A renewed interest in numerical optimization in the wave energy system can be seen in the last decade (Fig. 12.22).

Gato and Henriques [112] used Polak-Ribière's conjugate gradient algorithm to optimize the blade profile shape. The objective functions used are the minimization of the adverse pressure gradient of the inviscid flow and the minimization of the normalized load. By using the optimization technique, a new blade profile is derived from the standard NACA0015 profile. The new optimized profile significantly delays stall. Analysis of all the research work reported on optimization in wave energy shows that two major optimization techniques are used: surrogate-based optimization and an automated optimization algorithm (OPAL). Surrogate models are a type

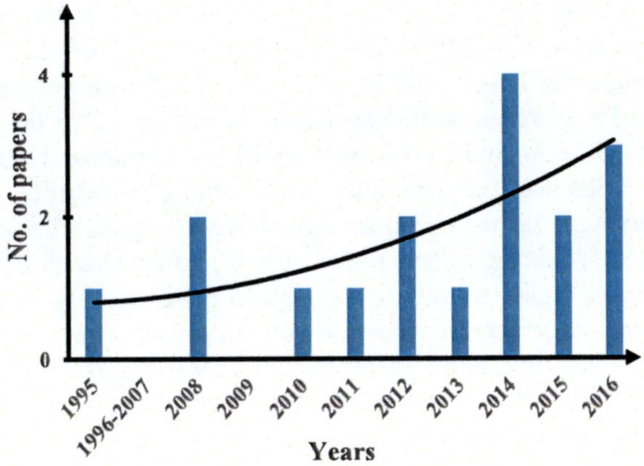

Fig. 12.22 Number of papers published in different years related to mathematical optimization applied to wave energy system

of approximation model which is built from the high-fidelity computationally expensive numerical codes such as computational fluid dynamics (CFD) [113]. Surrogate-based optimization techniques are effectively used in turbomachinery blade shape optimization [114, 115]. Badhurshah and Samad [116–119] used surrogate-based optimization to improve the performance of a bidirectional impulse turbine for wave energy conversion. In these works, response surface approximation, kriging, neural network, and weighted average surrogates were used to generate the population for multi-objective optimization. Halder and Samad [120] used surrogate modeling to optimize the operating range of the Wells turbine. The design variables considered are the angular velocity of the turbine and the axial air velocity. The objective function is to maximize efficiency. The optimization result indicated that the turbine has an efficiency of 63.2% at blade speed 2000 rpm for air velocity 7.85 m/s. This is better than the performance of the reference geometry for the same condition. Another study by Halder et al. [121] used the surrogate modeling technique to optimize the blade sweep angle. The optimized turbine blade has a backward sweep at the mid-section and a forward sweep at the tip section. For the optimized geometry, the torque coefficient increased by 28.28%, and the operating range increases by 18.18%.

OPAL is an automated optimization algorithm developed by researchers from Otto von Guericke University, Magdeburg, Germany [122]. This algorithm is used extensively to optimize the blade profile of the Wells turbine. OPAL is an optimization algorithm, which uses a master C program to couple the CAD model and CFD code. The detailed optimization procedure is discussed in [123, 124]. Optimized blade shape using this algorithm showed a considerable increase in power output (+11.3%) and efficiency (+1%) in the full operating range [125]. Mohamed et al. [126]carried out optimization using the OPAL algorithm taking three parameters: solidity, the thickness of the upper side of the blade, and thickness of the lower side of the blade. The optimized blade profile introduced an increase of efficiency by 2% and an increase in force coefficient by 4.5% compared to standard NACA2421 blades. Optimization of blade pitch angle was carried out by Mohamed and Shaaban[127, 128]. Two different blade profiles were used for the investigation: standard NACA0021 and an optimized aerofoil profile. The optimized blade pitch angle for NACA0021 is +0.3°, and for the optimized blade profile, it is +0.6°. A 2.3% increase in the efficiency was achieved for the NACA0021 turbine, while a 6.3% improvement was achieved for the optimized blade profile.

Shaaban [129] optimized the solidity distribution of the standard blade profile to come up with a novel variable chord blade. The rotor solidity at mid-span was kept constant while solidity at hub and tip was changed. The optimized geometry has a higher solidity at the tip and lower solidity at the hub. The optimized blade geometry increased the power output and the pressure drop across the turbine. Further optimization of air chamber geometry was carried out using a genetic algorithm to minimize the overall cost of electricity production. This study showed that a 1–1.5% reduction in the cost of electricity could be achieved by using an optimized rotor blade and air chamber geometry.

It is to be noted that all the optimization discussed here used direct optimization where the rotor blade shape is optimized starting from the reference blade geometry. Another way of doing the optimization is inverse design, where aerodynamic properties like blade loading were used to define the blade geometry. This is advantageous because it gave direct control of the flow field to optimize the rotor geometry [130]. However, inverse design often leads to rotor designs, which are not practically feasible. Also, formulating an inverse design problem is complicated because of the presence of unknown boundary conditions [125]. Gomes et al. [131] used a two-step inverse design method to optimize an axial flow impulse turbine used for wave energy conversion. A three-dimensional optimized geometry was developed using the inverse design technique, and it showed improved performance compared to the reference geometry.

12.11 Conclusion

A review of Wells turbine is presented here with an emphasis on optimization techniques applied to air turbines in the wave energy system. The review is not extensive; however, it gives the present state of the art of research in the field of Wells turbine and optimization techniques applied to wave energy systems. Different Wells turbine modifications carried out until date is represented in this paper with a comparative analysis of the different type of turbines. Some of the main conclusions from this review are mentioned here:

- Main geometric parameters are affecting the performance of Wells turbines, such as solidity, the hub to tip ratio, tip clearance, and blade thickness.
- Biplane turbine has a better overall performance than monoplane turbine. However, due to the mechanical complexity, the contra-rotating turbine is not feasible for practical use.
- Guide vanes help to reduce the exit kinetic energy losses. As a result, a turbine with a guide vane gives higher efficiency and a wider operating range.
- As the air velocity varies during inhalation and exhalation in the air chamber, turbines with self-pitch-controlled blades are more suitable for practical use.
- There is a renewed interest in numerical optimization applied to air turbines in wave energy systems. Mainly surrogate modeling and automated optimization algorithm (OPAL) were used in most of the works. Optimized rotor geometry showed better performance compared to standard turbines.

References

1. Clément, A., McCullen, P., Falcão, A., Fiorentino, A., Gardner, F., Hammarlund, K., Lemonis, G., Lewis, T., Nielsen, K., Petroncini, S., Pontes, M. T., Schild, P., Sjöström, B. O., Sørensen,

H. C., & Thorpe, T. (2002). Wave energy in Europe: Current status and perspectives. *Renewable and Sustainable Energy Reviews, 6*(5), 405–431.
2. Sannasiraj, S. A., & Sundar, V. (2016). Assessment of wave energy potential and its harvesting approach along the Indian coast. *Renewable Energy, 99*, 398–409.
3. Drew, B., Plummer, A. R., & Sahinkaya, M. N. (2009). A review of wave energy converter technology. *Proceedings of the Institution of Mechanical Engineers, Part A: Journal of Power and Energy, 223*(8), 887–902.
4. Sabzehgar, R., & Moallem, M. (2009). A review of ocean wave energy conversion systems. In *Proceedings of the IEEE Electrical Power & Energy Conference* (pp. 1–6).
5. Falcao, A. F. D. O. (2010). A review of wave energy extraction. *Renewable and Sustainable Energy Reviews, 14*, 899–918.
6. Guedes Soares, C., Bhattacharjee, J., & Karmakar, D. (2014). Overview and prospects for development of wave and offshore wind energy. *Brodogradnja/Shipbuilding, 65*(2), 87–109.
7. López, I., Andreu, J., Ceballos, S., Martínez de Alegría, I., & Kortabarria, I. (2013). Review of wave energy technologies and the necessary power-equipment. *Renewable and Sustainable Energy Reviews, 27*, 413–434.
8. Das, T. K., & Samad, A. (2020). Influence of stall fences on the performance of Wells turbine. *Energy, 194*, 116864.
9. Falcão, A. F. D. O., & Gato, L. M. C. (2012). Air turbines. In Sayigh A. A. (Ed.), *Comprehensive renewable energy, ocean energy* (Vol. 8, pp. 111–149). Elsevier.
10. Setoguchi, T., & Takao, M. (2006). Current status of self rectifying air turbines for wave energy conversion. *Energy Conversion and Management, 47*(15–16), 2382–2396.
11. Falcão, A. F. D. O., & Henriques, J. C. C. (2016). Oscillating-water-column wave energy converters and air turbines: A review. *Renewable Energy, 85*, 1391–1424.
12. Kaneko, K., Setoguchi, T., & Raghunathan, S. (1992). Self-rectifying turbines for wave energy conversion. *International Journal of Offshore and Polar Engineering, 2*(3), 238–240.
13. Masuda, Y. (1986). An experience of wave power generator through tests and improvement. In Falcao A. F. D. O. & Evans D. V. (Eds.), *Hydrodynamics of ocean wave-energy utilization* (pp. 445–452). Berlin: Springer.
14. Masuda, Y., & McCormick, M. (1986). Experiences in pneumatic wave energy conversion in Japan. In McCormick M. E. & Kim Y. C. (Eds.), *Utilization of ocean waves—Wave to energy conversion, American Society of Civil Engineers* (pp. 1–33).
15. Suzuki, M., Arakawa, C., & Takahashi, S. (2004). Performance of wave power generating system installed in breakwater at sakata port in Japan. In *Proceedings of The 14th International Offshore and Polar Engineering Conference, Toulon, France, May, 2004* (vol. 1, pp. 23–28).
16. Ravindran, M., & Koola, P. (1991). Energy from sea waves: The Indian wave energy program. *Current Science, 60*, 676–680.
17. Falcao, A. F. D. O. (2000). The shoreline OWC wave power plant at the Azores. In *Proceedings of the 4th European Wave Energy Conference, Denmark* (pp. 42–47).
18. Boake, C. B., Whittaker, T. J. T., Folley, M., & Ellen, H. (2002). Overview and initial operational experience of the LIMPET wave energy plant. In *Proceedings of The 12th International Offshore and Polar Engineering Conference, Japan, May 26–31* (pp. 586–594).
19. Zhang, D., Li, W., & Lin, Y. (2009). Wave energy in China: Current status and perspectives. *Renewable Energy, 34*(10), 2089–2092.
20. Ortubia, I., De Aguileta, L. I. L., & Marqués, J. (2009). Mutriku wave power plant: From the thinking out to the reality. In *8th European Wave and Tidal Energy Conference, Sweden* (pp. 319–329).
21. Wells, A. A. (1976). Fluid driven rotary transducer. British Patent Spec. No. 1595700.
22. Kaneko, K., & Setoguchi, T. (1988). Wells turbine for wave activated power generation, Patent No. JPS6385201 (A).
23. Dennis, T., Obeyesekera, B., Finnigan, T., & Alcorn, R. (2010). A blade pitch control mechanism. Patent No. US 2010/0290908 A1.
24. Carter, R. W. (2012). Bi directional axial flow turbine with self-pivoting blades for use in wave energy converter. Patent No. US2012/0124986 A1.

25. Arlitt, R., Carlous, T., & Tease, K. (2014). Rotor blade design for a Wells turbine, Patent No. US8678745 B2.
26. Arlitt, R., Banzhaf, H.-U., Starzmann, R., & Biskup, F. (2009). Air turbine for a wave power station, Patent No. WO2009089902.
27. Gato, L. M. C., & Falcão, A. F. D. O. (1988). Aerodynamics of the Wells turbine. *International Journal of Mechanical Science, 30*(6), 383–395.
28. Raghunathan, S., & Tan, C. P. (1983). Aerodynamic performance of a Wells air turbine. *Journal of Energy, 7*(January), 226–230.
29. Sheldahl, R. E., & Klimas, P. C. (1981). *Aerodynamic characteristics of seven symmetrical airfoil sections through 180-degree angle of attack for use in aerodynamic analysis of vertical axis wind turbines.* Sandia National Laboratories Energy Report, The United States of America.
30. Raghunathan, S., & Tan, C. P. (1996). The performance of biplane Wells turbine. *Journal of Energy, 118*(7), 210–215.
31. Gareev, A. (2011). *Analysis of variable pitch air turbines for oscillating water column (OWC) wave energy converters.* Australia: University of Woollongong.
32. Raghunathan, S. (1995). The Wells air turbine for wave energy conversion. *Progress in Aerospace Sciences, 31*(4), 335–386.
33. Raghunathan, S. (1995). A methodology for Wells turbine design for wave energy conversion. *Proceedings of the Institution of Mechanical Engineers, Part A: Journal of Power and Energy, 209*(31), 221–232.
34. Raghunathan, S., & Tan, C. P. (1981). Performance of the Wells turbine at starting. *Journal of Energy, 6*(6), 430–431.
35. Watterson, J. K., & Raghunathan, S. (1998). Computed effects of solidity on Wells turbine performance. *JSME International Journal, Series B, 41*(1), 177–183.
36. Raghunathan, S., Setoguchi, T., & Kaneko, K. (1994). Aerodynamics of monoplane Wells turbine—A review. *International Journal of Offshore and Polar Engineering, 4*(1), 68–75.
37. Halder, P., & Samad, A. (2015). Tip leakage effect on Wells turbine. In *Renewable Energies Offshore* (pp. 555–560).
38. Halder, P., Samad, A., Kim, J.-H., & Choi, Y.-S. (2015). High performance ocean energy harvesting turbine design—A new casing treatment scheme. *Energy, 86*, 219–231.
39. Taha, Z., Sugiyono, Tuan Ya, T. M. Y. S., & Sawada, T. (2011). Numerical investigation on the performance of Wells turbine with non-uniform tip clearance for wave energy conversion. *Applied Ocean Research, 33*(4), 321–331.
40. Kim, T. H., Setoguchi, T., Kinoue, Y., & Kaneko, K. (2001). Effects of blade geometry on performance of Wells turbine for wave power conversion. *Journal of Thermal Science, 10*(4), 293–300.
41. Webster, M., & Gato, L. M. C. (2003). The effect of rotor blade shape on the performance of Wells turbine. *International Journal of Offshore and Polar Engineering, 11*(3), 227–230.
42. Setoguchi, T. (2003). Effect of rotor geometry on the performance of Wells turbine. In *Proceedings of the thirteenth International Polar and Offshore Engineering Conference, Honolulu, Hawaii, USA May 25–30, 2003* (pp. 374–381).
43. Takao, M., Thakker, A., Abdulhadi, R., & Setoguchi, T. (2006). Effect of blade profile on the performance of a large-scale Wells turbine for wave-energy conversion. *International Journal of Sustainable Energy, 25*(1), 53–61.
44. Thakker, A., & Abdulhadi, R. (2007). Effect of blade profile on the performance of wells turbine under unidirectional sinusoidal and real sea flow conditions. *International Journal of Rotating Machinery,* 2007, 1–8, Article ID 51598.
45. Setoguchi, T., Kim, T. W., Takao, M., Thakker, A., & Raghunathan, S. (2004). The effect of rotor geometry on the performance of a Wells turbine for wave energy conversion. *International Journal of Ambient Energy, 25*(3), 137–150.
46. Suzuki, M., & Arakawa, C. (2008). Influence of blade profiles on flow around Wells turbine. *International Journal of Fluid Machinery and Systems, 1*(1), 148–154.

47. Webster, M., & Superior, I. (1998). The effect of rotor blade sweep on the performance of the Wells turbine. In *Proceeding of the Eighth International Polar and Offshore Engineering Conference, Montreal, Canada, May 24–29, 1998* (pp. 94–99).
48. Kim, T. H., Setoguchi, T., Kaneko, K., & Raghunathan, S. (2002). Numerical investigation on the effect of blade sweep on the performance of Wells turbine. *Renewable Energy, 25*(2), 235–248.
49. Starzmann, R., & Carolus, T. (2013). Effect of blade skew strategies on the operating range and aeroacoustic performance of the Wells turbine. *Journal of Turbomachinery, 136*(1), 11003.
50. Govardhan, M., & Chauhan, V. S. (2014). Numerical studies on performance improvement of self-rectifying air turbine for wave energy conversion. *Engineering Applications of Computational Fluid Mechanics, 1*(1), 57–70.
51. Soltanmohamadi, R., & Lakzian, E. (2015). Improved design of Wells turbine for wave energy conversion using entropy generation. *Meccanica, 51*(8), 1–10.
52. Takao, M., Takasaki, K., Okuhara, S., & Setoguchi, T. (2014). Wells turbine for wave energy conversion-improvement of stall characteristics by the use of 3-dimensional blades. *Journal of Fluid Science and Technology, 9*(3), 1–5.
53. Takao, M., Takasaki, K., Tsunematsu, T., Alam, M. M. A., & Setoguchi, T. (2015). Wells turbines with 3-dimensional blades-the performance under unsteady flow conditions. In *Proceedings of the ASME-JSME-KSME 2015 Joint Fluids Engineering Conference AJK Fluids 2015* (pp. 1–4).
54. Takasaki, K., Takao, M., & Setoguchi, T. (2014). Effect of blade shape on the performance of Wells turbine for wave energy conversion. *International Journal of Mechanical, Aerospace, Industrial and Mechatronics Engineering, 8*(12), 1999–2002.
55. Halder, P., & Samad, A. (2015). Wave energy harvesting turbine: Performanc enhancement. *Procedia Engineering, 116*, 97–102, Apac.
56. Takao, M., Setoguchi, T., Kinoue, Y., & Kaneko, K. (2007). Wells turbine with end plates for wave energy conversion. *Ocean Engineering, 34*(11–12), 1790–1795.
57. Takao, M., Setoguchi, T., Kinoue, Y., & Kaneko, K. (2006). Effect of end plates on the performence of a wells turbine for wave energy conversion. *Journal of Thermal Science, 15*(4), 319–323.
58. Takao, M., Setoguchi, T., Kinoue, Y., Kaneko, K., & Nagata, S. (2006). Improvement of Wells turbine performance by means of end plate. In *Proceedings of the Sixteenth (2006) International Offshore and Polar Engineering Conference* (Vol. 4, pp. 480–484).
59. Cui, Y., & Hyun, B.-S. (2016). Numerical study on Wells turbine with penetrating blade tip treatments for wave energy conversion. *International Journal of Naval Architecture and Ocean Engineering*, 2016.
60. Gato, L. M. C., & Curran, R. (1996). Performance of the biplane Wells turbine. *Journal of offshore Mechanics and Arctic Engineering, 118*, 210–215.
61. Kaneko, K., Setoguchi, T., Hamakawa, H., & Inoue, M. (1991). Biplane axial turbine for wave power generator. *International Journal of Offshore and Polar Engineering, 1*(2), 122–128.
62. Shaaban, S. (2012). Insight analysis of biplane Wells turbine performance. *Energy Conversion and Management, 59*, 50–57.
63. Gato, L. M. C., & Curran, R. (1996). Performance of the contrarotating Wells turbine. *International Journal of Offshore and Polar Engineering, 6*(1), 68–75.
64. Curran, B. R., Whittaker, T., Raghunathan, S., & Beattie, W. C. (1998). Performance prediction of contrarotating Wells turbine for wave energy converter design. *Journal of Energy Engineering, 124*(2), 35–53.
65. Folley, M., Curran, R., & Whittaker, T. (2006). Comparison of LIMPET contra-rotating Wells turbine with theoretical and model test predictions. *Ocean Engineering, 33*(8–9), 1056–1069.
66. Inoue, M., & Kaneko, K. (1985). Studies on Wells turbine for wave power generator (part 3: Effect of guide vanes). *Bulletin of JSME, 28*(243), 1986–1991.
67. Gato, L. M. C., & Falcao, A. F. D. O. (1990). Performance of the Wells turbine with a double row of guide vanes. *JSME International Journal, 33*(2), 265–271.

68. Gato, L. M. C., Warfield, V., & Thakker, A. (1996). Performance of a high-solidity wells turbine for an OWC wave power plant. *Journal of Energy Resources Technology, 118*(December), 263–268.
69. Takao, M., Setoguchi, T., & Kaneko, K. (1996). Performance of Wells turbine with guide vanes for wave energy conversion. *Journal of Thermal Science, 5*(2), 82–87.
70. Govardhan, M., Srikanth, J., & Gopalakrishna, G. (1997). Effect of guide vanes on the performance of the Wells turbine. *Journal of Energy, Heat and Mass Transfer, 19*, 187–197.
71. Setoguchi, T., Takao, M., & Kaneko, K. (1998). Effect of guide vanes on the performance of Wells turbine for wave energy conversion. *International Journal of Offshore and Polar Engineering, 8*(2), 155–160.
72. Govardhan, M., & Dhanasekaran, T. S. (1998). Effect of guide vanes on the performance of a variable chord self-rectifying air turbine. *Journal of Thermal Science, 7*(4), 209–217.
73. Govardhan, M., & Dhanasekaran, T. S. (2002). Effect of guide vanes on the performance of a self-rectifying air turbine with constant and variable chord rotors. *Renewable Energy, 26*(2), 201–219.
74. Suzuki, M., & Arakawa, C. (2000). Guide vanes effect of Wells turbine for wave power generator. *International Journal of Offshore and Polar Engineering, 10*(2), 153–159.
75. Suzuki, M. (2006). Design method of guide vane for Wells turbine. *Journal of Thermal Science, 15*(2), 126–131.
76. Setoguchi, T., Santhakumar, S., Takao, M., Kim, T. H., & Kaneko, K. (2001). Effect of guide vane shape on the performance of a Wells turbine. *Renewable Energy, 23*(1), 1–15.
77. Takao, M., Setoguchi, T., Kim, T. H., Kaneko, K., & Inoue, M. (2001). The performance of a Wells turbine with 3D guide vanes. *International Journal of Offshore and Polar Engineering, 11*(1), 72–76.
78. Halder, P., & Samad, A. (2014). Effect of guide vane angle on Wells turbine performance. In *Proceedings of the ASME 2014 Gas Turbine India Conference* (pp. 1–7).
79. Curran, R., & Gato, L. M. C. (1997). The energy conversion performance of several types of Wells turbine designs. *Proceedings of the Institution of Mechanical Engineers, Part A: Journal of Power and Energy, 211*(2), 133–145.
80. Gato, L. M. C., & Falcao, A. F. D. O. (1989). Aerodynamics of the Wells turbine: Control by swinging rotor blades. *International Journal of Mechanical Science, 31*(6), 425–434.
81. Setoguchi, T., Santhakumar, S., Takao, M., Kim, T. H., & Kaneko, K. (2003). A modified Wells turbine for wave energy conversion. *Renewable Energy, 28*(1), 79–91.
82. Raghunathan, S., & Tan, C. P. (1985). Effect of blade profile on the performance of the Wells self-rectifying air turbine. *International Journal of Heat and Fluid Flow, 6*(1), 17–22.
83. Inoue, M., Kaneko, K., Setoguchi, T., & Hamakawa, H. (1989). Air turbine with self-pitch controlled blades for wave power generator (estimation of performance by model testing). *JSME International Journal, Series II, 32*(1), 19–24.
84. Setoguchi, T., Kaneko, K., Takao, M., & Inoue, M. (1996). Air turbine with self-pitch-controlled blades for wave energy conversion. In *Proceeding of the sixth International Polar and Offshore Engineering Conference, Los Angeles, USA May 26–31, 1996* (Vol. I pp. 70–74).
85. Setoguchi, T., Raghunathan, S., Takao, M., & Kaneko, K. (1997). Air turbine with self pitch controlled blades for wave energy conversion (estimation of performances in periodically oscillating flow). *International Journal of rotating machinery, 3*(4), 233–238.
86. Takao, M., Setoguchi, T., Kaneko, K., & Inoue, M. (1997). Air turbine with self-pitch controlled blades for wave energy conversion. *International Journal of Offshore and Polar Engineering, 7*(4), 308–312.
87. Kim, T., Setoguchi, T., Kaneko, K., & Takao, M. (2001). The optimization of blade pitch settings of an air turbine using self-pitch-controlled blades for wave power conversion. *Journal of Solar Energy Engineering, 123*(4), 382.
88. Kinoue, Y., Setoguchi, T., Kaneko, K., Takao, M., & Lee, Y. W. (2002). Air turbine using self-pitch-controlled blades for wave energy conversion. In *Proceedings of the Twelfth (2002) International Offshore and Polar Engineering Conference* (Vol. 3, pp. 668–673).

89. Tease, W. K. (2009). Dynamic response of a variable pitch Wells turbine. *Acta Anaesthesiologica Scandinavica, 53*(6), 1341–1347.
90. Taylor, J. R. M., & Caldwell, N. J. (1998). Design and construction of the variable-pitch air turbine for the Azores wave energy plant. *Third European Wave Power Conference* (October, pp. 7–12).
91. Kinoue, Y., Setoguchi, T., Kim, T. H., Kaneko, K., & Inoue, M. (2003). Mechanism of hysteretic characteristics of Wells turbine for wave power conversion. *Transactions of the ASME, 125*(March), 302–307.
92. Thakker, A., & Abdulhadi, R. (2008). The performance of Wells turbine under bi-directional airflow. *Renewable Energy, 33*(11), 2467–2474.
93. Setoguchi, T., Kinoue, Y., Kim, T. H., Kaneko, K., & Inoue, M. (2003). Hysteretic characteristics of Wells turbine for wave power conversion. *Renewable Energy, 28*(13), 2113–2127.
94. Kim, T. H., Kinoue, Y., Setoguchi, T., & Kaneko, K. (2002). Effects of hub-to-tip ratio and tip clearance on hysteretic characteristics of wells turbine for wave power conversion. *Journal of Thermal Science, 11*(3), 207–213.
95. Paderi, M., & Puddu, P. (2013). Experimental investigation in a Wells turbine under bi-directional flow. *Renewable Energy, 57*, 570–576.
96. Puddu, P., Paderi, M., & Manca, C. (2014). Aerodynamic characterization of a Wells turbine under bi-directional airflow. *Energy Procedia, 45*, 278–287.
97. Setoguchi, T., Kinoue, Y., Moharnrnad, M., Kaneko, K., & Takao, M. (2004). Unsteady flow phenomena of Wells turbine in deep stall condition. In *Proceedings of The 14th International Offshore and Polar Engineering Conference, Toulon, France, May 23–28,2004* (pp. 266–271).
98. Kinoue, Y., Setoguchi, T., Kim, T. H., Mamum, M., Kaneko, K., & Inoue, M. (2004). Hysteretic characteristics of the Wells turbine in a deep stall condition. *Proceedings of the Institution of Mechanical Engineers, Part M: Journal of Engineering for the Maritime Environment, 218*(3), 167–173.
99. Mamun, M. (2006). *The study on the hysteretic characteristics of the Wells turbine in a deep stall condition.* Japan: Saga University.
100. Ghisu, T., Puddu, P., & Cambuli, F. (2016). Physical explanation of the hysteresis in Wells turbines: A critical reconsideration. *Journal of Fluids Engineering, 138*(11), 1–9.
101. Kinoue, Y., Kim, T. H., Setoguchi, T., Mohammad, M., Kaneko, K., & Inoue, M. (2004). Hysteretic characteristics of monoplane and biplane Wells turbine for wave power conversion. *Energy Conversion and Management, 45*(9–10), 1617–1629.
102. Inoue, M., Kaneko, K., Setoguchi, T., & Shimamoto, K. (1986). Studies on Wells turbine for wave power generator (part 4: Starting and running characteristics in periodically oscillating flow). *Bulletin of JSME, 29*(250), 1177–1182.
103. Inoue, M., Kaneko, K., Setoguchi, T., & Raghunathan, S. (1986). Simulation of starting characteristics of the Wells turbine. In *4th Joint Fluid Mechanics, Plasma Dynamics and Lasers Conference.*
104. Kim, T. H., Takao, M., Setoguchi, T., Kaneko, K., & Inoue, M. (2001). Performance comparison of turbines for wave power conversion. *International Journal of Thermal Sciences, 40*(7), 681–689.
105. Baños, R., Manzano-Agugliaro, F., Montoya, F. G., Gil, C., Alcayde, A., & Gómez, J. (2011). Optimization methods applied to renewable and sustainable energy: A review. *Renewable and Sustainable Energy Reviews, 15*(4), 1753–1766.
106. Chattopadhyay, A., Pagaldipti, N., & Chang, K. T. (1993). A design optimization procedure for efficient turbine airfoil design. *Computers and Mathematics with Applications, 26*(4), 21–31.
107. Chen, X., & Agarwal, R. K. (2013). Optimization of wind turbine blade airfoils using a multi-objective genetic algorithm. *Journal of Aircraft, 50*(2), 519–527.
108. Coutandin, D., Bucchieri, L., Brugali, L., & Galbiati, M. (2005). Optimisation techniques applied to the design of gas turbine blades cooling systems. In *International Congress on FEM Technology with ANSYS CFX & ICEM CFD Conference* (pp. 1–11).

109. Derksen, R. W., & Rogalsky, T. (2010). Bezier-PARSEC: An optimized aerofoil parameterization for design. *Advances in Engineering Software, 41*(7–8), 923–930.
110. Shahrokhi, A., & Jahangirian, A. (2007). Airfoil shape parameterization for optimum Navier-Stokes design with genetic algorithm. *Aerospace Science and Technology, 11*(6), 443–450.
111. Chehouri, A., Younes, R., Ilinca, A., & Perron, J. (2015). Review of performance optimization techniques applied to wind turbines. *Applied Energy, 142*, 361–388.
112. Gato, L. M. C., & Henriques, J. C. C. (1996). Optimization of symmetrical profiles for the Wells turbine rotor blades. In *ASME Fluids Engineering Division Summer Meeting*.
113. Han, Z.-H., & Zhang, K.-S. (2008). Surrogate-based optimization. *Real-World Applications of Genetic Algorithms*, pp. 343–362.
114. Samad, A., & Kim, K.-Y. (2009). Surrogate based optimization techniques for aerodynamic design of turbomachinery. *International Journal of Fluid Machinery and Systems*, 2(2), 179–188.
115. Samad, A., Kim, K.-Y., Goel, T., Haftka, R. T., & Shyy, W. (2008). Multiple surrogate modeling for axial compressor blade shape optimization. *Journal of Propulsion and Power, 24*(2), 301–310.
116. Badhurshah, R., & Samad, A. (2014). Efficiency enhancement of a bidirectional impulse turbine using artificial neural network. In *Proceedings of the ASME 2014 12th Biennial Conference on Engineering Systems Design and Analysis* (pp. 1–8).
117. Badhurshah, R., & Samad, A. (2014). Surrogate assisted design optimization of an air turbine. *International Journal of Rotating Machinery*, pp. 1–8. Article ID 563483.
118. Badhurshah, R., & Samad, A. (2015). Multi-objective optimization of a bidirectional impulse turbine. *Proceedings of the Institution of Mechanical Engineers, Part A: Journal of Power and Energy, 229*(6), 584–596.
119. Badhurshah, R., & Samad, A. (2015). Multiple surrogate based optimization of a bidirectional impulse turbine for wave energy conversion. *Renewable Energy, 74*, 749–760.
120. Halder, P., & Samad, A. (2016). Optimal Wells turbine speeds at different wave conditions. *International Journal of Marine Energy, 16*, 133–149.
121. Halder, P., Rhee, S. H., & Samad, A. (2016). Numerical optimization of Wells turbine for wave energy extraction. *International Journal of Naval Architecture and Ocean Engineering, 2016*.
122. Mohamed, M. H. (2011). *Design optimization of savonius and Wells turbines*. Germany: Otto-von-Guericke-Universität Magdeburg.
123. Thévenin, D., & Janiga, G. (2008). *Optimization and computational fluid dynamics*. Berlin, Heidelberg: Springer.
124. Mohamed, M. H., Janiga, G., Pap, E., & Thévenin, D. (2011). Optimal blade shape of a modified Savonius turbine using an obstacle shielding the returning blade. *Energy Conversion and Management, 52*(1), 236–242.
125. Mohamed, M. H., Janiga, G., Pap, E., & Thévenin, D. (2011). Multi-objective optimization of the airfoil shape of Wells turbine used for wave energy conversion. *Energy, 36*(1), 438–446.
126. Mohamed, M. H., Janiga, G., & Pap, E. (2008). Optimal shape of a modified Wells turbine considering mutual interaction between the blades. In *First International Conference of Energy Engineering ICEE-1, Aswan, Egypt* (pp. 1–15).
127. Mohamed, M. H., & Shaaban, S. (2014). Numerical optimization of axial turbine with self-pitch-controlled blades used for wave energy conversion. *International journal of energy research, 38*, 592–601.
128. Mohamed, M. H., & Shaaban, S. (2013). Optimization of blade pitch angle of an axial turbine used for wave energy conversion. *Energy, 56*, 229–239.
129. Shaaban, S. (2016). Aero-economical optimization of Wells turbine rotor geometry. *Energy Conversion and Management, 126*, 20–31.
130. Bonaiuti, D., & Zangeneh, M. (2009). On the coupling of inverse design and optimization techniques for the multiobjective, multipoint design of turbomachinery blades. *Journal of Turbomachinery, 131*(2), 21014.

131. Gomes, R. P. F., Henriques, J. C. C., Gato, L. M. C., & Falcão, A. F. D. O. (2012). Multi-point aerodynamic optimization of the rotor blade sections of an axial-flow impulse air turbine for wave energy conversion. *Energy, 45*(1), 570–580.

Chapter 13
Experimental Testing of Air Turbines for Wave Energy Conversion

Tapas K. Das, R. Suchithra, and Abdus Samad

Abstract The power take off device used in Oscillating Water Column (OWC) type wave energy converters is mainly a bidirectional turbine: either an impulse turbine or a reaction (Wells) turbine. Both these turbines have been studied numerically and experimentally to understand the influence of different parameters. However, an experimental analysis is more realistic in nature compared to the actual operation of the turbine. Such an experimental facility needs a system to create a bidirectional airflow similar to what happens in an OWC chamber. In this chapter, the experimental facility at the Wave Energy & Fluids Engineering Lab, IIT Madras, India, has been discussed in detail. The experimental procedure and the required instrumentations are explained meticulously. Finally, two case studies are presented: one with the impulse turbine and another with Wells turbine tested under sinusoidal flow conditions. The tests were carried out for no-load condition as well as with different resistive loading. The significant results from the experiments are mentioned with detailed explanation.

13.1 Introduction

The oscillating water column (OWC) type wave energy converter (WEC) is one of the most researched devices in the area of wave energy conversion [1]. The OWC type wave energy converters harness energy from the reciprocating nature of the wave into an oscillating air column; this airflow rotates a turbine coupled to an electrical generator. As the OWC system uses an air-turbine to transform the kinetic energy contained in the oscillating airflow into mechanical energy, the system has fewer moving parts inside the water and thus simple to implement. The air turbines used in the OWC devices have to rotate in one direction for bidirectional airflow.

T. K. Das (✉) · R. Suchithra · A. Samad
Wave Energy and Fluids Engineering Laboratory (WEFEL), Department of Ocean Engineering, Indian Institute of Technology Madras, Chennai, Tamil Nadu 600036, India

A. Samad
e-mail: samad@iitm.ac.in

A. Samad et al. (eds.), *Ocean Wave Energy Systems*, Ocean Engineering & Oceanography 14, https://doi.org/10.1007/978-3-030-78716-5_13

Initially, unidirectional turbines were used with a flow rectifying valve setup for such OWC devices. However, subsequently, the self-rectifying turbines substituted the unidirectional turbine setup [2]. The specialty of these self-rectifying air turbines is that it rotates in one direction for the bidirectional airflow. The common types of self-rectifying turbines used for OWC-WEC are axial flow turbines, in which the reaction and impulse turbines are the familiar types. Wells turbine is one type of reaction turbine that is commonly used in OWC prototypes.

The oscillating airflow rate and pressure interaction across the turbine plays a major role in the efficiency of the energy transfer from wave to wire of the OWC-WEC. Thus, both numerical and experimental analysis is equally substantial in order to comprehend the performance of such turbomachines. The experimental analysis is closer to actual operation and more realistic in nature. The impulse and Wells turbine were experimentally tested by different research groups to have a better understanding of their performance. Researchers at the National Institute of Ocean Technology (NIOT), India [3] designed an impulse turbine with a self-pitch-controlled guide vane. Sea trials were conducted for the turbine in a test site near the west coast of India. The test results showed that a self-pitch-controlled turbine performs better than the standard impulse turbine. However, the major drawback of the self-pitch-controlled turbine was the complex mechanical structure, which resulted in the increased maintenance cost of the system. Subsequently, the self-pitch-controlled guide vane was forced to be changed as a fixed guide vane arrangement. Maeda et al. [4] studied this fixed guide vane impulse turbine in detail and suggested the modification in the guide vane setting angles for improving the performance of the turbine. During this time, the concept of a twin unidirectional impulse turbine came up [5]. Such turbines have the benefit of running a pair of unidirectional turbines during the inhalation and exhalation phases of the OWC device. The unidirectional turbine has a higher efficiency during the exhalation phase of operation. Whereas in the inhalation phase, the efficiency drops down below the mean efficiency of bidirectional turbines [6]. Consequently, 1/3rd of the airflow is unused in the later phase of operation. This drop in efficiency is due to the high incidence losses occurring during the starting of the turbine, and as a result, the subsequent torque output is reduced as the turbine pairs are coupled in the same shaft.

The experimental [7] and numerical studies [8] on Wells turbine showed that there is a difference in energy transfer during the acceleration and deceleration phase of the operation. Such difference occurs mainly as a result of the compressibility effects inside the OWC chamber. Past investigators have demonstrated several experimental studies on both Wells turbine and impulse turbine to observe the flow characteristics under steady flow as well as bidirectional airflow. In the case of steady flow experiments, the inlet flow velocity is provided only in one direction. The flow velocity remains fixed with time for a particular rotation speed of the turbine. For bidirectional flow, the inlet airflow is provided from both directions, which is similar to the actual operation of the turbine-the flow velocity changes with time, usually in a sinusoidal manner. As a result, the turbine rotational speed also changes with time.

Many of the present experimental analyses of wave energy turbines were carried out only in steady-state conditions [4, 5, 9]. However, the steady-state conditions can only define the performance of the turbine in the accelerating regime. As discussed earlier, the performance of the turbine in the OWC varies between accelerating and decelerating zone as a result of the compressibility effect inside the OWC chamber. Thus, it is necessary to study the system in a bidirectional airflow.

The energy conversion in the OWC-WEC takes place in various stages; therefore, each subsection in the conversion process needs to be clearly analyzed. The various devices in the power take-off (PTO) system have equal importance in the wave-to-wire conversion of the system. The air turbine and electrical generator coupling can affect overall performance [10]. The wave to electrical power transfer is mostly dependent on the variable rotation of the turbine. This rotational speed of the turbine can be altered from the generator side, which in turn tunes the system for the incoming waves. For example, when using a direct current (DC) generator. The electrical system can act as an isolated system, and the power output can be delivered through a resistive load. The maximum power absorption can be achieved by varying the voltage and frequency (by changing the load resistance) of the generator. This type of speed control can be considered as the scalar control method.

This chapter reports the experimental analysis for bidirectional airflow conditions. The experimental setup produces a reciprocating airflow (pseudo sinusoidal) which simulate the airflow created by the action of the waves, with different height and period.

13.2 Experimental Setup

There are several facilities available around the world where the experimental work of the Wells and impulse turbines are carried out. Here we are going to explain the experimental test facility at the Wave Energy and Fluids Engineering Laboratory (WEFEL) of the Indian Institute of Technology Madras, India (Fig. 13.1). The test setup contains an air chamber of 1.2 m in diameter. One end of the chamber is connected to a crank piston arrangement (Fig. 13.1), which produces a reciprocating airflow. The other end of the chamber contains a duct through which the bidirectional airflow passes. A 15 kW induction motor acting as the prime mover drives the piston. Between the motor and piston, a slider-crank mechanism is used, which converts the angular motion of the induction motor into a translational motion of the piston. A variable frequency drive (VFD) controls the speed of the motor with the help of a proportional integral differential (PID) controller, which in turn controls the translational speed of the piston movement. The VFD frequency range for the test system can be set based on the generator speed rating. This range of frequencies can correspond to the different wave frequencies; inversely, it is the cycle time (CT) of the waves. Similarly, by changing the stroke length (SL) of the piston, the wave height can be varied. Table 13.1 gives the detailed components and specifications of the experimental setup.

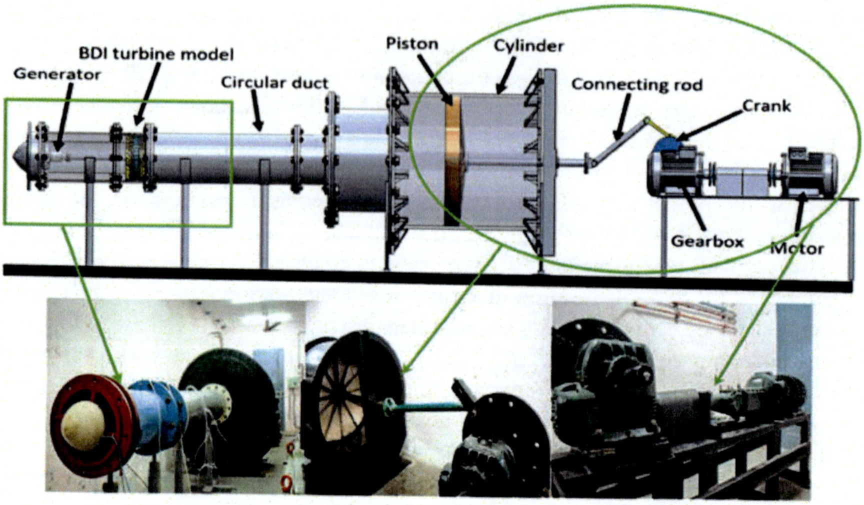

Fig. 13.1 Schematic and actual view of the experimental test rig at IIT Madras [11]

Table 13.1 Components and specification of the experimental setup [11]

Components	Specifications
3-phase squirrel induction motor	Rated power: 15 kW Rated speed: 2890 rpm Rated voltage: 440 V Rated current: 26 A
Gearbox	Gear ratio: 1:40
Crank	Diameter: 0.5 m
Connecting rod between piston and crank	Length: 1.4 m
Piston	Diameter: 1.2 m Seal: O-ring
Cylindrical chamber	Length: 1.5 m Diameter: 1.25 m
Duct containing turbine	Diameter: 300 mm Length: 2.5 m
VFD	Voltage: 440 V Power: 0.37–7.5 kW

13.3 Instrumentation: Sensors and Data Acquisition Systems

The experimental test setup is mainly focused on testing the air turbine, thus requires specific turbine parameters to be measured accurately. Each parameter needs different sensors and transducers in order to be measured in a feasible form. The various instruments used in the experimental process are airflow measuring flowmeter, differential pressure measuring transducer, rotational speed measuring device, torque transducer, current, and voltage measuring transducers. The flowmeter used for the airflow measurement was designed using two averaging type pitot tubes of diameter the same as the duct diameter, which was placed across inside the duct between the chamber and the turbine. This setup measured the inhalation airflow. The pitot tubes sense the static pressure at different points in the cross-section, and this output is connected to a multifunctional transducer. This airflow transducer then converts the static pressure points into airflow rate and airflow velocity as the output. The rotational speed of the turbine is measured using an induction-type speed sensor. This sensor is a non-contact type and is placed in close proximity to the turbine-generator rotor shaft. The rotor shaft is manufactured in the shape of a gear tooth, and the sensor measures the pulses with the number of gear teeth crossed. This conversion from gear tooth pulses to turbine angular velocity in rotation per minute is done using a frequency converter. Another important turbine parameter is the differential pressure measurement. The pressure differential across the turbine is measured using a differential pressure transducer. The transducer is a diaphragm-based instrument, which measures the difference in pressure between the two pressure inputs. Two pressure tapings are provided from the upstream and downstream of the test turbine; these tapings are provided on the duct wall. The torque output of the turbine is another critical parameter measured using a torque transducer placed in-between the turbine rotor and the generator. As the torque generated due to WECs are low in magnitude, the torque transducer should be very sensitive to measure the small variations. However, these instruments need to be precisely mounted on the shaft without any misalignment as misaligned mounting can also lead to unwanted noise, axial vibrations, uncertainty in the measurements. The final electrical outputs of the experiment system are the voltage drop across the load resistor and current flowing through the resistor. These electrical parameters are measured using two isolation amplifiers. The output from the two isolation amplifiers is used to calculate the total electrical power output of the system. The instrumentation of the experimental system is given in Table 13.2.

The measured instantaneous data from all the instruments are processed and recorded using a highly sophisticated data acquisition system (DAQ). Figure 13.2 shows the block diagram of the instrumentation system. The DAQ used in this experimental setup is based on the National Instruments (NI) controller platform compact-Rio (c-Rio). The c-Rio chassis contains the field-programmable gate array (FPGA) backplane, this FPGA links directly to the real-time controller and can be controlled using a host LabVIEW VI connected through Ethernet. A NI 9203 analog input module is used with this chassis, which is an 8-Channel, 16-Bit, ±20 mA current input–output (I/O) module. The different instruments are mapped to the corresponding channels of this module. This module is configured to a host computer, where the measured data are recorded. The input–output current range is configured for 0-20 mA for the analog module. Thus, all the instruments produce current in the range of 4–20 mA, which corresponds to the actual value of the measured parameter.

Table 13.2 Instrumentation specification [12]

Parameter	Instrument	Range	Output signal (m A)	Calibrated range
Rotational speed	Pepperl + Fuchs universal frequency converter	0–5000 rpm	4–20	0–5000 rpm
Differential pressure	ABB 266DSH	+10 kPa	4–20	+ 5 kPa
Airflow rate	Kimo CP-310 with averaging type pitot tube	0–100 m^3/s	4-20	0–0.9 m^3/s
Voltage	RISH Ducer TV 808	0–1000 V DC	4–20	0–300 V
Current	RISH Ducer TV 808 + shunt element	0–40 mA DC	4–20	0–5 A

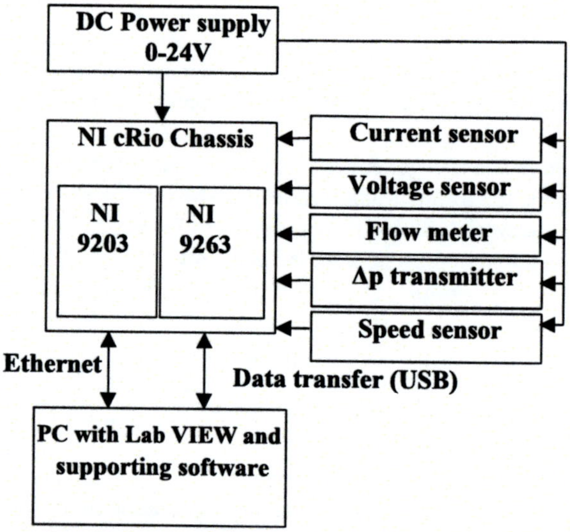

Fig. 13.2 Instrumentation of the experimental setup [12]

The DAQ system is capable of operating at a sampling frequency up to 1 kHz for measuring the current signals corresponding to turbine speed, torque, differential pressure, flow rate, and output electrical power.

The output recorder from the DAQ system was processed using the LabVIEW environment. A graphical user interface was created on the host computer to operate the experimental system for data logging and monitoring. This interface VI is designed for the transducers, and the I/O module mounted on the c-Rio chassis is connected via Ethernet.

Table 13.3 Generator specification [12]

Generator specification	Rating
Rated power	1/1.5 hp
Rated speed	1500/3000 rpm
Rated voltage	180 V
Rated current	5.2 A
Rated torque	4.707 Nm

13.4 Generator Selection

A permanent magnet direct current (PMDC) machine is used in this experimental setup. The machine is used because of its simplicity, ruggedness, and its ability to adapt to low-power applications efficiently. These machines do not need an external excitation circuit and also operate with low starting torque [13]. This machine can operate in both generating and motoring mode, depending upon the direction of the rotation of the rotor. The generator rotor shaft can be directly coupled to the turbine shaft without any requirement of gearing circuitry.

When the rotor is rotated by the action of the turbine, the turbine acts as the prime mover. This rotation of the rotor coils in the presence of the magnetic field created due to a permanent magnet stator induces an electromotive force (E_{emf}) across the armature coils. This generator can be directly connected to a battery or load resistor. This experimental process is mainly focused on determining the performance of the air turbine under different bidirectional flow conditions. Thus the complexities in other subsidiary devices are not investigated. The specifications of PMDC are given in Table 13.3.

13.5 Generator Characteristics

The PMDC can be represented by simple torque-speed characteristics [14], the induced electromagnetic force E_{emf} developed is given as,

$$E_{emf} = k_1 \psi_a \omega_t \tag{13.1}$$

where, ψ_a is the permanent magnet flux linkage. As the PMDC contains a permanent magnet and flux linkage cannot be changed with an external excitation circuitry, ψ_a is constant, i.e., $k_1 \psi_a = K_e =$ constant. This proportionality constant K_e, can be represented as the back emf constant, which is given as,

$$K_e = \frac{E_{emf}}{\omega_t} \tag{13.2}$$

where ω_t is the rotational speed of the turbine. similarly, the electromagnetic torque developed is given by,

$$T_e = k_2 \psi_a I_a \tag{13.3}$$

due to constant flux linkage, $k_2\psi_a$= constant $= K_t$, the Eq. (13.3) becomes,

$$K_t = \frac{T_e}{I_a} \tag{13.4}$$

thus, with the help of the armature current, we can calculate electromagnetic torque (T_e) produced by the generator. However, with the losses due to friction, the generator's electromagnetic torque is less than the turbine output torque (T_t). Equation 5 relates the turbine torque and electromagnetic torque is given as,

$$\frac{d\omega_t}{dt} = \frac{1}{J}\left(T_t - T_e - T_f\right) \tag{13.5}$$

where T_f is the frictional torque, which mainly corresponds to the mechanical losses along with the windage losses in the PMDC [13]. These mechanical losses are constant for a particular rotational speed and vary proportionally with speed.

The other losses in this generator are the electrical losses, which are the copper and iron loss. The copper loss is mainly due to the armature coil. The iron loss is due to the magnetizing and demagnetizing cycle of the permanent magnet. This cycle is nonlinear, and it creates a hysteresis loop known as the B/H curve (flux density/field intensity). This curve is purely based on the characteristics of the permeant magnet. However, in this chapter, iron loss is neglected, and the armature copper loss is estimated using the armature resistance. The measurement of armature resistance is a challenging task, as it is very sensitive to the position of the armature coils, temperature variations, and aging of the machine. Similarly, the measurement of the torque in the real sea test site is usually a cumbersome task of mounting the torque transducer in-between the turbine and generator shaft. Thus this chapter also provides an alternate approach to estimate the torque generated using the electrical power outputs [15]. This method can be considered as an indirect method of torque measurement. Usually, this method can be advantageous over transducer-based torque measurement, as it is simpler, faster, economical, and sufficiently accurate when utilized in industrial testing.

13.6 Experimental Procedure

The test turbines and the generator are coupled through the shaft. The various instruments stated in the above sections are connected to the system. In this setup, only the upstream airflow rate and airflow velocity are measured. The downstream measurements are not done. For measuring the downstream airflow rate and airflow velocity, the duct after the turbine needs to be extended further, and a similar flowmeter arrangement needs to be provided. However, the inhalation phase airflow rate is not measured experimentally; this is calculated from the relationship between the differential pressure and airflow rate obtained from the exhalation phase. The flowrate measurement recorded is for the total duct passage. However, the corresponding airflow rate for the annular area of the turbine needs to be calculated. The first step in this experimental procedure was to determine the unknown generator parameters [16], such as armature resistance (R_a), moment of inertia (J), and generator characteristics. In order to determine the R_a, V-I tests were conducted. The generator was operated in motoring mode, by applying a known voltage across the generator windings. The armature resistance can be theoretically considered as the series-connected resistances such as the armature winding resistance and brush contact resistance. The brush contact resistance behaves as a nonlinear resistance. Equation 13.6 shows the relation between terminal voltage, back emf, and voltage drop due to armature and brush contact [14],

$$V = E_{emf} + I_a R_a + \Delta V_{br} \tag{13.6}$$

where ΔV_{br} is the voltage drop across the brush contact, I_a is the armature current. To estimate the voltage drop due to armature resistance and segregate it from the brush contact drop, few tests were conducted for various excitation voltage sources. Similarly, no-load tests were performed to evaluate the back emf constant K_e, torque constant K_t Generator no-load efficiency. A retardation test provided the total moment of inertia J of the rotating element. For this test, the experimental setup was provided with the rated excitation voltage, and when the rotational speed reached the rated speed, the excitation power supply was switched off, and the corresponding decelerating speed was measured. This test was repeated for different speeds, and the machine parameter was estimated from the curves generated from this test.

The next step was to determine the stray losses and calibration errors; for this, a different set of measurements were conducted without any airflow variations or any excitation inputs to the system. This test provided the correction factors to be included during the analysis, as these errors also contribute to the uncertainty in the measurement.

After the initial and calibration tests are carried out, the bidirectional flow tests are performed. These tests are conducted for various combinations of SL and CT. The SL could be variated in a discrete manner from 0.2 to 0.6 m range with an increment step of 0.1 m. Whereas the CT can be varied with the help of the VFD. The speed of the induction motor connected to VFD was controlled through a PID controller. The PID is programmed in the host computer of the DAQ system. The CT can be varied from 3 to 12 s. The lower and upper limit was bound to the maximum operating range of the turbine-generator set up so that it operates safely and minimum airflow which could drive a turbine, respectively. These SL and CT variations also correspond to real-time wave action on the east coast of India. Table 13.4 shows the

Table 13.4 VFD frequencies in Hz [11]

Crank cycle time (s)	Stroke length (m)		
	0.3	0.4	0.5
4	22.4	22.5	22.8
5	18.5	18.5	18.7
6	15.5	15.5	15.7

VFD frequencies for a different combination of stroke lengths and cycle time used in the experiment.

13.7 Experimental Testing of an Impulse Turbine

13.7.1 Design and Fabrication

The preliminary process of analyzing a turbine is through numerical simulations using CFD codes. These codes predict the characteristics of the turbine rotor and guide the vane. The numerical simulations and optimization for a bidirectional impulse turbine were conducted by [17]. This study used the surrogate-based optimization technique along with CFD simulations to predict the performance of the turbine and reported the optimized turbine for better performance. The optimized rotor and guide vane geometry is given in Table 13.5

The optimized parameters, such as the number of rotor blades and the number of guide vane blades, were obtained from the above work. This optimized impulse turbine was designed and fabricated with a fixed guide vane. The design of the turbine rotor, guide vanes, casing, shaft, generator mounting flange, bullet nose was created using CATIA V5. The turbine guide vane assembly was fabricated in different phases. A Selective Laser Sintering (SLS) 3D printing technique was used to manufacture the turbine rotor and was made using polyamide plastic and consists of 38 blades

Table 13.5 Geometric parameters of the impulse turbine [12]

Parameter	Specifications
Rotor blade profile	Circular-Elliptic
Number of rotor blades	38
Tip diameter	298 mm
Hub/Tip	0.7
Blade inlet angle	60°
Guide vane profile	Plate type
Number of blades	24
Chord length	70 mm
Guide vane inlet angle	30°

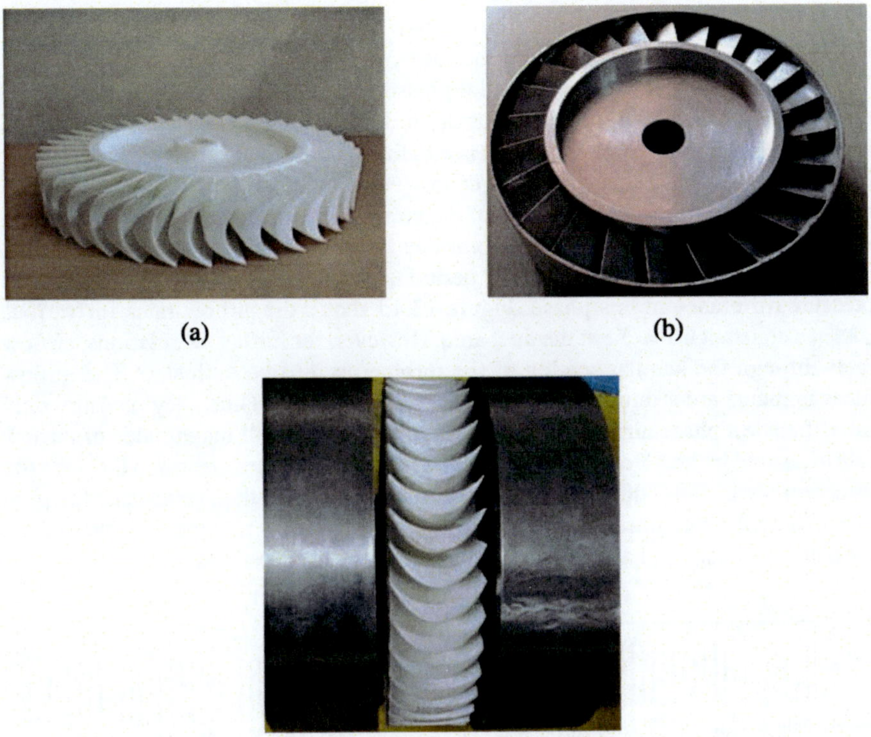

(a) (b) (c)

Fig. 13.3 **a** 3-D printed impulse turbine, **b** the guide vanes, **c** impulse turbine and guide vanes after assembly

as per the optimized design. The guide vanes were made of stainless-steel sheets through bending die to obtain the shape. The duct was made using mild steel with aluminum alloy flanges. The electrical generator was mounted on a mounting flange and was coupled to the turbine shaft made up of aluminum alloy through couplers. Figure 13.3 shows the picture of the impulse turbine and guide vanes used for the experiment.

13.7.2 No-Load Test

In this experimental process, the turbine is subjected to bidirectional airflow with no-load conditions. The experimental process variables are investigated for inflow and outflow phases. The airflow from the chamber flowing outwards through the turbine is the exhalation, and the suction from the outer atmosphere into the chamber through to the turbine is the inhalation. A flow asymmetry in the amplitude of every variable is observed between inflow and outflow phases.

Figure 13.4a, b show the differential pressure across the turbine and airflow rate through the duct section measured for the full cycle for a test case of SL = 0.5 m and CT = 5 s. Figure 13.4c shows the normalized differential pressure measured for inhalation and exhalation cycles recorded varying the SL with constant CT = 5 s. As per the observations, the magnitude of differential pressure in the exhalation phase is higher compared to that of the magnitude of the inhalation phase. Additionally, in the case of smaller SLs, the inflow period is very less which results in negligible pressure difference in this phase. Figure 13.4d shows the airflow rates for various SLs at a constant CT = 5 s at the duct area. However, for further calculations, airflow rates through the annular section of the turbine need to be evaluated. The airflow measurements are arranged only upstream of the turbine. Thus, they measure only the exhalation phase airflow rates. However, there are small magnitudes of airflow rate measured in the inflow phase, which is due to the diffracted airflow in the opposite direction (swirl flow) due to the obstruction caused due to the presence of the pitot tube. There is some filtration needed to analyze the airflow rate measurements. The measured inhalation phased airflow values need to be filtered out.

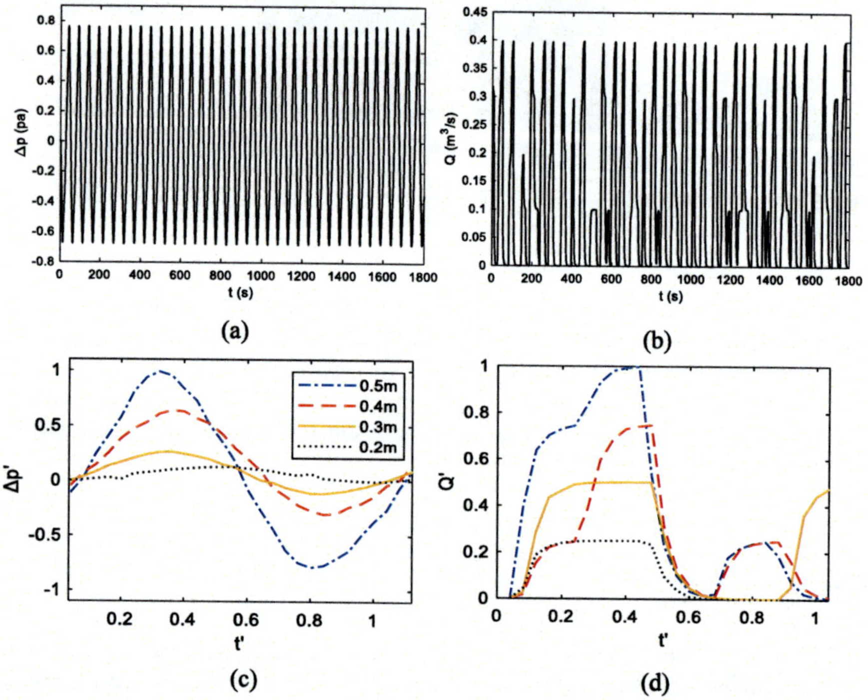

Fig. 13.4 Measured Time series for CT = 5 s **a** differential pressure for full-cycle, **b** flow rate for full-cycle, **c** differential pressure for a normalized cycle, **d** flow rate for a normalized cycle [12]

The experimental process is repeated by keeping the SL constant and varying the CT, which is shown in Fig. 13.5 (a, b). This process checks the effect of differential pressure across the turbine with the change of frequency of operation, which gives an understanding of the consequence of wave period on the air turbine operation.

From the no-load conditions, the rotational speed of the turbine rotor is measured. This test showed the starting and running characteristics of the turbine at different SL and CT. It can be observed that the starting time of the turbine is influenced by the variations of the SL. However, the maximum speed attained by the turbine is affected by the CT variations. Figure 13.6a shows the rotational speed observations for different CT and constant SL = 0.4 m. There is a fluctuation in the speed measurement around a set point for each CT. This oscillation around the setpoint is due to the inertia of the rotating part. As the moment of inertia creates a flywheel effect. It can be seen that the measured speed values for higher CT drops to zero for the inflow phase. The normalized voltage measurement for various CT at SL = 0.4 m with respect to normalized time is shown in Fig. 13.6b. It is observed that the voltage

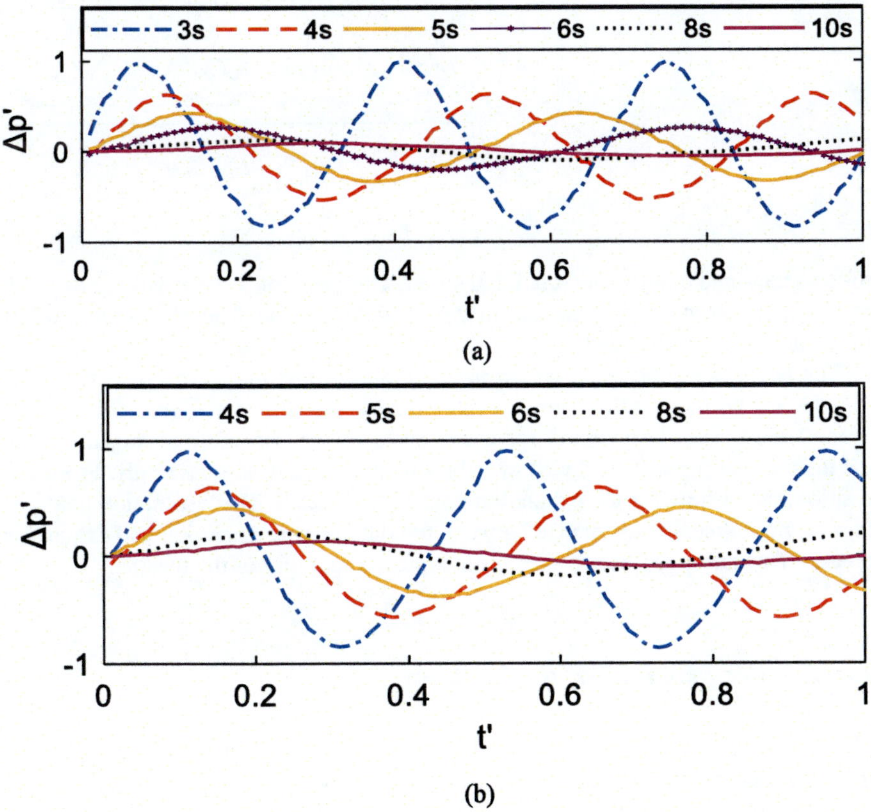

Fig. 13.5 Normalized differential pressure for different CT **a** SL = 0.4 m, **b** SL = 0.5 m [12]

Fig. 13.6 Normalized time series of turbine speed and voltage for SL = 0.4 m **a** Turbine speed, **b** generated voltage [12]

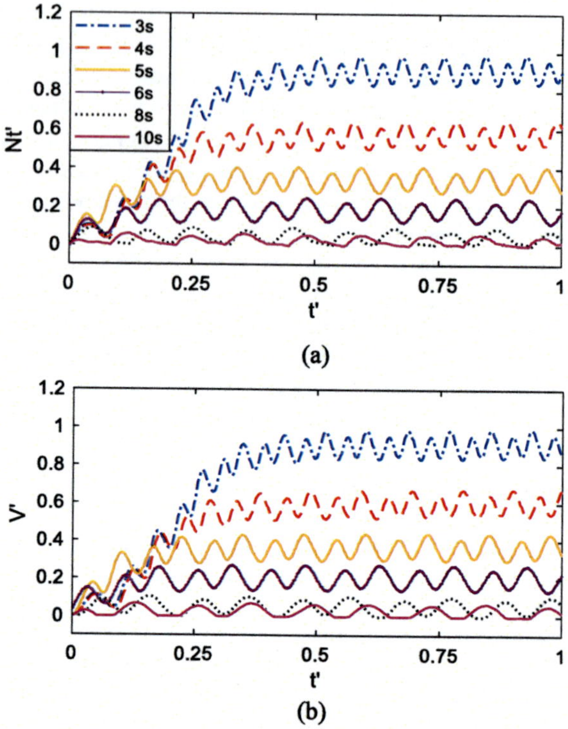

follows the same trend as the rotational speed. The magnitude of the voltage depends on the E_{emf} constant (K_e) defined in the previous section. The E_{emf} constant is a function of rotational speed.

The relationship between the airflow rate and differential pressure is shown in Fig. 13.7. The relationship shows a simple parabolic fitting for each SL and CT. This characteristic is obtained for every cycle of the ensemble-averaged data of the differential pressure and airflow values for the inhalation phase only. From this relation, the airflow for the exhalation phase is evaluated. This estimation holds, as there is no noticeable difference between the damping in inhalation and exhalation phases. This curve represents the power capture capability of the turbine.

13.7.3 Performance of the Turbine

The non-dimensional pressure coefficient with respect to flow coefficient is shown in Fig. 13.8a for different SL = 0.4 m and 0.5 m and CT = 4 s. For a single cycle of inhalation and exhalation, a loop is created. It can be observed that there is a phase lag observed in each cycle. The loop is in an anticlockwise direction for the

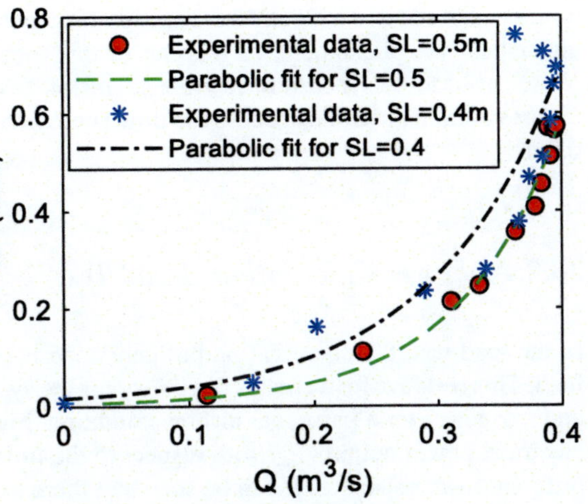

Fig. 13.7 Differential pressure versus flow rate for the exhalation phase [12]

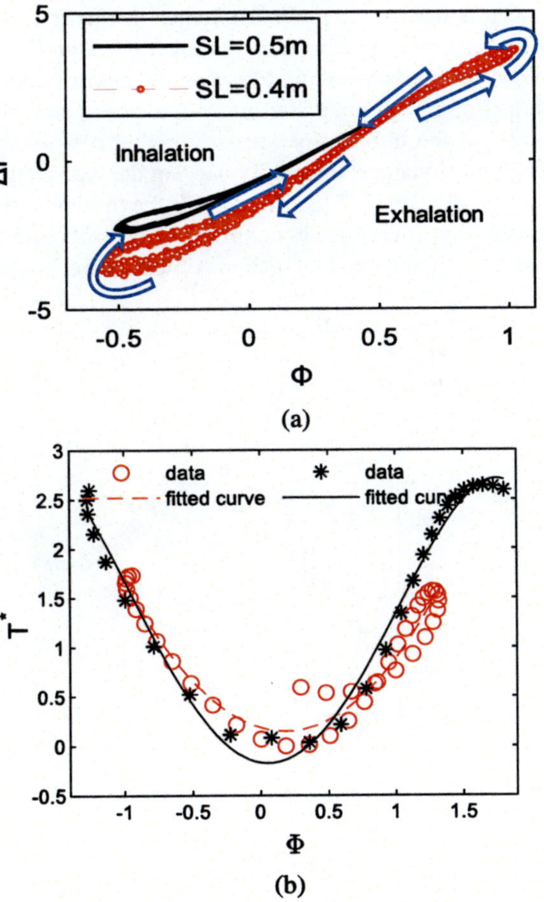

Fig. 13.8 Characteristics of the turbine **a** non-dimensional pressure coefficient for CT = 4 s [12], **b** torque coefficient for CT = 5 s

outflow phase and in a clockwise direction for the inflow phase. However, to clearly understand the phenomena of this lag in this loop, the inhalation and exhalation phases need to be synchronized with the piston frequency. Figure 13.8b shows the torque coefficient with respect to the flow coefficient for two different SL of 0.4 and 0.5 m.

13.7.4 Power Calculation: Load Test

In the load test, the electrical output generated is connected to a variable resistive load. This resistive load creates flexibility for the system to vary the terminal voltage and frequency for a particular airflow condition. Figure 13.9 shows the variation of electrical power output (P_e) with respect to the time for a single piston cycle (t') at different load resistance. It can be seen that there exists fluctuating electrical power around a set point for all the cases of resistive load. When the variable resistance value is decreased, there is a larger fluctuation in the voltage measurement, thus also results in reaching a higher magnitude. At this condition, a higher load current flows through the circuit. However, the nonzero power output is due to the flywheel effect of the system's inertia. On the other hand, when the value of load resistance is large, the turbine operates at a higher rotational speed, which in turn results in higher mechanical losses. By varying the load resistance, the speed of the generator is controlled, which in turn controls the turbine speed. This method of speed control can be considered scalar control. The control objective in this work is to operate the turbine at the speed at which maximum power is obtained. From the observations, it

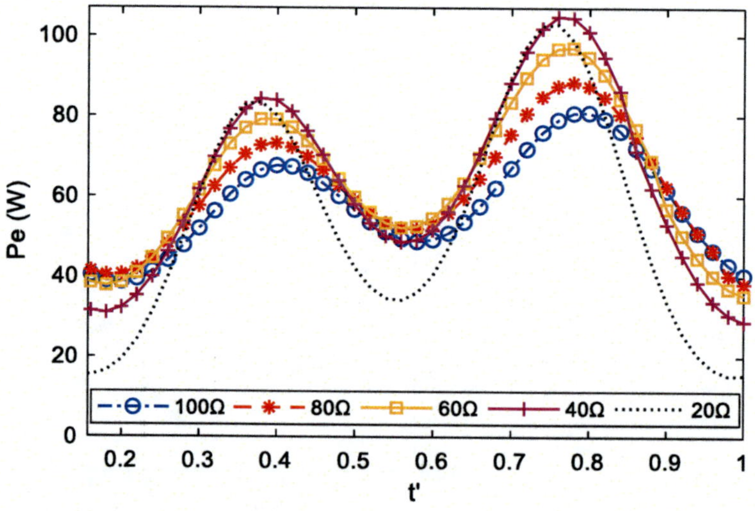

Fig. 13.9 Electrical power output [12]

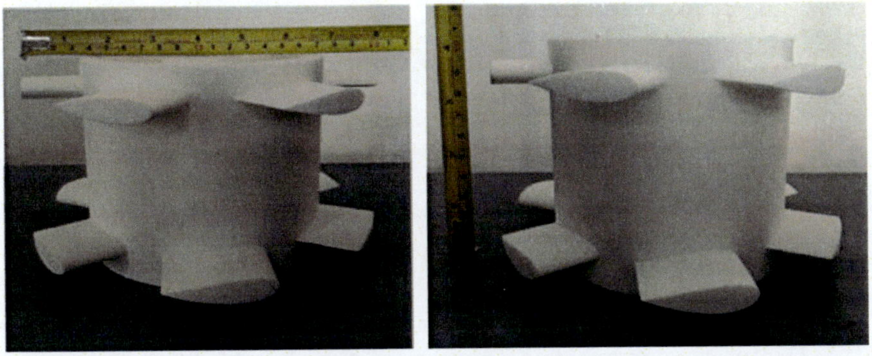

Fig. 13.10 The biplane Wells turbine used for the experiment [18]

can be seen that for the same input airflow setting, the turbine produces maximum power at $R_L = 40\Omega$, as during this condition, the turbine operates close to its design speed (~600 rpm).

13.8 Experimental Testing of a Wells Turbine

13.8.1 Design and Fabrication

A biplane Wells turbine of 196 mm diameter was tested at the Wave Energy and the Fluids Engineering Lab, IIT Madras, in the test set up as explained earlier. The Wells turbine has a 196 mm tip diameter, and the blades are placed in two planes (Fig. 13.10). The distance between the planes is 1.5 times the chord length. The chord length of the turbine is 50 mm. the blades are made of NACA0021 symmetrical profile. Each plane has six blades, and the stagger angle between the planes is 30°. There is a 2 mm tip clearance between the blade tip and the inner duct wall. The turbine is manufactured using a selective laser sintering method with polyamide material. The weight of the turbine is 1.1 kg after manufacturing.

13.8.2 Starting Characteristics

The experiment starts by setting the variable frequency drive at a certain frequency. Based on this frequency, the piston starts to reciprocate at a certain speed, which generates a sinusoidal velocity inside the duct. The turbine starts to rotate because of this inlet velocity. However, the turbine takes some time to overcome its inertia and reach a stable rotational speed. This is known as the starting characteristic of the turbine. Figure 13.11a shows the comparison of starting characteristics for two

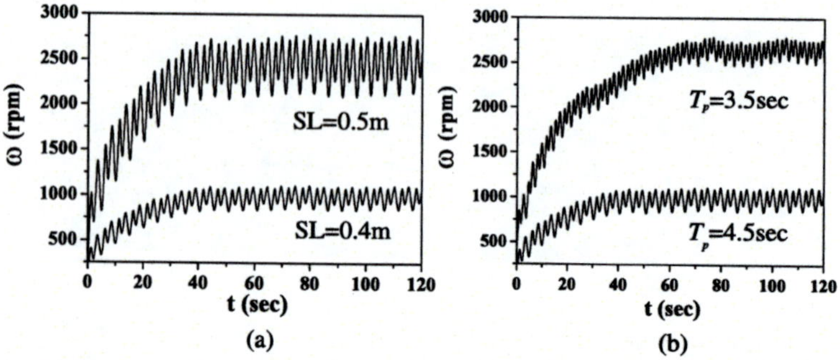

Fig. 13.11 Starting characteristics for **a** different stroke length at $T_p = 4.5$ s **b** different time period at SL = 0.4 m [18]

different piston stroke lengths at the same time period. Whereas, Fig. 13.11b shows the comparison of starting characteristics for two different piston time periods at the same stroke length. When the time period is fixed, the starting characteristic is almost the same for different piston stroke lengths. However, when the stroke length is fixed, if the time period less, the turbine takes more time to achieve a stable rotational speed.

13.8.3 No-Load Test

In the case of no-load tests, the turbine is left to run only with the generator without any external resistive loading. Figure 13.12 shows the comparison of rotational speed and differential pressure for different test cases. For the two test cases (Tp = 5.5 s, SL = 0.4 m and Tp = 4.5 s, SL = 0.3 m), the rotational speed of the turbine is less and almost touching zero when the transition from inhalation to exhalation stage. For Tp = 4.5 s, SL = 0.5 m and Tp = 3.5 s, SL = 0.4 m, the turbine rotational speed has similar values. For these two cases, the difference between the maximum and minimum rotational speed in one cycle is approximately 10% of the peak value. This shows that the turbine has a smooth operation for these two test cases.

The rotational speed of the turbine has a strong correlation with the pressure drop across the turbine. Similar to the rotational speed, for Tp = 4.5 s, SL = 0.5 m and Tp = 3.5 s, SL = 0.4 m, the pressure difference for these two cases is also in the similar range as seen from Fig. 13.12e, f. The peak value of pressure difference during exhalation is higher compared to that of the value during inhalation. The possible reason for this could be the presence of a torque transducer and generator, which may create a blockage for the incoming air. Along with the disturbances caused by the PTO elements, the downstream does not contain a duct to regulate the airflow at the inhalation phase, resulting in incident losses. However, this does not have any significant effect on the rotational speed of the turbine.

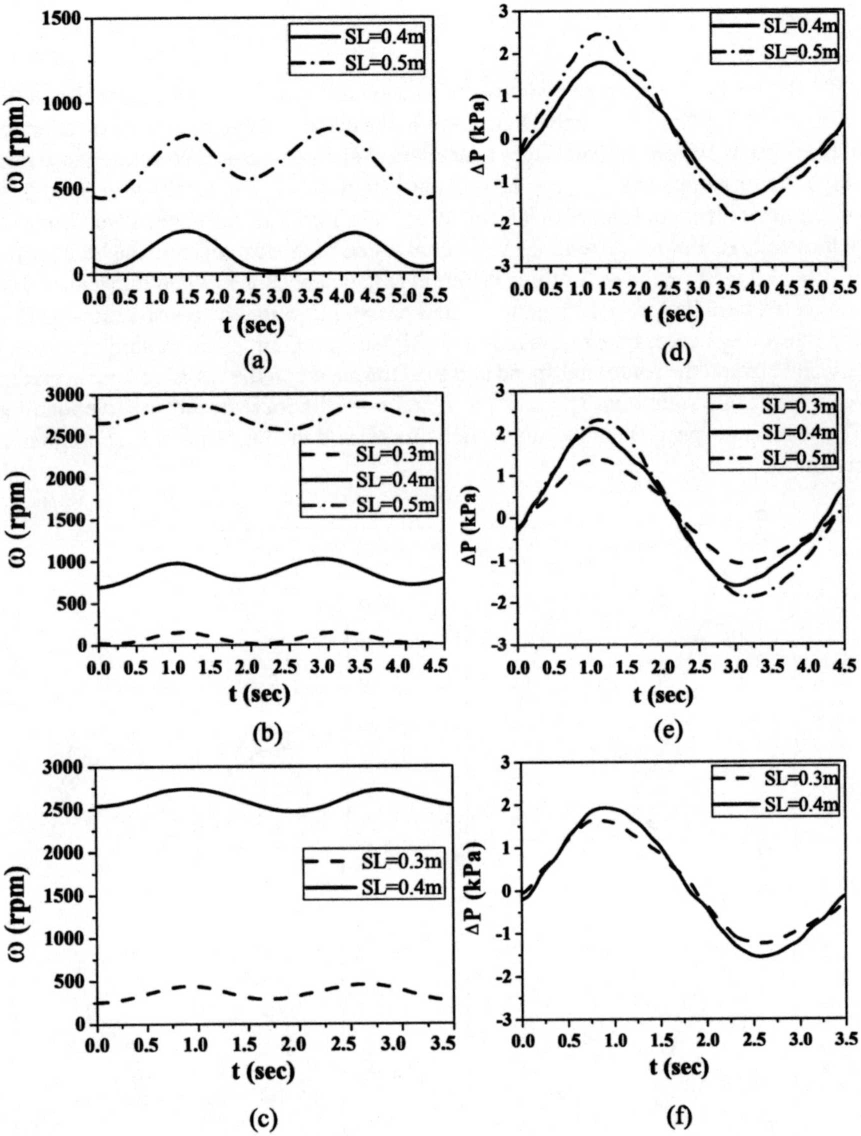

Fig. 13.12 Comparison of rotational speed (**a–c**) and differential pressure (**d–f**) for different test cases [18]

13.8.4 Test with Resistive Loading

After the no-load test, some resistive loads were attached to the generator to test the turbine with loading. A resistive load takes in the electrical energy from the generator in the form of current and produces heat energy. The generator is connected to a variable resistor so that the load can be manually controlled. The results presented here are for three different loads: 10, 50, and 100 Ω. Figure 13.13 shows the comparison of output voltage, output current, the rotational speed of the turbine, and the load torque at $T_p = 4.5$ s, SL = 0.4 m for three different resistance values. When the generator is connected with high resistance, the current passing through the resistive load is low, and the voltage across the resistive load is high as per Ohm's law. A similar analogy is seen between the rotational speed and load torque when the resistive load is varied. As a result, the rotational speed is increased with the increase in resistive loading. The rotational speed is analogous to the voltage, and the load torque is analogous to current.

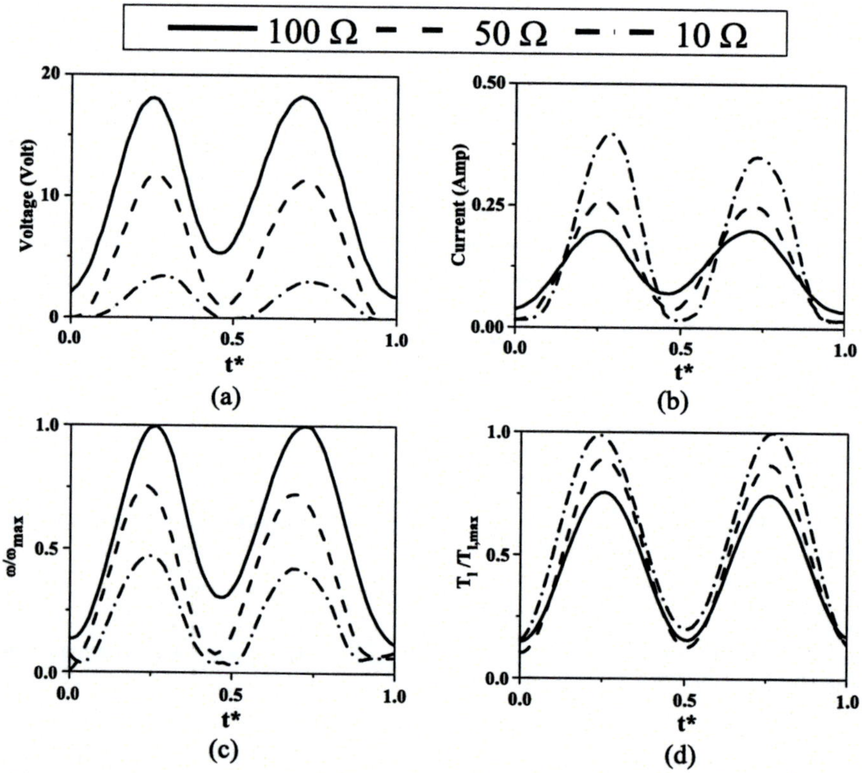

Fig. 13.13 Comparison of **a** voltage, **b** current, **c** rotational speed and **d** load-torque for different resistance at $T_p = 4.5$ s, SL = 0.4 m [18]

Table 13.6 Propagated uncertainty in the secondary parameter [18]

Secondary parameter	Uncertainty range (%)
Φ	±4.64
Δp^*	±2.52
P_e	±0.52

13.9 Uncertainty Analysis

The first step to determine the uncertainty of the experimental process is to determine the inaccuracies of the various measuring instruments. The values (primary variable) are obtained from the instrument datasheets. These values are then used to determine the uncertainty of each calculated variable (secondary variables). Thus, the uncertainty in each secondary variable will contain the propagated uncertainties of the primary variable. Table 13.6 provides the uncertainty of calculated secondary parameters.

13.10 Conclusions

The chapter discusses the experimental facility available at the Wave Energy and Fluids Engineering Lab, IIT Madras, India. The test facility contains a piston-cylinder assembly connected to a wind tunnel, which can generate bi-directional airflow similar to what is experienced in OWC type wave energy converters. A detailed description of the experimental procedure and the instruments used are mentioned here. Finally, two case studies are presented based on the experiment carried out at the laboratory. For the impulse turbine, the maximum rotational speed is about 1500 rpm. For the Wells turbine, the maximum rotational speed is very high, about 3000 rpm. The maximum rotational speed is limited by the rated speed of the generator. For experimental testing of any turbine, it is necessary to test both no-load and load conditions to understand the turbine characteristic completely.

References

1. Falcao, A. F. D. O., & Henriques, J. C. C. (2016). Oscillating-water-column wave energy converters and air turbines: A review. *Renewable Energy, 85*, 1391–1424.
2. Setoguchi, T., & Takao, M. (2006). Current status of self rectifying air turbines for wave energy conversion. *Energy Conversion and Management, 47*(15–16), 2382–2396.
3. Santhakumar, S., Jayashankar, V., Atmanand, M. A., Pathak, A. G., Ravindran, M., Setoguchi, T., Takao, M., & Kaneko, K. (1998). Performance of an impulse turbine based wave energy plant. In *Proceedings of Eighth (1998) International Offshore and Polar Engineering Conference, 1998* (pp. 75–80).

4. Maeda, H., Santhakumar, S., Setoguchi, T., Takao, M., Kinoue, Y., & Kaneko, K. (1999). Performance of an impulse turbine with fixed guide vanes for wave power conversion. *Renewable Energy, 17*(4), 533–547.
5. Jayashankar, V., Anand, S., Geetha, T., Santhakumar, S., Jagadeesh Kumar, V., Ravindran, M., Setoguchi, T., Takao, M., Toyota, K., & Nagata, S. (2009). A twin unidirectional impulse turbine topology for OWC based wave energy plants. *Renewable Energy, 34*(3), 692–698.
6. Pereiras, B., Valdez, P., & Castro, F. (2014). Numerical analysis of a unidirectional axial turbine for twin turbine configuration. *Applied Ocean Research, 47*, 1–8.
7. Setoguchi, T., Raghunathan, S., Takao, M., & Kaneko, K. (1997). Air-turbine with self-pitch-controlled blades for wave energy conversion (estimation of performances in periodically oscillating flow). *International Journal of Rotating Machinery, 3*(4), 233–238.
8. Ghisu, T., Puddu, P., & Cambuli, F. (2016). Physical explanation of the hysteresis in Wells turbines: A critical reconsideration. *Journal of Fluids Engineering, 138*(11), 111105.
9. Liu, Z., Cui, Y., Li, M., & Shi, H. (2017). Steady state performance of an axial impulse turbine for oscillating water column wave energy converters. *Energy, 141*, 1–10.
10. Pereiras, B., Montoya Andrade, D.-E. A., de la Villa Jaén, A., Castro, F., El Marjani, A., & Perez, P. (2011). Design of a turbine-generator joint for an oscillating water column. In *Proceedings of the 9th European Wave and Tidal Energy Conference, 2011* (pp. 1–8).
11. George, A., Anandanarayanan, R., Suchithra, R., Pattnaik, B., Dudhgaonkar, P., Jalihal, P., & Samad, A. (2018). Experimental analysis of turbine-chamber coupling for wave energy conversion. *International Journal of Energy Research, 42*(15), 4770–4782.
12. Ravikumar, S., Anandanarayanan, R., George, A., Pattanaik, B., Dudhgaonkar, P. V., Jalihal, P., & Samad, A. (2020). Experimental investigation of a bidirectional impulse turbine for oscillating flows at various resistive loads. *IEEE Journal of Oceanic Engineering*, pp. 1–17.
13. Krishnan, R. (2010). *Permanent magnet synchronous and brushless DC motor drives*. CRC Press Taylor & Francis Group.
14. Jacek, F. G., & Wing, M. (2002). *Permanent magnet technology: Design and applications* (2nd edn). New York: Marcel Dekker, Inc.
15. Filianoti, P., & Camporeale, S. M. (2009). In field measurements on a small scale OWC device. In *Proceedings of the 8th European Wave and Tidal Energy Conference, 2009* (pp. 284–294).
16. Wu, W. (2012). DC motor parameter identification using speed step responses. *Modelling and Simulation in Engineering, 2012*, 1–5.
17. Badhurshah, R., & Samad, A. (2015). Multiple surrogate based optimization of a bidirectional impulse turbine for wave energy conversion. *Renewable Energy, 74*, 749–760.
18. Das, T. K., Kumar, K., & Samad, A. (2020). Experimental analysis of a biplane wells turbine under different load conditions. *Energy, 206*, 118205.

Chapter 14
Passive Flow Control Methods for Performance Augmentation in Air Turbines Used for Wave Energy Conversion—A Review

P. Madhan Kumar, Paresh Halder, and Abdus Samad

Abstract The performance enhancement of air turbines used in wave energy extraction is quintessential to realize the commercialization of wave energy plants. This chapter discusses the various passive flow control methods used to enhance the performance of the Wells and axial impulse air turbines that are employed in the oscillating water column to harvest energy. The different passive flow control methods were elaborated with their effect on the performance parameters of the air turbines, and it was found that the implemented passive flow control devices addressed critical limitations such as narrow operating range, reduced efficiency, and low power output, etc. Moreover, the combined investigation of different passive flow control devices was suggested to achieve the overall performance enhancement of the Wells turbines. Lastly, only a handful of literature is available on the passive flow control of the axial impulse turbines, and dedicated research was recommended to analyze and implement compatible passive flow control devices to improve their performance.

List of Abbreviations

AOA	Angle of attack
CFD	Computational fluid dynamics
FC	Flow coefficient
FH	Flap height
GF	Gurney flap
LE	Leading-edge
LES	Large eddy simulation
OWC	Oscillating water column

P. M. Kumar · A. Samad
Wave Energy and Fluids Engineering Laboratory (WEFEL), Department of Ocean Engineering, Indian Institute of Technology Madras, Chennai 600036, TamilNadu, India
e-mail: samad@iitm.ac.in

P. Halder (✉)
Institute of Ocean Energy, Saga University, 1-Honjo-Machi, Saga-Shi, Saga 840-8502, Japan

A. Samad et al. (eds.), *Ocean Wave Energy Systems*, Ocean Engineering & Oceanography 14, https://doi.org/10.1007/978-3-030-78716-5_14

PS	Pressure side
PTO	Power take-off
RANS	Reynolds-averaged Navier-Stokes
RETB	Radiused edge tip blade
REF	Reference
SETB	Sharp-edged tip blade
SETE	The static extended trailing edge
SS	Suction side
SST	Shear stress transport
TE	Trailing edge
VTB	Variable thickness blade
WEC	Wave energy converter

Symbols

C	Chord length *(mm)*
L	Lift force (*N*)
D	drag force (*N*)
F_T	Tangential force (*N*)
F_A	Axial force (*N*)
V	Absolute velocity *(m/s)*
U_T	Blade tip velocity (*m/s*)
W	Relative velocity *(m/s)*
z_g	No. of guide vanes
z_r	No of rotor blades

Subscripts

1	Inlet
2	Outlet

14.1 Wave Energy

Energy generation from clean, non-polluting renewable sources is vital to cut back our carbon emissions. Currently, various types of renewable energy resources such as wind, hydro, solar, geothermal, biomass, and ocean energies, etc., have been installed around the world. Above all types, ocean energy constitutes a minor share of the entire renewable energy plants implemented across the globe [1]. Conversely, the ocean

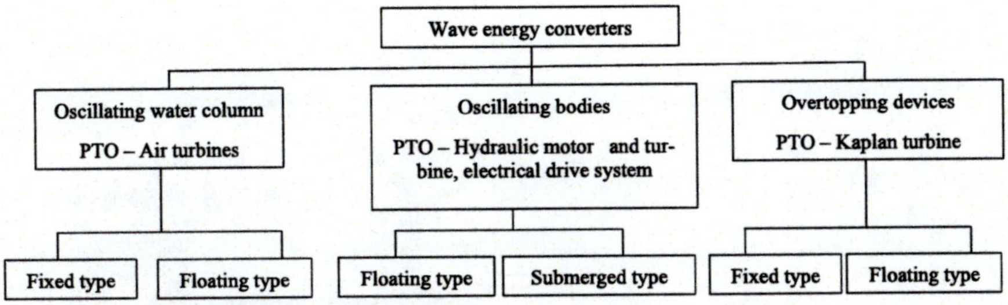

Fig. 14.1 Classification of wave energy converters. Reproduced from [3]

holds a vast amount of unexploited green energy, and it is adequate to meet our continuously increasing energy demands. Presently, ocean energy extraction occurs primarily through tidal power plants and wave energy conversion devices. The tidal energy is closer to commercial installation in contrast to the wave energy devices that are still in their development phases due to immature technology. On the other hand, wave energy has advantages such as high energy density in contrast to wind and solar energy, less pollution, higher availability, and it can travel longer distances with lesser energy losses, etc. [2]. Nevertheless, for an economically feasible wave energy extraction, rigorous research and development are needed to overcome the current technological barriers.

The annual mean power density along the Indian coastal areas is 80–90 kW/m, and the total wave energy potential is about 40 GW [4]. As explained earlier, despite its vast availability, energy production through waves is insignificant to date. Moreover, several demonstration plants and operational plants of wave energy converters (WEC) are installed in various countries such as Belgium, Canada, China, Denmark, India, Italy, Korea, Portugal, Spain, Sweden, and the United States of America [5]. There are various types of WECs available to harvest wave energy, and they are classified according to their operating principle (Fig. 14.1). Additionally, they use different power take-off (PTO) devices to convert the transformed energy from the primary converter to the electrical energy [6]. Among the different types of WECs available, the oscillating water column (OWC) is the common and actively researched WEC. The working principle of the OWC is discussed in the section below.

14.2 Oscillating Water Column

There are various types of WECs available to harvest wave energy; among them, the oscillating water column (OWC) is the common and actively researched WEC. The OWC with an air turbine for wave energy conversion is shown in Fig. 14.2. It consists of a partially submerged concrete structure with an opening below the water surface, which creates an air chamber. The incoming wave compresses and decompresses the entrapped air by an oscillating motion, and the air turbine inside the

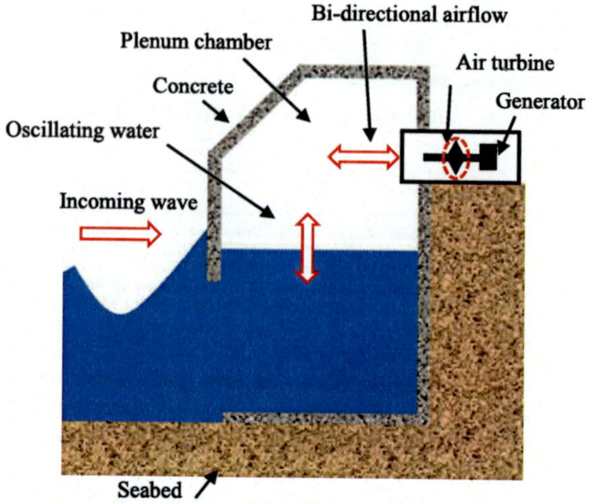

Fig. 14.2 Oscillating water column with air turbine for wave energy conversion [7]

OWC extracts energy from the bi-directional airflow. The OWC has many advantages such as relatively higher mean capture width ratio, easy maintenance, effective sea space usage, less moving parts, and the turbine placed above the seawater eliminates corrosion [8]. Additionally, the OWC can be integrated into the existing breakwater to reduce the cost of wave energy plants and can be modified to operate as a floating type in contrast to the fixed type structure.

14.3 Air Turbines for Wave Energy Conversion

The various types of air turbines used to extract wave energy in the OWC are listed below [9].

- Wells turbine
- Impulse turbine—Axial and radial flow types
- Dennis-Auld turbine
- HydroAir turbine

Of the above-mentioned turbines, the Wells and the axial impulse turbines have been studied extensively by various researchers. Recently, the studies on radial impulse turbines are gaining momentum, and various optimization works based on them were reported [10–12]. On the other hand, there is only less information is available on Dennis-Auld and HydroAir turbine in the open literature. Furthermore, this chapter deals with the passive flow control methods implemented in the Wells and the axial impulse turbines.

14.3.1 Wells Turbine

Wells turbine, a self-rectifying axial flow reaction turbine, is proposed by Dr. A. A. Wells at Queens University of Belfast to harvest wave energy from the OWC device. It comprises untwisted blades with NACA 00XX series symmetrical airfoils placed perpendicular to the incoming flow. It is insensitive to the incoming flow direction and produces unidirectional torque, owing to its blade symmetry. Figure 14.3 shows the velocity triangle for the Wells turbine experiencing bidirectional flow. The turbine blade rotates with peripheral velocity (U_T), whereas the incoming absolute and relative velocity are given by V_1 and W_1. Also, subscripts 1 and 2 imply the inlet and exit of the flow.

When the blade is stationary, the incoming flow hits the blade at a 90° angle of attack, which seems an impractical way to extract energy. However, when the blade is in rotation, the relative flow makes an angle of attack (AOA) in the range that is acceptable to produce lift force. The lift (L) and drag force (D) are generated perpendicular to the flow and along the flow direction, respectively. By resolving these force components, the total force acting on the axial and the tangential direction can be determined (Fig. 14.3). The same process is repeated for the decelerating flow, and it can be seen that the tangential force (F_T) acts in the same direction irrespective of flow direction, in contrast, the axial force (F_A) changes its orientation based on incoming flow direction. The increased F_T improves the turbine power output, whereas increased F_A reduces efficiency because of the higher pressure drop. The lift produced increases with an increase in AOA up to a critical value and drops rapidly

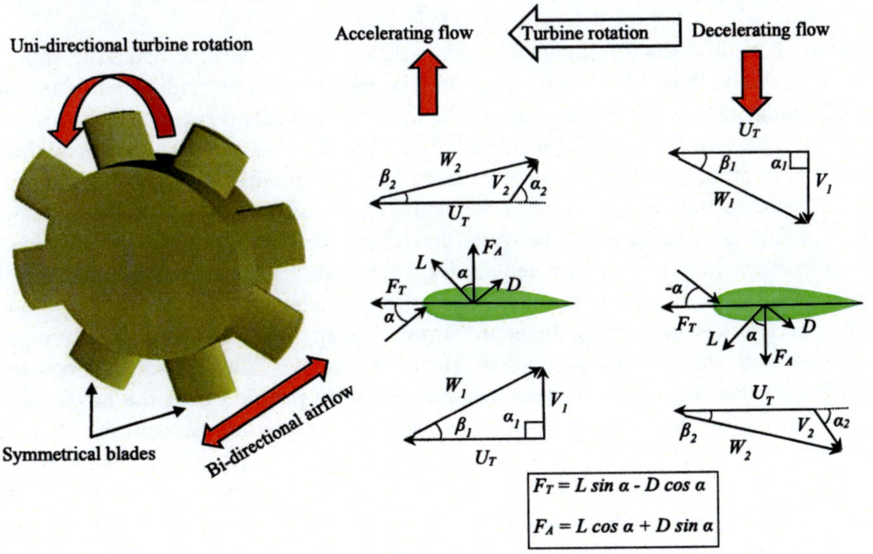

Fig. 14.3 Velocity triangle for Wells turbine in the bidirectional flow. Reproduced from [9] [13)

due to stalls caused by flow separation. The stall phenomenon reduces the operating range of the Wells turbine.

14.3.2 Axial Impulse Turbine

The axial impulse turbine offers solutions to various limitations of Wells turbine such as noisy operation, self-starting characteristics, and narrow operating range, etc. The axial impulse turbine consists of upstream and downstream guide vanes to guide the airflow to the rotor blade (Fig. 14.4). The downstream guide vane is kept as a mirror image of the upstream guide vane to maintain symmetry for the bi-directional flow. In contrast to the Wells turbine, the impulse turbine does not suffer from stall phenomenon due to flow separation and has a wider operating range. However, the aerodynamic efficiency of the axial impulse turbine is relatively less compared to the Wells turbine, mainly due to the losses at the entry of downstream guide vanes [14]. Luo et al. [15] discussed the various design parameters that affect the performance of the impulse turbine. Moreover, a detailed review of the numerical investigations performed on Wells and the axial impulse turbines is provided in the study of Cui et al. [16]. The passive flow control methods implemented in the Wells and the axial impulse turbines for performance enhancement are discussed below.

14.4 Flow Control Methods

Ludwig Prandtl is the forerunner of modern flow control, who elucidated the importance of the boundary layer and proposed various methods to modify the nature of the boundary layer, thereby managing the flow. There are two types of flow control methods, passive and active flow control (Fig. 14.5). The passive flow control method involves altering the geometry of an existing design to modify the flow behavior and thereby enhancing the performance. In contrast to the active flow control, the passive flow control devices cannot be activated and deactivated when required and continue to exist even when they are not required. On the other hand, the active flow control method requires energy input such as suction, blowing, etc., to control the flow. The passive flow control methods are simple, cheap, and require no energy input in contrast to the active flow control. Hence, the active flow control methods are preferred when their aerodynamic benefit margin is justified over the advantages of using passive flow control types [17]. Moreover, a detailed account of different methods to control flow separation is reported by Gad-el-Hak [18]. Owing to its simplicity and easy installation, various passive flow control methods are implemented in the Wells and the axial impulse turbines. The major objective is to enhance the operating range and efficiency in the Wells and the axial impulse turbines, respectively. They are discussed in the subsequent sections listed below.

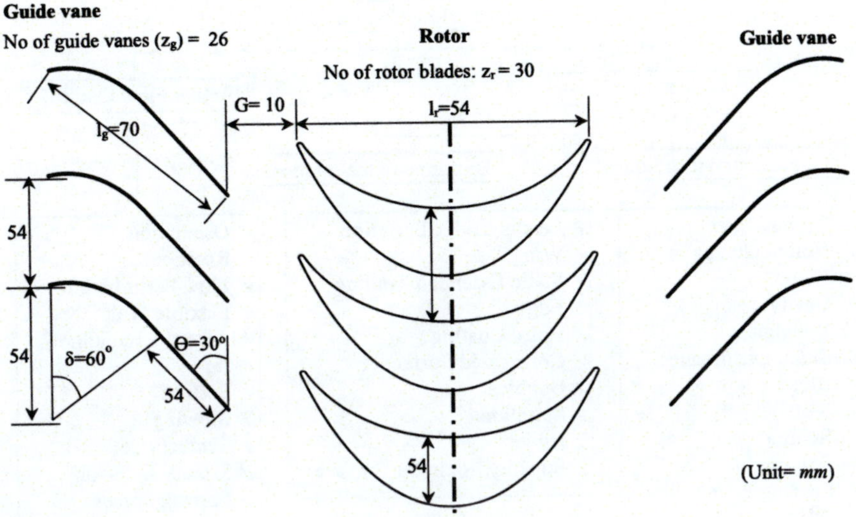

(a) 2-D schematic profile

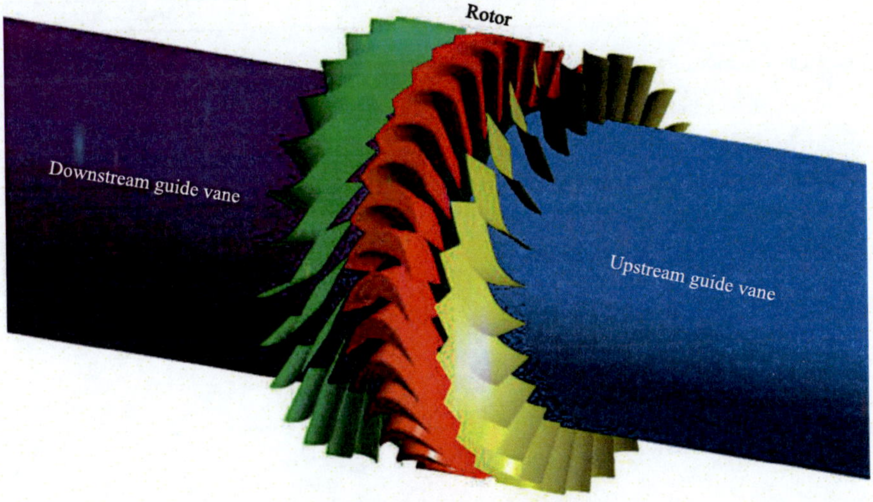

(b) 3-D Blade profile

Fig. 14.4 Axial impulse turbine geometry

14.5 Passive Flow Control Methods in Wells Turbine

The Wells turbine is an axial-flow reaction turbine that suffers from stall phenomenon due to flow separation at a higher flow rate. It is a well-researched area, and various

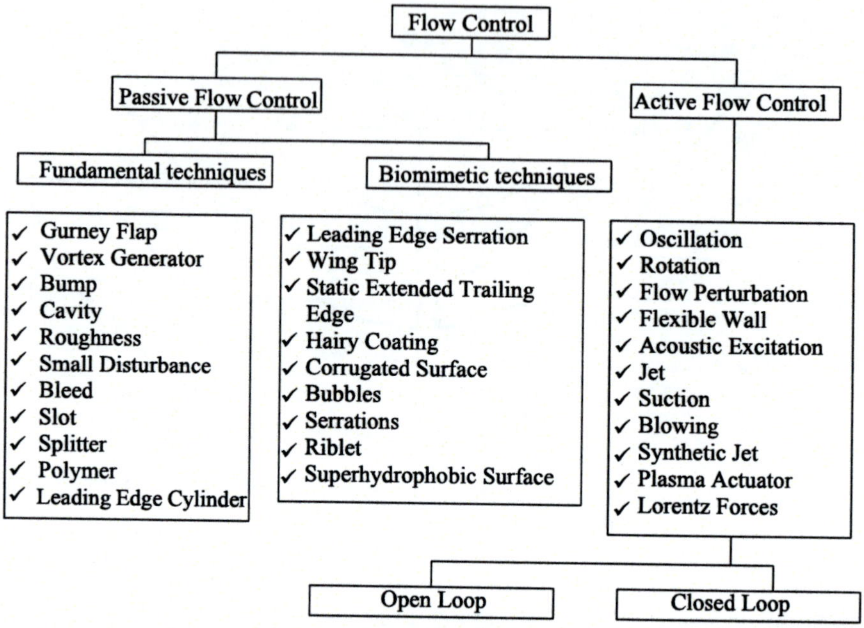

Fig. 14.5 Classification of flow control methods. Reproduce from [19]

researchers proposed different passive flow control methods to delay stall inception, thereby widening the operating range and enhancing efficiency. Owing to its operating conditions in the bidirectional flow, the blade symmetry of the Wells turbine should be preserved. Hence, the passive flow control methods have to be implemented in such a way that the blade symmetry is maintained. Some of the most prominent passive flow control methods applied on the Wells turbine to improve its performance are discussed below.

14.5.1 Blade Sweep

The blade sweep is one of the commonly applied design modifications in the Wells turbine to augment its performance. Webster and Gato [20] experimentally investigated the effect of the backward swept blade and reported enhanced operating range and delayed stall. The blade with a back sweep improves the stability of the variable pitch control operation by offsetting the blade pressure center [21]. Kim et al. [22] recommended the optimum blade sweep ratio as 0.35 for the NACA 0021 airfoil profile. Starzmann and Carolus [23] studied the Wells turbine with blade skew (sweep and dihedral) and reported that the skewed blades delayed stall and reduced the operating noise. The swept blade design used in the numerical study of Halder et al. [24] is shown in Fig. 14.6. They optimized the midspan and hub blade sweep angle and

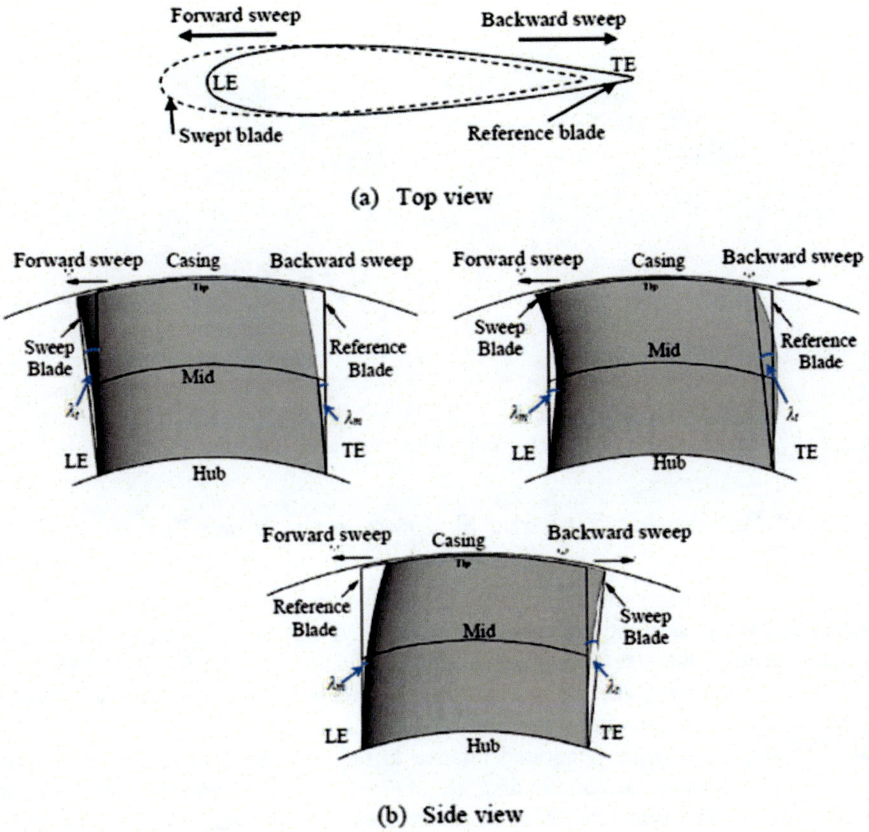

Fig. 14.6 Wells turbine blade with sweep [24]

reported that the optimized design enhanced the power and operating range by 29% and 18%, respectively.

14.5.2 Blade Setting Angle

The airflow data inside the Indian OWC plants showed that the outgoing flow velocity is higher than the incoming flow velocity [25], and it is asymmetrical. As the magnitude of the outgoing flow is relatively higher than the incoming flow, it indicates that the outgoing flow has more energy compared to the incoming flow, and with a proper blade modification, the turbine performance can be improved. Setoguchi et al. [25] proposed a Wells turbine with a non-zero blade setting angle (Fig. 14.7) to enhance its performance under asymmetrical airflow. They concluded that the 2° blade setting angle is the optimum value.

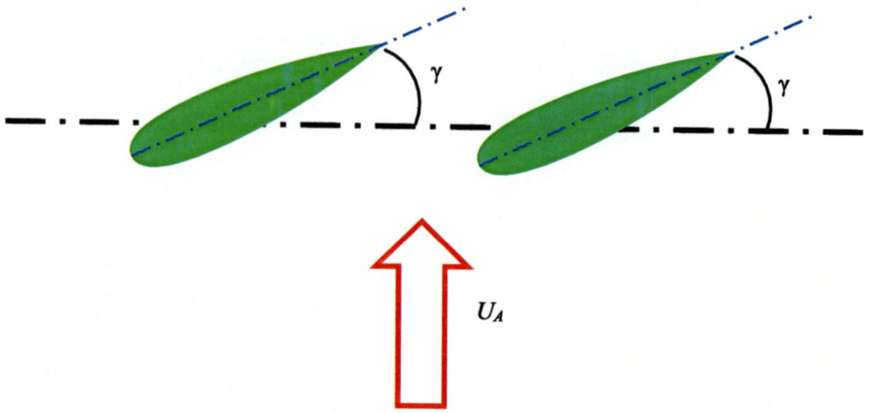

Fig. 14.7 Non-zero setting angle Wells turbine blade. Reproduced from [25]

14.5.3 Penetrating Ring and Endplate at the Blade Tip

Takao et al. [26] proposed end plates near the blade tip to mitigate the tip leakage losses and to increase the efficiency. They reported that performance enhancement is dependent on the size and location of the endplate. Cui and Hyun [27] studied the effect of various penetrating blade tip treatments for Wells turbine with fixed guide vanes and reported that the ring type Wells turbine performs better in terms of efficiency and torque generated relative to the reference and endplate model. Figure 14.8 shows the schematic diagram of the penetrating endplate and ring-type Wells turbine. The ring-type blade augmented the relative peak efficiency by 16% compared to the reference blade. Interestingly, in the work of Cui et al. [28], they reported that the endplates in Wells turbine without guide vanes outperformed the ring-type blade. It reduced the area of the separated flow on the blade suction side and enhanced the torque and operating range. The blade with an endplate enhanced the peak efficiency by 11% under steady-state flow conditions.

14.5.4 Non-uniform Tip Clearance

Watterson and Raghunathan [29] studied the effects of tip clearance on the Wells turbine performance. They reported that increased clearance reduced the efficiency and delayed stall, whereas the reduced clearance increased efficiency and advanced stall. The blade tip clearance plays a significant role in the stalling of the Wells turbine. Furthermore, they recommended that the tip clearance should not be higher than 2% of chord length (C) to avoid performance deterioration. The performance of the Wells turbine is highly dependent on tip clearance and tip leakage flow characteristics [30, 31]. Takao et al. [32] implemented non-uniform tip clearance such that the clearance

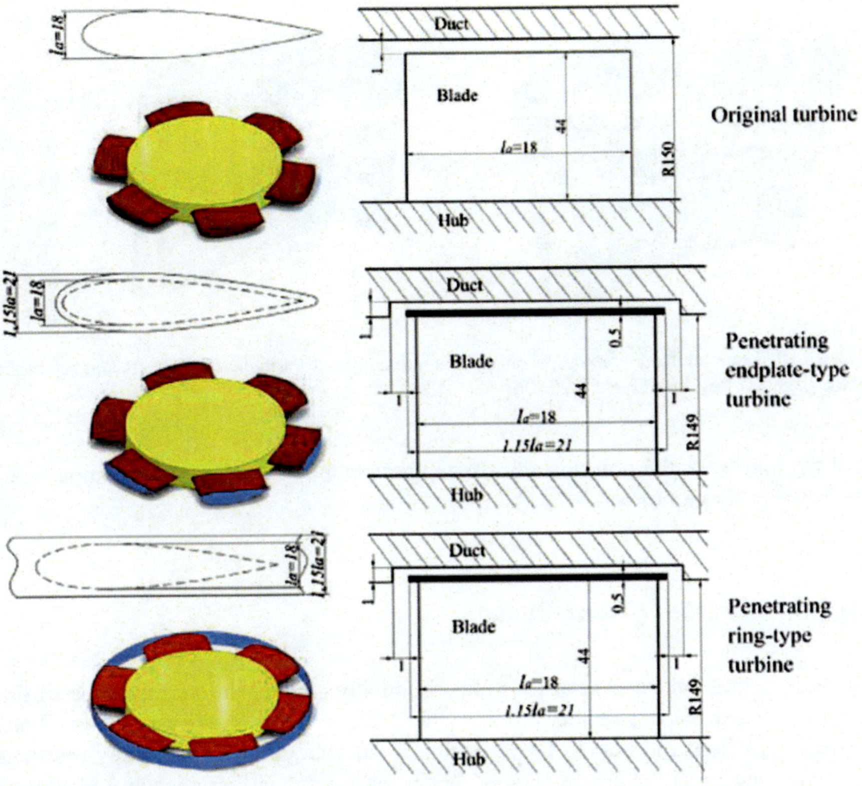

Fig. 14.8 Wells turbine blade with penetrating end plate and ring type turbine [27]

in the leading edge, mid chord, and trailing edge is 0.5 mm, 0.75 mm, and 1 mm, respectively. Figure 14.9 shows the implementation of non-uniform tip clearance in the Wells turbine. The non-uniform tip clearance blade enhanced the efficiency by 2% compared to the constant tip clearance blade. Likewise, Taha et al. [33] reported

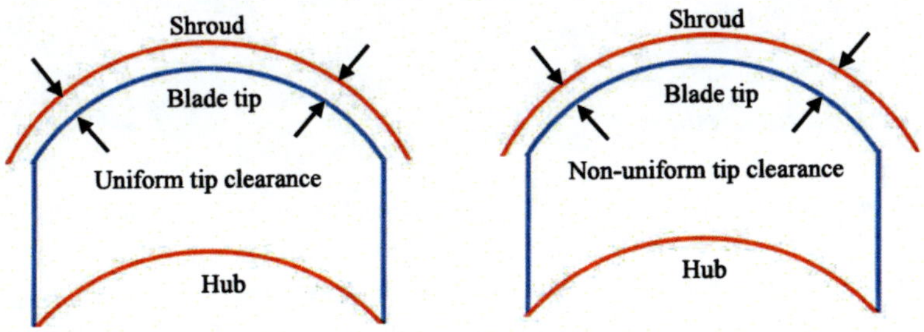

Fig. 14.9 Wells turbine blade with non-uniform tipclearance

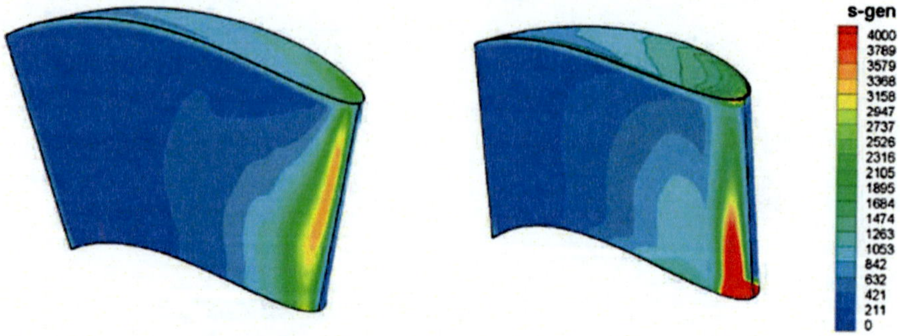

Fig. 14.10 Local entropy generation on blade suction side of variable chord (left) and reference blade (right) at flow coefficient 0.23 [36]

that the rotor with the non-uniform tip gap performed relatively better compared to the uniform tip gap rotor.

14.5.5 Variable Chord Blade

Govardhan and Dhanasekaran [34] experimentally studied the performance of the constant chord blade and the variable chord blade with guide vanes. The chord is increased from the hub to tip to design a variable chord blade. They reported that the variable chord blade showed better starting characteristics and produced relatively higher torque and delayed stall in contrast to the constant chord blade [35]. Soltanmohamadi and Lakzian [36] reported that the variable chord blade reduced the axial velocity at the tip, thereby decreasing the local angle of attack. This phenomenon is responsible for the delayed stall and improved operating range in the variable chord blade. Figure 14.10 displays the local entropy generation on the blade suction side for the variable and constant chord blade. It shows the entropy generation is relatively lesser in variable chord blade compared to the constant chord blade, and a relative decrement of 26,02% entropy is obtained in the variable chord blade case for the entire flow range.

14.5.6 Casing Groove

The casing groove is widely adopted in turbomachinery to improve performance. Halder et al. [37] proposed the implementation of a circumferential casing groove (Fig. 14.11) in the Wells turbine and studied its performance. The casing groove with 3% C depth produced an enhanced operating range, and the relative peak torque

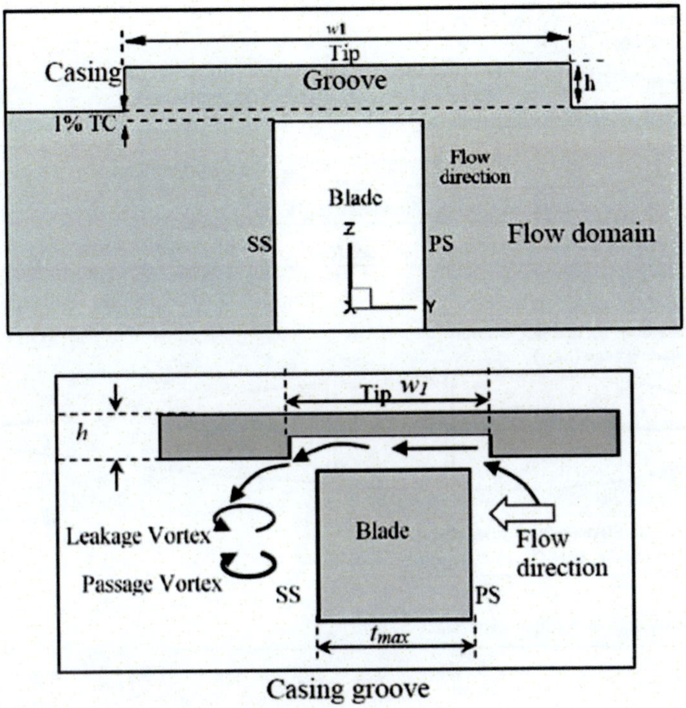

Fig. 14.11 Wells turbine blade with casing groove [37]

increased by 26%. They concluded that the tip vortices interacted weakly with the mainstream flow and delayed stall.

14.5.7 *Suction Slots*

The airfoil with a suction slot (Fig. 14.12) to enhance the performance is proposed by Shehata et al. [38]. They studied the effect of suction slots in 2-D NACA 0015 airfoil in a sinusoidal inlet velocity using the large eddy simulation (LES) method to replicate the working of Wells turbine, and they analyzed the entropy generation nature of the suction slots and reported enhanced torque and delayed stall. Shehata et al. [39] investigated a different variant of suction slots and reported the optimum position for the suction slots implemented. Shehata et al. [40] numerically examined the optimum angle for the suction slot and reported that a suction slot with a 10° angle in a clockwise direction enhanced the relative power by 21% in the unstalled regime and 44% in the stalled regime.

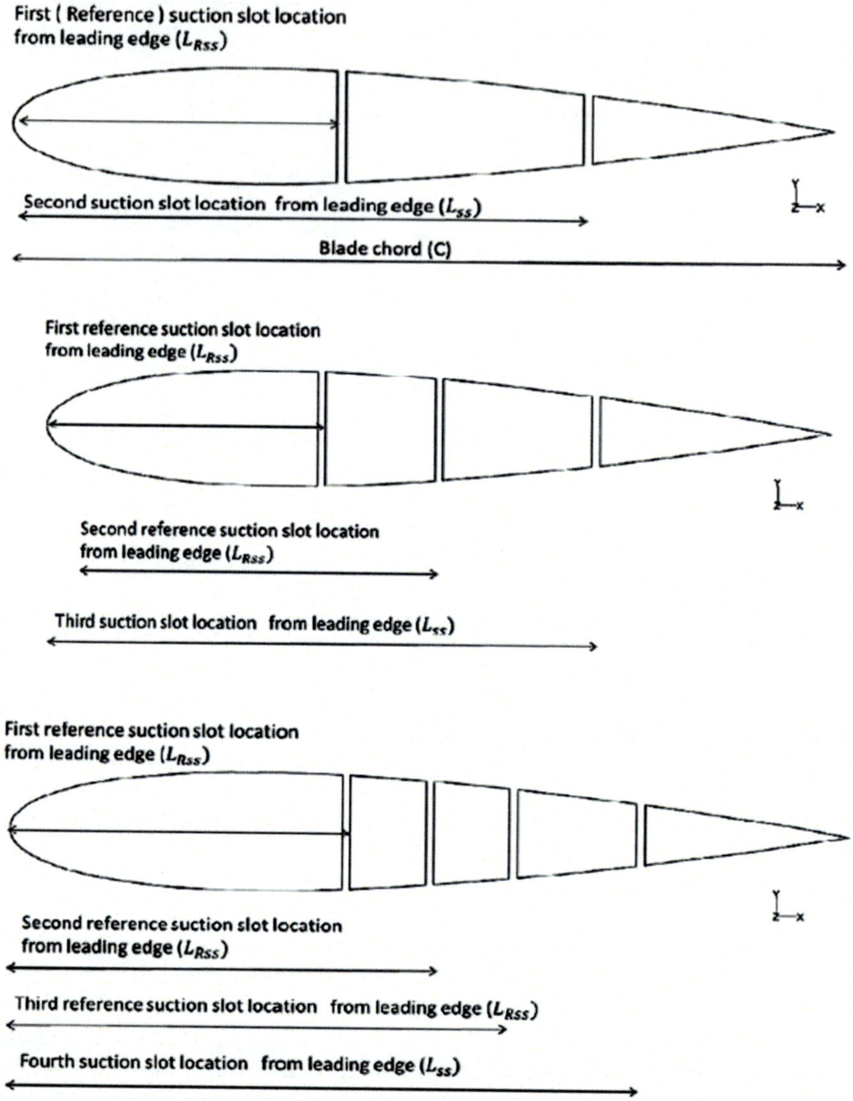

Fig. 14.12 Airfoils with different number of suction slots [39]

14.5.8 Variable Thickness Blade

Nazeryan and Lakzian [41] studied the effect of variable thickness rotor on the Wells turbine performance. Figure 14.13 illustrates the variable thickness blade used in the study. It comprises NACA 0010, NACA 0015, NACA 0020 profiles along with the hub, midspan, and tip, respectively. The entropy generation analysis indicated that the

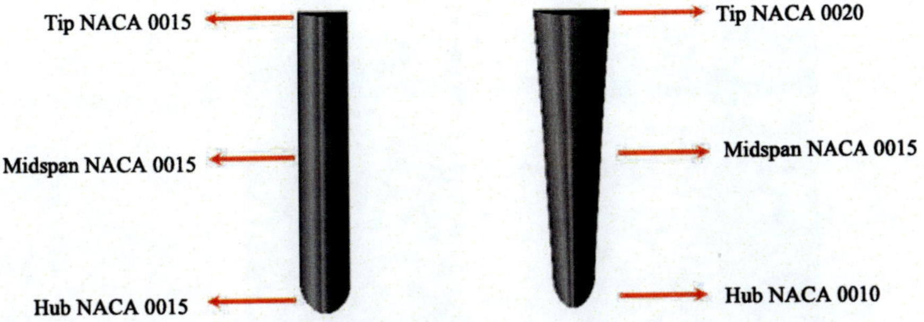

Fig. 14.13 Wells turbine blade with constant (left) and variable thickness (right). Reproduced from [41]

variable thickness blade minimized the mixing between the tip leakage flow and the primary flow, thus delaying stall condition. It augmented the torque and efficiency by 63.3% and 72.8%, respectively, in the post-stall regime. They also reported a relative decrement of 28.9% in the global entropy generation compared to the constant thickness blade in the post-stall regime. The variable thickness blade relatively enhanced the operating range and torque at higher flow coefficients compared to the constant thickness blade.

14.5.9 Leading-Edge Undulation

Various studies have shown that the leading-edge blade undulations inspired from the flippers of humpback whales improve the performance by suppressing the flow separation and delaying the stall inception [42–45]. Kumar et al. [46] introduced the concept of leading-edge undulation to the Wells turbine blade and numerically investigated the undulated blade with three amplitudes such as 1 mm, 2.5 mm, and 4 mm, respectively. Figure 14.14 displays the amplitude and wavelength description of the leading-edge undulated blade. The value of wavelength is fixed as 25 mm in their study, as it has a negligible effect on the performance of the blade [45]. The undulated blades generated streamwise vortices and modified the flow pattern on the blade suction side, thereby increasing the performance. They concluded that the undulated blade with amplitude 2.5 mm delayed stall and enhanced the relative operating range by 14.28%.

14.5.10 Radiused Edge Tip Blade

Several studies have shown that the performance of the Wells turbine depends principally on the tip leakage flow characteristics [29–32]. Kumar et al. [47] proposed a

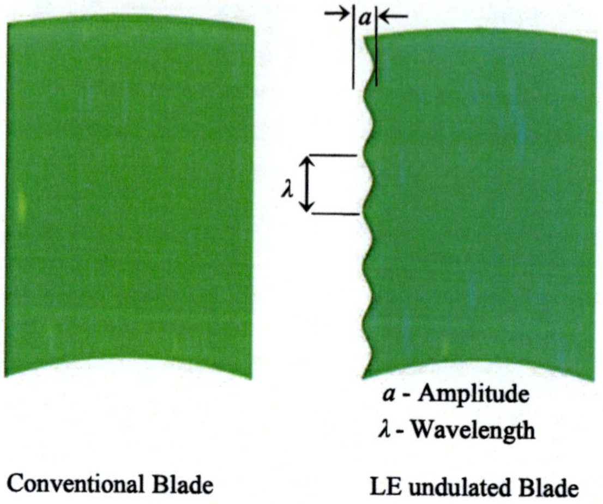

Fig. 14.14 Conventional (left) and LE undulated (right) Wells turbine blade

radiused edge tip blade (RETB) to alter the tip leakage flow features. Figure 14.15 displays the schematic of the sharp edge and the radiused edge tip blade. The fillet was created between the blade tip and span with a varying radius such as 1%, 2%, and 3% chord length (C), and their effect on performance on Wells turbine performance was studied. They concluded that the RETBs reduced the blade tip loading and weakened the interaction between the tip leakage flow and the primary flow. Owing to the reduced blade tip loading, the peak torque coefficient is shifted towards

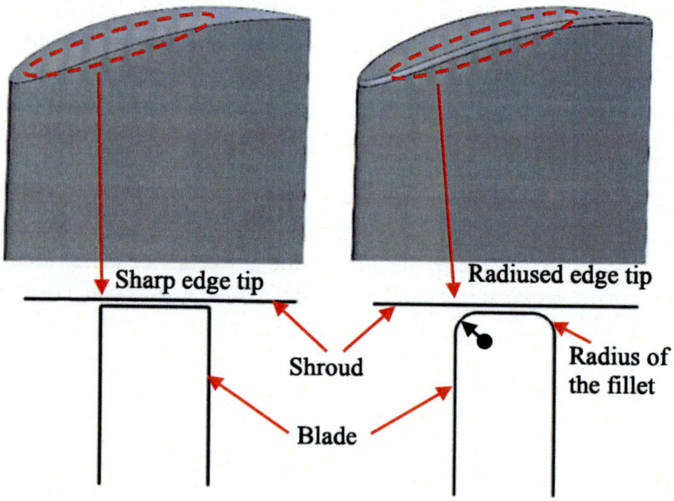

Fig. 14.15 Sharp edge tip and radiused edge tip blade [47]

a higher flow coefficient, and the stall inception was delayed. They reported that the relative peak torque and the operating range were enhanced by 37% and 22%, respectively.

14.5.11 Static Extended Trailing Edge (SETE)

Similar to the bio-mimicking inspiration of the leading-edge undulated blade, the concept of a static extended trailing edge was adopted from the wings of owl and merganser that consists of an extended trailing edge (Fig. 14.16). Kumar and Samad [48] analyzed blades with various extended lengths from 1 to 10% C with a fixed thickness of 0.25 mm. The SETE blades modified the pressure distribution across the blade and enhanced the torque generated. The torque enhancement increased with an increase in extended length up to 5% C, and for a length greater than 5% C, stall is advanced, and the operating range is reduced. The 5% C blade showed an improvement of 23.4% with an efficiency decrement of 5.4% because of the increased pressure drop. Additionally, they compared the performance of 5% C extended blade length with an increased chord length blade and concluded that the SETE blade performed better than the increased chord length blade.

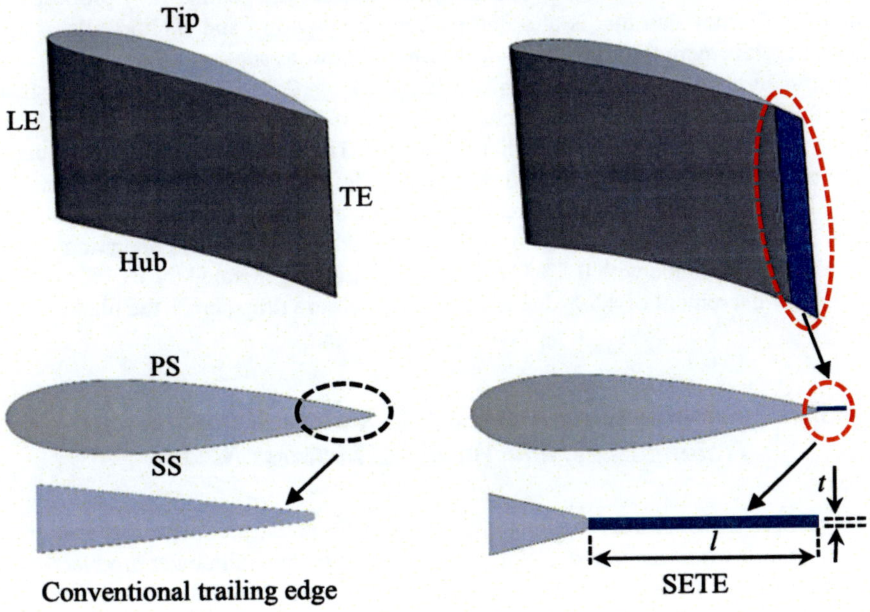

Fig. 14.16 Reference (left) and SETE (right) blade [48]

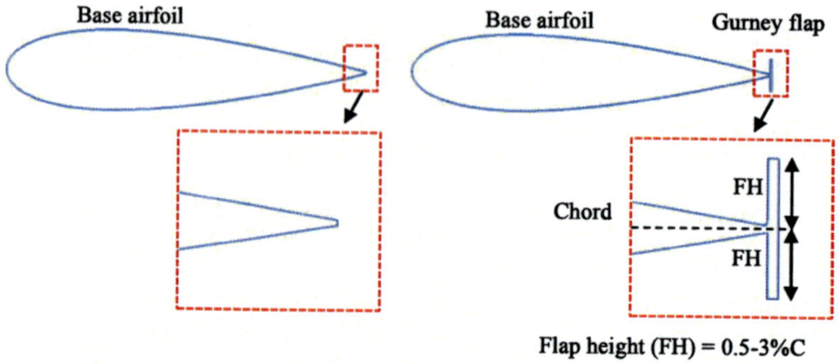

Fig. 14.17 Reference airfoil (left) and airfoil with Gurney flap on both sides (right) [49]

14.5.12 Gurney Flap

The Gurney flap is a commonly used passive flow control device to augment lift generation. It is a thin plate placed perpendicular to the pressure side of the blade trailing edge. Owing to its simplicity and efficacy, it finds applications in various fields such as aircraft, wind turbines, pumps, helicopter rotors, unmanned aerial vehicles, and axial fans, etc. Convectional Gurney flap design eliminates the blade symmetry of Wells turbine. To overcome this problem, Kumar and Samad [49] proposed a modified Gurney flap that is attached on both the pressure and suction side of the blade to retain symmetry for the bi-directional airflow, as shown in Fig. 14.17. They numerically studied the blade with flap heights from 0.5 to 3% C with a constant flap thickness of 0.5% C and concluded that the blades with Gurney modified the blade pressure distribution and enhanced torque. The Gurney flap blades produced counter-rotating vortex pair behind the blade trailing edge that modified the trailing edge Kutta condition, thereby increasing the circulation and lift. Also, for the blades with flap height greater than 1.5% C, the stall is advanced, and the operating range is reduced. The blade with 0.5% C enhanced the torque generated by 10.7% with an efficiency decrement of 4.7% due to increased pressure drop across the blade.

14.5.13 Combined Radiused Edge Tip, Static Extended Trailing Edge, and Variable Thickness Blade

The passive flow controls methods discussed so far enhanced either the operating range or the torque generated. However, to increase the overall power produced, the simultaneous enhancement of both the torque improvement throughout the flow range and a wider operating range is required. Kumar et al. [50] proposed and analyzed a combined design of various passive flow control methods such as a radiused edge tip, static extended trailing edge, and variable thickness blade to improve the operating

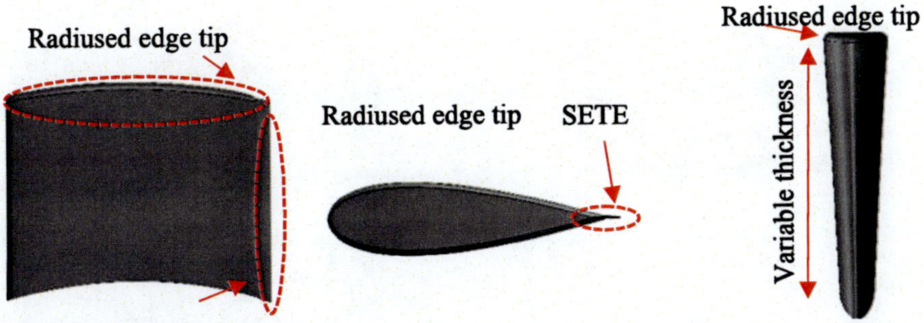

Fig. 14.18 Combined design Wells turbine blade with a radiused edge tip, static extended trailing edge, and variable thickness blade [50]

range and torque (Fig. 14.18). The radiused edge tip and variable thickness blade delayed the stall inception, whereas the static extended trailing edge is a lift enhancement device. They numerically investigated various design combinations and later concluded that the combined design with all three flow control devices performed better and enhanced the torque and operating range by 97% and 22%, respectively. They concluded that the combined design blade weakened the tip leakage and mainstream flow interaction due to reduced tip loading because of the radiused edge tip and enhanced the operating range. Moreover, the modified blade pressure distribution due to static extended trailing edge is responsible for the enhanced torque.

14.6 Passive Flow Control Methods in Axial Impulse Turbine

In contrast to the Wells turbine, the axial impulse turbine exhibits a wider operating range; however, the reduced aerodynamic efficiency is a prominent drawback of the axial impulse turbine. Various researchers studied several passive flow control methods to improve its efficiency; however, the impulse turbine has relatively lesser research works compared to the Wells turbine. The passive flow control works reported on the axial impulse turbine are discussed below.

14.6.1 Endplates

The endplates are beneficial to both Wells and impulse turbines as they modify the behavior of the tip leakage flow and enhance their performance. Figure 14.19 illustrates the endplate concept implemented in the impulse turbine. The concept of endplates in impulse turbine was introduced by Hyun et al. [51] and extended

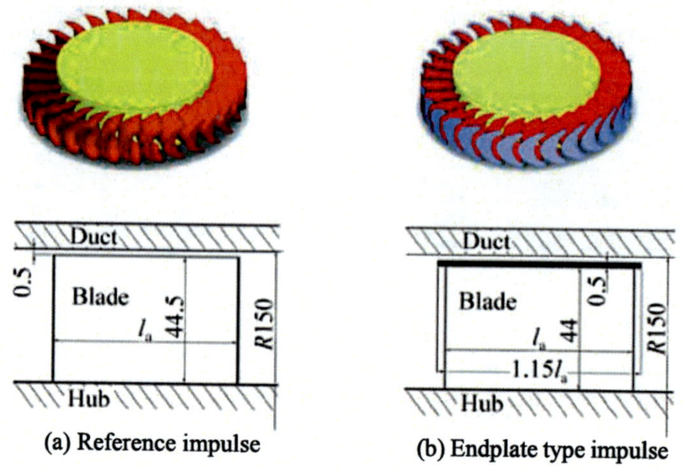

Fig. 14.19 Reference blade (left) and blade with endplate (right) [28]

by authors such as Takao et al. [52], Setoguchi et al. [53], and Cui et al. [28]. The endplates were proposed to minimize the adverse effects of tip leakage flow. They reported that the input coefficient was reduced in the blades with an endplate, and the efficiency was enhanced (up to 5%). They also investigated the endplates with different thickness, chord length, and eccentricity; besides they concluded that the thickness and chord length show minor influence on the blade performance. Additionally, they reported that the concentric endplates are superior to the eccentric endplates.

14.6.2 Blade Setting Angle

Liu et al. [54] extended the work of Setoguchi et al. [25] and implemented the impulse turbine blade with a set angle for the asymmetrical airflow to enhance its performance. The blade and guide vane with a non-zero setting angle are shown in Fig. 14.20. The blade offset increased the rotor blade pressure side area and helped in better energy extraction energy from the higher velocity available in the exhaled air flow. They varied the blade and guide vane setting angle from −5° to 7.5° and numerically investigated its effect on the impulse turbine performance. The blade with a set angle of 5° enhanced the relative mean efficiency by 9.2% at a higher flow coefficient.

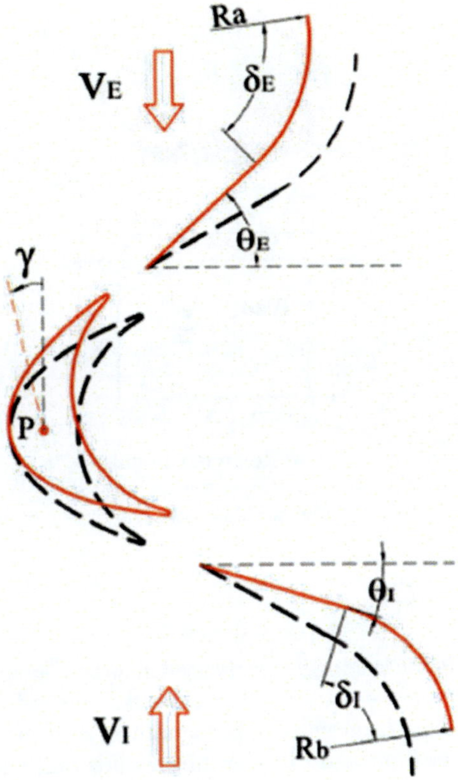

Fig. 14.20 Blade and guide vane setting angle in impulse turbine [54]

14.6.3 Penetration Ring

The ring-type impulse turbine was proposed by Hyun et al. [55]; they have adopted the concept from the ducted propellers used in marine applications (Fig. 14.21). Similar to the endplates, the ring-type blade is proposed to reduce the losses caused by the tip clearance flow. They reported that the ring-type blade relatively enhanced the efficiency up to a maximum value of 10% by minimizing the pressure drop. Likewise, Cui et al. [28] studied the ring-type impulse turbine blade with endplates and reported that the ring-type blade performs relatively better compared to the blades with an endplate. They reported that the ring-type blade relatively enhanced the peak efficiency by 4%.

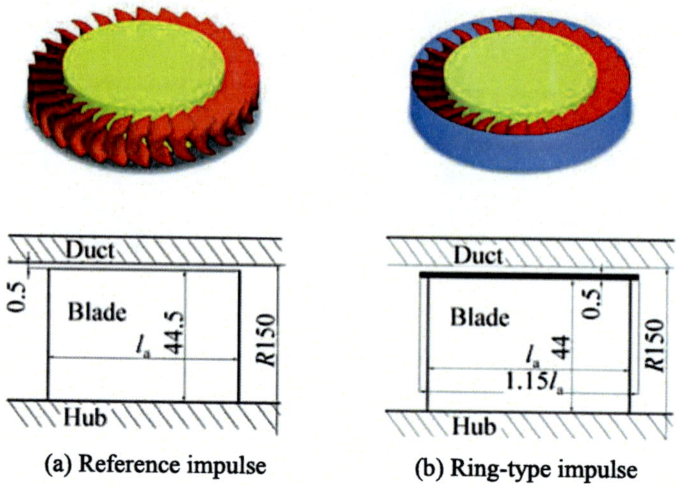

Fig. 14.21 Reference blade (left) and ring-type blade (right) [28]

14.6.4 Leaned Blade

Blade leaning is widely used in turbomachines to augment the performance by modifying the blade loading characteristics and. The concept of blade leaning in an axial impulse turbine was introduced by Ranjith et al. [56]. Figure 14.22 shows the 3D and the front view of the upstream guide vane with lean. The lean was implemented in the rotor blade and guide vanes in such a way that the blade symmetry is preserved for the bi-directional airflow. They investigated the effect of the combined rotor blade and

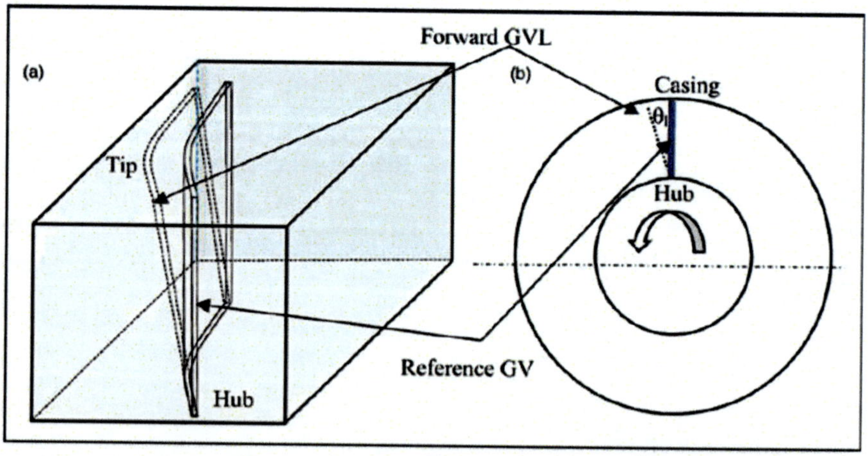

Fig. 14.22 Leaned guide vane 3D view (left) and front view of upstream guide vane (right) [56]

guide vane lean and reported that the efficiency is enhanced and the optimum guide vane lean angle is 10°. They concluded that the lean regulated the change in relative incidence angle at the blade leading edge and reduced the downstream recirculation region, thereby increasing the efficiency. The guide vane leans provided a higher efficiency increment compare to the rotor blade lean; however, it reduced the torque generated because of a decrease in the tangential velocity. The combined rotor blade and guide vane lean enhanced the relative efficiency by 3.4%, with an insignificant decrement in torque generated.

14.7 Conclusions

The present chapter broadly explores the passive flow control methods employed by various researchers to enhance the performance of air turbines used in wave energy conversion. The studies included are limited to the most commonly used air turbines in the oscillating water column, such as the Wells and the axial impulse turbines. The passive flow control methods were aimed at reducing the limitations of the above-mentioned turbine, such as narrow operating range, reduced efficiency, and lesser torque, etc., and thereby increasing the overall performance of the system. Among the several passive flow methods implemented in the Wells turbine, the enhancement of operating range served as the primary objective. Likewise, the augmentation of aerodynamic efficiency has been given more importance in the case of axial impulse turbines. Furthermore, the prominent conclusions from this detailed review were listed below.

- The important consideration in the implementation of the passive flow control methods in the air turbines is preserving the blade symmetry so that the turbine is suitable for the bi-directional airflow in the oscillating water column.
- The Wells turbine is a reaction turbine that has been realized with a higher number of passive flow control devices to enhance its performance compared to the axial impulse turbine.
- The characteristic of tip leakage flow is a critical parameter that is related to the Wells turbine stall, and several authors like Taha et al. [33], Halder et al. [37], Cui et al. [27], Kumar et al. [7] delayed the stall and enhanced the operating range by modifying the tip leakage flow behavior.
- Moreover, to achieve an overall better performance, the individual passive flow control methods alone are not sufficient. A combined study of passive flow control devices such as variable thickness blade, radiused edge tip, and static extended trailing edge provided simultaneous improvement of torque and operating range [50]. Likewise, a study of compatible passive flow control methods such as casing groove, blade sweep, etc., could provide a better turbine design for wave energy conversion.

- Furthermore, the implementation of passive flow control methods in the axial impulse turbines is still in infancy, and a dedicated research is recommended for their performance enhancement.

References

1. Renewables. (2019). *Global status report REN21*. REN21 Secretariat
2. Drew, B., Plummer, A., & Sahinkaya, M. N. (2009). A review of wave energy converter technology. *Journal of Power and Energy, 223*, 887–902. https://doi.org/10.1243/09576509JPE782
3. Falcão, A. F. O. (2010). Wave energy utilization: A review of the technologies. *Renewable and Sustainable Energy Reviews, 14*, 899–918https://doi.org/10.1016/j.rser.2009.11.003
4. Sannasiraj, S. A., & Sundar, V. (2016). Assessment of wave energy potential and its harvesting approach along the Indian coast. *Renewable Energy, 99*, 398–409. https://doi.org/10.1016/j.renene.2016.07.017.
5. Brito, E., Melo, A., Villate, J. L. (Eds.). (2017). *Annual report 2016: Ocean energy systems*. IEA-OES.
6. Pecher, A., & Kofoed, J. P. (Eds.). (2017). *Handbook of ocean wave energy, Ocean engineering & oceanography*. https://doi.org/10.1007/978-3-319-39889-1_8
7. Kumar, P. M., Halder, P., & Samad, A. (2021). Radiused Edge Blade Tip for a Wider Operating Range in Wells Turbine. *Arabian Journal for Science and Engineering, 46*(3), 2663–2676 https://doi.org/10.1007/s13369-020-05185-z
8. Heath, T. V. (2012). A review of oscillating water columns. *Philosophical Transactions of the Royal Society A: Mathematical, Physical and Engineering Sciences, 370*(1959), 235–245. https://doi.org/10.1098/rsta.2011.0164
9. Falcão, A. F. O., & Henriques, J. C. C. (2016). Oscillating-water-column wave energy converters and air turbines: A review. *Renewable and Sustainable Energy*, 85, 1391–1424. https://doi.org/10.1016/j.renene.2015.07.086
10. Ansarifard, N., Fleming, A., Henderson, A., Kianejad, S. S., & Chai, S. (2018). *Optimization study on the downstream section of a radial inflow turbine*. AWTEC 2018 Proc. 311.
11. Ansarifard, N., Fleming, A., Henderson, A., Kianejad, S. S., & Chai, S. (2019). Design optimisation of a unidirectional centrifugal radial-air-turbine for application in OWC wave energy converters. *Energies, 12*, 2791.
12. Ansarifard, N., Kianejad, S. S., Fleming, A., Henderson, A., & Chai, S. (2020). Design optimization of a purely radial turbine for operation in the inhalation mode of an oscillating water column. *Renewable Energy, 152*, 540–556.
13. Dixon, S. L., & Hall, C. A. (2010). Fluid mechanics and thermodynamics of turbomachinery (6th ed.). Butterworth–Heinemann.
14. Falcão, A. F., Henriques, J. C., & Gato, L. M. (2018). Self-rectifying air turbines for wave energy conversion: A comparative analysis. *Renewable and Sustainable Energy Reviews, 91*, 1231–1241.
15. Luo, Y., Presas, A., & Wang, Z. (2019). Numerical analysis of the influence of design parameters on the efficiency of an OWC axial impulse turbine for wave energy conversion. *Energies, 12*(5), 939.
16. Cui, Y., Liu, Z., Zhang, X., & Xu, C. (2019). Review of CFD studies on axial-flow self-rectifying turbines for OWC wave energy conversion. *Ocean Engineering, 175*, 80–102.
17. Greenblatt, D., Whalen, E. A., & Wygnanski, I. J. (2019). Introduction to the flow control virtual collection. *AIAA Journal, 57*, 3111–3114. https://doi.org/10.2514/1.J058507
18. Gad-el-Hak, M. (2007). *Flow control: Passive, active, and reactive flow management*. Cambridge university press.

19. Wang, J., & Feng, L. (2019). *Flow control techniques and applications*. Cambridge University Press.
20. Webster, M., & Gato, L. M. C. (1999). The Effect of rotor blade sweep on the performance of the Wells turbine. *International Journal of Offshore and Polar Engineering, 9*, 233–239.
21. Webster, M., & Gato, L. M. C. (2001). The effect of rotor blade shape on the performance of the Wells turbine. *International Journal Offshore Polar Engineering, 11*, 227–230. https://doi.org/10.1016/S0960-1481(00)00163-4
22. Kim, T. H., Setoguchi, T., Kaneko, K., & Raghunathan, S. (2002). Numerical investigation on the effect of blade sweep on the performance of Wells turbine. *Renewable Energy, 25*, 235–248. https://doi.org/10.1016/S0960-1481(00)00210-X
23. Starzmann, R., & Carolus, T. (2013). Effect of blade skew strategies on the operating range and aeroacoustic performance of the Wells turbine. *Journal of Turbomachinery, 136*, 011003. https://doi.org/10.1115/1.4025156
24. Halder, P., Samad, A., & Thevenin, D. (2017). Improved design of a Wells turbine for higher operating range. *Renewable Energy, 106*, 122–134. https://doi.org/10.1016/j.renene.2017.01.012
25. Setoguchi, T., Santhakumar, S., Takao, M., & Kim, K. K. (2003). A modified Wells turbine for wave energy conversion. *Renewable Energy, 28*, 79–91. https://doi.org/10.1016/S0960-1481(02)00006-X
26. Takao, M., Setoguchi, T., Kinoue, Y., & Kaneko, K. (2007). Wells turbine with end plates for wave energy conversion. *Ocean Engineering, 34*, 1790–1795. https://doi.org/10.1016/j.oceaneng.2006.10.009
27. Cui, Y., & Hyun, B. S. (2016). Numerical study on Wells turbine with penetrating blade tip treatments for wave energy conversion. *International Journal of Naval Architecture and Ocean Engineering, 8*(5), 456–465. https://doi.org/10.1016/j.ijnaoe.2016.05.009
28. Cui, Y., Hyun, B. S., & Kim, K. (2017). Numerical study on air turbines with enhanced techniques for OWC wave energy conversion. *China Ocean Engineering, 31*(5), 517–527.
29. Watterson, J. K., & Raghunathan, S. (1997). Computed effects of tip clearance on Wells turbine performance. 35th Aerospace Science Meeting Exhibition, AIAA Paper 97-0994.
30. Raghunathan, S. (1995). The Wells air turbine for wave energy conversion. *Progress in Aerospace Sciences, 31*, 335–386. https://doi.org/10.1016/0376-0421(95)00001-F
31. Taha, Z., & Sawada, T. (2010). A comparison of computational and experimental results of Wells turbine performance for wave energy conversion. *Applied Ocean Research, 32*, 83–90. https://doi.org/10.1016/j.apor.2010.04.002
32. Takao, M., Setoguchi, T., Nagata, S., Toyota, K. (2008). *A study on the effects of blade profile and non-uniform tip clearance of the Wells turbine*. Proceedings ASME 27th International Conference on Offshore Mechanics Arctic. Engineering OMAE2008. ASME, pp. 625–632. https://doi.org/10.1115/omae2008-57235
33. Taha, Z., Ya, T. T., & Sawada, T. (2011). Numerical investigation on the performance of Wells turbine with non-uniform tip clearance for wave energy conversion. *Applied ocean research, 33*(4), 321–331. https://doi.org/10.1016/j.apor.2011.07.002
34. Govardhan, M., & Dhanasekaran, T. S. (2002). Effect of guide vanes on the performance of a self-rectifying air turbine with constant and variable chord rotors. *Renewable Energy, 26*(2), 201–219. https://doi.org/10.1016/S0960-1481(01)00124-0
35. Govardhan, M., & Chauhan, V. S. (2007). Numerical studies on performance improvement of self-rectifying air turbine for wave energy conversion. *Engineering Applications of Computational Fluid Mechanics, 1*(1), 57–70. https://doi.org/10.1080/19942060.2007.11015182
36. Soltanmohamadi, R., & Lakzian, E. (2016). Improved design of Wells turbine for wave energy conversion using entropy generation. *Meccanica, 51*(8), 1713–1722. https://doi.org/10.1007/s11012-015-0330-x
37. Halder, P., Samad, A., Kim, J. H., & Choi, Y. S. (2015). High performance ocean energy harvesting turbine design—A new casing treatment scheme. *Energy, 86*, 219–231. https://doi.org/10.1016/j.energy.2015.03.131

38. Shehata, A. S., Xiao, Q., Saqr, K. M., Naguib, A., & Alexander, D. (2017). Passive flow control for aerodynamic performance enhancement of airfoil with its application in Wells turbine–under oscillating flow condition. *Ocean Engineering, 136*, 31–53. https://doi.org/10.1016/j.oceaneng.2017.03.010
39. Shehata, A. S., Xiao, Q., Selim, M. M., Elbatran, A. H., & Alexander, D. (2017). Enhancement of performance of wave turbine during stall using passive flow control: First and second law analysis. *Renewable Energy, 113*, 369–392. https://doi.org/10.1016/j.renene.2017.06.008
40. Shehata, A. S., Xiao, Q., Kotb, M. A., Selim, M. M., Elbatran, A. H., & Alexander, D. (2018). Effect of passive flow control on the aerodynamic performance, entropy generation and aeroacoustic noise of axial turbines for wave energy extractor. *Ocean Engineering, 157*, 262–300. https://doi.org/10.1016/j.oceaneng.2018.03.053
41. Nazeryan, M., & Lakzian, E. (2018). Detailed entropy generation analysis of a Wells turbine using the variation of the blade thickness. *Energy, 143*, 385–405. https://doi.org/10.1016/j.energy.2017.11.006
42. Fish, F. E., & Battle, J. M. (1995). Hydrodynamic design of the humpback whale flipper. *Journal of Morphology, 225*(1), 51–60.
43. Hansen, K. L., Kelso, R. M., & Dally, B. B. (2011). Performance variations of leading-edge tubercles for distinct airfoil profiles. *AIAA Journal, 49*(1), 185–194.
44. Skillen, A., Revell, A., Pinelli, A., Piomelli, U., & Favier, J. (2014). Flow over a wing with leading-edge undulations. *AIAA Journal, 53*(2), 464–472.
45. Johari, H., Henoch, C. W., Custodio, D., & Levshin, A. (2007). Effects of Leading-Edge Protuberances on Airfoil Performance. *AIAA journal, 45*(11), 2634–2642.
46. Kumar, P. M., Halder, P., Samad, A., & Rhee, S. H. (2018). Wave energy harvesting turbine: Effect of hub-to-tip profile modification. *International Journal of Fluid Machinery and Systems, 11*(1), 55–62.
47. Kumar, P. M., Halder, P., & Samad, A. (2018, June). *Performance analysis of wells turbine with radiused blade tip*. International Conference on Offshore Mechanics and Arctic Engineering (Vol. 51319, p. V010T09A049). American Society of Mechanical Engineers.
48. Kumar, P. M., & Samad, A. (2020). Nature-inspired design of a turbine blade harnessing wave energy. *Proceedings of the Institution of Mechanical Engineers, Part A: Journal of Power Energy, 234*(5), 670–689.
49. Kumar, P. M., & Samad, A. (2019). Introducing gurney flap to Wells turbine blade and performance analysis with OpenFOAM. *Ocean Engineering, 187*, 106212.
50. Kumar, P. M., Halder, P., Husain, A., & Samad, A. (2019). Performance enhancement of Wells turbine: Combined radiused edge blade tip, static extended trailing edge, and variable thickness modifications. *Ocean Engineering, 185*, 47–58.
51. Hyun, B. S., Moon, J. S., Hong, S. W., & Kim, K. S. (2005). Aerodynamic Characteristics of impulse turbine with an end plate for wave energy conversion. *Journal of Ocean Engineering and Technology, 19*(6), 1–7.
52. Takao, M., Setoguchi, T., Kinoue, Y., & Kaneko, K. (2007). Wells turbine with end plates for wave energy conversion. *Ocean Engineering, 34*(11–12), 1790–1795.
53. Setoguchi, T., Takao, M., Kaneko, K., Nagata, S., & Toyota, K. (2007, January). Effect of end plates on the performance of an impulse turbine for wave energy conversion. *Fluids Engineering Division Summer Meeting, 42894*, 1045–1049.
54. Liu, Z., Cui, Y., Kim, K. W., & Shi, H. D. (2016). Numerical study on a modified impulse turbine for OWC wave energy conversion. *Ocean Engineering, 111*, 533–542.
55. Hyun, B. S., Moon, J. S., Hong, S. W., & Kim, K. S. (2006). A Study on the performance of the ring-type impulse turbine for wave energy conversion. *Journal of Ocean Engineering and Technology, 20*(1), 20–25.
56. Ranjith, B., Halder, P., & Samad, A. (2019). High-performance ocean energy harvesting turbine design—Detailed flow analysis with blade leaning strategy. *Proceedings of the Institution of Mechanical Engineers, Part A: Journal of Power Energy, 233*(3), 379–396.

Chapter 15
Optimization of an Impulse Turbine for Efficient Wave Energy Extraction

Rameez Badhurshah and Abdus Samad

Abstract Wave energy extracting devices such as an oscillating water column (OWC) in general have lower efficiency compared to other energy extracting techniques due to non-optimal design of various components. OWC is an empty vertical column housing an Wells or an impulse turbine. This chapter presents a detailed methodology in optimizing few design variables such as the number of guide vanes, number of rotor blades, guide vane angle with an objective to maximize efficiency. This is achieved by using a combination of low fidelity method (surrogate modelling) and high fidelity method (CFD simulation) which utilizes computational resources effectively. The relative overall mean efficiency increment taken over a wider flow coefficient range was approximately 24% while for the peak to peak comparison, the relative enhancement was 28%. Flow dynamics have also been studied to understand the reason for enhancement in efficiency with the optimal design.

15.1 Introduction

The energy sector, which includes energy production, energy storage, energy management and energy transmission is one of the foremost factor driving the economy of a nation. Reasonably, most of the countries rely on fossil-fuel-based energy generation techniques to cater to the demand for energy generation. However, burning of fossil fuels reserves contributes to global warming and causes environmental pollution [1]. Thus, a considerable emphasis is being laid by several nations to move away from fossil-fuel-based energy generation and explore possibilities of alternative, cleaner and renewable sources of energy. Countries like Sweden, Denmark, Iceland, among several others, have already committed to being fossil-free by 2050 [2].

R. Badhurshah (✉)
IIT Bombay, Powai, Mumbai 400076, India

A. Samad
Wave Energy and Fluid Engineering Laboratory, Department of Ocean Engineering, Indian Institute of Technology Madras, Chennai 600036, India
e-mail: samad@iitm.ac.in

A. Samad et al. (eds.), *Ocean Wave Energy Systems*, Ocean Engineering & Oceanography 14, https://doi.org/10.1007/978-3-030-78716-5_15

Few of the popular renewable sources these countries rely on are solar energy, wind energy, wave energy, geothermal based energy and biomass-based energy [3]. However, with global awareness, many other countries have also taken steps towards renewable-based energy sources. Although, the measures and policies are still in initial developmental stage. While, solar, hydro-power and wind energy are gaining popularity and are also seen in the implementation and installation stages, the ocean-based resources are often being overlooked. Energy from the ocean can be extracted either from the waves, the underwater currents or through the tide formation. Possible factors for overlooking this tremendous potential of ocean energy generation could be a random distribution of wave and tides, extreme weather which could result in failure of the energy harvesting device owing to the increased structural loading, seasonal variation of sea states from month to month, and loss to aquatic flora and fauna. However, careful selection of location and studying the annual sea pattern can overcome the above factors. Some of the advantages of using ocean/marine-based energy harvesters over other renewable energy devices include the abundance and availability of ocean resources, high energy conversion factor and predictability in sea patterns. It has been estimated that the following various energy resources from ocean taken globally have the following potentials: tidal based resources 800 TWh per annum; osmotic 2000 TWh per annum; wave resources between 8000 and 80,000 TWh per annum; while thermal sources range between 10,000 and 87,600 TWh per annum, the total of potential results in overcasting global electricity demand which is approximately 16,000 TWh per annum (to be considered valid for the year the study was reported) [4].

Considering the immense potential the wave possess, it promises to be a resource that could compete against the fossil-based resources to fulfil the energy demand especially in offshore locations such as deep-sea oil exploration platforms or even at remote islands. Ocean waves are defined as the periodic undulation/oscillations of surface of the ocean caused due to blowing winds over it. Several methods have been explored to tap usable form of energy from the oscillating waves and can be broadly classified as oscillating water column (OWC), point absorbers, overtopping devices, oscillating bodies [5]. Based on the location of the energy harvesters, these devices can also be classified as onshore, near-shore and offshore devices.

15.2 Oscillating Water Column

Oscillating Water Column (OWC) is a hollow vertical column with one of its open end interacting with the waves while the other end being open to the atmosphere [6]. A turbine is housed inside the OWC near the atmosphere open end. A simple schematic is shown in Fig. 15.1. The periodic crests and troughs of the ocean waves cause the trapped air inside the cylindrical chamber to oscillate. Further, the reciprocating airflow causes the blades of the turbine to rotate. Notably, the rotation of turbine occurs in a single direction even though the air is reciprocating in nature; hence such turbines also derive their name as bidirectional air turbines. The turbine then transmits

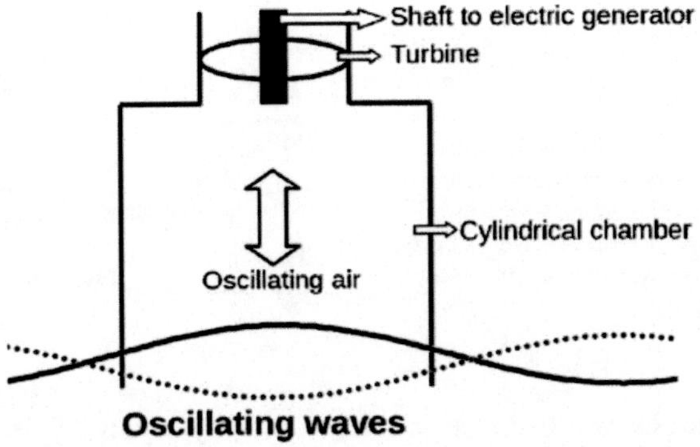

Fig. 15.1 A simple schematic showing the oscillating water column. The oscillating waves cause the air inside the chamber to oscillate which rotates the turbine attached near the other open end of the cylindrical chamber. The turbine shaft is attached to electric generator

the mechanical energy to an electric generator via a shaft, where the conversion of mechanical to electrical energy takes place and thus a usable form of energy is derived. To summarize, the complete conversion process of the wave to wire takes place in three stages listed as below:

- Wave hydrodynamic stage—Conversion of sea wave elevations to pressure gradients in the chamber
- Mechanical stage—Conversion of pressure gradient into mechanical energy via a turbine
- Electrical stage—Conversion of mechanical to electrical energy using an electric generator

Installation of OWCs can be onshore as well as offshore. Onshore OWCs installed near the shores are commonly fixed OWCs while for the offshore locations, more commonly floating type OWCs have been used. Onshore OWC has advantages such as ease of maintenance and repair, quick accessibility and cheaper energy transmission costs. While offshore devices cater to higher sea wave elevations, thereby having more considerable potential and relatively lower installation cost.

15.3 Turbine Selection for OWC

The turbine is the primary power fostering component of the OWC. The selection of turbine plays a critical role in the design and operation of the OWC [5]. Typically, axial flow turbines have been deployed since the perception of OWC began. However, there have been instances where radial type prototypes have been researched for their suitability for OWC applications. While, for the axial turbines, the inflow and the outflow are aligned parallel, for the radial turbine, the outflow and the inflow are perpendicular to each other. Based on the applications of OWC, two types of axial turbines have been in use, namely reaction turbine and impulse turbine. The reaction

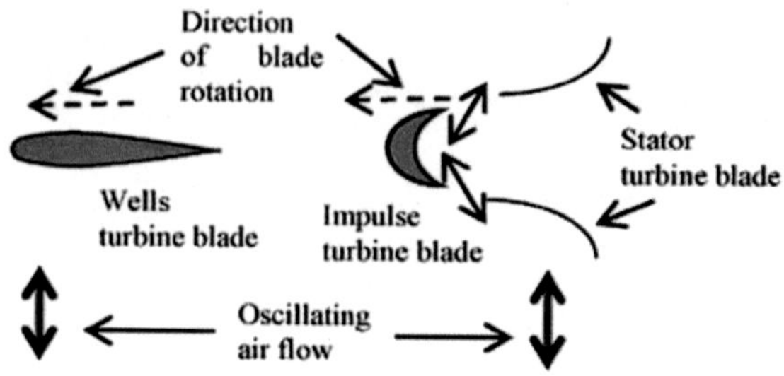

Fig. 15.2 A simple schematic showing the principle of operation of Wells turbine and impulse turbine. The Wells turbine works on the principle of aerodynamics where as the impulse turbine works on the principle of roto-dynamics [11]

turbine works on the principle of aerodynamics, while the impulse turbine works on principles of rotodynamic or jet dynamics.

Wells turbine, a reaction based turbine, was initially selected for drawing power in an OWC. Wells turbine uses aerofoil shaped blades mounted on to the hub and placed such that the inflow is perpendicular to the chord line of the blade. As the flow is essentially bidirectional (reciprocating) and hence, only symmetrical blade profiles are to be used. The pressure difference over the top and bottom of the blades generates a lift in the forward direction. A component of the lift force along the axis of the blade then helps in rotation of the hub.

Although efficiency produced from the reaction based turbines was appreciable owing to the aerodynamic design, the flow operating range is very narrow. Further, reaction turbines also had several other disadvantages such as poor starting characteristics, noisy operation due to high rotational speed and high axial thrust components. Thus, research for counteracting the problems as mentioned above began, and impulse turbine was found to be a better alternative with a wider flow operating regime. Impulse turbines thus gained immense popularity since the 2000s and slowly started replacing the reaction based turbine. A simple schematic is presented in Fig. 15.2 to highlight the principle of operation of Wells turbine and impulse turbine. Also, the peak rise and fall were found to be steady for impulse turbines as opposed to reaction-based turbines. In the reaction-based turbine, the jet of fluid is directly imposed on to the rotor blades mounted on the hub. Thus, the impulsive impact action of the jet force onto the surface of the rotor blades rotates the blades of the turbine. For successfully directing the flow on to the rotor blades, the impulse turbine needs an additional component "stator (guide vanes)" to guide the flow on to the rotor blades. Again, owing to the bidirectional nature of the flow, a set of guide vanes are needed each placed on either side of the rotor. As the stator blades are stationary, they are mounted on the shroud or the outer casing of the turbine. A provision for small gap is made between the guide vane and rotor to avoid the blades running onto each other and thereby damaging the turbine. Also, a tip clearance is needed between the tip of guide vane and the rotor hub as well as between the tip of rotor blades and the shroud to avoid hitting of blades with either the hub or the shroud. The guide vanes provided in the impulse turbine are either fixed type or movable type. In the movable type of guide vanes, the guide vanes pitch about its axis, thereby aligning itself with the flow. Although, the movable type of guide vane ensure lower flow bending losses, thereby improving efficiency as compared to the fixed type guide vane, but also adds maintenance costs owing to the moving parts. Also, design complexity is added, thereby increasing the designing and manufacturing costs. The primary role of guide vane is increasing the potential energy of the flow while the role of the rotor blade is to convert this potential energy into kinetic energy efficiently.

The restrictions of using a set of guide vanes and symmetrical blade profiles due to the bidirectional flow can be overcome using unidirectional flow turbines. Unidirectional flow turbines draw power only for one half-wave cycle, while, remaining idle for the other half of the wave cycle. For example, if the turbine draws power from the crest of the wave, the turbine remains idle during the trough. As bidirectional flow turbines work during both the half of the wave cycle, they are more efficient than the unidirectional flow turbines.

15.4 Numerical Studies on Impulse Turbine

Designing the turbine for harvesting wave energy efficiently is one of the major research work. Traditionally, a few prototype sets of turbines were manufactured, and the performance of each of the design was evaluated in the lab conditions to mimic the performance under real sea conditions [7]. The better performing design among the set was then selected for designing the actual turbine for real sea conditions, and the rest of the prototypes were discarded. This process is tedious and involves high manufacturing costs. Moreover, the design selected may be the best bet among the other prototypes but may not be the optimized and ideal selection.

15.4.1 Background Study

The traditional method of designing could be made easier, simpler as well as cost-effective using computational techniques. With the advent and technological development of computational resources and capabilities, the turbines can be designed as well as be simulated to mimic the real scenarios and can further be also optimized for better performance characteristics.

The earliest computational study performed on impulse turbine for wave energy extraction can be traced to the computational fluid dynamics (CFD) numerical simulations performed by Thakker et al. [8]. Two-dimensional simulations were performed on 0.6 m, 0.6 hub to tip diameter ratio impulse turbine with fixed guide vanes. The 2D CFD results were later extrapolated using numerical techniques to predict the three-dimensional performance characteristics of the turbine. Qualitatively, numerical results showed trends similar to experimental study, although magnitude-wise a huge discrepancy was observed. The error observed was attributed to the two-dimensional study, not capturing tip gap losses as well as the actual total pressure drop across the rotor. Also, the grid mesh size contained only 13,000 grid elements due to the limitations in computational resources during the early 2000s. Various turbulence models were also implemented in their models, and it was observed predictions by the $k - \epsilon$ turbulence model were closer to experimental results.

Thakker and Hourigan [9] performed the three-dimensional blade design and numerical simulations for the 0.6 m, 0.6 hub to tip diameter ratio 3-D blade model using commercial software Pro-E and Fluent 6.0. The 3-D flow domain was discretized into grid elements using hybrid meshing scheme wherein, hexahedral meshing near the blade region was used while for the rest of the fluid domain was meshed using tetrahedral and pyramidal elements. The numerical studies over-predicted the efficiency by approximately 10% compared to the experimental results. Again, the lack of computational resources was stated as the reason for the difference. Also, the numerical simulations were performed over a narrow range of flow coefficients.

15.4.2 Steps Involved in Performing CFD Simulations

Numerical techniques involved in performing CFD simulations are to convert the continuous differential equations into algebraic expressions and find a solution at discrete points in space or grids. The complete CFD process may involve the following steps [10]

1. Designing the geometry/blade design and flow fluid domain
2. Discretization/Grid generation: Splitting the geometry into smaller grid elements
3. Setting up the physical model, specifying initial and boundary conditions
4. Solving the equations for conservation of mass, momentum and energy
5. Checking the convergence of solution
6. Post processing and flow visualization.

15.4.3 Case Study on CFD Simulations for an Impulse Turbine for Wave Energy Extraction

In this section, a case study is done on the flow over an impulse turbine by performing numerical CFD simulations to mimic the real wave scenario. Instead of using a wavy input, steady-state uniform velocity is provided at the inlet to reduce the computational burden. The following subsection discusses this case study in minute details.

15.4.3.1 Impulse Turbine Geometry and Blade Profile

For the case study, we consider the design for a bidirectional flow impulse turbine chosen from [7]. The geometry profile for the rotor blade and the guide vane is shown in Fig. 15.3. A set of guide vanes prevail on either side of the rotor blade. The turbine has 30 number of rotor blades mounted on the hub, and 26 number of fixed type guide vanes along each side of the rotor blades mounted on the shroud or the casing. The guide vane is inclined at 30°. The rotor blade profile has a circular arc profile on the pressure side, and an elliptic arc profile on the suction side whereas the guide vane has a profile with a straight line camber followed by a circular arc for the smooth turning of flow. There were no criteria specified for the length of the straight path and radius for the curvature of the circular arc. Badhurshah et al. [11] deduced a relationship to co-relate the guide vane angle, the length of the straight path and the radius of the circular path. A simple schematic sketch is shown in Fig. 15.4 and the related equations (Eqs. 15.1 and 15.2) are as below

$$\alpha = l\sec(\theta) - \beta\tan(\theta)\,, \tag{15.1}$$

$$\beta = \frac{-lM^2 + \sqrt{l^2M^4 - 2M(l^2M^2 - L_{GV}^2\cos^2(\theta))}}{2M}\,, \tag{15.2}$$

where, $M = 1 - \sin(\theta)$, L_{GV} is the length of the guide vane fixed as 70 mm, l is the length of the straight profile while R is radius of the circular arc and θ is the guide vane angle.

A simple 3D CAD model of the complete assembly of an impulse turbine is shown in Fig. 15.5. For the chosen turbine geometry, the hub to tip diameter ratio was 0.7 with a hub diameter of 210 mm. To avoid hitting of the rotor blade with the external casing, a 1 mm tip clearance is provided between the rotor blades and shroud. Further, the gap between the rotor and the guide vane rows were 10 mm each on either side. The chord length for the rotor was 54 mm while for the guide vane, the chord length was 70 mm, and the blade thickness was 0.5 mm. Further to make computation cost-effective, instead of simulating the bidirectional flow over the entire turbine geometry, unidirectional flow over a single passage is modelled. Periodic boundary conditions are imposed on either side of the rotor blade, and guide vane domains as reported by Thakker and Hourigan [9]. The assembly was modelled using the frozen rotor approach in which the rotor blades are frozen the flow in the fluid flow domain is analyzed. The inlet and outlet flow domain are extended to 8.5 times the guide vane chord length to achieve a fully developed flow condition [12]. Reynolds averaged Navier Stokes equations (RANS) are solved along with additional equations for turbulence ($k - \epsilon$ turbulence model). The computational flow domain is shown in Fig. 15.6.

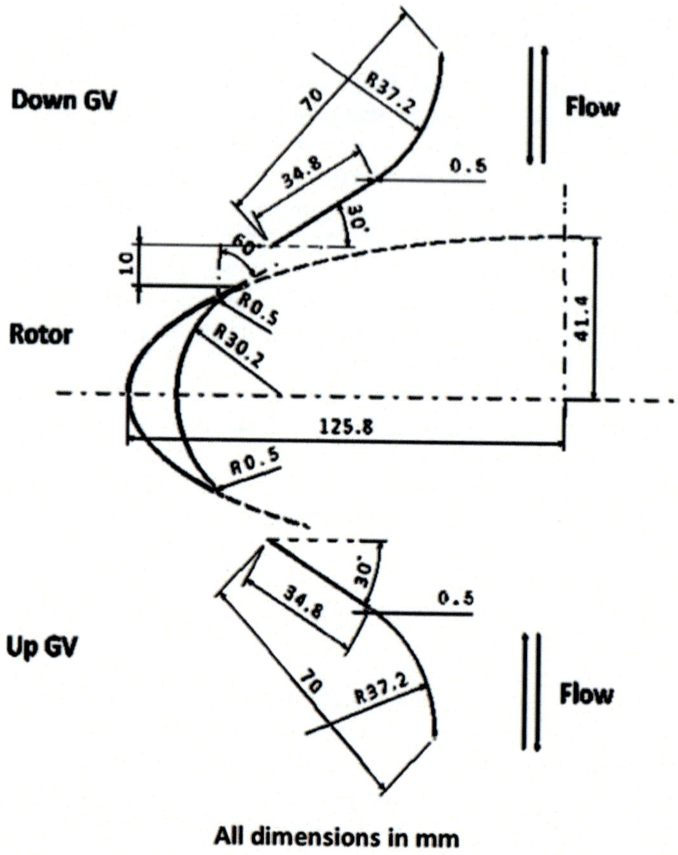

Fig. 15.3 Geometry profile for rotor blade and guide vane of an impulse turbine for wave energy applications. The flow is bidirectional in nature. Up GV and Down GV represent upstream guide vane and downstream guide vanes respectively [7]

15.4.3.2 Discretization/Grid Generation

The geometrically constructed flow domain is then discretized into small grid elements using commercial software ICEM-CFD. The boundary and the initial conditions are used to evaluate the fluid flow variables at the centre of each grid element. Hence, more the grid element, higher may be the accuracy, but the computational time increases exponentially with the number of grid elements. Grid elements adapted in the current study were non-uniform tetrahedral elements with very fine grid elements near the blade tip clearance. The mesh for the guide vane and rotor is shown in Fig. 15.7. The initial, as well as the boundary conditions, are set using commercial software Ansys CFX-Pre while the flow is solved using Ansys CFX-Solver. As

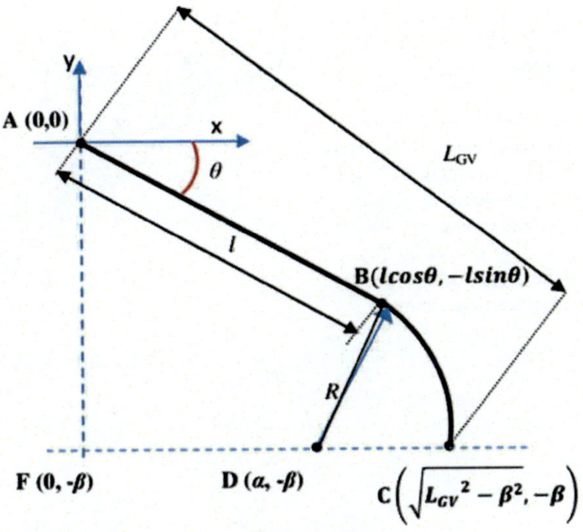

Fig. 15.4 Guide vane profile to derive a relationship to co-relate the guide vane angle, the length of the straight path and the radius of the circular path. Here, L_{GV} is the length of the guide vane which is fixed as 70 mm, l is the length of the straight profile while R is radius of the circular arc and θ is the guide vane angle [11]

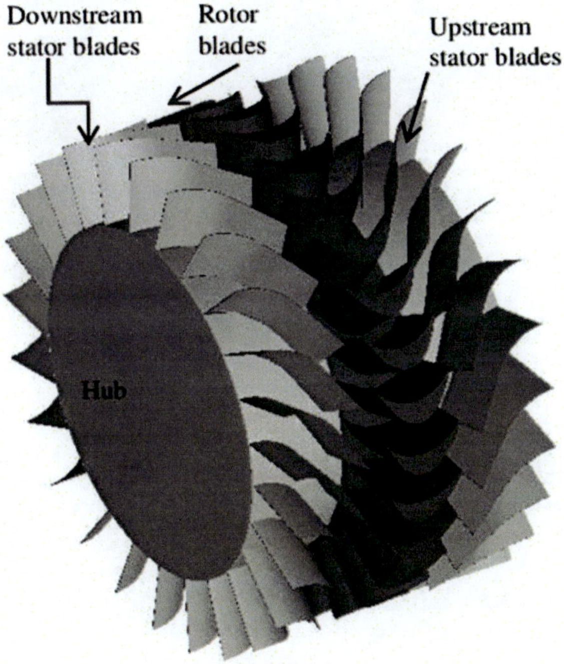

Fig. 15.5 A 3D CAD model of the impulse turbine showing the rotor blade sandwiched between guide vanes on either side. The rotor blades are mounted on the hub while the guide vanes are mounted on the outer casing or shroud [13]

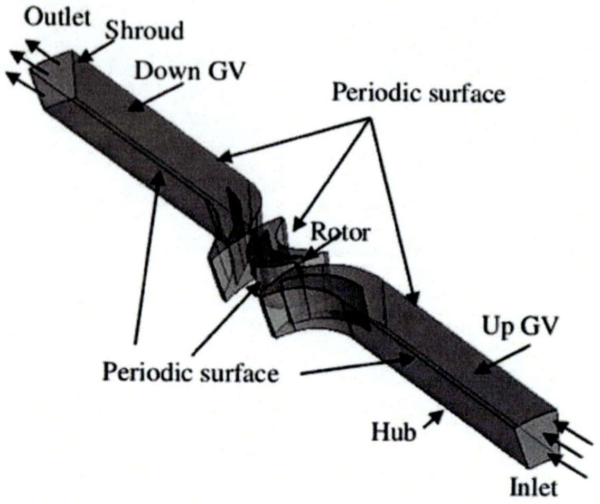

Fig. 15.6 The flow domain for a single passage of the turbine to reduce computational cost. Frozen rotor model is applied with the rotor being frozen. The inlet and outlet domain are extended to achieve fully developed flow conditions. On either side of the flow domain, periodic boundary conditions are imposed

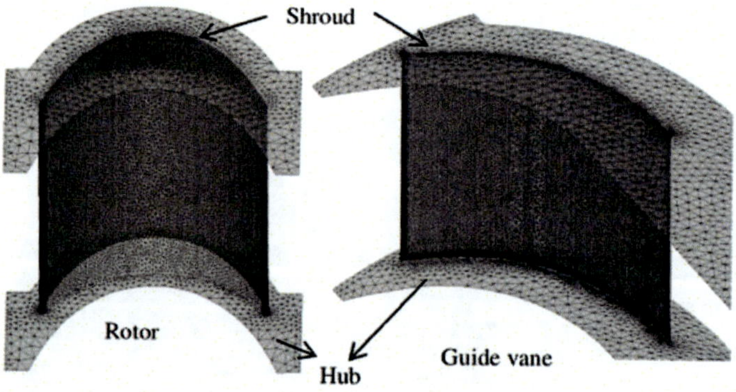

Fig. 15.7 Grid generation shown on the surface of rotor blade and the guide vane. The grids are finer near critical zones like the tip clearance and the blade edges and is coarser in the flow domain away from the blades [11]

there exist three different domains, the interfaces are coupled using the mixing plane model.

15.4.3.3 Initial and Boundary Conditions

Incoming flow velocity drives the turbine, thus the flow inlet velocity and the rotational velocity frame the basic input parameters. Further, flow coefficient (ϕ) (Eq. 15.3) defined as the ratio of flow inlet velocity (v) to the circumferential velocity (rotating velocity of the turbine (U) at mid-span) reduced the two input variables to one. To ensure wider regime of flow co-efficient, six different inlet velocity between 4–20 m/s were chosen while the circumferential velocity was fixed at 600 rpm (8.09

m/s). The inlet velocity was specified as the inlet boundary condition while the outlet boundary condition was specified as open to the atmosphere (1 atm). The turbulence model $k - \epsilon$ was chosen with medium (5%) turbulence intensity.

$$\phi = \frac{v}{U}, \tag{15.3}$$

15.4.3.4 Output Prameters

The turbine performance is evaluated in respect with certain parameters namely the torque (C_T), input power (C_A) and efficiency η. The expression for these variables are as below :

$$C_T = \frac{2T}{\rho(v^2 + U^2)bL_{RB}N_{RB}r}, \tag{15.4}$$

$$C_A = \frac{2\Delta pQ}{\rho(v^2 + U^2)bL_{RB}N_{RB}v}, \tag{15.5}$$

$$\eta = \frac{W}{\Delta pQ} = \frac{C_T}{\phi C_A}, \tag{15.6}$$

where, T is the torque, ρ is the density of air, b is blade height, L_{RB} is the chord length of the rotor blade, N_{RB} is the number of the rotor blade, r is mean radius of the rotor blade, Δp is the total pressure drop across the rotor blade, Q is the flow rate, and W is the shaft power.

15.4.3.5 Numerical Validation

The very first step is to validate the output CFD results obtained using the current methodology with the existing literature to determine the error between the two. Usually, an error up to 10% between computational and experimental results are acceptable depending on the severity of the problem statement. The current numerical results obtained via commercial software ANSYS-CFX are validated against experimental [7] as well as numerical results [14] from the literature. The outputs, torque coefficient (Eq. 15.4), input coefficient (Eq. 15.5) and efficiency (Eq. 15.6) for a wide range of flow coefficients (Eq. 15.3) were compared.

The variation in torque coefficient, input coefficient and efficiency plotted against the flow coefficient is shown in Fig. 15.8. For the torque coefficient, at lower flow coefficients, the CFD results of the current study match well the experimental value whereas, for the higher flow coefficients, magnitude is under predicted, the maximum relative error is approximately 8.25%. For input coefficient, at lower flow coefficients, CFD overpredicts the results while at higher flow coefficients, the value is underpredicted. As for the efficiency, for lower flow coefficients, there is a difference in magnitude observed at lower flow coefficients; however, the results match

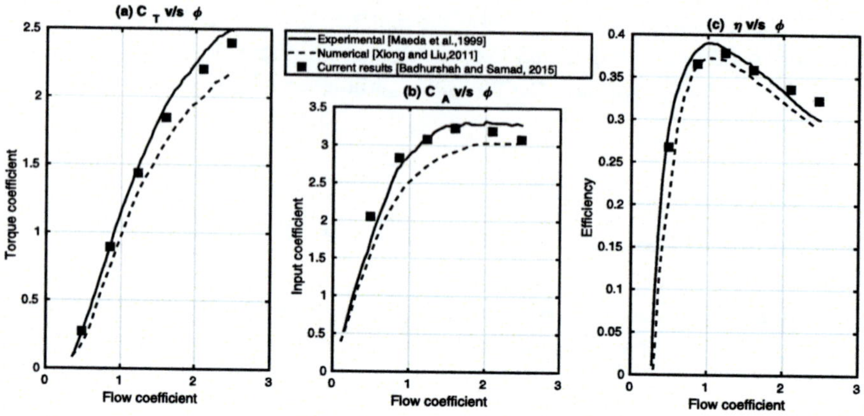

Fig. 15.8 Validation of current CFD results against existing literature [13]

very well with experimental data at high flow coefficients. Overall, the qualitative trend of the current study showed similarity with the experimental results. The difference in magnitude of results can be attributed to a number of factors such as the use of frozen rotor approximation, selection of grid, convergence criteria, selection of turbulence model, etc. Moreover, the experimental results also reported having uncertainty in results of ±2% [7]. Again, the various losses encountered during the experiments may not be accountable while performing numerical simulations. The efficiency initially rises, peaks at about 38% and then falls again with an increase in flow coefficients. The reason for the fall is due to the large pressure drop occurring at high flow coefficients. This large pressure drop is the measure of the unsuccessful conversion of potential energy into kinetic energy at higher flow coefficients.

15.5 Optimization

As observed in Sect. 15.4.3.5, the maximum efficiency of the turbine is limited to approximately 38%. With a strategic optimization methodology, the efficiency of the existing turbine can be improved by redesigning a few of the components of the turbine. An optimization is an efficient tool which also helps determine the problems associated with the existing design for poorer performance characteristics. Optimization should be studied as the science and art of selecting the best amongst available alternatives [15]. A general procedure followed for optimization is presented in Fig. 15.9.

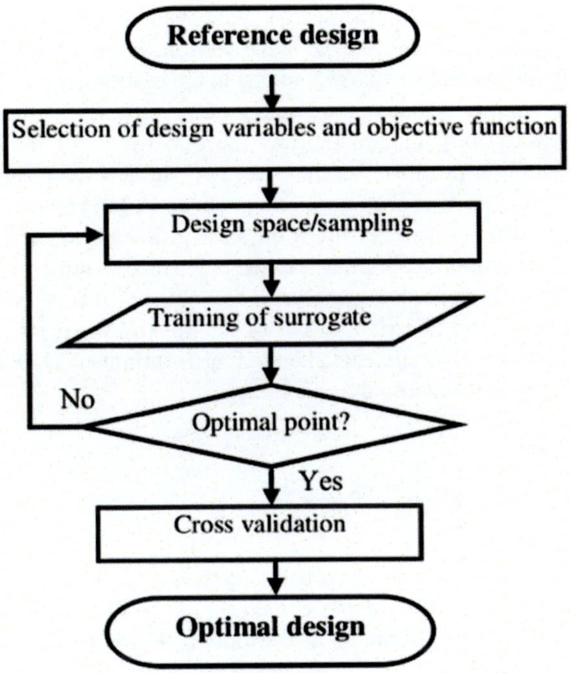

Fig. 15.9 General methodology adapted for optimization. Here, a reference design is chosen and fulfilling the procedural formalities, an optimal design is obtained. The optimized design has better performance characteristics as compared to the reference design [11]

15.5.1 Design Variable and Objective Function

The primary objective for any optimization process should be arriving at the best possible design. Certain output parameter would be needed to evaluate the performance of the available design. The best design may thus be chosen with the design with the best objective function. As discussed in Sect. 15.5, this objective functions can be arrived by modifying or redesigning certain input variables. Such input variables are referred to as the design variables.

Given the current study on impulse turbine, several factors exist which can be chosen design variables. Some of them include, number of rotor blades, number of guide vanes, guide vane angle, guide vane profile, tip clearance between blades and casing, hub to tip ratio, gap between the rotor and guide vane blades, rotor blade profile, flow blockage ratio, leaning of blades, etc. It would be a cumbersome task to optimize all the parameters at once. Hence, based on literature and instincts, a decision is to be made to narrow down variable. Additionally, inter-dependency of a variable over another should also be taken care of. In the current study, four design variables, namely number of guide vanes, number of rotor blades, guide vane angle and guide vane profile were chosen to optimize the turbine. It is very important to keep the variables other than the chosen four variables fixed to get accurate results. Following parameters were fixed for the current study, the hub to tip ratio kept as 0.7, hub diameter of 210 mm, the rotor profile and rest all the variable discussed above.

The selection of these design variables involved critical analysis of the pressure-flow contour for the reference blade section. Preliminary analysis revealed the flow over the blade had a wake region due to the section part of the successive blade. Thus, the motivation to select the number of guide vanes and rotor blades as design variables was to study the effect of flow width between the blades on the pressure contour on the blade profiles. For the other two variables, the selection was made on the basis of the study of literature. Setoguchi et al. [16] reported guide vane to be one of the critical parameters to turn the flow efficiently onto the rotor blades. However, the optimization reported was based on discrete point analysis, whereas the current study involved continuous analysis. The turning of flow is one of the critical parameters to make an effective impact of the flow on the rotor blade; the profile of the guide vane was thus chosen to be the fourth variable. The objective function of the current study was identified as efficiency.

15.5.2 Design Space

Design space involves organizing the design variables and the objective functions in a way to arrive at the effectiveness of the design variables on the chosen objective function. A design space distributes the set of sampling points chosen between the lower and upper limits of design variables. A number of sampling techniques exist, in the current study, Latin Hypercube sampling function was employed to distribute the design variables randomly, but in an organized manner [17]. The advantage of choosing Latin Hypercube sampling is the wider coverage of all portions of design space with the least possible number of sampling points. More number of sampling points may increase the number of design, thereby also increasing the number of CFD simulations and thus increasing the overall computational costs.

15.5.3 Use of Surrogates Method to Populate Sampling Points

The CFD simulations performed on the obtained sample points may not be adequate to fill the rest of the design space and choosing an optimized point from such a sample space may lead to erroneous results. The local optimal point achieved may not be the truly global optimized point. Hence, more number of sampling points may be needed. Choosing sampling points through high fidelity numerical simulation such as CFD may result in high computational costs. To counteract this, low fidelity surrogate techniques are used. Surrogates are adoptive methods, capable of producing the initial data and used mainly for populating a large data set using the initial set of data through various breeding methods. In the current study, the initial sampling points evaluated using CFD numerical simulation are used as a feeder to the surrogate methods. These surrogates are highly capable of replicating the results of actual simulations from the high-fidelity models such as CFD or finite element with an acceptable level of accuracy. Also, surrogates are problem dependent and thus using a single surrogate for a particular problem may lead to a greater probability of producing erroneous prediction. Hence, it is usually advised to avoid a single surrogate. Further, the use of multiple surrogates demands no extra cost or sampling points. The same set of initial sampling points may be used to feed the surrogate, amongst which, each surrogate may result in different optimal points [18]. This also increases the probability of finding the best optimal solution, thereby increasing the robustness of the surrogate methods.

The optimal point predicted by each of the surrogates needs to be then cross-validated with CFD results, and if there exists a higher error between the prediction of surrogate and CFD results, a sequential sampling technique is to implement. Sequential sampling is a method of adding additional new sample points, which are the data points predicted after the first run of surrogate on which CFD simulations were performed for calculating cross-validation error, to the initial sampling points. In the current study, four existing surrogates; namely Response surface approximation (RSA), Radial basis function (RBF), Kriging (KRG) and Weighted average surrogate (WAS) are used [18, 19]. A hybrid genetic algorithm is used as an optimizer which searches the optimal design point. Further, Weighted average surrogate (WAS) is a surrogate dependent on the other three surrogates. Using cross validation error, weights are assigned to the other individual surrogates (RSA, RBF and KRG), and the WAS surrogate is constructed. Fundamentally, lower weights are to be assigned to a surrogate with the highest CV error and vice versa. A brief description of each surrogate is presented below. For additional and detailed description of surrogates, readers may refer relevant literature [19, 20].

15.5.3.1 Response Surface Approximation (RSA)

RSA involves fitting a polynomial function. For a linear response, the fit may be expressed as

$$F(x) = \Sigma_{j=1}^{N} c_j x_j + e \,, \tag{15.7}$$

where, N is the number of design variables, x denotes the specific design variable while the terms c and e represents the regression coefficients independently distributed error with zero mean respectively.

In a similar fashion, second order polynomial may be fitted as

$$F(x) = c_0 \Sigma_{j=1}^{N} c_j x_j + \Sigma_{j=1}^{N} c_{jj} x_j^2 + \Sigma \Sigma_{i \neq j}^{N} c_{ij} x_i x_j \,, \tag{15.8}$$

In the current study, second order polynomial has been used and the number of regression coefficients were calculated using the expression $(N + 1) \times (N + 2)/2$.

15.5.3.2 Radial Basis Function (RBF)

RBF takes inspiration from human response functions acquired through the practice of prevailed experiences. Multiple processing nodes also called as neurons collect large inputs, process the output and store the processed information to retrieve information when recalled. In the current study, the linear nature of this activation function is used to reduce the computational burden. Two key critical parameters namely, the spread constant and error goal, are chosen as design parameters for the processing functions. The spread constant is chosen using a reiterative method such that the neurons do not respond similarly to all the input nor function becomes highly sensitive, thereby resulting in large variations for every input. Error goal is selected so as the network is not overstrained and also the accuracy remains largely unaffected.

15.5.3.3 Kriging (KRG)

KRG evaluates the response function as a sum value of two terms, first being a global model considered as $f(x)$ while other is systematic departure given as $Z(x)$

$$F(x) = f(x) + Z(x), \tag{15.9}$$

The global model usually evaluates the trend of the design space obtained by polynomial fit whereas the other term uses a Gaussian correlation with zero mean and non-zero covariance to quantify the local deviations in the design space points. They have been other variations to the Kriging model often referred to as Blind Kriging, Universal Kriging [21].

15.5.3.4 Weighted Average Surrogate (WAS)

WAS is a weight based surrogate which is constructed using averages of other surrogates [19]. The weights are thus assigned in a fashion such that the highest error is given the least weight and vice-versa. Thus,WAS may not give the best results but it certainly helps avoid selecting the worst surrogate.

$$F_{WAS}(x) = \Sigma_i^{N_{SM}} w_i(x) F_i(x) \ , \tag{15.10}$$

where, N_{SM} is the number of surrogates, and ith surrogate produces the respective weight $w_i(x)$.

15.5.4 Genetic Algorithm as an Optimizer

From the pool of the data points obtained using surrogate methods, selection of the best optimal design is made using genetic algorithm tool which searches for the global optimal design point. The genetic algorithm being a population-based algorithm may result in a different response value every time a run is performed. This problem is counteracted by clubbing the genetic algorithm with sequential quadratic programming, which is a local search algorithm, and it fine-tunes the optimal zone predicted by genetic algorithm. This combination of genetic algorithm and sequential quadratic programming is termed as a hybrid genetic algorithm [22]. Several optimization iterations are performed using hit, and trial method wherein the designer feeds initial guesses to search for a global optimal solution while using sequential quadratic programming alone. Thus, using the hybrid genetic algorithm, the designer may get riddance from the above-mentioned issue. Few terminologies in relation to the genetic algorithm have been described briefly as follows.

15.5.4.1 Population Size

The initial sampling points fed to the genetic algorithm is called as the initial population size. In the current study, the initial population was produced by performing CFD based numerical simulations. Typically, for the genetic algorithm to fetch the optimal point, large population size is needed. To reduce the computational resources, the remaining population is produced using surrogate models over several generations. This population being distributed randomly covers a wide range of feasible optimal solutions. Alternatively, one may even seed the population over certain critical areas which may have more probability of searching for the optimal solution.

15.5.4.2 Selection

The surrogate method usually produces a large pool of data points. Few of these points may produce poorer objective functions. It is important to filter such data points prior to the start of the next generation. Thus, the selection parameter allows choosing the data points from the current population size for the breeding process to produce next-generation data points. Thus, the selection strategy may permit the best individual in surviving and participating for the production of the next offspring, whereas the poorer individuals may be denied participation. There are a number of methods for selection like roulette wheel selection, rank selection, tournament selection, etc.

15.5.4.3 Crossover

Crossover process in the genetic algorithm is quite analogous to the actual biological reproduction system wherein crossover reproduces a child solution from the parent solutions, chosen by the selection strategy. Again, a number of different methods prevail for the crossover like a one-point crossover, two-point crossover, cut and slice, etc.

15.5.4.4 Mutation

The process of mutation allows for genetic diversity/variations from the current generation of the population to the successive generation, and so on. The process finds similarity with the biological mutation. Few gene values may be tailored to alter from its initial value and allow mutation in the upcoming generation. In our current study, the mutation probability has been set to very low.

15.6 Results and Discussions Based on Optimization of Impulse Turbine

This section presents the optimized turbine results obtained following the methodology discussed in Sect. 15.5. To summarize, a single objective function (efficiency) was chosen as the objective function and four design variables namely the number of guide vanes, number of rotor blades, guide vane angle and guide vane profile was selected.

15.6.1 Design Space

Using Hypercube sampling technique discussed in Sect. 15.5.2, a design space was created with fifteen sampling points randomly selected in order to cover the wider range. Along with the reference design taken as the sixteenth sampling point, CFD numerical simulations are performed. The sampling points were chosen as sixteen to accommodate the minimum number of regression coefficient needed for RSA surrogate to participate. For the four design variables, minimum of sixteen data points are required. Radial basis function, as well as Kriging, do not impose any such restriction, but with such a large initial population size, the pool of data points produced may be larger thereby assisting the genetic algorithm in choosing the optimal point from a wider design space.

The upper and lower limits or design boundary for the four design variables is shown in Table 15.1. The CAD design for the initial population size is modelled, and CFD numerical simulations are performed for each case. As the peak efficiency was observed at flow coefficient $\phi = 1.24$ (Fig.. 15.8c), the simulations for the remaining 15 sampling points were also performed at this fixed flow coefficient.

Table 15.1 Lower and upper limits of the design variables chosen for creating the design space

Design variable	Lower bound	Upper bound
Number of rotor blades	30	42
Number of guide vanes	16	32
Guide vane angle (°)	15	45
Guide vane profile (mm)	5	65

15.6.2 Sensitivity of Design Variable

These sampling points were populated using surrogate methods, and the hybrid genetic algorithm was run to search for the optimal solution from the different surrogates. The optimal solution was observed to be at the extreme boundaries and the objective function either under predicted or over predicted the results by a greater magnitude for all the surrogates except for RSA. Usually, such errors occur mainly due to either smaller initial population size or due to one of the chosen variable being insensitive to the objective function. To identify the cause, sensitivity analysis of each design variables was then performed using RSA. The sensitivity analysis curve of each design variable is plotted against relative efficiency, as shown in Fig. 15.10. While keeping the other three design variable fixed, the fourth design variable is varied by $\pm 10\%$ to assess it's effect on the objective function, namely the efficiency. In the plot, the abscissa plots the percentage deviation while the ordinate-axis plots the relative change in efficiency. From the plot, it can be clearly observed; the guide vane profile has no effect on the efficiency while the guide vane angle has a huge influence on the efficiency. Using this information and to increase the effectiveness and robustness of the genetic algorithm, the design variable guide vane profile is discarded for the next round. Thus, we use the same discussed sixteen sampling points, but feed only three design variables to the surrogates for replicating and producing more data points for the second iteration. The reduction in one design variable along with increasing the effectiveness of the surrogates also reduced computational time considerably. This claim of guide vane profile being insensitive is also cross-validated using CFD simulations and the discussion on same is done in Sect. 15.6.5

As discussed in Sect. 15.5.3.4, to construct WAS surrogate weights from each prior surrogate should be assigned. Thus, Table 15.2 shows the cross-validation error from each surrogate and the related weight assigned to each surrogate. It is observed that the cross-validation error is least for RSA model and thus higher weight is assigned to RSA as compared to other surrogate to construct WAS.

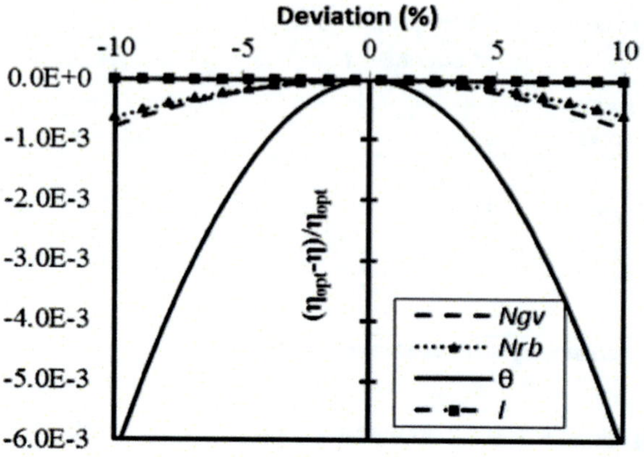

Fig. 15.10 Plot showing the sensitivity of each design variable to the objective function. Each design variable is changed by ±10% while keeping the other three fixed and variation of each of design variable against relative efficiency is plotted. Here, dashed black lines represent the number of rotor blades, dotted line with circle marker shows the number of guide vanes, the guide vane angle is shown by black solid line while the guide vane profile is shown using dashed line with square markers [11]

Table 15.2 Cross validation error and weights for construction of WAS [11]

Surrogate model	Cross validation error	Weight
RSA	0.045	0.362
RBF	0.059	0.282
KRG	0.046	0.355

15.6.3 Results from Optimization Algorithm

In this section, we discuss the optimal design point as predicted by the search algorithm from the pool of each of the surrogate. Each of these predictions is also cross-validated using CFD simulations to report the relative error between the surrogate-based prediction and actual CFD simulation.

Table 15.3 summarizes the results in the tabulated format. The first column represents the surrogate model with the first row showing the results of the reference design. The next three columns are the three chosen design variables. The fifth column plots the optimized efficiency as predicted by the genetic algorithm from the pool of data points each surrogate produced. Next, using each of these optimal design, CAD models are constructed again, and CFD based numerical simulations are performed. The efficiency from the CFD simulations is plotted in column six. The relative error between the surrogate predicted efficiency and CFD efficiency are listed next, and the last column shows the increment of efficiency compared to the base reference design. Maximum peak efficiency of 48.39% is observed for the CFD simulations performed at the RSA predicted design variables. Also, the relative error

Table 15.3 Optimization results as predicted by each pool of surrogate [11]

Surrogate model	Number of rotor blade	Number of guide vanes	Guide vane angle (°)	η_{sur}	η_{CFD}	Relative error (%)	Increment (%)
Reference	26	30	30.0°	–	37.74	–	–
RSA	22	35	39.6	50.41	48.39	−04.15	28.22
RBF	24	39	35.5	53.21	47.12	−12.91	24.84
KRG	21	38	29.7	47.16	42.98	−09.58	13.88
WAS	21	36	39.4	49.22	46.70	−05.43	25.23

is the least between the RSA prediction and actual CFD simulation. RBF surrogate over predicted the efficiency, and thus, a large discrepancy is observed between the surrogate prediction and actual CFD efficiency. Being a weighted model WAS neither performed the best nor the worst. The error in surrogate prediction and CFD results may be attributed to factors such as approximation errors associated with the surrogates, fractional round-up of the integer value for a number of rotor blade and guide vane and errors in general such as truncation and approximation associated with the CFD simulations.

15.6.4 Results for Optimized Turbine Over Wide Flow Range

Next, the turbine with the design variables predicted by RSA is chosen as the optimal design and hereby would be referred to as an optimized turbine. The turbine was run over different flow coefficient to cover a wider range of flow speeds and was compared with the base reference turbine, as shown in Fig. 15.11. It was very heartening to see the increment in efficiency was observed over the entire range of flow coefficient, which rightly justifies the use of surrogate models combines with CFD to help arrive at an optimized design. The maximum relative efficiency observed was approximately 32% at the flow coefficient of 2.2.

15.6.4.1 Pressure Contours at the Mid-Section Lamina

Next, a pressure distribution contour plot at the midsection lamina of the flow over the reference and the optimized blade is shown in Fig. 15.12. Quite convincingly, on observing the flow over each contour, the pressure drop is observed to be less in the optimized turbine, which invariably means the productive conversion of pressure energy into kinetic energy takes place in impulse turbine. Also, the role of flow bending by the guide vane can be understood by observing the pressure contours near the upstream guide vane. The flow is smooth over the optimized turbine which further reduces the pressure losses and thus results in higher efficiency.

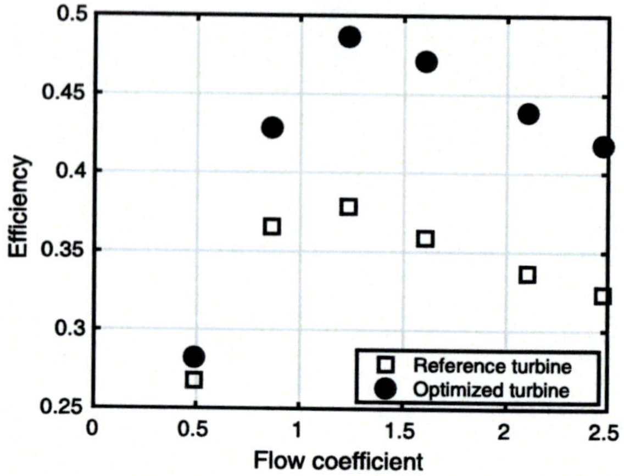

Fig. 15.11 Variation in efficiency over a wide flow coefficient range for the reference and the optimized impulse turbine. The increment in efficiency is observed over the entire range of flow coefficients [11]

15.6.4.2 Why the Optimized Turbine Has Better Efficiency?

The increment in efficiency observed for the impulse turbine can be primarily credited to the optimum guide vane angle. For the optimized impulse turbine, guide vane angle directs the flow over the rotor with lesser pressure loss and helps the flow pass smoothly, which reduces the flow bending losses. This design variable is very critical as very less guide vane angle may not turn the flow on to the rotor blade efficiently, whereas a higher guide vane angle will block the flow completely. Moreover, a higher number of guide vanes may also lead to flow blockage and hence an optimum number of guide vanes in co-ordination with optimum guide vane angle may help increase the efficiency of the turbine. The surrogate predicted an optimum number of the rotor blade as 35 and was higher than the reference design. A higher number of rotor blades may definitely imply increased energy transfer from the fluid to the turbine, however large number of rotor blades again causes flow blockage and thereby a substantial reduction in efficiency of the turbine.

15.6.5 Parametric Sensitivity Analysis

RSA predicted that the guide vane profile is insensitive to increment in efficiency (Sect. 15.6.3). In this section, the least sensitive parameter and the most sensitive parameters are validated by performing CFD simulations by varying the design variable. Needless to say, while performing the study of the sensitivity of one design variable, all the other design variables were kept at an optimized design variable. For the guide vane profile, two designs were considered, first in which the straight profile section of the guide vane was chosen as 20 mm (−50% of reference) while other being 50mm (+50% of reference). The efficiency at these extreme design point was

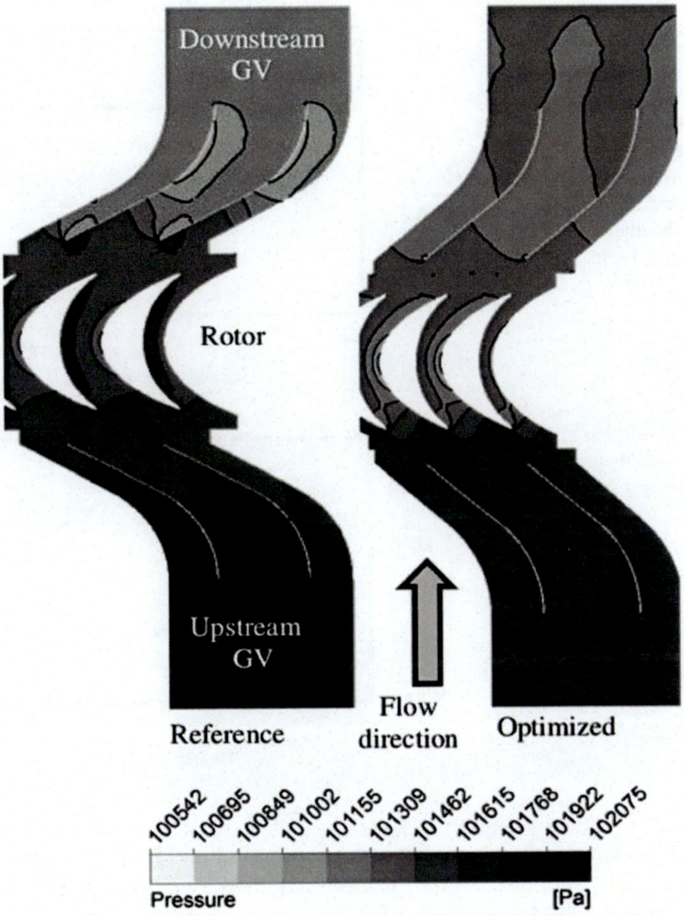

Fig. 15.12 Pressure contour over the mid-section lamina for the reference and optimized turbine. The pressure values corresponding to the contour shade are shown in the legend and both the turbine section were marked with the same pressure limits. The darker shades represents higher pressure [11]

then evaluated post performing CFD simulations and was plotted against the length of the straight section of guide vane profile as shown in Fig. 15.13. It is observed that the variation in the length of the straight section in the guide vane does not bring any change in efficiency. Hence, the prediction done by RSA can be considered to be valid.

Next, a similar exercise is carried to test the sensitivity of the guide vane angle on the efficiency of the turbine. Two turbines are designed, first with guide vane angle as 35.68° (−10% of optimum) and 43.61° (+10% of optimum), and the CFD simulations are performed with these two designs. The variation in efficiency is

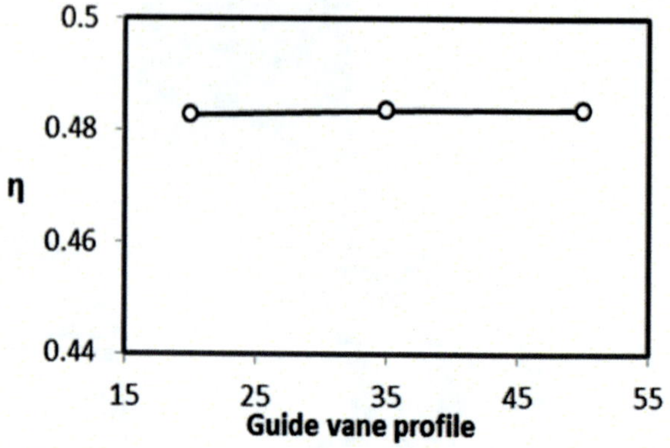

Fig. 15.13 Variation in efficiency with variation in the length of straight profile section of the guide vane. No change in efficiency is observed, thus guide vane profile is insensitive to the efficiency and thereby validating the claim predicted by RSA in Sect. 15.6.2 [11]

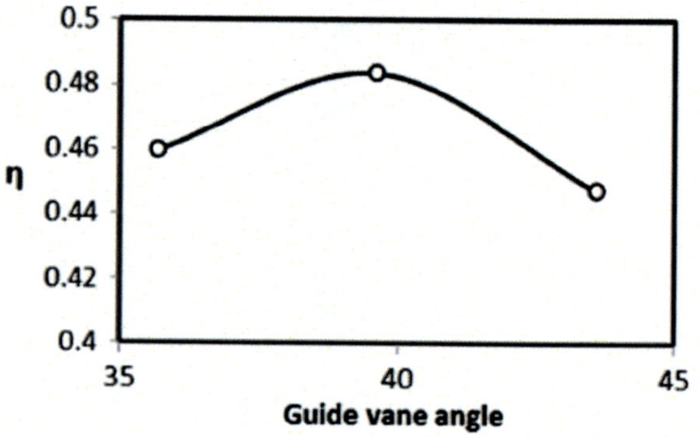

Fig. 15.14 Variation in efficiency plotted against variation in the guide vane angle [11]

plotted against the reference and above two designs in Fig. 15.14. On either side of the reference design, a fall in efficiency is observed, thereby revealing the sensitiveness of the parameter. The fall is steeper for the higher guide vane angle due to higher flow blockage.

15.7 Closure

The chapter describes the methodology used for optimizing a bidirectional flow impulse turbine used for wave energy extractions. Four design variables were chosen to be optimized in order to achieve the objective function of maximizing efficiency. Sixteen initial sample points, including the reference design, were randomly selected using a Latin Hypercube sampling technique. CFD simulations were performed on these sampling points, and the output results along with the design variables were fed to the surrogates as the initial population size. Four surrogate methods, namely response surface approximation, radial basis function, Kriging and weighted average surrogates, were constructed for populating the initial sample points. A hybrid genetic algorithm was used to search for the optimal point. The output predictions from the pool of each surrogate were then validated by CFD simulations using the CAD models generated for these design variables. A good match was observed between the surrogate predictions and actual CFD simulations and maximum efficiency of almost 48.24% was achieved. The relative increment in efficiency was approximately 28% for the peak value of efficiency reported at the flow coefficient of 1.24. The optimized design was reported to have 22 number of guide vanes, 35 number of rotor blades and guide vane angle of 39.64°. The fourth parameter, namely the guide vane profile, was found to remain insensitive to the objective efficiency and was thus eliminated as a design variable. Guide vane angle was reported to be a highly sensitive parameter. The optimized design overall contributed to better flow dynamics over the guide vanes and rotor blades of the impulse turbine, thereby increasing the efficiency effectively.

References

1. Shafiee, S., & Topal, E. (2009). When will fossil fuel reserves be diminished? *Energy Policy*, *37*(1), 181–9.
2. Sovacool, B. K., Noel, L., Kester, J., & de Rubens, G. Z. (2018). Reviewing Nordic transport challenges and climate policy priorities: Expert perceptions of decarbonisation in Denmark, Finland, Iceland, Norway, Sweden. *Energy*, *165*, 532–542.
3. Boyle, G. (2004). *Renewable energy*. Oxford University Press.
4. Khan, N., Kalair, A., Abas, N., & Haider. (2017). A review of ocean tidal, wave and thermal energy technologies. *Renewable and Sustainable Energy Reviews*, *72*, 590–604.
5. Antonio, F. D. (2010). Wave energy utilization: a review of the technologies. *Renewable and Sustainable Energy Reviews*, *14*(3), 899–918.
6. Evans, D. V. (1978). The oscillating water column wave-energy device. *IMA Journal of Applied Mathematics*, *22*(4), 423–33.
7. Maeda, H., Santhakumar, S., Setoguchi, T., Takao, M., Kinoue, Y., & Kaneko, K. (1999). Performance of an impulse turbine with fixed guide vanes for wave power conversion. *Renewable Energy*, *17*(4), 533–47.
8. Thakker, A., Khaleeq, HB., & Ansari, A. R. (2001). Numerical simulation of 0.6 m impulse turbine for wave power conversion under different flow conditions. In *The Eleventh International Offshore and Polar Engineering Conference International Society of Offshore and Polar Engineers*
9. Thakker, A., Hourigan, F. (2005). Computational fluid dynamics analysis of a 0.6 m, 0.6 hub-to-tip ratio impulse turbine with fixed guide vanes. *Renewable Energy*, *30*(9), 1387–1399.
10. Anderson, J. D., & Wendt, J. (1995). *Computational fluid dynamics*. New York: McGraw-Hill.
11. Badhurshah, R., Dudhgaonkar, P., Jalihal, P., & Samad, A. (2018). High efficiency design of an impulse turbine used in oscillating water column to harvest wave energy. *Renewable Energy*, *121*, 344–54.
12. Thakker, A., & Dhanasekaran, T. S. (2004). Computed effects of tip clearance on performance of impulse turbine for wave energy conversion. *Renewable Energy*, *29*(4), 529–47.
13. Badhurshah, R., & Samad, A. (2015). Multiple surrogate based optimization of a bidirectional impulse turbine for wave energy conversion. *Renewable Energy*, *74*, 749–60.
14. Xiong, C., & Liu, Z. (2011). Numerical analysis on impulse turbine for OWC wave energy conversion. In *2011 Asia-Pacific Power and Energy Engineering Conference* (pp. 1–5). IEEE
15. Deb, K. (2001). *Multi-objective optimization using evolutionary algorithms*. Wiley
16. Setoguchi, T., Santhakumar, S., Maeda, H., Takao, M., & Kaneko, K. (2001). A review of impulse turbines for wave energy conversion. *Renewable Energy*, *23*, 261–292.

17. McKay, M. D., Beckman, R. J., & Conover, W. J. (2000). A comparison of three methods for selecting values of input variables in the analysis of output from a computer code. *Technometrics*, *42*(1), 55–61.
18. Samad, A., & Kim, K. Y. (2008). Shape optimization of an axial compressor blade by multi-objective genetic algorithm. *Proceedings of the Institution of Mechanical Engineers, Part A: Journal of Power and Energy.*, *222*(6), 599–611.
19. Goel, T., Haftka, R. T., Shyy, W., & Queipo, N. V. (2007). Ensemble of surrogates. *Structural and Multidisciplinary Optimization*, *33*(3), 199–216.
20. Koziel, S., Ciaurri, D. E., & Leifsson, L. (2011). *Surrogate-based methods in Computational optimization, methods and algorithms* (pp. 33–59). Berlin, Heidelberg: Springer.
21. Badhurshah, R., & Samad, A. (2014). Surrogate assisted design optimization of an air turbine. *International Journal of Rotating Machinery*.
22. He, D. K., Wang, F. L., & Mao, Z. Z. (2008). Hybrid genetic algorithm for economic dispatch with valve-point effect. *Electric Power Systems Research.*, *78*(4), 626–33.

Chapter 16
Control of Wave Energy Converters

R. Suchithra and Abdus Samad

Abstract Energy from the ocean is one of the least exploited renewable and sustainable energy sources, and possesses an enormous amount of untapped energy. The harnessing of energy from the ocean can be accomplished by several methods such as salinity and temperature gradient, waves, currents, and tidal variations. Amidst all, the ocean waves are a highly promising form of energy. Ocean waves have a tremendous amount of energy that can be harvested to meet the increasing energy demand. There are several challenges in wave energy conversion, which include the corrosive environment, the constraints of the converters, untuned device performance for the incoming waves, and irregularities in waves and air velocity. These challenges affect the overall system performance and decrease the overall conversion efficiency. The solution for many of these problems is to tune the device according to the wave climate and a controlled operation of the system. As waves differ in height and period, an optimal control method can help improve the efficiency, performance, and power absorption of a wave energy converter (WEC). This chapter presents a review of various control schemes applied to wave energy devices to achieve higher efficiency. The actual efficiency of WEC in real-time is very less compared to theoretical efficiency which is obtained around 70–80%. This large efficiency gap is due to various stages involved in the wave energy conversion process. To achieve good overall efficiency an optimal design and control techniques need to be applied at different stages of WEC.

16.1 Introduction

The ocean waves are formed by the wind blowing over the surface of the ocean. These waves contain an enormous source of energy. Compared to the other renewable energy resources, the ocean waves have the highest energy density [1]. Approximately the entire ocean can generate an energy of about 8000–80,000 TW/year, i.e.

R. Suchithra · A. Samad (✉)
Wave Energy and Fluids Engineering Laboratory (WEFEL), Department of Ocean Engineering, Indian Institute of Technology Madras, Chennai, Tamil Nadu 600036, India
e-mail: samad@iitm.ac.in

A. Samad et al. (eds.), *Ocean Wave Energy Systems*, Ocean Engineering & Oceanography 14, https://doi.org/10.1007/978-3-030-78716-5_16

a single wave crest holds an power of approximately 10–50 kW/m [2]. These waves generate power constantly around the clock. The power generated by the waves provides a good balance between supply and demand. The energy generated from waves around the world is predicted nearly 1 TW or more [3]. A foremost part of the solution to the global energy emergency can be met by harnessing the enormous power of the waves.

Even though the waves contain this enormous power, the wave energy devices fit in the lower level of the technology readiness diagram [4, 5]. The main reason for this is that the wave energy device components do not operate and synchronize well in different wave climates. In recent years, a lot of attention is given to improve auxiliary subsystems such as gearing systems, control systems, electrical power take-off, power system protection, and grid connections of wave energy converte (WEC). An overall wave energy plant needs to be fully equipped with fault-tolerant devices for power system protection and the ability to withstand and counteract the unpredictable wave climate. The main issues in wave energy conversion include the harsh ocean environment, system limitations, mismatch of wave frequency to the resonant frequency, and irregular wave climates and air velocity. The solution for most of these problems is to control the process of the wave energy device. As waves fluctuate in height and period, a suitable control design would be able to improve the efficiency of a wave energy converter. There are several studies available on the application of control schemes on wave energy plants. The wave energy plant is dynamic in nature; it includes nonlinearities while operating. A control system must be designed such that it has a stable operation in all variable conditions. To extract maximum energy from waves, the hydrodynamic parameters or the device power take-off (PTO) must be tuned depending on the incoming waves. Various methods to control the operation of WECs are given below.

16.2 Control Techniques Applied to WECs

In recent years, a lot of importance is given to improve auxiliary subsystems such as gearing systems, control systems, electrical power take-off, power system protection, and grid connections of WEC. An overall wave energy plant needs to be fully equipped with fault-tolerant devices for power system protection and the ability to withstand and counteract the unpredictable wave phenomena. Challenges in wave energy conversion include the harsh ocean environment, system limitations, mismatch of wave frequency to the resonant frequency, and irregular wave climates and air velocity. Owing to all these difficulties, the complete device performance is drastically affected. The solution for many of these problems is to control the process of the wave energy device. An efficient control method can help tune the system corresponding to the input wave climate and improve the efficiency of a wave energy converter. There are several pieces of literature available on the application of control schemes on wave energy plants. The control strategies on WEC is broadly classified as:

1. Control of WEC primary parameters
2. Control of Secondary Converters.

16.2.1 Control of WEC Primary Parameters

The control system improves the cost-effectiveness of the WEC devices. The optimal hydrodynamic control increases energy production by a factor of 1.5–2.8 [6]. The control system adds an advantage in the reduction of initial cost due to the reduction in system dimensions for some specific amount of energy. Most of the WEC devices are based on relative oscillatory motion between two bodies or oscillating pressure distribution within a closed chamber. These oscillators have their resonance, which can be used to extract energy during certain wave conditions [7]. Due to the seasonal variation of wave spectra, the year-averaged efficiency of wave energy devices becomes very low. The solution to this problem is a proper selection of a control scheme. A control scheme can modify the dynamics of the device such that the efficiency of the WEC can be enhanced for a wide range of wave climates. Various kinds of active control mainly focus on structural control and control may be using passive loading, phase control by latching.

16.2.1.1 Passive Loading

To extract maximum energy from ocean waves, the device PTO has to be tuned depending on the incoming waves. Thus, control of the amplitude of the wave energy device motion is done with the help of tuning the damping of the device. Molinas et al. [8] developed a control scheme for primary conversion stages, which would increase the energy production of WEC. They developed a model using power electronic converters implemented on the power network. Passive loading refers to the implementation of a mechanism to alter dynamic resistance, and mechanical resistance experienced by the waves when they incident upon the device. Based on the wave time scale, this dynamic resistance may change constantly depending on the need. The load resistance is kept constant, provided by some machinery, which would maximize the converter energy. In this study, an oscillating water column (OWC) WEC system dynamic is studied, considering it as a cylinder with a hemispherical bottom subject to heave motion. Passive loading is a form of a linear control strategy, where the force exerted by the PTO is directly dependent on the velocity exerted by the air oscillation or the velocity of the buoy. In this method, the amplitude is only controlled, and no consideration is given to the phase. The comparison between passive loading with optimum control and its performance in irregular waves is done by Tedeschi et al. [9]. The peak power always doubles when passive loading is applied for sinusoidal wave conditions. The comparison of passive loading and optimal control shows that the maximum energy extraction is possible when the damping and the optimal loading are almost equal. Different passive loading schemes are classified in Table 16.1.

Table 16.1 Various passive loading schemes

Phase and amplitude	Discrete control	Latching control (sub-optimal)
		Declutching control
	Continuous control	Reactive control (optimal)
	Complex conjugate control (optimal)	
Airflow control	Valve in parallel with the turbine	
	Valve in series with the turbine	

16.2.1.2 Amplitude and Phase Control

Optimum interaction between the WEC device and the incoming wave is achieved by controlling the velocity of the device. The purpose of control is to use the optimum phase and amplitude to extract maximum energy. The optimum phase is obtained when the oscillating system is in resonance with the incoming wave. Falnes [10] described a system with optimal phase control of OWC-WEC, which is done by increasing the oscillation to attain an optimal interaction between the plant and wave. Maximum power absorption in the OWC device takes place under two conditions: the oscillation velocity should be in phase with the excitation force, and the amplitude of the oscillation must be tuned to the optimum value. This means the absorbed power must be equal to the power radiated into the sea. Two types of control strategies were stated in the above work; the first control strategy is used to maximize the converter power based on the measurement of the input wave. The second one is the amplitude of the oscillation needs to be tuned to the optimum value. This means the absorbed power to be equal to the power reradiated into the sea. Nunes et al. [11] applied phase and amplitude control on OWC-WEC containing variable pitch Wells turbine. The study showed that the response of the system could exactly match a particular frequency of the incoming wave, which is different from the device's natural frequency. A controller was developed to perform quality control, which could predict few more seconds into the future for a fixed time window and use the Fourier transform of the incoming wave inside the specific period to perform the spectral analysis. From the data of spectral analysis, a reference for the controller was constructed. The modeling of OWC-WEC in the above work did not consider the diffracted and radiated flow. However, for an OWC device exhibiting second order dynamic behavior, the optimal condition for energy extraction took place when the oscillating pressure inside the chamber and diffraction flow were in phase with each other. The excitation force is due to the action of water inside the chamber. To develop phase control on OWC-WEC, a Wells turbine with variable pitch was used in that work. As this turbine can act either as a compressor or as a turbine, there is a chance to control the pressure inside the chamber. The main factors behind implementing any control strategy are the quality of energy extracted and the cost of implementation. Any efficient control scheme must have a good correlation between these two factors.

16.2.1.3 Discrete Control

(a) **Latching control**

Latching control is a phase control scheme that was developed in the early 1970s by Budal and Falnes [12], which was used to demonstrate the application of control systems for optimizing the motion of WEC, which was oscillatory. In this method, the velocity of the absorber is put on to be in phase with the excitation forces, which is done by clutching the WEC system at an upper and lower position. In other words, the adequate delay in the release of the device in order to bring it to operate in phase with the diffraction force would make the system operate in a near-optimal condition. Thus, latching control is a sub-optimal control strategy as the amplitude is made sub-optimal, however, the phase is forced to be in optimal range by resisting the moving part of the body constant for a certain duration of the wave cycle.

Evaluation of latching control for heave oscillations was done by Korde [13]. He implemented latching control of a floating WEC implemented in deep water. The evaluation was done with a variation formulation of the control sequence in the time domain. Latching is applied to a hemispherical platform from a stationary onboard platform. The equipment includes a pneumatic cylinder with open/ close valves for latching. This time-domain calculation is carried out for small-amplitude regular and irregular waves [14]. Babarit et al. [15] described a latching control with the help of weak modeling, which is used to apply optimal control theory. It involves the generation of parametric resonance in the response of the mechanical system. Two types of WECs were studied. One was a heaving buoy with one degree of freedom. Next was a four-degree freedom WEC called SEAREV [16]. Latching applied to the above said sub-optimal weak formulation in regular waves, which could also be used to get an optimally controlled WEC in the time domain in a random sea. The analysis made is the maximum benefit of latching control on WEC [17]. A Comparative study of latching control for heaving point absorber is done in reference [18] were two strategies aiming at maximizing the absorbed energy and keeping the velocity and excitation force in phase. The result shows that irrespective of the control strategy of the WEC, the efficiency improves with latching control.

(b) **Declutching control**

Declutching or unlatching control, a dual latching control, involves the decoupling of the PTO. This infers that the PTO force is made equal to zero in the equation during certain portions of the power cycle. This type of control is mainly implemented in devices that have a longer natural period than the useful part of the spectrum. The primary moving part is permitted to adjust freely for half of the wave cycle, and the desired velocity PTO system is achieved. This method was listed in the paper [19], which showed that the correct PTO strategy can improve the power capture of wave plants from 12–15%. Unlatching is accomplished by bypassing the pumps at specified moments. These moments are established by the optimal command theory. Clement et al. [20] described discrete control strategy, which involved latching, unlatching, and combined latching-unlatching. Which showed that latching control needs mechanical breaks and locks, whereas declutching control needs simple elements such as valves in the case of hydraulic PTO and switches in the case of electrical PTO. Pico plant uses this method to prevent damage in the system, and SEAREV WEC uses it to increase energy production. Babarit et al. [15] described two control strategies such as pseudo- continuous control and declutching control shows the improvement of energy absorption with the implementation of the above-said methods.

16.2.1.4 Continuous Control: Reactive Control

The maximum energy from the WEC can be extracted when the amplitude and phase of the incident wave excite the resonant oscillations within the chamber. When the OWC is loaded, it can be expressed with a real part and an imaginary part. If at resonance, the real part equals the imaginary part of the resulting force from water column oscillation, 100% of the incident wave can be absorbed. However, when the load is applied to WEC, this load exhibits a reactive component, which is not equal to the resistive component. By controlling the resistive and reactive components, a maximum amount of energy extraction is possible. However, to achieve this maximum energy, optimum damping and resonance conditions should correlate with a particular wave period. Korde [21] has given a significant contribution to the reactive control of different wave energy devices. Reactive control is a frequency domain control used in the early days. The dynamic parameters such as spring coefficient and inertia of the WEC in primary conversion are mainly considered for control. Control is done to enable maximum energy absorption at a particular frequency. Korda's work [6, 21] formulated that, for a particular wave frequency, the wave energy device is resonated with the help of feedback signals of the displacement and acceleration commands. At the same time, for a particular frequency, the damping coefficient of the PTO mechanism can be tuned to balance the radiation damping of the system using velocity feedback. This allows the hydro-dynamically maximum energy absorption and can be done at each peak frequencies of the incoming wave spectrum. In order to achieve reactive control, the force should be in phase with displacement and acceleration. Many devices exploit the resonance phenomena to achieve maximum power, such as OWC-WECs and point absorbers. The importance of frequency matching considering dynamic variables is explained by Andrade et al. [22]. In principle, reactive control can be achieved by reversing the oil flow during part of the wave cycle. However, numerical analysis in [23] showed that reactive control is hard to achieve by reversing the oil flow direction during part of the wave cycle.

16.2.1.5 Complex Conjugate Control

In a point absorber system, the mechanical resonance appears when the imaginary part of the total impedance sums up to zero. For such a device the maximum energy absorption takes place when the impedance of the PTO mechanism becomes equal to its complex conjugate [22]. When such a control scheme is applied, the maximum energy transfer depends on the upcoming buoy's velocity values. This method is non-causal because it is needed to estimate an irregular wave for some instance into the future. Thus, the optimal PTO force cannot be implemented adequately in practice. To deal with non-causality, an approximate complex conjugate method is proposed. This involves causal functions in order to implement some alternative sub-optimal control strategies.

16.2.2 Airflow Control

A method to change the flow rate is by providing valves. There are two ways in which valves can be placed for an OWC duct, one is placing a bypass valve, and another method is to place a throttle valve. The enhancement of the OWC-WEC performance using bypass and throttle valve was studied by Falcão and Justino [24]. The study concluded that the use of a single or set of valves delayed the stall in the turbine and substantially increased the energy produced by the WEC, mainly during high wave incidence. The study also stated that the bypass valves have opted for a system where the turbine damping is much higher than the optimum damping, and throttle valves are mainly used for underdamped systems.

The performance of OWC-WEC coupled to a Wells turbine is significantly affected due to two reasons. Firstly, the unsteady bidirectional airflow across the turbine. The second reason is due to increasing the flow rate above critical limits. Due to these reasons, mechanical efficiency and power output drop rapidly. A method to change the flow rate is by providing valves (Fig. 16.1). There are two ways in which valves can be placed for an OWC duct, one is placing a bypass valve, and another method is to place a throttle valve. The next study using only bypass valves was given in the literature [25], where an increase in energy production was observed with the use of the valves. In the works of literature [26–28], the throttle valve is used in series with the device. A propotional integral derivative (PID) controller is used to control the position of the valve according to the incident wave along with the rotational speed control of the turbine. This complementary control action delayed the stall by altering the flow velocity across the turbine. However, their work used classical controllers, which are inefficient for robust tracking of the reference signals, and external disturbance rejection as the valve actuator is typically an uncertain system due to friction, external disturbances, and parametric uncertainties due to the cumbersome identification process. In the OWC device, the mass flow rate identification is a very tedious task. Thus, such systems have high parametric uncertainties, thus the conventional controller often fails to perform well.

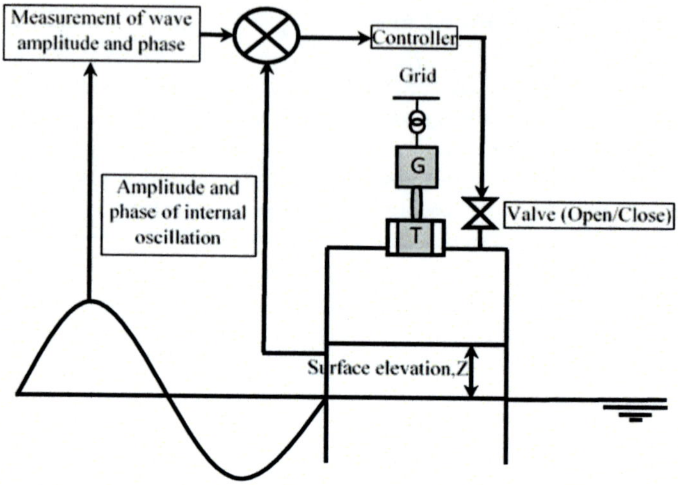

Fig. 16.1 Schematic of amplitude and phase control of OWC-WEC

Falcao et al. [24, 25] developed a model for efficiency improvement using a valve in series or parallel with the turbine shaft for airflow control the airflow. Numerical analysis shows that the turbine efficiency is particularly affected by two behaviors. Firstly, due to the unsteady flow of air displacement by free surface water oscillation. Secondly, due to the proportionality of airflow with turbine rotational speed. This model involves a throttle valve or bypass valve based on whether the WEC device is over-damped or under-damped by the turbine. The airflow control scheme also prevents losses due to aerodynamic stalls at the turbine rotor blade and prevents excessive air flow rate. This is of extreme importance when the performance of the turbine is greatly affected by rotor stalling. Study shows that throttle valve in series with the turbine is better compared to parallel mounted in enhancing the power and reduction of stalling.

16.2.3 Control of Secondary Converters

The control of the WEC device has to meet several criteria such as allowable speed range for optimal operation, the aerodynamic performance of the turbine, power quality, allowable fluctuation in power delivered to the grid, and efficiency of the wave to wire conversion. Control needs to be allied to various stages in order to obtain an efficient operation. Various control schemes applied to secondary are tabulated in Table 16.2.

Table 16.2 Control strategies for secondary converters

Rotational speed control	Voltage/frequency control on stator
	Voltage/frequency on rotor
Direct torque control	
Power quality control	
Real-time control of turbine blade pitch or geometry control [29, 30]	

16.2.3.1 Rotational Speed Control

Fixed rotational speed turbines are capable of operating only within a limited range of flow conditions around the best operating point. In particular, for the Well turbine during the low sea state, the power output is low were as during the most energetic sea state also its performance is poor which mainly due to aerodynamic losses. This loss is mainly caused by rotor stalling. Wells turbine operates only within a certain bandwidth. The restricted bandwidth is due to stalling. An efficient way to widen the response of the turbine is with the help of variable rotational speed. This enables the turbine to respond to a wide range of sea states. As the turbine flow rate is proportional to speed and power output is a proportional cube of speed. Thus, if the turbine speed is controlled, one can enhance the operation of the turbine by delaying the stall and maximizing the power absorption. The rotational speed should approximately match sea conditions. With just doubling the speed, the power output of the turbine is increased by 8 times; this increase of speed is done with the help of power electronic devices. The scheme of speed control was first addressed by Justino et al. [31] for the OWC plant. Their work considered three control strategies: (i) the electrical torque is controlled as a piecewise constant function of time. (ii) The electrical torque is controlled in such a way that differential torque is always a function of differential speed with respect to time. (iii) The electrical power is controlled as a continuous function of one or several state variables, as power is a function of pressure and speed. Voltage/frequency (V/F) control law is applied on stator and rotor as separate strategy and a comparison is made; his work shows that V/F control applied on the stator is optimal. There is always a trade-off between output power quality and the efficiency of a plant.

In the literature [32], a stochastic model was developed to design an optimal algorithm to control the turbine rotational speed of an OWC device coupled with the Wells turbine. The model is developed in the frequency domain with the assumption of known hydrodynamic coefficients and turbine performance curves. Linear control theory is applied on a random sea state, considering sea surface height as a normal probability density function. The relation between the instantaneous generator torque and rotational speed is considered for developing an optimal control law. Their work addresses several losses associated with a wave energy converting device, such as (i) viscous losses due to fluid effect; (ii) aerodynamic losses due to turbine stall; (iii) bearing frictional losses, and (iv) losses in the generator.

These losses can be minimized by successfully implementing control strategies to several stages of the plant. Stochastic modeling is mainly involved if the device is considering operating in random sea waves. The study concluded that a valve system for airflow control, when operated near-optimal point, could only increase the average power output by 3%. However, the control of rotational speed could increase the energy production by 20% to match the individual sea state.

Alberdi et al. [28] proposed a complementary control scheme in which both airflow and rotational speed are controlled. The rotational speed is adjusted instantaneously according to the slip of the generator in order to match the available sea state. The speed is regulated such that the stall region is avoided. This change of speed according to the input wave condition is done with the help of power electronic devices. A vector control scheme (field orientation control) is employed in such cases when alternating current (AC) machines are used in the systems. In this method, the three-phase current and voltage vectors are decoupled into two vector components each, one component is directly linked with the active power (a function of rotational speed), and the other is linked with the reactive power of the system. Thus, the systems can be controlled in terms of these vectors. The control block is modeled as a feed-forward controller where PI controllers are used for regulating the power output. It is observed that there is an improvement in the instantaneous power output, in particular, and there is a regulation of power quality. Portillo et al. [33] investigated the control problem on the OWC device, theoretically and experimentally. Their study was based on latching control and speed control of turbine using easily measurable variables by developing a causal control law. The rotational speed control was done using power and torque coefficients of the turbine to formulate a stochastic relationship to control the turbine and making it operates near the best operating point. This rotational scheme is also helpful in the case of an OWC device coupled with an impulse turbine. Even though this turbine is not affected by stalling and has a broad operating range compared to Wells turbine, there are certain losses due to an excessive incidence flow angle at the entry to the second row of guide vanes affecting the turbine performance. These losses are predominant during certain range of flow coefficient. Thus, by controlling the speed the turbine can be operated in the flow coefficient where the losses are minimum. The state of the art of the airflow control and rotational speed control is given in Table 16.3.

16.2.3.2 Direct Torque Control

Tedd et al. [44] modeled a direct torque control (DTC) technique for wave dragons. This device is a fully automated prototype that is deployed in Denmark. Wave dragon does not involve any active control to survive extreme sea conditions. The improvement in power absorption is greatly influenced by different control systems used in WEC. Control is split as slow, fast, and very fast. Very fast control ensures the turbine runs at the optimal speed, and the power quality is maintained good enough to deliver to the grid. Power electronics are used to give good performance and a stable system. Direct torque control is based on space vector modulation DTC- SVM as it is more

Table 16.3 State of the art of the control of OWC-WEC

Literature	Control technique	Turbine	Control variable	Inference
[34]	Turbine pitch control	Wells	Changing the pressure drop and flow rate across the turbine	Increase in the amount of energy extracted from sea state
[35][36]	Rotational speed and airflow	Wells	Control algorithm implemented in PLC	Single-speed operation insensitive to wave climate
[31] [37][38]	Rotational speed	Wells and bi-radial	Control of electromagnetic torque in PLC	Variable speed control operation
[28]	Rotational speed and airflow control	Wells	Complementary control: Throttle valve control using PI and control of rotational speed using PI	Variable speed control to delay the stall of the turbine
[33, 39]	Threshold latching control and rotational speed control	Bi-radial	Suboptimal latching control and theoretical formulation for rotational speed control	Numerical implementation of latching and speed control as a suboptimal optimization problem
[40–43]	Rotational speed control	Bi-radial	Hardware in-loop of back to back converter with PLC and a receding horizon latching control scheme is used	Increase in mean power of the spar buoy OWC device

advantageous than classical DTC. A review of various DTC for induction machines is given by Reza et al. [45]. This method combines DTC and FOC (field orientation control) for better operation. The proper implementation of torque controller is important concerning the power quality issue [46]. The DTC also ensures a power control for a wave energy device.

16.2.3.3 Power Quality Control

Considering the control of an electrical drive of WECs, there are two converters, namely rotor side converter, and grid side converter. The power control scheme helps in the protection of the rotor side converter to avoid overshoot in the rotor current.

The power quality issues in the wave energy converter need to rectify with a significant minimal error to connect it to a grid. Power quality issues are the total harmonic distortion, peak overshoot, sag, swell, and low power factor. These issues need appropriate attention in order to meet the energy demand. Lagoon et al. [47] worked on a model-based predictive power control scheme to ensure the active and reactive power to attain their reference values, which is designed for doubly fed induction generator (DFIG) using stator flux decoupling and deadbeat control loop. The control action was implemented on a model to be controlled, depending on the optimal reference point. The predictive control scheme optimizes the cost function and formulates the future control action. The model is assumed to be linear, and its state-space equations are used to calculate its active and reactive powers. For this analysis, information regarding leakage induction and resistance is needed for this control scheme. The paper addresses various controllers which can be chosen for power control such as internal model controller, fuzzy logic controller, direct torque controller its advantage and disadvantage are discussed. The internal mode controller has a good power response compared to the PI controller, but it is difficult to employ due to the predictive functional controller and internal mode controller scheme. When considering fuzzy logic controller, it contains a relatively complex transformation of voltages, currents, and the control outputs between the stationary, rotor, and synchronous reference frames. In the indirect torque controller, the switching states are chosen from an optimal switching table depending on the instantaneous error between the reference and predicted values of active and reactive power and angular position. The study concluded that the control scheme had better tracking performances for DFIG. Other control problems that arise on the grid side of the wave energy converter are active and reactive power control, DC-link voltage control, injected power quality control, grid synchronization, voltage harmonic compensation, and fault handling.

Xiang et al. [48] worked on developing power control for a direct drive wave energy converter. Power control stages require the following requirements: (i) Reactive power compensation in the active rectifier. Maximum energy transfer between the generator to the dc link is realized by means of a unity power factor controller, (ii) Regulation of the dc capacitor voltage, (iii) Current control of the inverter output current and consequently the inverter power into the power sink. Two control strategies addressed are frequency tuning control and unity power factor control. Both ensure a proper improvement in the power quality requirement of the grid. Garrido et al. [49] developed a control scheme for rotor side and grid side converter to control the power output of OWC coupled with a Wells turbine (Fig. 16.2). They used a sensor-less control scheme with an observer for induction rotor speed based on the disturbance model. For any given sea state, the control scheme could extract the maximum possible power using a point tracking technique. The performance curve is predefined for each turbine; the rotor speed is regulated so that the flow coefficient remains under desired bandwidth, resulting in a stall-free operation, yielding maximum power output.

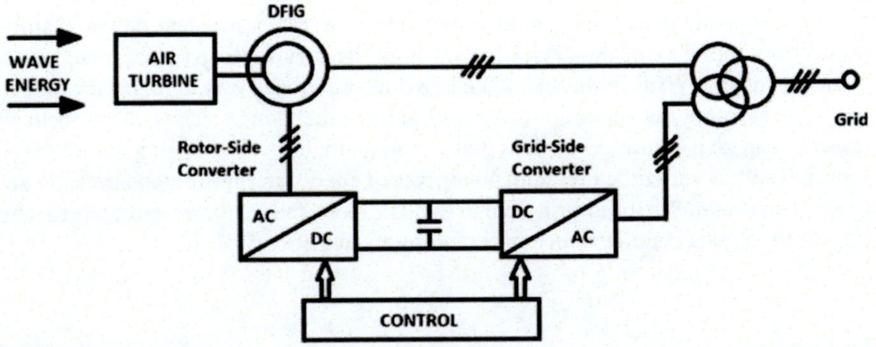

Fig. 16.2 Rotor side and grid side converter control of OWC-WEC

The performance of the OWC-WEC turbine can be significantly improved by the application of an appropriate control technique. The operating principle of OWC-WEC is typically different from other WECs, such as the point observers, AWS-WEC, overtopping devices. Thus, control strategies are also different for OWC based devices and have different control variables compared to other devices. For example, the reactive control variable for OWC-WEC is the air turbine blade angle. In the case of latching control, the control variable is the pressure inside the chamber; for airflow control, the valve area is altered, and for rotational speed control, the electromagnetic torque is controlled by appropriately switching the power converters.

16.2.3.4 Active Control of Turbine Blade Pitch or Geometry Control

The power extracted from the waves in the OWC device is limited mainly due to the stalling phenomena of the Wells turbine. In order to address this problem, a study was proposed to change the instantaneous angle of incidence of the turbine blade by controlling the blade pitch [30, 34]. However, this method faces technical difficulties, such as the implementation and maintenance of the actuators.

16.3 Conclusion

This chapter has summarized the recent research advances in control solutions applied to various wave energy converter technologies. Typical illustrations are addressed for OWC-WEC and AWS types of converters. With the increasing importance and complexity of energy systems, there is a need to focus much on the performance, scalability, reliability of the system. State-of-the-art gives good hope for developing wave energy converter technology. This chapter mainly insists on the importance of the involvement of each stage in energy conversion. Suitable selection of optimal control laws can significantly improve performance, efficiency, and power absorption. The complexity of the device is very high. Its survivability and cost are also a great concern for WEC developers. The initial installation is very high, but the potential of wave energy is very high compared to any other source. Thus, more focused research can give optimized devices that can adapt to the ever-changing wave energy source. There is very little literature on the role of energy storage in standalone power systems and complexity in grid connection. In these fields, future research can be improved to get a stabilized, better-performing wave energy plant.

References

1. Clément, A., McCullen, P., Falcão, A., Fiorentino, A., Gardner, F., Hammarlund, K., et al. (2002). Wave energy in Europe: Current status and perspectives. *Renewable and Sustainable Energy Reviews, 6*, 405–431. https://doi.org/10.1016/S1364-0321(02)00009-6
2. Sannasiraj, S. A., & Sundar, V. (2016). Assessment of wave energy potential and its harvesting approach along the Indian coast. *Renewable Energy, 99*, 398–409. https://doi.org/10.1016/j.renene.2016.07.017
3. Falnes, J. (2007). A review of wave-energy extraction. *Marine Structures, 20*, 185–201. https://doi.org/10.1016/j.marstruc.2007.09.001
4. Cruz, J. (2008). *Ocean wave energy*. Berlin, Heidelberg: Springer.
5. Masuda, Y. (1986). *Hydrodynamics of ocean wave-energy utilization*. Springer. https://doi.org/10.1007/978-3-642-82666-5
6. Korde, U. A. (2000). Control system applications in wave energy conversion. *Proceedings Oceans, 3*(2000), 1817–1824. https://doi.org/10.1109/OCEANS.2000.882202
7. Marsh, G. (2009). Maximising the power of waves. *Renewable Energy Focus, 10*, 80–84. https://doi.org/10.1016/S1755-0084(09)70242-X
8. Molinas, M., Skjervheim, O., Andreasen, P., Undeland, T., Hals, J., & Moan, T., et al. (2007). *Power electronics as grid interface for actively controlled wave energy converters.* In 2007 International Conference Clean Electrical Power, ICCEP '07, 2007 (pp. 188–195). https://doi.org/10.1109/ICCEP.2007.384210
9. Tedeschi, E., Carraro, M., Molinas, M., & Mattavelli, P. (2011). Effect of control strategies and power take-off efficiency on the power capture from sea waves. *IEEE Transactions on Energy Conversion, 26*, 1088–1098. https://doi.org/10.1109/TEC.2011.2164798
10. Falcão, A. F. D. O. (2008). Phase control through load control of oscillating-body wave energy converters with hydraulic PTO system. *Ocean Engineering, 35*, 358–366. https://doi.org/10.1016/j.oceaneng.2007.10.005
11. Nunes, G., Valério, D., Beirão, P., Sá da Costa, J. (2011). Modelling and control of a wave energy converter. *Renewable Energy, 36*, 1913–1921. https://doi.org/10.1016/j.renene.2010.12.018
12. Tedeschi, E., & Molinas, M. (2012). Tunable control strategy for wave energy converters with limited power takeoff rating. *IEEE Transactions on Industrial Electronics, 59*, 3838–3846. https://doi.org/10.1109/TIE.2011.2181131
13. Korde, U. A. (1991). On the control of wave energy devices in multi-frequency waves. *Applied Ocean Research, 13*, 132–144. https://doi.org/10.1016/S0141-1187(05)80060-4
14. Korde, U. A. (2002). Latching control of deep water wave energy devices using an active reference. *Ocean Engineering, 29*, 1343–1355. https://doi.org/10.1016/S0029-8018(01)00093-2
15. Babarit, A., Guglielmi, M., & Clément, A. H. (2009). Declutching control of a wave energy converter. *Ocean Engineering, 36*, 1015–1024. https://doi.org/10.1016/j.oceaneng.2009.05.006

16. Cordonnier, J., Gorintin, F., De Cagny, A., Clément, A. H., & Babarit, A. (2015). SEAREV: Case study of the development of a wave energy converter. *Renewable Energy, 80*, 40–52. https://doi.org/10.1016/j.renene.2015.01.061
17. Babarit, A., & Clément, A. H. (2006). Optimal latching control of a wave energy device in regular and irregular waves. *Applied Ocean Research, 28*, 77–91. https://doi.org/10.1016/j.apor.2006.05.002
18. Babarit, A., Duclos, G., & Clément, A. H. (2004). Comparison of latching control strategies for a heaving wave energy device in random sea. *Applied Ocean Research, 26*, 227–238. https://doi.org/10.1016/j.apor.2005.05.003
19. Salter, S. H., Taylor, J. R. M., & Caldwell, N. J. (2002). Power conversion mechanisms for wave energy. *Proceedings of the Institution of Mechanical Engineers Part M, 216*, 1–27. https://doi.org/10.1243/147509002320382112
20. Clement, A. H., & Babarit, A. (2012). Discrete control of resonant wave energy devices. *Philosophical Transactions of the Royal Society A—Mathematical Physical and Engineering Sciences, 370*, 288–314. https://doi.org/10.1098/rsta.2011.0132
21. Korde, U. A. (1991). Development of a reactive control apparatus for a fixed two-dimensional oscillating water column wave energy device. *Ocean Engineering, 18*, 465–483. https://doi.org/10.1016/0029-8018(91)90026-M
22. Montoya Andrade, D.-E. A., García Santana, A., de la Villa Jaén, A. (2012). Frequency-matching assessment under reactive control on wave energy converters. In *Proceedings of the 4th International Conference in Ocean Engineering (ICOE 2012)* (pp. 1–6).
23. Falcão, A. F. O., Justino, P. A. P., Henriques, J. C. C., & André, J. M. C. S. (2009). Reactive versus latching phase control of a two-body heaving wave energy converter. In *Proceeding of the European Control Conference 2009* (pp. 1–6).
24. Falcão, A. F. d. O., & Justino, P. A. P. OWC wave energy devices with air flow control. *Ocean Engineering, 26*, 1275–95. https://doi.org/10.1016/S0029-8018(98)00075-4
25. Falcão, a. F. D. O., Vieira, L. C., Justino, P. a. P., & André, J. M. C. S. (2003). By-pass air-valve control of an OWC wave power plant. *Journal of Offshore Mechanics and Arctic Engineering, 125*, 205. https://doi.org/10.1115/1.1576815
26. Amundarain, M., Alberdi, M., Garrido, A. J., Garrido, I., & Maseda, J. (2010). Wave energy plants: Control strategies for avoiding the stalling behaviour in the Wells turbine. *Renewable Energy, 35*, 2639–2648. https://doi.org/10.1016/j.renene.2010.04.009
27. Alberdi, M., Amundarain, M., Garrido, A. J., Garrido, I., & Maseda, F. J. (2011). Fault-ride-through capability of oscillating-water-column-based wave-power-generation plants equipped with doubly fed induction generator and airflow control. *IEEE Transactions on Industrial Electronics, 58*, 1501–1517. https://doi.org/10.1109/TIE.2010.2090831
28. Alberdi, M., Amundarain, M., Garrido, A. J., Garrido, I., Casquero, O., & De la Sen, M. (2011). Complementary control of oscillating water column-based wave energy conversion plants to improve the instantaneous power output. *IEEE Transactions on Energy Conversion, 26*, 1021–1032. https://doi.org/10.1109/TEC.2011.2167332
29. Kim, T. H., Setoguchi, T., Takao, M., Kaneko, K., & Santhakumar, S. (2002). Study of turbine with self-pitch-controlled blades for wave energy conversion. *International Journal of Thermal Sciences, 41*, 101–107. https://doi.org/10.1016/S1290-0729(01)01308-4
30. Setoguchi, T., Raghunathan, S., Takao, M., & Kaneko, K. (1997). Air-Turbine with self-pitch-controlled blades for wave energy conversion (Estimation of performances in periodically oscillating flow). *International Journal of Rotating Machinery, 3*, 233–238. https://doi.org/10.1155/S1023621X97000213
31. Justino, P. A. P., & Falcão, A. F. de O. (1999). Rotational speed control of an OWC wave power plant. *Journal of Offshore Mechanics and Arctic Engineering, 121*, 65. https://doi.org/10.1115/1.2830079.
32. Falcao, A. F. de O. (2004). Stochastic modelling in wave power-equipment optimization: Maximum energy production versus maximum profit. *Ocean Engineering, 31*, 1407–1421. https://doi.org/10.1016/j.oceaneng.2004.03.004

33. Portillo, J. C. C., Henriques, J. C. C., Falcão, A. F. O., Gomes, R. P. F., Gato, L. M. C. (2015). Theoretical and experimental investigation on latching and rotational speed control. *Renewable Energies Offshore*, 471–478.
34. Sarmento, A. J. N., Gato, L. M. C., & Falcão, A. F. O. (1990). Turbine-controlled wave energy absorption by oscillating water column devices. *Ocean Engineering*, *17*, 481–497. https://doi.org/10.1016/0029-8018(90)90040-D
35. Falcão, A. F. d. O. (2002). Control of an oscillating-water-column wave power plant for maximum energy production. *Applied Ocean Research*, *24*, 73–82. https://doi.org/10.1016/S0141-1187(02)00021-4
36. Falcão, A. F. O., & Rodrigues, R. J. (2002). Stochastic modelling of OWC wave power plant performance. *Applied Ocean Research*, *24*, 59–71. https://doi.org/10.1016/S0141-1187(02)00022-6
37. Falcão, A. F. O., & Henriques, J. C. C. (2014). Model-prototype similarity of oscillating-water-column wave energy converters. *International Journal of Marine Energy*, *6*, 18–34. https://doi.org/10.1016/j.ijome.2014.05.002
38. Falcão, A. F. O., Henriques, J. C. C., & Gato, L. M. C. (2016). Air turbine optimization for a bottom-standing oscillating-water-column wave energy converter. *Journal Ocean Engineering Marine Energy*, *2*, 459–472. https://doi.org/10.1007/s40722-016-0045-7
39. Henriques, J. C. C., Gato, L. M. C., Falcão, A. F. O., Robles, E., & Faÿ, F. X. (2016). Latching control of a floating oscillating-water-column wave energy converter. *Renewable Energy*, *90*, 229–241. https://doi.org/10.1016/j.renene.2015.12.065
40. Henriques, J. C. C., Gato, L. M. C., Lemos, J. M., Gomes, R. P. F., & Falcao, A. F. (2016). Peak-power control of a grid-integrated oscillating water column wave energy converter. *Energy*, *109*, 378–390. https://doi.org/10.1016/j.energy.2016.04.098
41. Henriques, C., Kelly, J., Mueller, M., Faÿ, F., Abusara, M., & Sheng, W., et al. (2020). Comparative assessment of control strategies for the biradial turbine in the Mutriku OWC plant. 146, 2766–2784. https://doi.org/10.1016/j.renene.2019.08.074
42. Falcão, A. F. O., Henriques, J. C. C., & Gato, L. M. C. (2017). Rotational speed control and electrical rated power of an oscillating-water-column wave energy converter. *Energy*, *120*, 253–261. https://doi.org/10.1016/j.energy.2016.11.078
43. Henriques, J. C. C., Portillo, J. C. C., Sheng, W., Gato, L. M. C., & Falcão, A. F. O. (2019). Dynamics and control of air turbines in oscillating-water-column wave energy converters: Analyses and case study. *Renewable and Sustainable Energy Reviews*, *112*, 571–589. https://doi.org/10.1016/j.rser.2019.05.010
44. Tedd, J., Kofoed, J. P., Jasinski, M., Morris, A., Friis-Madsen, E., & Wisniewski, R., et al. (2007). Advanced control techniques for WEC wave dragon. In *Proceeding of the 7th European Wave and Tidal Energy Conference* (pp. 1–7).
45. Reza, C. M. F. S., Islam, M. D., & Mekhilef, S. (2014). A review of reliable and energy efficient direct torque controlled induction motor drives. *Renewable and Sustainable Energy Reviews*, *37*, 919–932. https://doi.org/10.1016/j.rser.2014.05.067
46. Merabet Boulouiha, H., Allali, A., Laouer, M., Tahri, A., Denaï, M., & Draou, A. (2015). Direct torque control of multilevel SVPWM inverter in variable speed SCIG-based wind energy conversion system. *Renewable Energy*, *80*, 140–152. https://doi.org/10.1016/j.renene.2015.01.065.
47. Lagoun, M. S. (2014). A predictive power control of doubly fed induction generator for wave energy converter in irregular waves. In *International Conference on Green Energy ICGE* (pp. 26–31).
48. Xiang, J., Brooking, P. R. M., & Mueller, M. A. (2002). Control requirements of direct drive wave energy converters. In *IEEE Region 10 international conference on computers, communications, control power engineering TENCOM '02* (Vol. 3, pp. 2053–2056). https://doi.org/10.1109/TENCON.2002.1182746.
49. Garrido, I., Garrido, A. J., Alberdi, M., Amundarain, M., & Barambones, O. (2013). Performance of an ocean energy conversion system with DFIG sensorless control. *Mathematical Problems in Engineering*, *2013*. https://doi.org/10.1155/2013/260514.

Chapter 17
Recent Advances in Direct-Drive Power Take-Off (DDPTO) Systems for Wave Energy Converters Based on Switched Reluctance Machines (SRM)

Marcos Blanco, Jorge Torres, Miguel Santos-Herrán, Luis García-Tabarés, Gustavo Navarro, Jorge Nájera, Dionisio Ramírez, and Marcos Lafoz

Abstract This chapter is focused on Power Take-Off (PTO) systems for wave energy converters (WEC), being one of the most important elements since PTOs are responsible to transform the mechanical power captured from the waves into electricity. It presents Direct-Drive PTO (DDPTO) as one of the most reliable solutions to be adapted to some particular types of WEC, such as point absorbers. A discussion about modularity and adaptability, together with intrinsic characteristics of direct-drive PTOs, is also included. Among the different technologies of electric machines that can be used in direct-drive linear PTOs, switched reluctance machines (SRM) are described in further detail. In particular, the Azimuthal Multi-translator SRM is presented as a suitable solution in order to increase power density and reduce costs. Not only the electric machine, but also the associated power electronics are described in detail. The description includes the different configurations and topologies of power converters and the most appropriate control strategies. Finally, a superconducting linear generator solution is described, presenting it as a reliable alternative for the application of direct-drive PTOs. An example of concept and preliminary

M. Blanco · J. Torres · M. Santos-Herrán (✉) · L. García-Tabarés · G. Navarro · J. Nájera · M. Lafoz
CIEMAT, Avda. Complutense 40, 28045 Madrid, Spain
e-mail: miguel.santos@ciemat.es

M. Blanco
e-mail: marcos.blanco@ciemat.es; marcos.blanco.aguado@ciemat.es

J. Torres
e-mail: jorgejesus.torres@ciemat.es

J. Nájera
e-mail: jorge.najera@ciemat.es

M. Lafoz
e-mail: marcos.lafoz@ciemat.es

D. Ramírez
ETSII-UPM, Cale José Gutiérrez Abascal 2, 28006 Madrid, Spain
e-mail: Dionisio.ramirez@upm.es

A. Samad et al. (eds.), *Ocean Wave Energy Systems*, Ocean Engineering & Oceanography 14, https://doi.org/10.1007/978-3-030-78716-5_17

design is included in order to highlight the main challenges to be faced during this process.

17.1 Introduction

The Power Take-Off (PTO) system is the Wave Energy Converter (WEC), which is the component in charge of extracting the energy from the ocean resource. It is a key element at the core of the energy conversion process from the ocean waves to a final usable form of energy, e.g. electricity. From an economical perspective, it is forecasted that PTOs are going to become one of the main contributors to the wave energy arrays cost [1], representing up to 20–22% of the lifetime costs [2–4].

Accordingly, PTOs have always been at the focus of the research, development, and innovation in the wave energy sector. For instance, the Joint Research Centre (JRC), i.e. the science and knowledge service of European Commission (EC), has identified PTOs as one of the key technological priorities for wave energy since 2014 [5, 6]. In fact, the Strategic Research and Innovation Agenda (SRIA) for Ocean Energy developed under the project ETIP Ocean, which was released in 2020, still recognises "*improvement and demonstration of PTO and control systems*" as a challenge area and a priority topic for the design and validation of ocean energy devices [7]. Moreover, it is crucial for allowing ocean energy to become an essential contributor to the upcoming European Green Deal. As a consequence, in the recent years, EC has funded several projects addressing the development and optimisation of PTOs for wave energy conversion under the Horizon 2020 calls, such as OPERA, WaveBoost, WETFEET, IMAGINE and SEA-TITAN.[1]

The significance of PTO development as one of the key actions for an adequate progress in the sector was reinforced in 2015 with the creation of the Power Take-Off funding programme by Wave Energy Scotland (WES). To date, WES has supported 17 different projects under this PTO-specific call, providing GBP 17.5 million of funding. Furthermore, in 2018 the JRC held a workshop on the identification of future emerging technologies in the ocean energy sector [8], where innovative PTOs were found to be one of the nine main emerging technology families.

Given the aforementioned weight of the PTO cost shared in the total project costs, any potential cost reduction in this component might result in important savings in the final Levelised Cost of Energy (LCOE). This is especially relevant due to the current needs for cost reduction in the wave energy sector to foster its entry into the market. The Carbon Trust estimated that technical advances on PTOs for wave energy applications would lead to a cost centre progress rate of 93% [9], i.e. for each doubling of the installed capacity, a 7% reduction in PTO cost is achieved. For its part, WES identified PTO performance (in terms of conversion efficiency) and PTO reliability (in terms of availability) as the most relevant characteristics to drive potential LCOE cost reductions [10]. In this sense, a sensitivity analysis applied to

[1] Project Information can be consulted in https://cordis.europa.eu/project/.

a representative example found that a 10% improvement in efficiency or reliability could generate savings of 9% in the total LCOE. Similar results were found by the Partnership for Wave Energy in Denmark [11].

17.2 Power Take-Off (PTO) in Wave Energy Converters

17.2.1 Introduction to PTO-Concept

In 2007, the International Electrotechnical Commission (a worldwide standardisation organisation) established a Technical Committee dedicated to marine energy: TC114. Given the rapid growth of the emerging wave, tidal and water current energy industry, it became necessary to develop a Technical Specification (TS) presenting a common basis for language and nomenclature in the sector. *IEC TS 62600-1 Marine energy—Wave, tidal and other water current converters: Terminology* was released in 2011 (and amended in 2019). According to this TS, a PTO is defined as the "*mechanism that converts the motion of the prime mover into a useful form of energy such as electricity*", where "*prime mover*" is the "*physical component that acts as the interface between the marine resource and the energy converter from which energy is captured [Note 1 to entry: For wave energy converters the prime mover may be a heaving buoy, a hinged flap, an OWC runner,* etc.*]*".

It must be highlighted that the motion experienced by the prime mover will be determined not only by the excitation from the marine energy resource (i.e. the waves), but also by the characteristics of the PTO, which is able to create force/torque acting on the former. The main advantage of this force/torque lies in its controllability. Thus, the PTO can be considered equivalent to the "brain" of the WEC, since it is able to impose the necessary dynamics so that the prime mover-wave interaction maximises the energy extraction. This can only be achieved by means of a control system, which is responsible for adapting the WEC to a desirable behaviour under different load conditions (i.e. maximising energy generation during operation times; minimising the risk to the device in survivability mode; etc.).

17.2.2 Introduction to the Different Types of PTOs

As opposed to wind energy technology, or even tidal, where designs have focused on horizontal-axis machines, there is a lack of industrial convergence towards a common design or device in wave energy conversion. The European Marine Energy Centre (EMEC) in Scotland (UK) offers a comprehensive compilation of wave energy developers in the world. According to the latest update (in August 2020) [12], the most numerous device family—point absorbers—represents only 83 of the total 255 known devices, i.e. one third. Other relevant WEC families include attenuators,

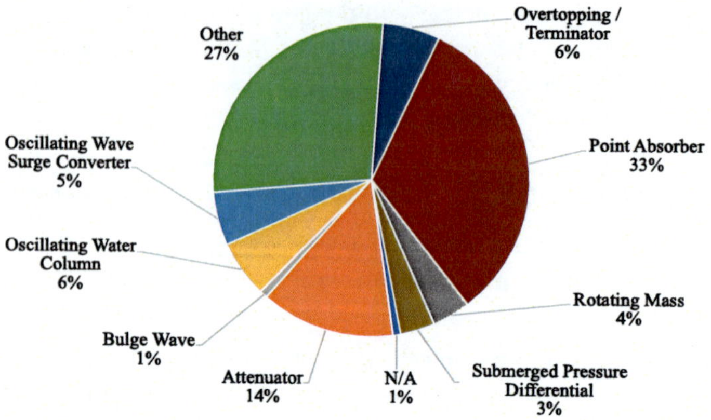

Fig. 17.1 Distribution of wave energy technologies according to EMEC's WEC-classification [12]

overtopping/terminators, oscillating water columns (OWC), or oscillating wave surge converters, as it can be observed in Fig. 17.1.

Wave energy PTOs are not oblivious to this technology diversity, and the aforementioned heterogeneity is transferred to their designs. Several PTO concepts have been investigated to date, given the existing connection between WEC and PTO typologies. For instance, on the one hand, OWCs will highly likely incorporate air turbines as their PTO, since their working principle consists of an oscillation in air pressure inside a chamber caused by the ocean waves. On the other hand, hydraulic PTOs will be probably utilised in WECs experiencing motions caused by high forces at low frequencies (e.g. attenuators or oscillating wave surge converters).

PTOs can be classified according to the type of energy conversion(s) in the path from ocean waves to electricity (see Fig. 17.2). Usually, the energy absorbed by the prime mover is firstly transformed into an intermediate form of energy, able to drive a rotary generator at which, secondly, this mechanical energy is converted into electricity. However, some PTO concepts, such as direct electrical drives or dielectric elastomers, can reduce this energy conversion process to a single step.

The vast majority of PTOs in wave energy conversion are dedicated to the generation of electricity. However, in some cases, the output is the production of pressurised water for desalination purposes (as proposed, for example, by Aquamarine Power with their Oyster device or by Carnegie Wave) [13]. Examples of the integration of PTOs, from the different families presented in Fig. 17.1, into wave energy converters can be seen in Fig. 17.3.

The main challenge for PTOs is the adaptation of the irregular high-power low-velocity wave energy resource to an acceptable form of electricity for the grid. Direct drive solutions are systems that directly couple the mechanical energy of the prime mover to the moving part of an electrical generator (the 'rotor' in rotary machines or 'translator' in linear ones). The coupling can be achieved either via a mechanical transmission system and gearbox (*direct mechanical drive systems*) or without

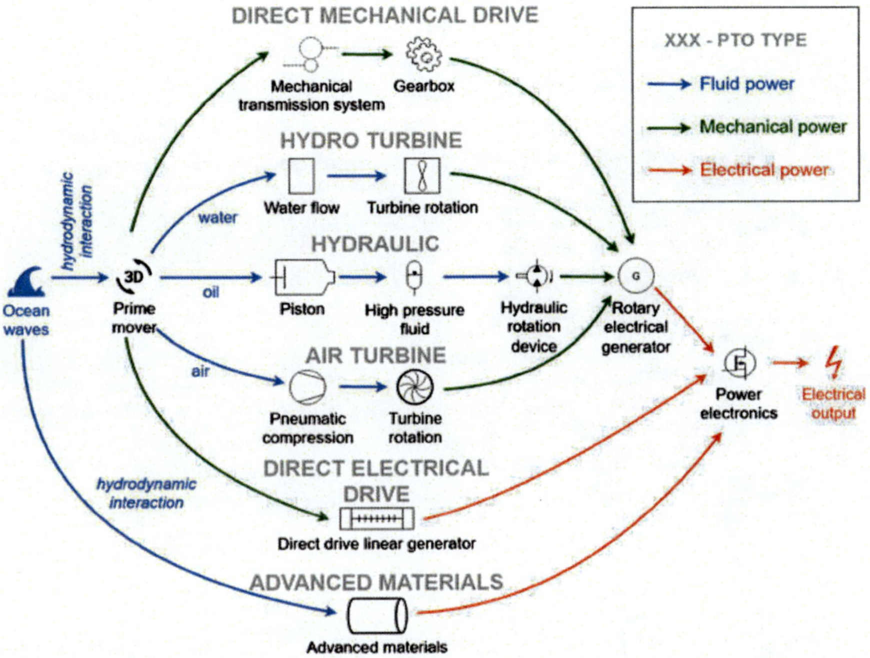

Fig. 17.2 Alternatives for wave energy to electricity conversion. Adapted from Magagna and Uihlein [6] and Têtu [11]

it (*direct electrical drive systems*). The benefit of employing direct drive systems lies in the removal of intermediate energy conversion stages, hence eliminating common failure sources (e.g. hydraulic components) and increasing its efficiency and reliability.

In the past, hydraulic systems used to represent the most common PTO type in wave energy conversion, accounting for around 40% of the total share in 2014 [6]. Nevertheless, in the recent years, the sector has witnessed an increased interest in direct-drive PTOs, moving from a 30% total PTO share representation in 2014 [6] to more than 40% in 2020 [19]. From the total of 98 active wave energy technology developers identified in [19], 31 companies are using direct mechanical drive systems, whereas in 13 cases direct electrical drives are considered.

Examples of real-sea deployments of WECs based on direct drive PTO systems and their characteristics [19–21] are listed in Table 17.1 (direct mechanical drive) and In the case of direct electrical drive systems, the absence of gearboxes allows to reduce noise levels and simplify drive train configurations [24], while the lower maintenance requirements led to increased availability and reduced lifetime costs. Furthermore, direct electrical drive machines increase the overall controllability of the WEC, since modern power electronics and control systems highly improve the performance of all electrical machines. This feature is of special relevance because it permits to implement more efficient strategies that maximise the energy harvesting

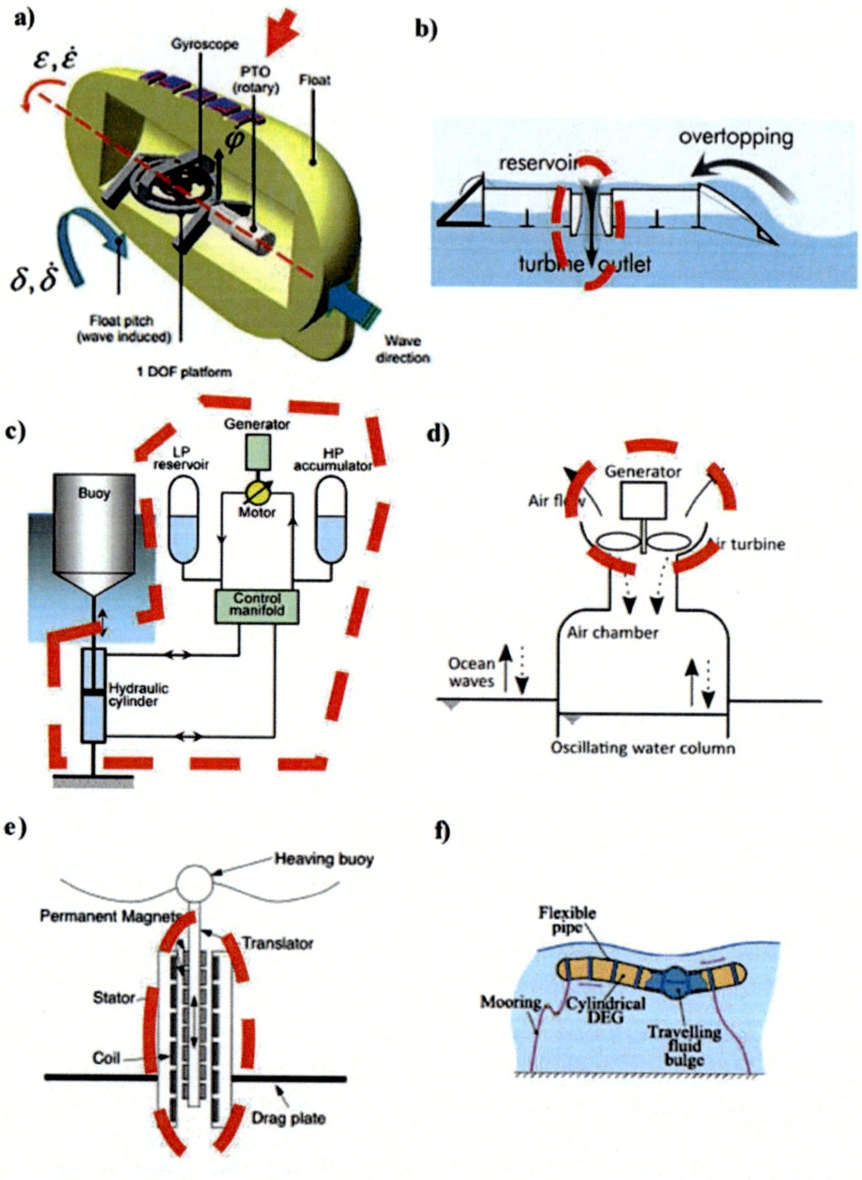

Fig. 17.3 Examples of PTO integration: **a** direct mechanical drive (taken from Liu et al. [14]); **b** hydro turbine (taken from [15]); **c** hydraulic (taken from Falcão [16]); **d** air turbine (taken from Têtu [11]; **e** direct electrical drive (taken from Drew et al. [17]); **f** advanced materials (taken from Moretti et al. [18]). The PTO components are indicated in red (Note that in the last case, the PTO and the prime mover are coincident, since the whole bulge-wave WEC is made of dielectric elastomer generators [DEGs])

Table 17.1 Direct mechanical drive PTO based WECs

WEC name	Technology developer	Website/references	Country of deployment	Year of deployment	Features	Rated power (kW)
Penguin	Wello	https://wello.eu/ [22]	UK	2012, 2017	Off-centre mass gyration drives a spinning flywheel	1000
Smart Power Buoy	Resen Waves	http://www.resenwaves.com/	Denmark	2017	Gearbox	0.6
C3	CorPower	https://www.corpowerocean.com/	UK	2018	Mechanical transmission by Cascade gearbox	25
EMERGE WEC	UmbraGroup	https://www.umbragroup.com/ [23]	UK	2019	Integration of a recirculating ballscrew and a permanent magnet generator	200
LAMWEC	Laminaria	http://www.laminaria.be/	UK	2020–2021	Drum pulley system	200

from the ocean wave resource. Therefore, the PTO concept cannot be limited to the electrical generator and must include the associated power electronics and control system. On the other hand, the main disadvantage of direct drives systems is that manages velocities—that are directly the velocities of the WEC prime mover—which are lower than conventional rotary generators equivalent speeds due to the low ocean waves frequencies. Consequently, the required forces are bigger. Since the force grows linearly with the current while the losses increase as the square of the current, these high force machines tend to be less efficient. These high forces give rise to very large machines and a heavy/robust structures due to the attractive forces between the stator and the translator [25].

Table 17.2 (direct electrical drive).

In the case of direct electrical drive systems, the absence of gearboxes allows to reduce noise levels and simplify drive train configurations [24], while the lower maintenance requirements led to increased availability and reduced lifetime costs. Furthermore, direct electrical drive machines increase the overall controllability of the WEC, since modern power electronics and control systems highly improve the performance of all electrical machines. This feature is of special relevance because it permits to implement more efficient strategies that maximise the energy harvesting

Table 17.2 Direct electrical drive PTO based WECs

WEC name	Technology developer	Website	Country of deployment	Year of deployment	Features	Rated power
AWS I	AWS Ocean Energy	http://www.awsocean.com/ [16]	Portugal	2004	Permanent-magnet linear synchronous generator	2000 kW
Wavebob	Wavebob	N/A [26]	Ireland		Linear electric generator	
L10	Oregon State University	https://oregonstate.edu/ [27]	USA	2008	Permanent-magnet tubular linear generator	10 kW
W1	Wedge Global	https://wedgeglobal.com/ [28]	Spain	2014	Switched Reluctance Machine	200 kW
SP WEC	SINN Power	https://www.sinnpower.com/ [13]	Greece	2015–2019		18 and 36 kW
Seabased	Seabased	https://seabased.com/ [22]	Sweden	2015	Longitudinal flux permanent magnet generator	30 kW
C-GEN Neptune	University of Edinburgh	https://www.cgen.eng.ed.ac.uk/ [29]	UK	2019	Transverse flux permanent magnet	150 kW
StingRAY H2	Columbia Power Technologies	https://cpower.co/ [30]	USA	2020–2021	LandRAY: rotary permanent magnet generator	500 kW

from the ocean wave resource. Therefore, the PTO concept cannot be limited to the electrical generator and must include the associated power electronics and control system. On the other hand, the main disadvantage of direct drives systems is that manages velocities—that are directly the velocities of the WEC prime mover—which are lower than conventional rotary generators equivalent speeds due to the low ocean waves frequencies. Consequently, the required forces are bigger. Since the force grows linearly with the current while the losses increase as the square of the current, these high force machines tend to be less efficient. These high forces give rise to very large machines and a heavy/robust structures due to the attractive forces between the stator and the translator [25].

The following section in this chapter will focus on the aforementioned direct electrical drive systems, hereinafter directly referred to as Direct-Drive Power Take-Offs (DDPTO).

17.3 Direct-Drive Power Take-Off: SRM Topology

17.3.1 Introductory Aspects

DDPTOs are very attractive due to the improvement in the overall efficiency and in the controllability of a WEC. In fact, they make possible the implementation of more efficient energy extraction strategies, but with the force capability limitation of the DDPTO. Since the velocity of the motions involved in WECs is small, 1 m/s as order of magnitude, the forces required to achieve typical renewable-energy power values are several orders of magnitude higher; e.g. a device which is designed to harvest 1 MW of energy, needs to produce a force in the order of MN.

From a comparative point of view, conventional rotating machines are very common to be found working in this range of power but at much higher velocity and much lower torques. Therefore, the best option to increase the relative velocity between the moving parts of the WEC is the combination of a DDPTO and a WEC working close to resonance, since the same power can be extracted from the waves with significantly less force than what is needed when operating far from resonance.

High power generators are long and heavy. The relatively high moving velocity which are involved limit the minimum airgap between both sides leading to stiff designs with strict tolerances. As explained before, big forces require big currents and consequently higher Joule losses, which will be associated to additional heat dissipation. Thermal calculations and design also become more relevant.

Therefore, all these points make a discussion about an attractive alternative (not exclusive) for increasing the force density of the PTO consisting of using superconducting coils. For the same level of machine power, the superconducting machine can be a solution to reduce size and weight (see Sect. 17.4). Excluding the superconducting machine, there are three possible main candidates of electric machines for using in a Power Take-Off (PTO) considering their operation principle. These three alternatives will be discussed in Sect. 17.3.3 [31, 32].

- Permanent Magnet Machines (PMM), including Permanent Magnet Synchronous Machines (PMSM)
- Variable Reluctance Machines (VRM), including Synchronous or Switched Reluctance Machines (SRM)
- Special machines based on a combination of the previous ones, like Vernier Machines or other Transverse Flux Permanent Magnet Machines.

One important issue in the linear electrical generators is the configuration of the machine, especially which part of the machine is the longer and the shorter one, since it affects to the size and weight of the system. Figure 17.4a shows an example of a linear machine configuration that balances transverse forces, in which all the developed DDPTOs are based on. In this case, the length of the stator (m) is smaller than the length of the translator (m) by twice the stroke of the PTO (s) [28, 33, 34]. Mechanical and electromagnetic considerations allow defining the geometrical requirements for the linear electrical machine to be used in the PTO:

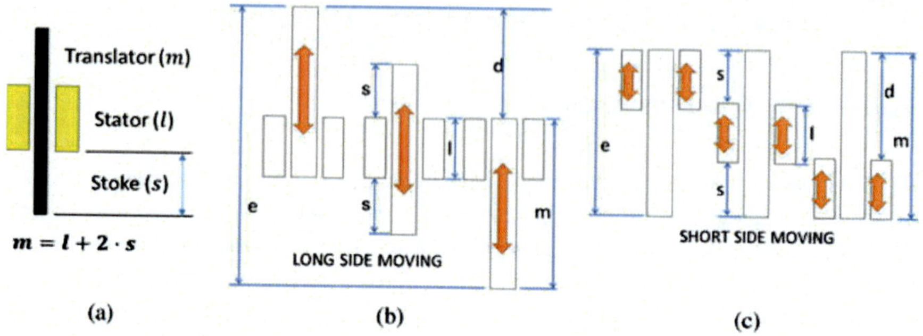

Fig. 17.4 **a** Example of short stator configuration; **b** Long side moving; **c** Short side moving

- *Stroke (s)*: This parameter is basically defined by its localization, the wave height and the operation mode (resonant or not). Even for small waves, the maximum amplitude of the translator motion (stroke) can be high if the point absorber works close to resonance. Combination of high waves and resonance usually lead to inadmissible values for the stroke, which must be limited.
- *Stator and Translator Dimensions*: The geometrical parameter constrains are defined based on the concept of Shear Stress (σ), which is the force that any electrical machine is able to produce by unit surface of its airgap. This variable is very useful in the analysis of an electrical machine for a PTO. The value of σ is proportional to the product of the electric load of the machine (expressed in kA/m) times the magnetic load of the machine (expressed in Tesla).

E.g. a typical value for PMSM spans from 20 to 40 kN/m^2 according to the magnetic flux density in the airgap (1–1.3 T) and the electric load (20–50 kA/m), which depends on the windings current density and their cooling system [24].

Based on that, each of the two sides on the machine can be classified according to four parameters:

- Its role: Which part of the machine is the active one and which is the passive one. The active part is where the coils are placed, and the passive one is the part which closes the magnetic circuit but does not include any coil that delivers active power to the system (e.g. in VRM, the part which only contains the magnetic steel; in PMSM, the part which includes the permanent magnets; in a conventional a synchronous machine, the part where the excitation coils are placed).
- Its length: Based on this parameter, there are two possibilities: to be a long or a short part. Making an analogy with a rotary machine, the 'short' part would represent 360 electrical degrees of it, therefore it would mean to a 'unit module' of the machine. On the other hand, the 'long' part is an integer multiple of the unit module, i.e. the overlapped part with the 'short' one. The difference between the length of the 'long' and the 'short' part is the total stroke of the linear machine.
- Its position: It could be external or internal. The part which surrounds the other one, is considered as external.

Table 17.3 Machine design parameters according to the configuration

	Long side moving	Short side moving
Stroke (***s***)	According to the PTO	According to the PTO
Short-side length (***l***)	According to the force	According to the force
Long-side length (***m***)	$m = l + s$	$m = l + s$
Displacement (***d***)	$d = s$	$d = s$
Machine envelope (***e***)	$e = 2 \cdot s + l$	$e = s + l$

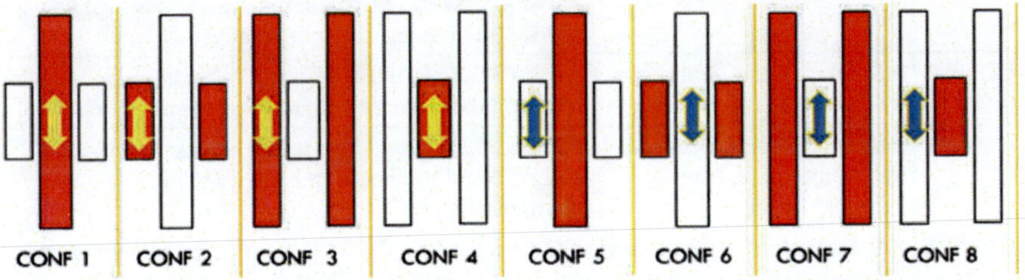

Fig. 17.5 Geometrical configurations of DDPTO. Active side (AS) in red I Passive side (PS) in white

- Its velocity: They could be defined as moving part or stationary part.

According to these parameters, there are different possible configurations. Figure 17.4b, c show two configurations, depending on whether the short part is moving or fixed, and for those possibilities, Table 17.3 shows the size of each part. The short-side length "l" is the length of the active side and depends on the rated generated force. Due to the linear machine envelope "e" is the length required by the structure to host the whole machine, it can be remarked that the power density for a generator with external translator is higher than for a generator with an internal one [35]. Based on this, Fig. 17.5 shows the possible configurations considering both possibilities (short or long moving) and the roles (active or passive). Additionally, a table with advantages and drawbacks for each alternative is included (Table 17.4).

17.3.2 Example of an Analysis of a WEC with a DDPTO

In this section, an example of a WEC with a linear electrical generator is analysed in order to show the impact of the DDPTO characteristics (efficiency, rated force, etc.) in the overall WEC energy extraction capacity. This study illustrates the need of a high-force PTO applied to a particular example, the Heaving Point-Absorber

Table 17.4 Advantages and drawbacks of the linear generator (DDPTO) geometrical configuration

CONF.	Geometrical Conf.	Advantages and drawbacks
CONF 1	**AS**: Long, Int. & Moving **PS**: Short, Ext. & Stationary	**UNFEASIBLE** I Cons.: long and moving active side
CONF 2	**AS**: Short, Ext. & Moving **PS**: Long, Int. & Stationary	**NOT PREFERABLE** I Cons.: The movement transmission is external and required flexible electrical connections
CONF 3	**AS**: Long, Ext. & Moving **PS**: Short, Int. & Stationary	**UNFEASIBLE** I Cons.: Similar to CONF 1 but even worse since the movement transmission must be external
CONF 4	**AS**: Short, Int. & Moving **PS**: Long, Ext. & Stationary	**FEASIBLE** I Cons.: Moving the active part and required flexible electrical connections/Adv.: The moving mass can be small, the passive side can be part of the structure and the spar shorter
CONF 5	**AS**: Long, Int. & Stationary **PS**: Short, Ext. & Moving	**NOT PREFERABLE** I Cons.: The active side must be long and the movement transmission must be external
CONF 6	**AS**: Short, Ext. & Stationary **PS**: Long, Int. & Moving	**FEASIBLE** I Cons.: The stroke is long, and the PS weight could be excessive I Adv.: Simplest configuration
CONF 7	**AS**: Long, Ext. & Stationary **PS**: Short, Int. & Moving	**NOT PREFERABLE** I Adv.: It could be considered for short strokes and high forces, where the length of the AS is not much smaller than the overall length
CONF 8	**AS**: Short, Int. & Stationary **PS**: Long, Ext. & Moving	**NOT PREFERABLE** I Cons.: Similar to CONFG 6 but the moving side can be even heavier, and the transmission must be external

(HPA) W1 prototype [36]. The results are obtained from a mathematical model, experimentally validated in the referred WEC (Fig. 17.6).

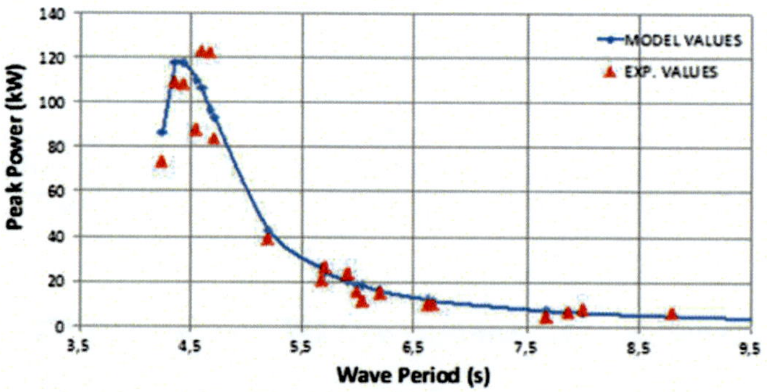

Fig. 17.6 Peak generated power predictions and measurements for the W1 WEC normalised to 1 m regular wave amplitude (fixed damping control strategy)

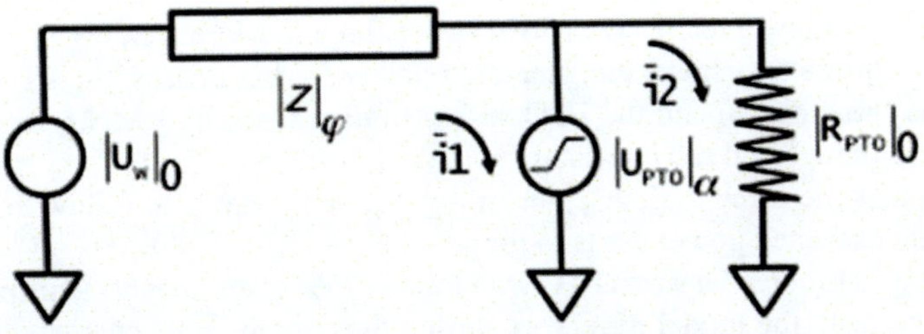

Fig. 17.7 Equivalent circuit of the PTO and WEC models [37, 38]

The model is based on the motion equations of two moving bodies of a HPA in the water, when they are excited by the presence of waves. The model is a simplified version where only the float movement is considered. Figure 17.7 shows the equivalent electrical circuit of this simplified model [37], including the following parameters:

- The two external forces acting on the HPA—wave excitation and PTO forces—are modelled with a voltage source U_w and U_{PTO}
- The float mechanical total impedance Z, which includes radiation, inertia (float mass plus added water mass) and buoyancy coefficients.
- The current equivalent to the velocities i_1 and i_2.
- The PTO losses in the form of a parallel resistance R_{PTO}.

Firstly, a benchmark ideal case is calculated in order to compare with the real PTO. Therefore, these results for this base case are obtained using the previous model with a control strategy which maximizes the generated electrical power [39, 40] for regular waves, assuming an ideal PTO (100% of efficiency) with and without a maximum force limitation of 1 MN. These results are shown in Fig. 17.8a where the red curve shows the maximum power that can be extracted for each wave period, (i.e. ideal case where there are no limitations), while the orange one represents the power that

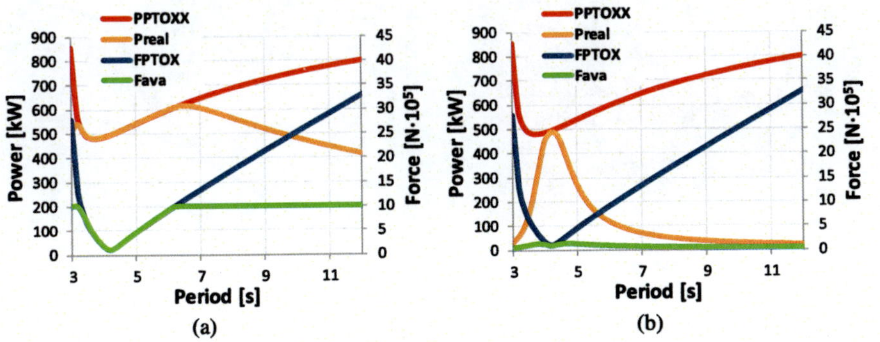

Fig. 17.8 Generated power and required force results obtained from the WEC & PTO models for different wave periods: **a** Ideal PTO; **b** Real PTO

is really extracted by a real PTO defined as benchmark. Moreover, Fig. 17.8a shows the required force to achieve the generated power for both cases, the PTO without limitations (blue curve) and the PTO with a limited force of 1 MN (green curve). Both curves orange and red one coincide in the range where the force is not limited, and as soon as the force limit acts, the orange curve is unable to follow the red one and the real extracted power starts to drop.

Next step is to consider a real PTO with losses (75% of efficiency). This parameter is estimated from the model developed during the experimental characterisation of the W1 prototype. These results are shown in Fig. 17.8b. The PTO has the same force limitation of 1 MN, nevertheless the results are totally different, since only in a narrow range of wave periods the PTO is able to extract the maximum available power. The rest of wave periods range, the extracted power by the real PTO (orange power) is smaller than the available power (red curve) and the force required by the PTO to extract that power (green curve) is well below the 1 MN limit. The reason is that the WEC control strategy prefers using less force than what is available to maximize energy production since losses for higher forces lead to a smaller energy production. Therefore, the main conclusion of the analysis is that the amount of energy which can be harvested from the waves in a wide range of periods for a real PTO with force restrictions and limited energy conversion efficiency drops dramatically with respect to what could be captured with an ideal lossless and force-unlimited PTO, as is shown in Fig. 17.9a.

Figure 17.9a shows an indicator to quantify the benefits of improving the PTO performance, especially the PTO force capability and efficiency. This parameter is quantified through the area enclosed by the electrical generated power curve, yellow area in the real PTO studied. This area will be named as Integrated Power Capture (IPC), expressed in kJ when power is expressed in kW. Moreover, the ratio between the IPC for a WEC with a real PTO and a WEC with an ideal one will be defined as the IPC Ratio (IPCR). The IPCR of a PTO could be improved augmenting the

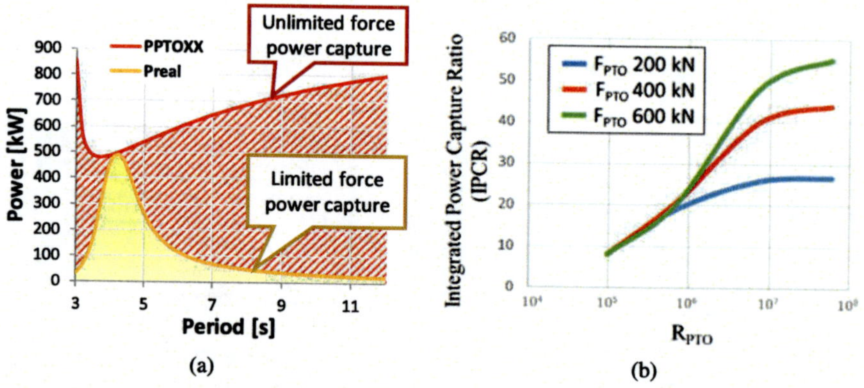

Fig. 17.9 **a** Integrated Power Capture Ratio (IPCR) concept; **b** IPCR versus PTO Resistance (R_{PTO}) for different PTO Forces (F_{PTO})

Force Capability, the Efficiency or both together. Figure 17.9b shows the effect of increasing the PTO force and efficiency, in terms of the parameter R_{PTO}, for the W1 PTO studied. In this study case, the dominant losses are the Joule losses, which are proportional to the square of the current and the force. The dependence of the R_{PTO} value in the IPCR is shown in Fig. 17.9b for three different force values (200, 400 and 600 kN), showing the mentioned benefits improving in the force level.

17.3.3 The Linear Switched Reluctance Machine (LSRM)

17.3.3.1 Introduction—Alternatives of Linear Electric Machines for Wave Energy Generation Application

Various types of linear electrical machines can be used for ocean wave energy conversion, as described in Sect. 17.3.1. The possible candidates for being used in a WEC are summarised in Fig. 17.10 and described in the following paragraphs [21].

Permanent Magnet Synchronous Machines (PMSM): The operating principle is based on the excitation from the magnets; the magnetic fields created move synchronously with the translator, at the same velocity, respect to the translator. As it occurs in other machines, force density could be increased using a double-sided configuration [41]. In this type of machines, two possible field configurations are used, longitudinal and transverse flux type. The implementation of a doubled-sided machine shows better performance than a single-sided machine according to the flux dispersion and bearing loads, since the lateral forces are fully compensated. One drawback of this type of machine is the cogging force produced by the interaction between rotor-stator magnetic fields. This effect reduces their efficiency in linear machines.

Lineal PM tubular machine: The operating principle of the PM tubular linear machine consists in the production of an electromotive force in a wire by means of changing the magnetic field (i.e. the force is generated by the interaction between the magnetic fields from the PMs and the currents on the coils). The translator must be perfectly aligned due to the fact that the force is perpendicular to the direction of the translator motion. The total cost of this type of linear machine is basically determined by the cost of the PM, usually Neodymium magnets due to their high

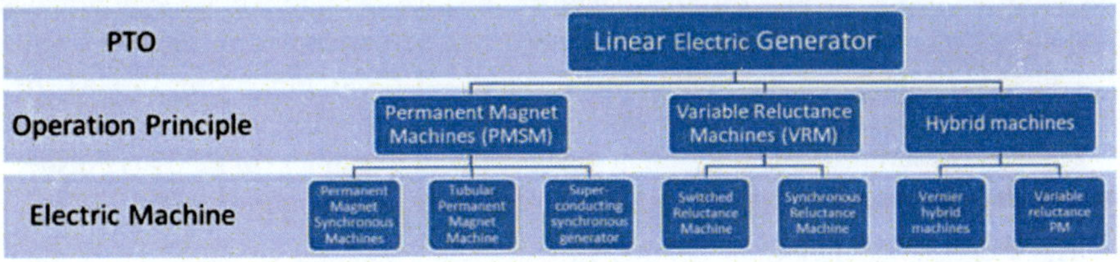

Fig. 17.10 Types of linear electric generators and assembly topologies for wave energy applications

magnetic field strength. Its main drawback is the de-magnetisation of the PMs after a certain time [42].

Linear superconducting synchronous machine: This topology has already been investigated for wind energy conversion. In such type of machine, superconducting magnets have been used instead of conventional ones in which a de-magnetisation effect appeared after a certain time. The typical wire used is the MgB2 [43], where its critical temperature is relatively high for the iron machines due to the fact that the magnetic field value is low.

- *Variable Reluctance Machines (VRM), including Switched Reluctance machines(SRM) and Synchronous Reluctance machines (SyRM) or*:

The operating principle of the linear SRM is similar to the rotating SRM and a stepper motor, it is based in the reluctance minimisation principle. The magnetic field is created from the stator current. The main advantages of these machines are their reliability, robustness, high efficiency and simple construction, lowest manufacturing cost due to the absence of magnets and high force. The operation of these machines is controlled by the translator position and a power electronic converter, similar to what occurs in a stepper motor. Due to its operating principle, its main drawback is the high force ripple compared to other topologies.

Regarding the SyRM, this topology has a similar structure as SRM since they have no permanent magnets mounted on their translators—based on salient poles—and the operating principle is the reluctance minimisation. Their stator structure is based on the induction machines, since the stator is wounded with a distributed winding with 3 phases, which creates a rotating magnetic field in the air gap of the electric machine. This field rotates at synchronous speed and can be considered sinusoidal. The main drawback of this topology is its low power factor.

- *Vernier hybrid machines*:

Vernier hybrid machines are a type of variable reluctance machines. These machines offer a low power factor, but it can be improved by auxiliary DC field excitation winding [44]. In Vernier hybrid machines, magnets are placed on the stator to create the magnetomotive force and the translator is slotted, as occurs in a reluctance machine, to modulate the magnetic field [45, 46].

- *Variable reluctance PM (VRPM) machine*:

The operating principle of these machines is the reluctance force which comes out to be the same as that for DC machines, PMSM or field wound synchronous machines [47]. Variable reluctance machines can be assembled in transverse flux machines (TFMs) or in longitudinal flux machines (LFMs), presenting TFMs better performance than LFMs for ocean wave applications [48–50].

Once the description of all different types of linear machines for PTOs is finished, we present the main differences, in order to select the best configuration:

- The PMSMs needs a high number of permanent magnets to produce the field. Modern rare-earth magnets allow generating magnetic fields up to 1.3 T with

masses in the order of 75 kg/m^2 of machine airgap [51]. In summary, a 500 kN permanent magnet machine would require between 12.5 and 25 m^2 of active air gap surface and between 950 and 1900 kg of permanent magnets.
- In the case of Saturated Variable Reluctance Machines (SVRMs), the Shear Stress is in a similar range as that for PMSMs. Although the magnetic field can be higher for SVRMs (up to 1.6 T), the electric load is usually smaller (<30 kA/m). Nevertheless, a great advantage of the SVRMs is its more efficient cooling and cheaper cost since their windings are concentrated.
- Another big difference is that SVRMs needs no permanent magnets that are expensive and delicate components for marine applications since they are prone to corrosion. Additionally, permanent magnets could be de-magnetised leading to a fatal breakdown because they will need to be magnetised again externally.
- The economic arguments are other point to consider. In a simple way, a conventional linear machine will be the sum, in different proportions, of three raw materials: Iron, copper and rare earth magnets. These elements need to be manipulated and transformed to its final state (cut, assembly, wound, protected, mounted, etc.). The approximate final prices for these components are in the following range [52]: Iron 3 €/kg, Copper 10 €/kg and Rare Earth Magnets 50 €/kg. Therefore, reducing the requirements of permanent magnets may also have apositive influence in the price of the machine.

Finally, these techno-economic considerations derive a conclusion: SRM can be considered a good compromised solution for a linear generator based PTO since it is simpler, more robust, more cost effective and can achieve the same controllability as any alternative machine with adequate power converters. Moreover, the linear SRM allows modular configurations. This means that once a polar structure is defined for both stator and translator, this structure can be repeated to achieve the required force and for any stroke by only modifying the length of the translator.

17.3.3.2 Introduction—Alternatives of Structure Configuration and Design of Linear Electric Machine Geometries

After the previous considerations on operation principle of the electric generator, two approaches arise to address the problem of increasing the force density, making more compact machines by means of structure configuration design:

- Increasing the airgap surface with limited impact on the machine volume
- Increasing the Shear Stress (σ).

The use of cylindrical designs is one of the possible options to increase the air gap surface in a Short-Stator configuration (CONF 6, see Fig. 17.4). Two examples of cylindrical designs are Oprea et al. [53] and Mendes et al. [54], based on PMSM and a Short-Stator SRM based on a cylindrical geometry, respectively. This structure configuration has two advantages, there are no coil ends, i.e. all the winding length produce force, and the lateral forces are fully compensated. Among the drawbacks

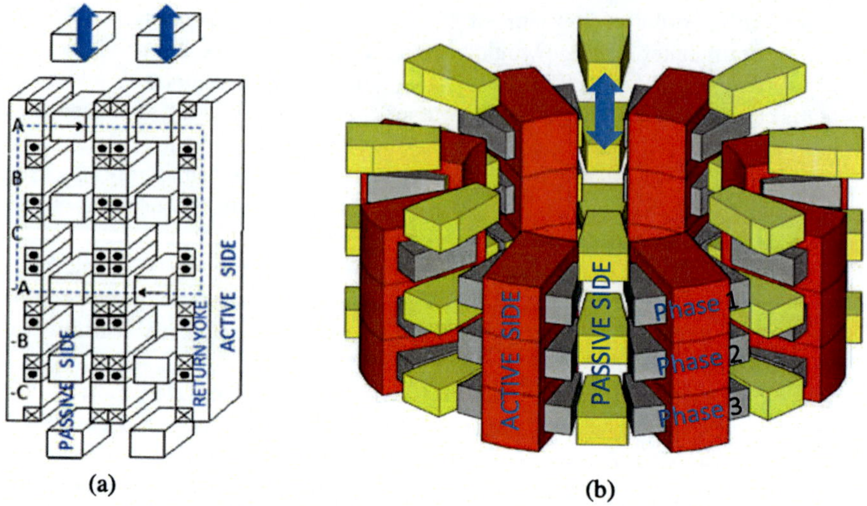

Fig. 17.11 **a** Multi-translator SRM [56]; **b** Azimuthal Multi-translator SRM [57]

can be highlighted the difficulty for placing the stator laminations in the required direction for minimising iron eddy currents, the additional iron in the axial direction of the stator to close the flux lines axially and the guidance of the cylindrical translator. Other way to increase the force density is using a multi-translator configuration (see Fig. 17.11a). This structure configuration increases by 30% the force density respect to a double-sided conventional one. It could be applied to SRM, Vernier or TFM. Finally, a new approach based on the existing multi-translator concept is shown in Fig. 17.11b. This azimuthal multi-translator SRM solution [55] aims to increase the airgap surface with reduced impact on the volume of the full machine.

The basic idea of the Rectangular Multi-translator Switched Reluctance Machine (RMSRM) shown in Fig. 17.11a is to introduce intermediate stators and translators in a Double-Sided Linear SRM. One intermediate double stator and one additional translator have been added to a conventional Double-Sided Machine with the corresponding two End Stators. Nevertheless, the Azimuthal Multi-translator SRM (AMSRM, see Fig. 17.11b) presents modifications respect to the RMSRM following the same operating principle but eliminating the return yoke and becoming all stators intermediate, without end stators. As explained before, the idea is to arrange the stator coils (active side) and the translator poles (passive side) circumferentially in a kind of toroidal geometry.

Table 17.5 shows a comparative study, which summarizes the Stator Force Density for the mentioned topologies respect to the coil depth. The benchmark case is set as the Double-Sided SRM. The results for the Single-side LSRM shows a smaller overall force density when the translator is included. Moreover, the results for the RMSRM are obtained for one intermediate stator. If this number is increased, the force density will be augmented.

Table 17.5 Relative stator force density for different LSRMs referred to the double-sided solution

	Single-sided LSRM	Double-sided LSRM	RMSRM	AMSRM
Relative stator force density	1.00	1.00	1.33	2.00

Table 17.6 Comparison of the RMSRM solution versus the AMSRM configuration

Parameter	RMSRM	AMSRM
Electric Load (A/m)	26,000	30,000
Magnetic load (T)	1.5	1.5
Shear Stress (σ) (kN/m²)	19	27
Air gap surface (m^{-2})	8.8	37.5

Finally, the Azimuthal Multi-translator Switched Reluctance Machine has several advantages over the rest of SRM topologies:

- There are no lateral stators, and the required amount of iron is minimum, increasing the force density. This idea is to create an azimuthal flux path, closing the flux through identical Intermediate Stators, eliminating the end stators.
- The leakage flux is smaller than in the RMSRM configuration, so more force per ampere-turn can be achieved than in the equivalent RMSRM.
- Its circular arrangement is more suitable for fitting in the typical configuration of HPA, since the PTO is usually hosted in a cylindrical vessel. The azimuthal geometry has higher filling factor than the rectangular one.
- The benefits of an azimuthal configuration are increased with the square of the external diameter since air gap surface is proportional to this geometrical value.

These advantages conclude that the AMSRM configuration increases the Shear Stress (σ), which comes from a higher current density and a better magnetic circuit design. In addition, the AMSRM configuration increases the air gap surface, which is a direct consequence of the configuration of the new machine and the adaptation to the PTO shape. Table 17.6 presents a comparison between a RMSRM and an AMSRM pre-designed for a wave energy application.[2] These parameters summarize the improvement in the overall force and force density. Both machines have equal overall stator length, pole pitch (λ) and magnetic load. The comparison extracts that the AMSRM topology requires smaller pole depth as RMSRM, showing also better thermal performance. Therefore, its better cooling allows increasing the current density from 3.5 A/m^2 of the RMSRM to 4.5 A/m^2 in the AMSRM topology.

[2] The results are based on a preliminary electromagnetic design applying the methodology carried out in H2020 project SEATITAN (ID: 764014).

17.3.3.3 AMSRM Working Principle

Once AMSRM has been selected as a promising linear generator candidate, this section describes in detail its operating principle. As described before, the SRM operating principle is based on a stepper motor, where the stator phases are energising sequentially to obtain energy in generator mode or to produce energy when motoring. When the energized stator poles and the corresponding translator poles try to align, producing a net force which can act in the same direction of motion (motor operation) or opposite to it (generation). This sequential process is shown in Fig. 17.12. For a linear SRM.

Figure 17.12a describes the motor mode, in which the green phase is activated in the fully misaligned position. In this case, the pole (P) tries to align the green phase in order to minimize the reluctance of the circuit. This process generates an electromagnetic force in the same direction of the translator movement. When the pole achieves the fully alignment position, the phase is switched off and the next one is activated, in this case the yellow one. The generator mode is analogous; the phase is activated when the pole (P) is in the fully alignment position. The translator is moving towards the fully misaligned position due to an external force that pulls the translator. In this movement, an electromagnetic force due to the phase activation acts in the opposite direction of the velocity trying to brake the motion. When misalignment is achieved, the phase is switched off and the next one is switched on, in this case the yellow one.

The energisation of a SRM phase is related to the inductance evolution for each phase. The inductance changes as the translator is moving, being maximum at the fully alignment position and minimum at the fully misalignment situation. The evolution of the inductance respect to the SRM translator pole position, is shown in Fig. 17.13a. Based on this curve, two working zones are deduced related to the two operating modes. The motor mode is applied in the zone with positive slope ($dL/dx > 0$) of the energised phase, accelerating the translator since the force is positive, as Fig. 17.13b shows. However, in the generator mode, the objective is to brake the translator transforming mechanical energy into electrical one. Therefore, the phase which has a translator pole near fully alignment position and moving to a fully misalignment one is energised, corresponding to the zone with $dL/dx < 0$.

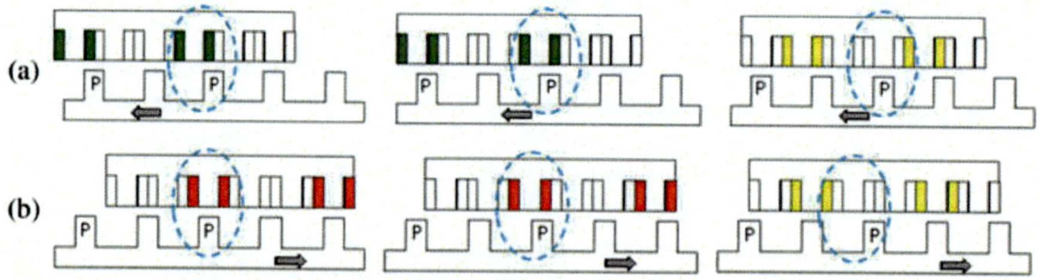

Fig. 17.12 Working principle of the LSRM: **a** Motor mode **b** Generator mode

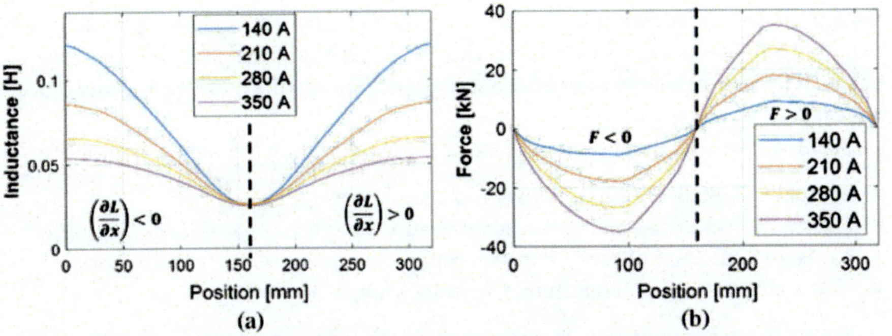

Fig. 17.13 Example of LSRM characteristics as a function of the position: **a** Inductance and **b** Force

The legend of the images shown in Fig. 17.13, shows the variations of the force and inductance as a function of the current supplied on the phases.

According to that two zones, the phase activation sequence of the SRM is showed in Fig. 17.14. In motor operation mode, phases are activated with positive slope of the inductance as a function of the position. However, the phases are switched on with negative slope when the machine works in a generator mode. Figure 17.14 represents an ideal situation since the current waveform shown is rectangular, settling and cancelling instantaneously. In a real case, the current cannot be established

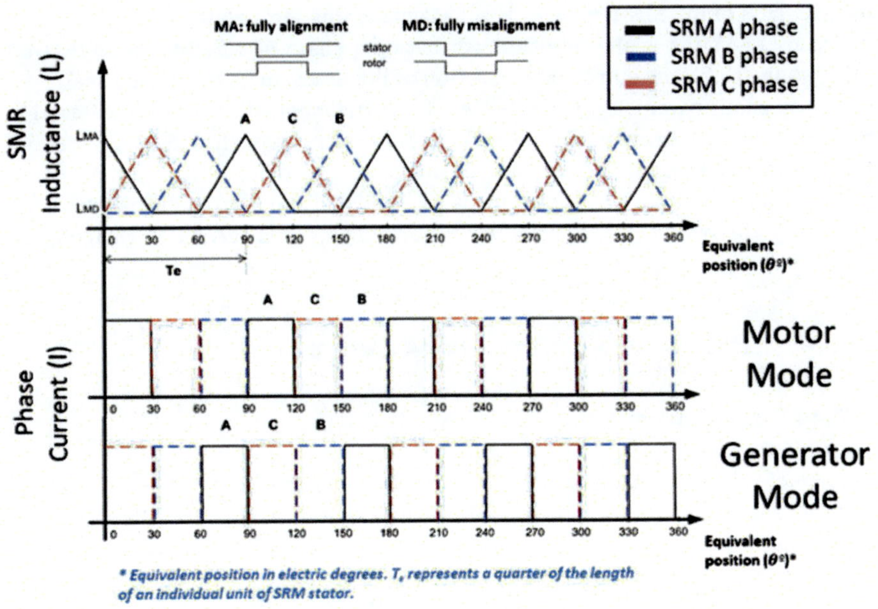

Fig. 17.14 SRM phase activation sequence

instantaneously, and therefore, currents of phases will overlap, being necessary to carry out a forward and delay to the activation positions of the phases respectively. A constant force is achieved in permanent regime thanks to the overlapping and current control.

A SRM electromagnetic model can be set out based on the next assumptions:

- Mutual magnetic coupling between phases is negligible.
- Losses by hysteresis and parasitic currents in magnetic circuit are negligible.
- Resistance of each phase is constant and independent of current frequency.
- Translator moves with negligible viscous friction coefficient.

Under these assumptions, the operation of the SRM can be described by a set of equations, one for each SRM phase. The instantaneous voltage in the phase terminals can be obtained from Faraday's law like it is shown in (1).

$$u = R \cdot i + \frac{d\phi(x,i)}{dt} = R \cdot i + \frac{\partial \phi(x,i)}{\partial i} \cdot \frac{\partial i}{\partial t} + \frac{\partial \phi(x,i)}{\partial x} \cdot \frac{\partial x}{\partial t}; \left(v = \frac{\partial x}{\partial t} \right) \tag{17.1}$$

$$\frac{\mathrm{d}i}{\mathrm{d}t} = \frac{u - R \cdot i - \frac{\partial \phi(x,i)}{\partial x} \cdot v}{\frac{\partial \phi(x,i)}{\partial i}}; \tag{17.2}$$

where, $R[\Omega]$: electric resistance of the coils of one phase; t[s]: time; i[A]: electric current flowing in a SRM phase; x[m]: relative position between the active and passive parts of the SRM; $\phi(x,i)$[Wb]: Magnetic flux through the coils.

From (17.1), the evolution of the current on a phase of a SRM can be derived. Integrating (17.2), the current as a function of the time can be obtained. Finally, the electromagnetic parameters of the SRM—force and magnetic flux—are obtained by means of a finite element tool. Figure 17.15 shows an example for a SRM, presenting force, magnetic flux and the derivative terms of magnetic flux as function of current and translator position. Those data maps are tabulated and used in the mathematical model together with (17.2) to calculate the evolution of the current on each phase.

17.3.4 SRM Power Electronics and Control

17.3.4.1 Introduction

A complete DDPTO encompasses the linear electric machine, the power electronics (PE, which interconnect the electric grid with the electric machine), and the associated control [24]. In addition, it is worth mentioning that the SRM only works driven by a PE converter in order to activate/deactivate its concentrated coils in the proper sequence [58]. The PE and the control of a LSRM is described in this section.

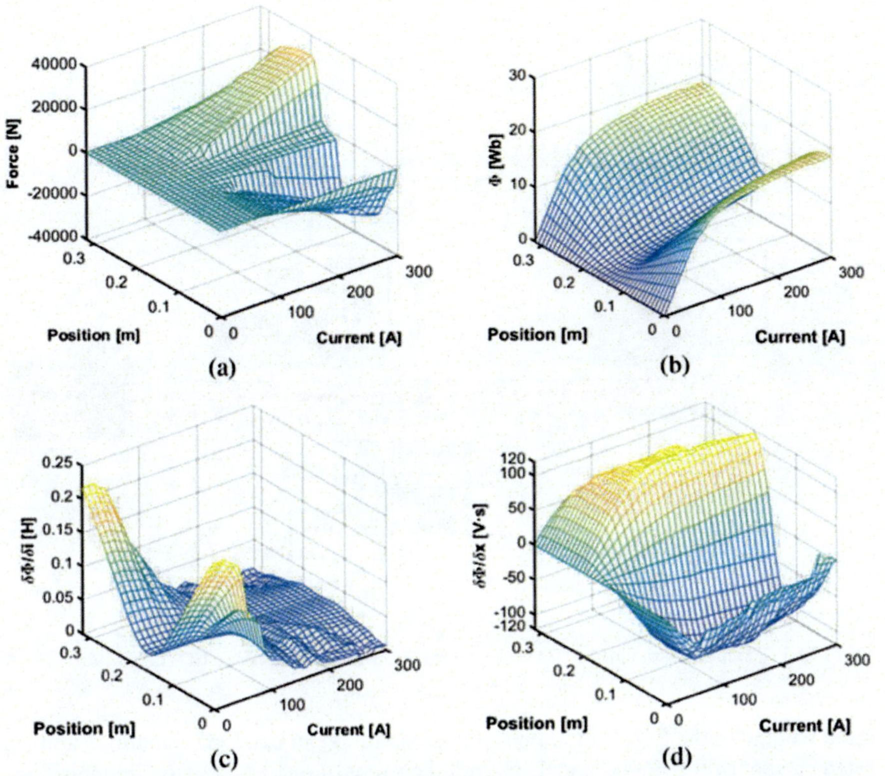

Fig. 17.15 Electromagnetic parameters of a SRM as a function of the position and current: **a** Force; **b** Magnetic flux; **c** and **d** derivative terms of the magnetic flux

One of the priority topics highlighted in the Strategic Research and Innovation Agenda for Ocean Energy [7] is the "*Improvement and demonstration of PTO and control systems*", and one of the remarked actions is "*Optimisation and simplification through standardisation, modularity and scalability of key PTO components*". The proposed LMSR, PE and control match this requirement.

The PE can be differentiated in two types of converters: the Linear Generator Side Converter (LGSC), which controls the energy extracted from the waves by means of the control over the linear generator; and the Electric Grid-Tie Converter (EGTC), which injects the generated electric energy in the electric grid. Nowadays, EGTC can be equipped with advanced grid functionalities such as synthetic inertia or frequency fast response, being called grid-forming converters [59].

In terms of PTO modularity (see Fig. 17.16), one LGSC module can drive serial associations of several LSRM modules. The rated current of the LGSC module is the rated current of the LSRM module, and the rated voltage is the multiplication of the LSRM module rated voltage by the number of modules in the LSRM association. In turn, one EGTC can manage parallel associations of LGSC modules. Therefore, the

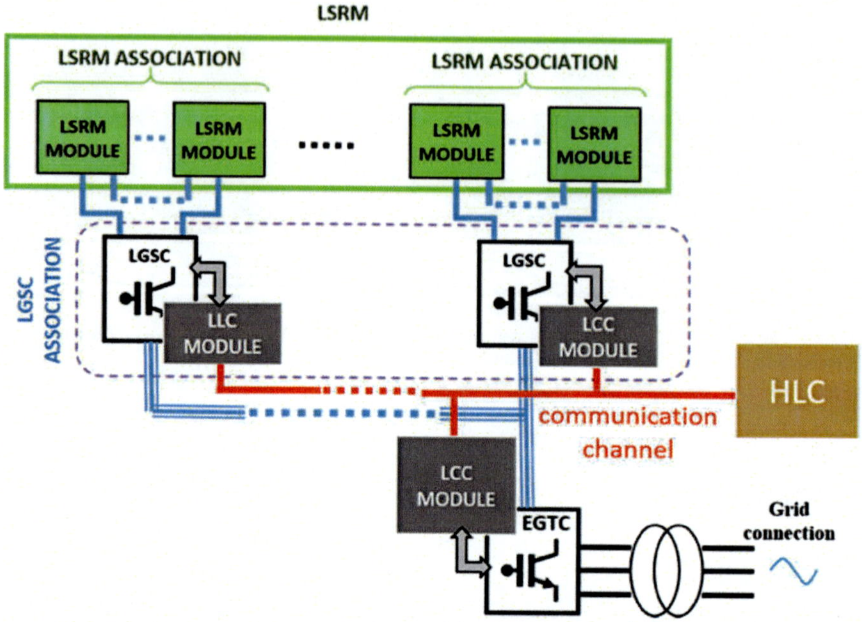

Fig. 17.16 Example of modular PTO scheme (one LGSC association/one EGTC)

rated voltage of the EGTC module is the rated voltage of one LGSC module, and the rated current of the EGTC equals the rated current of one LCSG module multiplied by the number of modules in the LGSC association. The way to associate the LSRM modules and the LGSC modules is a compromise solution between the voltage and current to be managed by the PE converters.

Two types of control can be identified in the control loop (see Fig. 17.17):

- High Level Control (HLC), which controls the mechanic variables of the WEC, defining a force command. The aim of this control loop is to maximise the energy extracted from the waves.

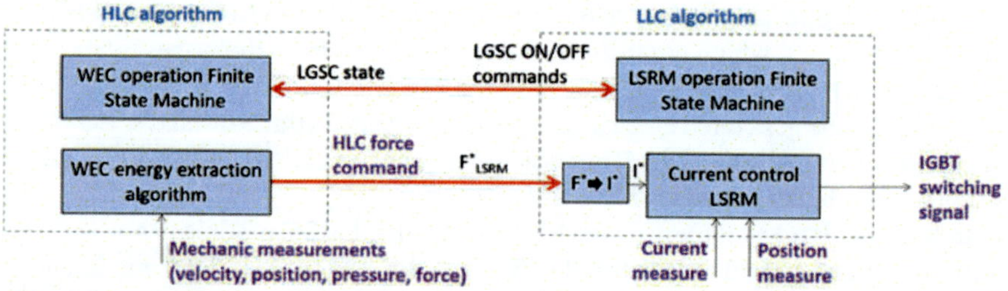

Fig. 17.17 Example of LLC/HLC scheme for the PTO

- Low Level Control (LLC), which drives the LSRM by controlling its electric variables by means of the state of semiconductors. The objective of this control loop is to impose the command force defined by the HLC.

The proposed LLC is modular, one for each PE converter module since the LLC drives the PE. The proposed HLC is unique, driving the global mechanic variables of the WEC. The modular distributed control would be interconnected by a common communication channel with an industrial field bus such as CAN o MODBUS.

The EGTC would be a common tree-phase voltage source inverter [60]. The LGSC is specific, in terms of power electronic converter topology and LLC, for a LSRM. The complete electric drive, encompassing the LSRM, the power electronics and the LLC is described in Sect. 17.3.4.2.

17.3.4.2 SRM Electrical Drive: Power Electronics Topology, Control Algorithms and Operation

The LGSC must achieve a series of requirements to operate the LSRM, namely:

- Each phase of the LSRM must be independently controllable.
- The LGSC must be able to limit the current at low velocities by switching the voltage applied to the phases (i.e. hysteresis band control).
- The LGSC must be equally capable of energising the LSMR phases at rated current before entering its region of motor force.
- In the same way, the LGSC must be able to energize the next phase before the phase that has just been deactivated is completely de-energised, so that power can be extracted from several phases at the same time, reducing the force ripple.
- For the switching periods to be small, the converter must be able to energize and de-energize each phase in a very short time. This requires sufficiently high phase activation and de-activation voltages. However, the phase activation frequency in the wave energy case is not as high as in high-velocity applications [61].
- The converter must allow the free flow (freewheeling) of the current in order to reduce the switching frequency. A path of free circulating current is a so-called “zero voltage” current loop, which does not include the DC bus and applies a zero to the generator phase to rise its current with a smaller time derivate (di/dt).

To extract the energy from waves it is necessary to regulate the current at each one of the generator phases. In this case, the current through the phases of the electrical machine is unidirectional, so, by applying different voltages in the phases, it is possible to regulate this current value [58]. There are different strategies to control the current through the phases of a LSRM: hysteresis band strategy, closed-loop pulse width modulation (PWM) and direct torque control [62]. In this case, a hysteresis band strategy is proposed, which will be described in detail in Sect. ‘Switching Strategy for Current Regulation’.

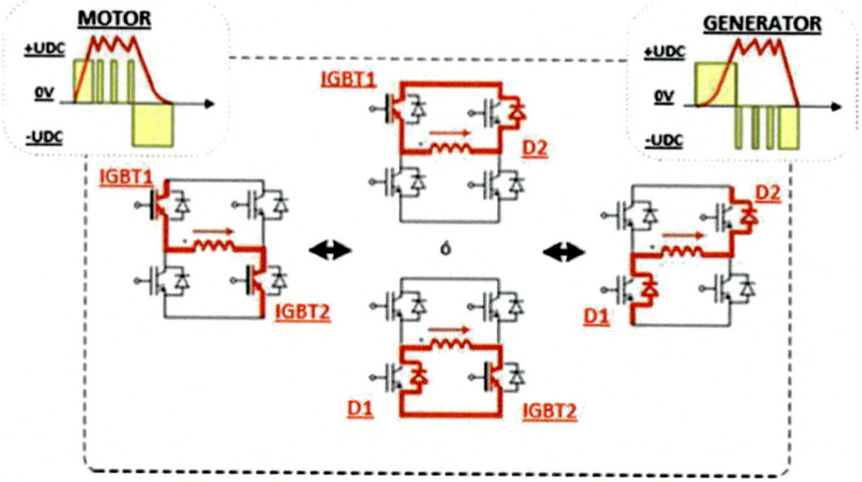

Fig. 17.18 State of the semiconductors and current flow through them in motor and generator mode for a phase of a switched reluctance machine

Switching Strategy for Current Regulation

The LGSC, which connects a DC link (U_{DC}) with an electrical machine, must be able to regulate the current through the phases of the machine. Figure 17.18 shows the status of the switches for the different voltages applied in the case of selecting an asymmetric half-bridge converter (see Sect. Topologies for the Linear Generator Side Converter (LGSC)).

At the beginning of the activation period of one phase $+U_{DC}$ is applied to rapidly increase the current from zero to the reference value and *0 V* and $-U_{DC}$ is used alternately to keep the current within the reference band. Finally, $-U_{DC}$ is applied to extinguish the current at the end of the activation period of each phase. At low and medium velocities, when the counter-electromotive force (EMF) is not very high, current regulation is achieved by soft switching applying 0 V to increase the current within the band. This switching mode is known as 'soft switching' since, when the current is within the limits of the band, 0 V is applied to raise the current.

An alternative to increase the current in this situation is to apply $+Udc$ instead of 0 V. This switching mode is known as 'hard switching'. By applying $+Udc$, the current increases faster than with 0 V. From the point of view of the converter losses, this mode is worse than soft switching because the switching frequency of the power converter increases. In general, hard switching is only suitable when a small hysteresis band width is required to minimize the current ripple and, therefore, when it is necessary to make the current evolve very fast.

From the point of view of the state of the switches, there are two possible combinations to apply 0 V. The two possibilities are closing either IGBT1 or IGBT2. In case of closing IGBT1, the current will circulate through IGBT1 and D1, and if

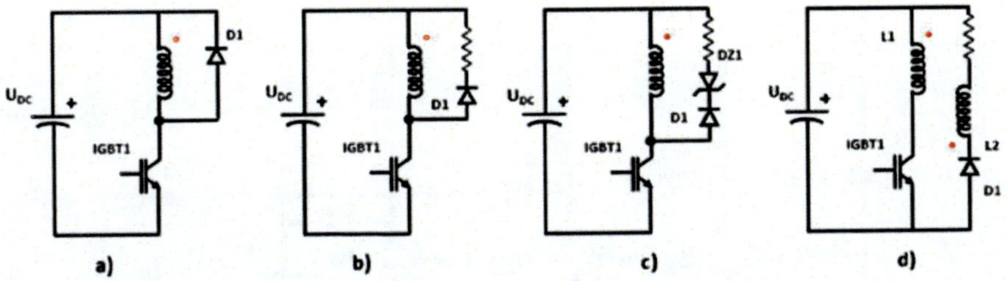

Fig. 17.19 Different topologies of LGSC converters with one switch per phase

IGBT2 is closed, the current will circulate through IGBT2 and D2 (see Fig. 17.18). A possible switching strategy consists of toggling IGBTT1 and IGBT2 in each activation period to apply 0 V, so that one cycle closes IGBTT1 and IGBT2 is used in the next period (cyclic soft switching). An alternative is to change IGBT1 and IGBT2 in each switching event within an activation period (alternating soft-switching).

Topologies for the Linear Generator Side Converter (LGSC)

Three main topologies of converters can be identified to drive an SRM:

(a) Power converters with one switch per phase

One switch per generator phase is enough to provide unidirectional current to the phases of the generator. Four different configurations are shown in Fig. 17.19. In the case of the topologies in Figs. 17.19a–c, the energy supplied to the magnetic field during phase activation is not returned to the source (DC link), but rather dissipated in the elements of the free circulation circuit, resulting in an inefficient option. This does not occur for the topology in Fig. 17.19d, where each phase of the machine is formed by a double winding (two windings per phase, L1 and L2), so, when the switch is open, the current through the primary winding is transferred to the secondary winding and flows through the diode. However, the drawbacks of this topology are the losses in copper associated to the auxiliary winding, and also the blocking of $2{\cdot}U_{DC}$ voltage by the phase switch during the off state (double value than in the previous cases).

An extended family of power converters with a single switch per phase is represented by the "C-dump" converter shown in Fig. 17.20 [63]. The main idea of this family is to create a second voltage bus (capacitor C in Fig. 17.20) to reverse bias the phases during de-energisation after de-activation. The voltage level in this second bus is regulated on $2{\cdot}U_{DC}$ by means of a DC/DC converter implemented through the switch IGBT4 and the inductance L4. This value is chosen to be able to apply a voltage of $-U_{DC}$ to the de-activated phase for rapid demagnetisation. In this sense, the main switches must be sized for $2{\cdot}U_{DC}$ since it is the voltage to be blocked. In this case, the energy supplied to the magnetic field is returned to the source. The great disadvantage of this type of converters lies in the need for an additional power converter, which increases the difficult of the PE control.

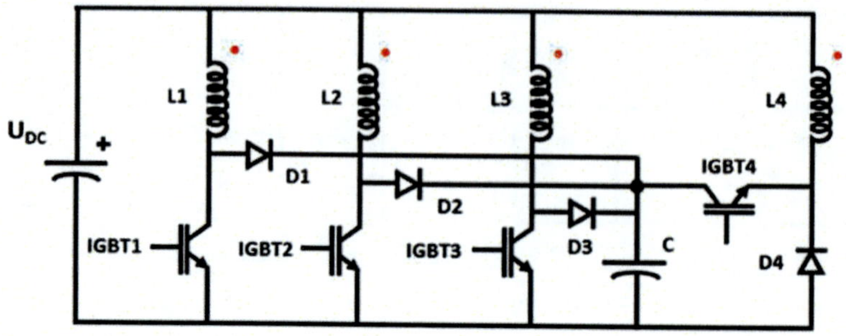

Fig. 17.20 C-dump power converter for three phases LSRM

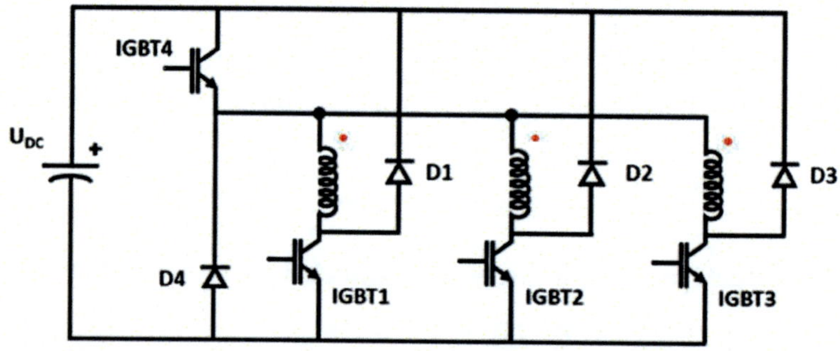

Fig. 17.21 Miller power converter for 3 phases LSRM

(b) Power converters with shared switches

There are other topologies which share switches between one or more phases, thus reducing the number of components and thus the cost of the converter. This is the reason why converters such as the one in Fig. 17.21, called "Miller" converter or "n + 1" switch converter, have emerged [64].

In this circuit, the number of switches is one more than the number of phases of the machine. One of the switches, the upper IGBT4, is shared between all phases. To activate one of the phases, this upper switch and the lower switch are closed in series with the corresponding phase. The regulation of its current can be done either by hard-switching using both switches, or by soft-switching since various possibilities of free flow of current are offered. To de-energize the phase after de-activation, both switches are opened applying $-U_{DC}$ to the phase, thus returning the magnetic energy to the source. The limitation of this PE absence of independency in the control of the different LSRM phases, since the next phase to be activated cannot be energised until the phase that has just been de-activated is completely de-energised (since the common switch is being used to apply reverse voltage to the off phase).

There are variations of this converter, such as the Pollock converter [63], whose topology is based on the block in Fig. 17.22. As extracted from Fig. 17.22, adjacent phases of the converter share one of the switches; this considerably reduces the

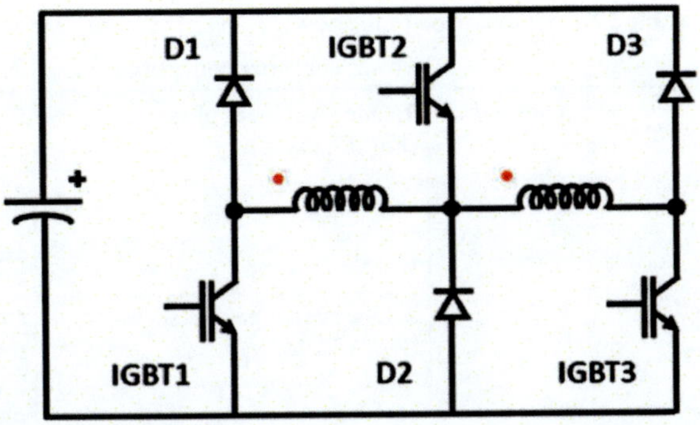

Fig. 17.22 Pollock power converter for LSRM

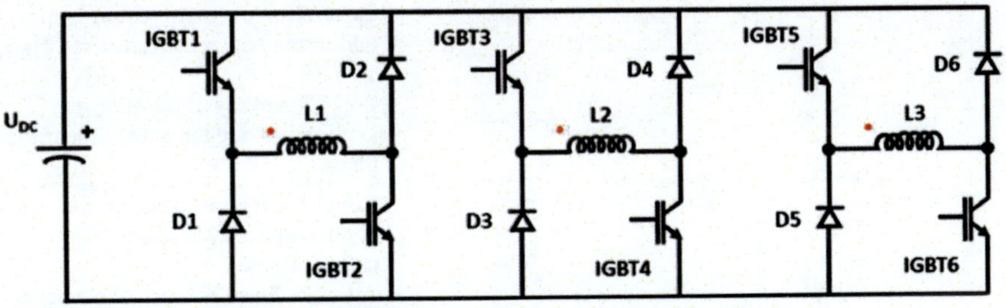

Fig. 17.23 Power converter topology for three phases switched reluctance machine

number of semiconductors, however limits the independent control of the phases. If the two SRM phases that share a switch are activated consecutively, the next phase cannot be energised until the current in the previous phase is extinguished (since the shared switch is open to de-energize). Therefore, this topology is more convenient for SRM with a high number of phases, where the distribution of the windings in the converter can be established so that consecutive phases do not share elements. It should also be mentioned that shared switches have a higher average switching frequency and therefore will present higher losses than in other topologies.

(c) Asymmetric half bridge power converter

In this type of PE converter, each LSRM phase is in series with the switches [60]. To increase reliability and apply optimised strategies, more than one switch per phase is used. In this way, there is a full bridge converter for each LSRM phase (composed of two switches and two free-wheeling diodes), as shown in Fig. 17.23.

The upper and lower switch of each phase can be controlled independently to give rise to different switching schemes. To apply $+U_{DC}$ to each phase, IGBT1 and IGBT2 switches are closed; to apply 0 V, either IGBT1 or IGBT2 are closed; and for $-U_{DC}$, no switch is closed. This converter offers great control flexibility since the current can be treated completely independently in each phase. It also provides extra reliability to the LSRM PTO, since it avoids a short circuit of the DC bus through the main switches, as the phase winding is connected in series between both semiconductors.

Table 17.7 Comparison of the power converter topologies for LGSC

Topology	Power converter configurations	Advantages and drawbacks
One switch per phase	**1**: One switch per phase (Fig. 17.19a–c)	Cons.: It is not possible to recover the energy supplied during the activation of the phase, low efficiency, and low versatility Adv.: Reduced number of components and low cost
	2: One switch per phase (Fig. 17.19d)	Cons.: Low efficiency due to power losses in the secondary winding, the blocking of 2·UDC voltage by the phase switch and higher complexity Adv: Low cost
	3: C-dump power converter	Cons.: Complexity in control, an additional power converter and high cost Adv.: Recovery of the energy supplied during the activation of the phase
Shared switches	**4**: Miller power converter	Cons.: No independence in the control of each phase and high current ripple Adv.: Reduced number of elements and low cost
	5: Pollock power converter	Cons.: Higher average switching frequency and no independence in the control of each phase in machines with low number of phases Adv.: Reduced number of components and low cost
Half bridge	**6**: Asymmetric half bridge power converter	Cons.: High number of elements and higher cost. Adv.: Independence in the control of each phase, higher control flexibility and lower current ripple

Furthermore, all switches and diodes must be sized for the same U_{DC} voltage. The main drawback of this topology is the greater number of switches required. Also, for low voltage applications, the voltage drop across the two semiconductors can be significant relative to the DC bus voltage. However, this converter is optimal for medium/high voltage applications where performance is critical and, for all its advantages, it is widely used for LSRM applications.

(d) Summary

As a summary, the following comparative table is proposed (Table 17.7). From the topologies presented in the table, a particularly suitable topology is PCT 5 because, despite the greater number of semiconductors, it allows independent control of the current of each phase.

Usually, the linear electric machines are designed with wider airgaps that conventional rotary machines due to the long displacement strokes, and this issue leads to a limitation in the maximum slope of the current evolutions. In order to a proper control of the phase currents, it is necessary overlap the activations of the phases, and hence, a power electronic converter capable to control independently the current in each phase.

Control Techniques to Reduce the Torque Ripple in LSRM

As mentioned in Sect. 0, the SRM has a high ripple in force when compared to other types of electrical machines. There are two types of techniques to minimise the force ripple, related either to modifications in the machine topology or to the power converter control strategy. This section describes different existing techniques based on control strategies. The main existing techniques are listed below:

(a) Current and angle modulations

Traditionally, there have been two different ways of reducing torque in switched reluctance machines from a control point of view: the on-off angle optimisation method and the method for following a certain current profile. The first one consists in optimising the angles to achieve a certain torque. However, this method fails to accurately master the behaviour of the torque. The second of the methods consists of following a certain current reference that achieves a certain torque reduction. The drawback of this method is that it is necessary to store a large amount of data to know the exact current value in each situation to give a certain torque value. This method is proposed in [65]. Another additional method derived from the previous one consists of the independent control of the current in each of the poles [66].

(b) Average torque control and Direct torque control

A new scheme for the control of the mean torque in reluctance machines is proposed in [67]. The average torque control (ATC) technique can be implemented using the online average torque and energy ratio estimation method for closed loop control. The torque reference varies constantly as a function of the commutation angles and the reference current to achieve a constant average torque given as the reference value. Due to the adaptive adjustment of the system parameters, precise torque control and acceptable reduction in torque ripple are obtained.

The technique known as direct torque control (DTC) has been employed to minimize torque ripple in SRM [68]. The value of the reference current is obtained from a look-up table in which the position of the moving part of the machine, the torque and the current are related. Semiconductor trigger signals are obtained by comparing the current reference value and the measured value. There is another control technique derived from the previous one called instantaneous torque control (DITC) that achieves a faster response in reducing the torque error [69]. The torque measured is estimated online and the semiconductor trigger signals are the result of a hysteresis band control of the torque. Another technique of instantaneous torque control regulating the reluctance co-energy of the machine is proposed in [70].

(c) Torque sharing function

Another control technique to reduce the torque ripple is the torque sharing function (TSF) with the hysteresis band current control. The reference torque can be defined by a TSF, where the sum of the torque references of each of the phases is equal to the reference of the output torque. The torque of each phase is obtained from the predefined torque profile in a look-up table that relates current, position of the moving part of the machine and torque. The total reference torque of the machine is divided by the number of phases of the machine depending on the position of the machine and as many torque references as phases of the machine are generated. Once the position of the machine and the torque reference are known, the reference currents through the phases of the machine are generated. The reference currents are compared to the measured currents and the switching signals are generated by a hysteresis band control. This method allows to smooth the torque shape, avoiding torque peaks that cause a high ripple. Torque peaks occur during the activation switching of each phase, so there are advanced TSF techniques to minimize torque ripple in these transitions and obtain a better response [71].

(d) Feedback linearisation control, Iterative learning control and Intelligent control

Other techniques used to reduce torque ripple are: feedback linearisation control (FBL), iterative learning control (ILC) and intelligent control (IC). FBL uses a feedback linearisation controller to achieve a closed-loop linearised system. This loop makes it possible to compensate for the non-linear behaviour of the reluctance machine [72]. ILC is an appropriate control technique for non-linear systems with many coupled variables and in which a very precise analysis of the model is required without the need to identify the machine parameters. This method consists of an iterative process in which the necessary waveform to give a certain torque and the voltage that must be applied to the machine phase to achieve that waveform are determined [73]. Finally, IC is based on the use of techniques such as neural networks, or fuzzy logic. This method has a great adaptive capacity and robust self-learning. It is based on the use of a precise estimation of the torque from the voltages and currents measured in the phases of the machine. The estimated torque is compared with the reference torque and the controller based on one of the optimisation techniques discussed generates the reference currents in the machine phases [74].

Multimode Hysteresis Band for LSRM Current Control

As explained in the previous Sect. 'Switching Strategy for Current Regulation', the proposed current control for the LSRM consist in a hysteresis-band soft-switching strategy. However, at low velocities, the low values of EMF (electromotive force) cannot drive the LSRM current. In particular, according to the SRM phase Eq. (17.2),

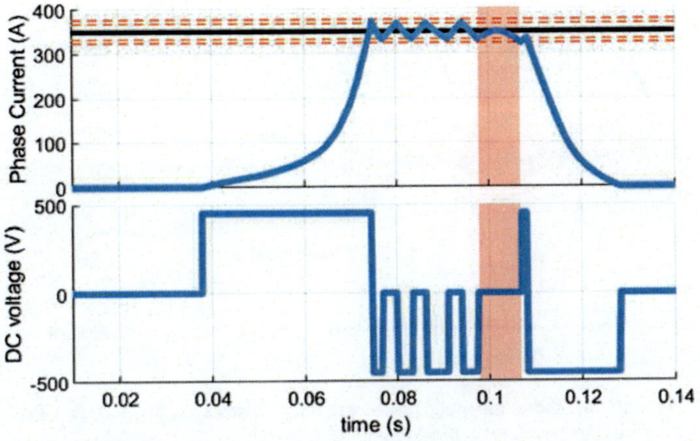

Fig. 17.24 Example of LSRM current with a bad estimation of the velocity limit

If 0 V is imposed when the LSRM is working as a generator, the current evolution depends on the value of the velocity. The current slope is positive at high velocities, as expected in a generator mode. The current slope is negative at low velocities, as it occurs in motor mode due to the LSRM is operating in a non-regenerative generator working point.

A velocity value delimits each LSRM current behaviour, and this limit value depends on the current. In the case of wave energy extraction, the velocity of the electric drive changes of sign in every wave cycle and, hence, the two LRMS current behaviours occur and they must be taken into account. One solution is the use of a 'multimode soft-switching hysteresis band' [75]. This hysteresis-band soft-switching strategy drives the current by applying:

- A generator mode commutation when the velocity is upper than the velocity limit: [0 V/$-U_{DC}$] pair to increase/decrease the current respectively.
- A motor commutation mode when the velocity is lower than the velocity limit: [+U_{DC}/0 V] to increase/decrease the current respectively.

An accurate calculation of the velocity limit value for each current would be required for the multimode operation. For this reason, an external hard-switching hysteresis band is establishing in order to manage incorrect estimations of the velocity limit values at each current. Figure 17.24 shows an example of the LSRM current behaviour with a bad estimation of the velocity limit. In the last part of the phase current evolution, the slope of the current changes sign without a change in the switching state. Figure 17.25 shows a scheme of the current behaviour when a multimode soft-switching band is applied, including the hard-switching external band.

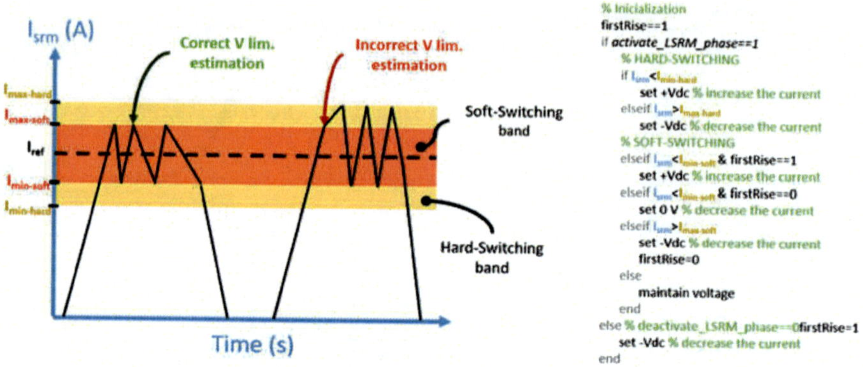

Fig. 17.25 Scheme of the multimode soft-switching hysteresis band (left). Pseudocode of multimode soft-switching hysteresis band (right)

17.4 Superconducting Linear Switched Reluctance Machine (LSRM)

17.4.1 Introduction

As explained in previous sections, there are a number of different approaches to harvest energy from the ocean waves. As shown in Sect. 17.2.2 the approach can be summarised on 10 types of different technologies, with the Point Absorber as the most common one, being present in 33% of those devices. The Oscillating Water Column and the Oscillating Wave Surge Converter Pressure Differential were selected in the 6% and 3% of the devices, respectively. One of the characteristics that these three technologies have in common (since they are based on a linear displacement between two bodies) is that they are very suitable for using DDPTOs to perform the energy conversion process as they are linear displacement devices with a single conversion stage of energy, which makes them very efficient.

At the same time, it is quite intuitive that the forces required in this conversion process are very high. For instance, for a 10 s period wave of 2 m in amplitude the peak vertical velocity would be 0.4 m/s. If a peak power of 1 MW is required, a peak force of 2.5 MN would be needed (only considering the active force component, although in some devices an additional reactive force is required). When talking about such big forces for an electrical machine, superconductivity becomes an interesting option to be considered. Ultimately, a high force machine means a high current machine for what a superconducting option can be almost indispensable. Besides, there is an additional consideration in favour of a superconducting version which is the higher efficiency, something which is also indispensable to harvest energy when high reactive forces are involved.

17.4.2 The Importance of High Force/High Efficiency PTOs

To analyse the convenience of a superconducting PTO let us consider the case of the Heaving Point Absorber (HPA), which, as it has been already explained, is based on the relative movement of two bodies: One is stationary or pseudo-stationary and it is called the Spar, while the other is moving and it is called the Float. To analyse the impact on the requirements of the PTO force and efficiency, we will consider the real WEC example already presented in Sect. 17.3.2 (UNDIGEN W1 prototype [36]). The results were presented in Fig. 17.8a (100% of efficiency, 1 MN PTO) and Fig. 17.8b (75% of efficiency, 1 MN PTO). There are two main conclusions that can be extracted from Fig. 17.8b (75% of efficiency PTO):

- Only for a very narrow window of wave periods, the maximum available power (red curve) can be extracted. For the rest of periods, the power that can be harvested (yellow curve) is very small.
- The force that the PTO uses to produce power (green curve) is below the maximum available force to avoid increasing the losses above the extracted power and, in any case, it is much below the required force to extract the maximum power (blue curve).

If we now use an ideal superconducting PTO with the same force (1 MN) but an efficiency of 100%, alternative results are obtained (see Fig. 17.8a) and another two important conclusions can be extracted:

- The window for which the maximum theoretical limit can be extracted becomes much broader.
- All the available force is used until the required force is bigger than what the PTO can provide, and the extracted power is unable to follow the maximum power path. If we make a more powerful PTO, we should be able to follow the maximum power curve up to the required wave period.

Superconducting PTOs become then a very convenient alternative to increase the area of the Power versus Wave Period curve, making it closer to the maximum magnitude that can be theoretically achieved. Both, the increasing of force capability and the improvement of the efficiency contribute to this increment.

17.4.3 Some Simple Concepts on Superconductivity and Cryogenics

Superconductivity is the property that some materials present under certain conditions to transport electrical current without losses. In other words, the absence of electrical resistance. Besides this fact, two other phenomena characterize the superconducting state: Diamagnetism (*Meissner Effect*) and a finite leap in the specific heat curve. The explanation of superconductivity as well as its physical details are far away from the scope of this introduction, but some useful concepts should be mentioned here. As previously said, the superconducting state only occurs under certain conditions which can be expressed in a simple manner [76]: The point defined by the following three coordinates: the current density flowing through the superconductor (J), the magnetic flux density (B) in which it is immersed and the temperature (T) at which it is working, must be inside a volume which is defined by the intersection of the so-called critical surface with the coordinate planes. The intersection of the critical surface with the J, B & T axes define the critical values: Critical J (J_c), Critical B (B_c) and Critical T (T_c).

There are two big groups of superconductors attending, basically, to the value of their critical temperature. The first group constitute the Low T_c Superconductors with critical temperatures below 20 K, which are mostly intermetallic alloys and obey to the BCS theory to explain the mechanism of superconductivity. There is an exception within this group which is the MgB2 (Magnesium Diboride) [5] with a critical temperature of 39 K, a non-metallic compound which also obeys to the BCS theory. Figure 17.26 shows the critical surfaces for the three most relevant superconductors of the previous group. The other group is called High T_c Superconductors with critical temperatures above 70 K. They are ceramic materials based on copper oxides and a rare earth element and their superconductivity is not explained by the BCS theory. Figure 17.26 also shows the area of operation for different types of superconducting magnets, including conventional ones, high field ones and those working at higher temperatures but moderated fields. For each type, a different type of superconductor becomes more suitable.

Superconducting wires are able to transport current densities which can be easily two orders of magnitude higher than those for normal conducting wires and this implies a significant reduction in the size of the devices but also in their power

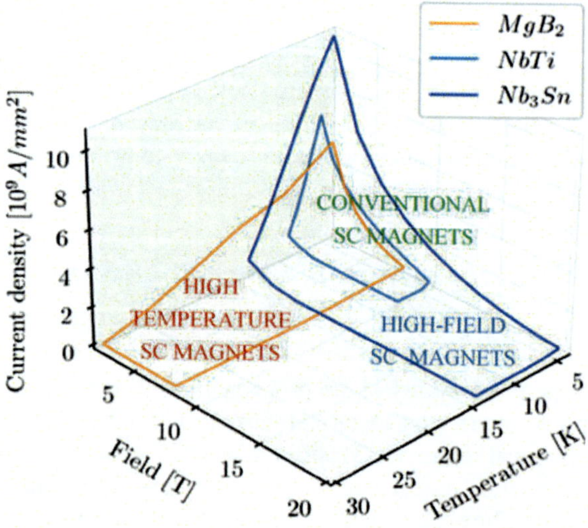

Fig. 17.26 Critical surfaces for NbTi, Nb_3Sn & MgB_2

consumption. In this manner, they are especially adequate for magnet manufacturing in general and for electrical machine windings in particular, which can be considered as a special case of magnets. Nevertheless, superconductivity has two big issues to overcome:

- While the resistance is null for direct current (DC), when transporting alternate current (AC) there are losses associated to the induced electric fields in the superconductor. Besides, there are also dissipative currents induced in the metal matrix in which the SC material is allocated. The consequence is that both terms heat up the material driving it to the normal state. This fact limits the application of superconductors to DC or to low frequency ones and requires the use of special wires based on very small filaments of SC material and high electrical resistance matrices [77]
- The magnet has to operate at cryogenic temperatures (in the order of 10 K or less) and this requires a complex cryogenic facility to cool the magnet down but also to keep it cold.

Usually, cryogenic aspects of superconductivity are even more complex than the superconducting aspects themselves. To cool the magnet down there are different possibilities spanning from submerging it into a liquid coolant as helium, circulating a coolant liquid or a gas though it or extracting the heat from the magnet by conduction using a cryocooler [78]. To keep it cool, the magnet must be thermally insulated from outside, minimising the heat which is conducted and radiated. For the first case the magnet must be in a vacuum atmosphere (typically 10–6 mbar) while for the second one, the solution is to shield the magnet with a screen that reduces the radiated heat to the outer wall of the vacuum chamber. In any case, it becomes essential to reduce the losses associated to the magnet to the minimum possible value since, extracting 1 W at 4.2 K implies a real power consumption of some kW at room temperature (significantly higher than the theoretical Carnot limit). In this sense, increasing the operational temperature of the superconductor has a great impact on the overall design, either because the refrigeration power is reduced or because the requirements of thermal insulation are not so strict.

17.4.4 Options for Superconducting Machines: A Superconducting PTO

Theoretically, every conventional machine has its corresponding superconducting version, nevertheless, the limitations associated to superconductivity, restricts the real number of types of superconducting (SC) machines that can be performed. Particularly, there are two severe restrictions to consider: (1) AC losses must be reduced below admissible values below the refrigeration system capabilities. For this reason, historically, superconducting machines only introduce SC windings associated to the DC side (2) powering SC windings that are moving and must work at cold is very complex and this is the reason why, normally, only stationary windings are superconducting, although there are exceptions. In other words, in most of the existing realisations concerning SC machines only the excitation DC field wingdings are stationary and superconducting. Alternatively, solutions for using bulk superconductors as high field permanent magnets to create the excitation field in the electrical machine have also been proposed, even in linear machines [79, 80]. There are few realisations integrating superconductivity in linear machines for PTO applications. One interesting case was the one proposed by Keysan and Mueller [81, 82] which is based on a Linear Synchronous Machine with SC excitation and normal conducting stator, to avoid AC losses. The SC excitation may have homopolar or bipolar configurations and it is based on using a superconducting coil and an iron yoke to create a dipole magnetic field which crosses the copper armature windings powered by electric current. Interaction between the magnetic field and the current, creates a net thrust (action & reaction) on each side of the machine.

Generally speaking, there are three possible configurations for a linear SC machine, regarding the operational temperature of its main components. In principle there is only one condition to fulfil: SC coils must be at cryogenic temperatures; the rest can be either at cold or warm temperatures. Figure 17.27a shows the configuration where only the coils are at cold, Fig. 17.27b the configuration where all the stator is at cold (iron & coils) while the translator is warm and Fig. 17.27c shows a configuration where all the machine is cold. The big difference between them concerns the size of the air gap.

While for the first case the airgap can be as small as in any conventional machine (say in the order of 1 mm), for the second one there is a minimum airgap given by the necessary width of the cryostat which can easily be in the order of 50 mm. Finally, the third case requires an airgap of around 3–5 mm, big enough to allocate a radiation screen but much smaller than what is needed for the previous one. Apparently, the first case should be preferable to the other two cases since the airgap is the smallest of them, but there is a big issue associated to it: since each coil requires its own cryostat,

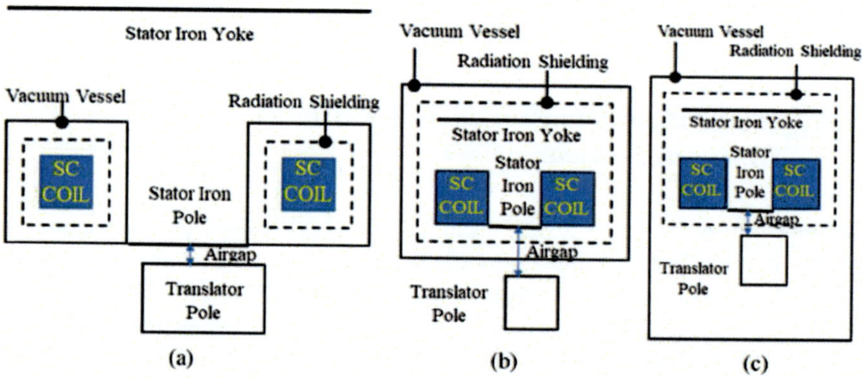

Fig. 17.27 **a** Only superconducting (SC) coils working at cold; **b** Full stator working at cold; **c** Full stator & translator working at cold

the equivalent coil size (including the coil and the cryostat) can be of the same order of magnitude as for a conventional non-SC coil. Besides this fact, the mechanical support of the coil in this solution becomes very complicated. Solution presented in Fig. 17.27b is extremely compact, and coils are directly wounded around the iron making very easy their structural support; the problem now is the huge airgap which is needed and the corresponding associated stray field. Even if the ampere-turns are increased to compensate the bigger gap, the machine becomes very inefficient from a magnetic point of view. Solution depicted in Fig. 17.27c includes the advantages of the two previous ones, avoiding their drawbacks: the coil can be easy supported, retaining the big electromagnetic forces that appear, while the airgap can be small enough to guarantee an efficient magnetic design.

Figure 17.28 shows a proposal for a SC linear displacement PTO based on the third solution [83]. All the machine is allocated inside a flexible cryostat consisting of two upper and a lower bellows joined to a central stationary body where the stator is allocated and surrounded by a radiation screen. The translator is fixed to the moving parts of the cryostat but thermally insulated from them. A guiding system guarantees a rectilinear displacement. The only limitation of the third solution in general and in this proposal in particular, is the cryostat length which reduces the applications of

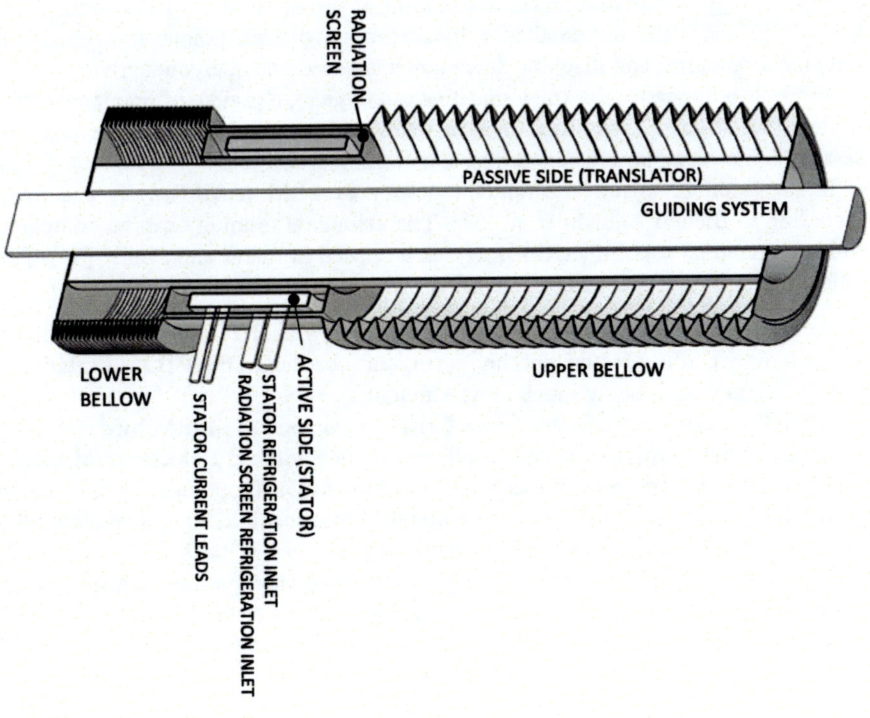

Fig. 17.28 Superconducting PTO based on a flexible cryostat

the machine to those involving a reciprocating short-medium stroke movement but not a long linear displacement.

Concerning the type of electrical machine than can be used, there are several options with the only restriction of avoiding translators with windings. Machine represented in Fig. 17.27 corresponds to a cylindrical SRM type [54] but also synchronous or permanent magnet machines can be used. In any case, the important contribution of superconductivity is the increasement of current density and, consequently, the strong growth of the ampere-turns with a similar machine size. Since, for a saturated machine, the force is approximately proportional to the ampere-turns, SC machines can easily provide forces one order of magnitude bigger than the conventional ones with similar sizes, when all the different aspect, including cryogenics, are taken into account.

17.5 Concluding Remarks

This chapter analyses the use of linear electrical switched reluctance machines (SRM) as PTO in wave energy systems. After an introduction about general concepts of wave energy conversion and a description of the direct-drive PTO concepts in Sects. 17.1 and 17.2, the chapter is focused on two main points: new electrical machine topologies; and machine developments based on superconductivity.

Respect to the first point (new machine topologies), a review of previous works related with linear electric machines in wave energy conversion are presented in Sect. 17.3 including power electronics and control issues.

In this section, a new azimuthal topology of SRM (AMSRM) is described according to Blanco Aguado et al. [57]. The azimuthal topology can be compared with a rectangular topology (RMSRM), as proposed in Lucas and Pinilla [84]. The following improvements can be remarked:

- Azimuthal configuration has an advantage for a WEC PTO application, since it has a higher filling factor that the rectangular one due to the PTO is hosted in a cylindrical vessel, being much more efficient.
- AMSRM increases the Stator Force Density compared to MSRM by developing an azimuthal configuration of a Multi-translator Switched Reluctance Machine. AMSRM fully eliminates the End Stators of the MSRM topology, by creating an azimuthal flux path, closing through identical Intermediate Stators, avoiding the need of lateral iron to close the magnetic circuit.
- AMSRM reduces the machine weight since this configuration eliminates the lateral end stators, increasing the force density respect to the MSRM. Moreover, azimuthal configuration presents an airgap surface growth by a consequence of its configuration and its adaptation to the PTO shape.

The future perspectives of the AMSRMs are great since their performance increasing with their outer diameter, i.e. the force developed by the machine grows higher than the square of their surface, as shown in Table 17.8. Therefore, the higher

Table 17.8 Comparison of the force developed by three AMSRM configurations

Parameter	AMSRM Conf 1	AMSRM Conf 2	AMSRM Conf 3
No. of active sides	8	16	32
No. of passive sides	8	16	32
No. of gaps per phase	16(2mmeach)	32(2mmeach)	64(2mmeach)
Diameter (D) (mm)	1000	2000	3000
Axial length (l) (mm)	280	620	800
Height (h) (mm)	800	1056	1303
Average force @ nominal current (N)	34,293	207,424	574,529

the outer diameter, the larger the air gap surface, and with it, the force developed by the machine.

Respect to the second point (superconducting machines), it has been justified the superconductivity solution in Sect. 17.4, in addition to presenting a specific design. A Power Take-Off for wave energy extraction converters is one of the applications where superconductivity becomes more adequate in the coming years for two main reasons which are closely related: on the one hand, big forces are required to achieve significant power levels, since the moving speed is very low. Furthermore, if the PTO capability of supplying reactive force to increase the power capture ability is required, the level of needed force is even higher and superconductivity is almost indispensable if an all-electric PTO is preferred. On the other hand, big forces imply big currents which mean big losses in resistive generators that must be oversized. Besides, these losses can make the use of reactive force not efficient which, again limits extracting all the available energy for a given sea state.

In this regard superconductivity allows transporting very big current without losses which, at the end, allows increasing the operational magnetic field. Since the force is the product of the current times the magnetic field, a significant force in-crease can be achieved and what is even more important; a dramatic force density (force per generator unit mass) increment will happen.

A brief list and description of superconducting technologies has been presented in the chapter (Sect. 17.4) as well as the main issues that superconductivity must overcome like the A.C. losses or the cryogenic configuration to cool the magnet down to operational temperatures. It has also been shown how the way the coils are insulated in their vessels affects enormously to the magnetic efficiency and in this sense a very compact design based on cold Active and Passive Sides has been presented as an example of what could be a compact superconducting generator.

Acknowledgements This research, developed under the Proyect SETITAN (ID: 764014), has been supported and has received funding from European Union's Horizon 2020 research and innovation programme under H2020-EU.3.3.2.—Low-cost, low-carbon energy supply (LCE-07-2016-2017).

References

1. (2015) *International Levelised Cost of Energy (LCOE) for ocean energy technologies.*
2. Low Carbon Innovation Coordination Group. (2012). *Technology Innovation Needs Assessment (TINA), Marine Energy, Summary Report.*
3. (2013) *Ocean energy: Cost of energy and cost reduction opportunities.*
4. Iglesias, G., Astariz, S., & Vazquez, A. (2018). The economics of wave and tidal energy. In *Wave and tidal energy* (pp. 513–532). Wiley.
5. Magagna, D., Monfardini, R., & Uihlein, A. (2016). *JRC ocean energy status report 2016 edition.* Publications Office of the European Union.
6. Magagna, D., & Uihlein, A. (2015). *2014 JRC ocean energy status report.* Luxembourg: Publications Office of the European Union.
7. Villate, J. L. (2020). *Strategic research and innovation agenda for ocean energy.*
8. Magagna, D., Margheritini, L., Alessi, A., Bannon, E., Boelman, E., Bould, D., & et al. (2018). *Workshop on identification of future emerging technologies in the ocean energy sector.* European Commission JRC.
9. (2011) Accelerating marine energy: The potential for cost reduction—Insights from the carbon trust marine energy accelerator. *Technical report for carbon trust.*
10. *WES|Programmes: Power Take Off.* https://www.waveenergyscotland.co.uk/programmes/details/power-take-off/. Accessed May 27, 2020.
11. Têtu, A. (2017). *Power take-off systems for WECs* (pp. 203–220).
12. *Wave devices: EMEC: European Marine Energy Centre.* http://www.emec.org.uk/marine-energy/wave-devices/. Accessed October 12, 2015.
13. LiVecchi, A., Copping, A., Jenne, D., Gorton, A., Preus, R., Gill, G., Robichaud, R., Green, R., Geerlofs, S., Gore, S., Hume, D., McShane, W., Schmaus, C., & Spence, H. (2019). *Exploring opportunities for marine renewable energy in maritime markets.* U.S. Department of Energy (DOE).
14. Liu, Z., Zhang, R., Xiao, H., & Wang, X. (2020). Survey of the mechanisms of power take-off (PTO) devices of wave energy converters. *Acta Mechanica Sinica, 36*, 644–658. https://doi.org/10.1007/s10409-020-00958-z
15. (2012) *Wave dragon.* https://en.wikipedia.org/wiki/Wave_Dragon. Accessed March 11, 2021.
16. Falcão, A. F de O. (2010). Wave energy utilization: A review of the technologies. *Renewable and Sustainable Energy Reviews, 14*, 899–918. https://doi.org/10.1016/j.rser.2009.11.003
17. Drew, B., Plummer, A. R., & Sahinkaya, M. N. (2009). A review of wave energy converter technology. *Proceedings of the Institution of Mechanical Engineers, Part A: Journal of Power and Energy, 223*, 887–902. https://doi.org/10.1243/09576509JPE782
18. Moretti, G., Santos Herran, M., Forehand, D., Alves, M., Jeffrey, H., Vertechy, R., & Fontana, M. (2020). Advances in the development of dielectric elastomer generators for wave energy conversion. *Renewable and Sustainable Energy Reviews, 117*, 109430. https://doi.org/10.1016/j.rser.2019.109430
19. Ahamed, R., McKee, K., & Howard, I. (2020). Advancements of wave energy converters based on power take off (PTO) systems: A review. *Ocean Engineering, 204*, 107248. https://doi.org/10.1016/J.OCEANENG.2020.107248
20. Magagna, D. (2019). *Ocean energy technology development report.*

21. Khatri, P., & Wang, X. (2020). Comprehensive review of a linear electrical generator for ocean wave energy conversion. *IET Renewable Power Generation, 14*, 949–958.
22. Joubert, J., van Niekerk, J., Rienecke, J., & Meyer, I. (2016). *Wave energy convertors (WEC): A review of the technology and power generation.*
23. S.p.A., U. C. (2017). *ReBaS generator (Reciprocating Ball Screw Generator)—PTO Stage 2—Public Report*. UMBRA Cuscinetti S.p.A.
24. Mueller, M., & Polinder, H. (2013). *Electrical drives for direct drive renewable energy systems.* Woodhead Publishing.
25. Penalba, M., & Ringwood, J. (2016). A review of wave-to-wire models for wave energy converters. *Energies, 9*, 506. https://doi.org/10.3390/en9070506
26. Lin, Y., Bao, J., Liu, H., Li, W., Tu, L., & Zhang, D. (2015). Review of hydraulic transmission technologies for wave power generation. *Renewable and Sustainable Energy Reviews, 50*, 194–203. https://doi.org/10.1016/j.rser.2015.04.141
27. Waters, R., Stålberg, M., Danielsson, O., Svensson, O., Gustafsson, S., Strömstedt, E., Eriksson, M., Sundberg, J., & Leijon, M. (2007). Experimental results from sea trials of an offshore wave energy system. *Applied Physics Letters, 90*, 034105. https://doi.org/10.1063/1.2432168
28. Santos, M., Lafoz, M., Blanco, M., García-Tabarés, L., García, F., Echeandía, A., & Gavela, L. (2012). Testing of a full-scale PTO based on a switched reluctance linear generator for wave energy conversion. In *4th International Conference on Ocean Energy (ICOE2012).*
29. Edinburgh U of C-GEN Direct Drive PTO. (2017). *PTO Stage 2—Public Report*. University of Edinburgh
30. Keller, J. (2020). *Test of a novel, commercial-scale wave energy direct-drive rotary power take-off at the NREL National Wind Technology Center: Cooperative Research and Development Final Report, CRADA Number CRD-14-571.*
31. Mueller, M. A. (2002). Electrical generators for direct drive wave energy converters. In *Proceedings of the generation, transmission and distribution* (pp 446–456). IEE.
32. Polinder, H., Mueller, M., & Scuotto, M. (2007). Linear generator systems for wave energy conversion. In *Proceedings of the 7th European wave and tidal energy conference.*
33. Elwood, D., Yim, S. C., Prudell, J., Stillinger, C., von Jouanne, A., Brekken, T., Brown, A., & Paasch, R. (2010). Design, construction, and ocean testing of a taut-moored dual-body wave energy converter with a linear generator power take-off. *Renewable Energy, 35*, 348–354. https://doi.org/10.1016/j.renene.2009.04.028
34. Polinder, H., Damen, M. E. C. E. C., & Gardner, F. (2004). Linear PM Generator system for wave energy conversion in the AWS. *IEEE Transactions on Energy Conversion, 19*, 583–589. https://doi.org/10.1109/TEC.2004.827717
35. Liu, C., Yu, H., Hu, M., Liu, Q., Zhou, S., & Huang, L. (2014). Research on a permanent magnet tubular linear generator for direct drive wave energy conversion. *IET Renewable Power Generation, 8*, 281–288. https://doi.org/10.1049/iet-rpg.2012.0364
36. Blanco, M., Lafoz, M., Pinilla, G., Gavela, L., García-Tabarés, L., & Echeandia, A. (2010). Electric linear generator to optimize a point absorber wave energy converter. In *3rd International conference and exhibition on Ocean Energy, ICOE 2010.*
37. Falnes, J. (2002). *Ocean waves and oscillating systems, linear interactions including wave-energy extraction.* Cambridge University Press.
38. Hai, L., Goteman, M., & Leijon, M. (2016). A methodology of modelling a wave power system via an equivalent RLC circuit. *IEEE Transaction on Sustainable Energy, 3029*, 1–1. https://doi.org/10.1109/TSTE.2016.2538803
39. Falnes, J. (2007). A review of wave-energy extraction. *Marine Structures, 20*, 185–201. https://doi.org/10.1016/j.marstruc.2007.09.001
40. Hals, J., Falnes, J., & Moan, T. (2011). A comparison of selected strategies for adaptive control of wave energy converters. *Journal of Offshore Mechanics and Arctic Engineering, 133*, 31101. https://doi.org/10.1115/1.4002735
41. Joseph, D. M., & Cronje, W. A. (2007). Design and analysis of a double-sided tubular linear synchronous generator with particular application to wave-energy conversion. In *2007 IEEE power engineering society conference and exposition in Africa—PowerAfrica* (pp. 1–8).

42. Lu, Q., & Ye, Y. (2009). Design and analysis of tubular linear PM generator. *IEEE Transactions on Magnetics, 45*, 4716–4719. https://doi.org/10.1109/TMAG.2009.2022184
43. Nakamura, T., Nishio, H., Amemiya, N., Kajikawa, K., & Wakuda, T. (2012). Performance of induction/synchronous motor having ${\rm MgB}_{2}$ cage windings for liquid hydrogen circulation pump. *IEEE Transactions on Applied Superconductivity, 22*, 5200404. https://doi.org/10.1109/TASC.2011.2179398
44. Ching, T. W., Chau, K. T., & Li, W. (2016). Power factor improvement of a linear vernier permanent-magnet machine using auxiliary DC field excitation. *IEEE Transactions on Magnetics, 52*, 1–4. https://doi.org/10.1109/TMAG.2016.2524533
45. Toba, A., & Lipo, T. A. (2000). Generic torque-maximizing design methodology of surface permanent-magnet vernier machine. *IEEE Transactions on Industry Applications, 36*, 1539–1546. https://doi.org/10.1109/28.887204
46. Li, J., Chau, K. T., Jiang, J. Z., Liu, C., & Li, W. (2010). A new efficient permanent-magnet vernier machine for wind power generation. *IEEE Transactions on Magnetics, 46*, 1475–1478. https://doi.org/10.1109/TMAG.2010.2044636
47. Lomonova, E. A., Tang, Y., Paulides, J. J. H. (2015). Hybrid-excited variable reluctance machines. In *2015 International Aegean Conference on Electrical Machines Power Electronics (ACEMP), 2015 International Conference on Optimization of Electrical Electronic Equipment (OPTIM) 2015 International Symposium on Advanced Electromechanical Motion Systems (ELECTROMOTION)* (pp. 1–10).
48. Chen, A., Nilssen, R., & Nysveen, A. (2010). Performance comparisons among radial-flux, multistage axial-flux, and three-phase transverse-flux PM machines for downhole applications. *IEEE Transactions on Industry Applications, 46*, 779–789. https://doi.org/10.1109/TIA.2009.2039914
49. Guo, YouGuang, Zhu, J. G., Watterson, P. A., & Wei, Wu. (2006). Development of a PM transverse flux motor with soft magnetic composite core. *IEEE Transactions on Energy Conversion, 21*, 426–434. https://doi.org/10.1109/TEC.2005.860403
50. Ahn, H., Jang, G., Chang, J., Chung, S., & Kang, D. (2008). Reduction of the torque ripple and magnetic force of a rotatory two-phase transverse flux machine using herringbone teeth. *IEEE Transactions on Magnetics, 44*, 4066–4069. https://doi.org/10.1109/TMAG.2008.2001586
51. Rodewald, W., Katter, M., & Vacuumschmelze. (2004). *Properties and applications of high performance magnets*.
52. Polinder, H., Mecrow, B. C., Jack, A. G., Dickinson, P. G., & Mueller, M. A. (2005). Conventional and TFPM linear generators for direct-drive wave energy conversion. *IEEE Transactions on Energy Conversion, 20*, 260–267. https://doi.org/10.1109/TEC.2005.845522
53. Oprea, CA, Martis, CS, Jurca, FN, Fodorean, D, Szabó, L, & Szabó, L (2011) Permanent magnet linear generator for renewable energy applications: Tubular vs. four-sided structures. In *2011 International conference on clean electrical power (ICCEP)* (pp. 588–592).
54. Mendes, R. G., Calado, M., & Mariano, S. (2014). *Identification of some tubular topologies of linear switched reluctance generator for direct drive applications in ocean wave energy conversion*.
55. Garcia-Tabares Rodriguez, L., Lafoz, M., Torres, J. J., Obradors, D., Blanco, M., Najera, J., Navarro, G., Garcia, F., & Sanchez, A. (2020). New type of linear switched reluctance generator for wave energy applications. *IEEE Transactions on Applied Superconductivity, 1*.https://doi.org/10.1109/TASC.2020.2981900
56. Lucas, J., & Pinilla, M. (2012). *Switched reluctance linear motor/generator*.
57. Blanco Aguado, M., Garcia Lorenzo, F. E., Garcia-Tabares Rodriguez, L., Lafoz Pastor, M., Obradors Campos, D., Torres Miranda, J. J., Jesus, M. O. C. D. T. M. J. (2019). *Switched reluctance machine, 14*.
58. Krishnan, R. (2001). *Switched reluctance motor drives: Modeling, simulation, analysis, design, and applications*. CRC Press.
59. Arghir, C., Jouini, T., & Dörfler, F. (2018). Grid-forming control for power converters based on matching of synchronous machines. *Automatica, 95*, 273–282. https://doi.org/10.1016/j.automatica.2018.05.037

60. Mohan, N., Undeland, T. M., & Robbins, W. P. (2003). *Power electronics: Converters, applications, and design*. Wiley.
61. Vukosavic, S., & Stefanovic, V. R. (1991). SRM inverter topologies: A comparative evaluation. *IEEE Transactions on Industry Applications, 27*, 1034–1047. https://doi.org/10.1109/28.108453
62. Bilgin, B., Jiang, J. W., & Emadi, A., *Switched reluctance motor drives: Fundamentals to applications*.
63. Barnes, M., & Pollock, C. (1998). Power electronic converters for switched reluctance drives. *IEEE Transactions on Power Electronics, 13*, 1100–1111. https://doi.org/10.1109/63.728337
64. Miller, T. J. E. (1989). *Brushless permanent-magnet and reluctance motor drives*. Clarendon Press.
65. Mikail, R., Husain, I., Sozer, Y., Islam, M. S., & Sebastian, T. (2013). Torque-ripple minimization of switched reluctance machines through current profiling. *IEEE Transactions on Industry Applications, 49*, 1258–1267. https://doi.org/10.1109/TIA.2013.2252592
66. Lin, F., & Yang, S. (2007). An approach to producing controlled radial force in a switched reluctance motor. *IEEE Transactions on Industrial Electronics, 54*, 2137–2146. https://doi.org/10.1109/TIE.2007.895129
67. Inderka, R. B., & De Doncker, R. W. A. A. (2003). High-dynamic direct average torque control for switched reluctance drives. *IEEE Transactions on Industry Applications, 39*, 1040–1045. https://doi.org/10.1109/TIA.2003.814579
68. Inderka, R. B., & De Doncker, R. W. A. A. (2003). DITC-direct instantaneous torque control of switched reluctance drives. *IEEE Transactions on Industry Applications, 39*, 1046–1051. https://doi.org/10.1109/TIA.2003.814578
69. Klein-Hessling, A., Hofmann, A., & De Doncker, R. W. (2017). Direct instantaneous torque and force control: A control approach for switched reluctance machines. *IET Electric Power Applications, 11*, 935–943. https://doi.org/10.1049/iet-epa.2016.0515
70. Wong, K. F., Cheng, K. W. E., & Ho, S. L. (2009). On-line instantaneous torque control of a switched reluctance motor based on co-energy control. *IET Electric Power Applications, 3*, 257–264. https://doi.org/10.1049/iet-epa.2008.0124
71. Vujičić, V. P. (2012). Minimization of torque ripple and copper losses in switched reluctance drive. *IEEE Transactions on Power Electronics, 27*, 388–399. https://doi.org/10.1109/TPEL.2011.2158447
72. Ilic'-Spong, M., Marino, R., Peresada, S., & Taylor, D. (1987) . Feedback linearizing control of switched reluctance motors. *IEEE Transaction on Automatic Control, 32*, 371–379. https://doi.org/10.1109/TAC.1987.1104616
73. Sahoo, N. C., Xu, J. X., & Panda, S. K. (2001). Low torque ripple control of switched reluctance motors using iterative learning. *IEEE Transactions on Energy Conversion, 16*, 318–326. https://doi.org/10.1109/60.969470
74. Mir, S., Elbuluk, M. E., & Husain, I. (1999). Torque-ripple minimization in switched reluctance motors using adaptive fuzzy control. *IEEE Transactions on Industry Applications, 35*, 461–468. https://doi.org/10.1109/28.753642
75. Blanco, M., Navarro, G., & Lafoz, M. (2009). Control of power electronics driving a switched reluctance linear generator in wave energy applications. In *13th European Conference on Power Electronics and Applications, 2009. EPE '09* (pp 1–9). IEEE.
76. Hull, J. R., Wilson, M. N., Bottura, L., Rossi, L., Green, M. A., Iwasa, Y., Hahn, S., Duchateau, J.-L., & Kalsi, S. S. (2015). Superconducting magnets. *Applied superconductivity* (pp. 403–602). Weinheim, Germany: Wiley-VCH Verlag GmbH & Co. KGaA.
77. Iwasa, Y. (2009). *Case studies in superconducting magnets: Design and operational issues* (2nd ed.). US: Springer.
78. Dolan, T. J. (2013). Cryogenic systems. In *Lecture Notes in Energy* (pp 491–511). Springer.
79. Terao, Y., Sekino, M., & Ohsaki, H. (2010). Design study of linear synchronous motors using superconducting coils and bulks. In *The 2010 international power electronics conference—ECCE ASIA* (pp. 1760–1765).

80. Jin, J. X., Zheng, L. H., Guo, Y. G., & Zhu, J. G. (2011). Performance characteristics of an HTS linear synchronous motor with HTS bulk magnet secondary. *IEEE Transactions on Industry Applications, 47*, 2469–2477. https://doi.org/10.1109/TIA.2011.2168934
81. Hodgins, N., Keysan, O., McDonald, A. S., & Mueller, M. A. (2012). Design and testing of a linear generator for wave-energy applications. *IEEE Transactions on Industrial Electronics, 59*, 2094–2103. https://doi.org/10.1109/TIE.2011.2141103
82. Keysan, O., & Mueller, M. A. (2012). A linear superconducting generator for wave energy converters. In *6th IET International conference on power electronics, machines and drives (PEMD 2012)* (pp. 1–6).
83. García-Tabarés, L., Munilla, J., López de Toledo, C., García, F., Sarmiento, G., & Sanz, S. (2020). *Máquina Eléctrica Lineal Superconductora de Movimiento Alternativo.*
84. Lucas, T. J., & Pinilla, M. M. (2006). *Switched reluctance linear motor/generator*.

Chapter 18
Grid Integration of Wave Energy Devices

Marcos Blanco, Jorge Torres, Miguel Santos-Herrán, Álvaro Santiago, Gustavo Navarro, Jorge Nájera, Juan Ignacio Pérez, and Marcos Lafoz

Abstract This chapter is focused on the integration in the electric grid of wave energy devices. In particular, the problem of the output power oscillations in wave farms (WF) is presented and analysed, since these oscillations imply frequency oscillations in the electric grid. The chapter begins with Sect. 18.1, which draws the big picture of the integration of renewable energies, and specifically wave energy in the electric system. Section 18.2 is devoted to the study of the power oscillations presented in wave energy generation. Firstly, the nature of the wave oscillations is analysed, and several wave models such as regular and irregular waves are presented through their energy spectrum, spectral parameters, statistics, etc. Then, the effect of the waves in the electric power oscillations is studied and some stochastic models are presented as well as several types of wave energy converters (WEC). In the last part of this section, the impact of power oscillations in the power grid and the grid codes that wave generation must meet are assessed. Next, Sect. 18.3 discusses power smoothing solutions and focuses on storage, available storage technologies,

M. Blanco (✉) · J. Torres · M. Santos-Herrán · Á. Santiago · G. Navarro · J. Nájera · M. Lafoz
CIEMAT, Avda. Complutense 40, 28045 Madrid, Spain
e-mail: marcos.blanco@ciemat.es

J. Torres
e-mail: jorgejesus.torres@ciemat.es

M. Santos-Herrán
e-mail: miguel.santos@ciemat.es

Á. Santiago
e-mail: alvaro.santiago@ciemat.es

G. Navarro
e-mail: marcos.blanco.aguado@ciemat.es

J. Nájera
e-mail: jorge.najera@ciemat.es

M. Lafoz
e-mail: marcos.lafoz@ciemat.es

J. I. Pérez
ETSICCP-UPM, Campus Ciudad Universitaria, C/Profesor Aranguren 3, 28040 Madrid, Spain
e-mail: ji.perez@upm.es

A. Samad et al. (eds.), *Ocean Wave Energy Systems*, Ocean Engineering & Oceanography 14, https://doi.org/10.1007/978-3-030-78716-5_18

and dimensions. Finally, Sect. 18.4 summarizes energy storage systems used in wave energy devices in real projects.

18.1 Introduction: Wave Energy Impacts in Electric Grid

18.1.1 Impact in the Power Grid of Renewable Energy

Plans to transit to a 100% renewable electric power system are fully firm. The list of countries all over the world that have committed to drastically reduce the share of fossil-based electricity generation and/or achieve a 100% renewable electric power system is endless.

In the European Union (EU), the Directive (EU) 2018/2001 establishes a binding target of 32% for the share of renewable energy consumed in the EU by 2030, which translates into a rather higher share target when referred to the electricity consumption. According to the International Renewable Energy Agency (IRENA), the share of renewable energy in electricity production will reach between 41 and 50% by 2030 in the EU [1].

Not only the EU has initiated a firm transition to a fully renewable electric power system but also other regions of the world. Just as an example, California state has committed to producing half of the State's electricity from renewable energy by 2030 [2], as have Victoria and Queensland States in Australia done as well [3].

This transition seems a one-way path though which is not exempt from difficulties. Dozens of scientific articles have discussed the difficulties in the operation of an electric power system with a large (or even a 100%) share of renewable generation. The impacts on the system operation that have been discussed in the literature are diverse, but most of them are due to the variable and unpredictable nature of renewable generation sources, and perhaps more importantly, to the fact that most renewable generation units are connected to the grid through power electronic devices.

Spreading renewable generation units throughout large regions can help mitigate the impact of variability and unpredictability of renewable resources [4]. However, a large share of converter-interfaced generation (CIG) may render the system more vulnerable to sudden generation-demand imbalances due to the lack of sufficient synchronous inertia. High penetration of CIG may therefore affect the stability of the system's frequency [5]. Even though CIG can provide synthetic inertia to the system and hence contribute to mitigating the reduction in the system's synchronous inertia, the quality of the inertia provided by CIG is still somewhat questionable, mainly due to the dependency on "fallible" measurement and communication systems [6]. CIG does not use the rotor speed as frequency measurement but rather utilizes the Phase-Locked Loop (PLL) to estimate the local bus frequency. PLL can show stability issues [7] and introduce a relevant delay [6]. On the other hand, important difficulties have been encountered when using Phasor Measurement Units (PMUs) to assess the system's Rate of Change of Frequency [8, 9].

When a fault occurs, synchronous generators provide around six times rated current (fault current). This fault current is used to trigger protection relays. A similar response cannot be provided by CIG due to the converter's limits on the admissible current overshoots. The fault level contribution of CIG ranges only from 1.1 to 1.5 times their rated current [10] In electric power systems with a large penetration of CIG protective relays may lose the ability to sense the fault conditions because the available fault current is drastically reduced [11].

As opposed to synchronous generation, the physical dynamics of CIG are mainly dictated by their control loops, which are significantly faster than those of synchronous generation. The physical dynamics of CIG are of the same order of magnitude similar to the dynamics of the transmission lines. The dynamic coupling of CIG with the grid must be carefully considered for the design of CIG controls, especially in low to medium voltage grids [12]. In power systems with a high penetration of distributed CIG, the risk of unintentional islanding increases. Conventional anti-islanding techniques aim to make some generators trip as fast as possible. These techniques have proved to be effective in electric power systems with moderate levels of CIG. However, other techniques will likely be necessary for fully renewable power systems [11]. As it was mentioned above, geographically spreading renewable generation sources mitigate the impact of resources' variability from a global perspective. However, locally the variability of renewable resources may cause voltage flicker which can propagate locally and affect the power quality [13, 14]. Another local impact of CIG on power quality is the well-known harmonics which are caused by the switching of the grid-side inverter and contain high-frequency harmonic components at multiples of the switching frequency of the inverter [15, 16].

Of course, the deployment of renewable generation sources brings very valuable societal benefits that by far overcome the above-described impacts or difficulties in the operation of the power system. We are still learning how to properly operate electric power systems with a large or full share of renewable CIG sources and have already implemented a few technological solutions that mitigate the above-described impacts. Interconnections are said to be key to the successful operation of electric power systems with large shares of renewable CIG. Such is the case of Denmark where instantaneous penetrations of renewable CIG have reached 100% on certain occasions [4]. However, it is important to bear in mind that as discussed in [17], when areas with different instantaneous penetrations of CIG are interconnected, there is inertia heterogeneity between the areas, and significant frequency swings can appear between the areas [18]. The use of retired thermal generators as synchronous condensers can also help mitigate some of the above-mentioned impacts by the provision of synchronous inertia, voltage-reactive power control, and fault current [19].

With a proper design of the converter control, renewable CIG can provide frequency control ancillary services with the same or better quality than a synchronous generation [11]. Several demonstration projects have been recently carried out which have unveiled the extraordinary ability of renewable CIG to provide frequency control [20, 21]. All technological solutions proposed so far to mitigate the impacts of renewable CIG pass through a careful design of the grid codes [22], the system's operational procedures and ancillary services, and the electricity and ancillary services markets rules. Interesting high-level actions that have been taken to mitigate the impacts of renewable CIG are the minimum synchronous inertia requirement imposed by the Australian Energy Market Operator [23], and the implementation of a new synchronous inertial response ancillary service by EirGrid [24].

Last but not least, it is important to highlight here the huge potential of energy storage to mitigate the impacts of renewable CIG from different perspectives. Among others, energy storage can contribute to reduce renewable energy curtailments, using excess renewable energy in moments of renewable energy deficit, smooth the fluctuations of the output power of renewable CIG and enhance the system's frequency control [25, 26]. An application of energy storage to smoothing power oscillations of a wave farm (WF) will be later discussed in this chapter.

18.1.2 Current Status and Perspectives of Ocean Energy

Despite being an established industry, ocean energy is still some steps away from reaching commercialisation and bankability. In recent years, and in order to aid in this transition, the sector has made an effort in outlining the necessary steps to be taken (in the form of the so-called technology and commercialisation roadmaps) in order to reach that market stage. In the latest sector roadmap, the Strategic Research and Innovation Agenda (SRIA) for Ocean Energy (developed under the project ETIP Ocean), released in 2020, the industry notes the demonstration of wave energy pilot farms (WF) as a high-priority challenge area [27].

In addition, integration of storage is remarked as a key action to consider in the deployment of the aforementioned demonstration projects. Furthermore, the demonstration of storage solutions is again highlighted as a key action in another challenge area of the roadmap: the improvement and demonstration of power take-off (PTO) solutions. In this case, the focus is set on the delivery of grid-compliant power by the wave energy device or device, and the corresponding power output smoothing when required. In this latter case, the need for storage capacity is estimated at 5–10 s at the device's rated power.

Another challenging area in the aforementioned SRIA is the integration in the energy system. The wave energy resource has a correlated variability, and it is not linked to wind or solar power, which becomes a significant advantage in enhancing the penetration of renewable energy in the grid. Smoother daily generation profiles can be achieved, reducing the requirements for back-up systems for power balancing (such as spinning reserves or energy storage), transmission, and demand-response, thus allowing for relevant cost reductions [27]. Furthermore, wave and tidal energy are more predictable than wind and solar, which make the former particularly attractive for renewable generation in smaller grids such as islands.

In the case of isolated systems and islands, where energy can reach high prices ($\approx$0.35 €/kWh) due to the fuel, transport, and storage costs associated with the fossil-fuelled generation, wave power might become an attractive potential competitor.

Moreover, in the case of islands, opportunities are present beyond utility-scale power generation. With the expected growth of the so-called Blue Economy, wave energy positions itself as a potential alternative to decarbonise activities such as desalination, space cooling, shipping, and tourism [28]. Wave energy alternative applications also include aquaculture, green hydrogen generation, oceanography (power supply for environmental monitoring equipment), and underwater vehicle charging (e.g. gliders) [28, 29]. Nowadays, these Blue Economy opportunities for wave energy are present in the development agendas of key players from leading countries around the world, such as the European Commission in the EU or the Department of Energy in the USA. Figure 18.1 summarises the main synergies between islands and ocean or wave energy.

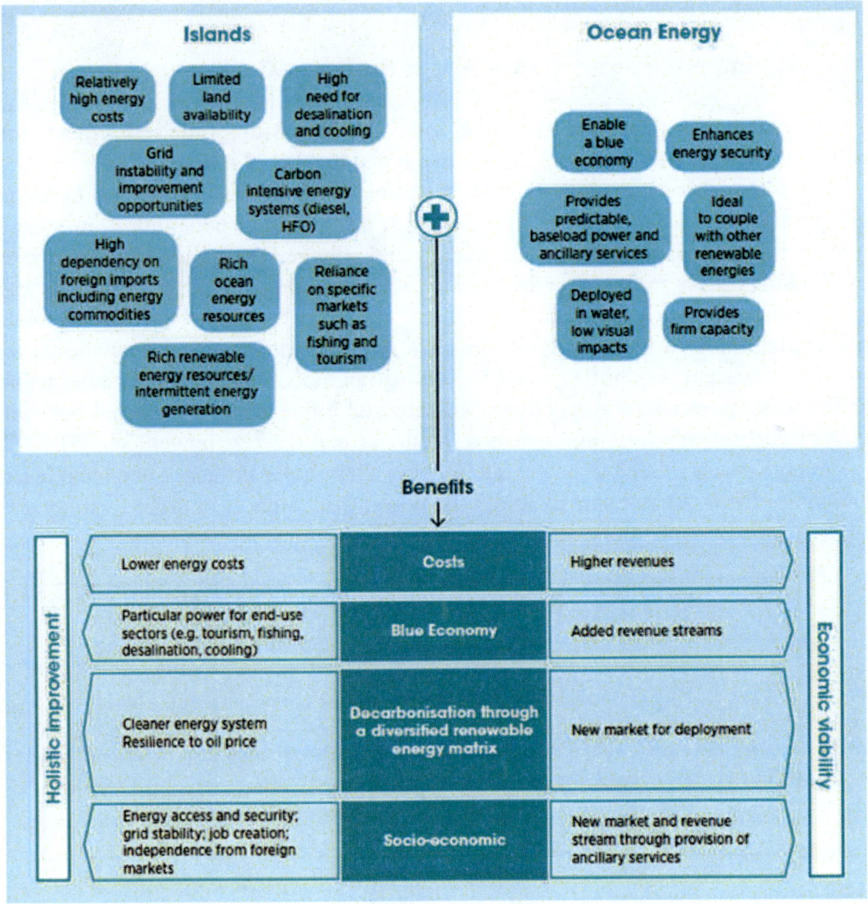

Fig. 18.1 Synergies between islands and ocean/wave energy. *Source* IRENA [30]

In terms of the storage system requirements, it must be noted that, in electrically-isolated grids, the storage should be designed for both long-term (balancing of generation and consumption) and short-term operation (power oscillation smoothing). The isolated grids themselves should incorporate such short-term operations in their designs so that variations in generation or demand do not have any impact on the potential weakness of the grid. Long-term operation of energy storage systems should meet the needs of the whole electric grid (and should presumably be operated by the Transmission or Distribution System Operator (TSO or DSO).

18.1.3 Problems with the Grid Integration of Wave Energy Converters

18.1.3.1 Impact of Power Oscillations in the Power Grid

The electric power generated by a WEC follows an oscillating time series due to the nature of the renewable source, similar to the one displayed in Fig. 18.2.

The intermittency associated with it causes a continuous imbalance between generation and demand which, for grids with a high WEC penetration, may lead to power quality issues related to frequency and voltage. The period of those power oscillations occurs in the range of 5–25 s, causing sudden frequency variations and voltage distortion such as flicker and harmonics [31–33]. Moreover, introducing those power oscillations in weak grids with low X/R ratios may lead to severe voltage excursions, since instantaneous frequency and voltage are not only dependent on the active and reactive power flow respectively, and crossed influences are expected between voltage and active power, and between frequency and reactive power [34, 35].

Voltage issues related to power quality have their major influence at a local level, and their effects are not spread across the power grid. Hence, in order to study and

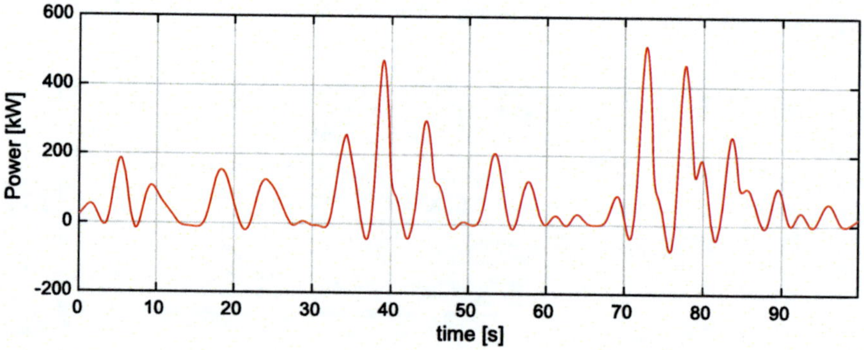

Fig. 18.2 Typical electric power time-series generated by a WEC [31]

analyse them, detailed power grid models including non-aggregated loads, distributions and transmission lines, power generation and storage units, and the rest of the equipment connected at the nodes nearby a WF are required [36]. On the other hand, frequency variations are a global problem for the power grid, and sudden power variations introduced by a WF induce frequency oscillations far from the WEC connection node. Thus, frequency issues can be studied and assessed with power balancing models, which account for aggregated generation and demand, and no-load, generator, or equipment needs to be modelled in detail. Considering the power quality issues described above, voltage issues related to WECs integration tend to be minor, given that WF is commonly connected to transmission lines with high X/R ratio, and minor voltage excursions can be effectively solved with commercial equipment such as STATCOMs [34, 37, 38]. Moreover, conclusions extracted from voltage analyses highly depend on the studied location, making them non-generalizable [35]. Thereby, this chapter focuses on assessing and mitigating frequency variations when integrating WF on the power grid.

Frequency deviation thresholds are, in the vast majority of the cases, strictly defined by the grid codes of the different electric regions. A common threshold categorization includes three levels of operation: normal, alert, and emergency. For the Continental Europe region, normal operation corresponds to frequencies within 50 Hz $\pm$ 50 MHz and frequencies within the alert operation thresholds for less than 5 min, alert operation corresponds to frequencies between 50 Hz $\pm$ 200 MHz for periods longer than 5 min, and emergency operation corresponds to frequencies out of the 50 Hz $\pm$ 200 MHz range [39]. Note to the reader that these thresholds are not universal, and there might be slight variations for other electric regions. In order to keep the frequency within the normal operation range, an extensive regulation is described in the different grid codes, including primary, secondary, and tertiary frequency control. According to the frequency regulation for a system with conventional generation units, primary control acts in the range of 1–20 s, and corresponds with a fast power response from the generators, based on their statism. The secondary regulation acts in the range of 2–20 min and corresponds to a new power generation set point for few generation units, which restores both the frequency to the base value (50 Hz for Continental Europe) and the programmed power dispatched by the rest of generation units. At last, tertiary regulation restores the power available in a specific area for performing secondary regulation, in case there is a further frequency event [40].

This frequency control is effectively implemented by conventional generation units, but it is not straightforward how to provide this kind of control with RES units, such as WECs. However, with the increasing penetration of RES to detriment of conventional generation units, this frequency control should not be delegated to the remaining conventional generators, and RES units should also contribute to power quality control. In case they remain aside, frequency restoration mechanisms that are usually demanded in fatal events would have a major role, making the power grid unsafety and prone to suffer a cascade accident that ends in a region blackout.

18.1.3.2 Introduction of Grid Codes Compliance

The high penetration of renewable energies into the electric power system has led grid codes to become progressively more restrictive to all generation facilities. As solar and wind energy used to be exempted from most of the interconnection requirements and currently should meet them, the same may be demanded of wave energy.

Although wave energy, as emergent technology, is exempt from meeting most of the grid codes' restrictions, the technological solutions should be analysed regarding future deployments. Considering that the intermittency of renewable generation causes problems to the grid, especially for islands and weak grids and that some countries or regions such as *Puerto Rico, Hawaii*, or *Ireland* have already imposed power ramp limitations, it should be necessary a system that allows to achieving power ramp-rate control, as Energy Storage Systems (ESS) integration or active power curtailment with forecasting systems [41–44]. In addition, although the power oscillations of WF may not be constrained in the grid codes, the resultant voltage and frequency oscillations are constrained to narrow margins around their rated values.

The existence of international standardization committees, as IEC TC 8A, for grid integration of variable power generation from renewable energies, such as PV and wind energy, could also be taken into consideration as grid compliance assessment methods for wave energy future integration considering the similar resource variability problems and the inexistence of more specificity due to its novelty.

18.2 Power Oscillations in Wave Energy Generation

18.2.1 Wave Energy Resource: Characterization of Oscillations

The ocean wave energy resource has an intrinsic oscillatory nature. In a general way, the characterization of the ocean waves or sea-states is carried out by means of spectral analysis [45, 46]. Usually, energy wave spectrums (S_f) are calculated by means of a Fast Fourier Transform (FTT) applied on the time profiles of the ocean-free surface elevation with durations from 15 min to 1 h. S_f is represented in Energy [m^2/Hz], for each frequency bin. From the energy wave spectrum, it is possible to calculate different spectral parameters which give information about the periods and heights of the oscillations of each sea-state [47], such as:

1) Significant wave height (H_s). H_s is traditionally defined as the average value of the wave height (trough to crest) of the highest third of the waves. H_s is evaluated by means of the following equation:

$$H_s = 4 \cdot \sqrt{m_0} = 4 \cdot \sqrt{\int_0^{\infty} S_f df} \tag{18.1}$$

where m0 is the spectral moment of order 0 (defined as $m_n = \int_0^{\infty} f^n \cdot S_f df$), and f is the frequency of oscillation (being the period of oscillation T = 1/f.

2) Wave energy period (T_e). T_e represents the period of oscillation of the energy in the ocean wave. T_e is evaluated by means of the following equation:

$$T_e = T_{-1,0} = \frac{m_{-1}}{m_0} = \left(\int_0^{\infty} f^{-1} \cdot S_f df\right) / \left(\int_0^{\infty} S_f df\right) \tag{18.2}$$

3) Wave peak period (T_p). T_p is evaluated as the period of oscillation where S_f reach its maximum value.

Figure 18.3 shows one-year information about the hourly sea-states in a particular location. The figure shows a matrix scatter diagram of the values of H_s, T_e, and the main direction of the propagation of the waves for each analysed sea-state.

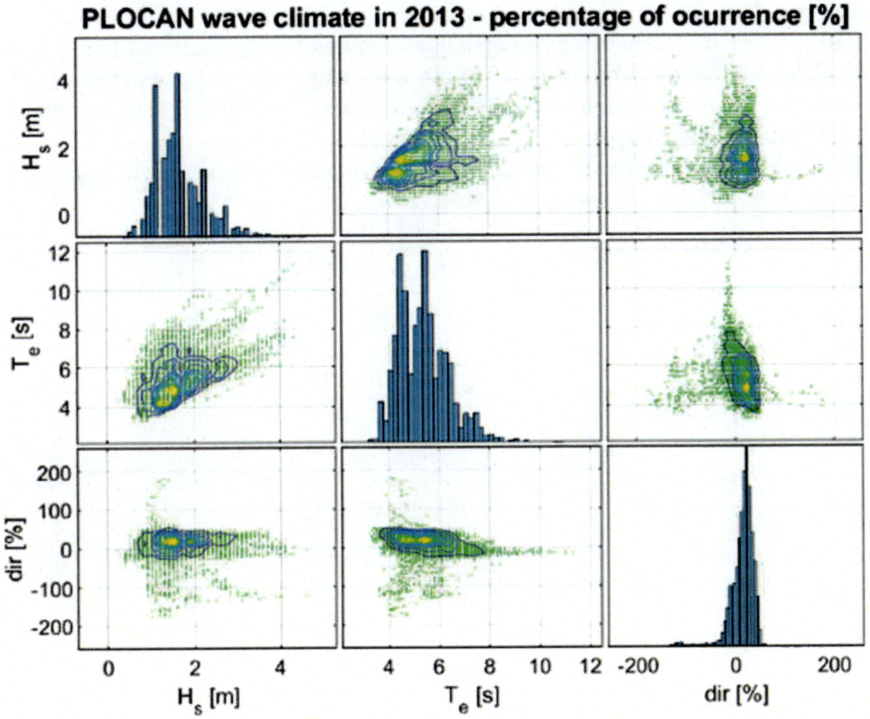

Fig. 18.3 Annual dataset of sea-state characteristic (H_s, T_e and wave direction) of the open-sea test facility of WECs PLOCAN [48] (Canary Islands, Spain)

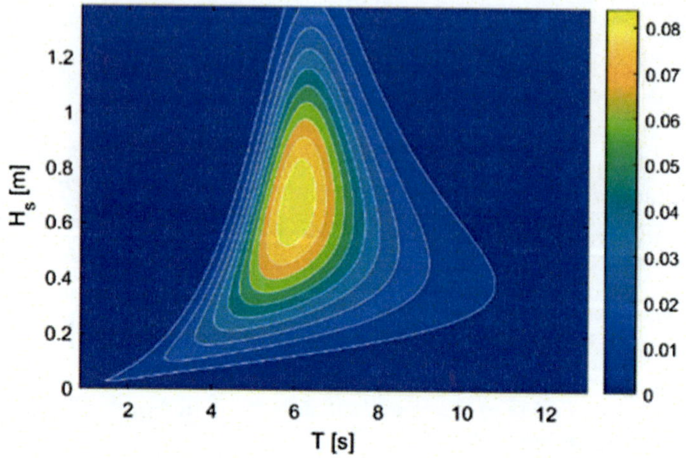

Fig. 18.4 Wave amplitude and period probability of a sea state of H_s of 2 m and a T_p of 8 s

Another way to characterise the oscillatory nature of the ocean waves is by means of statistical analysis. This type of analysis gives information about the probability of occurrence of a certain single oscillation, represented by its period and height. The joint distribution probability (p) of the wave amplitudes (A) and periods (T) is obtained through the *Longuet-Higgins* approximation [49, 50]. This approximation is shown in its non-dimensional form by means of the Eqs. (18.3)–(18.7). Figure 18.4 shows the probability of each possible single oscillation for a sea-state characterised by means of a H_s of 2 m and a T_p of 8 s.where

$$\mathrm{p(R,}\,\tau) = \frac{2}{\pi^{\frac{1}{2}}\cdot\nu}\frac{\mathrm{R}^2}{\tau^2}\mathrm{e}^{-\mathrm{R}^2\cdot\left(1+\frac{\left(1-\frac{1}{\tau}\right)^2}{\mathrm{v}^2}\right)}\mathrm{L}(\nu) \tag{18.3}$$

$$\mathrm{L(v)} \approx 1+\frac{1}{4}\cdot\mathrm{v}^2;\ \mathrm{v}=\left(\frac{\mathrm{m}_0\cdot\mathrm{m}_2}{\mathrm{m}_1^2}-1\right)^{\frac{1}{2}} \tag{18.4}$$

$$\mathrm{R}=\frac{\mathrm{A}}{(2\cdot\mathrm{m}_0)^{\frac{1}{2}}} \tag{18.5}$$

$$\tau=\frac{\mathrm{T}\cdot\mathrm{m}_1}{2\cdot\pi\cdot\mathrm{m}_0} \tag{18.6}$$

$$m_n(\omega)=\int_0^\infty \omega^{\mathrm{n}}\cdot\mathrm{E}(\omega)\cdot\mathrm{d}\omega \tag{18.7}$$

18.2.2 Evaluation of Electric Power Oscillations in Wave Energy Converters

The oscillatory behaviour of the energy source is directly translated into an oscillating electric power generation by the WECs. Owing to the special characteristics of wave power conversion, there is an important impact of low-frequency oscillation in the amplitude of the electric power, with periods of oscillation ranging from a few seconds to a few dozens of seconds (typically, the peak of the spectrum of the output power of a WF occurs at a period value twice the period of the peak of the sea-state [51]). The power oscillations are more pronounced in the point absorber WEC, especially when using direct-drive systems as power take-off (PTO) [52, 53]. In this case, there is not any intermediate system capable of storing energy (except, perhaps, for the capacitors of the DC-link of the power electronics of the PTO, as explained in Sect. 3.1). There are some other WEC technologies and PTOs capable to store energy in intermediate stages, which present reduced power oscillations, such as [54]:

Overtopping WEC technology, which working principle implies a water reservoir. The water of the reservoir stores energy as potential energy.

Hydraulic PTOs, which in some configurations have pressure reservoirs in other to smooth the power oscillations.

Oscillating Water Columns, which can include flywheel directly in the shaft of the air turbine.

Nevertheless, the technology of the point absorbers is considered by a large number of developers [55] and has hence been selected as well in one of the two example studies presented in Sect. 18.3. This power fluctuation leads to frequency variations [31, 56] and voltage distortions such as flicker [33] and harmonics [57], which become more acute in weak or isolated grids (such as the case of islands), where the system inertia is low (see Sect. 2.3.2). For the sake of example, the power generated by a single point-absorber with a direct-drive linear generator is shown in Fig. 18.5. In addition, the real measurement of the electric power generated by one turbine of the *Mutriku* OWC power plant is shown in Fig. 18.6.

It is worth to be mentioned that, if the dynamics of the WEC are considered linear, it is possible to set a WEC stochastic model [56, 58]. This model allows analysing the WEC dynamics (velocities, forces, etc.) in an analogous way as the sea-states statistical analysis presented in Sect. 2.1. Therefore, an analysis of the probability of the amplitude and period of each single power oscillation is possible. For the sake of example, Fig. 18.7 shows the evolution of the position, velocity, and force in a 2-body point absorber with respect to the amplitude and period of each oscillation.

When a certain number of single WECs are associated in a WF, the amplitude of the WF output power oscillations is lower than the sum of amplitudes of the individual WEC power output oscillations. This mitigation/cancellation effect of the power oscillations depends—mainly—on the relative position of each WEC, on the number of WECs, and on the direction and dispersion of the waves of each sea-state [51, 60]. For the sake of example, Fig. 18.8 shows this cancelation effect in a

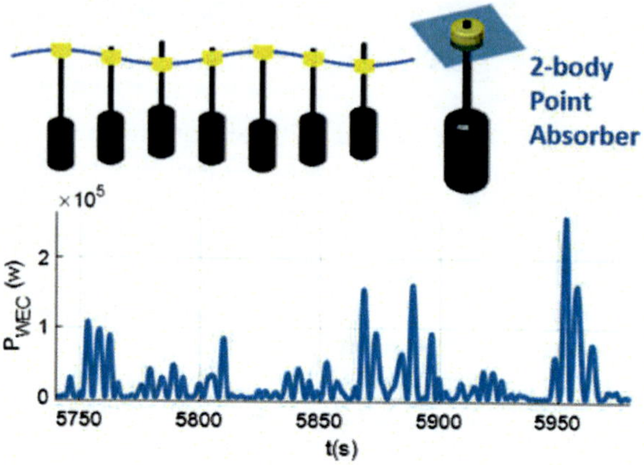

Fig. 18.5 Simulation of the power of a 2-body point absorber with a direct-drive linear generator

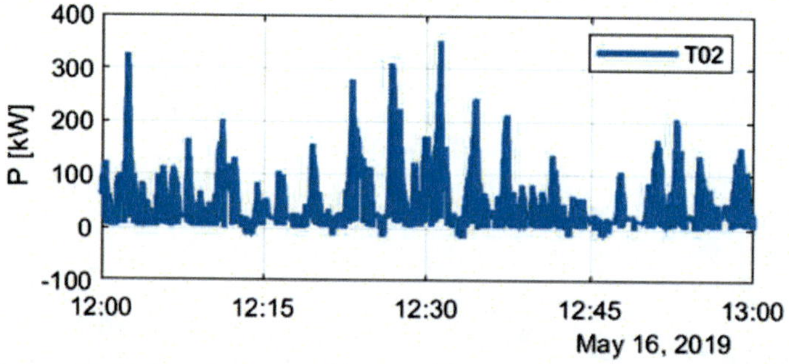

Fig. 18.6 Real measurement of the power output of one turbine of the Oscillating Water Column Power Plant of ***Mutriku*** [59]

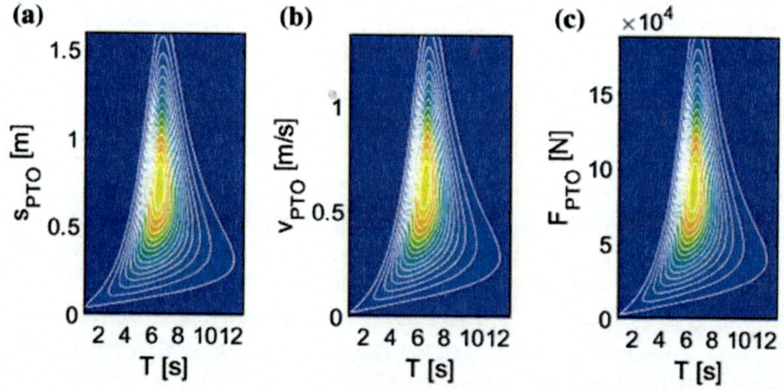

Fig. 18.7 Probability of amplitude and period of each single oscillations of the mechanic variables heave position (**a**), heave velocity (**b**), and force in the PTO (**c**) in the case of a sea state of H_s of 2 m and a T_p of 8 s

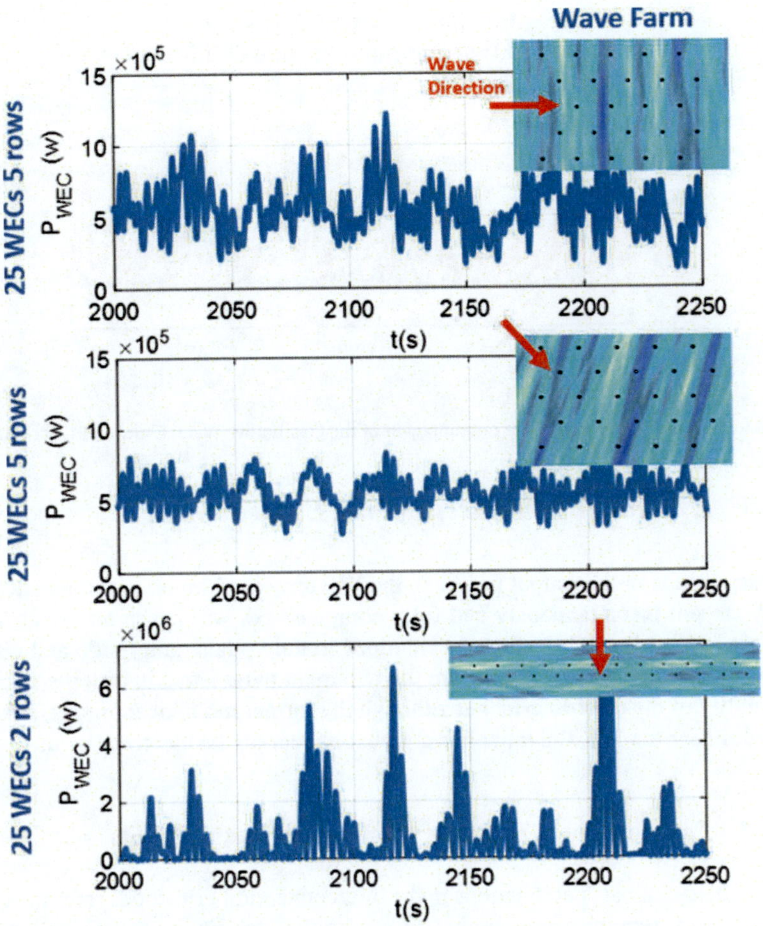

Fig. 18.8 Simulation of the output power of a Wave Farm of 2-body point absorbers

simulated WF of point absorbers, and Fig. 18.9 shows the power oscillations in a real OWC power plant.

It worth to be mentioned that, in other to evaluate the power oscillations in a WF, it is necessary to evaluate the free ocean surface in the location of each WEC. The directional wave energy spectrum of the sea-state is used to reproduce free ocean surface in the points taken up by each WEC [46, 61], but other approximations based on propagation delay times can be used [62].

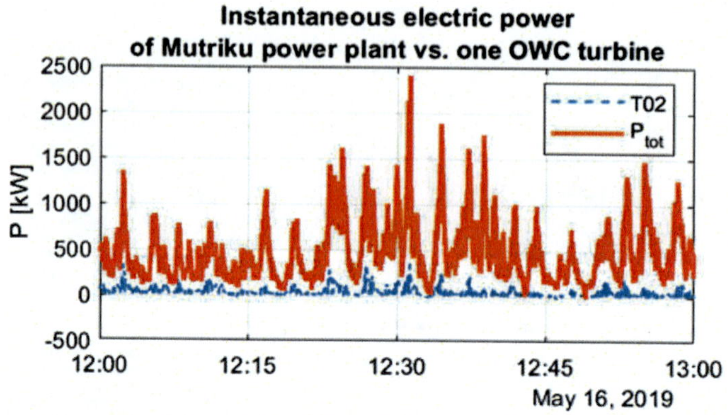

Fig. 18.9 Real measurement of the output power of the Oscillating Water Column Plant of *Mutriku*

18.2.3 Evaluation of Impacts on the Electric Grid

The oscillations in the output power in the WF presented in the previous Sect. 2.2, if they are not instantaneously and fully compensated, will cause an imbalance in the electrical grid which leads to oscillations in the voltage amplitude and electric frequency. Voltage and frequency are the two main parameters in order to maintain the stability of the electric grid, but although the former has a local impact, affecting the nodes near the WF, the latter has a global impact on the electrical grid.

18.2.3.1 Introduction to Frequency Grid Codes Requirements

Most of the technical specifications in the interconnection grid codes refer to voltage and frequency requirements, aiming to ensure the stability of the electric system against oscillations and disturbances, as well as seeking to maintain values within specific ranges both in normal operation and in the event of disturbances in the electric system. However, as the grid is something dynamic, the power system could be off-balance even when production and demand have been coupling in the day-ahead electricity market since production and demand must match so perfectly and instantaneously.

This is directly related to the idea of complementary services for eventually using them to keep the power system in balance and enable the Transmission or Distribution System Operator (TSO) to maintain the integrity of the electric power system, as well as the power quality [63, 64].

While there are various types of ancillary services and can be classified as related to frequency control, voltage control, or system restart, it is important to highlight the main importance of frequency-related services (primary and secondary regulations) as they are directly associated with the system balance and essential to maintaining the frequency command value. For this purpose, the power-frequency regulation capability must be guaranteed in most of the interconnection grid code regulations and most countries point out that large generators must have a form of ancillary services agreement to be able to provide basic services considered essential for the operation of the transmission system.

Thereby, operation requirements for generators as primary frequency control (Frequency Response) are mandatory for large generators, compelling them to guarantee a local control that responds autonomously to changes in frequency by raising or lowering the output of a unit in order to counteract the change in frequency, all coordinated with the TSO. This controller droop, out of the defined dead-band, shall be adjustable according to the specifications of the TSO, being in the range of 2 and 12% the output active power at the moment of the event (according to European Operation Handbook [63]).

Considering WF, it should be noted the Commission Regulation (UE) 2016/631, establishing a Network Code on Requirements for Generators (RfG) in Europe [65]. This regulation firstly specifies the requirements for Offshore Power Park Modules, which must comply with most of the frequency and voltage stability requirements governing power park modules situated onshore and participate in the primary regulation.

Taking into account that intermittency of renewable generation causes problems to the grid, especially for islands or weak grids, some regulations also establish output power ramp-rate restrictions in some countries or regions such as Puerto Rico [41], Ireland [66], Germany, or Hawaii [42] trying to avoid or limit the problems related to the intermittency of renewable generation. For example, Puerto Rico requires a maximum ramp-rate of 10% per minute of the rated power both to the increase and decrease of power output (a value also required by the German Transmission System Operator for positive ramps). In other countries as Ireland, the TSO (EirGrid) establishes the ramp-rate and limits its deviation due to resource availability up to 3% per minute of the registered capacity or ±0.5 MW the demanded rate. In the case of the Hawaiian grid code (HECO) a limit on the ramps of ±2 MW/min is demanded during most of the hours of the day, with some exceptions from (00:00–4:00) and (18:00–20:00) when ramps are limited to ±1 MW/min.

Although there are no clear restrictions on output power ramp-rates at present in most of the grid codes, its limits are commonly set by the TSOs as 10–15% of the rated power per minute, which can be regarded as a reference considering the interconnection of WF when anticipating future restrictions on power oscillations.

18.2.3.2 Models and Response of the Electric Grid

In this sub-section, a power balancing model with primary and secondary frequency regulation is described, so frequency events introduced by a WF can be assessed, and mitigation techniques can be analyzed. This model can be further expanded with a local voltage model. Moreover, a case example for *Tenerife* island is provided, where a real WF with a real generation profile is integrated into the power grid, so the frequency events associated with it can be observed.

A power balancing model should include, from a general point of view, a wave farm model (WFM) and a power grid model (PGM) [31]. The WFM consists of a model of a single WEC, which can be parameterized based on specific WEC characteristics and on the WF layout. Results of WF output power obtained from models are afore presented in Sect. 2.2 by means of the use of dynamic models or stochastic models. The WFM inputs the ocean-wave resource conditions and outputs the total electric power generated by the WF, the latter being one of the inputs of the PGM.

In order to evaluate the frequency events associated with the WF power generation, a PGM based on a single-bus inertia simulation model is proposed in this section. The PGM, which can be parameterized based on the grid frequency characteristics (inertia, load sensitivity, etc.) and the generation condition (generation from conventional and renewable power plants), includes both the primary and secondary frequency response of the power grid. Hence, the PGM outputs a frequency time-profile based on the WFM power output, which inputs the model together with the power profile consumed by the loads.

The PGM, depicted in Fig. 18.10, includes the primary frequency control (R_1), which is represented by the statism of the spinning power generation units (R) and the time response of the speed governors (T_{R1}). Besides, an automatic generation control (AGC) is included in order to perform the secondary frequency regulation. The secondary frequency control contribution (R_2), which restores the frequency to its nominal value once the primary control has stabilized it, is calculated based on the ACG set-point signal (P^*_{sp}) and a transfer function that replicates the R_2 dynamics [67, 68]. At last, the aggregated inertia of the system is represented by the constant M, and the load sensitivity to frequency changes by D_{net}, being f_{grid} the output.

The equations that govern the PGM are listed as follows [31]:

$$M \cdot \frac{df_{grid}}{dt} = P_{WEC} - P_{LOAD} - D_{net} \cdot f_{grid} + R_1 + R_2 \tag{18.8}$$

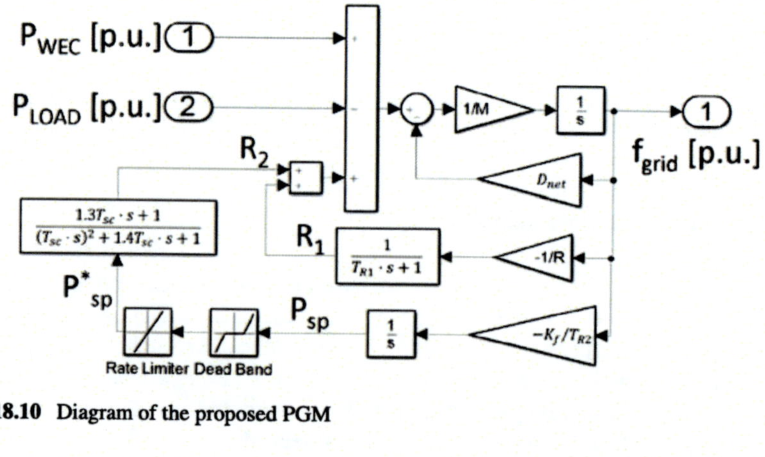

Fig. 18.10 Diagram of the proposed PGM

$$\frac{dR_1}{dt} = \frac{1}{T_{R_1}} \cdot \left(-\frac{f_{grid}}{R} - R\right) \tag{18.9}$$

$$\frac{dP_{sp}}{dt} = -\frac{K_f}{T_{R_2}} \cdot f_{grid} \tag{18.10}$$

$$1.3 \cdot T_{sc} \cdot \frac{dP_{sp}^*}{dt} + P_{sp}^* = T_{sc}^2 \cdot \frac{d^2R_2}{dt^2} + 2 \cdot 0.7 \cdot T_{sc} \cdot \frac{dR_2}{dt} + R_2 \tag{18.11}$$

In order to evaluate the impact of the WF power generation on the electric power grid, the resulting frequency profile, together with its average value, its variance, and its maximum and minimum values should be recorded. A common approach to evaluating this impact, which is the one proposed in this chapter and defined by European grid codes [69, 70], is based on categorizing the frequency events as normal, alert, or emergency state. European grid codes consider that the grid is at a normal state when the frequency deviations remain within a narrow range (±50 MHz), or when further deviations of ±200 MHz are corrected in less than 5 min. An alert state is reached for deviations within ±100 MHz and ±200 MHz for periods longer than 5 min, or deviations within ±50 MHz for more than 50 min. Any other frequency situation experienced by the power grid is considered an emergency state.

Once the PGM and the frequency evaluation methodology have been detailed, a further improvement to the PGM regarding voltage quality assessment can be included. In this sense, and given that voltage quality issues such as voltage excursions out of thresholds, flicker, or harmonics have a local effect without spreading their effects across the grid, a simplified but rigorous approach should include a Thevenin equivalent with the following elements: a three-phase line that connects the node where the WFM is plugged to the point of common coupling (PCC), and a three-phase voltage source connected at the PCC, whose frequency variation is modelled as the single-bus inertia described above. In order to evaluate voltage quality, the new PGM elements should be parameterized with the short circuit ratio at the PCC (SCR), defined as [71]:

$$SCR = \frac{S_{sc}}{S_n} \tag{18.12}$$

where S_{sc} is the short circuit ratio at the PCC, and S_n is the power connection at the PCC, i.e., the WF-rated power. Moreover, X/R ratio is also needed to parameterize the three-phase line. This is not a constant parameter and can be varied depending on the constructive elements of the line and on the distance between the WF connection node and the PCC [71].

As mentioned in previous sections, minor voltage excursions in transmission lines can be effectively solved with STATCOMs [34]. This equipment, which allows for a fast reactive power supply, is effective in lines with a high X/R ratio, where there is a direct relation between voltage levels and reactive power (the higher the reactive power generation, the higher the voltages). However, in lines with lower

X/R ratios, usually associated with low voltage and/or isolated power grids, this direct relationship is no longer valid, and a crossed influence between active power and voltage quality is also presented, as well as a crossed influence between reactive power and frequency variations [34, 35]. This situation may leave some room for WF connected at weaker networks to contribute to voltage quality regulation.

When parameterizing the PGM elements related to voltage behavior, common values for SCR and X/R ratios can be obtained from [71–75].

As mentioned in previous sections, given the local character of the voltage issues and that WF are commonly connected to transmission lines, this section is focused on frequency quality.

For sake of example, a case study with a frequency deviation assessment in *Tenerife* island (Spain) where a WF is integrated with the power grid is presented in this section. The WFM has been parameterized for 54 two-body point absorbers, and for ocean-wave and electric conditions during a complete year (Fig. 18.11). The PGM parameters related to frequency behavior have been parameterized based on

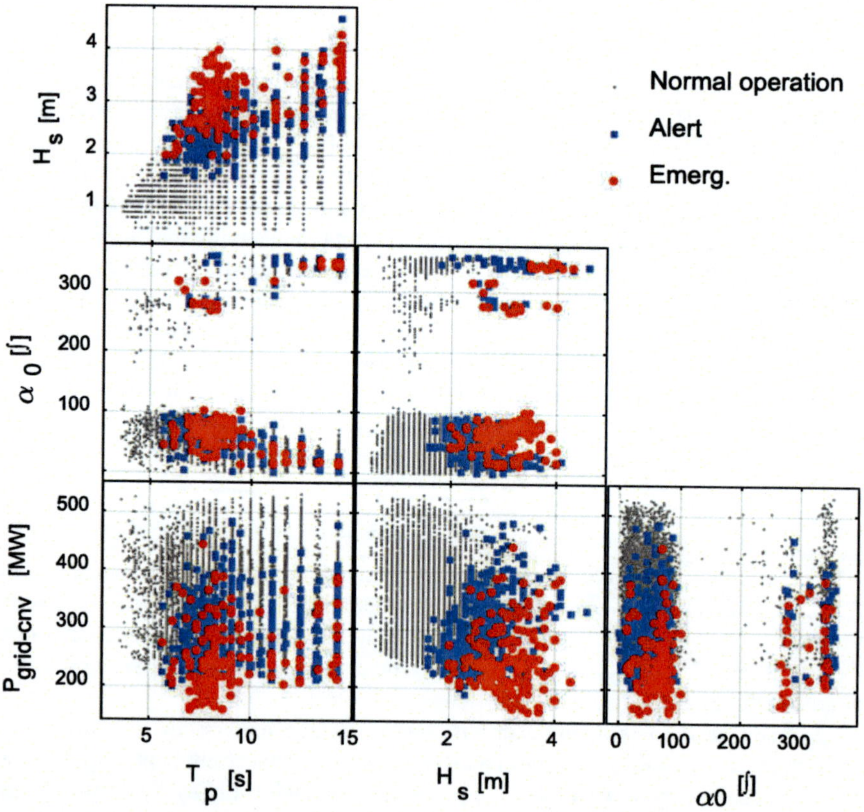

Fig. 18.11 Scatter diagram for parameters H_s, T_p, α_w. and $P_{grid-cnv}$ for the case study presented in this sub-section

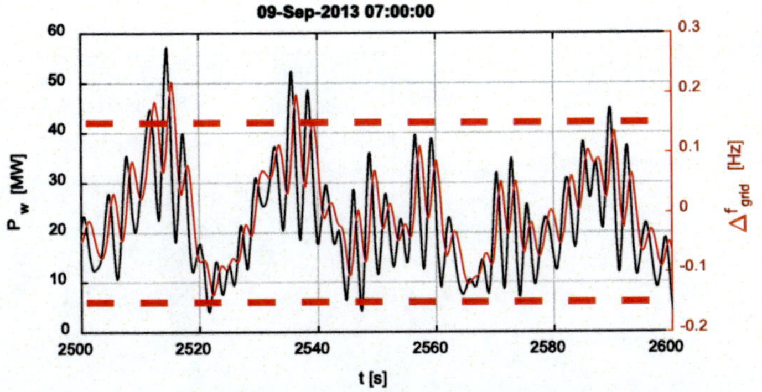

Fig. 18.12 Power profile generated by the WFM versus frequency profile obtained from the PGM

the data provided at [67, 68, 70]. The power generated by conventional power plants ($P_{grid-cnv}$) is also shown in Fig. 18.11, which is directly related to the inertia of the grid. Moreover, the criteria for evaluating the WF integration in the power grid has been taken from the Spanish Insular Grid Code [76], which establishes the alert and emergency threshold at 50 Hz ± 15 MHz and 50 Hz ± 25 MHz, respectively. Once both models are parameterized, the output of the WFM is obtained (P_{WEC}), i.e., the power generated by the WF for one year. Figure 18.12. shows a one-hour detailed profile for P_{WEC}. This power inputs the PGM, obtaining the frequency profile (f_{grid}). The frequency profile corresponding to the previous P_{WEC} profile is shown in the same Fig. 18.12 for sake of example.

As aforementioned, the hourly average value of both P_{wec} and f_{grid}, as well as their variance and peak values are recorded. As seen in Fig. 18.12, a WF integration generating a pulsing power profile with peaks up to 90 MW can lead to a frequency profile that accounts for a high number of events out of the normal operation range (50 Hz ± 15 MHz). These frequency deviations are the ones located in the red-colored area in Fig. 18.12. These frequency events are also highlighted in Fig. 18.11, where the different parameters of the WFM are marked in blue, red, and yellow according to the frequency value obtained under that specific parameter condition. This information is quite relevant since it confirms that alert and emergency frequency events are most likely to occur during off-peak periods and high WEC penetration (high H_s and T_p).

The histogram presented in Fig. 18.13 indicates the number of oscillations along a year that the grid faces an alert or emergency situation. According to the values presented in the histograms, it is clear that a high penetration coming from the WECs originates an unacceptable number of alert and emergency events in one year.

It is noticeable that the average WF power represents 4% of the synchronized power of the grid, being the WEC power generation equal or greater than 21.2% only 0.5% of the time. Hence, a scenario with unsmoothed WFs is absolutely unacceptable, since the power oscillations introduced in the power grid are directly related to the

Fig. 18.13 Histogram for the frequency peak values

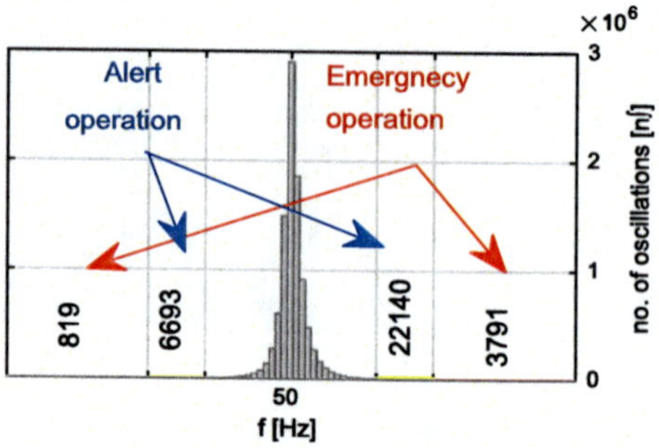

frequency deviations, which could potentially trigger emergency measures to maintain the power quality, and could eventually end in a system failure or blackout. The direct relationship between power oscillations and frequency deviations is detailed in Fig. 18.14.

The linear regression of P_{WEC} and number of events in order to evaluate if the two variables are correlated is plotted in Fig. 18.14. The analysis shows a R^2 factor higher than 0.98. This high correlation between penetration of P_{WEC} and number of alert events establishes a limit in the penetration of the wave energy in a particular electric grid. In the presented example, the penetration limit is 2.5%. However, diminishing the power peak oscillations introduced by the WF in the grid immediately reduces the number of alert and frequency events experienced by the latter. Hence, smoothing the power generated by WF becomes crucial when aiming for higher penetration of these types of generation units.

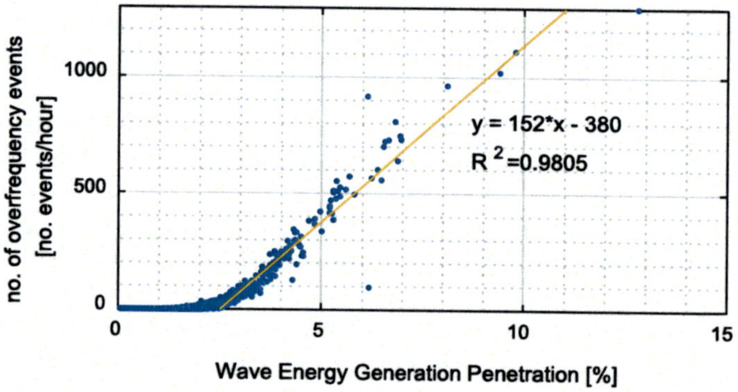

Fig. 18.14 Scatter diagram relating the penetration of P_{WEC} and the number of alert or emergency overfrequency events

18.3 Analysis of Smoothing Power Solutions

18.3.1 Introduction: Relevant Solutions for Wave Power Smoothing

As previously stated, there is a significant challenge in being able to physically connect the systems that extract energy from the waves to the electrical grid. Although this fact has important similarities with offshore wind farms [77]—mainly in terms of offshore equipment and interconnection with the grid, alternative problems should be faced in wave energy related to power quality [57, 78] and grid stability due to the oscillating nature of extracted power [79, 80] (see Figs. 18.2, 18.5, 18.6, 18.8 or 18.9). In this sub-section, solutions to smooth the extracted power by a wave farm (WF) are introduced. The main objective is the mitigation of the impact in the grid of low-frequency oscillation on the amplitude of the injected electrical power. If an oscillatory electrical power such as shown in Fig. 18.2 were injected directly into the grid, it would involve a power grid quality or stability problem that must be considered in the future development of this technology.

The four possible solutions introduced in this sub-section [52] are focused on the mitigation of power oscillation from a few seconds to a few dozen seconds:

- **Solution 1**: planning the distribution of each wave energy device in the WF [60, 81] to maximize the cancelation effect in the output power oscillations of the whole WF by the sum of the output power of each WEC.
  This first solution is based on a suitable WEC distribution that minimizes the power oscillation. Due to the irregularity of the real waves, a more in-depth analysis of the height, period, and direction of wave propagation is necessary. For example, when the layout is located in the same direction, the greatest oscillating power occurs. However, a WF with properly planned layouts can significantly reduce the power oscillations. [45, 46] (see Fig. 18.8).
- **Solution 2**: modification of the energy extraction control of the WECs by controlling the entire wave park so that the oscillations of each WEC are cancelled. In this solution, a control of the reactive mechanical power in WEC energy extraction to extricate the maximum power from the waves is supposed [82]. If the apparent power delivered by the PTO does not exceed its rated power, extra reactive mechanical power should be considered to manage the delay between the oscillations of each individual. The delay capacity depends on the WEC mechanical characteristic, and the maximum extra reactive mechanical power available [52].
- **Solution 3**: controlling the power electronics of the WEC, in particular the Grid-Tied Converter [83, 84], with a hybrid strategy that includes a voltage control in the DC-link and a direct power control [85].
  The solution is based on the idea of taking advantage of the capacity to store extra energy in the DC-link capacitors (which links the converter connected to the generator and the converter connected to the electrical grid). The energy stored in

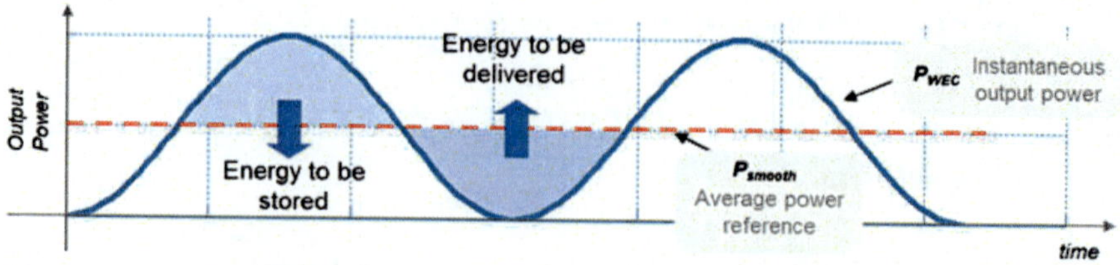

Fig. 18.15 Energy is stored and delivered to maintain a certain average output power as constant

them depends on the capacity and the voltage squared, therefore, an oscillation of the voltage amplitude in the DC bus must be allowed to reduce the oscillation of the total power generated. This oscillation is limited by the nominal voltage level of the capacitors and the minimum voltage required for PTO to work properly.
To control the oscillation of the voltage on the DC bus and reduce the amplitude of the power oscillation, it is necessary to implement a direct power control [6] with a mixed reference, depending on the voltage level of the DC bus. If the voltage oscillation is between the limits of the set band, the reference for the control is a constant power. When the voltage is outside these limits the reference is the voltage [85].

- **Solution 4**: Use of adding a set of additional energy storage systems to keep the electrical power output at a constant average value. This solution is explained in Sect. 3.2.4.
The previously mentioned solutions do not include additional systems to the PTO and are implemented from the control point of view. Sometimes these solutions are difficult to implement and the expected results are not obtained. The use of storage systems can greatly improve the power oscillation problem and maybe a more viable solution. There are different possibilities to calculate the required energy of the storage system based on how the system is being operated [52]. One possibility is to consider that both the velocity and the force generated are sinusoidal. This implies that the generated power ($P = F{\cdot}v$) also has a sinusoidal shape as shown in Fig. 18.15. To size the energy storage system, a certain level of average power is calculated during a certain period. This would be the power value that you want to deliver to the electrical grid. When the power generated by the point absorbers is greater than the average power, the storage system will be charged. On the contrary, if the generated power is less than the average value, the storage system will provide the necessary power to reach the target value (calculated average power).

Some of these solutions can be implemented together to increase the benefits of the resulting system. Solutions 2 and 3 just imply modifications in control algorithms to use the inherited energy storage capacity of each WEC: the energy storage capacity of the WEC mass and the DC-link capacitors of the power electronics. However, these solutions have a limited capacity to smooth the power oscillations. Solution 1 (WF planning) has a great capacity to smooth the power oscillations directly related to the number of WECs, but the smoothing capacity is strongly dependent on the direction and period of the waves. Solutions 1, 2, and 3 cannot completely solve the problem of the oscillations but can reduce the requirements of the implementation of solution 4 (an energy storage device).

In general terms, the storage system for wave energy application must have the following characteristics:

- The power requirement is greater than the power needs, so a high-power density, low energy density system is preferable. A suitable sizing allows maximizing the cost/benefit of the storage system.
- The storage system must withstand a large number of charge/discharge cycles since the system must work continuously.
- The response time must be very high.
- It must be a system with high power density to reduce the occupied space. In this case, the volume is more important than the mass.
- It must be a system that requires low maintenance since the environment in which these systems operate is not very accessible.

From the existing storage systems on the market, batteries and supercapacitors are the ones that would best suit the needs of the application. Perhaps, supercapacitors would be better suited since they have a higher power density than batteries, the electrical response is faster, the maintenance is lower, and aging is slower when the cycling frequency is high. An alternative to supercapacitors is flywheels. On the other hand, the cost of batteries on the market is currently lower than that of supercapacitors. In any case, some advantages can be found in each one, so the final choice will depend on the particular scenario considered. There is also the possibility of hybridizing several storage systems to take advantage of each of them.

This solution 4 (additional energy storage system) is detailed in the following Sects. 3.2.2, 3.2.3, and 3.2.4.

18.3.2 Analysis of Energy Storage as Power Smoothing Solution

18.3.2.1 Introduction: Control and Integration Configuration

Energy storage is a recommended solution to smooth the intrinsic power oscillations of the ocean-wave energy generation. One of the remarked actions by the European ocean-energy developing roadmap ([27]) is to "[...] *Demonstrate delivery of grid-compliant power including short-term energy storage solutions to smooth power output when needed [...]*", i.e. one of the main challenges in the wave energy development is to deliver electric power that complies the electric grid-codes and, in addition, one of the proposed solutions to face this task is the integration of short-term energy storage system in the wave energy power plants. This energy storage could be integrated in each WEC or could be set as a single onshore device connected to the

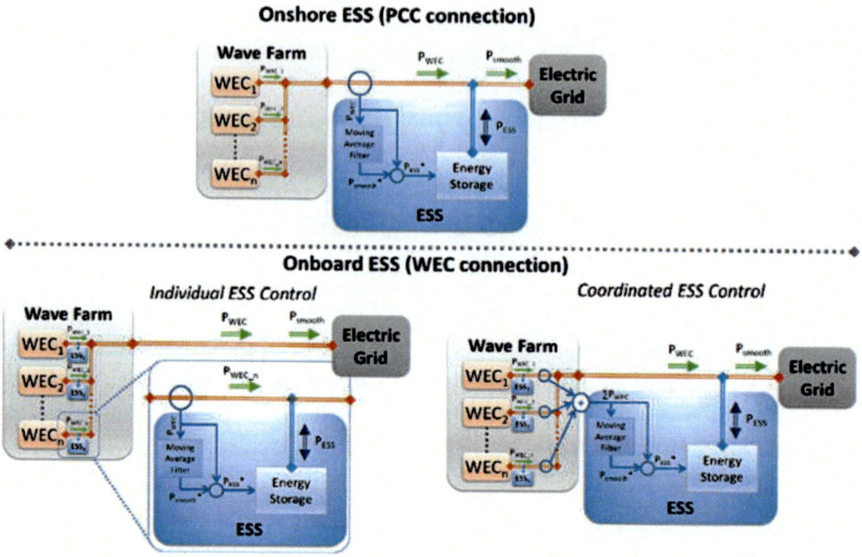

Fig. 18.16 Configurations of the integration of Energy Storage System (ESS) in Wave Farms (WF)

Point of Common Coupling (PCC) of the WF. Figure 18.16 shows different combinations of ESS location -offshore integrated in each WEC, or onshore connected to the PCC of the WF- and ESS controls -individual or coordinated.

The onshore location presents the advantage of reducing the complexity of the system (just one energy storage device/system) and reduce the OPEX (better and easier access for maintenance). On the other hand, the offshore location can reduce the power requirement of the power umbilical cables. Regarding the control options, the coordinated configuration takes the advantage of reducing the ESS requirements (75% less of ESS required energy). However, this option increases the power oscillations at the output of each WEC—maintaining the power oscillation level at the output of the WF [86].

The objective of the ESS control is the smoothing of the generated power and, for this, the control is usually set as Fig. 18.16 shows:

- An objective value of the output power is set (P_{smooth}* according to Fig. 18.16). This objective value is the power injected into the system by the ensemble of the WEC—or WF—and the ESS.
- The command value of the power that the ESS storages or delivers (P_{ESS}* according to Fig. 18.16), is obtained as the difference between P_{smooth}* and the ocean-wave power generated by the WEC or the WF (P_{WEC} according with Fig. 18.16).
- Different approaches can be carried out to obtain the P_{smooth}* value [87]–[95]. In this chapter, we are using an average value of the P_{WEC}, which is calculated with a Moving Average Filter (MAF).

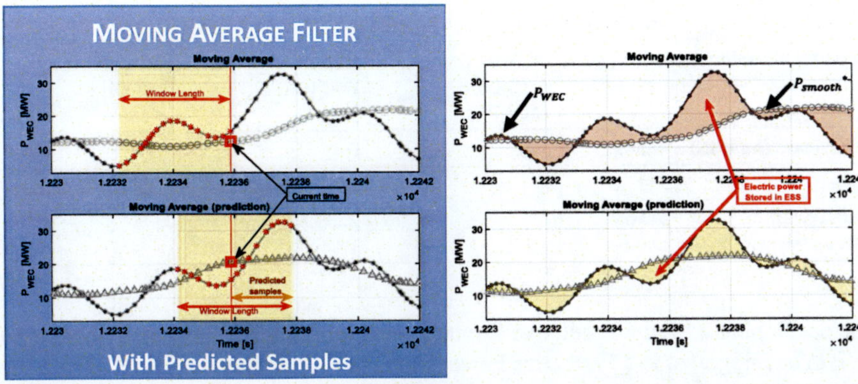

Fig. 18.17 Alternatives of the calculation of the power reference: with/without predicted samples

The calculation of the P_{smooth}* value is a key part of the ESS control algorithm used to smooth the WEC power oscillations. The ESS control proposed in this section is the MAF previously mentioned [see (18.13)]. An interesting upgrade of this control is the use of predicted values in the MAF as it is shown in Fig. 18.17 -left graph- [see (18.14)].

$$P_{smooth}{}^{*}(t_o) = \frac{1}{N}\sum_{i=-N}^{0} P_{WEC}(t_i) \tag{18.13}$$

$$P_{smooth}{}^{*}(t_o) = \frac{1}{N}\sum_{i=-\frac{N}{2}-1}^{N/2} P_{WEC}(t_i) \tag{18.14}$$

where N is the number of samples of the Moving Average Filter (MAF); i is the index of the time sample; t_i is the time instant i, where positive index represents future values, negative index represents past values and t_0 is the current instant.

As it is shown in Fig. 18.17 (right graph), the energy managed by the ESS is the area between the two power profiles, and it can be reduced through the use of predicted values. In a simple example,[1] the reduction of the ESS requirements is approx. a 60% as it is shown in Table 18.1.

Other approaches to the control of ESS can be considered, such as the generation of the reference P_{smooth}* through a Low-Pass Filter [87–89]; the use of Moving Average Filters (as is proposed in this section) to generate the reference P_{smooth}* [90, 91]; the consideration of the Flicker limits to define the reference P_{smooth}* [92]; the use of the ESS to stabilizes the DC-link voltage of the power electronics [93, 94]; or

[1] A WF located at the north of *Tenerife* Island (Canary Islands, Spain) is considered. The WF is composed of 25 point-absorbers of 150 kW of rated power. The calculation of ESS has been carried out just in one sea-state of 1.5 m of significant height (H_s) and 10.5 s of peak period (T_p) which is the predominant sea-state of the location.

Table 18.1 Energy storage requirements depend on the P_{smooth} calculation algorithm

Control algorithm	σ (P_{WEC})	N° Alert events Δ (f_{grid}) > 150 MHz	ESS rated power [kW]	ESS rated energy [kW · h]
Base case (no ESS)	4854.6	340	0	0
MAF	3117.8	7	2647.0	2.257
MAF with predicted samples	3116.6	7	2239.8	0.900

the developing a Model Predictive Control which predicts the generated power of a WF to be integrated in a Predictive Power Management System which optimizes the ESS operation [95].

18.3.2.2 Methodology for Dimensioning Energy Storage Systems (Study Case 1)

Description of the Study Case

The results presented in this sub-section are based on an analysis of an example case of a WF with the following characteristics [31]:

- The WF is located in the north of *Tenerife* Island, Spain (Fig. 18.18a)
- The WF has a rated power of 15 MW and is composed of 25 WECs (Fig. 18.18c).
- The considered WECs are 2-body point absorber as shown in Fig. 18.18b.
- The ESS is connected in the PCC of the WF (Fig. 18.18c). The ESS control is based on a MAF filter.

The complete system has been simulated employing a single node electric grid model (see Sect. 2.3.2) and a dynamic model of the WF. It has been simulated each hour of one year, considering the characteristics of the electric grid (electric inertia, renewable penetration, consumption, etc.) and the ocean-wave state (height, period, main direction, etc.).

Calculation of the Power Smoothing Level—Parametrization of the Control Algorithm

The **first step** of the dimensioning process is to determine the level of power smoothing which is necessary to solve a particular problem. In this case, the identified problem is the frequency oscillations produced in the electric grid due to the fluctuating power output of the WF. In this step, the ESS control algorithm is parameterized.

The objective of the ESS is to smooth the power output to ensure the grid compliance of the WF. In some grid codes such as Ireland, Germany, or Puerto Rico, the

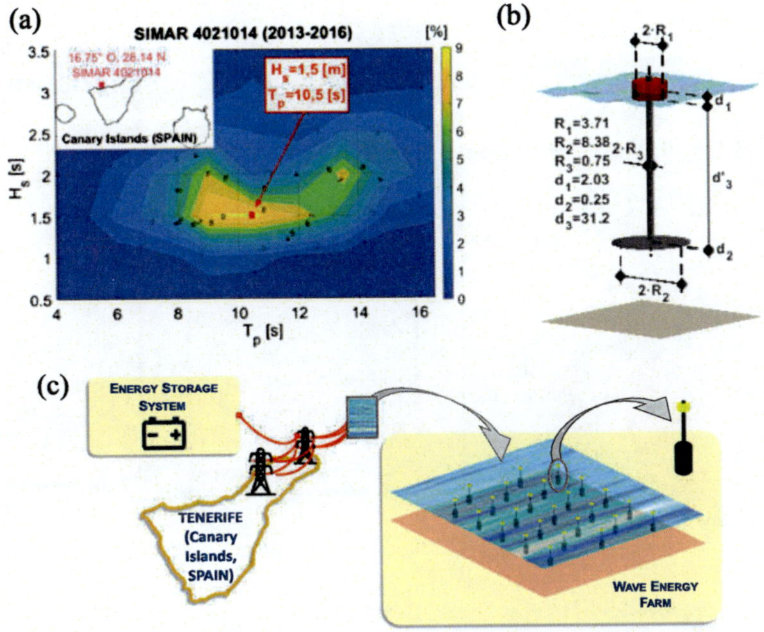

Fig. 18.18 Example case: 15 MW Wave Farm (WF) of point absorbers in *Tenerife* Island (Spain)

power oscillations are limited through the limit of the power ramp, as presented in Sect. 2.3.1. However, there is no reference to power oscillations neither in the European Grid Code [65, 96], nor in the specific Spanish Grid Code for insular electric grid systems [76]. Therefore, the approach of this analysis is the quantification of the problem caused by the power oscillations: the electric frequency oscillations.

In Spain, a normal operating state in an insular electric system is defined when the frequency is between its normal operational security limits of 49.85 and 50.15 Hz [76]. Also, it is accepted intervals of <5 min with values outside the mentioned range and between 49.75 and 50.25 Hz (alert operating state). The frequency values beyond these limits lead to an emergency operational state. Considering that the electricity generation by means of ocean-wave energy should not disturb the electrical grid beyond its normal operational state, frequency deviations of the grid produced by the WF oscillating active power must be within the ranges mentioned above. Thus, the activation of the load-frequency control is avoided [63]. In summary, regarding frequency values, the ESS should be dimensioned to smooth the WF power oscillations, maintaining the frequency deviations within their operating limits (49.85 and 50.15 Hz).

The power and frequency oscillations are correlated due to the frequency response has a relevant linear component [see Eq. (18.8) in Sect. 2.3.2]. Hence, the standard deviation—as a variable to represent the oscillations—of frequency and power is directly correlated. Figure 18.19 represents the standard deviation of the power

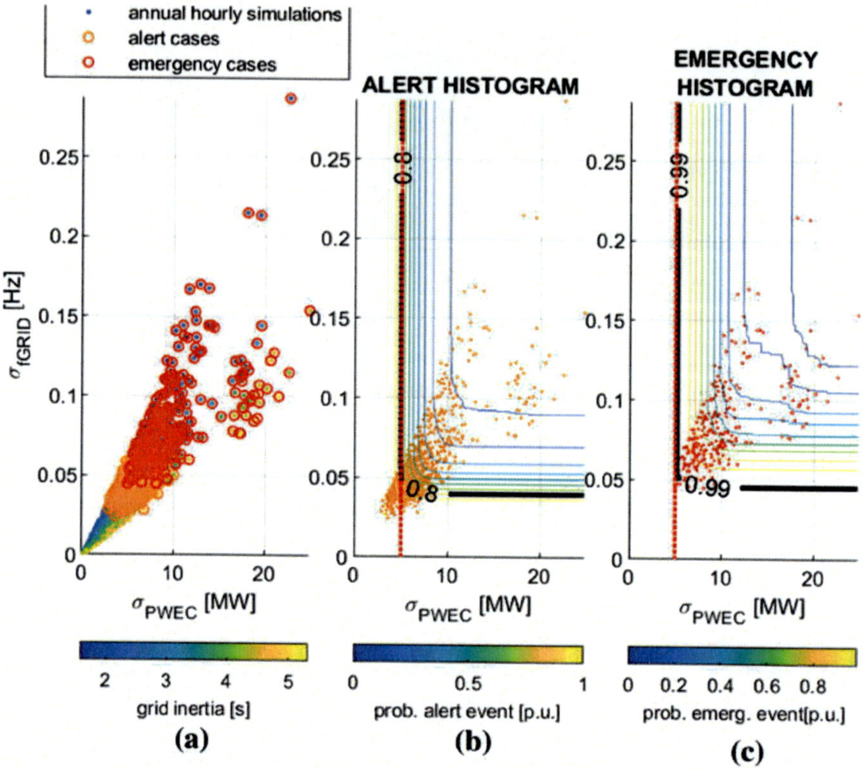

Fig. 18.19 Standard deviation of the output power versus standard deviation of the frequency of hourly periods during one year of the example case

injected by the WF concerning the standard deviation of the electric frequency of the grid in the example case. The standard deviation has been calculated for hourly periods. Also, if the alert and emergency events occur, the mark of the hourly period is remarked with a yellow and red circle, respectively. Figure 18.19a shows a direct relationship between the standard deviation of power and that of frequency, a relationship that depends on the inertia of the hourly period. The inertia value is remarked with the colour of each mark. Figure 18.19b, c represent the bivariate histogram of the alert and emergency events, respectively.

From Fig. 18.19 graphs, it is possible to define the maximum standard deviation of the WF power to minimize the alert and emergency events. The objective is to ensure that the normal operational mode is maintained along the longest part of the considered year. For sake of example, the criteria to eliminate 80% of the alert events and 100% of emergency events are considered. This leads to a standard deviation value of the WF power of 5 MW.

Considering the theoretical P_{smooth} equal to the power output command ($P_{smooth}*$), P_{smooth} can be calculated from P_{WEC} [see (18.13)]. Through the variance properties,

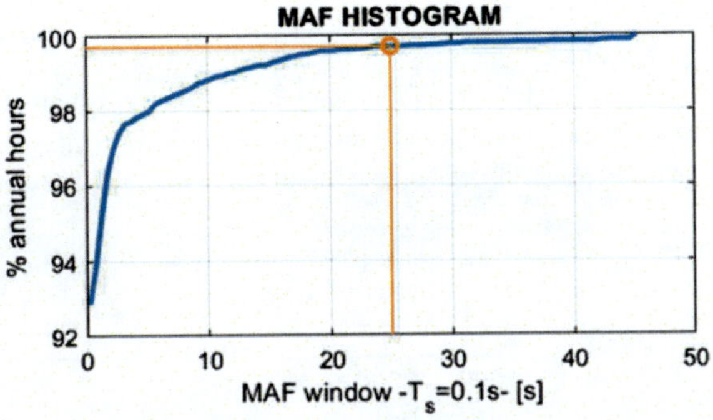

Fig. 18.20 Histogram of the theoretical MAF window values (N) -according to (18.16)- in each hourly simulation to reduce the standard deviation value of the WF power to the objective value (5 MW)

in particular the variance of a linear combination, the variance (σ^2) of the P_{smooth} can be calculated from the covariances (cov) of P_{WEC} as shown in (18.1) shows. Also, the covariances of P_{WEC} can be calculated from its frequency spectrum (S_f) by the application of the Weiner-Khinchin theorem [97] [see (18.16)].

$$\sigma^2\left(P_{smooth}{}^*\right) = \sum_{j=1}^{N}\sum_{k=1}^{N} cov(P_{WEC}(t_0 - (k-1)T_s), P_{WEC}(t_0 - (j-1)T_s)) \tag{18.15}$$

$$cov(P_{WEC}(t_0 - (k-1)T_s), P_{WEC}(t_0 - (j-1)T_s)) = \int_{-\infty}^{\infty} S_f(f) e^{2\pi i (j-k) T_s f} df \tag{18.16}$$

In summary, using (18.1) and (18.16) and from the frequency spectrum, each hourly, P_{WEC} can be calculated as the minimum value of the MAF window N which satisfies the condition of the P_{smooth} variance value. From Fig. 18.20, it is possible to select the MAF window as 25 s. The MAF window was selected as lower as possible due to the size of the MAF window is directly related to the rated energy value of the ESS; i.e., as lower is the power output variance, as higher is the energy storage requirements.

Calculation of the Rated Values of the ESS—Power and Energy

In a **second step** of the dimensioning process, the rated power and energy values of the Energy Storage System are defined.

Each hourly period of the example case is re-evaluated but including in the simulation an ESS theoretical model (no power/energy limited, and no losses). The ESS power output command is calculated with a MAF with a window of 30 s. From this analysis, it is possible to obtain a bi-variate histogram of the hourly rated energy and power requirements of the ESS (see Fig. 18.22). The rated ESS power requirement

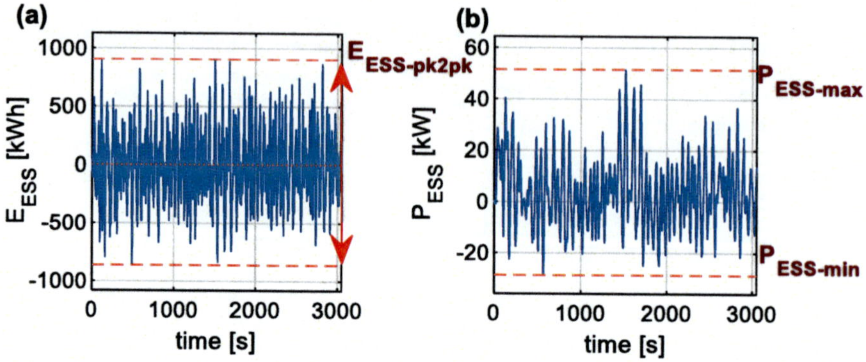

Fig. 18.21 Example of one hourly E_{ESS} (**a**) and P_{ESS} (**b**) profiles

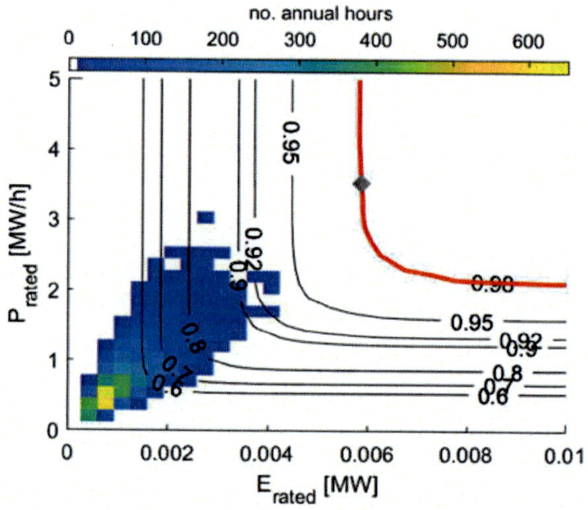

Fig. 18.22 Time (% of the year) in which power oscillations are completely compensated as a function of the ESS energy and power. ESS power losses are not considered

($P_{Rated\text{-}ESS}$) is calculated as the maximum absolute value of the hourly profile of P_{ESS} ($P_{ESS} = P_{smooth} - P_{WEC}$). The rated ESS energy requirement ($E_{Rated\text{-}ESS}$) is the peak-to-peak value of the hourly profile of E_{ESS}, as it is shown in Fig. 18.21.

In a **third step**, the technology of energy storage should be selected. The histogram (Fig. 18.22) is commonly used to select the proper energy storage technology, selecting a % of power oscillation to be compensated. For sake of example, the compensation selection of the 98% of the oscillations leads to a minimum rated power of 2.39 MW and a minimum rated energy of 4 kWh at the knee point. Another important parameter to select a proper energy storage technology is the oscillations number to be compensated, which are equal to the number of charge/discharge cycles in the ESS. In the example case, the number of ESS cycles per year is 9,651,044.

Fig. 18.23 *Mutriku* OWC power plant integrated in the breakwater [98, 100] (courtesy of BIMEP)

A further explanation of the energy storage technology selection is presented in Sect. 3.2.4.

18.3.2.3 Influence of Using Prediction in the MAF Calculation (Study Case 2)

Study Case Description

In this example case, the analysis has been carried out over a real ocean-wave energy generation plant. Figure 18.23a shows a photograph of the Oscillating Water Column (OWC) located in the harbor of *Mutriku* (Spain). This ocean-wave energy plant is suppling energy since 2011 and it encompasses 16 modules of OWC with a total rated power of 296 kW [98]. It is remarkable the relevance of the OWC power plant of *Mutriku*, since it is the ocean-wave system that has generated the most electrical energy in Europe [99].

The characteristics of the considered ESS are the following:

- The ESS is considered connected to the PCC of the OWC power plant.
- The ESS control is based on a MAF filter that uses predicted values [see (18.14)].
- The ESS is considered a Kinetic Energy Storage System (KESS) of 125 kW and 0.5kWh of rated power and energy, respectively.

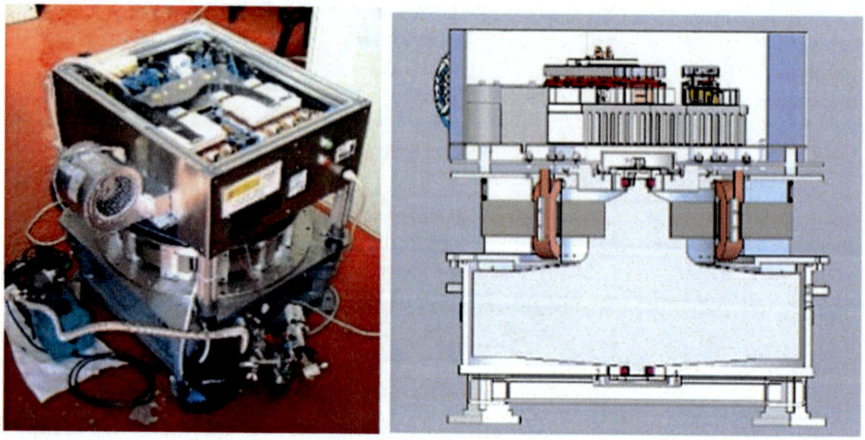

Fig. 18.24 Kinetic Energy Storage System unit considered in this example case

Fig. 18.25 KESS efficiency map [102]

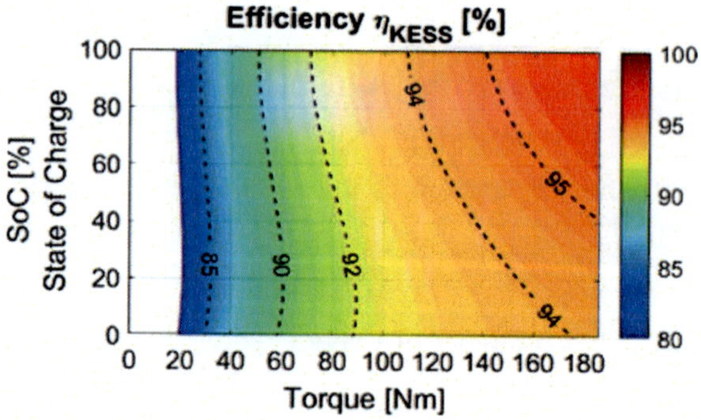

The KESS has modeled using a Loss Model, which allows evaluating the stored energy from the energy management at the connection point of the KESS. The KESS to be modeled is shown in Fig. 18.24, and the system encompasses a metallic flywheel, a Switched Reluctance Machine (SRM), and its associated power electronics [101]. The loss model considers the losses of the SRM (copper and iron losses), the mechanical losses in the flywheel (friction losses in bearings, and windage losses), and the losses in the power electronics (switching and conduction losses) [102]. Considering the whole KESS losses of the KESS, it is possible to obtain the ESS efficiency map showed in Fig. 18.25.

The complete system has been simulated by means of the use of real output power profiles of *Mutriku*, the efficiency map of the KESS, and an algorithm based on MAF with predicted values (see Fig. 18.26). In this example case, 2 representative days of generation of Mutriku has been analyzed (completed 24-h power profile). An example of this power profile has been shown in Figs. 18.6 and 18.9.

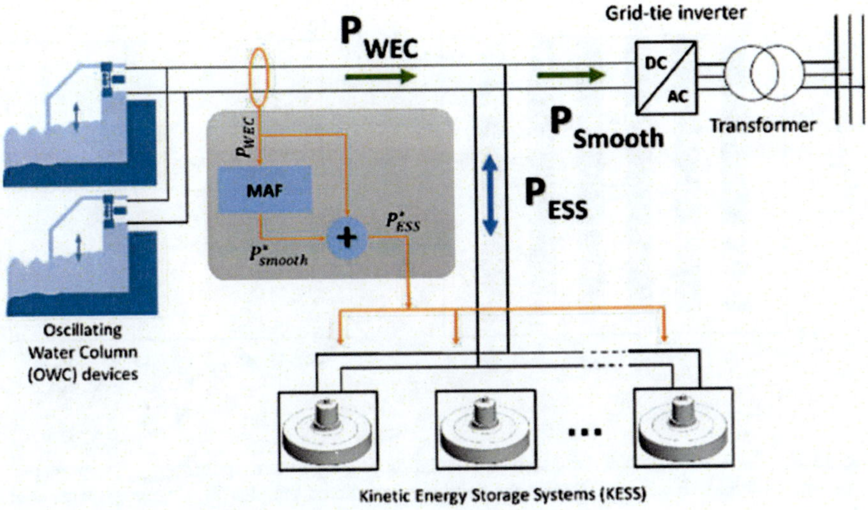

Fig. 18.26 Scheme of the ESS integrated in the OWC plant

Power Smoothing Level Calculation—Control Algorithm Parametrization

In this **first step**, the MAF window size is set to accomplish a certain power output smoothing level. The variable used to evaluate the smoothing level accomplished is the standard deviation of the output power ($\sigma_{P_{smooth}}$). In a first simple analysis, the correlation between the size of the MAF window (N) -considering a sample period (T_s) of 0.1 s- and the $\sigma_{P_{smooth}}$ is calculated. Besides, the correlation between the number of predicted values used by the MAF (N_p) and the $\sigma_{P_{smooth}}$ is calculated too. The conclusions that emerge from Fig. 18.27 are that N_p does not influence in $\sigma_{P_{smooth}}$, and hence the selection of N is the only necessary parameter to define the smoothing level [103].

Figure 18.28 shows the analysis of the 2 representative 24-h profiles, representing the $\sigma_{P_{smooth}}$ with respect to N. The figure shows as, from the value of 500 samples of MAF window size (N), no further reduction in the standard deviation (samples) is achieved. For sake of example, the value of N is set in 500.

To complete the parametrization of the control, the influence of the predicted sample number (N_p) in the ESS definition should be analyzed, and the value of N_p should be set. Figure 18.29 shows the Cumulative Distribution Functions (CDF) of the required energy for each case of N_P simulated. The figure shows as, for the value of N set in 500, an increment in the value of N_P implies a reduction of the Energy requirements due to the same level of cumulative percentage (y-axis value), the energy requirement is lower. The N_p value is selected from the information of Fig. 18.30; the value of 250 of predicted samples—the half of the MAF windows size—is a minimum in the ESS energy requirements.

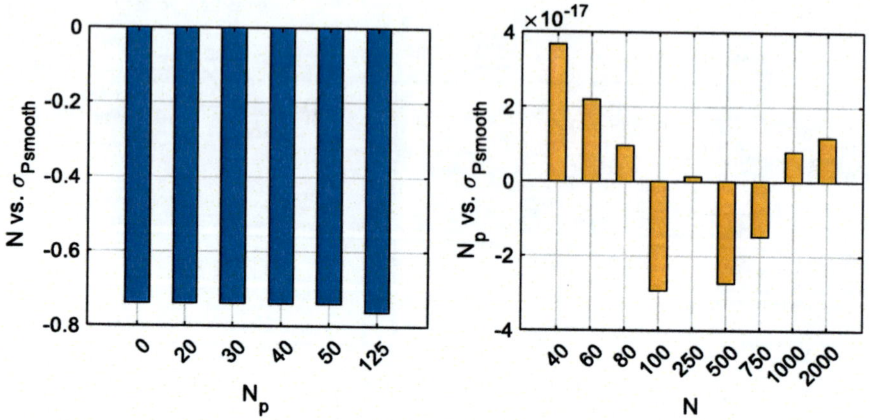

Fig. 18.27 Correlation between the standard deviation of the power output ($\sigma_{P_{smooth}}$) with respect to the size of the MAF window (N) and the number of predicted samples (N_p). *Source* Torres et al. [103]

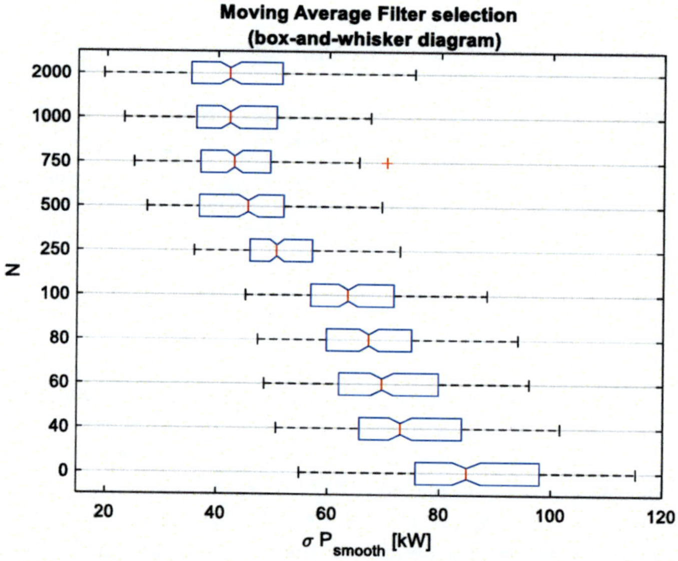

Fig. 18.28 Standard deviation of the power output ($\sigma_{P_{smooth}}$) with respect to the size of the MAF window (N). *Source* Torres et al. [103]

ESS Size Calculation—Number of KESS Units

Once the control is parametrized, in a **second step** it is necessary to define the size of the ESS (rated values of power and energy). In this example case, a particular

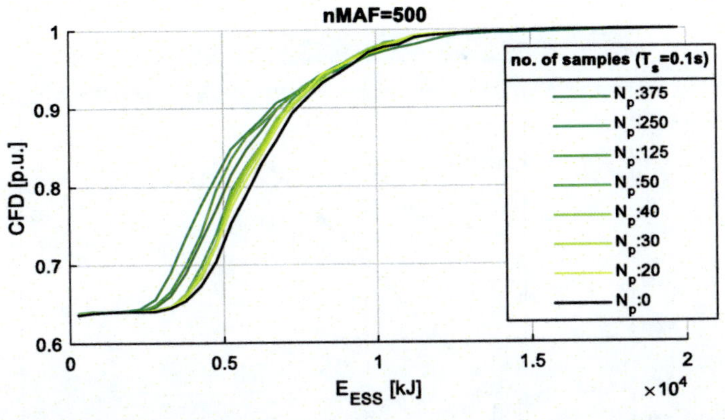

Fig. 18.29 CDF (Cumulative Distribution Function) of the required energy in the ESS. There are shown the cases of a MAF window (N) of 500 samples and different numbers of predicted values (N_p). *Source* Torres et al. [103]

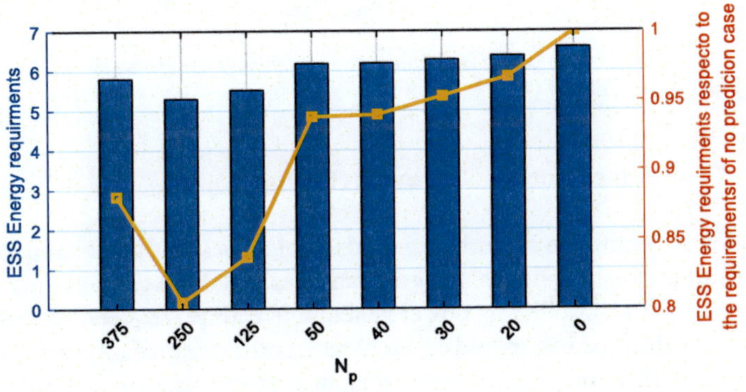

Fig. 18.30 Energy requirements for different values of N_p (number of predicted values) for the case of a MAF window size of 500 samples ($T_s = 0.1$)

ESS is considered, and in consequence, it is necessary to define the number of KESS to overcome the energy and power requirements. From the system simulation for the case of the particular parametrization of the control, it is possible to obtain a bi-variate histogram of energy and power requirements (see Fig. 18.31) which is analogous to the histogram shown in Fig. 18.22 for the previous example case.

Over the histogram, the black-line plotted represent the energy-power curve of the considered KESS, where each squared mark represents several KESS units. To cover 99% of the cases and smooth the power generated 99% of the time, the ESS would be composed of 5 KESS units.

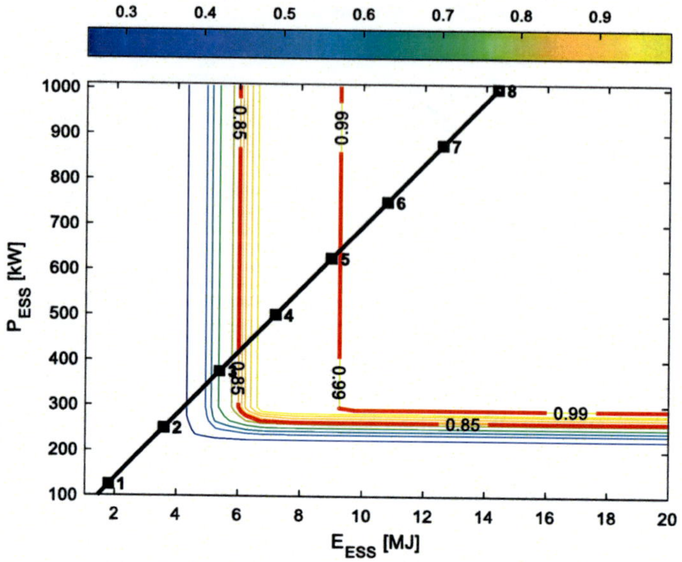

Fig. 18.31 Time (%) in which power oscillations are completely compensated as a function of the ESS energy and power. *Source* [103]

18.3.2.4 Selection of Energy Storage Technology

Once created the complete electric grid model which includes power generation and power consumption at the nodes, the grid variables evolution can be analyzed along the measuring period. Oscillating power generation leads to frequency excursions out of the limits established by the grid codes. Among other methods, such as generation or consumption curtailment or control strategies, energy storage is one of the most effective solutions to maintain the frequency stability of the grid. It was previously described how to obtain the energy and power parameters of the energy storage system required to achieve that goal.

However, reducing 100% of frequency excursions would lead to large energy storage, most of the time underused and therefore unviable. As a consequence, the selection of power and energy parameters needs to consider a reliability factor. Considering this reliability factor, Fig. 18.32 presents a family of curves related to the power and energy requirements for different reliability factors of Study Case 1 (see Sect. 3.2.2). Although it depends highly on the type of electric grid and it is mainly a qualitative decision that considers many factors, 95–97% could be a good first approach for the selection of power and energy parameters. In this case, the curve of application requirements is selected in 98% of the cases and highlighted as a red line.

The next step is to present the suitable technologies of energy storage for that application on that figure. Since the storage technologies are characterized by their power and energy and it can be considered a certain modularity, they can be depicted

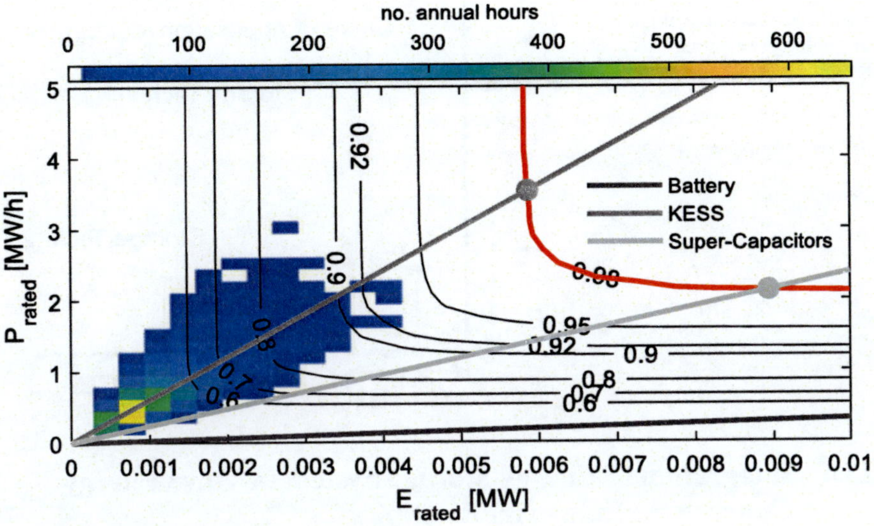

Fig. 18.32 ESS power and energy requirements

as straight lines. It is important to highlight that the closer to the knee of the curves is the line representing the storage technology, the more appropriate is to accomplish the requirements of the present application. Table 18.2 shows the characteristics of 3 different ESS and the necessary units to cover 98% of the annual hours.

In this study case, the line corresponding to the flywheel is the most appropriate, and therefore, it defines the most suitable energy storage technology, as well as the power and energy parameters. In other words, the number of modules of flywheels of this type is required.

Sometimes, the option of combining several energy storage technologies, comprising a hybrid energy storage system, is economically more viable and should be analyzed. In that case, the methodology to select and define the parameters of the system is similar to the already described. The way to reach the most convenient point, for instance, the knee of the 98% curve is through two-line sections, corresponding to two storage technologies, respectively. Figure 18.33 shows a scheme of this method, getting the power and energy parameters of each technology: (P1, E1) and (P2, E2).

Table 18.2 Electrical characteristics of different energy storage systems

Technology	P rated [kW]	E rated [kWh]	Cost [k$]	Ud	E_nom/E_req (%)
'LiMn2O4'	2219.2	69.65	4160	17	1720
'Flywheel'	2426.3	4.03	2620	5	100
'SuperCaps'	2219.2	9.33	2060	17	229

Source Mongird et al. [104]

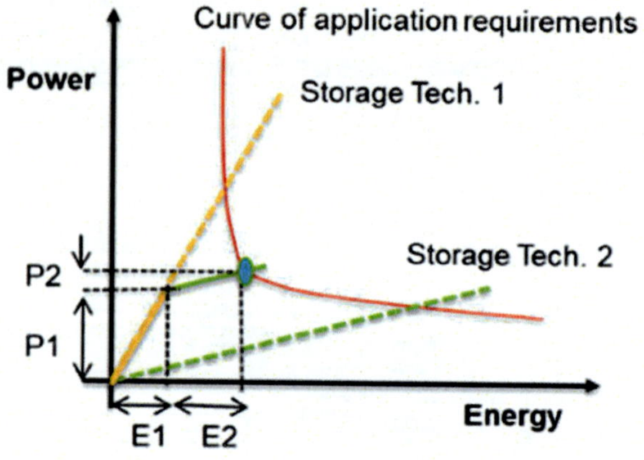

Fig. 18.33 Example of ESS technologies selection

18.4 Integration of Energy Storage System in Wave Energy Converters: Industrial Examples

This section lists the companies that market systems that extract energy from waves and include some type of onboard storage system. The purpose of this storage system and its characteristics are also described. The main companies of this type that currently exist in the market are:

- AW-Energy
- Ocean Power Technologies (OPT)
- Ocean Harvesting Technologies (OHT).

The company **AW-Energy** [105] offers a robust and predictable source of energy from waves to complement other types of generation systems with renewables such as solar and wind. It is a system that transforms wave energy into electrical energy with devices that operate in areas near the coast (below 2 km from the shore) and at depths that are around 8–20 m. Depending on the wave conditions in the operating area, these systems will be almost or completely submerged and anchored to the seabed. The power of these systems together with the PTO is in the range of 350–1000 kW with a capacity factor of 25–50% depending on the swell condition of the location where the system is installed. The wave energy is extracted utilizing hydraulic pumps attached to the moving part (Waveroller) and others placed in the fixed part in a closed circuit. The pressurized fluid is led to a supply and smoothing system that connects a hydraulic motor that controls an electric generator. The extracted electrical power is sent to the electrical grid. From the point of view of energy storage, the peculiarity of this system is that it integrates an on-board power system and a hydraulic system that allow the smoothing of the output power that is delivered to the power inverters located on the coast. Four prototypes developed by the company from 2002 to 2014.

AW-Energy has participated or is participating in four projects [106, 107]: on the coast of Portugal (FOAK and SURGE three units of 100 kW), on the French

Table 18.3 Electrical characteristics and dimensions of prototype PB3

WEC dimensions	Electrical characteristics
Height: 13.3 m	Continuous Average Power based on 8.4 kWh/day
Draft: 9.28 m	Payload Peak Power: up to 3 kW Power to load: 7.5 kW (custom)
Spar diameter: 1 m	Nominal battery Capacity: 50 kWh; Modular and scalable to 100, 150 kWh
Float diameter: 2.65 m	Zero wave day capacity: 100 W load for 1 + weeks
Weight: 8.3 kg	DC output: 24 and 300 V (standard) 5–600 V (custom)

Datasource from PB3 PowerBuoy® [109]

coast (WATTMOR 1 MW project), another on the east coast of Mexico (10 MW Waveroller wave farm) and another on the coast of Southeast Asia.

Another company that markets a system that extracts energy from waves with integrated storage is **OPT** [108]. Among the different products offered to convert wave energy into electrical energy is PB3 Powerbuoy [109]. The device is floating and in turn moored to the bottom of the sea and can operate in ocean depths in the range of 20–3000 m. The PB3 device supplies power to loads on board or to equipment on the seabed. The system is constantly sending data and communicating with the facilities that are located on the coast. The energy is extracted through an onboard direct-drive generator that charges a battery pack, also on-board. Batteries supply the necessary power to meet the end user's power needs when the power extracted from the waves is not enough. The storage system allows this device to be used in applications with variable power demand and short periods of peak demand.

The PB3 technology was validated in a US Navy Project to feed coastal safety net systems and was subjected to adverse atmospheric conditions such as hurricanes. The main characteristics for the PB3 are shown in Table 18.3.

Finally, the Swedish company **OHT** has developed a hydraulic collection system from the WEC's to a central tower with an energy storage system that allows less fluctuating electrical power [110]. The hydraulic system discussed above is connected to an input shaft of a planetary box. The output shaft is connected to an electric generator and a floating ring gear is connected to a rack and pinion drive. A weight running on a linear guide gives a constant torque in a gearbox and a constant pressure to the hydraulic system. The electric generator is controlled to work at a constant speed that gives an average electrical output power corresponding to the average energy of the sea state at that time. If the speed of the hydraulic motor increases a rear gear is forced to rotate forward to lift the weight on the accumulator. On the other hand, when the input power is less than the desired power to keep the generator speed constant, the ring gear rotates in the opposite direction to keep the electrical output power constant.

The five phases from concept development to the demonstration WF were scheduled to start in 2014 and end in 2020 [111]. These five phases consist of:

- Phase 1: Development and evaluation of the concept through simulations of a 1 MW collection tower with 5 buoys. The collaborating companies in this stage were Waves4Power and CorPower.
- Phase 2: Design of a system based on standard components for a 10 MW power output demonstration array that integrates 40–50 buoys and a collection tower with the gravity storage system. The buoys were supplied by CorPower and the Waves4Power company connected the tower to the storage system.
- Phase 3: Half scale proof of design test-rig. A collection tower dimensioned to be used with 3 buoys for dry tests was built. The peak power of the hydraulic motor was 750 kW to output an electrical power of 100 kW.
- Phase 4: A Collection tower connected to 3 weeks to carry out tests at sea. PTO in WEC's modified with radial piston pumps.
- Phase 5: Tests in a sea environment with an array of 10 MW wave energy converters. The functionality and design validated in the 100 kW prototype (stage 3 and 4) will be scaled up to a 10 MW collection tower connected to 40–50 buoys.

Acknowledgements This research has been developed in the framework of the CYTED Thematic Network REMAR (OPORTUNIDADES DE INTEGRACIÓN EN REDES ELÉCTRICAS IBEROAMERICANAS DE LAS ENERGÍAS DEL MAR; http://www.cyted.org/?q=es/detalle_proyecto&un=1024)—proposal Number P720RT0246 | Thematic Networks Call—CYTED 2020.

In addition, the authors would like to thank EVE (Ente Vasco de la Energía) and BiMEP for the data of the output power of the Mutriku OWC plant provided in other to carry out the study case presented in Sect. 3.2.3.

References

1. California Energy Office, "California's. 2030. *Climate commitment renewable resources for half of the state's electricity by 2030 benefits from renewables for half of electricity use by 2030* (p. 2013).
2. Renewable Energy Prospects for the European Union. 2018. Accessed January 13, 2021 [Online]. Available https://www.irena.org/publications/2018/Feb/Renewable-energy-prospects-for-the-EU
3. Australian Energy Market Operator. (2019). Draft 2020 integrated system plan, no. December (pp. 1–83) [Online]. Available https://aemo.com.au/energy-systems/major-publications/integrated-system-plan-isp/2020-integrated-system-plan-isp
4. Bloom, A., et al. (2017, November). It's Indisputable: Five facts about planning and operating modern power systems. *IEEE Power and Energy Magazine, 15*(6), 22–30. https://doi.org/10.1109/MPE.2017.2729079
5. Flynn, D., et al. (2017). Technical impacts of high penetration levels of wind power on power system stability. *Wiley Interdisciplinary Reviews: Energy and Environment, 6*(2). https://doi.org/10.1002/wene.216
6. Milano, F., Dorfler, F., Hug, G., Hill, D. J., & Verbič, G. (20182018).Foundations and challenges of low-inertia systems (Invited Paper). In 20th Power systems computation conference. PSCC 2018. https://doi.org/10.23919/PSCC.2018.8450880
7. Huang, L., Xin, H., Dong, W., & Dörfler, F. (2019).Impacts of grid structure on PLL-synchronization stability of converter-integrated power systems. *arXiv*.

8. Riepnieks, A., & Kirkham, H. (2016). Rate of change of frequency measurement. In *2016 57th International scientific conference power and electrical engineering of RIGA Technical University, RTUCON 2016*. https://doi.org/10.1109/RTUCON.2016.7763110
9. Kirkham, H., & Pandey, S. (2018, February). Is ROCOF measureable? In *2018 IEEE power energy society innovative smart grid technologies conference (ISGT)* (pp. 1–5). https://doi.org/10.1109/ISGT.2018.8403334
10. Turcotte, D., & Katiraei, F. (2009). Fault contribution of grid-connected inverters. In *2009 IEEE electrical power and energy conference, EPEC 2009* (pp. 1–5). https://doi.org/10.1109/EPEC.2009.5420365
11. Kroposki, B., et al. (2017). Achieving a 100% renewable grid: Operating electric power systems with extremely high levels of variable renewable energy. *IEEE Power and Energy Magazine, 15*(2), 61–73. https://doi.org/10.1109/MPE.2016.2637122
12. Pogaku, N., Prodanovic, M., & Green, T. C. (2007). Modeling, analysis and testing of autonomous operation of an inverter-based microgrid. *IEEE Transactions on Power Electronics, 22*(2), 613–625. https://doi.org/10.1109/TPEL.2006.890003
13. Rahman, S., Moghaddami, M., Sarwat, A. I., Olowu, T., & Jafaritalarposhti, M. (2018). Flicker estimation associated with PV integrated distribution network. In *Proceedings of IEEE* SOUTHEASTCON (Vol. 2018-April, no. 1553494, pp. 1–6). https://doi.org/10.1109/SECON.2018.8479058.
14. Larsson, A. (2002). Flicker emission of wind turbines during continuous operation. *IEEE Transactions on Energy Conversion, 17*(1), 114–118. https://doi.org/10.1109/60.986447
15. Chen, Z., Zhang, L., & Yeung, R. W. (2019). Analysis and optimization of a Dual Mass-Spring-Damper (DMSD) wave-energy convertor with variable resonance capability. *Renewable Energy, 131*, 1060–1072. https://doi.org/10.1016/J.RENENE.2018.07.006
16. Liang, X., & Andalib-Bin-Karim, C. (2018). Harmonics and mitigation techniques through advanced control in grid-connected renewable energy sources: A review. *IEEE Transactions on Industry Applications, 54*(4), 3100–3111. https://doi.org/10.1109/TIA.2018.2823680
17. Adrees, A., Milanović, J. V., & Mancarella, P. (2019). Effect of inertia heterogeneity on frequency dynamics of low-inertia power systems. *IET Generation, Transmission and Distribution, 13*(14), 2951–2958. https://doi.org/10.1049/iet-gtd.2018.6814
18. Ulbig, A., Borsche, T. S., & Andersson, G. (2014). *Impact of low rotational inertia on power system stability and operation. IFAC, 19*(3).
19. Al Masood, N., Yan, R., Saha, T. K., & Bartlett, S. (2016). Post-retirement utilisation of synchronous generators to enhance security performances in a wind dominated power system. *IET Generation, Transmission & Distribution, 10*(13), 3314–3321. https://doi.org/10.1049/iet-gtd.2016.0267
20. Gevorgian, V., & O'Neill, B. (2016). *Advanced grid-friendly controls demonstration project for utility-scale PV power plants*. National Renewable Energy Laboratory, no. January (p. 102) [Online]. Available https://www.nrel.gov/docs/fy16osti/65368.pdf
21. Loutan, C., et al. (2017, March). *Demonstration of essential reliability services by a 300-MW solar photovoltaic power plant*. https://doi.org/10.2172/1349211
22. Al-Shetwi, A. Q., Hannan, M. A., Jern, K. P., Mansur, M., & Mahlia, T. M. I. (2020). Grid-connected renewable energy sources: Review of the recent integration requirements and control methods. *Journal of Cleaner Production, 253*, 119831. https://doi.org/10.1016/j.jclepro.2019.119831
23. Gu, H., Member, S., Yan, R., & Saha, T. (2019). Review of system strength and inertia requirements for the national electricity market of Australia. *CSEE Journal of Power and Energy System, 5*(3), 295–305. https://doi.org/10.17775/cseejpes.2019.00230
24. Fernandez-Munoz, D., Guisandez, I., Perez-Diaz, J. I., Chazarra, M., Femandez-Espina, A., & Burke, F. (2018, June). Fast frequency control services in Europe. In International conference on the European energy market (Vol. 2018). https://doi.org/10.1109/EEM.2018.8469973.
25. Castillo, A., & Gayme, D. F. (2014). Grid-scale energy storage applications in renewable energy integration: A survey. *Energy Conversion and Management, 87*, 885–894. https://doi.org/10.1016/j.enconman.2014.07.063

26. Stenclik, D., Denholm, P., & Chalamala, B. (2017). Maintaining balance: The increasing role of energy storage for renewable integration. *IEEE Power and Energy Magazine, 15*(6), 31–39. https://doi.org/10.1109/MPE.2017.2729098
27. Villate, J. L. (2020). Strategic Research and Innovation Agenda for Ocean Energy [Online]. Available https://www.etipocean.eu/news/targeted-innovation-to-drive-large-scale-deployment-of-ocean-energy-says-new-etip-ocean-report/
28. IRENA. (2020). Fostering a blue economy: Offshore renewable energy. *International Renewable*, Abu Dhabi. 978-92-9260-288-8.
29. LiVecchi, A., et al. (2019). *Exploring opportunities for marine renewable energy in maritime markets*. U.S. Department of Energy (DOE).
30. IRENA. (2020). *Innovation outlook: Ocean energy technologies*. Abu Dhabi (UAE).
31. Villalba, I., et al. (2019, October). Wave farms grid code compliance in isolated small power systems. *IET Renewable Power Generation, 13*(1), 171–179. https://doi.org/10.1049/iet-rpg.2018.5351
32. Eltigani, D., & Masri, S. (2015). Challenges of integrating renewable energy sources to smart grids: A review. *Renewable and Sustainable Energy Reviews, 52*, 770–780. https://doi.org/10.1016/j.rser.2015.07.140
33. Blavette, A., O'Sullivan, D. L., Alcorn, R., Lewis, T. W., & Egan, M. G. (2014). Impact of a medium-size wave farm on grids of different strength levels. *IEEE Transactions on Power Systems, 29*(2), 917–923. https://doi.org/10.1109/TPWRS.2013.2284513
34. Mendonça, H., de Castro, R., Martínez, S., & Montalbán, D. (2017). Voltage impact of a wave energy converter on an unbalanced distribution grid and corrective actions. *Sustainability, 9*(10), 1844. https://doi.org/10.3390/su9101844
35. Nájera, J., Mendonça, H., de Castro, R., & Arribas, J. (2019). Strategies comparison for voltage unbalance mitigation in LV distribution networks using EV chargers. *Electronics, 8*(3), 289. https://doi.org/10.3390/electronics8030289
36. Johnson, B., & Cotilla-Sanchez, E. (2020). Estimating the impact of ocean wave energy on power system reliability with a well-being approach. *IET Renewable Power Generation, 14*(4), 608–615. https://doi.org/10.1049/iet-rpg.2019.0567
37. Naresh, M., & Tripathi, R. K. (2017, April). Power flow control of DFIG generators for wind turbine variable speed using STATCOM. In *2017 Innovations in power and advanced computing technologies (i-PACT)* (pp. 1–5). https://doi.org/10.1109/IPACT.2017.8245039.
38. Jalali, A., & Aldeen, M. (2016, November).Placement and operation of STATCOM-storage for voltage stability enhancement of power systems with embedded wind farms. In *2016 IEEE innovative smart grid technologies—Asia (ISGT-Asia)* (pp. 948–953). https://doi.org/10.1109/ISGT-Asia.2016.7796513
39. C. R. (EU) 2017/1485. (2017). *Guideline on electricity transmission system operation* [Online]. Available https://eur-lex.europa.eu/legal-content/EN/TXT/?uri=CELEX%3A32017R1485
40. Kundur, P. (1994). *Power system stability and control*. McGraw-Hill Education.
41. PREPA. (2012). *Puerto Rico Electric Power Authority minimum technical requirements for (PV) generation projects*.
42. HECO. (2014). *Hawaiian Electric Company: State of the system, summary for ESSS RFP*.
43. Gevorgian, V., & Booth, S. (2013). Review of PREPA technical requirements for interconnecting wind and solar generation. Accessed November 24, 2020 [Online]. Available www.nrel.gov/publications
44. Piwko, R. (2012, September). *Special assessment interconnection requirements for variable generation*.
45. Goda, Y. (2010). *Random seas and design of maritime structures—Advanced series on ocean engineering* (3rd ed.). World Scientific Publishing Company.
46. Falnes, J. (2002). *Ocean waves and oscillating systems, linear interactions including wave-energy extraction*. Cambridge: Cambridge University Press.
47. Falnes, J. (2007). A review of wave-energy extraction. *Marine Structures, 20*, 185–201. https://doi.org/10.1016/j.marstruc.2007.09.001

48. PLOCAN (PLataforma Oceania de CANarias) Homepage. http://www.plocan.eu/index.php/es/. Accessed July 03, 2015.
49. Longuet-Higgins, M. S. (1983). On the joint distribution of wave periods and amplitudes in a random wave field. *Proceedings of the Royal Society of London. Series A: Mathematical and Physical Sciences, 389*(1797), 241–258. https://doi.org/10.2307/2397713
50. Falcão, A. F. de O., & Rodrigues, R. J. (2002, April). Stochastic modelling of OWC wave power plant performance. *Applied Ocean Research, 24(2), 59–71.* https://doi.org/10.1016/S0141-1187(02)00022-6
51. Blanco, M., Navarro, G., Lafoz, M., & Pérez, J. I. (2017). How harmful is the wave energy penetration in an electric grid? In *Proceedings of the Twelfth European wave and tidal energy conference* (pp. 942–942).
52. Lafoz, M., Blanco, M., & Ramirez, D. (2011). Grid connection for wave power farms. In *Proceedings of the 2011 14th European conference on power electronics and applications (EPE 2011)* (pp. 1–10).
53. Zhou, Z., Benbouzid, M., Frédéric Charpentier, J., Scuiller, F., & Tang, T. (2013). A review of energy storage technologies for marine current energy systems. *Renewable and Sustainable Energy Reviews, 18*, 390–400. https://doi.org/10.1016/j.rser.2012.10.006
54. Falcão, A. F. de O. (2010, April). Wave energy utilization: A review of the technologies. *Renewable and Sustainable Energy Reviews,* 14(3), 899–918. https://doi.org/10.1016/j.rser.2009.11.003
55. Wave developers: EMEC: European Marine Energy Centre. http://www.emec.org.uk/marine-energy/wave-developers/. Accessed May 14, 2020.
56. Lafoz, M., Blanco, M., Beloqui, L., Navarro, G., & Moreno-Torres, P. (2016). Dimensioning methodology for energy storage devices and wave energy converters supplying isolated loads. *IET Renewable Power Generation, 10*(10), 1468–1476. https://doi.org/10.1049/iet-rpg.2016.0074
57. Polinder, H., & Scuotto, M. (2005). Wave energy converters and their impact on power systems. In *2005 International conference on future power systems* (p. 9). https://doi.org/10.1109/FPS.2005.204210
58. Falcão, A. F. de O., Cândido, J. J., Justino, P. A. P., & Henriques, J. C. C. (2012). Hydrodynamics of the IPS buoy wave energy converter including the effect of non-uniform acceleration tube cross section. *Renewable Energy, 41*, 105–114. https://doi.org/10.1016/j.renene.2011.10.005
59. Bimep - Área de ensayos de energía marina de Armintza. http://bimep.com/. Accessed October 12, 2015.
60. Göteman, M., Engström, J., Eriksson, M., & Isberg, J. (2015). Optimizing wave energy parks with over 1000 interacting point-absorbers using an approximate analytical method. *International Journal of Marine Energy, 10*, 113–126. https://doi.org/10.1016/j.ijome.2015.02.001
61. Blavette, A., O'Sullivan, D. L., Lewis, T. W., & Egan, M. G. (2015). Dimensioning the equipment of a wave farm: Energy storage and cables. *IEEE Transactions on Industry Applications, 51*(3), 2470–2478. https://doi.org/10.1109/TIA.2014.2382754
62. Blavette, A., et al. (2017). Influence of the wave dispersion phenomenon on the flicker generated by a wave farm. In *Proceedings of the 12th European wave tidal energy conference (EWTEC 2017)* [Online]. Available https://hal.archives-ouvertes.fr/hal-01754055/file/Blavette--Influenceofthewavedispersion_EWTEC2017-revised_v4.pdf
63. ENTSO-E. (2013). *Network code on load-frequency control and reserves* [Online]. Available http://networkcodes.entsoe.eu/wp-content/uploads/2013/08/130628-NC_LFCR-Issue1.pdf%5Cn; http://networkcodes.entsoe.eu/wp-content/uploads/2013/08/130628-NC%7B_%7DLFCR-Issue1.pdf
64. SOGL. (2020). *Nordic system operation agreement (SOA)—Annex load-frequency control & reserves (LFCR).*
65. Comisión Europea. (2016). *Reglamento (UE) 2016/631 de la Comisión de 14 de abril de 2016 que establece un código de red sobre requisitos de conexión de generadores a la red*. D. Of. la Unión Eur., no. 112, 27 de abril (pp. 1–68).

66. EirGrid. (2015). *EirGrid Grid Code.*
67. E.-E. ENTSO-E. (2009). *Continental Europe operation handbook.*
68. Egido, I., Fernández-Bernal, F., Rouco, L., Porras, E., & Sáiz-Chicharro, Á. (2004). Modeling of thermal generating units for automatic generation control purposes. *IEEE Transactions on Control Systems Technology, 12*(1), 205–210.
69. TSOs, N. (2018). *Explanatory document for the Nordic synchronous area Proposal for the dimensioning rules for FCR in accordance with Article 153 of the Commission Regulation (EU) 2017/1485 of 2 August 2017 establishing a guideline on electricity transmission system operation.*
70. A1-Appendix 1: Load-frequency control and performance [E]. Accessed November 24, 2020 [Online]. Available https://eepublicdownloads.entsoe.eu/clean-documents/pre2015/publications/entsoe/Operation_Handbook/Policy_1_Appendix_final.pdf.
71. Etxegarai, A., Eguia, P., Torres, E., Iturregi, A., & Valverde, V. (2015). Review of grid connection requirements for generation assets in weak power grids. *Renewable and Sustainable Energy Reviews, 41*, 1501–1514.
72. Farias, M. F., Battaiotto, P. E., & Cendoya, M. G. (2010). Wind farm to weak-grid connection using UPQC custom power device. In *2010 IEEE international conference on industrial technology* (pp. 1745–1750).
73. Tande, J. O., Di Marzio, G., & Uhlen, K. (2007). *System requirements for wind power plants.* SINTEF Energy Research.
74. Chang, Y.-P., & Wu, C.-J. (2003). Effect of utility system X/R ratio on harmonic amplification characteristics of a plastic plant. In *2003 IEEE PES transmission and distribution conference and exposition (IEEE Cat. No. 03CH37495)* (Vol. 1, pp. 68–72).
75. Alizadeh, S. M., Ozansoy, C., & Alpcan, T. (2016).The impact of X/R ratio on voltage stability in a distribution network penetrated by wind farms. In *2016 Australasian universities power engineering conference (AUPEC)* (pp. 1–6).
76. Ministerio de Industria energía y Turismo. (2012). Procedimiento Operación 3.1, 3.2, 9, 14.4, SEIE 1, SEIE 2.2, SEIE 2.3, SEIE 3.1, SEIE 7.1, SEIE 7.2, SEIE 8.2, SEIE 9 (p. 57263).
77. Strachan, N., & Jovcic, D. (2009). Offshore wind farm with centralised power conversion and DC interconnection. *IET Generation, Transmission and Distribution, 3*(6), 586–595. https://doi.org/10.1049/iet-gtd.2008.0372
78. Boström, C., Lejerskog, E., Stålberg, M., Thorburn, K., & Leijon, M. (2009). Experimental results of rectification and filtration from an offshore wave energy system. *Renewable Energy, 34*(5), 1381–1387. https://doi.org/10.1016/j.renene.2008.09.010
79. Murray, B., Egan, M. G., Hayes, J. G., & Sullivan, D. L. (2009). Applications of supercapacitor energy storage for a wave energy converter system. In *Proceedings of the 7th European wave tidal energy conference (EWTEC 2009).*
80. Aubry, J. J., et al. (2015, October). Energy storage system sizing for smoothing power generation of direct wave energy converters. In *Proceedings of the 3rd international conference on ocean energy (ICOE 2010)*, no. October. ISBN 978-84-693-5467-4, Accessed October 26, 2015 [Online]. Available http://hal.archives-ouvertes.fr/hal-00526435/
81. Villalba, I., Greiner, D., Díaz, F., Blanco, M., Lafoz, M., & Pérez, J. I. (2018). Multiobjective optimization of a wave energy farm. In *6th European seminar on computing (ESCO 2018).*
82. Hals, J., Falnes, J., & Moan, T. (2011). A comparison of selected strategies for adaptive control of wave energy converters. *Journal of Offshore Mechanics and Arctic Engineering, 133*, 31101. https://doi.org/10.1115/1.4002735
83. Ndreko, M., Rüberg, S., & Winter, W. (2018). Grid forming control for stable power systems with up to 100% inverter based generation: A paradigm scenario using the IEEE 118-Bus System.
84. Mohan, N., Undeland, T. M., & Robbins, W. P. (2003). *Power electronics: Converters, applications, and design.* Wiley.
85. Malinowski, M., Kazmierkowski, M. P., & Trzynadlowski, A. (2000). Direct power control with virtual flux estimation for three-phase PWM rectifiers. In *ISIE'2000. Proceedings of the 2000 IEEE international symposium on industrial electronics (Cat. No. 00TH8543)* (Vol. 2, pp. 442–447). https://doi.org/10.1109/ISIE.2000.930338

86. Kovaltchouk, T., Blavette, A., Aubry, J., Ben Ahmed, H., & Multon, B. (2016, September). Comparison between centralized and decentralized storage energy management for direct wave energy converter farm. *IEEE Transactions on Energy Conversion, 31*(3), 1051–1058. https://doi.org/10.1109/TEC.2016.2547462
87. Rojas-Delgado, B., Alonso, M., Amaris, H., & de Santiago, J. (2019). Wave power output smoothing through the use of a high-speed kinetic buffer. *Energies, 12*(11), 2196. https://doi.org/10.3390/en12112196
88. Agarwal, A., Iyer, V. M., Anurag, A., & Bhattacharya, S. (2019, September). Adaptive control of a hybrid energy storage system for wave energy conversion application. In *2019 IEEE energy conversion congress and exposition (ECCE)* (pp. 4994–5001). https://doi.org/10.1109/ECCE.2019.8912897
89. Hazra, S., & Bhattacharya, S. (2016, September). Hybrid energy storage system comprising of battery and ultra-capacitor for smoothing of oscillating wave energy. In *2016 IEEE energy conversion congress and exposition (ECCE)* (pp. 1–8). https://doi.org/10.1109/ECCE.2016.7855172
90. Nie, Z., et al. (2017). Designing and testing composite energy storage systems for regulating the outputs of linear wave energy converters. *Energies, 10*(1), 114. https://doi.org/10.3390/en10010114
91. Brando, G., Dannier, A., Del Pizzo, A., Di Noia, L. P., & Pisani, C. (2016). Grid connection of wave energy converter in heaving mode operation by supercapacitor storage technology. *IET Renewable Power Generation, 10*(1), 88–97. https://doi.org/10.1049/iet-rpg.2015.0093
92. Kovaltchouk, T., Multon, B., Ben Ahmed, H., Aubry, J., & Venet, P. (2015, May). Enhanced aging model for supercapacitors taking into account power cycling: application to the sizing of an energy storage system in a direct wave energy converter. *IEEE Transactions on Industry Applications, 51*(3), 2405–2414. https://doi.org/10.1109/TIA.2014.2369817.
93. Zou, Y., & Cheng, K. W. E. (2017). A vertical flux-switching permanent magnet based oscillating wave power generator with energy storage. *Energies, 10*(7), 887. https://doi.org/10.3390/en10070887
94. Kamat, P., Hazra, S., & Bhattacharya, S. (2017, March). Stand-alone low-cost wave energy generation with energy storage integration. In *2017 IEEE applied power electronics conference and exposition (APEC)* (pp. 1550–1557). https://doi.org/10.1109/APEC.2017.7930905
95. Rajapakse, G., Jayasinghe, S., Fleming, A., & Negnevitsky, M. (2018). Grid integration and power smoothing of an oscillating water column wave energy converter. *Energies, 11*(7), 1871. https://doi.org/10.3390/en11071871
96. European Commission. (2017). *Regulation (EU) 2017/1485—Guideline on electricity transmission system operation.*
97. Wiener, N. (1930). Generalized harmonic analysis. *Acta Mathematica, 55*, 117–258. https://doi.org/10.1007/BF02546511
98. BiMEP, Mutriku technical characteristics—BiMEP. https://www.bimep.com/en/mutriku-area/technical-characteristics/. Accessed January 15, 2021.
99. Magagna, D. (2019).*Ocean energy technology development report*. https://doi.org/10.2760/158132
100. Faÿ, F.-X., Robles, E., Marcos, M., Aldaiturriaga, E., & Camacho, E. F. (2020). Sea trial results of a predictive algorithm at the Mutriku Wave power plant and controllers assessment based on a detailed plant model. *Renewable Energy, 146*, 1725–1745. https://doi.org/10.1016/j.renene.2019.07.129
101. Navarro, G., Torres, J., Moreno-Torres, P., Blanco, M., & Lafoz, M. (2015, September). Technology description and characterization of a low-cost flywheel for energy management in microgrids. In *2015 17th European conference on power electronics and applications (EPE'15 ECCE-Europe)* (pp. 1–10). https://doi.org/10.1109/EPE.2015.7309324
102. Torres, J., Moreno-Torres, P., Navarro, G., Blanco, M., & Lafoz, M. (2018). Fast energy storage systems comparison in terms of energy efficiency for a specific application. *IEEE Access, 6*(2), 40656–40672. https://doi.org/10.1109/ACCESS.2018.2854915

103. Torres, J., Blanco, M., Lafoz, M., Navarro, G., Nájera, J., & Santos-Herran, M. (2020). Dimensioning methodology of energy storage systems for power smoothing in a wave energy conversion plant considering efficiency maps and filtering control techniques. *Energies, 13*(13), 3380. https://doi.org/10.3390/en13133380
104. Mongird, K., et al. (2019). *Energy storage technology and cost characterization report | Department of Energy*. no. July [Online]. Available https://www.energy.gov/eere/water/downloads/energy-storage-technology-and-cost-characterization-report
105. (2015) AW Energy Homepage (WAVEROLLER developer). http://aw-energy.com/. Accessed June 14, 2015.
106. Kasanen, E. (2015). AW-energy—Positive experiences of the Waveroller in Portugal and France. In *WavEC conference*.
107. Mäki, T. (2017). Wave energy, waveroller technology & aw-energy. In *EU Green Week* [Online]. Available http://www.eugreenweek.eu/sites/default/files/speakers/documents/session_2.1_tuula_maki_0.pdf
108. http://www.oceanpowertechnologies.com/. (2015). *Ocean Power Technologies Homepage*. http://www.oceanpowertechnologies.com/home. Accessed June 14, 2015.
109. *PB3 PowerBuoy®—Ocean Power Technologies*. https://oceanpowertechnologies.com/pb3-powerbuoy/. Accessed January 14, 2021.
110. *Ocean harvesting Technologies*. https://www.oceanharvesting.com/. Accessed January 14, 2021.
111. OHT. (2021). *H2020-SMEINST-1-2016-2017—A hydraulic collection tower, with a novel energy storage device for wave energy arrays*. https://cordis.europa.eu/project/id/775250. Accessed February 14, 2021.